Quasistatic Contact Problems in Viscoelasticity and Viscoplasticity

AMS/IP
Studies in
Advanced
Mathematics

Volume 30

Quasistatic Contact Problems in Viscoelasticity and Viscoplasticity

Weimin Han and Mircea Sofonea

American Mathematical Society • International Press

Shing-Tung Yau, General Editor

2000 *Mathematics Subject Classification.* Primary 74–02, 74M15, 74M10, 74S05, 74B05, 74B20, 74Cxx, 74Dxx;
Secondary 65M06, 65M12, 65M15, 65M60, 65N12, 65N15, 65N30, 49J40.

Library of Congress Cataloging-in-Publication Data

Han, Weimin.
Quasistatic contact problems in viscoelasticity and viscoplasticity / Weimin Han and Mircea Sofonea.
p. cm. — (AMS/IP studies in advanced mathematics; v. 30)
Includes bibliographical references and index.
ISBN 0-8218-3192-5 (alk. paper)
1. Viscoelasticity. 2. Viscoplasticity. 3. Contact mechanics—Mathematical models. I. Sofonea, Mircea. II. Title. III. Series.

QA931.H26 2002
620.1′1232—dc21 2002027716

Visit the AMS home page at http://www.ams.org/
Visit the International Press home page at URL: http://www.intlpress.com/

10 9 8 7 6 5 4 3 2 1 07 06 05 04 03 02

Dedicated to

DAQING HAN, SUZHEN QIN
HUIDI TANG, ELIZABETH JING HAN
MICHAEL YUE HAN

and

In memory of
SEMINICA SOFONEA

Contents

Preface

Phenomena of contact between deformable bodies or between deformable and rigid bodies abound in industry and everyday life. Contact of braking pads with wheels, tires with roads, pistons with skirts are just a few simple examples. Common industrial processes such as metal forming, metal extrusion, involve contact evolutions. Because of the importance of contact processes in structural and mechanical systems, a considerable effort has been made in its modeling and numerical simulations. The engineering literature concerning this topic is extensive. Owing to their inherent complexity, contact phenomena are modeled by nonlinear evolutionary problems that are difficult to analyze.

In early mathematical publications it was invariably assumed that the deformable bodies were linearly elastic and the processes were static. However, a long time ago it was recognized the need to consider contact problems involving viscoelastic and viscoplastic materials as well as a large variety of contact and frictional boundary conditions, which lead to time dependent models. The mathematical theory of contact problems, that can predict reliably the evolution of the contact process in different situations and under various conditions, is emerging currently. It deals with rigorous modeling of the contact phenomena, based on the fundamental physical principles, as well as with the variational and numerical analysis of the models. A thorough treatment of contact problems require knowledge from functional analysis, modern partial differential equations, numerical approximations and error analysis.

The purpose of this book is to introduce the reader to a mathematical theory of contact problems involving deformable bodies. The contents cover the mechanical modeling, mathematical formulations, variational analysis, and the numerical solution of the associated formulations. Our intention is to give a complete treatment of some contact problems by presenting arguments and results in modeling, analysis, and numerical simulations.

In the book we treat quasistatic contact processes in the infinitesimal strain theory. Quasistatic processes arise when the applied forces vary slowly in time, and therefore the system response is relatively slow so that the inertial terms in the equations of motion can be neglected. We model the material behavior with elastic, viscoelastic or viscoplastic constitutive laws; some of our results are extended to materials with internal state variables and to perfectly plastic materials. The contact is modeled with various conditions, including Signorini nonpenetration condition, normal compliance and normal damped response conditions. The friction is modeled with versions of Coulomb's and Tresca's friction laws or with laws involving a dissipative frictional potential. We also consider problems with friction and wear and we use a version of the Archard law to model the evolution of wear.

Variational analysis of the models includes existence and uniqueness results of weak solutions as well as results of continuous dependence of the solution on the data and parameters. Links between different mechanical models are discussed; for example, elasticity as a limiting case of viscoelasticity, perfect plasticity as a limiting case of viscoplasticity, Signorini nonpenetration condition as a limiting case of the normal compliance contact condition. In carrying out the variational analysis we systematically use results on elliptic and evolutionary variational inequalities, convex analysis, nonlinear equations with monotone operators and fixed points of operators.

Two kinds of approximation schemes are introduced and analyzed. When only the spatial variables are discretized, we obtain semi-discrete schemes. If both the spatial and temporal variables are discretized, we arrive at fully discrete schemes. For both kinds of schemes we prove existence and uniqueness results. We show convergence of the discrete solutions under the basic solution regularity available from the well-posedness results of the variational problems. We also present optimal order error estimates under additional regularity assumption on the solution.

To demonstrate the performance of the numerical schemes, a number of numerical simulations are discussed. The test problems range from one to three dimensional geometries. The finite element method is used to discretize the spatial domain and finite differences are used for the time derivatives. We describe in the book numerical results in the study of some model contact problems for elastic, viscoelastic and viscoplastic materials.

The book is intended to be self-contained and accessible to a large number of readers. It is divided into four parts, as described in the following.

Part I is devoted to the basic notions and results which are fundamental to the development later in this book. We review here the background on functional analysis, function spaces, finite difference approximations and finite element method. Then we apply these results in the study of elliptic and evolutionary variational inequalities. The material presented in this part is self-contained. It does not need any knowledge in Contact Mechanics and could be aimed at graduate students and researchers interested in a general treatment of variational problems and their numerical approximations.

Part II presents preliminary material in Contact Mechanics. We summarize here basic notions and general principles of Mechanics of Continua. Then we introduce contact boundary conditions with or without friction as well as constitutive laws which are used in the rest of the book. We also present preliminary results on variational and numerical analysis in contact problems and we apply these results in the study of models involving elastic bodies. The material in Part II is aimed at those readers who are interested in the mechanical background of contact problems and mathematical theory of some contact problems in elasticity.

Parts III and IV represent the main parts of the book. They deal with the study of quasistatic problems for Kelvin-Voigt viscoelastic materials and rate-type viscoplastic materials, respectively. These parts are written based on our original research. We consider here a number of problems with various contact and frictional or frictionless boundary conditions, for which we provide variational analysis and numerical approximations. For some of the contact problems, we include results of numerical simulations to show the performance of the numerical schemes. These two parts of the book would interest mainly researchers for an in-depth knowledge of the mathematical theory of quasistatic contact problems.

The list of the references at the end of the book is by no means exhaustive. It only includes papers or books that were used for or are directly connected with the subjects treated in this book. Each part is concluded with a section entitled *Bibliographical Notes* that discusses references on the principal results treated, as well as information on important topics related to, but not included in, the body of the text.

Each of the four parts of the book is divided into several chapters. All the chapters are numbered consecutively. Mathematical relations (equalities or inequalities) are numbered by chapter and their order of occurrence. For example, (5.3) is the third numbered mathematical relation in Chapter 5. Definitions, examples, problems, theorems, lemmas, corollaries, propositions and remarks are numbered consecutively within each chapter. For example, in Chapter 10, Problem 10.1 is followed by Theorem 10.2.

The present book is the result of three years of cooperation between the two authors. In writing it we have drawn on the results of our joint collaboration with numerous colleagues and friends to whom we address our thanks. We express our gratitude to Professor Meir Shillor for our beneficial cooperation as well as for the interesting discussions on the models treated in the book. We particularly thank Dr. J.R. Fernández-García who provided the numerical simulations included in Parts III and IV of the book. We extend our thanks to Dr. M. Barboteu who realized the numerical results in Part II of the book. We especially thank Professor Shing-Tung Yau for his support and encouragement of our work.

The work of W.H. was supported by NSF/DARPA under Grant DMS-9874015, James Van Allen Natural Science Fellowship, Faculty Development Award at the University of Iowa, and K.C. Wong Education Foundation. Part of the book manuscript was prepared when W.H. was visiting the State Key Laboratory of Scientific and Engineering Computing, Chinese Academy of Sciences, during the summer of 2000 through the support of the K.C. Wong Education Foundation. He particularly thanks Professor Lie-heng Wang for the warm hospitality.

W.H. — M.S.
Iowa City — Perpignan

List of Symbols

Sets

$\mathbb{N}$: the set of positive integers

$\mathbb{Z}_+$: the set of non-negative integers

$\mathbb{R}$: the real line

$\overline{\mathbb{R}} = \mathbb{R} \cup \{\pm\infty\}$: extended real line

$\mathbb{R}_+$: the set of non-negative numbers

$\mathbb{R}^d$: the d-dimensional Euclidean space

$\mathbb{S}^d$: the space of second order symmetric tensors on $\mathbb{R}^d$

$\mathbb{Q}_+^d$: the set of orthogonal matrices of order d with determinant 1

S_d: the unit sphere in $\mathbb{R}^d$

Ω: an open, bounded, connected set in $\mathbb{R}^d$ with a Lipschitz boundary Γ

Γ: the boundary of the domain Ω, that is decomposed as $\Gamma = \overline{\Gamma}_1 \cup \overline{\Gamma}_2 \cup \overline{\Gamma}_3$ with Γ_1, Γ_2 and Γ_3 relatively open with respect to Γ

Γ_1: the part of the boundary where displacement condition is specified; $\operatorname{meas}(\Gamma_1) > 0$ is assumed throughout the book

Γ_2: the part of the boundary where traction condition is specified

Γ_3: the part of the boundary for contact

$\overline{I} = [0, T]$: time interval of interest

$I = (0, T)$

Operators

ε: deformation operator, i.e. $\boldsymbol{\varepsilon}(\boldsymbol{u}) = (\varepsilon_{ij}(\boldsymbol{u}))$, $\varepsilon_{ij}(\boldsymbol{u}) = \frac{1}{2}(u_{i,j} + u_{j,i})$ (page 97)

Div: divergence operator, i.e. $\operatorname{Div}\boldsymbol{\sigma} = (\sigma_{ij,j})$ (page 100)

Π^h: finite element interpolation operator (page 52)

$\mathcal{P}_K$: projection operator onto a set K (page 16)

$\mathcal{P}^h$: finite element projection operator (pages 77, 181)

$\mathcal{P}_{Q^h}$: finite element projection on Q^h (page 154)

Function spaces

$L^p(\Omega)$: the Lebesgue space of p-integrable functions, with the usual modification if $p = \infty$ (page 28)

$C^m(\overline{\Omega})$: the space of functions whose derivatives up to and including order m are continuous up to the boundary Γ (page 26)

$C_0^\infty(\Omega)$: the space of infinitely differentiable functions with compact support in Ω (page 26)

$W^{k,p}(\Omega)$: the Sobolev space of functions whose weak derivatives of orders less than or equal to k are p-integrable on Ω (page 30)

$H^k(\Omega) \equiv W^{k,2}(\Omega)$ (page 30)

$W_0^{k,p}(\Omega)$: the closure of $C_0^\infty(\Omega)$ in $W^{k,p}(\Omega)$ (page 30)

$H_0^k(\Omega) \equiv W_0^{k,2}(\Omega)$ (page 30)

$H^{-1}(\Omega)$: the dual of $H_0^1(\Omega)$ (page 34)

$H^{\frac{1}{2}}(\Gamma)$: a Sobolev space on Γ, defined as the range of the trace operator on $H^1(\Omega)$ (page 36)

$H^{-\frac{1}{2}}(\Gamma)$: the dual of $H^{\frac{1}{2}}(\Gamma)$ (page 36)

$H_\Gamma = H^{\frac{1}{2}}(\Gamma)^d$ (page 143)

H'_Γ: dual of H_Γ (page 144)

$H = \{\, \boldsymbol{v} = (v_1, \dots, v_d)^T : v_i \in L^2(\Omega),\ 1 \le i \le d \,\} = L^2(\Omega)^d$, inner product $(\boldsymbol{u}, \boldsymbol{v})_H = \int_\Omega u_i(\boldsymbol{x})\, v_i(\boldsymbol{x})\, dx$ (page 142)

$Q = \{\, \boldsymbol{\tau} = (\tau_{ij}) : \tau_{ij} = \tau_{ji} \in L^2(\Omega),\ 1 \le i, j \le d \,\} = L^2(\Omega)_s^{d\times d}$, inner product $(\boldsymbol{\sigma}, \boldsymbol{\tau})_Q = \int_\Omega \sigma_{ij}(\boldsymbol{x})\, \tau_{ij}(\boldsymbol{x})\, dx$ (page 142)

$Q_1 = \{\, \boldsymbol{\tau} \in Q : \operatorname{Div}\boldsymbol{\tau} \in H \,\}$, inner product $(\boldsymbol{\sigma}, \boldsymbol{\tau})_{Q_1} = (\boldsymbol{\sigma}, \boldsymbol{\tau})_Q + (\operatorname{Div}\boldsymbol{\sigma}, \operatorname{Div}\boldsymbol{\tau})_H$ (page 145)

$V = \{\, \boldsymbol{v} \in H^1(\Omega)^d : \boldsymbol{v} = \boldsymbol{0}$ a.e. on $\Gamma_1 \,\}$, inner product $(\boldsymbol{u}, \boldsymbol{v})_V = (\boldsymbol{\varepsilon}(\boldsymbol{u}), \boldsymbol{\varepsilon}(\boldsymbol{v}))_Q$ (page 144)

$V_1 = \{\boldsymbol{v} \in V : v_\nu = 0$ a.e. on $\Gamma_3\}$ with the inner product $(\boldsymbol{u}, \boldsymbol{v})_V$ (page 144)

$V_2 = \{\boldsymbol{v} \in V : v_\nu \le 0$ a.e. on $\Gamma_3\}$ with the inner product $(\boldsymbol{u}, \boldsymbol{v})_V$ (page 144)

X: a Hilbert space or its subset with inner product $(\cdot,\cdot)_X$, or a Banach space or its subset with norm $\|\cdot\|_X$

$C^m(\overline{I}; X) = \{ v \in C(\overline{I}; X) : v^{(j)} \in C(\overline{I}; X),\ j = 1, \dots, m \}$ (page 37)

$L^p(I; X) = \{ v : \overline{I} \to X$ measurable $: \|v\|_{L^p(I;X)} < \infty \}$ (page 38)

$W^{k,p}(I; X) = \{ v \in L^p(0, T; X) : \|v^{(j)}\|_{L^p(I;X)} < \infty\ \forall j \le k \}$ (page 39)

$H^k(I; X) \equiv W^{k,2}(I; X)$ (page 40)

Other symbols

d: a positive integer taking the values $1, 2, 3$ in applications

c: a generic positive constant

$r_+ = \max\{0, r\}$: positive part of r

$\boldsymbol{\nu}$: unit outward normal on the boundary of Ω

$\langle\cdot,\cdot\rangle$: the duality pairing between $H^{\frac{1}{2}}(\Gamma)$ and $H^{-\frac{1}{2}}(\Gamma)$ (page 36)

$\langle\cdot,\cdot\rangle_\Gamma$: the duality pairing between H_Γ and H'_Γ (page 144)

$\forall$: for any

$\exists$: there exist(s)

$\overline{A}$: closure of the set A

int A or $\mathring{A}$: interior of the set A

∂A: boundary of the set A

δ_{ij}: the Kronecker delta

s.t.: such that

a.e.: almost everywhere

iff: if and only if

$O(h)$: for some constant $c > 0$ independent of h such that $|O(h)| \le c\,h$

$\Delta w_n = w_n - w_{n-1}$: backward difference

$\delta_n w_n = \Delta w_n / k_n$: backward divided difference

Part I

Nonlinear Variational Problems and Numerical Approximation

Chapter 1

Preliminaries of Functional Analysis

In this chapter we review some basic notions and results of functional analysis that are fundamental to the development later in this work. Most of the material is standard and can be found in many books on functional analysis. For this reason some important results have been given without proof while we refer to the bibliography for proofs as well as for more information in topics presented here. Basic results on linear algebra and general topology are assumed to be familiar to the reader.

1.1 Normed Spaces and Banach Spaces

Spaces provide an adequate setting in which we make calculations and perform mathematical analysis. Often we need two kinds of structure in our spaces: algebraic structure and topological structure. The algebraic structure allows us to perform arithmetic operations while the topological structure allows us to define such useful concepts as convergence. We assume the reader is familiar with the definition of a linear space. Recall that a *linear space* is a set of elements (called vectors) over which a vector addition and a scalar-vector multiplication are defined such that the ordinary arithmetic rules of vector addition and scalar-vector multiplication in the plane $\mathbb{R}^2$ are valid. In this work, we only deal with real linear spaces so that all the scalars are real. Simple examples of linear spaces are the d-dimensional vector space $\mathbb{R}^d = \{\, \boldsymbol{x} = (x_1, \ldots, x_d)^T : x_1, \ldots, x_d \in \mathbb{R}\,\}$ and the space of d by d matrices, $\mathbb{R}^{d\times d}$. Over the space $\mathbb{R}^{d\times d}$, we have an additional algebraic operation: matrix multiplication. A linear space used often in this work is $\mathbb{S}^d$, the set of all the real symmetric matrices of order $d \times d$. This is a linear subspace of $\mathbb{R}^{d\times d}$.

The notion of a norm in a general linear space is an extension of the ordinary length of a vector in $\mathbb{R}^2$ or $\mathbb{R}^3$.

Definition 1.1 *Given a linear space X, a* norm $\|\cdot\|_X$ *is a function from X to $\mathbb{R}$ with the following properties.*

1. $\|v\|_X \geq 0$ *for any* $v \in X$, *and* $\|v\|_X = 0$ *iff* $v = 0$.

2. $\|\alpha v\|_X = |\alpha|\,\|v\|_X$ *for any* $v \in X$ *and any* $\alpha \in \mathbb{R}$.

3. $\|u+v\|_X \le \|u\|_X + \|v\|_X$ *for any* $u, v \in X$.

The pair $(X, \|\cdot\|_X)$ *is called a* normed space*; we will simply say* X *is a normed space when the definition of the norm is understood from the context.*

Example 1.2 *In* $\mathbb{R}^d$*, the usual norm is*

$$\|\boldsymbol{x}\|_2 = \Big(\sum_{i=1}^d |x_i|^2\Big)^{1/2},$$

called the Euclidean norm. *Generally, we can define the p-norm* $\|\cdot\|_p$ *by the formulas*

$$\|\boldsymbol{x}\|_p = \Big(\sum_{i=1}^d |x_i|^p\Big)^{1/p}, \quad 1 \le p < \infty,$$
$$\|\boldsymbol{x}\|_\infty = \max_{1\le i\le d} |x_i|.$$

If it is not specified explicitly, then the norm in the space $\mathbb{R}^d$ *is taken to be the Euclidean norm. The space* $\mathbb{R}^d$ *endowed with the Euclidean norm is usually called a Euclidean space.*

Example 1.3 *Over the space* $\mathbb{R}^{d\times d}$ *of the matrices of order d, the quantity*

$$\|A\|_F = \Big(\sum_{i,j=1}^d |a_{ij}|^2\Big)^{1/2} \text{ for } A = (a_{ij})_{d\times d}$$

defines a norm, called the Frobenius norm.

Let X be a normed space with norm $\|\cdot\|_X$. If $Y \subset X$ is a subspace, then $(Y, \|\cdot\|_X)$ is a normed space too. If it is not explicitly stated otherwise, the norm over a subspace is taken to be the norm of the original normed space.

Let X_1 and X_2 be two normed spaces with norms $\|\cdot\|_{X_1}$ and $\|\cdot\|_{X_2}$. The standard norm over the Cartesian product space $Y = X_1 \times X_2$ is

$$\|w\|_Y = \|u\|_{X_1} + \|v\|_{X_2} \quad \forall\, w = (u, v) \in Y.$$

We have seen from Example 1.2 that over a linear space, it is possible to define infinite many norms. Sometimes it is desirable to know if two norms are related.

Definition 1.4 *Let* $\|\cdot\|^{(1)}$ *and* $\|\cdot\|^{(2)}$ *be two norms over a linear space* X. *We say the norm* $\|\cdot\|^{(1)}$ *is* stronger *than the norm* $\|\cdot\|^{(2)}$ *if there exists a constant* $c > 0$ *such that*

$$\|v\|^{(2)} \le c\,\|v\|^{(1)} \quad \forall\, v \in X.$$

The two norms are said to be equivalent *if there are two constants $c_1, c_2 > 0$ such that*

$$c_1 \|v\|^{(1)} \le \|v\|^{(2)} \le c_2 \|v\|^{(1)} \quad \forall\, v \in X.$$

Example 1.5 *Over the space $\mathbb{R}^d$, any two norms $\|\cdot\|_p$ and $\|\cdot\|_q$, $p, q \in [1, \infty]$, are equivalent. We can easily show*

$$\|\boldsymbol{x}\|_\infty \le \|\boldsymbol{x}\|_p \le d^{1/p} \|\boldsymbol{x}\|_\infty \quad \forall\, \boldsymbol{x} \in \mathbb{R}^d.$$

In fact, we have the following well-known general result.

Theorem 1.6 *Over a finite dimensional space, any two norms are equivalent.*

Over an infinite dimensional space, not any two norms are equivalent.

Example 1.7 *Consider the space $X = C[0,1]$ of real-valued continuous functions on $[0,1]$. Both*

$$\|v\|^{(1)} = \max_{0 \le x \le 1} |v(x)|$$

and

$$\|v\|^{(2)} = \int_0^1 |v(x)|\, dx$$

are norms. We have

$$\|v\|^{(2)} \le \|v\|^{(1)} \quad \forall\, v \in X.$$

But the two norms are not equivalent as there is no constant $c > 0$ such that $\|v\|^{(1)} \le c\, \|v\|^{(2)}$ for all $v \in X$. For example, for any $n \in \mathbb{N}$, let

$$v_n(x) = \begin{cases} 1 - nx & \text{if} \quad 0 \le x \le 1/n, \\ 0 & \text{if} \quad 1/n < x \le 1. \end{cases}$$

Then $\|v_n\|^{(1)} = 1$, $\|v_n\|^{(2)} = 1/(2\,n)$ and therefore inequality $\|v_n\|^{(1)} \le c\, \|v_n\|^{(2)}$ for all $n \in \mathbb{N}$ leads to a contradiction. The conventional norm on the space $C[0,1]$ is $\|\cdot\|^{(1)}$, which is usually written as $\|\cdot\|_{C[0,1]}$.

The notion of a semi-norm is useful in error estimates of some numerical approximations.

Definition 1.8 *Given a linear space X, a* semi-norm *$|\cdot|_X$ is a function from X to $\mathbb{R}$ satisfying the properties of a norm except that $|v|_X = 0$ does not necessarily imply $v = 0$.*

Example 1.9 *Over the space $X = C^1[0,1]$ of real-valued continuously differentiable functions on $[0,1]$,*

$$\|v\|_{C^1[0,1]} = \max_{0\le x\le 1} |v(x)| + \max_{0\le x\le 1} |v'(x)|$$

is the standard norm, while

$$|v|_{C^1[0,1]} = \max_{0\le x\le 1} |v'(x)|$$

is a semi-norm. Here v' denotes the derivative of the function v.

With a norm at our disposal, we use the quantity $\|u - v\|_X$ to measure the distance between u and v. Consequently, the norm is used to define the convergence of a sequence in the space X.

Definition 1.10 *Let X be a normed space. A sequence $\{u_n\} \subset X$ is said to converge (strongly) to $u \in X$ if*

$$\|u_n - u\|_X \to 0 \quad \text{as } n \to \infty.$$

In this case u is called the limit *of the sequence $\{u_n\}$ and we write*

$$u = \lim_{n\to\infty} u_n \quad \text{or} \quad u_n \to u \quad \text{in } X.$$

It is easy to verify that a limit of a sequence, if it exists, is unique.

The adjective "strong" is introduced in the previous definition to distinguish this convergence from other types of convergence which will be introduced in the next section. For two equivalent norms, convergence in one norm is equivalent to convergence in the other norm.

Example 1.11 *In the space $\mathbb{R}^d$, for any $p \in [1,\infty]$, the convergence of a sequence of vectors in $\|\cdot\|_p$ is equivalent to componentwise convergence.*

The convergence of sequences is used to define closed sets in a normed space.

Definition 1.12 *Let X be a normed space. A subset $X_1 \subset X$ is said to be (strongly)* closed *if it contains the limits of all convergent sequences $\{u_n\} \subset X_1$.*

To test the convergence of a sequence without knowing the limiting element, it is usually convenient to refer to the notion of a Cauchy sequence.

Definition 1.13 *Let X be a normed space. A sequence $\{u_n\} \subset X$ is called a* Cauchy sequence *if $\|u_m - u_n\|_X \to 0$ as $m, n \to \infty$.*

Obviously, a convergent sequence is a Cauchy sequence. In a finite dimensional space, say $\mathbb{R}^d$, a sequence is convergent iff it is a Cauchy sequence. However, in a general infinite dimensional space, a Cauchy sequence may fail to converge.

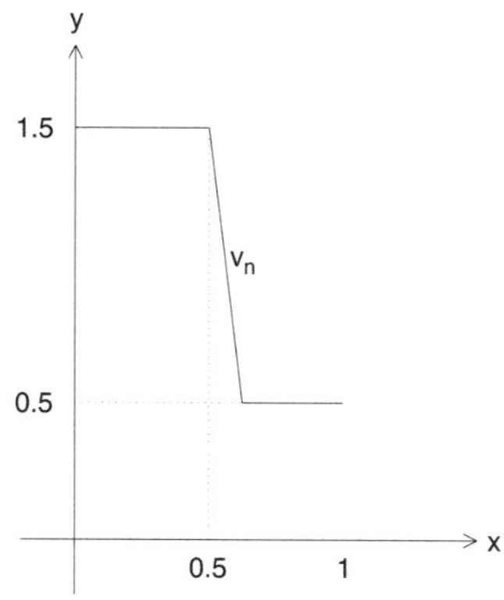

Figure 1.1: The function v_n in Example 1.14

Example 1.14 *Consider the space $X = C[0,1]$ endowed with the norm*

$$\|v\| = \int_0^1 |v(x)|\,dx.$$

We define a sequence $\{v_n\} \subset X$:

$$v_n(x) = \begin{cases} 1.5 & \text{if } 0 \le x \le \frac{1}{2}, \\ 1.5 - n\,(x - \frac{1}{2}) & \text{if } \frac{1}{2} < x < \frac{1}{2} + \frac{1}{n}, \\ 0.5 & \text{if } \frac{1}{2} + \frac{1}{n} \le x \le 1 \end{cases}$$

(cf. Figure 1.1) and let v denote the function

$$v(x) = \begin{cases} 1.5 & \text{if } 0 \le x < \frac{1}{2}, \\ 0.5 & \text{if } \frac{1}{2} < x \le 1. \end{cases}$$

Then $\|v_n - v\| \to 0$ as $n \to \infty$, but the limit function v does not belong to X no matter how we define the value $v(\frac{1}{2})$.

The previous example justifies the following definition.

Definition 1.15 *A normed space is said to be* complete *if every Cauchy sequence from the space converges to an element in the space. A complete normed space is called a* Banach space.

Example 1.16 *Any finite-dimensional normed space is a Banach space. In particular, the space $\mathbb{R}^d$ with any p-norm $\|\cdot\|_p$ is a Banach space.*

Definition 1.17 *Let A be a set of a normed space. The* closure $\overline{A}$ *of A is the union of A and the set of the limits of all the convergent sequences from A.*

Figure 1.2: Some convex sets in $\mathbb{R}^2$

Figure 1.3: Some non-convex sets in $\mathbb{R}^2$

Definition 1.18 *Let X be a normed space. A subset $A \subset X$ is said to be* dense *in X if $\overline{A} = X$.*

Example 1.19 *It is a classical result that $C[0,1]$ is dense in $L^p(0,1)$, $p \in [1,\infty)$; i.e., for any $v \in L^p(0,1)$, there exists a sequence $\{v_n\} \subset C[0,1]$ such that*

$$\|v_n - v\|_{L^p(0,1)} \to 0 \quad \text{as } n \to \infty.$$

Definition 1.20 *Let X be a linear space. A subset $K \subset X$ is said to be* convex *if it has the property*

$$u, v \in K \;\Rightarrow\; (1-t)\,u + t\,v \in K \quad \forall\, t \in [0,1].$$

For $t \in [0,1]$, the expression $(1-t)\,u + t\,v$ is said to be a *convex combination* of u and v. The set $\{(1-t)\,u + t\,v : t \in [0,1]\}$ consists of all the points on the line segment connecting u and v. We see that if K is convex and $u, v \in K$, then the line segment connecting u and v is contained in K. Let K be convex. By an induction argument, for any $u_1, \dots, u_n \in K$ and any nonnegative numbers $t_1, \dots, t_n$ with $\sum_{i=1}^n t_i = 1$, we have $\sum_{i=1}^n t_i u_i \in K$. The quantity $\sum_{i=1}^n t_i u_i$ with nonnegative numbers $t_1, \dots, t_n$ satisfying $\sum_{i=1}^n t_i = 1$ is called a *convex combination* of the elements $u_1, \dots, u_n$. Some convex and non-convex sets in $\mathbb{R}^2$ are shown in Figures 1.2 and 1.3, respectively.

1.2 Linear Operators and Linear Functionals

Given two sets X and Y, an *operator* T from X to Y is a rule which assigns to each element in a subset of X a unique element in Y. The domain $\mathcal{D}(T)$ of T is the subset of X where T is defined,

$$\mathcal{D}(T) = \{\, v \in X : T(v) \text{ is defined} \,\},$$

and the range $\mathcal{R}(T)$ of T is the set of the elements in Y generated by T,

$$\mathcal{R}(T) = \{\, w \in Y : w = T(v) \text{ for some } v \in \mathcal{D}(T) \,\}.$$

Usually the domain $\mathcal{D}(T)$ is understood to be the whole set X, unless it is explicitly stated otherwise. A real-valued operator defined on a linear space X is called a *functional.*

If both X and Y are normed spaces, we can consider the continuity and boundedness of the operators.

Definition 1.21 *Let $(X, \|\cdot\|_X)$ and $(Y, \|\cdot\|_Y)$ be two normed spaces. An operator $T : X \to Y$ is said to be*

1. continuous *at $v \in X$ if*

$$v_n \to v \text{ in } X \;\Rightarrow\; T(v_n) \to T(v) \text{ in } Y;$$

2. continuous *if it is continuous at any element of the space X;*

3. Lipschitz continuous *if there exists $L_T > 0$ such that*

$$\|T(u) - T(v)\|_Y \le L_T \|u - v\|_X \quad \forall\, u, v \in V;$$

4. bounded *if for any $r > 0$ there is an $R > 0$ such that*

$$\|v\|_X \le r \;\Rightarrow\; \|T(v)\|_Y \le R.$$

Clearly, if T is Lipschitz continuous then it is a continuous operator. We observe that an alternative definition of the boundedness is that for any set $B \subset X$,

$$\sup_{v \in B} \|v\|_X < \infty \;\Rightarrow\; \sup_{v \in B} \|T(v)\|_Y < \infty.$$

We now consider a particular yet important type of operators called linear operators. From now on, when we write $T : X \to Y$, we will implicitly assume $\mathcal{D}(T) = X$, unless stated otherwise.

Definition 1.22 *Let X and Y be two linear spaces. An operator $L : X \to Y$ is called* linear *if*

$$L(\alpha_1 v_1 + \alpha_2 v_2) = \alpha_1 L(v_1) + \alpha_2 L(v_2) \quad \forall\, v_1, v_2 \in X,\ \alpha_1, \alpha_2 \in \mathbb{R},$$

or equivalently,

$$\begin{aligned} L(v_1 + v_2) &= L(v_1) + L(v_2) \quad \forall\, v_1, v_2 \in X, \\ L(\alpha\, v) &= \alpha\, L(v) \quad \forall\, v \in X,\ \alpha \in \mathbb{R}. \end{aligned}$$

For a linear operator L, we usually write $L(v)$ as Lv. For simplicity, we sometimes write Lv even when L is not linear. A well-known important property of a linear operator is that the continuity and boundedness are equivalent.

Theorem 1.23 *Let X and Y be normed spaces, $L : X \to Y$ a linear operator. Then L is continuous on X iff it is bounded on X.*

Because of Theorem 1.23, we will use the words *continuity* and *boundedness* interchangeably for linear operators. Also from Theorem 1.23, we conclude that for a linear operator, continuity and Lipschitz continuity are equivalent.

We will use the notation $\mathcal{L}(X,Y)$ for the set of all the continuous linear operators from a normed space X to another normed space Y. In the special case $Y = X$, we use $\mathcal{L}(X)$ to replace $\mathcal{L}(X,X)$. Since boundedness is equivalent to continuity for a linear operator, for $L \in \mathcal{L}(X,Y)$ it is meaningful to define the quantity

$$\|L\|_{\mathcal{L}(X,Y)} = \sup_{0 \neq v \in X} \frac{\|Lv\|_Y}{\|v\|_X}. \tag{1.1}$$

It can be shown that the set $\mathcal{L}(X,Y)$ is a linear space under the regular addition and scalar multiplication of operators, and (1.1) defines a norm over the space. The norm (1.1) is called the *operator norm*. The operator norm enjoys the following compatibility property

$$\|Lv\|_Y \leq \|L\|_{\mathcal{L}(X,Y)} \|v\|_X \quad \forall\, v \in X.$$

This property is very useful in some context of error analysis for numerical methods.

Example 1.24 *A matrix $A = (a_{ij})$ from the space $\mathbb{R}^{d\times d}$ can be viewed as a linear continuous operator from $\mathbb{R}^d$ to $\mathbb{R}^d$ through the matrix/vector multiplication:*

$$\boldsymbol{v} \in \mathbb{R}^d \mapsto A\boldsymbol{v} \in \mathbb{R}^d.$$

So $\mathbb{R}^{d\times d}$ may be identified with $\mathcal{L}(\mathbb{R}^d, \mathbb{R}^d)$. The formula (1.1) provides a way to define a family of matrix operator norms:

$$\|A\|_{p,q} = \sup_{\mathbf{0} \neq \boldsymbol{x} \in \mathbb{R}^d} \frac{\|A\boldsymbol{x}\|_q}{\|\boldsymbol{x}\|_p}.$$

In particular, when $q = p$, we use the symbol $\|A\|_p$ to replace $\|A\|_{p,p}$. The most important matrix operator norms are

$$\|A\|_1 = \max_{1\leq j\leq d} \sum_{i=1}^{d} |a_{ij}|,$$

$$\|A\|_2 = \max\{|\lambda|^{1/2} : \lambda \text{ is an eigenvalue of } A^T A\},$$

$$\|A\|_\infty = \max_{1\leq i\leq d} \sum_{j=1}^{d} |a_{ij}|.$$

The matrix norm $\|\cdot\|_2$ is also called the spectral norm.

As for the completeness of the space $\mathcal{L}(X, Y)$, we have the next result.

Theorem 1.25 *Let X be a normed space, and Y be a Banach space. Then $\mathcal{L}(X, Y)$ is a Banach space.*

For a normed space X, the space $\mathcal{L}(X, \mathbb{R})$ is called the *dual space* of X and is denoted by X'. The elements of X' are *linear continuous functionals* on X. The duality pairing between X' and X is usually denoted by $\ell(v)$ or $\langle v', v\rangle$ for $\ell, v' \in X'$ and $v \in X$. As it follows from (1.1), the norm on X' is given by

$$\|\ell\|_{X'} = \sup_{0 \neq v \in X} \frac{|\ell(v)|}{\|v\|_X}. \tag{1.2}$$

By Theorem 1.25 we know that $(X', \|\cdot\|_{X'})$ is a Banach space.

Example 1.26 *Let $\Omega \subset \mathbb{R}^d$ be an open set. For $p \in [1, \infty]$, its* conjugate exponent *$p' \in [1, \infty]$ is defined by the relation*

$$\frac{1}{p} + \frac{1}{p'} = 1.$$

Here by convention, $1/\infty = 0$. Assume $p \in [1, \infty)$. Then a classical result is $(L^p(\Omega))' = L^{p'}(\Omega)$, i.e., the dual space of $L^p(\Omega)$ can be identified with $L^{p'}(\Omega)$.

We can now introduce another kind of convergence in a normed space.

Definition 1.27 *Let X be a normed space. A sequence $\{u_n\} \subset X$ is said to* converge weakly *to $u \in X$ if for any $\ell \in X'$,*

$$\ell(u_n) \to \ell(u) \quad \text{as } n \to \infty.$$

In this case u is called the weak limit *of $\{u_n\}$ and we write $u_n \rightharpoonup u$ as $n \to \infty$.*

It follows from Hahn-Banach Theorem that the weak limit of a sequence, if it exists, is unique. It is easy to see that the strong convergence implies the weak convergence, i.e. if $u_n \to u$ in X, then $u_n \rightharpoonup u$ in X. The converse of this property is not true in general.

Example 1.28 *Consider the space $L^2(0, \pi)$. For any $f \in L^2(0, \pi)$, we have*

$$\int_0^\pi f(x) \sin nx \, dx \to 0 \quad \text{as } n \to \infty.$$

Thus, $\sin nx \rightharpoonup 0$. However, $\sin nx \not\to 0$ since

$$\int_0^\pi |\sin nx|^2 dx = \frac{\pi}{2} \not\to 0.$$

The weak convergence of sequences is used to define weakly closed sets in normed space.

Definition 1.29 *Let X be a normed space. A subset $X_1 \subset X$ is said to be* weakly closed *if it contains the limits of all weakly convergent sequences $\{u_n\} \subset X_1$.*

Clearly, every weakly closed subset of X is closed but the converse of this property is not true in general. However, we have the following result.

Theorem 1.30 *A convex closed set in a normed space is weakly closed.*

Let X be a normed space. We denote by $X'' = (X')'$ the dual of the Banach space X', which will be called the *bidual* of X. The bidual X'' is a Banach space. Each element $u \in X$ induces a linear continuous functional $\ell_u \in X''$ by the relation $\ell_u(v') = \langle v', u\rangle$ for any $v' \in X'$. The mapping $u \mapsto \ell_u$ from X into X'' is linear and isometric, i.e. $\|\ell_u\|_{X''} = \|u\|_X$ for all $u \in X$. Therefore, the normed space X may be viewed as a linear subspace of the Banach space X'' by the embedding $u \mapsto \ell_u = \chi(u)$. We introduce the following definition.

Definition 1.31 *A normed space X is said to be* reflexive, *if X may be identified with X'' by the canonical embedding χ (i.e. if $\chi(X) = X''$).*

Example 1.32 *Continuing Example 1.26, we see that for $p \in (1, \infty)$, the space $L^p(\Omega)$ is reflexive.*

A reflexive space must be complete and is hence a Banach space. We have the following important property of a reflexive space.

Theorem 1.33 *If X is a reflexive Banach space, then any bounded sequence in X has a weakly convergent subsequence.*

It follows that if X is a reflexive Banach space and the sequence $\{u_n\} \subset X$ is bounded (i.e. $\sup_n \|u_n\|_X < \infty$), then we can find a subsequence $\{u_{n'}\} \subset \{u_n\}$ and an element $u \in X$ such that $u_n \rightharpoonup u$ in X. Furthermore, if the limit u is independent of the subsequence extracted, then the whole sequence $\{u_n\}$ converges weakly to u.

Definition 1.34 *Let X be a normed space and let X' denote its dual. A sequence $\{u'_n\} \subset X'$ is said to* converge weakly* *to $u' \in X'$ if*

$$\langle u'_n, v\rangle \to \langle u', v\rangle \quad \text{as } n \to \infty, \quad \forall v \in X.$$

In this case u' is called the weak* limit *of $\{u'_n\}$ and we write $u'_n \rightharpoonup^* u'$ as $n \to \infty$.*

It is easy to verify that the weak* limit of a sequence, if it exists, is unique. Moreover, it can be shown that the weak convergence in X' implies the weak* convergence, i.e. if $u'_n \rightharpoonup u'$ in X', then $u'_n \rightharpoonup^* u'$ in X'.

Definition 1.35 *A normed space X is said to be* separable *if there exists an enumerable set $\{w_1, w_2, \dots\} \subset X$ such that finite linear combinations of $\{w_i\}$ are dense in X, i.e. any element in X can be approximated by a sequence of finite linear combinations of $\{w_i\}$.*

In the definition for a separable space, any finite number of elements from the set $\{w_i\}$ can be assumed to be linearly independent. We have the following important result on separable Banach spaces.

Theorem 1.36 *If X is a separable Banach space, then any bounded sequence in X' has a weakly * convergent subsequence.*

It follows from the previous theorem that if X is a separable Banach space and the sequence $\{u'_n\} \subset X'$ is such that $\sup_n \|u'_n\|_{X'} < \infty$, then we can find a subsequence $\{u'_{n_k}\} \subset \{u'_n\}$ and an element $u' \in X'$ such $u'_{n_k} \rightharpoonup^* u'$ in X'.

Example 1.37 *Let $\Omega \subset \mathbb{R}^d$ be an open set. It is known that $L^1(\Omega)$ is a separable Banach space and $(L^1(\Omega))' = L^\infty(\Omega)$. Thus from Theorem 1.36, we conclude that if $\{u_n\}$ is a bounded sequence in $L^\infty(\Omega)$, then we can find a subsequence $\{u_{n_k}\}$ and an element $u \in L^\infty(\Omega)$ such that*

$$\int_\Omega u_{n_k}(\boldsymbol{x})\, v(\boldsymbol{x})\, dx \to \int_\Omega u(\boldsymbol{x})\, v(\boldsymbol{x})\, dx \text{ as } k \to \infty, \ \forall\, v \in L^1(\Omega).$$

We will need the concept of compact operators.

Definition 1.38 *Let X and Y be two normed spaces and $T : X \to Y$. The operator T is said to be* compact *if for any bounded sequence $\{v_n\} \subset X$, the sequence $\{T(v_n)\} \subset Y$ has a subsequence converging in Y.*

1.3 Hilbert Spaces

In a large number of applications, inner product spaces are usually used. These are the spaces where a norm can be defined through an inner product and the notion of orthogonality of two elements can be introduced. The inner product in a general space is a generalization of the usual scalar or dot product in the plane $\mathbb{R}^2$ or the space $\mathbb{R}^3$.

Definition 1.39 *Let X be a real linear space. An* inner product $(\cdot,\cdot)_X$ *is a function from $X \times X$ to $\mathbb{R}$ with the following properties.*

1. $(v, v)_X \ge 0\ \forall\, v \in X$, *and* $(v, v)_X = 0$ *iff* $v = 0$.
2. $(u, v)_X = (v, u)_X\ \forall\, u, v \in X$.
3. $(\alpha\, u + \beta\, v, w)_X = \alpha\, (u, w)_X + \beta\, (v, w)_X\ \forall\, u, v, w \in X,\ \forall\, \alpha, \beta \in \mathbb{R}$.

The pair $(X, (\cdot,\cdot)_X)$ is called an inner product space. *When the definition of the inner product $(\cdot,\cdot)_X$ is clear from the context, we simply say X is an inner product space.*

For an inner product, there is an important property called the *Cauchy–Schwartz inequality*:

$$|(u,v)_X| \le \sqrt{(u,u)_X\,(v,v)_X} \quad \forall\, u,v \in X, \tag{1.3}$$

with the equality holding iff u and v being linearly dependent. An inner product $(\cdot,\cdot)_X$ induces a norm through the formula

$$\|v\|_X = \sqrt{(v,v)_X} \quad \forall\, v \in X.$$

Unless stated otherwise, the norm in an inner product space is the one induced by the inner product through the above formula.

Among the inner product spaces, of particular importance are the Hilbert spaces.

Definition 1.40 *A complete inner product space is called a* Hilbert space.

From the definition, we see that an inner product space X is a Hilbert space if X is a Banach space under the norm induced by the inner product.

Example 1.41 *The Euclidean space $\mathbb{R}^d$ is a Hilbert space with the canonical inner product*

$$\boldsymbol{x}\cdot\boldsymbol{y} = x_i y_i \quad \forall\, \boldsymbol{x} = (x_i), \boldsymbol{y} = (y_i) \in \mathbb{R}^d.$$

Note that everywhere in this book we adopt the summation convention on the repeated index. The Euclidean norm is related to the inner product by the relation

$$\|\boldsymbol{x}\|_2 = \sqrt{\boldsymbol{x}\cdot\boldsymbol{x}}.$$

Example 1.42 *The space $\mathbb{R}^{d\times d}$ of the matrices of order d is a Hilbert space with the canonical inner product*

$$\boldsymbol{\sigma}\cdot\boldsymbol{\tau} = \sigma_{ij}\tau_{ij} \quad \forall\, \boldsymbol{\sigma} = (\sigma_{ij})_{d\times d},\ \boldsymbol{\tau} = (\tau_{ij})_{d\times d} \in \mathbb{R}^{d\times d}.$$

The induced norm is the Frobenius norm. In this book, we use the above formula for the inner product of two matrices or two tensors of order two.

With the notion of an inner product at our disposal, we can then define the angle between two vectors as following

$$\theta = \arccos\left[\frac{(u,v)_X}{\|u\|_X\,\|v\|_X}\right].$$

This definition makes sense if $u, v \ne 0$ since the argument of arccos is between -1 and 1 by the Cauchy-Schwartz inequality. The case of a right angle is particularly important.

Definition 1.43 *We say two vectors u and v are* orthogonal *if* $(u, v)_X = 0$. *An element u is said to be orthogonal to a subset $X_1 \subset X$ if $(u, v)_X = 0$ for any $v \in X_1$.*

On Hilbert spaces, continuous linear functionals are limited in the forms they can take. The following theorem makes this more precise. The result is an important tool in developing solvability theory for various equations.

Theorem 1.44 (Riesz representation theorem) *Let X be a Hilbert space and let $\ell \in X'$. Then there is a unique $u \in X$ for which*

$$\ell(v) = (u, v)_X \quad \forall\, v \in X.$$

Moreover,

$$\|\ell\|_{X'} = \|u\|_X.$$

By the Riesz representation theorem we may identify a Hilbert space with its dual. Consequently a Hilbert space can also be identified with its bidual. Therefore, every Hilbert space is reflexive.

We introduce the important concept of monotone operators in inner product spaces.

Definition 1.45 *Let X be a space with inner product $(\cdot, \cdot)_X$ and norm $\|\cdot\|_X$. An operator $A : X \to X$ is said to be a* monotone operator *if*

$$(A(u) - A(v), u - v)_X \geq 0 \quad \forall\, u, v \in X.$$

The operator A is strictly monotone *if*

$$(A(u) - A(v), u - v)_X > 0 \quad \forall\, u, v \in X,\ u \neq v,$$

and strongly monotone *if there exists a constant $m > 0$ such that*

$$(A(u) - A(v), u - v)_X \geq m\, \|u - v\|_X^2 \quad \forall\, u, v \in X.$$

For a linear operator $A : X \to X$, the monotonicity, strict monotonicity and strong monotonicity are characterized by

$$(Av, v)_X \geq 0 \quad \forall\, v \in X,$$
$$(Av, v)_X > 0 \quad \forall\, 0 \neq v \in X$$

and

$$(Av, v)_X \geq m \|v\|_X^2 \quad \forall\, v \in X,$$

respectively. Here $m > 0$ is a constant.

An important class of nonlinear operators defined in a Hilbert space is provided by the following result.

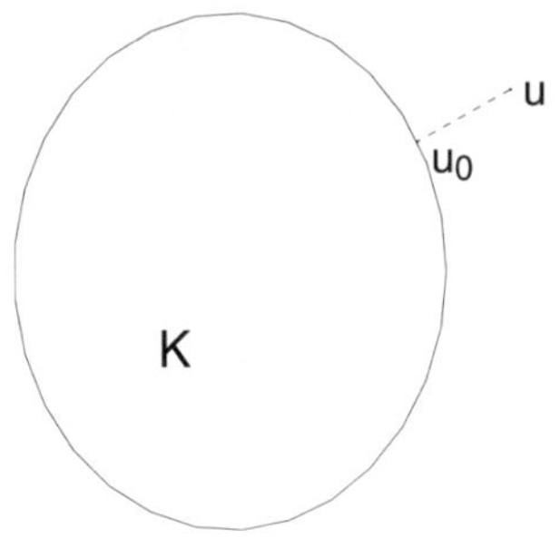

Figure 1.4: Projection on a non-empty closed convex set

Theorem 1.46 (Projection theorem) *Let K be a nonempty closed convex subset in a Hilbert space X. Then for any $u \in X$ there is a unique element $u_0 = \mathcal{P}_K u \in K$ such that*

$$\|u - u_0\|_X = \min_{v \in K} \|u - v\|_X.$$

The operator $\mathcal{P}_K : X \to K$ is called the *projection operator* onto K. The element $u_0 = \mathcal{P}_K u$ is called the *projection* of u on K (see Figure 1.4) and is characterized by the inequality

$$u_0 \in K, \quad (u_0 - u, v - u_0)_X \ge 0 \quad \forall\, v \in K. \tag{1.4}$$

The geometrical interpretation of the characterization is that the angle between the two vectors $u - u_0$ and $v - u_0$ is not smaller than $\pi/2$ (see Figure 1.5). Using inequality (1.4) it is easy to verify that the projection operator is nonexpansive:

$$\|\mathcal{P}_K u - \mathcal{P}_K v\|_X \le \|u - v\|_X \quad \forall\, u, v \in X, \tag{1.5}$$

and monotone:

$$(\mathcal{P}_K u - \mathcal{P}_K v, u - v)_X \ge 0 \quad \forall\, u, v \in X. \tag{1.6}$$

In general, the projection operator $\mathcal{P}_K : X \to K$ is nonlinear. However, if in particular $K = M$ is a closed subspace of X, then $\mathcal{P}_M$ is a linear continuous operator on X and $\|\mathcal{P}_M\|_{\mathcal{L}(X)} = 1$. In this case, every element $u \in X$ can be written as $u = u_1 + u_2$, where $u_1 = \mathcal{P}_M u \in M$ (cf. Figure 1.6), $u_2 = u - \mathcal{P}_M u \in M^\perp$, $M^\perp$ being the orthogonal complement of M defined by

$$M^\perp = \{\, v_2 \in X : (v_1, v_2)_X = 0 \quad \forall\, v_1 \in M \,\}.$$

Since $M \cap M^\perp = \{0\}$, it follows that X is a direct sum of the closed subspace M and its orthogonal complement $M^\perp$: $X = M \oplus M^\perp$.

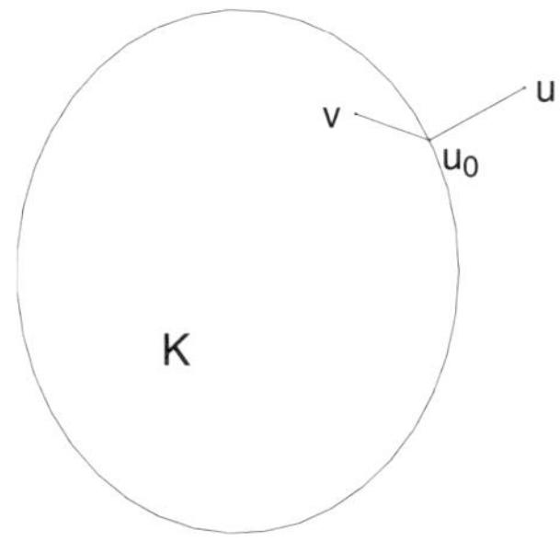

Figure 1.5: Characterization of the projection

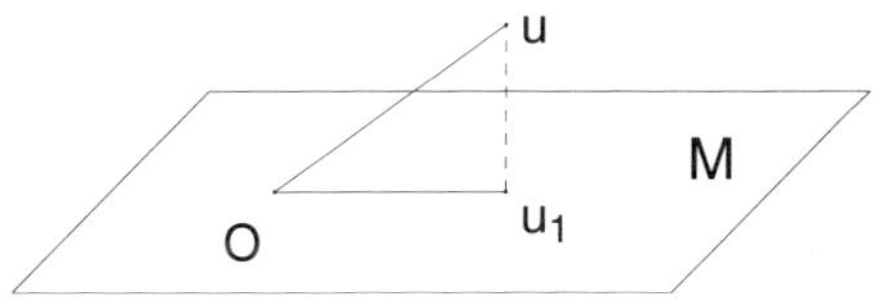

Figure 1.6: Orthogonal projection onto a closed subspace

1.4 Convex Functions

Let X be a linear space. In the study of convex functions, it is convenient to consider functions that take on values on the extended real line $\overline{\mathbb{R}} = \mathbb{R} \cup \{\pm\infty\}$. Here we follow the convention that $\infty + \infty = \infty$ and $(-\infty) + (-\infty) = -\infty$ while an expression of the form $\infty + (-\infty)$ is undefined.

Definition 1.47 *Let $f : X \to \overline{\mathbb{R}}$. The function f is said to be* proper *if $f(v) > -\infty$ for all $v \in X$ and $f(u) < \infty$ for some $u \in X$. The function $f : X \to \overline{\mathbb{R}}$ is* convex *if*

$$f((1-t)\,u + t\,v) \le (1-t)\,f(u) + t\,f(v) \tag{1.7}$$

for any $u, v \in X$ and $t \in (0,1)$ for which the right hand side is meaningful, i.e., $f(u)$ and $f(v)$ are not simultaneously infinite with opposite signs. The function f is strictly convex *if the strict inequality in (1.7) holds for $u \ne v$ and $t \in (0,1)$.*

An inductive argument shows that if f is convex then

$$f\Big(\sum_{i=1}^{n} t_i u_i\Big) \le \sum_{i=1}^{n} t_i f(u_i)$$

for any $u_1, \dots, u_n \in X$ and any nonnegative numbers $t_1, \dots, t_n$ with

$$\sum_{i=1}^{n} t_i = 1.$$

We observe that if f is strictly convex, then $f(v) > -\infty$ for any $v \in X$. Moreover, it follows easily from the definition of a convex function that if $f, g : X \to \overline{\mathbb{R}}$ are convex and $\lambda > 0$ then the functions $f + g$, λf and $\sup\{f, g\}$ are all convex. Here we let $f(v) + g(v) = \infty$ if $f(v) = -g(v) = \pm\infty$.

Given a function $f : X \to \overline{\mathbb{R}}$, we define its *effective domain*

$$\operatorname{dom}(f) = \{ v \in X : f(v) < \infty \}$$

and its *epigraph*

$$\operatorname{epi}(f) = \{ (v, t) \in X \times \mathbb{R} : f(v) \le t \}.$$

It is easy to show that for a convex function the effective domain is a convex set in X and the epigraph is a convex set in $X \times \mathbb{R}$.

From now on, we assume that X is a normed space.

Definition 1.48 *A function $f : X \to \overline{\mathbb{R}}$ is said to be* lower semicontinuous *(l.s.c.) at $u \in X$ if*

$$\liminf_{n\to\infty} f(u_n) \ge f(u) \tag{1.8}$$

for any sequence $\{u_n\} \subset X$ converging to u in X. The function f is l.s.c. on a subset of X if it is l.s.c. at any point of the subset. We say f is l.s.c. if it is l.s.c. on X. When the inequality (1.8) *holds for any sequence $\{u_n\} \subset X$ converging weakly to u, the function f is said to be* weakly lower semicontinuous *(weakly l.s.c.) at u, weakly l.s.c. on the subset, and weakly l.s.c., respectively.*

A continuous function is l.s.c. However, lower semicontinuity does not imply continuity. For example, the function $f : \mathbb{R} \to \mathbb{R}$ given by

$$f(x) = \begin{cases} 0 & \text{if} \quad x \le 0, \\ 1 & \text{if} \quad x > 0 \end{cases}$$

is l.s.c., but it is discontinuous at $x = 0$.

There is a useful characterization of the lower semicontinuity that provides an alternative definition of l.s.c. (weakly l.s.c.) in many references.

Proposition 1.49 *Let $f : X \to \overline{\mathbb{R}}$. Then the following statements are equivalent:*

1. *f is (weakly) lower semicontinuous;*
2. *for any $r \in \mathbb{R}$ the set $\{v \in X : f(v) \le r\}$ is (weakly) closed;*
3. $\operatorname{epi}(f)$ *is (weakly) closed in $X \times \mathbb{R}$.*

Keeping in mind Theorem 1.30 and Proposition 1.49 we deduce the following result.

Corollary 1.50 *A proper convex function $f : X \to \overline{\mathbb{R}}$ is lower semicontinuous if and only if it is weakly lower semicontinuous.*

Example 1.51 (Indicator function) *Let X be a real normed space and let $K \subset X$. Consider the indicator function*

$$I_K(v) = \begin{cases} 0 & \text{if} \quad v \in K, \\ +\infty & \text{if} \quad v \notin K. \end{cases}$$

Using Proposition 1.49 we can prove that the set K is a non-empty closed convex set of X if and only if its indicator function I_K is a proper convex lower semicontinuous function.

By a consequence of the Hahn-Banach theorem, we have the following property of convex lower semicontinuous functions.

Proposition 1.52 *Let X be a normed space and $f : X \to \overline{\mathbb{R}}$ a proper convex lower semicontinuous function. Then f is bounded from below by an affine function, i.e. there exists $u' \in X'$ and $\alpha \in \mathbb{R}$ such that $f(u) \geq \langle u', u \rangle + \alpha$ for all $u \in X$.*

An important property of lower semicontinuous functions is given by the following well-known theorem of the Weierstrass type.

Theorem 1.53 *Let X be a reflexive Banach space and $f : X \to \overline{\mathbb{R}}$ a proper convex lower semicontinuous function such that $f(u) \to +\infty$ when $\|u\|_X \to +\infty$ (i.e. f is coercive). Then f is bounded from below and attains its infimum on X. The set of minimizers of f is closed and is a convex set in X. Moreover, if f is a strictly convex function, then f attains its infimum on X at only one point.*

Now we introduce the definition of subdifferential and subgradient. The notion of subdifferential is useful in describing various mechanical laws including some arising in contact problems.

Definition 1.54 *Let X be a real normed space with the dual X', and $f : X \to \overline{\mathbb{R}}$. Let $u \in X$ be such that $f(u) \neq \pm\infty$. Then the* subdifferential *of f at u is defined to be the set*

$$\partial f(u) = \{\, u' \in X' : \ f(v) \geq f(u) + \langle u', v - u \rangle \quad \forall\, v \in X \,\}.$$

Any $u' \in \partial f(u)$ is called a subgradient *of f at u. The function f is said to be* subdifferentiable *at $u \in X$ if $\partial f(u) \neq \emptyset$. The function f is said to be* subdifferentiable *if it is subdifferentiable at any point $u \in X$.*

We see that if $\partial f(u) \neq \emptyset$, then $f(v) > -\infty$ for any $v \in X$.

Example 1.55 *For a real-valued real-variable function $f : \mathbb{R} \to \mathbb{R}$, its subdifferential at $u \in \mathbb{R}$ is the set of slopes of the straight lines through the point $(u, f(u))$ lying below the curve of f. In particular, if $f'(u)$ exists, then $\partial f(u) = \{f'(u)\}$.*

Example 1.56 *The absolute value function $f(x) = |x|$ is not differentiable at $x = 0$ in the classical sense, yet it is subdifferentiable at $x = 0$ and we have*

$$\partial f(0) = [-1, 1].$$

Example 1.57 (Support functional) *Let X be a real normed space, and let $K \subset X$ be a convex set. Consider the subdifferential of the indicator function $I_K(v)$ defined in Example 1.51. If $u \notin K$, then $\partial I_K(u) = \emptyset$. Assume $u \in K$. Then $u' \in \partial I_K(u)$ iff*

$$I_K(v) \geq \langle u', v - u \rangle \quad \forall\, v \in X,$$

i.e.,

$$\langle u', v - u \rangle \leq 0 \quad \forall\, v \in X.$$

Thus we have the characterization

$$\partial I_K(u) = \{\, u' \in X' : \langle u', v - u \rangle \leq 0 \quad \forall\, v \in K \,\}.$$

Any subgradient $u' \in \partial I_K(u)$ is called a support functional *to K at u. We always have $0 \in \partial I_K(u)$ for $u \in K$. It is easily seen that if $u \in \text{int}\,(K)$, then $\partial I_K(u) = \{0\}$. For $u \in \partial K$ and the case $\text{int}\,(K) \neq \emptyset$, it can be shown the existence of a nonzero subgradient $u' \in \partial I_K(u)$.*

In the previous example and below, by $\text{int}\,(X_1)$ we denote the interior of the set $X_1 \subset X$. More generally, we have the following result.

Proposition 1.58 *Let X be a real normed space and let $f : X \to \overline{\mathbb{R}}$ be a proper convex lower semicontinuous functional. Then $\partial f(u) \neq \emptyset$ for all $u \in \text{int}\,(\text{dom}\,(f))$, i.e. f is subdifferentiable on $\text{int}\,(\text{dom}\,(f))$.*

It is easy to verify the following relations from the definition of the subdifferential:

$$\partial(\lambda\, f(u)) = \lambda\, \partial f(u) \quad \forall\, \lambda > 0,$$
$$\partial(f_1 + f_2)(u) \supset \partial(f_1)(u) + \partial(f_2)(u).$$

We now recall the definition of Gâteaux derivative.

Definition 1.59 *Let $f : X \to \overline{\mathbb{R}}$ and let $u \in X$ such that $f(u) \in \mathbb{R}$. Then f is Gâteaux differentiable at u if there exists $u' \in X'$ such that*

$$\lim_{t \to 0} \frac{f(u + t\, v) - f(u)}{t} = \langle u', v \rangle \quad \forall\, v \in X.$$

The element u' is unique and is called the Gâteaux derivative *of f at u, usually denoted by $f'(u)$.*

The next result provides a relation between the Gâteaux derivative and the subdifferential.

Proposition 1.60 *Let X be a real normed space and $f : X \to \overline{\mathbb{R}}$ be convex. Assume $u \in X$ and $f(u) \in \mathbb{R}$. Then the following statements hold.*

(a) *If $f'(u)$ exists as a Gâteaux derivative, then $\partial f(u) = \{f'(u)\}$.*

(b) *If f is continuous at u and $\partial f(u)$ contains exactly one element, then $f'(u)$ exists as a Gâteaux derivative.*

The notion of Gâteaux derivative can be used to characterize the convexity of Gâteaux-differentiable functionals.

Theorem 1.61 *Let X be a normed space and $K \subset X$ be a non-empty convex subset. Assume $f : X \to \mathbb{R}$ is Gâteaux differentiable. Then the following three statements are equivalent.*

(a) *f is convex.*

(b) $f(v) \geq f(u) + \langle f'(u), v - u\rangle \quad \forall\, u, v \in X$.

(c) $\langle f'(v) - f'(u), v - u\rangle \geq 0 \quad \forall\, u, v \in X$.

1.5 Banach Fixed-point Theorem

Banach fixed-point theorem will be used repeatedly later in this work in proving existence of solutions to variational problems in contact mechanics.

Let X be a Banach space with the norm $\|\cdot\|_X$, and K a subset of X. Let $T : K \to X$ be an operator defined on K. We are interested in the existence of a solution of the operator equation

$$u = T(u), \quad u \in K \tag{1.9}$$

and the possibility of approximating the solution u by the iterative method: $u_0 \in K$ and

$$u_{n+1} = T(u_n), \quad n = 0, 1, \ldots. \tag{1.10}$$

A solution u of the equation (1.9) is called a *fixed-point* of the operator T on K.

Theorem 1.62 (Banach fixed-point theorem) *Let K be a nonempty closed set in a Banach space X. Assume $T : K \to K$ is a contractive mapping with contractivity constant $\alpha \in [0, 1)$, i.e.,*

$$\|T(u) - T(v)\| \leq \alpha\, \|u - v\| \quad \forall\, u, v \in K.$$

Then:

(a) (Existence and uniqueness) *There exists a unique* $u \in K$ *such that* $u = T(u)$.

(b) (Convergence and error estimates of the iteration) *For any* $u_0 \in K$, *the formula* (1.10) *defines a sequence* $\{u_n\} \subset K$ *that converges to* u:

$$\|u_n - u\|_X \to 0 \quad \text{as } n \to \infty.$$

For the error, there hold the estimates

$$\|u_n - u\|_X \le \frac{\alpha^n}{1-\alpha} \|u_0 - u_1\|_X, \tag{1.11}$$

$$\|u_n - u\|_X \le \alpha \|u_{n-1} - u\|_X, \tag{1.12}$$

$$\|u_n - u\|_X \le \frac{\alpha}{1-\alpha} \|u_{n-1} - u_n\|_X. \tag{1.13}$$

Proof. (a) Since $T : K \to K$, the sequence $\{u_n\}$ is well-defined. Let us first prove that $\{u_n\}$ is a Cauchy sequence. Using the contractivity of the mapping T, we have

$$\|u_{n+1} - u_n\|_X \le \alpha \|u_n - u_{n-1}\|_X \le \cdots \le \alpha^n \|u_1 - u_0\|_X.$$

Then for any $m > n \ge 1$,

$$\begin{aligned} \|u_m - u_n\|_X &\le \sum_{j=0}^{m-n-1} \|u_{n+j+1} - u_{n+j}\|_X \\ &\le \sum_{j=0}^{m-n-1} \alpha^{n+j} \|u_1 - u_0\|_X \\ &\le \frac{\alpha^n}{1-\alpha} \|u_1 - u_0\|_X. \end{aligned} \tag{1.14}$$

Since $\alpha \in [0, 1)$, $\|u_m - u_n\|_X \to 0$ as $m, n \to \infty$. Thus $\{u_n\}$ is a Cauchy sequence, and has a limit $u \in K$ since K is a closed set in the Banach space X. We take the limit $n \to \infty$ in $u_{n+1} = T(u_n)$ to see that $u = T(u)$ by the continuity of T, i.e., u is a fixed-point of T.

Suppose $u_1, u_2 \in K$ are fixed-points of T. Then from $u_1 = T(u_1)$ and $u_2 = T(u_2)$, we obtain

$$u_1 - u_2 = T(u_1) - T(u_2).$$

Hence

$$\|u_1 - u_2\|_X = \|T(u_1) - T(u_2)\|_X \le \alpha \|u_1 - u_2\|_X$$

which implies $\|u_1 - u_2\|_X = 0$ since $\alpha \in [0, 1)$. So a fixed-point is unique.

(b) We prove now the error estimates. Let $m \to \infty$ in (1.14), we get the estimate (1.11). From

$$\|u_n - u\|_X = \|T(u_{n-1}) - T(u)\|_X \le \alpha \|u_{n-1} - u\|_X$$

we obtain the estimate (1.12). This estimate together with

$$\|u_{n-1} - u\|_X \le \|u_{n-1} - u_n\|_X + \|u_n - u\|_X$$

implies the estimate (1.13). ■

Later in this book, we also need a variant of the Banach fixed-point theorem. For an operator T, we define its powers inductively by the formula $T^m = T(T^{m-1})$ for $m \ge 2$.

Corollary 1.63 *Assume K is a nonempty closed set in a Banach space X, and that $T : K \to K$. Suppose T^m is a contraction for some positive integer m. Then T has a unique fixed-point in K. Moreover, for any $u_0 \in K$, the iteration method*

$$u_{n+1} = T(u_n), \quad n = 0, 1, 2, \ldots.$$

converges.

Proof. By Theorem 1.62, the mapping T^m has a unique fixed point $u \in K$. From

$$T^m(u) = u,$$

we obtain

$$T^m(T(u)) = T(u).$$

Thus $T(u) \in K$ is also a fixed-point of T^m. Since T^m has a unique fixed point, we must have

$$T(u) = u,$$

i.e. u is a fixed-point of T. Uniqueness of a fixed-point of T follows easily from that of T^m.

For any $j = 0, 1, \ldots, m-1$, the subsequence

$$\{u_{mn+j}\}_{n\ge 0} = \{(T^m)^n(u_j)\}_{n\ge 0}$$

converges to u. Therefore, the whole sequence $\{u_n\}_{n\ge 0}$ converges to u. ■

Chapter 2

Function Spaces and Their Properties

In this chapter we review definitions and properties of some function spaces, including spaces of continuous, continuously differentiable and p-integrable functions, Sobolev spaces, and spaces of vector-valued functions. All of the function spaces used in the book are real spaces. To restrict the chapter to a reasonable length, most of the basic theorems are stated without proofs. The material presented in this chapter will be used throughout the rest of the book.

2.1 The Spaces $C^m(\overline{\Omega})$ and $L^p(\Omega)$

We review definitions of some spaces involving continuous, continuously differentiable and p-integrable functions in this section. Let Ω be an open bounded subset of $\mathbb{R}^d$. We denote by Γ the boundary of Ω, and $\overline{\Omega} = \Omega \cup \Gamma$ the closure of Ω. A typical point in $\mathbb{R}^d$ is denoted by $\boldsymbol{x} = (x_1, \dots, x_d)^T$. It is convenient to use the multi-index notation for partial derivatives. An ordered collection of d non-negative integers, $\alpha = (\alpha_1, \dots, \alpha_d)$, is called a *multi-index*. The quantity $|\alpha| = \sum_{i=1}^d \alpha_i$ is said to be the *length* of α. If v is an m-times differentiable function, then for any α with $|\alpha| \le m$,

$$D^\alpha v(\boldsymbol{x}) = \frac{\partial^{|\alpha|} v(\boldsymbol{x})}{\partial x_1^{\alpha_1} \cdots \partial x_d^{\alpha_d}}$$

is the αth partial derivative. This is a handy notation for partial derivatives. Some examples are

$$\begin{aligned} \frac{\partial v}{\partial x_1} &= D^\alpha v, \quad \alpha = (1, 0, \dots, 0), \\ \frac{\partial^d v}{\partial x_1 \cdots \partial x_d} &= D^\alpha v, \quad \alpha = (1, 1, \dots, 1). \end{aligned}$$

The set of all the partial derivatives of order m of a function v can be expressed as $\{ D^\alpha v : |\alpha| = m \}$. There are other notations commonly used for partial derivatives; e.g., the partial derivative $\partial v / \partial x_i$ is also written as $\partial_{x_i} v$, or $\partial_i v$, or

v_{x_i}, or $v_{,i}$.

Spaces of continuous and continuously differentiable functions. Let $C(\overline{\Omega})$ be the space of functions which are *uniformly continuous* on Ω. Any function in $C(\overline{\Omega})$ is bounded. The notation $C(\overline{\Omega})$ is consistent with the fact that a uniformly continuous function on Ω has a unique continuous extension to $\overline{\Omega}$. For $v \in C(\overline{\Omega})$, we always understand its boundary values on Γ to be the continuous extension of the values of v in Ω. The space $C(\overline{\Omega})$ is a *Banach space* with the norm

$$\|v\|_{C(\overline{\Omega})} = \sup\{\, |v(\boldsymbol{x})| : \boldsymbol{x} \in \Omega\} \equiv \max\{|v(\boldsymbol{x})| : \boldsymbol{x} \in \overline{\Omega}\,\}.$$

Denote $\mathbb{Z}_+$ the set of non-negative integers. For any $m \in \mathbb{Z}_+$, $C^m(\overline{\Omega})$ is the space of functions which, together with their derivatives of order less than or equal to m, are continuous up to the boundary,

$$C^m(\overline{\Omega}) = \{\, v \in C(\overline{\Omega}) : D^\alpha v \in C(\overline{\Omega}) \text{ for } |\alpha| \le m\,\}.$$

When $m = 0$, we usually write $C(\overline{\Omega})$ instead of $C^0(\overline{\Omega})$. The space $C^m(\overline{\Omega})$ is a Banach space with the norm

$$\|v\|_{C^m(\overline{\Omega})} = \sum_{|\alpha| \le m} \|D^\alpha v\|_{C(\overline{\Omega})}.$$

We also set

$$C^\infty(\overline{\Omega}) = \bigcap_{m=0}^{\infty} C^m(\overline{\Omega}) \equiv \{\, v \in C(\overline{\Omega}) : v \in C^m(\overline{\Omega}) \ \ \forall\, m \in \mathbb{Z}_+\,\},$$

the space of infinitely differentiable functions.

Given a function v on Ω, its *support* is defined to be

$$\operatorname{supp} v = \overline{\{\, \boldsymbol{x} \in \Omega : v(\boldsymbol{x}) \ne 0\,\}}.$$

We say that v has a *compact support* if $\operatorname{supp} v$ is a proper subset of Ω: $\operatorname{supp} v \subset \Omega$. Thus, if v has a compact support, then there is a neighboring open strip about the boundary Γ such that v is zero on the part of the strip that lies inside Ω. Later on, we need the space

$$C_0^\infty(\Omega) = \{\, v \in C^\infty(\Omega) : \operatorname{supp} v \subset \Omega\,\}.$$

Obviously, $C_0^\infty(\Omega) \subset C^\infty(\overline{\Omega})$. In case Ω is an interval with $\Omega \supset (-1,1)$, a well-known example of a non-analytic $C_0^\infty(\Omega)$ function is (cf. Figure 2.1)

$$v(x) = \begin{cases} e^{1/(x^2-1)} & \text{if } \ |x| < 1, \\ 0 & \text{otherwise.} \end{cases} \tag{2.1}$$

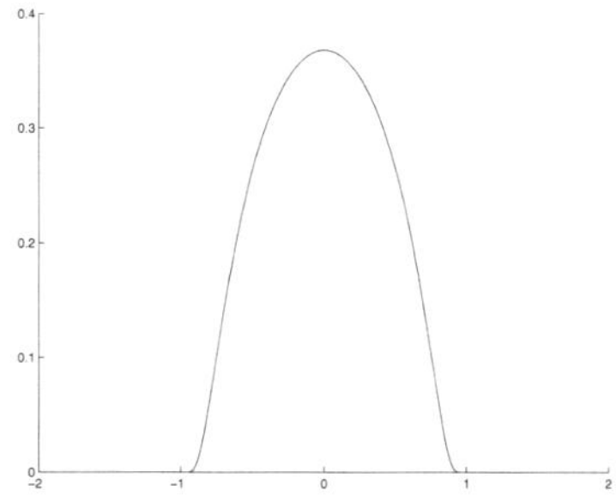

Figure 2.1: The function (2.1)

A function v defined on Ω is said to be *Lipschitz continuous* if for some constant c, there holds the inequality

$$|v(\boldsymbol{x}) - v(\boldsymbol{y})| \le c\,\|\boldsymbol{x}-\boldsymbol{y}\| \quad \forall\, \boldsymbol{x}, \boldsymbol{y} \in \Omega.$$

In this relation, $\|\boldsymbol{x}-\boldsymbol{y}\|$ denotes the standard Euclidean distance between $\boldsymbol{x}$ and $\boldsymbol{y}$: $\|\boldsymbol{x}-\boldsymbol{y}\| = (\sum_{i=1}^d |x_i - y_i|^2)^{1/2}$. The smallest possible constant in the above inequality is called the Lipschitz constant of v, and is denoted by $\mathrm{Lip}(v)$. The Lipschitz constant is characterized by the formula

$$\mathrm{Lip}(v) = \sup\left\{ \frac{|v(\boldsymbol{x}) - v(\boldsymbol{y})|}{\|\boldsymbol{x}-\boldsymbol{y}\|} : \boldsymbol{x}, \boldsymbol{y} \in \Omega,\ \boldsymbol{x} \ne \boldsymbol{y} \right\}.$$

Certainly, a Lipschitz continuous function belongs to $C(\overline{\Omega})$. More generally, a function v is said to be *Hölder continuous* with exponent $\beta \in (0,1]$ if for some constant c,

$$|v(\boldsymbol{x}) - v(\boldsymbol{y})| \le c\,\|\boldsymbol{x}-\boldsymbol{y}\|^\beta \quad \forall\, \boldsymbol{x}, \boldsymbol{y} \in \Omega.$$

The Hölder space $C^{0,\beta}(\overline{\Omega})$ is defined to be the subspace of $C(\overline{\Omega})$ which consists of functions that are Hölder continuous with the exponent β. With the norm

$$\|v\|_{C^{0,\beta}(\overline{\Omega})} = \|v\|_{C(\overline{\Omega})} + \sup\left\{ \frac{|v(\boldsymbol{x}) - v(\boldsymbol{y})|}{\|\boldsymbol{x}-\boldsymbol{y}\|^\beta} : \boldsymbol{x}, \boldsymbol{y} \in \Omega,\ \boldsymbol{x} \ne \boldsymbol{y} \right\}$$

the space $C^{0,\beta}(\overline{\Omega})$ becomes a Banach space. In particular, $C^{0,1}(\overline{\Omega})$ is the space of all the Lipschitz continuous functions on Ω.

For a general nonnegative integer m and $\beta \in (0,1]$, we similarly define the Hölder space

$$C^{m,\beta}(\overline{\Omega}) = \left\{ v \in C^m(\overline{\Omega}) : D^\alpha v \in C^{0,\beta}(\overline{\Omega}) \text{ for all } \alpha \text{ with } |\alpha| = m \right\};$$

this is a Banach space with the norm

$$\|v\|_{C^{m,\beta}(\overline{\Omega})} = \|v\|_{C^m(\overline{\Omega})} + \sum_{|\alpha|=m} \sup\left\{ \frac{|D^\alpha v(\boldsymbol{x}) - D^\alpha v(\boldsymbol{y})|}{\|\boldsymbol{x}-\boldsymbol{y}\|^\beta} : \boldsymbol{x}, \boldsymbol{y} \in \Omega,\ \boldsymbol{x} \ne \boldsymbol{y} \right\}.$$

L^p spaces. In $L^p(\Omega)$ spaces, we identify functions which are equal a.e. on Ω, i.e. two functions are considered identical if they are equal a.e. on Ω. For $p \in [1, \infty)$, $L^p(\Omega)$ is the linear space of Lebesgue measurable functions $v : \Omega \to \mathbb{R}$ such that

$$\|v\|_{L^p(\Omega)} = \left\{ \int_\Omega |v(\boldsymbol{x})|^p dx \right\}^{1/p} < \infty. \tag{2.2}$$

The space $L^\infty(\Omega)$ consists of all essentially bounded measurable functions $v : \Omega \to \mathbb{R}$,

$$\|v\|_{L^\infty(\Omega)} = \operatorname*{ess\,sup}_{\boldsymbol{x} \in \Omega} |v(\boldsymbol{x})| < \infty. \tag{2.3}$$

Some basic properties of the L^p spaces are summarized in the following theorem. For $p \in [1, \infty]$, its conjugate exponent $p' \in [1, \infty]$ is defined by the relations

$$\frac{1}{p} + \frac{1}{p'} = 1 \quad \text{if } p \neq 1,$$
$$p' = \infty \quad \text{if } p = 1.$$

Here, we adopt the convention $1/\infty = 0$ and therefore $p' = 1$ if $p = \infty$.

Theorem 2.1 *Let Ω be an open bounded set in $\mathbb{R}^d$, $p \in [1, \infty]$. Then:*

(a) *$L^p(\Omega)$ is a Banach space.*

(b) *Every Cauchy sequence in $L^p(\Omega)$ has a subsequence which converges pointwise a.e. on Ω.*

(c) (HÖLDER'S INEQUALITY) *Let $u \in L^p(\Omega)$, $v \in L^{p'}(\Omega)$. Then*

$$\int_\Omega |u(\boldsymbol{x})\, v(\boldsymbol{x})|\, dx \le \|u\|_{L^p(\Omega)} \|v\|_{L^{p'}(\Omega)}.$$

In the particular case $p = 2$, we have the Cauchy–Schwartz inequality

$$\int_\Omega |u(\boldsymbol{x})\, v(\boldsymbol{x})|\, dx \le \|u\|_{L^2(\Omega)} \|v\|_{L^2(\Omega)} \quad \forall\, u, v \in L^2(\Omega).$$

(d) *For $1 \le p < \infty$, $(L^p(\Omega))' = L^{p'}(\Omega)$. Hence if $p \in (1, \infty)$, the space $L^p(\Omega)$ is reflexive.*

(e) *For $1 \le p < \infty$, $L^p(\Omega)$ is a separable space.*

A simple consequence of the theorem is the following result.

Corollary 2.2 *The space $L^2(\Omega)$ is a Hilbert space with the inner product*

$$(u, v)_{L^2(\Omega)} = \int_\Omega u(\boldsymbol{x})\, v(\boldsymbol{x})\, dx \quad \forall\, u, v \in L^2(\Omega).$$

Definition 2.3 *Let $1 \le p < \infty$. A function $v : \Omega \subset \mathbb{R}^d \to \mathbb{R}$ is said to be* locally p-integrable, *$v \in L^p_{\text{loc}}(\Omega)$, if for every $\boldsymbol{x} \in \Omega$, there is an open neighborhood Ω' of $\boldsymbol{x}$ such that $\Omega' \subset\subset \Omega$ (i.e., $\overline{\Omega'} \subset \Omega$) and $v \in L^p(\Omega')$.*

We have the following useful result ([218, p. 18]).

Lemma 2.4 (GENERALIZED VARIATIONAL LEMMA) *Let $v \in L^1_{\text{loc}}(\Omega)$ with Ω a nonempty open set in $\mathbb{R}^d$. If*

$$\int_\Omega v(\boldsymbol{x})\, \phi(\boldsymbol{x})\, dx = 0 \quad \forall\, \phi \in C_0^\infty(\Omega),$$

then $v = 0$ a.e. on Ω.

The following theorem of Lebesgue is useful when we localize a global variational relation (for a proof, see e.g., [214, p. 108]).

Theorem 2.5 *Assume $v \in L^1_{\text{loc}}(\Omega)$. Then*

$$\lim_{h \to 0+} \frac{1}{\text{meas}\,(Q_h)} \int_{Q_h} v(\boldsymbol{y})\, dy = v(\boldsymbol{x}) \quad \text{for almost all } \boldsymbol{x} \in \Omega,$$

where Q_h is a sequence of closed sets with $\boldsymbol{x}$ as an interior point such that meas $(Q_h) \to 0$ *as $h \to 0+$.*

2.2 Sobolev Spaces

Sobolev spaces are indispensable tools in the study of boundary value problems. To introduce Sobolev spaces, we first need to extend the definition of derivatives. The starting point is the classical "integration by parts" formula

$$\int_\Omega v(\boldsymbol{x})\, D^\alpha \phi(\boldsymbol{x})\, dx = (-1)^{|\alpha|} \int_\Omega D^\alpha v(\boldsymbol{x})\, \phi(\boldsymbol{x})\, dx \tag{2.4}$$

which holds for $v \in C^k(\Omega)$, $\phi \in C_0^\infty(\Omega)$ and $|\alpha| \le k$. This formula, relating differentiation and integration, is a most important formula in calculus. The *weak derivative* is an extension of the classical derivative, i.e., if the classical derivative exists then the two derivatives coincide. To ensure that the weak derivative is useful, we require that the integration by parts formula (2.4) holds. A more general approach for the extension of the classical derivatives is to first introduce the derivatives in the distributional sense. A detailed discussion of distributions and the derivatives in the distributional sense can be found in several monographs, e.g., [198]. Here we choose to introduce the concept of the weak derivatives directly, which is sufficient for this work.

Definition 2.6 *Let Ω be a nonempty open set in $\mathbb{R}^d$, $v, w \in L^1_{\text{loc}}(\Omega)$. Then w is called an αth* weak derivative *of v if*

$$\int_\Omega v(\boldsymbol{x})\, D^\alpha \phi(\boldsymbol{x})\, dx = (-1)^{|\alpha|} \int_\Omega w(\boldsymbol{x})\, \phi(\boldsymbol{x})\, dx \quad \forall\, \phi \in C_0^\infty(\Omega). \tag{2.5}$$

Applying Lemma 2.4, we see that when a weak derivative exists, it is uniquely defined up to a set of measure zero. Also from the definition of the weak derivative, we see that if v is k-times continuously differentiable on Ω, then for each α with $|\alpha| \le k$, the classical partial derivative $D^\alpha v$ is also the αth weak derivative of v. For this reason, we use the notation $D^\alpha v$ also for the αth weak derivative of v.

Sobolev spaces of integer order. We start with the following definition.

Definition 2.7 *Let k be a nonnegative integer, $p \in [1, \infty]$. The Sobolev space $W^{k,p}(\Omega)$ is the set of all the functions $v \in L^1_{\rm loc}(\Omega)$ such that for each multi-index α with $|\alpha| \le k$, the αth weak derivative $D^\alpha v$ exists and $D^\alpha v \in L^p(\Omega)$. The norm in the space $W^{k,p}(\Omega)$ is defined as*

$$\|v\|_{W^{k,p}(\Omega)} = \begin{cases} \Big(\sum\limits_{|\alpha| \le k} \|D^\alpha v\|^p_{L^p(\Omega)} \Big)^{1/p}, & 1 \le p < \infty, \\ \max\limits_{|\alpha| \le k} \|D^\alpha v\|_{L^\infty(\Omega)}, & p = \infty. \end{cases}$$

When $p = 2$, we write $H^k(\Omega) \equiv W^{k,2}(\Omega)$.

Usually we replace $\|v\|_{W^{k,p}(\Omega)}$ by the simpler notations $\|v\|_{k,p,\Omega}$, or even $\|v\|_{k,p}$ when no confusion will be caused. When $p = 2$, we use $\|v\|_{k,\Omega}$ or $\|v\|_k$ for $\|v\|_{H^k(\Omega)}$. The semi-norm over the space $W^{k,p}(\Omega)$ is

$$|v|_{W^{k,p}(\Omega)} = \begin{cases} \Big(\sum\limits_{|\alpha| = k} \|D^\alpha v\|^p_{L^p(\Omega)} \Big)^{1/p}, & 1 \le p < \infty, \\ \max\limits_{|\alpha| = k} \|D^\alpha v\|_{L^\infty(\Omega)}, & p = \infty. \end{cases}$$

It is not difficult to see that $W^{k,p}(\Omega)$ is a normed space. Moreover, we have the following result.

Theorem 2.8 *The Sobolev space $W^{k,p}(\Omega)$ is a Banach space.*

A simple consequence of the theorem is the following result.

Corollary 2.9 *The Sobolev space $H^k(\Omega)$ is a Hilbert space with the canonical inner product*

$$(u, v)_k = \int_\Omega \sum_{|\alpha| \le k} D^\alpha u(\boldsymbol{x})\, D^\alpha v(\boldsymbol{x})\, dx, \quad u, v \in H^k(\Omega).$$

The closure of the space $C_0^\infty(\Omega)$ with respect to the norm $\|\cdot\|_{k,p}$ gives a closed subspace of $W^{k,p}(\Omega)$, denoted $W_0^{k,p}(\Omega)$. When $p = 2$ we use the notation $H_0^k(\Omega) \equiv W_0^{k,2}(\Omega)$. It follows from Theorem 2.8 and Corollary 2.9 that $W_0^{k,p}(\Omega)$

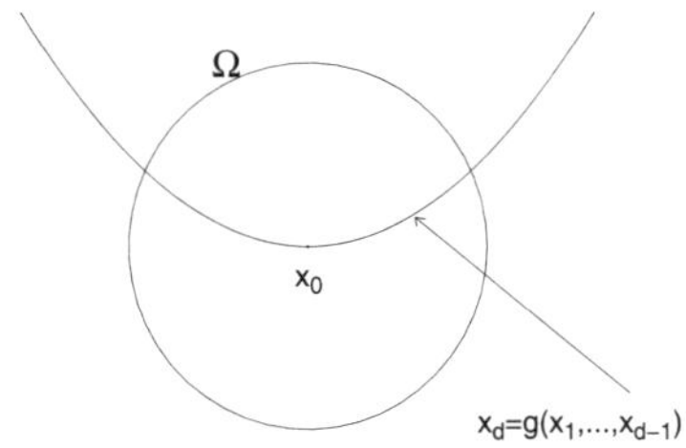

Figure 2.2: Smoothness of the boundary

is a Banach space and $H_0^k(\Omega)$ is a Hilbert space. It can be shown that the semi-norm $|\cdot|_{k,p}$ is a norm on $W_0^{k,p}(\Omega)$ and there exists a constant $c > 0$ such that

$$|v|_{k,p} \le \|v\|_{k,p} \le c\,|v|_{k,p} \quad \forall\, v \in W_0^{k,p}(\Omega).$$

Usually, we will use

$$\|v\|_{W_0^{k,p}(\Omega)} \equiv |v|_{W^{k,p}(\Omega)}$$

as the norm over the space $W_0^{k,p}(\Omega)$. For example,

$$\|v\|^2_{H_0^1(\Omega)} = \int_\Omega |\nabla v|^2 dx$$

is used to define the norm on $H_0^1(\Omega)$.

We now collect some important properties of the Sobolev spaces. Some of them require a certain degree of regularity on the boundary Γ of the domain Ω, that is assumed to be open and bounded.

Definition 2.10 *Let Ω be open and bounded in $\mathbb{R}^d$, and let X denote a space of real valued functions on $\mathbb{R}^{d-1}$. We say Γ is* of class X *if for each point $\boldsymbol{x}_0 \in \Gamma$, there exist an $r > 0$ and a function $g \in X$ such that upon relabelling the coordinate axes if necessary, we have (cf. Figure 2.2)*

$$\Omega \cap B(\boldsymbol{x}_0, r) = \{\, \boldsymbol{x} \in B(\boldsymbol{x}_0, r) : x_d > g(x_1, \ldots, x_{d-1})\,\}.$$

Here, $B(\boldsymbol{x}_0, r)$ denotes the ball centered at $\boldsymbol{x}_0$ with radius r.

In particular, when X consists of Lipschitz continuous functions, we say Ω is a Lipschitz domain. *When X consists of C^m functions, we say Ω is a C^m domain. Finally, when X consists of $C^{m,\beta}$ functions, we say that Ω is a $C^{m,\beta}$ domain.*

Some smooth domains are shown in Figure 2.3. We remark that in engineering applications, most domains are Lipschitz continuous. A well-known non-Lipschitz domain is one with cracks, see Figure 2.4.

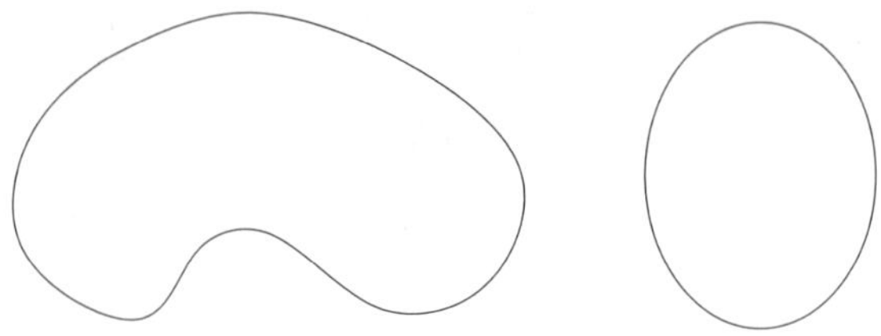

Figure 2.3: Smooth domains

(a) (b)

Figure 2.4: (a) A Lipschitz domain; (b) A non-Lipschitz domain

Since Γ is a compact set in $\mathbb{R}^d$, we can actually find a finite number of points $\{\boldsymbol{x}_i\}_{i=1}^I$ on the boundary, positive numbers $\{r_i\}_{i=1}^I$ and functions $\{g_i\}_{i=1}^I \subset X$ such that Γ is covered by the union of the balls $B(\boldsymbol{x}_i, r_i)$, $1 \le i \le I$, and

$$\Omega \cap B(\boldsymbol{x}_i, r_i) = \{\, \boldsymbol{x} \in B(\boldsymbol{x}_i, r_i) : x_d > g_i(x_1, \dots, x_{d-1}) \,\}$$

upon relabelling the coordinate axes if necessary.

We have the following result on the approximation of Sobolev functions by smooth functions.

Theorem 2.11 *Assume Ω is a Lipschitz domain, $k \in \mathbb{Z}_+$, $1 \le p < \infty$. Then for any $v \in W^{k,p}(\Omega)$, there exists a sequence $\{v_n\} \subset C^\infty(\overline{\Omega})$ such that*

$$\|v_n - v\|_{k,p} \to 0 \quad \text{as } n \to \infty.$$

From the definition of the space $W_0^{k,p}(\Omega)$, we immediately get the following density result.

Theorem 2.12 *For any $v \in W_0^{k,p}(\Omega)$, there exists a sequence $\{v_n\} \subset C_0^\infty(\Omega)$ such that*

$$\|v_n - v\|_{k,p} \to 0 \quad \text{as } n \to \infty.$$

Sobolev embedding theorems are important, e.g., in analyzing the regularity of a weak solution of a boundary value problem.

Definition 2.13 *Let X and Y be two Banach spaces with $X \subset Y$. We say the space X is* continuously embedded *in Y and write $X \hookrightarrow Y$, if there exists $c > 0$ such that*

$$\|v\|_Y \le c\,\|v\|_X \quad \forall\, v \in X. \tag{2.6}$$

If X and Y are function spaces and $X \hookrightarrow Y$, then functions in X are smoother than other functions in Y.

Theorem 2.14 *Let $\Omega \subset \mathbb{R}^d$ be a nonempty open bounded Lipschitz domain. Then the following statements are valid.*

(a) *If $k < \frac{d}{p}$, then $W^{k,p}(\Omega) \hookrightarrow L^q(\Omega)$ for any $q \le p^*$, where $\frac{1}{p^*} = \frac{1}{p} - \frac{k}{d}$.*

(b) *If $k = \frac{d}{p}$, then $W^{k,p}(\Omega) \hookrightarrow L^q(\Omega)$ for any $q < \infty$.*

(c) *If $k > \frac{d}{p}$, then*

$$W^{k,p}(\Omega) \hookrightarrow C^{k-[\frac{d}{p}]-1,\beta}(\Omega),$$

where

$$\beta = \begin{cases} \left[\frac{d}{p}\right] + 1 - \frac{d}{p}, & \text{if } \frac{d}{p} \ne \text{integer}, \\ \text{any positive number } < 1, & \text{if } \frac{d}{p} = \text{integer}. \end{cases}$$

In the theorem, $[x]$ denotes the integer part of x. In this work, we will in particular use the embedding result

$$H^2(\Omega) \hookrightarrow C(\overline{\Omega})$$

when the spatial dimension $d \le 3$. More precisely, for any $v \in H^2(\Omega)$, there is $\tilde{v} \in H^2(\Omega) \cap C(\overline{\Omega})$ such that $\tilde{v} = v$ a.e. in Ω and $\|\tilde{v}\|_{C(\overline{\Omega})} \le c\,\|v\|_{H^2(\Omega)}$ for some constant c independent of v. We identify v with $\tilde{v}$, and still use the symbol v. Thus, in this sense, it is meaningful to use pointwise values of an $H^2(\Omega)$ function.

Sobolev spaces are defined through $L^p(\Omega)$ spaces. Hence Sobolev functions are uniquely defined only a.e. in Ω. Now that the boundary Γ has measure zero in $\mathbb{R}^d$, it seems the boundary value of a Sobolev function is not well-defined. Nevertheless it is possible to define the trace of a Sobolev function on the boundary in such a way that for a Sobolev function that is continuous up to the boundary, its trace coincides with its boundary value.

Theorem 2.15 *Assume Ω is an open bounded Lipschitz domain in $\mathbb{R}^d$ with boundary Γ and $1 \le p < \infty$. Then there exists a continuous linear operator $\gamma : W^{1,p}(\Omega) \to L^p(\Gamma)$ such that $\gamma v = v|_\Gamma$ if $v \in W^{1,p}(\Omega) \cap C(\overline{\Omega})$. Moreover, the mapping $\gamma : W^{1,p}(\Omega) \to L^p(\Gamma)$ is compact, i.e., for any bounded sequence $\{v_n\}$ in $W^{1,p}(\Omega)$, there is a subsequence $\{v_{n'}\} \subset \{v_n\}$ such that $\{\gamma v_{n'}\}$ is convergent in $L^p(\Gamma)$.*

The operator γ is called the *trace operator*, and γv can be termed the *generalized boundary value* of $v \in W^{1,p}(\Omega)$. For simplicity, when no ambiguity occurs, we usually write v instead of γv. Note that the continuity of γ implies that there exists a constant $c > 0$ such that

$$\|\gamma v\|_{L^p(\Gamma)} \le c \, \|v\|_{W^{1,p}(\Omega)} \quad \forall \, v \in W^{1,p}(\Omega). \tag{2.7}$$

Generally, the trace operator is neither an injection nor a surjection from $W^{1,p}(\Omega)$ to $L^p(\Gamma)$ (the only exception is when $p = 1$: the trace operator is surjective from $W^{1,1}(\Omega)$ to $L^1(\Gamma)$). The range $\gamma(W^{1,p}(\Omega))$ is a space smaller than $L^p(\Gamma)$.

In studying boundary value problems, necessarily we need to be able to impose essential boundary conditions properly in formulations. For second-order boundary value problems, essential boundary conditions involve only function values on the boundary, so Theorem 2.15 is enough for the purpose. For higher-order boundary value problems, we need to use the traces of partial derivatives on the boundary. For example, for fourth-order boundary value problems, any boundary conditions involving derivatives of order at most one are treated as essential boundary conditions. Since a tangential derivative of a function on the boundary can be obtained by taking a differentiation of the boundary value of the function, we only need to use traces of a function and its normal derivative.

Let $\boldsymbol{\nu} = (\nu_1, \ldots, \nu_d)^T$ denote the outward unit normal to the boundary Γ of Ω. Recall that if $v \in C^1(\overline{\Omega})$, then its classical normal derivative on the boundary is

$$\frac{\partial v}{\partial \nu} = \sum_{i=1}^{d} \frac{\partial v}{\partial x_i} \nu_i.$$

From Theorem 2.15 it is straightforward to obtain the next result that expresses the fact that the classical normal derivative can be extended to a generalized normal derivative.

Theorem 2.16 *Assume Ω is an open bounded Lipschitz domain in $\mathbb{R}^d$ with boundary Γ and $1 \le p < \infty$. Then there exists a continuous linear operator $\gamma_1 : W^{2,p}(\Omega) \to L^p(\Gamma)$ such that $\gamma_1 v = \partial v / \partial \nu|_\Gamma$ a.e. on Γ when $v \in W^{2,p}(\Omega) \cap C^1(\overline{\Omega})$.*

The negative order Sobolev space $H^{-1}(\Omega)$ is defined as the dual of $H_0^1(\Omega)$. Thus, any $\ell \in H^{-1}(\Omega)$ is a bounded linear functional on $H_0^1(\Omega)$, i.e. there exists $M > 0$ such that

$$|\ell(v)| \le M \, \|v\|_{H_0^1(\Omega)} \quad \forall \, v \in H_0^1(\Omega).$$

The norm of ℓ is

$$\|\ell\|_{H^{-1}(\Omega)} = \sup_{v \in H_0^1(\Omega)} \frac{\ell(v)}{\|v\|_{H_0^1(\Omega)}}.$$

Any function $f \in L^2(\Omega)$ naturally induces a bounded linear functional $f \in H^{-1}(\Omega)$ by the relation

$$\langle f, v\rangle = \int_\Omega f\, v\, dx \quad \forall\, v \in H_0^1(\Omega).$$

Sometimes even when $f \in H^{-1}(\Omega)\backslash L^2(\Omega)$, we write $\int_\Omega f\, v\, dx$ for the duality pairing $\langle f, v\rangle$ between $H^{-1}(\Omega)$ and $H_0^1(\Omega)$.

Sobolev spaces of non-integer order. It is possible to extend the definition of Sobolev spaces with non-negative integer order to any real order. We first introduce Sobolev spaces of positive non-integer order. Assume $p \in [1, \infty)$.

Definition 2.17 *Let $s = k + \sigma$ with $k \geq 0$ an integer and $\sigma \in (0,1)$. We define the Sobolev space*

$$W^{s,p}(\Omega) = \Big\{ v \in W^{k,p}(\Omega) : \frac{|D^\alpha v(\boldsymbol{x}) - D^\alpha v(\boldsymbol{y})|}{\|\boldsymbol{x} - \boldsymbol{y}\|^{\sigma + d/p}} \in L^p(\Omega \times \Omega)\ \ \forall\, \alpha : |\alpha| = k \Big\}$$

with the norm

$$\|v\|_{s,p,\Omega} = \Big(\|v\|_{k,p,\Omega}^p + \sum_{|\alpha|=k} \int_{\Omega\times\Omega} \frac{|D^\alpha v(\boldsymbol{x}) - D^\alpha v(\boldsymbol{y})|^p}{\|\boldsymbol{x} - \boldsymbol{y}\|^{\sigma\, p + d}}\, dx\, dy \Big)^{1/p}.$$

It can be shown that the space $W^{s,p}(\Omega)$ is a Banach space. It is reflexive if and only if $p \in (1, \infty)$. When $p = 2$, $H^s(\Omega) \equiv W^{s,2}(\Omega)$ is a Hilbert space with the inner product

$$(u, v)_{s,\Omega} = (u, v)_{k,\Omega} + \sum_{|\alpha|=k} \int_{\Omega\times\Omega} \frac{(D^\alpha u(\boldsymbol{x}) - D^\alpha u(\boldsymbol{y}))\,(D^\alpha v(\boldsymbol{x}) - D^\alpha v(\boldsymbol{y}))}{\|\boldsymbol{x} - \boldsymbol{y}\|^{2\sigma + d}}\, dx\, dy.$$

Most properties of Sobolev spaces of integer order, such as density of smooth functions and Sobolev embedding theorems discussed above, carry over to Sobolev spaces of positive non-integer order introduced here. The introduction of the spaces $W^{s,p}(\Omega)$ in this work serves two purposes: for the definition of Sobolev spaces over boundaries and for a more precise statement of Sobolev trace theorems. Therefore, we do not give detailed discussions of the properties of the spaces $W^{s,p}(\Omega)$. An interested reader can consult [131, Chapter 4, Part I].

The space $C_0^\infty(\Omega)$ does not need to be dense in $W^{s,p}(\Omega)$. So it makes sense to define $W_0^{s,p}(\Omega)$ to be the closure of the space $C_0^\infty(\Omega)$ in $W^{s,p}(\Omega)$. When $p = 2$ we use the notation $H_0^s(\Omega) \equiv W_0^{s,2}(\Omega)$. Clearly, $W_0^{s,p}(\Omega)$ is a Banach space and $H_0^s(\Omega)$ is a Hilbert space.

With the spaces $W_0^{s,p}(\Omega)$, we can then define Sobolev spaces with negative order.

Definition 2.18 *Let $s \geq 0$, either an integer or a non-integer. Let $p \in [1, \infty)$ and p' be its conjugate exponent. Then we define $W^{-s,p'}(\Omega)$ to be the dual space of $W_0^{s,p}(\Omega)$. In particular, $H^{-s}(\Omega) \equiv W^{-s,2}(\Omega)$.*

The space $H^{\frac{1}{2}}(\Gamma)$. It is possible to define general Sobolev spaces over the boundary Γ. Properties of the Sobolev spaces over boundaries are summarized in [131, Chapter 4, Part I]. In this work, we will only need to use the space $H^{\frac{1}{2}}(\Gamma)$. This space is defined to be the range of the trace operator γ over the space $H^1(\Omega)$. It may be organized as a real Hilbert space with an inner product denoted $(\cdot, \cdot)_{1/2}$.

We have the following result.

Theorem 2.19 *Assume Ω is a bounded Lipschitz domain in $\mathbb{R}^d$ with boundary Γ. Then $\gamma : H^1(\Omega) \to H^{\frac{1}{2}}(\Gamma)$ is a linear, continuous and surjective operator. Moreover, there exists a linear continuous operator $z : H^{\frac{1}{2}}(\Gamma) \to H^1(\Omega)$ such that*

$$\gamma(z(\xi)) = \xi \quad \forall\, \xi \in H^{\frac{1}{2}}(\Gamma).$$

We denote by $H^{-\frac{1}{2}}(\Gamma)$ the dual space of $H^{\frac{1}{2}}(\Gamma)$ and by $\langle \cdot, \cdot \rangle_{1/2}$ the duality pairing mapping between $H^{-\frac{1}{2}}(\Gamma)$ and $H^{\frac{1}{2}}(\Gamma)$. The duality pairing $\langle \cdot, \cdot \rangle_{1/2}$ is an extension of the $L^2(\Gamma)$ inner product: If $\xi' \in L^2(\Gamma)$, then $\xi' \in H^{-\frac{1}{2}}(\Gamma)$ and

$$\langle \xi', \xi \rangle_{1/2} = \int_\Gamma \xi' \xi \, da \quad \forall\, \xi \in H^{\frac{1}{2}}(\Gamma).$$

Finally, we recall that if Ω is an open bounded Lipschitz domain, then the kernel of the trace operator on $H^1(\Omega)$ is $H_0^1(\Omega)$.

In Theorem 2.16, the generalized normal derivative is defined for $W^{2,p}(\Omega)$ functions. This normal derivative can be defined for less regular Sobolev functions. Assume $u \in H^1(\Omega)$ with $\Delta u \in L^2(\Omega)$. Then, given $v \in H^{\frac{1}{2}}(\Gamma)$, for any $w \in H^1(\Omega)$ with $\gamma(w) = v$ on Γ, the value of the quantity

$$\int_\Omega (\Delta u \, w + \nabla u \cdot \nabla w) \, dx$$

depends only on u and v (i.e., it does not depend on the particular choice of the function w) and can be bounded by

$$\left(\|\Delta u\|_{L^2(\Omega)} + \|\nabla u\|_{L^2(\Omega)^d}\right) \|v\|_{H^{\frac{1}{2}}(\Gamma)}.$$

Based on this result, the normal derivative of u is defined as an element of $H^{-\frac{1}{2}}(\Gamma)$ such that

$$\left\langle \frac{\partial u}{\partial \nu}, v \right\rangle_{1/2} = \int_\Omega (\Delta u \, v + \nabla u \cdot \nabla v) \, dx \quad \forall\, v \in H^1(\Omega).$$

Detailed discussion can be found in, e.g., [140].

2.3 Spaces of Vector-valued Functions

In this section we consider vector-valued functions. If it is not specified otherwise, $(X, \|\cdot\|_X)$ will denote a real Banach space. Let $T > 0$. In quasistatic contact problems to be discussed later in this book, $[0, T]$ is the time interval of interest. To simplify the notation, we write $I = (0, T)$ and $\overline{I} = [0, T]$.

The spaces $C^m(\overline{I}; X)$. We define $C(\overline{I}; X)$ to be the space of functions $v : \overline{I} \to X$ that are continuous on the closed interval $\overline{I}$. With the norm

$$\|v\|_{C(\overline{I};X)} = \max_{t\in\overline{I}} \|v(t)\|_X,$$

the space $C(\overline{I}; X)$ becomes a Banach space.

Definition 2.20 *A function $v : \overline{I} \to X$ is said to be (strongly)* differentiable *at $t_0 \in \overline{I}$ if there exists an element in X, denoted as $v'(t_0)$ and called the derivative of v at t_0, such that*

$$\lim_{h\to 0} \left\| \frac{1}{h}(v(t_0 + h) - v(t_0)) - v'(t_0) \right\|_X = 0,$$

where the limit is taken with respect to h with $t_0 + h \in \overline{I}$. The derivative at $t_0 = 0$ is defined as a right-sided limit, and that at $t_0 = T$ as a left-sided limit. The function $v(t)$ is said to be differentiable on $\overline{I}$ if it is differentiable at any $t_0 \in \overline{I}$. It is differentiable a.e. if it is differentiable a.e. on $\overline{I}$. In this case the function v' is called the (strong) derivative *of v. Higher derivatives $v^{(j)}(t_0)$, $j \ge 2$, are defined recursively by $v^{(j)} = (v^{(j-1)})'$. Usually we use the notation $\dot{v}(t_0) = v'(t_0)$ and we understand $v^{(0)}$ to be v.*

For an integer $m \ge 0$, we define the space

$$C^m(\overline{I}; X) = \{\, v \in C(\overline{I}; X) : v^{(j)} \in C(\overline{I}; X),\ j = 1, \dots, m \,\}.$$

This is a Banach space with the norm

$$\|v\|_{C^m(\overline{I};X)} = \sum_{j=0}^{m} \max_{t\in\overline{I}} \|v^{(j)}(t)\|_X.$$

In particular, $C^1(\overline{I}; X)$ denotes the space of continuously differentiable functions on $\overline{I}$ with values in X. This is a Banach space with the norm

$$\|v\|_{C^1(\overline{I};X)} = \max_{t\in\overline{I}} \|v(t)\|_X + \max_{t\in\overline{I}} \|\dot{v}(t)\|_X.$$

We also set

$$C^\infty(\overline{I}; X) = \bigcap_{m=0}^{\infty} C^m(\overline{I}; X) \equiv \{\, v \in C(\overline{I}; X) : v \in C^m(\overline{I}; X)\ \ \forall\, m \in \mathbb{Z}_+ \,\},$$

the space of infinitely differentiable functions defined on $\overline{I}$ with values in X.

The spaces $L^p(I; X)$. Let us first bring in some definitions.

Definition 2.21 *A function $v : \overline{I} \to X$ is said to be (strongly)* measurable *if there exists a sequence $\{v_n\}$ of piecewise constant functions $v_n : \overline{I} \to X$ such that*

$$\lim_{n\to\infty} \|v_n(t) - v(t)\|_X = 0 \quad \text{a.e. } t \in I.$$

The function is Bochner integrable *if there exists a sequence $\{v_n\}$ of piecewise constant functions $v_n : \overline{I} \to X$ such that*

$$\lim_{n\to\infty} \int_I \|v(t) - v_n(t)\|_X dt = 0.$$

When $v : \overline{I} \to X$ is Bochner integrable, its Bochner integral over any measurable set $B \subset \overline{I}$ is defined as

$$\int_B v(t)\, dt = \lim_{n\to\infty} \int_B v_n(t)\, dt \quad \text{in } X.$$

The following result can be proved.

Theorem 2.22 (Bochner) *Let $v : \overline{I} \to X$. Then v is Bochner integrable iff v is measurable and $\int_I \|v(t)\|_X dt < \infty$.*

For $p \in [1, \infty)$, we define $L^p(I; X)$ to be the space of all measurable functions $v : \overline{I} \to X$ such that $\int_I \|v(t)\|_X^p dt < \infty$. With the norm

$$\|v\|_{L^p(I;X)} = \left(\int_I \|v(t)\|_X^p dt \right)^{1/p},$$

the space $L^p(I; X)$ becomes a Banach space. We define $L^\infty(I; X)$ to be the space of all measurable functions $v : \overline{I} \to X$ such that $t \mapsto \|v(t)\|_X$ is essentially bounded on I. The space $L^\infty(I; X)$ is a Banach space with the norm

$$\|v\|_{L^\infty(I;X)} = \operatorname*{ess\,sup}_{t\in I} \|v(t)\|_X.$$

When $(X, (\cdot,\cdot)_X)$ is a Hilbert space, $L^2(I; X)$ is also a Hilbert space with the inner product given by

$$(u, v)_{L^2(I;X)} = \int_I u(t)\, v(t)\, dt.$$

Similar to Theorem 2.5 (page 29), the following result of Lebesgue is useful when we localize a global relation.

Theorem 2.23 *Assume that $v \in L^1(I; X)$. Then*

$$\lim_{h\to 0} \frac{1}{h} \int_{t_0}^{t_0+h} \|v(t) - v(t_0)\|_X\, dt = 0 \quad \text{for almost all } t_0 \in I.$$

We see that Theorem 2.23 implies that if $v \in L^1(I;X)$, then

$$\lim_{h\to 0} \frac{1}{h} \int_{t_0}^{t_0+h} v(t)\,dt = v(t_0) \quad \text{for almost all } t_0 \in I.$$

Here the limit is understood in the sense of the norm of X, that is,

$$\lim_{h\to 0} \left\| \frac{1}{h} \int_{t_0}^{t_0+h} v(t)\,dt - v(t_0) \right\|_X = 0 \quad \text{for almost all } t_0 \in I.$$

The next result describes the dual of the space $L^p(I;X)$, in the Hilbertian case.

Theorem 2.24 *Let $p \in [1,\infty)$ and let $p' \in (1,\infty]$ be its conjugate. Assume X is a Hilbert space. Then the dual of the space $L^p(I;X)$ is $L^{p'}(I;X)$.*

The duality pairing between the space $L^p(I;X)$ and its dual is given by

$$\langle u', u\rangle = \int_I (u'(t), u(t))_X dt.$$

The spaces $W^{k,p}(I;X)$. We start with the following definitions.

Definition 2.25 *A function $v : \overline{I} \to X$ is said to be* locally-integrable, *$v \in L^1_{\text{loc}}(I;X)$, if for any closed interval $B \subset I$, we have*

$$\int_B \|v(t)\|_X dt < \infty.$$

Definition 2.26 *A function $v : \overline{I} \to X$ is weakly differentiable if $v \in L^1_{\text{loc}}(I;X)$ and there exists $w \in L^1_{\text{loc}}(I;X)$ such that*

$$\int_I v(t)\,\phi'(t)\,dt = -\int_I w(t)\,\phi(t)\,dt \quad \forall\,\phi \in C_0^\infty(I).$$

We write $w = v'$, and call it the weak derivative *of v. Higher order weak derivatives $v^{(j)}$ ($j \geq 2$) are defined recursively by $v^{(j)} = (v^{(j-1)})'$. The first two weak derivatives are usually also denoted as $\dot{v}$, $\ddot{v}$.*

For $k \in \mathbb{N}$ and $1 \leq p \leq \infty$, we introduce the space

$$W^{k,p}(I;X) = \{\, v \in L^p(I;X) : \|v^{(j)}\|_{L^p(I;X)} < \infty \;\forall j \leq k \,\}.$$

When $p < \infty$, we define the norm in the space $W^{k,p}(I;X)$ by

$$\|v\|_{W^{k,p}(I;X)} = \Big(\int_I \sum_{0\leq j\leq k} \|v^{(j)}(t)\|_X^p dt \Big)^{1/p}.$$

When $p = \infty$, the norm is defined as

$$\|v\|_{W^{k,\infty}(I;X)} = \max_{0\le j\le k} \operatorname*{ess\,sup}_{t\in I} \|v^{(j)}\|_X.$$

If X is a Hilbert space and $p = 2$, then the space

$$H^k(I;X) \equiv W^{k,2}(I;X)$$

is a Hilbert space with the inner product

$$(u,v)_{H^k(I;X)} = \sum_{0\le j\le k} \int_I (u^{(j)}(t), v^{(j)}(t))_X \, dt.$$

The following result is needed later.

Lemma 2.27 *Assume X is a Banach space or a closed set of a Banach space. Let $p \in [1,\infty]$. Then $W^{1,p}(I;X) \hookrightarrow C(\overline{I};X)$.*

The next result can help us to understand the nature of the functions from the spaces $W^{k,p}(I;X)$. For this, we recall the definition of absolute continuity.

Definition 2.28 *A function $v : \overline{I} \to X$ is* absolutely continuous *if for any $\varepsilon > 0$, there exists a $\delta = \delta(\varepsilon) > 0$ such that $\sum_n \|v(b_n) - v(a_n)\|_X < \varepsilon$ whenever $\{(a_n,b_n)\}_n \subset \overline{I}$ is a denumerable family of pairwisely disjoint intervals with total length $\sum_n (b_n - a_n) < \delta$.*

We have the following characterization for the space $W^{k,p}(I;X)$.

Theorem 2.29 *Let $k > 0$ be an integer, $p \in [1,\infty]$, and $v \in L^p(I;X)$. Then $v \in W^{k,p}(I;X)$ if and only if $v(t) = w(t)$ a.e. on I for some $w \in C^{k-1}(\overline{I};X)$ such that w, w', ..., $w^{(k-1)}$ are absolutely continuous on $\overline{I}$, the k-th strong derivative $w^{(k)}$ exists a.e. and $w^{(k)} \in L^p(I;X)$.*

Because of Theorem 2.29, we usually simply say $v \in W^{k,p}(I;X)$ iff v, v', ..., $v^{(k-1)}$ are absolutely continuous on $\overline{I}$ and $v^{(k)} \in L^p(I;X)$.

It is well known that every real-valued absolutely continuous function $v : \overline{I} \to \mathbb{R}$ is almost differentiable on I and it can be expressed as the indefinite integral of the derivative. Some simple examples (see for instance [26, p. 15]) show that this fails when v is absolutely continuous from $[0,T]$ to a general Banach space X. However, the following theorem, obtained in [139], holds.

Theorem 2.30 *Let X be a reflexive Banach space and let $v : \overline{I} \to X$ be an absolutely continuous function. Then v is a.e. differentiable on I, its derivative $\dot{v}$ belongs to $L^1(I;X)$ and*

$$v(s) = v(0) + \int_0^s \dot{v}(t)\, dt. \tag{2.8}$$

Using Theorems 2.29 and 2.30 we can show the following results.

Lemma 2.31 *Let X be a reflexive Banach space and let $v : \overline{I} \to X$. Then:*

(a) *v belongs to $W^{1,1}(I;X)$ iff v is an absolutely continuous function on $\overline{I}$ to X;*

(b) *v belongs to $W^{1,\infty}(I;X)$ iff v is a Lipschitz continuous function on $\overline{I}$ to X.*

It is easy to prove the next result.

Lemma 2.32 *Let X be a reflexive Banach space and $u \in W^{1,p}(I;X)$ for some $p \in [1,\infty]$. Then*

$$\|u(t) - u(s)\|_X \le \int_s^t \|\dot{u}(\tau)\|_X d\tau, \quad 0 \le s \le t \le T.$$

We also have

$$\|u(t) - u(s)\|_X^p \le (t-s)^{p-1} \int_s^t \|\dot{u}(\tau)\|_X^p d\tau, \quad 0 \le s \le t \le T,$$

if $p < \infty$, and

$$\|u(t) - u(s)\|_X \le (t-s)\|\dot{u}\|_{L^\infty(0,T;X)}, \quad 0 \le s \le t \le T,$$

if $p = \infty$.

Lemma 2.33 *If $(X, (\cdot,\cdot)_X)$ is a Hilbert space and $u \in H^1(I;X)$, then the real-valued function $t \mapsto \frac{1}{2}\|u(t)\|_X^2$ is absolutely continuous on $\overline{I}$,*

$$\frac{d}{dt}\left(\frac{1}{2}\|u(t)\|_X^2\right) = (\dot{u}(t), u(t))_X \quad \text{a.e. on } I$$

and the following equality holds:

$$\frac{1}{2}\|u(t)\|_X^2 = \frac{1}{2}\|u(0)\|_X^2 + \int_0^t (\dot{u}(s), u(s))_X dt$$

for all $t \in \overline{I}$.

Let X be a Banach space with the norm $\|\cdot\|_X$, or a Hilbert space with the inner product $(\cdot,\cdot)_X$. For a subset $S \subset X$, we will use the symbols $C^m(\overline{I};S)$ and $W^{k,p}(I;S)$. For example, $C^m(\overline{I};S)$ is the set of the continuous functions $v : \overline{I} \to S$ such that $v^{(j)} : \overline{I} \to S$, $1 \le j \le m$, are continuous. We write

$$\|v\|_{C^m(\overline{I};S)} = \sum_{j=0}^m \max_{t\in\overline{I}} \|v^{(j)}(t)\|_X.$$

Chapter 3

Introduction to Finite Difference and Finite Element Approximations

The finite difference method is a universally applicable numerical method for the solution of differential equations. The finite element method has been the dominant numerical method in computational mechanics in the last several decades. We review the basic notions of the two methods and the theory of standard finite element interpolation error estimation in this chapter.

3.1 Finite Difference Approximations

The derivatives (in the classical sense) are defined through limits of divided differences. The basic idea of the finite difference approximation is to approximate derivatives by appropriate difference quotients. There are a variety of ways to do the approximation.

Suppose f is a differentiable real-valued function on $\mathbb{R}$ and $t \in \mathbb{R}$. We are interested in approximating the derivative $f'(t)$ by combinations of values of f. For this purpose, choose a small parameter $k > 0$, called the *step-size*. Then we have the following three popular difference approximations:

$$\begin{aligned} f'(t) &\approx \frac{f(t+k) - f(t)}{k} \\ &\quad \frac{f(t) - f(t-k)}{k} \\ &\quad \frac{f(t+k) - f(t-k)}{2k}. \end{aligned}$$

These differences are called a *forward difference*, a *backward difference* and a *centered difference*, respectively. Supposing f has a bounded second derivative, it is easy to verify, using Taylor's theorem, that the approximation errors for all the three differences are $O(k)$. If the third derivative of f exists and is bounded, then the approximation error for the centered difference is $O(k^2)$. We see that if the function is smooth, the centered difference is a more accurate approximation to the derivative.

Let us use these difference formulas to formulate some difference schemes for a sample initial value problem:

$$\dot{u} = f(u,t), \quad t \in \overline{I} = [0,T], \tag{3.1}$$

$$u(0) = u_0. \tag{3.2}$$

Here, $T > 0$, $f : \mathbb{R} \times \overline{I} \to \mathbb{R}$ and $u_0 \in \mathbb{R}$ are given. We assume f is uniformly Lipschitz continuous with respect to its first argument, i.e. for some constant $c > 0$,

$$|f(u,t) - f(v,t)| \le c\,|u - v| \quad \forall\, u, v \in \mathbb{R},\ \forall\, t \in \overline{I}.$$

Under this assumption, the initial value problem has a unique solution $u : \overline{I} \to \mathbb{R}$. To develop finite difference schemes, we first introduce a partition of the interval $\overline{I}$: $0 = t_0 < t_1 < \cdots < t_N = T$, and denote $k_n = t_n - t_{n-1}$ for $n = 1, \ldots, N$, $k = \max_{1\le n\le N} k_n$. The points t_n, $n = 0, 1, \ldots, N$, are called the node points, and k_n, $n = 0, 1, \ldots, N$, are the step-sizes. We will use the notation $u_n = u(t_n)$, $n = 0, 1, \ldots, N$. A difference scheme is to find a sequence $\boldsymbol{u}^k = \{u_n^k\}_{n=0}^N$ such that

$$u_0^k = u_0$$

and a system of difference equations is satisfied. We have many choices for the difference scheme. The *forward scheme* is

$$\frac{u_n^k - u_{n-1}^k}{k_n} = f(u_{n-1}^k, t_{n-1}), \quad n = 1, \ldots, N. \tag{3.3}$$

The *backward scheme* is

$$\frac{u_n^k - u_{n-1}^k}{k_n} = f(u_n^k, t_n), \quad n = 1, \ldots, N. \tag{3.4}$$

These schemes are first order approximations of the differential equation (3.1) since the divided difference $(u_n^k - u_{n-1}^k)/k_n$ is a first order approximation of the derivative $\dot{u}(t_n)$ and $\dot{u}(t_{n-1})$, respectively.

We can also consider the differential equation at a half grid point $t = t_{n-1/2} = (t_n + t_{n-1})/2$ and approximate the derivative $\dot{u}(t_{n-1/2})$ by a centered difference:

$$\dot{u}(t_{n-1/2}) \approx \frac{u_n - u_{n-1}}{k_n}.$$

As a result we obtain the *Crank-Nicolson scheme*:

$$\frac{u_n^k - u_{n-1}^k}{k_n} = f\Big(\frac{u_n^k + u_{n-1}^k}{2}, t_{n-1/2}\Big), \quad n = 1, \ldots, N. \tag{3.5}$$

This scheme is of second order since the divided difference $(u_n^k - u_{n-1}^k)/k_n$ is a second order approximation of the derivative $\dot{u}(t_{n-1})$. A variant of the Crank-Nicolson scheme is

$$\frac{u_n^k - u_{n-1}^k}{k_n} = \frac{1}{2}\left(f(u_n^k, t_n) + f(u_{n-1}^k, t_{n-1})\right), \quad n = 1, \dots, N, \tag{3.6}$$

which is also a second order approximation.

The schemes (3.3), (3.4) and (3.5) are special cases of a *generalized mid-point scheme*:

$$\frac{u_n^k - u_{n-1}^k}{k_n} = f(\theta\, u_n^k + (1-\theta)\, u_{n-1}^k, t_{n-1+\theta}), \quad n = 1, \dots, N. \tag{3.7}$$

Here $\theta \in [0,1]$ and $t_{n-1+\theta} = \theta\, t_n + (1-\theta)\, t_{n-1}$. When $\theta = 0, 1, 1/2$, the generalized mid-point scheme reduces to (3.3), (3.4) and (3.5), respectively. The scheme (3.7) is of second order when $\theta = 1/2$, and is of first order for all other choices of θ.

A variant of the generalized mid-point scheme is

$$\frac{u_n^k - u_{n-1}^k}{k_n} = \theta\, f(u_n^k, t_n) + (1-\theta)\, f(u_{n-1}^k, t_{n-1}), \quad n = 1, \dots, N. \tag{3.8}$$

This scheme includes the schemes (3.3), (3.4) and (3.6) as special cases.

The forward scheme (3.3) is an example of an explicit method since we can compute the approximate value u_n^k directly from (3.3). In contrast, the generalized mid-point scheme with $\theta \neq 0$ is called an implicit scheme, since at each step, we need to solve an algebraic equation to get u_n^k. For the implicit schemes, with an application of the Banach fixed-point theorem, it is not difficult to show that the value u_n^k exists and is unique, as long as the (maximal) step-size k is sufficiently small.

In this work, we use finite differences to approximate the time derivative.

3.2 Basis of the Finite Element Approximation

In contrast to the finite difference method of approximating derivatives directly by finite differences, the finite element method for solving a differential equation problem is based on a weak formulation of the problem and the use of finite element approximations. There are some basic aspects in the construction of finite element approximations. First we need a partition (or triangulation) of the domain of the differential equation into sub-domains called elements. Associated with the partition, we define a finite element space. Finally, we choose basis functions for the finite element space. The basis functions should have small supports. For simplicity, we assume Ω is a polygonal (polyhedral) domain. The reader is referred to [47, 49] for some detailed discussion on the use of the finite element method when Ω is not polygonal.

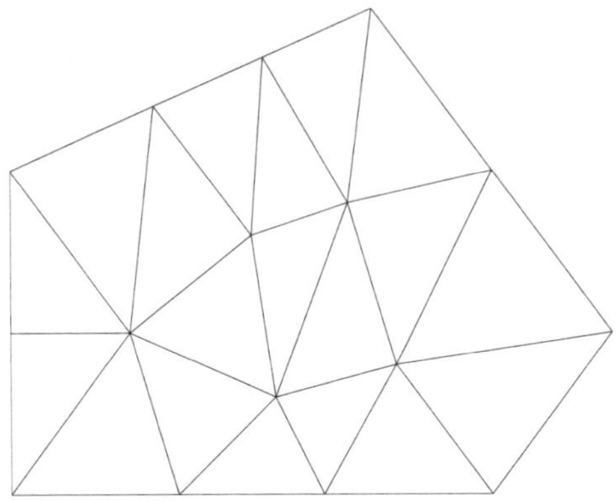

Figure 3.0: A finite element mesh

Triangulation. A *triangulation*, *partition*, or *mesh* is a partition $\mathcal{T}_h = \{K\}$ of the polygonal domain $\overline{\Omega}$ into a finite number of subsets K, called *elements*, with the following properties:

1. $\overline{\Omega} = \cup_{K \in \mathcal{T}_h} K$.

2. Each K is a polyhedron with a nonempty interior $\overset{\circ}{K}$.

3. For distinct $K_1, K_2 \in \mathcal{T}_h$, $\overset{\circ}{K}_1 \cap \overset{\circ}{K}_2 = \emptyset$.

4. For distinct $K_1, K_2 \in \mathcal{T}_h$, $K_1 \cap K_2$ is either empty, or a common vertex, a common side or a common face of K_1 and K_2.

5. $\text{diam}\,(K) \leq h$ for all $K \in \mathcal{T}_h$.

From now on, we focus our discussion to two-dimensional domains; most of the discussion can be extended in a straightforward fashion to higher-dimensional domains. We assume the domain Ω is a polygon so that it can be partitioned into straight-sided triangles and quadrilaterals. For convenience in practical implementation as well as in theoretical analysis, it is assumed that there exist a finite number of fixed polyhedra, called *reference elements* or *master elements*, ambiguously represented by one symbol $\hat{K}$, such that for each element K, there is an invertible affine mapping function F_K with $K = F_K(\hat{K})$. In the two-dimensional case, it is customary to use an equilateral or right isosceles triangle for triangular elements, and squares with side length 1 or 2 for quadrilateral elements.

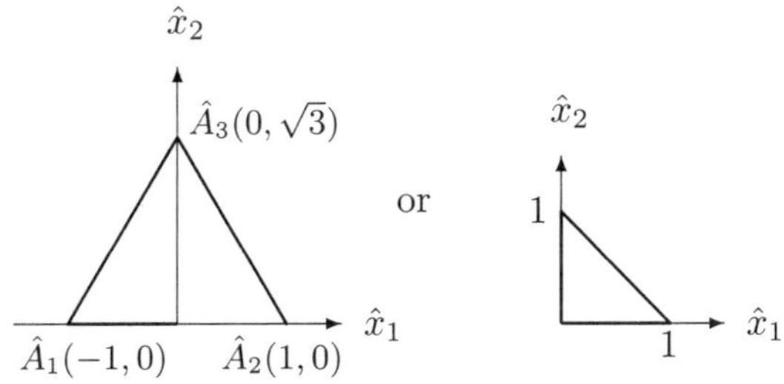

Figure 3.1. Reference triangular elements in $\mathbb{R}^2$

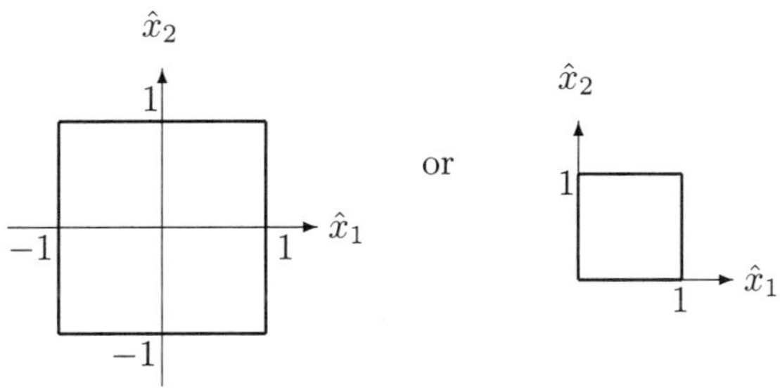

Figure 3.2. Reference rectangular elements in $\mathbb{R}^2$

For an arbitrary element K, we denote

$$h_K = \operatorname{diam}(K) = \max\{\, \|\boldsymbol{x} - \boldsymbol{y}\| : \boldsymbol{x}, \boldsymbol{y} \in K \,\}$$

and

$$\rho_K = \text{diameter of the largest sphere } S_K \text{ inscribed in } K.$$

When dealing with the reference element $\hat{K}$ we denote the corresponding quantities by $\hat{h}$ and $\hat{\rho}$. The quantity h_K describes the size of K, while the ratio h_K/ρ_K is an indication whether the element is flat: the larger the ratio h_K/ρ_K, the flatter the element K. For a triangulation $\mathcal{T}_h$, we denote

$$h = \max_{K \in \mathcal{T}_h} h_K, \tag{3.9}$$

often called the *mesh parameter* . The quantity h is a measure of how refined the mesh is. The smaller h is, the finer the mesh. We say a family of triangulations $\{\mathcal{T}_h\}_h$ is *regular* if

1. there exists a constant $\sigma \geq 1$ such that

$$\frac{h_K}{\rho_K} \leq \sigma \quad \forall\, K \in \mathcal{T}_h,\ \forall\, h;$$

2. the discretization parameter $h \to 0$.

A condition equivalent to the first requirement is the minimal angle condition, i.e. the minimal angles of all the elements are bounded below away from 0.

Polynomial spaces on the reference elements. Function spaces over a general element can be constructed from those on the reference element. Thus we first introduce a polynomial space $\hat{X}$ on $\hat{K}$. Although it is possible to choose any finite dimensional function space as $\hat{X}$, the overwhelming choice for $\hat{X}$ for practical use is a polynomial space, due to the simplicity in constructing a polynomial and the good approximation property of polynomial spaces.

Example 3.1 *For definiteness, we choose the equilateral triangle $\hat{K}$ to be the reference triangle with the vertices $\hat{A}_1(-1,0)$, $\hat{A}_2(1,0)$ and $\hat{A}_3(0,\sqrt{3})$. It is convenient to use the* barycentric coordinates $\hat{\lambda}_i$, $i = 1,2,3$, *associated with the triangle $\hat{K}$. These functions are linear and satisfy the relations*

$$\hat{\lambda}_i(\hat{A}_j) = \delta_{ij}.$$

It is easy to verify that

$$\begin{aligned}
\hat{\lambda}_1(\hat{\boldsymbol{x}}) &= \frac{1}{2}\left(1 - \hat{x}_1 - \frac{\hat{x}_2}{\sqrt{3}}\right),\\
\hat{\lambda}_2(\hat{\boldsymbol{x}}) &= \frac{1}{2}\left(1 + \hat{x}_1 - \frac{\hat{x}_2}{\sqrt{3}}\right),\\
\hat{\lambda}_3(\hat{\boldsymbol{x}}) &= \frac{\hat{x}_2}{\sqrt{3}}.
\end{aligned}$$

Then for any linear function $\hat{v} \in \mathcal{P}_1(\hat{K})$, we have the formula

$$\hat{v}(\hat{\boldsymbol{x}}) = \sum_{i=1}^{3} \hat{v}(\hat{A}_i)\,\hat{\lambda}_i(\hat{\boldsymbol{x}}).$$

Because of this formula, the vertices $\{\hat{A}_i\}_{i=1}^3$ are called the nodes, the function values $\{\hat{v}(\hat{A}_i)\}_{i=1}^3$ are the parameters used to determine the linear function, and $\{\hat{\lambda}_i\}_{i=1}^3$ are the basis functions of the space $\mathcal{P}_1(\hat{K})$.

Now consider the space of quadratic functions on $\hat{K}$. A quadratic function has six coefficients and we need six interpolation conditions to determine it. For this, we can use the function values at the three vertices $\{\hat{A}_i\}_{i=1}^3$ and the three side mid-points,

$$\hat{A}_{ij} = \frac{1}{2}\,(\hat{A}_i + \hat{A}_j), \quad 1 \le i < j \le 3.$$

Then any quadratic function $\hat{v} \in \mathcal{P}_2(\hat{K})$ can be uniquely expressed as

$$\hat{v}(\hat{\boldsymbol{x}}) = \sum_{i=1}^{3} \hat{v}(\hat{A}_i)\,\hat{\lambda}_i(\hat{\boldsymbol{x}})\,(2\,\hat{\lambda}_i(\hat{\boldsymbol{x}}) - 1) + \sum_{1\le i<j\le 3} 4\,\hat{v}(\hat{A}_{ij})\,\hat{\lambda}_i(\hat{\boldsymbol{x}})\,\hat{\lambda}_j(\hat{\boldsymbol{x}}).$$

In this case, the vertices and the side mid-points are called the nodes, the function values at the nodes are the parameters used to determine the quadratic function.

In the above example, the parameters are function values at the nodes. The corresponding finite elements to be constructed later are called Lagrange finite elements. Lagrange finite elements are the natural choice in solving second-order boundary value problems. We focus on the discussion of Lagrange finite elements only.

Generally, let $\hat{X}$ be a polynomial space over $\hat{K}$, $\dim \hat{X} = N$, and we choose a set of nodal points $\{\hat{\boldsymbol{x}}_i\}_{i=1}^N$ in $\hat{K}$ such that any function $\hat{v} \in \hat{X}$ is uniquely determined by its values at the nodes $\{\hat{\boldsymbol{x}}_i\}_{i=1}^N$. Then we have the formula

$$\hat{v}(\hat{\boldsymbol{x}}) = \sum_{i=1}^{N} \hat{v}(\hat{\boldsymbol{x}}_i)\, \hat{\phi}_i(\hat{\boldsymbol{x}}),$$

where the functions $\{\hat{\phi}_i\}_{i=1}^N$ form a basis for the space $\hat{X}$ with the property

$$\hat{\phi}_i(\hat{\boldsymbol{x}}_j) = \delta_{ij}, \quad 1 \le i, j \le N.$$

Affine-equivalent finite elements. We then define a function space over a general element K that is the image of the reference element $\hat{K}$ under an invertible affine mapping

$$F_K : \hat{K} \to K, \quad F_K(\hat{\boldsymbol{x}}) = \boldsymbol{T}_K \hat{\boldsymbol{x}} + \boldsymbol{b}_K. \tag{3.10}$$

The mapping F_K is a bijection between $\hat{K}$ and K.

Over the element K, we define a finite dimensional function space X_K by

$$X_K = \hat{X} \circ F_K^{-1} \equiv \{\, v : v = \hat{v} \circ F_K^{-1},\ \hat{v} \in \hat{X}\,\}. \tag{3.11}$$

Since F_K is an invertible affine mapping, if $\hat{X}$ is a polynomial space of certain degree, then X_K is a polynomial space of the same degree. Any function v defined on K is associated with a function $\hat{v}$ defined on $\hat{K}$ through $\hat{v} = v \circ F_K$. We see that $v = \hat{v} \circ F_K^{-1}$. Thus we have the relation

$$v(\boldsymbol{x}) = \hat{v}(\hat{\boldsymbol{x}}) \quad \forall\, \boldsymbol{x} \in K,\ \hat{\boldsymbol{x}} \in \hat{K},\ \text{with } \boldsymbol{x} = F_K(\hat{\boldsymbol{x}}).$$

Using the nodal points $\hat{\boldsymbol{x}}_i$, $1 \le i \le N$, of $\hat{K}$, we can define the nodal points on K:

$$\boldsymbol{x}_i^K = F_K(\hat{\boldsymbol{x}}_i), \quad i = 1, \dots, N. \tag{3.12}$$

Recall that $\{\hat{\phi}_i\}_{i=1}^N$ are the basis functions of the space $\hat{X}$ associated with the nodal points $\{\hat{\boldsymbol{x}}_i\}_{i=1}^N$ with the property that

$$\hat{\phi}_i(\hat{\boldsymbol{x}}_j) = \delta_{ij}, \quad 1 \le i, j \le N.$$

We define

$$\phi_i^K = \hat{\phi}_i \circ F_K^{-1}, \quad i = 1, \dots, N.$$

Then the functions $\{\phi_i^K\}_{i=1}^N$ have the property that

$$\phi_i^K(\boldsymbol{x}_j^K) = \delta_{ij}, \quad 1 \le i, j \le N.$$

Hence, $\{\phi_i^K\}_{i=1}^N$ form a set of basis functions for the space X_K.

The next result on the affine transformation (3.10) plays an important role in estimating finite element interpolation errors. The matrix norm used here is the spectral norm, i.e. the operator matrix norm induced by the Euclidean vector norm.

Lemma 3.2 *For the invertible affine map $F_K : \hat{K} \to K$ of (3.10), we have the estimates*

$$\|\boldsymbol{T}_K\| \le \frac{h_K}{\hat{\rho}} \quad \text{and} \quad \|\boldsymbol{T}_K^{-1}\| \le \frac{\hat{h}}{\rho_K}.$$

Finite element spaces. A global finite element function v^h is defined piecewise by the formula

$$v^h|_K \in X_K \quad \forall\, K \in \mathcal{T}_h.$$

A natural question is whether such finite element functions can be used to solve a boundary value problem. For a linear second-order elliptic boundary value problem, the space V for the corresponding weak formulation is a subspace of $H^1(\Omega)$. Since the restriction of v^h on each element K is a smooth function, a necessary and sufficient condition for $v^h \in H^1(\Omega)$ is $v^h \in C(\overline{\Omega})$ (cf. [14]). Thus we need to check if $v^h \in C(\overline{\Omega})$ holds. We then define a finite element space corresponding to the triangulation $\mathcal{T}_h$,

$$X^h = \{\, v^h \in C(\overline{\Omega}) : v^h|_K \in X_K \ \forall\, K \in \mathcal{T}_h \,\}.$$

We observe that if $\hat{X}$ consists of polynomials, then a function from the space X^h is a piecewise image of polynomials. In our special case of an affine family of finite elements, F_K is an affine mapping, and $v^h|_K$ is a polynomial.

We can use the finite element space X^h to approximate the space $H^1(\Omega)$. Some boundary value problems involve essential boundary conditions. Then we need finite element sets to approximate subsets of $H^1(\Omega)$. As an example, suppose $\Gamma_1 \subset \Gamma$ is a relatively closed subset of the boundary Γ such that it is the union of some polyhedra. Consider the space

$$V = \{\, v \in H^1(\Omega) : v = 0 \text{ a.e. on } \Gamma_1 \,\}.$$

We construct finite element partitions of Ω that are compatible to the subset Γ_1, i.e., if K is a finite element, then either $K \cap \Gamma_1 = \emptyset$ or $K \cap \Gamma \subset \Gamma_1$. Then we define the finite element space to be $V^h = V \cap X^h$. In other words,

$$V^h = \{\, v^h \in X^h : v^h = 0 \text{ on } \Gamma_1 \,\}.$$

For an affine family of finite elements, the mesh is completely determined by the reference element $\hat{K}$ and the family of affine mappings $\{F_K : K \in \mathcal{T}_h\}$.

The finite element space X^h is completely described by the space $\hat{X}$ on $\hat{K}$ and the family of affine mappings.

Local interpolation. We first introduce an interpolation operator $\hat{\Pi}$ for continuous functions on $\hat{K}$. Recall that $\{\hat{\boldsymbol{x}}_i\}_{i=1}^N$ are the nodal points while $\{\hat{\phi}_i\}_{i=1}^N$ are the associated basis functions of the polynomial space $\hat{X}$ in the sense that $\hat{\phi}_i(\hat{\boldsymbol{x}}_j) = \delta_{ij}$ is valid. We define

$$\hat{\Pi} : C(\hat{K}) \to \hat{X}, \quad \hat{\Pi}\hat{v} = \sum_{i=1}^N \hat{v}(\hat{\boldsymbol{x}}_i)\hat{\phi}_i. \tag{3.13}$$

Evidently, $\hat{\Pi}\hat{v} \in \hat{X}$ is uniquely determined by the interpolation conditions

$$\hat{\Pi}\hat{v}(\hat{\boldsymbol{x}}_i) = \hat{v}(\hat{\boldsymbol{x}}_i), \quad i = 1, \ldots, N.$$

On any element K, we define similarly the interpolation operator Π_K by

$$\Pi_K : C(K) \to X_K, \quad \Pi_K v = \sum_{i=1}^N v(\boldsymbol{x}_i^K)\phi_i^K. \tag{3.14}$$

We see that $\Pi_K v \in X_K$ is uniquely determined by the interpolation conditions

$$\Pi_K v(\boldsymbol{x}_i^K) = v(\boldsymbol{x}_i^K), \quad i = 1, \ldots, N.$$

The following result explores the relation between the two interpolation operators. The result is of fundamental importance in error analysis of finite element interpolation.

Theorem 3.3 *For the above interpolation operators $\hat{\Pi}$ and Π_K, we have $\hat{\Pi}(\hat{v}) = (\Pi_K v) \circ F_K^{-1}$, i.e., $\hat{\Pi}\hat{v} = \widehat{\Pi_K v}$.*

Example 3.4 *Let us consider a planar polygonal domain Ω which is partitioned into triangles $\{K\}$. For a generic element K, again use $\boldsymbol{a}_1$, $\boldsymbol{a}_2$ and $\boldsymbol{a}_3$ to denote its vertices. Then with linear elements, for a continuous function v defined on K, its linear interpolant is*

$$\Pi_K v(\boldsymbol{x}) = \sum_{i=1}^3 v(\boldsymbol{a}_i)\,\lambda_i(\boldsymbol{x}), \quad \boldsymbol{x} \in K.$$

For the function $\hat{v} = v \circ F_K^{-1}$ defined on the reference element $\hat{K}$, its linear interpolant is

$$\hat{\Pi}\hat{v}(\hat{\boldsymbol{x}}) = \sum_{i=1}^3 \hat{v}(\hat{\boldsymbol{a}}_i)\,\hat{\lambda}_i(\hat{\boldsymbol{x}}), \quad \hat{\boldsymbol{x}} \in \hat{K}.$$

Here, $\hat{\boldsymbol{a}}_i = F_K^{-1}(\boldsymbol{a}_i)$, $1 \le i \le 3$, are the vertices of $\hat{K}$. Since $v(\boldsymbol{a}_i) = \hat{v}(\hat{\boldsymbol{a}}_i)$ by definition and $\lambda_i(\boldsymbol{x}) = \hat{\lambda}_i(\hat{\boldsymbol{x}})$ for $\boldsymbol{x} = F_K(\hat{\boldsymbol{x}})$, obviously the relation $\hat{\Pi}\hat{v} = \widehat{\Pi_K v}$ holds.

By using the mid-points of the sides, we can give a similar discussion of quadratic elements.

Global interpolation. For a function $v \in C(\overline{\Omega})$, we construct its global interpolant $\Pi^h v$ in the finite element space X^h by the formula

$$\Pi^h v|_K = \Pi_K v \quad \forall\, K \in \mathcal{T}_h.$$

Let $\{\boldsymbol{x}_i\}_{i=1}^{N_h} \subset \overline{\Omega}$ be the set of the nodes collected from the nodes of all the elements $K \in \mathcal{T}_h$. We have the representation formula

$$\Pi^h v = \sum_{i=1}^{N_h} v(\boldsymbol{x}_i)\phi_i \tag{3.15}$$

for the global finite element interpolant. Here ϕ_i, $i = 1, \dots, N_h$, are the global basis functions that span X^h. The basis function ϕ_i is associated with the node $\boldsymbol{x}_i$, i.e., ϕ_i is a piecewise polynomial, $\phi_i|_K \in X_K$, and $\phi_i(\boldsymbol{x}_j) = \delta_{ij}$. If the node $\boldsymbol{x}_i$ is a vertex $\boldsymbol{x}_l^K$ of the element K, then $\phi_i|_K = \phi_l^K$. If $\boldsymbol{x}_i$ is not a node of K, then $\phi_i|_K = 0$. Thus the functions ϕ_i are constructed from local basis functions ϕ_i^K.

Example 3.5 *We examine an example of linear elements. Assume $\Omega \subset \mathbb{R}^2$ is a polygonal domain, which is triangulated into triangles K, $K \in \mathcal{T}^h$. Denote $\{\boldsymbol{x}_i\}_{i=1}^{N_{\text{int}}}$ the set of the interior nodes, i.e. the vertices of the triangulation that lie in Ω; and $\{\boldsymbol{x}_i\}_{i=N_{\text{int}}+1}^{N_h}$ the set of the boundary nodes, i.e. the vertices of the triangulation that lie on $\partial\Omega$. From each vertex $\boldsymbol{x}_i$, denote $\tilde{K}_i$ the patch of the elements K which contain $\boldsymbol{x}_i$ as a vertex. The basis function ϕ_i associated with the node $\boldsymbol{x}_i$ is a continuous function on $\overline{\Omega}$, which is linear on each K and is non-zero only on $\tilde{K}_i$. The corresponding piecewise linear function space is then*

$$X^h = \operatorname{span}\{\, \phi_i,\ 1 \le i \le N_h \,\}.$$

Suppose we need to solve a linear elliptic boundary value problem with Neumann boundary condition. Then the function space is $V = H^1(\Omega)$, and we choose the linear element space to be $V^h = X^h$. Now suppose the homogeneous Dirichlet boundary condition is specified. Then the function space is $V = H_0^1(\Omega)$, while the linear element space is

$$V^h = \operatorname{span}\{\, \phi_i,\ 1 \le i \le N_{\text{int}} \,\}.$$

In other words, we only use those basis functions associated with the interior nodes, so that again the finite element space is a subspace of the space for the boundary value problem. In case of a mixed boundary value problem, where we have a homogeneous Dirichlet boundary condition on a relatively closed part of the boundary Γ_1 and a Neumann boundary condition on the remaining part of the boundary Γ_2, then the triangulation should be compatible with the splitting $\partial\Omega = \Gamma_1 \cup \Gamma_2$, i.e., if an element K has one side on the boundary, then that side must belong entirely to either Γ_1 or $\overline{\Gamma}_2$. The corresponding linear element space is then constructed from the basis functions associated with the interior nodes together with those boundary nodes located on Γ_1.

3.3 Finite Element Interpolation Error Estimates

In this section we present some estimates for the finite element interpolation error. The interpolation error estimates can be derived through the use of the reference element technique; i.e., error estimates are first derived on the reference element, which are then translated to a general finite element. The results discussed in this section can be extended to the case of a general d-dimensional domain. Definitions of triangulation and finite elements in d-dimensional case are similar to those for the 2-dimensional case, cf. e.g. [47, 49].

Interpolation error estimates on the reference element. We first present some interpolation error estimates over the reference element.

Theorem 3.6 *Let k and m be nonnegative integers with $k > 0$, $k+1 \geq m$, and $\mathcal{P}_k(\hat{K}) \subset \hat{X}$. Let $\hat{\Pi}$ be the operators defined in* (3.13). *Then there exists a constant c such that*

$$|\hat{v} - \hat{\Pi}\hat{v}|_{m,\hat{K}} \leq c\,|\hat{v}|_{k+1,\hat{K}} \quad \forall\, \hat{v} \in H^{k+1}(\hat{K}). \tag{3.16}$$

In Theorem 3.6, the assumption $k > 0$ is made to warrant the continuity of an $H^{k+1}(\hat{K})$ function. In the d-dimensional case, this assumption is replaced by $k+1 > d/2$. A crucial ingredient in the usual proof of Theorem 3.6 is the *polynomial invariance property*:

$$\hat{\Pi}\hat{v} = \hat{v} \quad \forall\, \hat{v} \in \mathcal{P}_k(\hat{K}).$$

This property follows from the definition (3.13) and the inclusion $\mathcal{P}_k(\hat{K}) \subset \hat{X}$.

Interpolation error estimates. We first consider the finite element interpolation error over each element K. As in Theorem 3.6, we assume $k > 0$, ensuring $H^{k+1}(K) \hookrightarrow C(K)$. Then for $v \in H^{k+1}(K)$, pointwise values $v(\boldsymbol{x})$ are meaningful. Let $\Pi_K : H^{k+1}(K) \to X_K \subset H^m(K)$ be the interpolation operator defined by (3.14).

To translate the result of Theorem 3.6 from the reference element $\hat{K}$ to the element K, we need to discuss the relations between Sobolev norms over the reference element and a general element.

Theorem 3.7 *Assume $\boldsymbol{x} = \boldsymbol{T}_K\hat{\boldsymbol{x}} + \boldsymbol{b}_K$ is a bijection from $\hat{K}$ to K. Then $v \in H^m(K)$ if and only if $\hat{v} \in H^m(\hat{K})$. Furthermore, for some constant c independent of K and $\hat{K}$, the estimates*

$$|\hat{v}|_{m,\hat{K}} \leq c\,\|\boldsymbol{T}_K\|^m |\det \boldsymbol{T}_K|^{-1/2} |v|_{m,K},$$
$$|v|_{m,K} \leq c\,\|\boldsymbol{T}_K^{-1}\|^m |\det \boldsymbol{T}_K|^{1/2} |\hat{v}|_{m,\hat{K}}$$

hold.

We can combine the preceding results, including Lemma 3.2, to obtain an estimate for the interpolation error in the semi-norm $|v - \Pi_K v|_{m,K}$.

Theorem 3.8 *Let k and m be nonnegative integers with $k > 0$, $k+1 \geq m$, and $\mathcal{P}_k(\hat{K}) \subset \hat{X}$. Let Π_K be the operators defined in* (3.14). *Then there is a constant c depending only on $\hat{K}$ and $\hat{\Pi}$ such that*

$$|v - \Pi_K v|_{m,K} \leq c\, \frac{h_K^{k+1}}{\rho_K^m} |v|_{k+1,K} \quad \forall\, v \in H^{k+1}(K). \tag{3.17}$$

The error bound in (3.17) depends on two parameters h_K and ρ_K. In the case of a regular family of affine finite elements, we can deduce the following error estimate from Theorem 3.8.

Corollary 3.9 *We keep the assumptions stated in Theorem 3.8. Furthermore, assume $\{\mathcal{T}_h\}$ is a regular family of finite element partitions. Then there is a constant c such that*

$$\|v - \Pi_K v\|_{m,K} \leq c\, h_K^{k+1-m} |v|_{k+1,K} \quad \forall\, v \in H^{k+1}(K),\ \forall\, K \in \mathcal{T}_h. \tag{3.18}$$

Example 3.10 *Let K be a triangle in a regular family of affine finite elements. We take the three vertices of K to be the nodal points. The local function space X_K is $\mathcal{P}_1(K)$. Assume $v \in H^2(K)$. Applying the estimate* (3.18) *with $k = 1$, we have*

$$\|v - \Pi_K v\|_{m,K} \leq c\, h_K^{2-m} |v|_{2,K} \quad \forall\, v \in H^2(K). \tag{3.19}$$

This estimate holds for $m = 0, 1$.

Since the global finite element interpolant is defined piecewise in terms of local finite element interpolants, we have the next result for global finite element interpolation error estimates.

Theorem 3.11 *Assume that all the conditions of Corollary 3.9 hold. Then there exists a constant c independent of h such that*

$$\|v - \Pi^h v\|_{m,\Omega} \leq c\, h^{k+1-m} |v|_{k+1,\Omega} \quad \forall\, v \in H^{k+1}(\Omega),\ m = 0, 1. \tag{3.20}$$

3.4 Finite Element Analysis of Linear Elliptic Boundary Value Problems

In this section, we briefly discuss the application of the finite element method in solving linear elliptic boundary value problems.

We first consider a model problem. Let $f \in L^2(\Omega)$ and consider the homogeneous Dirichlet boundary value problem of the Poisson equation

$$-\Delta u = f \quad \text{in } \Omega, \tag{3.21}$$

$$u = 0 \quad \text{on } \partial\Omega. \tag{3.22}$$

Let us derive a weak formulation of the problem. Assume u is a smooth function satisfying (3.21)–(3.22). We multiply the equation (3.21) by a smooth test function $v \in C_0^\infty(\Omega)$ and integrate over Ω to find

$$-\int_\Omega \Delta u\, v\, dx = \int_\Omega f\, v\, dx.$$

We then perform an integration by parts on the term of the left hand side and use the boundary condition $v = 0$ on $\Gamma = \partial\Omega$ to obtain

$$\int_\Omega \nabla u \cdot \nabla v\, dx = \int_\Omega f\, v\, dx. \tag{3.23}$$

We proved this relation under the assumptions that u is smooth and $v \in C_0^\infty(\Omega)$. Observe that the integrals in the relation (3.23) are well defined for $u, v \in H^1(\Omega)$. Recalling the boundary condition (3.22), we thus require $u \in H_0^1(\Omega)$. Since the closure of $C_0^\infty(\Omega)$ in $H^1(\Omega)$ is $H_0^1(\Omega)$, we derive from (3.23) that

$$u \in H_0^1(\Omega), \quad \int_\Omega \nabla u \cdot \nabla v\, dx = \int_\Omega f\, v\, dx \quad \forall\, v \in H_0^1(\Omega). \tag{3.24}$$

This is the *weak formulation* of the boundary value problem (3.21)–(3.22).

Now suppose $u \in H_0^1(\Omega)$ is a solution of the problem (3.24). Let us explore in what sense it solves the boundary value problem (3.21)–(3.22). We let $v \in C_0^\infty(\Omega)$ in (3.24) to obtain

$$-\Delta u = f \quad \text{in the sense of distributions.}$$

Since $f \in L^2(\Omega)$, we find that the equation (3.21) holds a.e. in Ω. The condition $u \in H_0^1(\Omega)$ implies that the boundary condition (3.22) holds a.e. on Γ.

The problem (3.24) is an example of a general abstract framework for linear elliptic boundary value problems of the form

$$u \in X, \quad a(u, v) = \ell(v) \quad \forall\, v \in X. \tag{3.25}$$

Here

$$\begin{aligned} X &= H_0^1(\Omega), \\ a(u, v) &= \int_\Omega \nabla u \cdot \nabla v\, dx, \\ \ell(v) &= \int_\Omega f\, v\, dx. \end{aligned}$$

The norm of the space X is chosen to be $\|v\|_{H_0^1(\Omega)} = \|\nabla v\|_{L^2(\Omega)^d}$, which is equivalent to the norm $\|v\|_{H^1(\Omega)}$. We see that

$$|a(u, v)| \le \|u\|_{H_0^1(\Omega)} \|v\|_{H_0^1(\Omega)} \quad \forall\, u, v \in H_0^1(\Omega), \tag{3.26}$$

$$a(v, v) = \|v\|_{H_0^1(\Omega)}^2 \quad \forall\, v \in H_0^1(\Omega). \tag{3.27}$$

By the well-known Lax-Milgram Lemma (cf. Corollary 4.8 on page 64), the weak formulation (3.24) has a unique solution.

In general, it is impossible to find the exact solution of the problem (3.25) because the space $H^1(\Omega)$ is infinite dimensional. A natural approach to constructing an approximate solution is to solve a finite dimensional analog of the problem (3.25). For this purpose, let $\mathcal{T}_h$ be a finite element partition of $\overline{\Omega}$ and let $X^h \subset X$ be a corresponding finite element space. We project the problem (3.25) onto X^h:

$$u^h \in X^h, \quad a(u^h, v^h) = \ell(v^h) \quad \forall\, v^h \in X^h. \tag{3.28}$$

The discrete problem has a unique solution, again by the Lax-Milgram Lemma.

We can express the problem (3.28) in the form of a linear system. Indeed, let $\{\phi_i\}_{i=1}^{N_h}$ be a basis of the finite dimensional space X^h, $N_h = \dim(X^h)$. We write

$$u^h = \sum_{j=1}^{N_h} \xi_j \phi_j$$

and take $v \in X^h$ in (3.28) each of the basis functions ϕ_i. As a result, (3.28) is equivalent to a linear system

$$A\boldsymbol{\xi} = \boldsymbol{b}.$$

Here, $\boldsymbol{\xi} = (\xi_j) \in \mathbb{R}^{N_h}$ is the unknown vector, $A = (a(\phi_j, \phi_i)) \in \mathbb{R}^{N_h \times N_h}$ is called the *stiffness matrix* and $\boldsymbol{b} = (\ell(\phi_i)) \in \mathbb{R}^{N_h}$ is the *load vector*. So the solution of the problem (3.28) can be found by solving a linear system.

The approximate solution u^h is, in general, different from the exact solution u. To increase the accuracy, it is natural to seek the approximate solution u^h in a larger finite element space X^h. Thus, for a sequence of finite element subspaces $X^{h_1} \subset X^{h_2} \subset \cdots \subset X$, we compute a corresponding sequence of approximate solutions $u^{h_i} \in X^{h_i}$, $i = 1, 2, \ldots$.

For convergence and error analysis, we apply *Céa's inequality*

$$\|u - u^h\|_{H_0^1(\Omega)} \le \inf_{v^h \in X^h} \|u - v^h\|_{H_0^1(\Omega)}. \tag{3.29}$$

A proof of Céa's inequality is rather simple. From (3.25) and (3.28) we obtain the error relation

$$a(u - u^h, v^h) = 0 \quad \forall\, v^h \in X^h. \tag{3.30}$$

Now for any $v^h \in X^h$, using the conditions (3.27), (3.28) and the error relation (3.30), we have

$$\begin{aligned}
\|u - u^h\|^2_{H_0^1(\Omega)} &\le a(u - u^h, u - u^h) \\
&= a(u - u^h, u - v^h) \\
&\le \|u - u^h\|_{H_0^1(\Omega)} \|u - v^h\|_{H_0^1(\Omega)}.
\end{aligned}$$

Then

$$\|u - u^h\|_{H_0^1(\Omega)} \le \|u - v^h\|_{H_0^1(\Omega)} \quad \forall\, v^h \in X^h.$$

We then obtain (3.29).

Based on (3.29), we have the following convergence result and error estimate.

Theorem 3.12 *Let $\{X^h\} \subset H_0^1(\Omega)$ be a family of affine-equivalent finite element spaces of piecewise polynomials of degree less than or equal to k, $k \ge 1$, corresponding to a regular family of finite element partitions of $\overline{\Omega}$. Then the finite element method converges:*

$$\|u - u^h\|_{H_0^1(\Omega)} \to 0 \quad \text{as } h \to 0.$$

Assume $u \in H^{k+1}(\Omega)$. Then there exists a constant c such that the following error estimate holds

$$\|u - u^h\|_{H_0^1(\Omega)} \le c\, h^k |u|_{H^{k+1}(\Omega)}. \tag{3.31}$$

Proof. It is well-known that $C_0^\infty(\Omega)$ is dense in $H_0^1(\Omega)$ (Theorem 2.12 on page 32). For any $\varepsilon > 0$, we can find $\tilde{u} \in C_0^\infty(\Omega)$ such that

$$\|u - \tilde{u}\|_{H_0^1(\Omega)} < \frac{\varepsilon}{2}.$$

Let $\Pi^h \tilde{u} \in X^h$ be the finite element interpolant of $\tilde{u}$. Then by Theorem 3.11,

$$\|\tilde{u} - \Pi^h \tilde{u}\|_{H_0^1(\Omega)} \le c\, h^k |\tilde{u}|_{H^{k+1}(\Omega)}.$$

So for h small enough,

$$\|\tilde{u} - \Pi^h \tilde{u}\|_{H_0^1(\Omega)} < \frac{\varepsilon}{2}.$$

By Céa's inequality (3.29),

$$\|u - u^h\|_{H_0^1(\Omega)} \le \|u - \Pi^h \tilde{u}\|_{H_0^1(\Omega)} \le \|u - \tilde{u}\|_{H_0^1(\Omega)} + \|\tilde{u} - \Pi^h \tilde{u}\|_{H_0^1(\Omega)}.$$

Hence, if h is small enough,

$$\|u - u^h\|_{H_0^1(\Omega)} < \varepsilon,$$

and we conclude the convergence.

Under the assumption $u \in H^{k+1}(\Omega)$, we have by Céa's inequality (3.29) that

$$\|u - u^h\|_{H_0^1(\Omega)} \le \|u - \Pi^h u\|_{H_0^1(\Omega)}.$$

An application of Theorem 3.11 yields the estimate (3.31). ■

The above discussion can be extended to the weak formulation (3.25) and its finite element approximation (3.28) under more general assumptions with

X a Hilbert space with the inner product $(\cdot,\cdot)_X$ and associated norm $\|\cdot\|_X$, $a : X \times X \to \mathbb{R}$ a bilinear form, and $\ell \in X'$. Under the assumptions of boundedness and X-ellipticity of the form a, i.e.

$$\begin{aligned} |a(u,v)| &\le M\,\|u\|_X \|v\|_X \quad \forall\, u,v \in X, \\ a(v,v) &\ge m\,\|v\|_X^2 \quad \forall\, v \in X \end{aligned}$$

for some constants $M, m > 0$, it follows from the Lax-Milgram Lemma that both (3.25) and (3.28) have unique solutions. For error analysis, Céa's inequality holds:

$$\|u - u^h\|_X \le c \inf_{v^h \in X^h} \|u - v^h\|_X \tag{3.32}$$

with $c = M/m$. The Céa inequality and finite element interpolation error estimates altogether yield error estimates for the finite element solution.

Chapter 4

Variational Inequalities

The theory of variational inequalities plays an important role in the study of both the qualitative and numerical analysis of some nonlinear boundary value problems arising in mechanics and other applications. At the heart of this theory is the intrinsic inclusion of free boundaries. The aim of this chapter is to present an introduction to the theory of variational inequalities on Hilbert spaces as well as their numerical approximations. The results presented here serve as a preparation for later study of contact problems.

4.1 Elliptic Variational Inequalities

In this section we present some theorems on the solvability of elliptic variational inequalities. Most of the concepts and results presented here can be extended to more general cases. Here, for simplicity, we restrict ourselves to the framework of Hilbert spaces, as this will be sufficient for later development. Everywhere in this section X denotes a real Hilbert space with inner product $(\cdot,\cdot)_X$ and norm $\|\cdot\|_X$. Let there be given an operator $A : X \to X$, a nonempty set $K \subset X$ and a proper functional $j : X \to \overline{\mathbb{R}}$. Then an *elliptic variational inequality of the first kind* is a problem of the form: Given $f \in X$, find u such that

$$u \in K, \quad (Au, v-u)_X \geq (f, v-u)_X \quad \forall\, v \in K. \tag{4.1}$$

An *elliptic variational inequality of the second kind* is a problem of the form: Given $f \in X$, find u such that

$$u \in X, \quad (Au, v-u)_X + j(v) - j(u) \geq (f, v-u)_X \quad \forall\, v \in X. \tag{4.2}$$

We are interested in sufficient conditions for the existence and uniqueness of solutions to the inequalities (4.1) and (4.2). We remark that if we take $j = I_K$, the indicator function of the set K, then the variational inequality (4.2) reduces to the variational inequality (4.1). Therefore, elliptic variational inequalities of the first kind are special cases of elliptic variational inequalities of the second kind. For this reason, we start with an existence and uniqueness result in the study of variational inequalities of the form (4.2). To this end we make the following assumptions.

$$\left.\begin{array}{l} A : X \to X \text{ is a strongly monotone Lipschitz} \\ \text{continuous operator on } X, \text{ i.e.} \\ \text{(a) there exists } M > 0 \text{ such that} \\ \qquad \|Au - Av\|_X \le M\|u - v\|_X \quad \forall\, u, v \in X; \\ \text{(b) there exists } m > 0 \text{ such that} \\ \qquad (Au - Av, u - v)_X \ge m\,\|u - v\|_X^2 \quad \forall\, u, v \in X. \end{array}\right\} \tag{4.3}$$

$$j : X \to \overline{\mathbb{R}} \text{ is a proper, convex and l.s.c. functional.} \tag{4.4}$$

We have the following existence and uniqueness result.

Theorem 4.1 *Let X be a Hilbert space. Assume* (4.3) *and* (4.4). *Then for any $f \in X$, the variational inequality* (4.2) *has a unique solution. Moreover, the solution depends Lipschitz continuously on f.*

The proof of Theorem 4.1 will be carried in two steps. It is based on arguments of minimization of convex functions and an application of the Banach fixed-point theorem. The first step consists of solving (4.2) in the case when A is the identity operator on X.

Lemma 4.2 *Assume* (4.4). *Then for any $f \in X$, there exists a unique element u such that*

$$u \in X, \quad (u, v - u)_X + j(v) - j(u) \ge (f, v - u)_X \quad \forall\, v \in X. \tag{4.5}$$

Moreover, if u_1 and u_2 denote the solutions of the inequality (4.5) *for $f = f_1, f_2 \in X$, respectively, then*

$$\|u_1 - u_2\|_X \le \|f_1 - f_2\|_X. \tag{4.6}$$

Proof. Define a functional $J : X \to \overline{\mathbb{R}}$ by the formula

$$J(v) = \frac{1}{2}\,\|v\|_X^2 + j(v) - (f, v)_X.$$

Due to (4.4) we see that J is proper, strictly convex and lower semicontinuous. From Proposition 1.52 (page 19) it follows that $J(v) \to \infty$ when $\|v\|_X \to \infty$, i.e. J is coercive. Therefore, applying Theorem 1.53 we deduce that there exists a unique element $u \in X$ such that

$$J(v) \ge J(u) \quad \forall\, v \in X. \tag{4.7}$$

In particular, this implies $J(u) < \infty$ and $j(u) < \infty$.

Let us show that $u \in X$ is a minimizer of $J(\cdot)$ iff u satisfies the variational inequality (4.5). Let $v \in X$ and $t \in (0, 1)$. It follows from (4.7) that

$$J(u + t(v - u)) \ge J(u).$$

Using the convexity of j we deduce that

$$t\,(u, v-u)_X + \frac{t^2}{2}\|v-u\|_X^2 + t\,(j(v)-j(u)) \geq t\,(f, v-u)_X.$$

Dividing both sides of the inequality by $t > 0$ and passing to the limit $t \to 0+$, we conclude that u satisfies the variational inequality (4.5). Conversely, suppose u is a solution of the variational inequality (4.5). Since j is a proper functional we know that $j(u) < \infty$. Then for any $v \in X$, we have

$$\begin{aligned} J(v) - J(u) &= (u, v-u)_X + j(v) - j(u) - (f, v-u)_X + \frac{1}{2}\|v-u\|_X^2 \\ &\geq \frac{1}{2}\|v-u\|_X^2 \\ &\geq 0. \end{aligned}$$

Thus, u is a solution of the minimization problem (4.7).

Hence, the minimization problem (4.7) and the variational inequality (4.5) are equivalent. Since $J(\cdot)$ has a unique minimizer, we infer that the variational inequality (4.5) has a unique solution.

Let now $f_1, f_2 \in X$ and let u_1, $u_2 \in X$ be such that

$$(u_1, v-u_1)_X + j(v) - j(u_1) \geq (f_1, v-u_1)_X \quad \forall\, v \in X, \tag{4.8}$$

$$(u_2, v-u_2)_X + j(v) - j(u_2) \geq (f_2, v-u_2)_X \quad \forall\, v \in X. \tag{4.9}$$

It follows that $j(u_1) < \infty$ and $j(u_2) < \infty$. Therefore, taking $v = u_2$ in (4.8) and $v = u_1$ in (4.9) and adding the obtained inequalities, we find

$$\|u_1 - u_2\|_X^2 \leq (f_1 - f_2, u_1 - u_2)_X, \tag{4.10}$$

which implies (4.6). ■

When $j : X \to \overline{\mathbb{R}}$ is a proper, convex, lower semicontinuous functional and $f \in X$, the solution u of the variational inequality (4.5) is called the *proximal element* of f with respect to the functional j and it is usually denoted $\mathrm{Prox}_j f$. The operator $\mathrm{Prox}_j : X \to X$ defined by $f \mapsto \mathrm{Prox}_j f$ is called the *proximity operator*, first introduced in [155]. Lemma 4.2 states that Prox_j is a non-expansive operator:

$$\|\mathrm{Prox}_j f_1 - \mathrm{Prox}_j f_2\|_X \leq \|f_1 - f_2\|_X \quad \forall\, f_1,\, f_2 \in X. \tag{4.11}$$

Moreover, from (4.10) it follows that Prox_j is a monotone operator,

$$(\mathrm{Prox}_j f_1 - \mathrm{Prox}_j f_2, f_1 - f_2)_X \geq 0 \quad \forall\, f_1, f_2 \in X. \tag{4.12}$$

Finally, we remark that if K is a nonempty closed convex subset of X, then by Lemma 4.2 we may consider the operator Prox_{I_K} where I_K denotes the indicator function of K. By the characterization (1.4) of the projection (page 16), it follows that $\mathrm{Prox}_{I_K} = \mathcal{P}_K$, the projection operator on K. Clearly, inequalities (4.11) and (4.12) generalize the properties (1.5) and (1.6), respectively.

We now provide a proof of Theorem 4.1.

Proof of Theorem 4.1. Let $f \in H$ and let $\rho > 0$ be a parameter to be chosen later. Since $\rho j : X \to \overline{\mathbb{R}}$ is again a proper, convex, lower semicontinuous functional, we can define an operator $T : X \to X$ by

$$T(v) = \text{Prox}_{\rho j}(\rho f - \rho Av + v) \quad \forall\, v \in X. \tag{4.13}$$

We will show that with a suitable choice of ρ the operator T is a contractive mapping on X. To this end, let $u,\, v \in X$. Using (4.13) and (4.11) we find

$$\begin{aligned} \|T(u) - T(v)\|_X^2 &\le \|u - v - \rho(Au - Av)\|_X^2 \\ &= \|u - v\|_X^2 - 2\rho\,(Au - Av, u - v)_X + \rho^2\|Au - Av\|_X^2. \end{aligned}$$

Using the assumption (4.3) we obtain

$$\|T(u) - T(v)\|_X^2 \le (1 - 2\rho m + \rho^2 M^2)\|u - v\|_X^2.$$

It follows now that if $0 < \rho < \frac{2m}{M^2}$, then $0 < 1 - 2\rho m + \rho^2 M^2 < 1$. Thus, taking $\alpha = (1 - 2\rho m + \rho^2 M^2)^{\frac{1}{2}}$ we deduce that $\alpha \in (0,1)$ and

$$\|T(u) - T(v)\|_X \le \alpha\,\|u - v\|_X,$$

i.e. $T : X \to X$ is a contraction. Therefore, from Banach's fixed point theorem, it follows that T has a fixed point u:

$$u = \text{Prox}_{\rho j}(\rho f - \rho Au + u).$$

By the definition of the proximal element (see (4.5)) we then obtain

$$(u, v - u)_X + \rho j(v) - \rho j(u) \ge (\rho f - \rho Au + u, v - u)_X \quad \forall\, v \in X,$$

i.e.,

$$\rho\,[(Au, v - u)_X + j(v) - j(u)] \ge \rho\,(f, v - u)_X \quad \forall\, v \in X.$$

Since $\rho > 0$, we deduce from the above inequality that u is a solution of the variational inequality (4.2).

To show the uniqueness, assume there are two solutions $u_1,\, u_2 \in X$ to the variational inequality (4.2). Then,

$$\begin{aligned} (Au_1, v - u_1)_X + j(v) - j(u_1) \ge (f, v - u_1)_X \quad \forall\, v \in X, \\ (Au_2, v - u_2)_X + j(v) - j(u_2) \ge (f, v - u_2)_X \quad \forall\, v \in X. \end{aligned}$$

Since j is proper, we have $j(u_1) < \infty$, $j(u_2) < \infty$. Taking $v = u_2$ in the first inequality, $v = u_1$ in the second one and adding the corresponding inequalities, we find

$$(Au_1 - Au_2, u_1 - u_2)_X \le 0.$$

Using now (4.3)(b) we obtain $u_1 = u_2$, which proves the uniqueness part of Theorem 4.1.

A similar argument show that if u_1 and u_2 are the solutions of the inequality (4.2) for $f = f_1$ and f_2, then

$$(Au_1 - Au_2, u_1 - u_2)_X \le (f_1 - f_2, u_1 - u_2)_X.$$

Using again (4.3)(b) we find

$$\|u_1 - u_2\|_X \le \frac{1}{m} \|f_1 - f_2\|_X,$$

which shows that the solution of (4.2) depends Lipschitz continuously on f. ∎

We can specialize Theorem 4.1 to the case of elliptic variational inequalities of the first kind.

Corollary 4.3 *Let X be a Hilbert space and $K \subset X$ be a nonempty convex closed subset. Assume* (4.3). *Then for any $f \in X$, there exists a unique solution to the variational inequality* (4.1). *Moreover, the solution depends Lipschitz continuously on f.*

Proof. From Example 1.51 (page 19) we know that the indicator function I_K is convex, proper and l.s.c. Thus the result follows from Theorem 4.1 with the choice $j \equiv I_K$. ∎

In the special case $j \equiv 0$ we obtain the following result from the theory of monotone operators.

Corollary 4.4 *Assume that X is a Hilbert space and $A : X \to X$ is a strongly monotone Lipschitz continuous operator. Then for any $f \in X$ the equation $Au = f$ has a unique solution $u \in X$, which depends Lipschitz continuously on f.*

A large number of applications lead to elliptic variational inequalities in which the operator A and the element f need to be specified by Riesz representation theorem (Theorem 1.44). For this reason, we present some variants of the previous results in a form useful for some applications. Since we will assume the operator A is linear, it is convenient to introduce the notion of a bilinear form.

Definition 4.5 *Let $a : X \times X \to \mathbb{R}$. We say $a(\cdot,\cdot)$ is a* bilinear form *on X if it is linear with respect to each argument. We say the bilinear form $a(\cdot,\cdot)$ is* continuous *or* bounded *if there exists a number $M > 0$ such that*

$$|a(u, v)| \le M \|u\|_X \|v\|_X \quad \forall\, u, v \in X. \tag{4.14}$$

*The bilinear form $a(\cdot,\cdot)$ is X-*elliptic *if there is a constant $m > 0$ such that*

$$a(v, v) \ge m \|v\|_X^2 \quad \forall\, v \in X \tag{4.15}$$

and $a(\cdot,\cdot)$ is symmetric *if*

$$a(u, v) = a(v, u) \quad \forall\, u,\, v \in X.$$

Given a linear operator $A : X \to X$, a bilinear form $a : X \times X \to \mathbb{R}$ can be defined through the relation

$$a(u,v) = (Au,v)_X \quad \forall\, u,v \in X. \tag{4.16}$$

Conversely, it follows from the Riesz representation theorem that if $a : X \times X \to \mathbb{R}$ is a continuous bilinear form, then there exists a linear operator $A : X \to X$ such that (4.16) holds. It is easy to see that the boundedness of a is equivalent to that of A, and the X-ellipticity of a is equivalent to the strong monotonicity of A.

We now specialize Theorem 4.1 to obtain the following result.

Corollary 4.6 *Let X be a Hilbert space. Assume $a : X \times X \to \mathbb{R}$ is a continuous, X-elliptic bilinear form, $j : X \to \overline{\mathbb{R}}$ is proper, convex and l.s.c. on X and $\ell : X \to \mathbb{R}$ is a linear continuous functional. Then there exists a unique solution to the elliptic variational inequality of the second kind*

$$u \in X, \quad a(u, v-u) + j(v) - j(u) \geq \ell(v-u) \quad \forall\, v \in X. \tag{4.17}$$

Proof. We define an operator $A : X \to X$ by the relation (4.16). Then A is a linear operator. The continuity of the bilinear form a implies that the operator A is bounded and is hence Lipschitz continuous. The X-ellipticity of a implies the strongly monotonicity of A. Then the result follows from Theorem 4.1. ■

Using the same arguments, as a consequence of Corollary 4.3, we obtain:

Corollary 4.7 *Let X be a Hilbert space and let $K \subset X$ be a non-empty, convex, closed subset. Assume $a : X \times X \to \mathbb{R}$ is a continuous, X-elliptic bilinear form and $\ell : X \to \mathbb{R}$ is a linear continuous functional. Then there exists a unique solution to the elliptic variational inequality of the first kind*

$$u \in K, \quad a(u, v-u) \geq \ell(v-u) \quad \forall\, v \in K. \tag{4.18}$$

As a consequence, we have the well-known Lax-Milgram Lemma for linear elliptic boundary value problems.

Corollary 4.8 (Lax-Milgram Lemma) *Let X be a Hilbert space, $a : X \times X \to \mathbb{R}$ be a bounded, X-elliptic bilinear form and let $\ell : X \to \mathbb{R}$ be a linear continuous functional. Then there exists a unique solution to the variational equation*

$$u \in X, \quad a(u,v) = \ell(v) \quad \forall\, v \in X. \tag{4.19}$$

Corollary 4.8 follows from Corollary 4.6 in the special case $j \equiv 0$. It can also be recovered from Corollary 4.7 with $K = X$.

The term "variational" in the description of a problem of the form (4.19) derives from its association with a minimization problem, in the case in which the bilinear form a is symmetric (i.e. $a(u,v) = a(v,u)$ for $u, v \in X$). Indeed, we can consider the problem

$$\min_{v \in X} J(v) \tag{4.20}$$

where $J : X \to \mathbb{R}$ is the energy functional defined by

$$J(v) = \frac{1}{2} a(v, v) - \ell(v). \tag{4.21}$$

Following the proof of Lemma 4.2 we see that if a is symmetric and X-elliptic, then $u \in X$ is a minimizer of J if and only if it is a solution of (4.19). Thus, under the stated assumptions, the variational equation (4.19) and the minimization problem (4.20) are equivalent. The unique solvability of these problems may be inferred by the Lax-Milgram Lemma. Similar results can be proved for elliptic variational inequalities: if a is symmetric and X-elliptic, then solving the variational inequality (4.17) is equivalent to solving the minimum problem (4.20) with the functional

$$J(v) = \frac{1}{2} a(v, v) + j(v) - \ell(v),$$

and solving the variational inequality (4.18) is equivalent to solving the minimum problem

$$\min_{v \in K} J(v)$$

with the energy functional (4.21).

4.2 Approximation of Elliptic Variational Inequalities

In this section we discuss numerical approximations of elliptic variational inequalities and their error analysis. Everywhere in this section X denotes a real Hilbert space with inner product $(\cdot, \cdot)_X$ and norm $\|\cdot\|_X$, a is a continuous, X-elliptic bilinear form on X and ℓ is a continuous linear functional on X. Let $K \subset X$ be a non-empty, convex and closed set. Then the following elliptic variational inequality of the first kind

$$u \in K, \quad a(u, v - u) \geq \ell(v - u) \quad \forall\, v \in K \tag{4.22}$$

has a unique solution.

A general framework for numerical solution of the variational inequality (4.22) can be described as follows. Let $X^h \subset X$ be a finite element space, and let $K^h \subset X^h$ be non-empty, convex and closed. Then the finite element approximation of the problem (4.22) is

$$u^h \in K^h, \quad a(u^h, v^h - u^h) \geq \ell(v^h - u^h) \quad \forall\, v^h \in K^h. \tag{4.23}$$

The discrete problem (4.23) has a unique solution under the stated assumptions on the given data.

A general convergence result of the finite element method can be found in [90]. Here our emphasis is on error estimation for the finite element solution u^h. We follow [75] and first give an abstract error analysis.

Theorem 4.9 *There is a constant $c > 0$ independent of h and u, such that*

$$\begin{aligned}\|u - u^h\|_X \le c \Big\{ &\inf_{v^h \in K^h} \big[\|u - v^h\|_X + |R(v^h - u)|^{\frac{1}{2}}\big] \\ &+ \inf_{v \in K} |R(v - u^h)|^{\frac{1}{2}} \Big\},\end{aligned} \tag{4.24}$$

where

$$R(v) = a(u, v) - \ell(v)$$

represents a residual quantity.

The inequality (4.24) is a generalization of Céa's inequality to the finite element approximation of elliptic variational inequalities of the first kind. It is easy to see that the inequality (4.24) reduces to Céa's inequality (3.32) in the case of finite element approximation of a variational equation problem.

Remark 4.10 *In the case $K^h \subset K$, we have the so-called* internal approximation *of the elliptic variational inequality of the first kind. Since now $u^h \in K$, the second term on the right hand side of* (4.24) *vanishes, and the error inequality* (4.24) *reduces to*

$$\|u - u^h\|_X \le c \inf_{v^h \in K^h} \left[\|u - v^h\|_X + |R(v^h - u)|^{1/2}\right].$$

Now we consider a model problem.

Example 4.11 *Let $\Omega \subset \mathbb{R}^2$ be a polygon of boundary Γ, $f \in L^2(\Omega)$ and let $\psi \in H^2(\Omega)$ such that $\psi \le 0$ on Γ. We take*

$$\begin{aligned} X &= H_0^1(\Omega), \\ a(u, v) &= \int_\Omega \nabla u \, \nabla v \, dx, \\ \ell(v) &= \int_\Omega f \, v \, dx, \\ K &= \{ v \in X : v \ge \psi \text{ a.e. in } \Omega \}. \end{aligned}$$

The set K is nonempty, because the function $\max\{0, \psi\}$ *belongs to K. It is easy to verify that the set K is closed and convex, the bilinear form a is bounded and X-elliptic, and the linear form ℓ is continuous on X. Applying Corollary 4.7 we see that there exists a unique solution of the variational inequality*

$$u \in K, \quad \int_\Omega \nabla u \cdot \nabla (v - u) \, dx \ge \int_\Omega f (v - u) \, dx \quad \forall \, v \in K. \tag{4.25}$$

This problem is called the obstacle problem. *It models the equilibrium of an elastic membrane which passes through a closed curve Γ, the boundary of a planar domain Ω, lies above an obstacle of height ψ and is subject to the action* *of a vertical force of density f. Here, the unknown u represents the vertical displacement of the membrane.*

The inequality (4.24) can be used in a derivation of error estimates for the approximation of the obstacle problem. Such an estimate was first proved in [75]. Assume $u, \psi \in H^2(\Omega)$ and Ω is a polygon. We use linear elements on a regular family of triangulations for the approximation, and take the discrete admissible set to be

$$K^h = \{ v^h \in H_0^1(\Omega) : v^h \text{ is piecewise linear, } v^h(\boldsymbol{x}) \geq \psi(\boldsymbol{x}) \ \forall \text{node } \boldsymbol{x} \}.$$

We see that any function in K^h is a continuous piecewise linear function, vanishing on the boundary and dominating the obstacle function at the interior nodes of the mesh. In general, $K^h \not\subset K$. We have the optimal order error estimate

$$\|u - u^h\|_{H^1(\Omega)} \leq c\, h$$

for some constant $c > 0$ depending only on $|u|_2$, $\|f\|_0$ and $|\psi|_2$.

We now consider the approximation of elliptic variational inequalities of second kind. In addition to the previous notation and assumptions, we denote by $j : X \to \overline{\mathbb{R}}$ a proper, convex and l.s.c. functional. Under these assumptions, there exists a unique solution of the elliptic variational inequality

$$u \in X, \quad a(u, v - u) + j(v) - j(u) \geq \ell(v - u) \quad \forall v \in X. \tag{4.26}$$

Let $X^h \subset X$ be a finite element space. Then the finite element approximation of the problem (4.26) is

$$u^h \in X^h, \quad a(u^h, v^h - u^h) + j(v^h) - j(u^h) \geq \ell(v^h - u^h) \quad \forall v^h \in X^h. \tag{4.27}$$

Assuming additionally that j is proper also on X^h, as is always the case in applications, we conclude that the discrete problem (4.27) has a unique solution u^h and $j(u^h) \in \mathbb{R}$. We now present an abstract error estimate for $u - u^h$, a derivation of which can be found in, e.g. [96].

Theorem 4.12 *There is a constant $c > 0$ independent of h and u such that*

$$\|u - u^h\|_X \leq c \inf_{v^h \in X^h} \left\{ \|u - v^h\| + |R(v^h, u)|^{1/2} \right\}, \tag{4.28}$$

where

$$R(v^h, u) = a(u, v^h - u) + j(v^h) - j(u) - \ell(v^h - u)$$

represents a residual quantity.

We observe that (4.28) is a generalization of Céa's inequality (3.32) (page 58) to the finite element approximation of elliptic variational inequalities of the second kind. The inequality (4.28) is the basis for deriving error estimates of finite element solutions of various application problems.

Now we consider a model problem.

Example 4.13 *Let $\Omega \subset \mathbb{R}^2$ be an open bounded set, with a Lipschitz boundary Γ. We take*

$$\begin{aligned} X &= H^1(\Omega), \\ a(u,v) &= \int_\Omega (\nabla u \cdot \nabla v + u\,v)\,dx, \\ \ell(v) &= \int_\Omega f\,v\,dx, \\ j(v) &= g \int_\Gamma |v|\,ds. \end{aligned}$$

This problem is a simplified version of the friction problem in elasticity (cf. [70]). It models the antiplane shear deformation of a linearly elastic cylinder of cross section Ω that is in frictional contact on its lateral surface. Here $f \in L^2(\Omega)$ denotes the axial component of the body forces, $g > 0$ represents the given friction bound and u is the displacement field.

We choose this model problem for its simplicity while at the same time it contains the main feature of an elliptic variational inequality of the second kind. Applying Corollary 4.6, we see that the corresponding variational inequality problem

$$u \in X, \quad a(u, v-u) + j(v) - j(u) \geq \ell(v-u) \quad \forall\, v \in X \tag{4.29}$$

has a unique solution. Given a finite element space X^h, let u^h denote the corresponding finite element solution defined in (4.27). Assume Ω is a polygonal domain and write $\Gamma = \cup_{i=1}^{i_0} \Gamma_i$, where each Γ_i is a line segment. The following error estimate can be derived based on the inequality (4.28).

Theorem 4.14 *For the above model problem, we assume the solution $u \in H^2(\Omega)$ and $u|_{\Gamma_i} \in H^2(\Gamma_i)$, $1 \leq i \leq i_0$. Let X^h be a piecewise linear finite element space constructed from a regular partition of the domain Ω. Let $u^h \in X^h$ be the finite element solution defined by (4.27). Then we have the optimal order error estimate*

$$\|u - u^h\|_{H^1(\Omega)} \leq c(u)\,h. \tag{4.30}$$

A major issue in solving the discrete system (4.27) is the treatment of the non-differentiable term j. In practice, several approaches have been proposed, e.g., regularization technique, method of Lagrangian multipliers, method of numerical integration. Some discussions of these approaches are found in [14] and the references listed there.

4.3 An Evolutionary Variational Inequality

In this section we present an existence and uniqueness result in the study of a class of evolutionary variational inequalities which arise in the study of quasistatic contact problems as well as in the primal formulation of some plasticity problems, see e.g. [96].

Let $T > 0$ and $1 < p \leq \infty$. Recall that $I = (0, T)$, $\overline{I} = [0, T]$. Everywhere in this section X denotes a real Hilbert space with inner product $(\cdot, \cdot)_X$ and norm $\|\cdot\|_X$, $a : X \times X \to \mathbb{R}$, $j : X \to \mathbb{R}$ and $f : \overline{I} \to X$. The problem to be discussed in this section is the following.

Problem 4.15 *Find $u : \overline{I} \to X$ such that for almost any $t \in I$,*

$$a(u(t), v - \dot{u}(t)) + j(v) - j(\dot{u}(t)) \geq (f(t), v - \dot{u}(t))_X \quad \forall\, v \in X, \tag{4.31}$$

and

$$u(0) = u_0. \tag{4.32}$$

In the study of (4.31)–(4.32) we make the following assumptions.

$$\left.\begin{array}{l} a : X \times X \to \mathbb{R} \text{ is a symmetric bilinear form and} \\ \text{(a) there exists } M > 0 \text{ such that} \\ \qquad |a(u, v)| \leq M \|u\|_X \|v\|_X \quad \forall\, u, v \in X; \\ \text{(b) there exists } m > 0 \text{ such that } a(v, v) \geq m\, \|v\|_X^2 \quad \forall\, v \in X. \end{array}\right\} \tag{4.33}$$

$$j \;\text{ is a continuous semi-norm on } X. \tag{4.34}$$

$$f \in W^{1,p}(I; X). \tag{4.35}$$

$$u_0 \in X. \tag{4.36}$$

$$a(u_0, v) + j(v) \geq (f(0), v)_X \quad \forall\, v \in X. \tag{4.37}$$

We remark that the assumption (4.34) requires the functional j to be a semi-norm on X such that it is continuous with respect to the norm $\|\cdot\|_X$.

Well-posedness of Problem 4.15 is given by the following result.

Theorem 4.16 *Assume* (4.33)–(4.37). *Then there exists a unique solution $u \in W^{1,p}(I; X)$ to Problem 4.15. Moreover, the mapping $(f, u_0) \mapsto u$ is Lipschitz continuous from $W^{1,1}(I; X) \times X$ to $L^\infty(I; X)$.*

The proof of Theorem 4.16 is carried out in several steps and is based on a time discretization method, compactness and lower semicontinuity arguments. We assume in what follows that assumptions in Theorem 4.16 hold. In the first step we introduce an implicit scheme for solving Problem 4.15. To this end let $N \geq 1$ be a sequence of increasing integers and let $k = T/N$ represent the stepsizes. For each N we denote the grid points $t_n = k\,n$ for $n = 0, 1, \ldots, N$. To simplify the notation, we write $f_n = f(t_n) \in X$, $0 \leq n \leq N$. Notice that since $W^{1,p}(I; X) \hookrightarrow C(\overline{I}; X)$ (Lemma 2.27 on page 40), the function $f : \overline{I} \to X$ is continuous in t. So it is meaningful to write pointwise values $f(t)$, for any $t \in \overline{I}$. For a sequence $\{w_n\}$ we define the difference $\Delta w_n = w_n - w_{n-1}$.

Let us consider a discrete analogue of Problem 4.15. We define a sequence $\{w_n\}_{n=0}^N \subset X$ by the relations:

$$w_0 = u_0 \tag{4.38}$$

and for $n = 1, \ldots, N$,

$$a(w_n, v - \Delta w_n) + j(v) - j(\Delta w_n) \geq (f_n, v - \Delta w_n)_X \quad \forall v \in X. \tag{4.39}$$

Lemma 4.17 *The sequence* $\{w_n\}_{n=0}^N \subset X$ *is well-defined by* (4.38)–(4.39) *and the following estimates hold:*

$$\|\Delta w_n\|_X \leq c\,\|\Delta f_n\|_X, \quad 1 \leq n \leq N, \tag{4.40}$$

$$\max_{0 \leq n \leq N} \|w_n\|_X \leq c\left(\|u_0\|_X + \int_I \|\dot{f}(t)\|_X dt\right). \tag{4.41}$$

Proof. For $n = 1, \ldots, N$, we rewrite (4.39) in the form

$$\begin{aligned} &a(\Delta w_n, v - \Delta w_n) + j(v) - j(\Delta w_n) \\ &\qquad \geq (f_n, v - \Delta w_n)_X - a(w_{n-1}, v - \Delta w_n) \quad \forall v \in X. \end{aligned} \tag{4.42}$$

This elliptic variational inequality has a unique solution $\Delta w_n \in X$ by Corollary 4.6. This result, together with the initial value condition (4.38), implies that the sequence $\{w_n\}_{n=0}^N \subset X$ is well-defined.

To derive the estimates (4.40) and (4.41), we take $v = 0$ in (4.42) to obtain

$$a(\Delta w_n, \Delta w_n) \leq (f_n, \Delta w_n)_X - j(\Delta w_n) - a(w_{n-1}, \Delta w_n). \tag{4.43}$$

For $n \geq 2$, let $v = \Delta w_{n-1} + \Delta w_n$ in (4.39) with n replaced by $n-1$,

$$a(w_{n-1}, \Delta w_n) + j(\Delta w_{n-1} + \Delta w_n) - j(\Delta w_{n-1}) \geq (f_{n-1}, \Delta w_n)_X.$$

So

$$-a(w_{n-1}, \Delta w_n) \leq -(f_{n-1}, \Delta w_n)_X + j(\Delta w_{n-1} + \Delta w_n) - j(\Delta w_{n-1}).$$

Using this inequality in (4.43) it follows that

$$\begin{aligned} a(\Delta w_n, \Delta w_n) &\leq (\Delta f_n, \Delta w_n)_X + j(\Delta w_{n-1} + \Delta w_n) - j(\Delta w_{n-1}) - j(\Delta w_n) \\ &\leq (\Delta f_n, \Delta w_n)_X, \end{aligned}$$

where in the last step we used the triangle inequality for the functional j. Thus,

$$\|\Delta w_n\|_X \leq c\,\|\Delta f_n\|_X \quad \text{for } n \geq 2.$$

We still need to show that

$$\|\Delta w_1\|_X \leq c\,\|\Delta f_1\|_X.$$

For this, we first get, from (4.38) and (4.43),

$$a(\Delta w_1, \Delta w_1) \leq (f_1, \Delta w_1)_X - j(\Delta w_1) - a(u_0, \Delta w_1). \tag{4.44}$$

Using the assumption (4.37), we have

$$(f_0, \Delta w_1)_X \leq j(\Delta w_1) + a(u_0, \Delta w_1),$$

i.e.,

$$-j(\Delta w_1) - a(u_0, \Delta w_1) \le -(f_0, \Delta w_1)_X.$$

Add this inequality to (4.44) to obtain

$$a(\Delta w_1, \Delta w_1) \le (f_1 - f_0, \Delta w_1)_X,$$

and then

$$\|\Delta w_1\|_X \le c\, \|\Delta f_1\|_X.$$

Hence the estimates (4.40) are proved.

Now write

$$w_n = u_0 + \sum_{j=1}^{n} \Delta w_j.$$

We have

$$\begin{aligned} \|w_n\|_X &\le \|u_0\|_X + c \sum_{j=1}^{n} \|\Delta w_j\|_X \\ &\le \|u_0\|_X + c \sum_{j=1}^{n} \|\Delta f_j\|_X. \end{aligned}$$

Applying Lemma 2.32, we obtain

$$\|\Delta f_j\|_X \le \int_{t_{j-1}}^{t_j} \|\dot{f}(t)\|_X dt.$$

Thus,

$$\|w_n\|_X \le \|u_0\|_X + c \int_0^{t_n} \|\dot{f}(t)\|_X dt,$$

from which, we conclude the estimate (4.41). ∎

We then define a piecewise linear interpolant w^k from the sequence $\{w_n\}_{n=0}^N \subset X$ by the formula

$$w^k(t) = w_{n-1} + \frac{t - t_{n-1}}{k}\, \Delta w_n, \quad t \in [t_{n-1}, t_n], \quad 1 \le n \le N.$$

In the next step we provide convergence results involving the sequence $\{w^k\}$.

Lemma 4.18 *There is a function $u \in W^{1,p}(I; X)$ and a subsequence of $\{w^k\}$, still denoted by $\{w^k\}$, such that*

$$w^k \rightharpoonup^* u \quad \text{in } L^\infty(I; X),$$

and

$$w^k \rightharpoonup u \quad \text{in } W^{1,p}(I;X), \quad \text{if } p < \infty$$

or

$$w^k \rightharpoonup^* u \quad \text{in } W^{1,\infty}(I;X), \quad \text{if } p = \infty.$$

Proof. We have

$$\dot{w}^k(t) = \frac{1}{k}\Delta w_n \quad \text{for } t \in (t_{n-1}, t_n).$$

Thus, if $p < \infty$, then

$$\int_{t_{n-1}}^{t_n} \|\dot{w}^k(t)\|_X^p dt = \frac{1}{k^p}\int_{t_{n-1}}^{t_n} \|\Delta w_n\|_X^p dt \le \frac{c}{k^p}\int_{t_{n-1}}^{t_n} \|\Delta f_n\|_X^p dt.$$

By applying Lemma 2.32,

$$\int_{t_{n-1}}^{t_n} \|\dot{w}^k(t)\|_X^p dt \le c\int_{t_{n-1}}^{t_n} \|\dot{f}(t)\|_X^p dt$$

and so,

$$\int_I \|\dot{w}^k(t)\|_X^p dt \le c\int_I \|\dot{f}(t)\|_X^p dt.$$

When $p = \infty$, we have

$$\|\dot{w}^k\|_{L^\infty(I;X)} \le c \max_n \|\Delta f_n\|_X \le c\|\dot{f}\|_{L^\infty(I;X)}.$$

Hence, in any case, $\{w^k\}$ is a bounded sequence in the space $W^{1,p}(I;X)$, which completes the proof of the lemma. ■

Introduce two piecewise constant functions $\tilde{w}^k : \overline{I} \to X$ and $\tilde{f}^k : \overline{I} \to X$ by the formulas

$$\tilde{w}^k(t) = w_n, \ \tilde{f}^k(t) = f_n, \ t \in (t_{n-1}, t_n], \quad n = 1, \ldots, N.$$

Here and below, when $n = 1$, we make the following modification:

$$\tilde{w}^k(t) = w_1, \ \tilde{f}^k(t) = f_1, \ t \in [t_0, t_1].$$

In what follows u will denote the element of $W^{1,p}(I;X)$ provided by Lemma 4.18 and $\{w^k\}$, $\{\tilde{w}^k\}$, $\{\tilde{f}^k\}$ will represent appropriate subsequences of the sequences $\{w^k\}$, $\{\tilde{w}^k\}$, $\{\tilde{f}^k\}$, respectively.

In the next two steps we provide some convergence and semicontinuity results.

Lemma 4.19 *We have the following convergence as $k \to 0$:*

$$\int_I a(\tilde{w}^k(t), v(t))\, dt \to \int_I a(u(t), v(t))\, dt, \tag{4.45}$$

$$\int_I (\tilde{f}^k(t), v(t))_X\, dt \to \int_I (f(t), v(t))_X\, dt. \tag{4.46}$$

Proof. For all $t \in (t_{n-1}, t_n]$ we have

$$\tilde{w}^k(t) - w^k(t) = w_n - \Big(w_{n-1} + \frac{t - t_{n-1}}{k}\, \Delta w_n\Big) = \frac{t_k - t}{k}\, \Delta w_n.$$

We take the norm in $L^\infty(I; X)$ to obtain

$$\|\tilde{w}^k - w^k\|_{L^\infty(I;X)} \le \max_n \|\Delta w_n\|_X \le c \max_n \int_{t_{n-1}}^{t_n} \|\dot{f}(t)\|_X dt.$$

By the absolute continuity of the integral, we have

$$\|\tilde{w}^k - w^k\|_{L^\infty(I;X)} \to 0 \quad \text{as } k \to 0$$

which implies

$$\|\tilde{w}^k - u\|_{L^\infty(I;X)} \to 0 \quad \text{as } k \to 0. \tag{4.47}$$

A similar argument leads to

$$\|\tilde{f}^k - f\|_{L^\infty(I;X)} \to 0 \quad \text{as } k \to 0. \tag{4.48}$$

Then the convergence properties (4.45) and (4.46) follow from (4.47) and (4.48), by the assumption (4.33). ∎

Lemma 4.20 *We have the following convergence as $k \to 0$:*

$$\int_I a(\tilde{w}^k(t), \dot{w}^k(t))\, dt \to \int_I a(u(t), \dot{u}(t))\, dt, \tag{4.49}$$

$$\int_I (\tilde{f}^k(t), \dot{w}^k(t))_X\, dt \to \int_I (f(t), \dot{u}(t))_X\, dt, \tag{4.50}$$

$$\liminf \int_I j(\dot{w}^k(t))\, dt \ge \int_I j(\dot{w}(t))\, dt. \tag{4.51}$$

Proof. Write

$$\begin{aligned}\int_I a(\tilde{w}^k(t), \dot{w}^k(t))\, dt &- \int_I a(u(t), \dot{u}(t))\, dt \\ &= \int_I a(u(t), \dot{w}^k(t) - \dot{u}(t))\, dt \\ &\quad + \int_I a(\tilde{w}^k(t) - u(t), \dot{w}^k(t))\, dt. \end{aligned} \tag{4.52}$$

Since $\dot{w}^k \rightharpoonup \dot{u}$ or $\dot{w}^k \rightharpoonup^* \dot{u}$ in $L^p(I;X)$, we have

$$\int_I a(u(t), \dot{w}^k(t) - \dot{u}(t))\,dt \to 0 \quad \text{as } k \to 0. \tag{4.53}$$

Also,

$$\begin{aligned}\left|\int_I a(\tilde{w}^k(t) - u(t), \dot{w}^k(t))\,dt\right| &\leq c \int_I \|\tilde{w}^k(t) - u(t)\|_X \|\dot{w}^k(t)\|_X\,dt \\ &\leq c\,\|\tilde{w}^k - u\|_{L^\infty(I;X)} \|\dot{w}^k\|_{L^1(I;X)} \\ &\to 0 \quad \text{as } k \to 0. \end{aligned} \tag{4.54}$$

The convergence (4.49) is now a consequence of (4.52)–(4.54). The convergence (4.50) follows from a similar argument. Finally, the convergence (4.51) follows from (4.34), standard semicontinuity arguments and Lemma 4.18. ∎

We have now all the ingredients to prove the theorem.

Proof of Theorem 4.16. From (4.39), we obtain

$$a(\tilde{w}^k(t), v - \dot{w}^k(t)) + j(v) - j(\dot{w}^k(t)) \geq (\tilde{f}^k(t), v - \dot{w}^k(t))_X \quad \forall\, v \in X,$$

for almost any $t \in I$. By integrating this inequality on $\overline{I}$ we find

$$\begin{aligned}\int_I a(\tilde{w}^k(t), v(t) - \dot{w}^k(t))\,dt + \int_I j(v(t))\,dt - \int_I j(\dot{w}^k(t))\,dt \\ \geq \int_I (\tilde{f}^k(t), v(t) - \dot{w}^k(t))_X\,dt \quad \forall\, v \in L^p(I;X).\end{aligned}$$

Using now Lemmas 4.19 and 4.20 we deduce

$$\begin{aligned}\int_I a(u(t), v(t) - \dot{u}(t))\,dt + \int_I j(v(t))\,dt - \int_I j(\dot{u}(t))\,dt \\ \geq \int_I (f(t), v(t) - \dot{u}(t))_X\,dt \quad \forall\, v \in L^p(I;X).\end{aligned}$$

Then performing a localization argument by applying Theorem 2.23 (cf. [96, p. 165]) we obtain that the element $u \in W^{1,p}(I;X)$ satisfies (4.31). Moreover, since $w_k(0) = w_0 = u_0$ (see (4.38)), we deduce from Lemma 4.18 that $u(0) = u_0$, which proves that u is a solution of Problem 4.15.

Assume $u_1, u_2 \in W^{1,p}(I;X)$ are solutions of Problem 4.15. The inequalities below hold for all $v \in X$ and a.e. $t \in I$:

$$\begin{aligned} a(u_1(t), v - \dot{u}_1(t)) + j(v) - j(\dot{u}_1(t)) &\geq (f(t), v - \dot{u}_1(t))_X, \\ a(u_2(t), v - \dot{u}_2(t)) + j(v) - j(\dot{u}_2(t)) &\geq (f(t), v - \dot{u}_2(t))_X. \end{aligned}$$

We take $v = \dot{u}_2(t)$ in the first inequality, $v = \dot{u}_1(t)$ in the second inequality. Adding the corresponding inequalities and using the symmetry of the bilinear form, we obtain

$$\frac{1}{2}\frac{d}{dt}\,a(u_1(t) - u_2(t), u_1(t) - u_2(t)) \leq 0. \tag{4.55}$$

Integrating this inequality from 0 to s and using the initial conditions $u_1(0) = u_2(0) = u_0$, we obtain

$$a(u_1(s) - u_2(s), u_1(s) - u_2(s)) \le 0 \quad \forall\, s \in \overline{I}.$$

Since a is X-elliptic, we deduce that $u_1(s) = u_2(s)$ for all $s \in \overline{I}$, which proves the uniqueness part in Theorem 4.16.

Let now $f_i \in W^{1,p}(I;X)$ and $u_{i,0} \in X$ be such that (4.37) holds for $f = f_i$ and $u_0 = u_{i,0}$, $i = 1, 2$. We denote by $u_i \in W^{1,p}(I;X)$ the solution of the Cauchy problem (4.31)–(4.32) for the data f_i and $u_{i,0}$. A computation similar to the one for (4.55) leads to the inequality

$$\frac{1}{2}\frac{d}{dt}\, a(u_1 - u_2, u_1 - u_2) \le (f_1 - f_2, \dot{u}_1 - \dot{u}_2)_X \quad \text{a.e. on } I.$$

Integrating the inequality from 0 to t, we obtain

$$\begin{aligned}\|u_1(t) - u_2(t)\|_a^2 &\le \|u_{1,0} - u_{2,0}\|_a^2 + 2\int_0^t (f_1(s) - f_2(s), \dot{u}_1(s) - \dot{u}_2(s))_X\, ds \\ &= \|u_{1,0} - u_{2,0}\|_a^2 + 2\,(f_1(0) - f_2(0), u_{1,0} - u_{2,0})_X \\ &\quad - 2\int_0^t (\dot{f}_1(s) - \dot{f}_2(s), u_1(s) - u_2(s))_X\, ds\end{aligned}$$

for all $t \in \overline{I}$, where

$$\|v\|_a = a(v,v)^{1/2} \tag{4.56}$$

is a norm on X, equivalent to $\|\cdot\|_X$. Thus,

$$\begin{aligned}\|u_1(t) - u_2(t)\|_a^2 \le c\Big\{&\|u_{1,0} - u_{2,0}\|_X^2 + \|f_1(0) - f_2(0)\|_X^2 \\ &+ \int_0^t \|\dot{f}_1(s) - \dot{f}_2(s)\|_X \|u_1(s) - u_2(s)\|_a ds\Big\}.\end{aligned}$$

Then,

$$\begin{aligned}\|u_1 - u_2\|_{L^\infty(I;X)}^2 \le c\Big\{&\|u_{1,0} - u_{2,0}\|_X^2 + \|f_1(0) - f_2(0)\|_X^2 \\ &+ \|\dot{f}_1 - \dot{f}_2\|_{L^1(I;X)} \|u_1 - u_2\|_{L^\infty(I;X)}\Big\}.\end{aligned}$$

Therefore,

$$\begin{aligned}\|u_1 - u_2\|_{L^\infty(I;X)} \le c\Big\{&\|u_{1,0} - u_{2,0}\|_X + \|f_1(0) - f_2(0)\|_X \\ &+ \|\dot{f}_1 - \dot{f}_2\|_{L^1(I;X)}\Big\}.\end{aligned} \tag{4.57}$$

Since $W^{1,1}(I;X) \hookrightarrow C(\overline{I};X)$, from (4.57) we deduce

$$\|u_1 - u_2\|_{L^\infty(I;X)} \le c\Big\{\|u_{1,0} - u_{2,0}\|_X + \|f_1 - f_2\|_{W^{1,1}(I;X)}\Big\},$$

which concludes the proof of Theorem 4.16. ■

We observe from the above proof that the symmetry of the bilinear form is not needed in the existence part.

4.4 Semi-discrete Approximations of the Evolutionary Variational Inequality

Assume all the conditions stated in Theorem 4.16 hold so that Problem 4.15 has a unique solution $u \in W^{1,p}(I;X)$. We now turn to numerical approximations of Problem 4.15. We will study semi-discrete approximation schemes in this section and fully discrete approximation schemes in the next section. The analysis given in this and the next sections represents a slight extension of that discussed in [96] in two aspects: First, the initial value u_0 is not assumed zero, and second, the time stepsize can vary. Everywhere in this and the next sections, c will denote positive constants which are independent of the discretization parameters h and k defined later.

Two types of semi-discrete approximation schemes can be developed. In the first type, we replace the infinite-dimensional space X by a finite-dimensional subspace X^h, leading to a finite-dimensional system of ordinary differential equations and inequalities. Such approximation schemes can be termed as *spatially semi-discrete schemes.* In the second type, we replace the time derivatives by finite differences that lead to elliptic variational inequalities over infinite-dimensional space X at each time step. Such approximation schemes can be termed as *temporally semi-discrete schemes.* In this work when we study semi-discrete schemes, we only consider spatially semi-discrete schemes. One reason behind this selection of the presentation materials is that after a spatially semi-discrete discretization, we can employ some ODE solvers for the numerical solution, and the development of ODE solvers has reached some maturity. Another reason we do not consider temporally semi-discrete schemes is that analysis of these semi-discrete schemes is simpler than that of fully discrete schemes. One is referred to [96] for a detailed discussion on error analysis of temporally semi-discrete schemes.

Now we approximate Problem 4.15 with the infinite-dimensional space X being replaced by a family of finite-dimensional subspaces $\{X^h\}$. The subspaces $\{X^h\}$ are intended to be finite element spaces, though much of the analysis applies to more general situations.

Let $\{X^h\}$ be a family of finite-dimensional subspaces of X, with the property that

$$\lim_{h\to 0} \inf_{v^h \in X^h} \|v - v^h\|_X = 0 \quad \forall\, v \in X.$$

Then a semi-discrete internal approximation of Problem 4.15 is:

Problem 4.21 *Find $u^h : \overline{I} \to X^h$ such that*

$$u^h(0) = \mathcal{P}^h u_0 \tag{4.58}$$

and for almost all $t \in I$,

$$a(u^h(t), v^h - \dot{u}^h(t)) + j(v^h) - j(\dot{u}^h(t)) \geq (f(t), v^h - \dot{u}^h(t))_X \quad \forall\, v^h \in X^h. \tag{4.59}$$

Above, $\mathcal{P}^h u_0 \in X^h$ is the *finite element projection* of u_0 on X^h defined by

$$a(\mathcal{P}^h u_0, v^h) = a(u_0, v^h) \quad \forall\, v^h \in X^h. \tag{4.60}$$

By the properties of a, $\mathcal{P}^h u_0 \in X^h$ exists and is unique. By Céa's inequality (3.32) (page 58), we have

$$\|u_0 - \mathcal{P}^h u_0\|_X \le c \inf_{v^h \in X^h} \|u_0 - v^h\|_X.$$

Using the condition (4.37), we obtain its discrete analog:

$$a(\mathcal{P}^h u_0, v^h) + j(v^h) \ge (f(0), v^h)_X \quad \forall\, v^h \in X^h. \tag{4.61}$$

Thus the existence of a unique solution u^h to Problem 4.21 follows from Theorem 4.16 with X replaced by X^h. We note from Theorem 4.16 that $u^h \in W^{1,p}(I; X)$. This regularity result implies that $u^h \in C(\overline{I}; X)$; in particular, the value $u^h(0)$ is well-defined. We also have the stability estimate

$$\begin{aligned} \|u_1^h - u_2^h\|_{L^\infty(I;X)} \le c \big\{ &\|\mathcal{P}^h u_{1,0} - \mathcal{P}^h u_{2,0}\|_X + \|f_1(0) - f_2(0)\|_X \\ &+ \|\dot{f}_1 - \dot{f}_2\|_{L^1(I;X)} \big\} \end{aligned}$$

for the semidiscrete solutions u_1^h and u_2^h corresponding to two initial values $\mathcal{P}^h u_{1,0}$, $\mathcal{P}^h u_{2,0}$, and two right-hand sides $f_1, f_2 \in W^{1,p}(I; X)$, see (4.57).

We turn to an estimate for the semi-discrete approximation error $u - u^h$. For convenience we use the equivalent norm $\|\cdot\|_a$ defined in (4.56). The equalities and inequalities below hold for almost any $t \in I$. We begin by setting $v = \dot{u}^h(t)$ in (4.31) to obtain

$$a(u(t), \dot{u}^h(t) - \dot{u}(t)) + j(\dot{u}^h(t)) - j(\dot{u}(t)) \ge (f(t), \dot{u}^h(t) - \dot{u}(t))_X. \tag{4.62}$$

We now add (4.62) to (4.59) and obtain

$$\begin{aligned} &a(u(t), \dot{u}^h(t) - \dot{u}(t)) + a(u^h(t), v^h - \dot{u}^h(t)) + j(v^h) - j(\dot{u}(t)) \\ &\quad \ge (f(t), v^h - \dot{u}(t))_X. \end{aligned} \tag{4.63}$$

For any $v^h \in X^h$, we write

$$\begin{aligned} &\frac{1}{2}\frac{d}{dt}\|u(t) - u^h(t)\|_a^2 \\ &\quad = a(u(t) - u^h(t), \dot{u}(t) - \dot{u}^h(t)) \\ &\quad = a(u(t) - u^h(t), \dot{u}(t) - v^h) + a(u(t) - u^h(t), v^h - \dot{u}^h(t)). \end{aligned}$$

Then using (4.63), we have

$$\frac{1}{2}\frac{d}{dt}\|u(t) - u^h(t)\|_a^2 \le a(u(t) - u^h(t), \dot{u}(t) - v^h) + R(t; \dot{u}(t), v^h), \tag{4.64}$$

where

$$R(t;\dot{u}(t),v^h) = a(u(t),v^h-\dot{u}(t)) + j(v^h) - j(\dot{u}(t)) \\ - (f(t),v^h-\dot{u}(t))_X \tag{4.65}$$

is a residual quantity. Since

$$\begin{aligned} &a(u(t)-u^h(t),\dot{u}(t)-v^h) \\ &\quad \le a(u(t)-u^h(t),u(t)-u^h(t))^{1/2} a(\dot{u}(t)-v^h,\dot{u}(t)-v^h)^{1/2} \\ &\quad \le c\left(\|u(t)-u^h(t)\|_a^2 + \|\dot{u}(t)-v^h\|_X^2\right), \end{aligned}$$

from (4.64) we find that for any $v^h = v^h(t) \in X^h$,

$$\begin{aligned} &\frac{d}{dt}\|u(t)-u^h(t)\|_a^2 \\ &\quad \le c\left(\|u(t)-u^h(t)\|_a^2 + \|\dot{u}(t)-v^h(t)\|_X^2 + R(t;\dot{u}(t),v^h(t))\right). \end{aligned} \tag{4.66}$$

We multiply the inequality (4.66) by e^{-ct} and integrate it from 0 to s to obtain

$$\begin{aligned} \|u(s)-u^h(s)\|_a^2 &\le c\,\|u_0-\mathcal{P}^h u_0\|_a^2 \\ &\quad + c\,e^{cs}\int_0^s e^{-ct}\left(\|\dot{u}(t)-v^h(t)\|_X^2 + R(t;\dot{u}(t),v^h(t))\right)dt, \end{aligned}$$

for all $s \in \overline{I}$. This in turn leads to the Céa-type inequality

$$\begin{aligned} \|u-u^h\|_{L^\infty(I;X)} &\le c\,\|u_0-\mathcal{P}^h u_0\|_X^2 + c \inf_{v^h\in L^2(I;X^h)}\Big(\|\dot{u}-v^h\|_{L^2(I;X)} \\ &\quad + \|R(\cdot;\dot{u}(\cdot),v^h(\cdot))\|_{L^1(I)}^{1/2}\Big). \end{aligned} \tag{4.67}$$

To conclude, we have proved the following result.

Theorem 4.22 *Suppose the assumptions stated in Theorem 4.16 are satisfied. Assume that X^h is a finite-dimensional subspace of X. Let $u \in W^{1,p}(I;X)$ and $u^h \in W^{1,p}(I;X^h)$ be the solutions of Problems 4.15 and 4.21, respectively. Then the estimate* (4.67) *holds.*

The inequality (4.67) is the basis for various asymptotic error estimates. See Section 8.4 in the context of solving a quasistatic contact problem in this direction.

4.5 Fully Discrete Approximations of the Evolutionary Variational Inequality

We now consider a family of *fully discrete approximation schemes.* In addition to the finite-dimensional subspace X^h, we need a partition of the time interval: $\overline{I} = \cup_{n=1}^N [t_{n-1},t_n]$ with $0 = t_0 < t_1 < \cdots < t_N = T$. We denote $k_n = t_n - t_{n-1}$

for the length of the sub-interval $[t_{n-1}, t_n]$, and $k = \max_n k_n$ for the maximal step-size. In [96], the discussion of the approximation scheme is given for the case of uniform partition of $\overline{I}$. Here we allow the more general case of an arbitrary partition. For a parameter $\theta \in [\frac{1}{2}, 1]$, we denote $t_{n-1+\theta} = \theta\, t_n + (1-\theta)\, t_{n-1}$ and $f_{n-1+\theta} = f(t_{n-1+\theta})$. For the given linear functional $f \in W^{1,p}(I; X)$ and the solution $u \in W^{1,p}(I; X)$, we use the notation $f_n = f(t_n)$ and $u_n = u(t_n)$, which are well-defined. The symbol Δu_n is used to denote the backward difference $u_n - u_{n-1}$, and $\delta_n u_n = \Delta u_n / k_n$ for the backward divided difference. No summation is implied over the repeated index n.

A fully discrete internal approximation of Problem 4.15 is the following.

Problem 4.23 *Find $u^{hk} = \{u_n^{hk}\}_{n=0}^N \subset X^h$ such that*

$$u_0^{hk} = \mathcal{P}^h u_0 \tag{4.68}$$

and for $n = 1, 2, \ldots, N$,

$$\begin{aligned} a(\theta\, u_n^{hk} + (1-\theta)\, u_{n-1}^{hk}, v^h - \delta_n u_n^{hk}) + j(v^h) - j(\delta_n u_n^{hk}) \\ \geq (f_{n-1+\theta}, v^h - \delta_n u_n^{hk})_X \quad \forall\, v^h \in X^h. \end{aligned} \tag{4.69}$$

Here $\mathcal{P}^h u_0$ is defined in (4.60) and hence (4.61) holds.

The reason why the parameter θ is restricted in the range $\left[\frac{1}{2}, 1\right]$ is explained in [96]. For completeness, we repeat it here. The choice $\theta > 1$ is not good, since the scheme (4.69) requires a value of f outside the time interval $\overline{I}$. The scheme with the choice $\theta = 0$ is meaningless. For $0 \neq \theta < \frac{1}{2}$, the scheme is unstable. To see this, consider the extreme case when $j = 0$ and $u_0 = \mathcal{P}^h u_0 = 0$. Then the continuous problem is to find u such that $u(0) = 0$ and at any time $t \in \overline{I}$,

$$a(u(t), v) = (f(t), v)_X \quad \forall\, v \in X.$$

The discrete approximation is $u_0^{hk} = 0$ and for $n = 1, \ldots, N$,

$$a(\theta\, u_n^{hk} + (1-\theta)\, u_{n-1}^{hk}, v^h) = (f_{n-1+\theta}, v^h)_X \quad \forall\, v^h \in X^h.$$

Now consider a perturbed problem for u^{hk}: $\hat{u}_0^{hk} = \varepsilon$, and for $n = 1, \ldots, N$,

$$a(\theta\, \hat{u}_n^{hk} + (1-\theta)\, \hat{u}_{n-1}^{hk}, v^h) = (f_{n-1+\theta}, v^h)_X \quad \forall\, v^h \in X^h.$$

For the difference $e_n = \hat{u}_n^{hk} - u_n^{hk}$ we have $e_0 = \varepsilon$ and for $n = 1, \ldots, N$,

$$a(\theta\, e_n + (1-\theta)\, e_{n-1}, v) = 0 \quad \forall\, v^h \in X^h.$$

Thus, for $n \geq 1$,

$$e_n = -\frac{1-\theta}{\theta}\, e_{n-1},$$

and as a consequence,

$$e_n = (-1)^n \left(\frac{1-\theta}{\theta}\right)^n e_0 = (-1)^n \left(\frac{1-\theta}{\theta}\right)^n \varepsilon.$$

Since $0 \neq \theta < \frac{1}{2}$, it follows that $|(1-\theta)/\theta| > 1$. Therefore, a small perturbation in the initial value may cause arbitrarily large errors in the numerical approximation.

We now show the well-posedness of Problem 4.23.

Theorem 4.24 *Suppose the assumptions stated in Theorem 4.16 are satisfied. Assume that X^h is a finite-dimensional subspace of X. Then Problem 4.23 admits a unique solution u^{hk} and the solution is stable in the sense that for given $u_{1,0}, u_{2,0} \in X$ and $f_1, f_2 \in W^{1,p}(I;X)$, the corresponding solutions $u_{1,n}^{hk}$ and $u_{2,n}^{hk}$, $0 \le n \le N$ satisfy the inequality*

$$\max_{0\le n\le N} \|u_{1,n}^{hk} - u_{2,n}^{hk}\|_X \le c\left(\|\mathcal{P}^h u_{1,0} - \mathcal{P}^h u_{2,0}\|_X + \|f_1(0) - f_2(0)\|_X \right.$$
$$\left. + \|f_1 - f_2\|_{L^1(I;X)}\right). \tag{4.70}$$

Proof. Because of the positive homogeneity of j, the inequality (4.69) can be rewritten as

$$\begin{aligned}&\theta\, a(\Delta u_n^{hk}, v^h - \Delta u_n^{hk}) + j(v^h) - j(\Delta u_n^{hk})\\&\quad \ge (f_{n-1+\theta}, v^h - \Delta u_n^{hk})_X - a(u_{n-1}^{hk}, v^h - \Delta u_n^{hk}) \quad \forall\, v^h \in X^h,\end{aligned} \tag{4.71}$$

which admits a unique solution $\Delta u_n^{hk} \in X^h$.

We then derive the stability inequality (4.70). For $n = 1, 2, \ldots, N$, we have $\delta_n u_{1,n}^{hk}, \delta_n u_{2,n}^{hk} \in X^h$, and

$$\begin{aligned}&a(\theta\, u_{1,n}^{hk} + (1-\theta)\, u_{1,n-1}^{hk}, v^h - \delta_n u_{1,n}^{hk}) + j(v^h) - j(\delta_n u_{1,n}^{hk})\\&\qquad \ge (f_{1,n-1+\theta}, v^h - \delta_n u_{1,n}^{hk})_X \qquad \forall\, v^h \in X^h,\end{aligned} \tag{4.72}$$

$$\begin{aligned}&a(\theta\, u_{2,n}^{hk} + (1-\theta)\, u_{2,n-1}^{hk}, v^h - \delta_n u_{2,n}^{hk}) + j(v^h) - j(\delta_n u_{2,n}^{hk})\\&\qquad \ge (f_{2,n-1+\theta}, v^h - \delta_n u_{2,n}^{hk})_X \qquad \forall\, v^h \in X^h.\end{aligned} \tag{4.73}$$

Let $e_n = u_{1,n}^{hk} - u_{2,n}^{hk}$ denote the difference between the two solutions. Taking $v^h = \delta_n u_{2,n}^{hk}$ in (4.72) and $v^h = \delta_n u_{1,n}^{hk}$ in (4.73), and adding the two resultant inequalities, we obtain

$$a(\theta\, e_n + (1-\theta)\, e_{n-1}, \delta_n e_n) \le (f_{1,n-1+\theta} - f_{2,n-1+\theta}, \delta_n e_n)_X.$$

Let us bound the quantity

$$A_n = a(\theta\, e_n + (1-\theta)\, e_{n-1}, \delta_n e_n) \tag{4.74}$$

from below. From

$$\begin{aligned}A_n &= \frac{1}{k_n}\,[\theta\, a(e_n, e_n) - (2\theta - 1)\, a(e_n, e_{n-1}) - (1-\theta)\, a(e_{n-1}, e_{n-1})]\\&\ge \frac{1}{k_n}\,\left[\theta\, \|e_n\|_a^2 - (2\theta-1)\, \|e_n\|_a \|e_{n-1}\|_a - (1-\theta)\, \|e_{n-1}\|_a^2\right]\\&\ge \frac{1}{k_n}\,\left[\theta\, \|e_n\|_a^2 - (2\theta-1)\, \tfrac{1}{2}\,(\|e_n\|_a^2 + \|e_{n-1}\|_a^2) - (1-\theta)\, \|e_{n-1}\|_a^2\right],\end{aligned}$$

we obtain the lower bound

$$A_n \geq \frac{1}{2k_n} \left(\|e_n\|_a^2 - \|e_{n-1}\|_a^2\right). \tag{4.75}$$

Then for $n = 1, \ldots, N$,

$$\|e_n\|_a^2 - \|e_{n-1}\|_a^2 \leq 2\,(f_{1,n-1+\theta} - f_{2,n-1+\theta}, e_n - e_{n-1})_X.$$

A simple induction shows that

$$\begin{aligned}\|e_n\|_a^2 &\leq \|e_0\|_a^2 + 2\sum_{j=1}^{n}(f_{1,j-1+\theta} - f_{2,j-1+\theta}, e_j - e_{j-1})_X \\ &= \|e_0\|_a^2 + 2\,(f_{1,n-1+\theta} - f_{2,n-1+\theta}, e_n)_X \\ &\quad + 2\sum_{j=1}^{n-1}((f_{1,j-1+\theta} - f_{1,j+\theta}) - (f_{2,j-1+\theta} - f_{2,j+\theta}), e_j)_X.\end{aligned}$$

With $M = \max_n \|e_n\|_a$, we then find that for $n = 1, \ldots, N$,

$$\begin{aligned}\|e_n\|_a^2 \leq \|e_0\|_a^2 + c\Big(&\|f_{1,n-1+\theta} - f_{2,n-1+\theta}\|_X \\ &+ \sum_{j=1}^{n-1}\|(f_{1,j-1+\theta} - f_{1,j+\theta}) - (f_{2,j-1+\theta} - f_{2,j+\theta})\|_X\Big)\,M.\end{aligned}$$

Therefore,

$$\begin{aligned}M^2 \leq \|e_0\|_a^2 + c\Big(&\|f_{1,N-1+\theta} - f_{2,N-1+\theta}\|_X \\ &+ \sum_{j=1}^{N-1}\|(f_{1,j-1+\theta} - f_{1,j+\theta}) - (f_{2,j-1+\theta} - f_{2,j+\theta})\|_X\Big)\,M,\end{aligned}$$

that is,

$$\begin{aligned}\max_n \|e_n\|_a \leq \|e_0\|_a + c\Big(&\|f_{1,N-1+\theta} - f_{2,N-1+\theta}\|_X \\ &+ \sum_{j=1}^{N-1}\|(f_{1,j-1+\theta} - f_{1,j+\theta}) - (f_{2,j-1+\theta} - f_{2,j+\theta})\|_X\Big). \end{aligned}\tag{4.76}$$

We then get the estimate (4.70). ■

Now we turn our attention to an error analysis for the fully discrete scheme. The quantity of interest is the error $e_n = u_n - u_n^{hk}$, $0 \leq n \leq N$. We consider the quantities

$$A_n = a(\theta\, e_n + (1-\theta)\, e_{n-1}, \delta_n e_n), \quad n = 1, \ldots, N. \tag{4.77}$$

As in (4.75), we have a lower bound for A_n:

$$A_n \geq \frac{1}{2k_n} \left(\|e_n\|_a^2 - \|e_{n-1}\|_a^2 \right). \tag{4.78}$$

To derive an upper bound for A_n, we write

$$\begin{aligned} A_n &= a(\theta\, u_n + (1-\theta)\, u_{n-1}, \delta_n u_n - \delta_n u_n^{hk}) \\ &\quad - a(\theta\, u_n^{hk} + (1-\theta)\, u_{n-1}^{hk}, \delta_n u_n - v_n^h) \\ &\quad - a(\theta\, u_n^{hk} + (1-\theta)\, u_{n-1}^{hk}, v_n^h - \delta_n u_n^{hk}), \end{aligned}$$

where $v_n^h \in X^h$ is arbitrary. Using (4.69) to bound the last term, we obtain

$$\begin{aligned} A_n &\leq a(\theta\, u_n + (1-\theta)\, u_{n-1}, \delta_n u_n - \delta_n u_n^{hk}) \\ &\quad - a(\theta\, u_n^{hk} + (1-\theta)\, u_{n-1}^{hk}, \delta_n u_n - v_n^h) \\ &\quad + j(v_n^h) - j(\delta_n u_n^{hk}) - (f_{n-1+\theta}, v_n^h - \delta_n u_n^{hk})_X. \end{aligned} \tag{4.79}$$

Now we take $v = \delta_n u_n^{hk} \in K$ in (4.31) at $t = t_{n-1+\theta}$ to get

$$\begin{aligned} 0 &\leq a(u_{n-1+\theta}, \delta_n u_n^{hk} - \dot{u}_{n-1+\theta}) + j(\delta_n u_n^{hk}) - j(\dot{u}_{n-1+\theta}) \\ &\quad - (f_{n-1+\theta}, \delta_n u_n^{hk} - \dot{u}_{n-1+\theta})_X. \end{aligned} \tag{4.80}$$

Adding (4.79) and (4.80), we have

$$\begin{aligned} A_n &\leq a(\theta\, u_n + (1-\theta)\, u_{n-1}, \delta_n u_n - \delta_n u_n^{hk}) \\ &\quad - a(\theta\, u_n^{hk} + (1-\theta)\, u_{n-1}^{hk}, \delta_n u_n - z_n^h) \\ &\quad + a(u_{n-1+\theta}, \delta_n u_n^{hk} - \dot{u}_{n-1+\theta}) \\ &\quad + j(v_n^h) - j(\dot{u}_{n-1+\theta}) - (f_{n-1+\theta}, v_n^h - \dot{u}_{n-1+\theta})_X, \end{aligned}$$

which is rewritten as

$$\begin{aligned} A_n &\leq \frac{1}{k_n}\, a(E_{n,\theta}(u), e_n - e_{n-1}) \\ &\quad + a(\theta\, e_n + (1-\theta)\, e_{n-1}, \delta_n u_n - v_n^h) \\ &\quad - a(E_{n,\theta}(u), \delta_n u_n - v_n^h) + R_n(\dot{u}_{n-1+\theta}, v_n^h), \end{aligned} \tag{4.81}$$

where

$$E_{n,\theta}(u) = \theta\, u_n + (1-\theta)\, u_{n-1} - u_{n-1+\theta} \tag{4.82}$$

and

$$\begin{aligned} R_n(\dot{u}_{n-1+\theta}, v_n^h) &= a(u_{n-1+\theta}, v_n^h - \dot{u}_{n-1+\theta}) + j(v_n^h) \\ &\quad - j(\dot{u}_{n-1+\theta}) - (f_{n-1+\theta}, v_n^h - \dot{u}_{n-1+\theta})_X. \end{aligned}$$

To proceed further, we need estimates for several quantities involving $E_{n,\theta}(u)$ defined in (4.82). These estimates are proved in [96] in the case of uniform step size. The first two results follow immediately from the proofs in [96] for the same results with uniform step size.

Lemma 4.25 *Under the assumption $\ddot{w} \in L^1(I;X)$, we have*

$$\|E_{n,\theta}(w)\|_X \le 2\,\theta\,(1-\theta)\,k_n\,\|\ddot{w}\|_{L^1(t_{n-1},t_n;X)}.$$

If we further assume that $\ddot{w} \in L^\infty(I;X)$, then

$$\|E_{n,\theta}(w)\|_X \le \frac{\theta\,(1-\theta)}{2}\,k_n^2\,\|\ddot{w}\|_{L^\infty(I;X)}.$$

Lemma 4.26 *Assume that $\ddot{w} \in L^1(I;X)$. Then*

$$\|\delta w_n - \dot{w}_{n-1+\theta}\|_X \le \|\ddot{w}\|_{L^1(t_{n-1},t_n;X)}.$$

Furthermore, if $w^{(3)} \in L^1(I;X)$, then

$$\|\delta w_n - \dot{w}_{n-1/2}\|_X \le \frac{k_n}{8}\,\|w^{(3)}\|_{L^1(t_{n-1},t_n;X)}.$$

For the next lemma, we need to modify slightly the proof in [96] for a similar result. For completeness, we include its proof here.

Lemma 4.27 *Under the assumption $\ddot{w} \in L^1(I;X)$, we have*

$$\|E_{n,\theta}(w) - E_{n+1,\theta}(w)\|_X \le c\,(k_n + k_{n+1})\,\|\ddot{w}\|_{L^1(t_{n-1},t_{n+1};X)}. \tag{4.83}$$

If we further assume that $w^{(3)} \in L^1(I;X)$ and, $\theta = 1$ or the step size change is of second order in k from one time sub-interval to the next, i.e.,

$$|k_n - k_{n+1}| = O(k^2), \quad n = 1,\dots,N-1, \tag{4.84}$$

then

$$\begin{aligned}\|E_{n,\theta}(w) - E_{n+1,\theta}(w)\|_X &\le c\,k^2(k_n + k_{n+1})\,\|\ddot{w}\|_{L^\infty(I;X)}\\ &\quad + c\,k^2\|w^{(3)}\|_{L^1(t_{n-1},t_{n+1};X)}.\end{aligned} \tag{4.85}$$

Proof. Write

$$\begin{aligned}
w_{n+1} &= w_n + k_{n+1}\dot{w}_n + \int_{t_n}^{t_{n+1}} (t_{n+1} - t)\,\ddot{w}(t)\,dt,\\
w_{n-1} &= w_n - k_n\dot{w}_n + \int_{t_n}^{t_{n-1}} (t_{n-1} - t)\,\ddot{w}(t)\,dt,\\
w_{n-1+\theta} &= w_n - (1-\theta)\,k_n\dot{w}_n + \int_{t_n}^{t_{n-1+\theta}} (t_{n-1+\theta} - t)\,\ddot{w}(t)\,dt,\\
w_{n+\theta} &= w_n + \theta\,k_{n+1}\dot{w}_n + \int_{t_n}^{t_{n+\theta}} (t_{n+\theta} - t)\,\ddot{w}(t)\,dt.
\end{aligned}$$

Then

$$
\begin{aligned}
&E_{n,\theta}(w) - E_{n+1,\theta}(w) \\
&\quad = -\theta \int_{t_n}^{t_{n+1}} (t_{n+1} - t)\, \ddot{w}(t)\, dt + (1-\theta) \int_{t_n}^{t_{n-1}} (t_{n-1} - t)\, \ddot{w}(t)\, dt \\
&\quad\quad + \int_{t_n}^{t_{n+\theta}} (t_{n+\theta} - t)\, \ddot{w}(t)\, dt - \int_{t_n}^{t_{n-1+\theta}} (t_{n-1+\theta} - t)\, \ddot{w}(t)\, dt.
\end{aligned}
$$

Thus the estimate (4.83) holds.

Assume now $w^{(3)} \in L^1(I;X)$. Then

$$
\ddot{w}(t) = \ddot{w}_n + \int_{t_n}^{t} w^{(3)}(s)\, ds,
$$

and we have

$$
\begin{aligned}
E_{n,\theta}(w) - E_{n+1,\theta}(w) = {} & \frac{1}{2}\, \theta\, (1-\theta)\, (k_n - k_{n+1})\, (k_n + k_{n+1})\, \ddot{w}_n \\
& - \theta \int_{t_n}^{t_{n+1}} (t_{n+1} - t) \int_{t_n}^{t} w^{(3)}(s)\, ds\, dt \\
& + (1-\theta) \int_{t_n}^{t_{n-1}} (t_{n-1} - t) \int_{t_n}^{t} w^{(3)}(s)\, ds\, dt \\
& + \int_{t_n}^{t_{n+\theta}} (t_{n+\theta} - t) \int_{t_n}^{t} w^{(3)}(s)\, ds\, dt \\
& - \int_{t_n}^{t_{n-1+\theta}} (t_{n-1+\theta} - t) \int_{t_n}^{t} w^{(3)}(s)\, ds\, dt.
\end{aligned}
$$

Then if $\theta = 1$ or (4.84) holds, we have the estimate (4.85). ∎

From the upper bound (4.81) and the lower bound (4.78), we obtain the inequality

$$
\begin{aligned}
\frac{1}{2k_n} \left(\|e_n\|_a^2 - \|e_{n-1}\|_a^2 \right) \le {} & \frac{1}{k_n}\, a(E_{n,\theta}(u), e_n - e_{n-1}) + c\, M\, \|\delta_n u_n - v_n^h\|_X \\
& + c\, \|E_{n,\theta}(u)\|_X\, \|\delta_n u_n - v_n^h\|_X \\
& + |R_n(\dot{u}_{n-1+\theta}, v_n^h)|, \qquad (4.86)
\end{aligned}
$$

where $M = \max_n \|e_n\|_a$. Thus

$$
\begin{aligned}
\|e_n\|_a^2 - \|e_{n-1}\|_a^2 \le {} & 2\, a(E_{n,\theta}(u), e_n - e_{n-1}) + c\, M\, k_n\, \|\delta_n u_n - v_n^h\|_X \\
& + c\, k_n\, \|E_{n,\theta}(u)\|_X\, \|\delta_n u_n - v_n^h\|_X \\
& + c\, k_n\, |R_n(\dot{u}_{n-1+\theta}, v_n^h)|.
\end{aligned}
$$

A simple induction argument yields, for $1 \le n \le N$,

$$\begin{aligned}\|e_n\|_a^2 \le 2\sum_{j=1}^{n} a(E_{j,\theta}(u), e_j - e_{j-1}) + c\,M\sum_{j=1}^{n} k_j\|\delta_j u_j - v_j^h\|_X \\ + c\,(\max_n \|E_{n,\theta}(u)\|_X)\sum_{j=1}^{n} k_j\|\delta_j u_j - v_j^h\|_X \\ + c\sum_{j=1}^{n} k_j|R_j(\dot{u}_{j-1+\theta}, v_j^h)| + \|e_0\|_a^2.\end{aligned}$$

Notice that

$$\begin{aligned}\sum_{j=1}^{n} a(E_{j,\theta}(u), e_j - e_{j-1}) = a(E_{n,\theta}(u), e_n) \\ + \sum_{j=1}^{n-1} a(E_{j,\theta}(u) - E_{j+1,\theta}(u), e_j).\end{aligned}$$

Hence

$$\begin{aligned}M^2 \le c\,M\Big(\|E_{N,\theta}(u)\|_X + \sum_{j=1}^{N-1}\|E_{j,\theta}(u) - E_{j+1,\theta}(u)\|_X \\ + \sum_{j=1}^{N} k_j\|\delta_j u_j - v_j^h\|_X\Big) \\ + c\,\max_n \|E_{n,\theta}(u)\|_X \sum_{j=1}^{N} k_j\|\delta_j u_j - v_j^h\|_X \\ + c\sum_{j=1}^{N} k_j|R_j(\dot{u}_{j-1+\theta}, v_j^h)| + \|e_0\|_a^2.\end{aligned} \tag{4.87}$$

Using the relation

$$x, a, b \ge 0 \text{ and } x^2 \le a\,x + b \Longrightarrow x^2 \le a^2 + 2\,b,$$

we find from (4.87) that

$$\max_n \|u_n - u_n^{hk}\|_X \le c\Big(\|E_{N,\theta}(u)\|_X + \sum_{j=1}^{N-1} \|E_{j,\theta}(u) - E_{j+1,\theta}(u)\|_X$$
$$+ \sum_{j=1}^{N} k_j \|\delta_j u_j - v_j^h\|_X + \|u_0 - \mathcal{P}^h u_0\|_X\Big)$$
$$+ c\Big\{\max_n \|E_{n,\theta}(u)\|_X \sum_{j=1}^{N} k_j \|\delta_j u_j - v_j^h\|_X\Big\}^{1/2}$$
$$+ c\Big\{\sum_{j=1}^{N} k_j |R_j(\dot u_{j-1+\theta}, v_j^h)|\Big\}^{1/2}. \tag{4.88}$$

Concrete error estimates follow from (4.88) if we apply the results given in Lemmas 4.25, 4.27, and 4.26. Let us focus on the orders of the schemes by assuming that the solution is smooth. Specifically, we assume that $u \in W^{3,1}(I;X)$. Then we also have $\ddot u \in L^\infty(I;X)$. From Lemma 4.25,

$$\max_n \|E_{n,\theta}(u)\|_X \le c\,k^2.$$

From Lemma 4.27,

$$\sum_{j=1}^{N-1} \|E_{j,\theta}(u) - E_{j+1,\theta}(u)\|_X \le c\,k^2,$$

and from Lemma 4.26,

$$\sum_{j=1}^{N} \|\delta_j u_j - \dot u_{j-1+\theta}\|_X \le c \quad \text{if } \theta \ne \frac{1}{2},$$
$$\sum_{j=1}^{N} \|\delta_j u_j - \dot u_{j-1/2}\|_X \le c\,k.$$

With these inequalities, together with the triangle inequality

$$\|\delta_j u_j - v_j^h\|_H \le \|\delta_j u_j - \dot u_{j-1+\theta}\|_X + \|\dot u_{j-1+\theta} - v_j^h\|_X,$$

we obtain the following error estimates assuming $u^{(3)} \in L^1(I;X)$:

$$\max_n \|u_n - u_n^{hk}\|_X \le c\,k + c\,\|u_0 - \mathcal{P}^h u_0\|_X + c\sum_{j=1}^{N} k_j \|\dot u_{j-1+\theta} - v_j^h\|_X$$
$$+ c\Big\{\sum_{j=1}^{N} k_j |R_j(\dot u_{j-1+\theta}, v_j^h)|\Big\}^{1/2} \quad \text{if } \theta \ne \frac{1}{2} \tag{4.89}$$

and

$$\max_n \|u_n - u_n^{hk}\|_X \le c\,k^2 + c\,\|u_0 - \mathcal{P}^h u_0\|_X + c\sum_{j=1}^N k_j \|\dot{u}_{j-1/2} - v_j^h\|_X$$

$$+\,c\Big\{\sum_{j=1}^N k_j |R_j(\dot{u}_{j-1/2}, v_j^h)|\Big\}^{1/2} \quad \text{if } \theta = \frac{1}{2}, \qquad (4.90)$$

where $v_j^h \in X^h$, $1 \le j \le N$, are arbitrary. These estimates are applied in Section 8.4 in the context of solving a quasistatic contact problem.

We now summarize the results of the section in the form of a theorem.

Theorem 4.28 *Suppose the assumptions stated in Theorem 4.16 are satisfied. Let $u \in W^{1,p}(I;X)$ and u^{hk} be the solutions of Problems 4.15 and 4.23, respectively. Then if $u \in W^{3,1}(I;X)$, we have the estimates* (4.89) *and* (4.90).

We observe that the orders are optimal with respect to the time step-size in the error estimates (4.89) and (4.90). In particular, when $\theta = 1$, we have a backward Euler scheme, and it is a first-order method with respect to the temporal step-size. When $\theta = \frac{1}{2}$, we have the second-order accurate Crank–Nicolson-type scheme.

Bibliographical Notes

The material presented in Chapter 1 is standard and can be found in many books on functional analysis. For more information in the field we refer the reader to the books [36, 220, 221]. Some recent textbooks on basic notions and results of functional analysis and their applications in numerical analysis are [14, 175]. A complete treatment on general theory of convex functions as well as proofs of the results presented in Section 1.4 can be found in the works [73, 111, 216], for example.

For a comprehensive treatment of basic aspects of spaces introduced in Sections 2.1 and 2.2 as well as for useful generalizations, the reader is referred to [1, 148, 160]. For more detailed information on the spaces of vector-valued functions presented in Section 2.3 we refer to the references [26] and [35].

More details on theoretical analysis of the finite difference method presented in Chapter 3 can be found in several books on the topic, e.g. [150, 200]; see also in-depth survey articles [149, 212]. Standard references on mathematical analysis of the finite element method include [19, 21, 31, 32, 47, 49, 128, 165, 199, 206].

Interest in variational inequalities originates in mechanical problems, see for instance [70], where many problems in mechanics and physics are formulated in the framework of variational inequalities. More recent references in the mathematical analysis of variational inequalities include [22, 86, 136, 167, 185]. Details concerning the numerical analysis of variational inequalities can be found in [90, 91, 109, 112, 134]. A recent reference in the study of mathematical and numerical analysis of variational inequalities arising in hardening plasticity is [96]. The proof of Theorem 4.1 based on the properties of the proximity maps follows [33]. The results presented in Sections 4.3 and 4.4 in the study of evolutionary variational inequalities are based on and represent a slight extension of the results presented in [96]. The study of a more general class of evolutionary variational inequalities of the form (4.31)–(4.32) in which the functional j depends on the solution can be found in [158].

Part II

Mathematical Modelling in Contact Mechanics

Chapter 5

Preliminaries of Contact Mechanics of Continua

In this chapter we summarize some basic notions and general principles of Mechanics of Continua with emphasis in the study of contact problems involving deformable bodies. We first introduce the main concepts of kinematics of continua (the deformation gradient, deformation tensors, and strain tensors), then present elements of dynamics of continua including the balance laws, stress tensors and their principal properties. Most of these notions and results are standard and can be found in many books on mechanics and therefore we skip the details. We then describe the general physical setting of contact problems and summarize a number of contact and friction boundary conditions which will be employed in this book. In this chapter, all variables are assumed to have sufficient degree of smoothness consistent with developments they are involved in.

5.1 Kinematics of Continua

Following the convention in the literature, we understand a *continuous medium* to be a body composed of material that is continuously distributed, i.e. without gaps or empty spaces. By this it is meant, first, that such a body occupies and completely fills a region of the three-dimensional Euclidean space which will vary with time as the body deforms. Second, it is assumed that properties and behavior of such a body can be described by continuous functions. This statement implies that derivatives of the functions exist and are continuous too if they enter the theory, since all functions entering the theory are assumed continuous. This hypothetical continuous body will be called a *continuous medium* or *continuum*.

The study of the behavior of continuous media conveniently begins with development of a suitable framework within which the motion of the body can be described. This framework is independent of any agents acting on the body, and it is also independent of the constitution of the body. In other words, it concerns solely with the geometry of motion. This is known as *kinematics of continua* and we now proceed to review its main concepts. For generality our discussion is given for the d-dimensional case, $d = 1, 2, 3$.

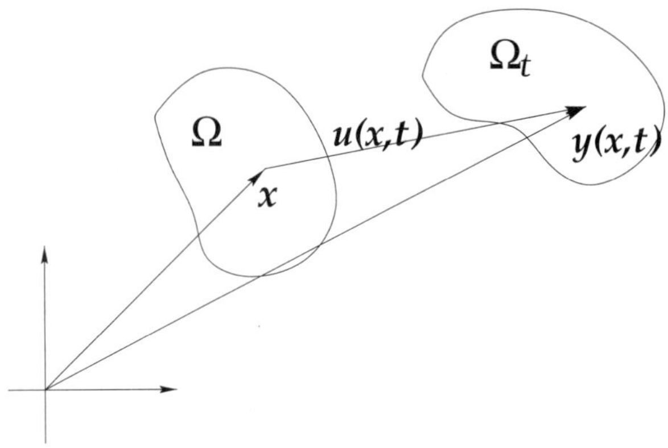

Figure 5.1: Reference and current configurations of a continuous medium

Configuration, motion, displacement field. We assume in what follows that an origin O and an orthonormal basis $\{\boldsymbol{e}_1, \dots, \boldsymbol{e}_d\}$ have been chosen in the d-dimensional Euclidian space which will then be identified with the space $\mathbb{R}^d$. We denote by Ω the subset of $\mathbb{R}^d$ occupied by a continuous medium at time $t = 0$ and by Ω_t the subset of $\mathbb{R}^d$ occupied by the medium at the time $t > 0$. We assume that Ω and Ω_t are bounded, open, and connected subsets of $\mathbb{R}^d$. The set Ω is called the *reference* or *undeformed configuration* of the body and, for every $t > 0$, Ω_t is called the *current* or *deformed configuration* of the body. We refer to a point $\boldsymbol{x} \in \Omega$ as a *material point.*

The *motion* of the body may be described mathematically by introducing a vector-valued function $\boldsymbol{y}$ of position and time. Thus, a material particle initially located at $\boldsymbol{x}$ will have position $\boldsymbol{y}(\boldsymbol{x}, t)$ at time t, see Figure 5.1. We say that the motion is described by the equation

$$\boldsymbol{y} = \boldsymbol{y}(\boldsymbol{x}, t) \tag{5.1}$$

or, in component form,

$$y_i = y_i(x_1, \dots, x_d, t), \quad i = 1, \dots, d,$$

for $\boldsymbol{x} = (x_1, \dots, x_d) \in \Omega$ and $t > 0$.

Here and below the indices i, j run from 1 to d, and the summation convention over repeated indices is adopted. Recall that the index following a comma indicates a partial derivative with respect to the corresponding component of the independent variable x_i.

We assume that the motion (5.1) is continuous and possesses continuous partial derivatives with respect to their arguments. Moreover, we assume the Jacobian

$$J = \det\left(\frac{\partial y_i}{\partial x_j}\right) \neq 0 \tag{5.2}$$

so that the motion is one-to-one. Condition (5.2) is the mathematical form of the axiom of impenetrability, i.e. the motion can only carry every region into a region, every surface into a surface and every curve into a curve. No one portion of a matter can penetrate into another. Indeed, as usual in the literature on mechanics, we will assume the orientation-preserving condition $J > 0$.

Together with the motion $\boldsymbol{y}$, it is often convenient to introduce the *displacement* field $\boldsymbol{u}$ defined by the relation

$$\boldsymbol{u}(\boldsymbol{x}, t) = \boldsymbol{y}(\boldsymbol{x}, t) - \boldsymbol{x} \tag{5.3}$$

for all $\boldsymbol{x} \in \Omega$ and $t > 0$. In the study of most problems arising in solid mechanics, the displacement field plays the role of a primary unknown.

The *velocity* field $\dot{\boldsymbol{u}}$ and the *acceleration* field $\ddot{\boldsymbol{u}}$ corresponding to the displacement field $\boldsymbol{u}$ are defined by

$$\dot{\boldsymbol{u}}(\boldsymbol{x}, t) = \frac{\partial^2 \boldsymbol{u}(\boldsymbol{x}, t)}{\partial t}, \qquad \ddot{\boldsymbol{u}}(\boldsymbol{x}, t) = \frac{\partial^2 \boldsymbol{u}(\boldsymbol{x}, t)}{\partial t^2},$$

for all $\boldsymbol{x} \in \Omega$ and $t > 0$.

Deformation gradient and strain tensors. The displacement alone does not give complete information about the deformation of the body. We need to be able to distinguish, for example, between a simple *rigid body motion*, in which the body is rotated and translated to a new position without deformation, and a situation in which the body indeed assumes a new shape. To this end, we start by defining the matrix

$$\boldsymbol{F} = (F_{ij}), \qquad F_{ij} = \frac{\partial y_i}{\partial x_j} \tag{5.4}$$

at each point $\boldsymbol{x} \in \Omega$ and for any $t > 0$. $\boldsymbol{F}$ is called the *deformation gradient*; it determines a mapping from the reference configuration to the current configuration up to an arbitrary constant vector. We also introduce the symmetric tensor

$$\boldsymbol{C} = \boldsymbol{F}^T \boldsymbol{F}, \qquad C_{ij} = F_{ki} F_{kj}, \tag{5.5}$$

called the *right Cauchy-Green tensor*. Here and below we use $\boldsymbol{A}^T$ to define the transpose of the matrix $\boldsymbol{A}$. In order to show the significance of tensor $\boldsymbol{C}$ let us take the spatial differential of both sides of (5.1) for fixed t. It follows that

$$\boldsymbol{dy} = \boldsymbol{F}\boldsymbol{dx} \qquad (dy_i = F_{ij}\, dx_j).$$

Therefore,

$$\|\boldsymbol{dy}\|^2 = \boldsymbol{dy} \cdot \boldsymbol{dy} = (\boldsymbol{F}\boldsymbol{dx}) \cdot (\boldsymbol{F}\boldsymbol{dx}) = (F_{ki} dx_i)(F_{kj} dx_j) = F_{ki} F_{kj} dx_i dx_j.$$

Here and below we use "$\cdot$" and $\|\cdot\|$ to denote the canonical inner product and norm on spaces $\mathbb{R}^d$ and $\mathbb{R}^{d \times d}$, defined in Examples 1.41 and 1.42. By (5.5) we find

$$\|\boldsymbol{dy}\|^2 = C_{ij} dx_i dx_j = \boldsymbol{dx} \cdot \boldsymbol{C}\boldsymbol{dx}. \tag{5.6}$$

It follows that the knowledge of the tensor $\boldsymbol{C}$ allows us to compute the lengths of the infinitesimal vectors in the deformed configuration.

With the intention of showing that the tensor $\boldsymbol{C}$ is indeed a good measure of "strain", understood here in its intuitive sense of "change in form or size", let us consider a rigid motion in which for all $t > 0$ the current configuration Ω_t is obtained by rotating the reference configuration Ω around the origin with a rotation $\boldsymbol{Q}(t)$ and by translating it with a vector $\boldsymbol{a}(t)$. We have

$$\boldsymbol{y}(\boldsymbol{x}, t) = \boldsymbol{a}(t) + \boldsymbol{Q}(t)\,\boldsymbol{x} \quad \forall \boldsymbol{x} \in \Omega,\ t > 0 \tag{5.7}$$

where $\boldsymbol{a}(t) \in \mathbb{R}^d$, $\boldsymbol{Q}(t) \in \mathbb{Q}^d_+$, $\mathbb{Q}^d_+$ being the set of orthogonal matrices of order d whose determinants are $+1$. For the motion (5.7) we have

$$\boldsymbol{F}(\boldsymbol{x}, t) = \boldsymbol{Q}(t) \quad \forall \boldsymbol{x} \in \Omega,\ t > 0.$$

Therefore from (5.5) it follows that

$$\boldsymbol{C} = \boldsymbol{I} \quad \forall \boldsymbol{x} \in \Omega,\ t > 0, \tag{5.8}$$

where $\boldsymbol{I}$ denotes the identity matrix. It is remarkable that the converse is also true: If (5.8) is satisfied then the corresponding motion is a rigid motion, i.e. it is of the form (5.7) with $\boldsymbol{a}(t) \in \mathbb{R}^d$, $\boldsymbol{Q}(t) \in \mathbb{Q}^d_+$ for $t > 0$. Consequently, the condition (5.8) characterizes the rigid motion, i.e. an arbitrary motion is a rigid one iff (5.8) is satisfied.

In order to compare the deformed configuration with the reference configuration we introduce a relative measure $\boldsymbol{E}$ of the deformation by

$$\boldsymbol{E} = \frac{1}{2}(\boldsymbol{C} - \boldsymbol{I}) = \frac{1}{2}(\boldsymbol{F}^T\boldsymbol{F} - \boldsymbol{I}). \tag{5.9}$$

The tensor $\boldsymbol{E}$ is called the *finite strain tensor* or the *Green-Saint Venant tensor*. It follows from (5.6) that

$$\|\boldsymbol{dy}\|^2 - \|\boldsymbol{dx}\|^2 = C_{ij}\,dx_i\,dx_j - \delta_{ij}\,dx_i\,dx_j = (C_{ij} - \delta_{ij})\,dx_i\,dx_j.$$

Using the definition (5.9), we can write

$$\|\boldsymbol{dy}\|^2 - \|\boldsymbol{dx}\|^2 = 2E_{ij}\,dx_i\,dx_j, \tag{5.10}$$

i.e. the tensor $\boldsymbol{E}$ characterizes the variation of the length square of the material element. Using the definition of the strain tensor $\boldsymbol{E}$ we can rephrase the previously mentioned result: $\boldsymbol{y}$ is a rigid motion iff $\boldsymbol{E} = \boldsymbol{0}$, the zero matrix.

In what follows we express the tensors $\boldsymbol{F}$, $\boldsymbol{C}$ and $\boldsymbol{E}$ in terms of the displacement gradient $\nabla\boldsymbol{u}$, given by

$$\nabla\boldsymbol{u} = \Big(\frac{\partial u_i}{\partial x_j}\Big). \tag{5.11}$$

From (5.3)–(5.5) it follows that

$$\boldsymbol{F} = \boldsymbol{I} + \nabla\boldsymbol{u}, \quad \boldsymbol{C} = \boldsymbol{I} + \nabla\boldsymbol{u} + (\nabla\boldsymbol{u})^T + (\nabla\boldsymbol{u})^T\nabla\boldsymbol{u}.$$

By (5.9) we obtain

$$\boldsymbol{E} = \frac{1}{2}\Big(\nabla\boldsymbol{u} + (\nabla\boldsymbol{u})^T + (\nabla\boldsymbol{u})^T\nabla\boldsymbol{u}\Big). \tag{5.12}$$

In component form, from (5.11) and (5.12) we have

$$E_{ij} = \frac{1}{2}(u_{i,j} + u_{j,i} + u_{k,i}u_{k,j})$$

which shows that the finite strain tensor $\boldsymbol{E} = (E_{ij})$ is a nonlinear function of the displacement field $\boldsymbol{u}$.

Infinitesimal theory. There are many problems of practical interest for which the deformations can be regarded as "small" in some sense, and under such circumstances it is natural to consider whether the formulation of the problem might be simplified by exploiting this feature. Of course, first it is necessary to formalize and to qualify what is meant by the word "small". For the purposes of this work the following definition suffices: A body is said to undergo *infinitesimal deformation* if the displacement gradient $\nabla\boldsymbol{u}$ is sufficiently small so that the nonlinear term in (5.12) can be neglected. In other words, in the infinitesimal deformation theory we suppose that $\eta = \|\nabla\boldsymbol{u}\| << 1$ and we neglect all the quantities of order 2 or more in η. In this case we may replace the finite strain tensor $\boldsymbol{E}$ by the tensor $\boldsymbol{\varepsilon}$ defined by

$$\boldsymbol{\varepsilon} = \frac{1}{2}\Big(\nabla\boldsymbol{u} + (\nabla\boldsymbol{u})^T\Big). \tag{5.13}$$

The tensor $\boldsymbol{\varepsilon}$ is called the *small strain tensor*, the *infinitesimal strain tensor*, or the *linearized strain tensor*. Using the above approximation we have $\boldsymbol{\varepsilon} \approx \boldsymbol{E}$ and $\boldsymbol{\varepsilon}$ will have the same physical meaning as $\boldsymbol{E}$. Since everywhere above it was implicitly assumed that the strain is defined for a particular displacement field, we shall use sometimes the notation $\boldsymbol{\varepsilon}(\boldsymbol{u})$ instead of $\boldsymbol{\varepsilon}$, in order to underline the dependence of the small strain tensor on the displacement field $\boldsymbol{u}$. Thus, by (5.11) and (5.13) we have

$$\boldsymbol{\varepsilon}(\boldsymbol{u}) = (\varepsilon_{ij}(\boldsymbol{u})), \quad \varepsilon_{ij}(\boldsymbol{u}) = \frac{1}{2}(u_{i,j} + u_{j,i}). \tag{5.14}$$

We have seen earlier that the finite strain tensor $\boldsymbol{E}$ vanishes if and only if the body undergoes a rigid body motion. Since we are concerned in the sequel with the infinitesimal deformation theory, it is necessary to characterize a rigid body motion in the context of this theory. To this end, suppose that the body undergoes an infinitesimal rigid body motion: $\boldsymbol{\varepsilon}(\boldsymbol{u}) = \boldsymbol{0}$, i.e.

$$\nabla\boldsymbol{u} = -(\nabla\boldsymbol{u})^T. \tag{5.15}$$

Thus the displacement gradient is skew. It can be shown that condition (5.15) is satisfied iff the displacement field $\boldsymbol{u}$ is given by

$$\boldsymbol{u}(\boldsymbol{x}, t) = \boldsymbol{u}_0(t) + \boldsymbol{\omega}(t) \wedge \boldsymbol{x} \quad \forall\, \boldsymbol{x} \in \Omega,\ t > 0 \tag{5.16}$$

where $\boldsymbol{u}_0(t)$, $\boldsymbol{\omega}(t) \in \mathbb{R}^d$ for all $t > 0$. Here $\wedge$ denotes the exterior product on $\mathbb{R}^d$. We conclude that in the linearized strain theory the motion of the body is rigid iff the corresponding displacement field is of the form (5.16). The set of displacements $\boldsymbol{u}$ of the form (5.16) is called the set of *rigid displacements*.

5.2 Dynamics of Continua

In this section we move away from the purely geometric nature of kinematics and discuss some features of external and internal forces, formulate the mass conservation law and the principles of linear angular momenta. A further development is the introduction of the notion of stress, as a quantity that characterizes the state of internal forces acting in a body and which satisfies the equation of motion.

The principal concepts in the dynamic of continua are *mass* and *forces*. We do not attempt to provide precise definitions for them here. Mass is treated as a measure of material which the body contains. A force is considered as a measure of interactions between the body and its environment or between different parts of the body.

Balance law of mass. Let $\Omega \subset \mathbb{R}^d$ represent the reference configuration of a deformable body and, as before, let $\Omega_t \subset \mathbb{R}^d$ denote the current configuration at some moment $t > 0$. Consider a part of the body, $\mathcal{P}$, which in the reference configuration occupies an open subset $\omega \subset \Omega$ and in the current configuration occupies the set $\omega_t \subset \Omega_t$.

A mass distribution is characterized by a function $\rho : \Omega \to \mathbb{R}_+$, called *the density of mass* in the reference configuration such that the mass of $\mathcal{P}$ is given by

$$m(\mathcal{P}) = \int_\omega \rho(\boldsymbol{x})\, dx.$$

We denote by $\rho_t : \Omega_t \to \mathbb{R}_+$ the density of mass in the current configuration. Thus, the mass of $\mathcal{P}$ in the current configuration is given by

$$m_t(\mathcal{P}) = \int_{\omega_t} \rho_t(\boldsymbol{y})\, dy.$$

The mass conservation law states that the mass does not change during the motion, i.e. $m_t(\mathcal{P}) = m(\mathcal{P})$. Therefore

$$\int_{\omega_t} \rho_t(\boldsymbol{y})\, dy = \int_\omega \rho(\boldsymbol{x})\, dx. \tag{5.17}$$

Since $\omega_t = \boldsymbol{y}(\omega, t)$, we may perform a change of variables to write

$$\int_{\omega_t} \rho_t\, dy = \int_\omega \rho_t J\, dx \tag{5.18}$$

where J is the Jacobian defined in (5.2). It follows from (5.17) and (5.18) that

$$\int_\omega \rho_t J \, dx = \int_\omega \rho \, dx.$$

Since ω is an arbitrary part of Ω we must have

$$\rho = \rho_t J, \tag{5.19}$$

that is

$$\rho(\boldsymbol{x}) = \rho_t(\boldsymbol{y}(\boldsymbol{x},t)) \, J(\boldsymbol{x},t) \quad \forall \, \boldsymbol{x} \in \Omega, \ t > 0.$$

Equation (5.19) represents the *balance law of mass.* It is useful in deriving relations between quantities defined in reference and current configurations, respectively.

System of forces. The body is subject to *external* and *internal* forces. The external forces are also called *applied* forces and are of two kinds. They model the action exerted by agencies external to the body. The internal forces model the interactions between different parts of the deformable body.

We start with a description of the applied forces. First, there are *applied body forces* defined by a vector field $\boldsymbol{f}_0(\cdot,t) : \Omega \to \mathbb{R}^d$, called the *density of the applied body forces per unit reference volume.* An elementary force $\boldsymbol{f}_0(\boldsymbol{x},t)$ is exerted at each material point $\boldsymbol{x}$ at time t such that the resultant of the applied body force action at time t is $\int_\Omega \boldsymbol{f}_0(\boldsymbol{x},t)\,dx$. The density $\boldsymbol{f}_0(\boldsymbol{x},t)$ is assumed to be given. Gravity is a canonical example; in this case $\boldsymbol{f}_0(\boldsymbol{x}\cdot,t) = g\rho(\boldsymbol{x})\,\boldsymbol{e}$ where g is the gravitational acceleration and $\boldsymbol{e}$ is the unit vector pointing in the downward vertical direction.

The second kind of applied forces are the *applied surface forces.* They are defined by a vector field $\boldsymbol{f}_2(\cdot,t) : \Gamma \to \mathbb{R}^d$, called the *density of the applied surface forces per unit reference area.* An elementary force $\boldsymbol{f}_2(\boldsymbol{x},t)$ is exerted at each material point $\boldsymbol{x}$ at time t such that the resultant of the applied surface force action at time t is $\int_\Gamma \boldsymbol{f}_2(\boldsymbol{x},t)\,da$. The density $\boldsymbol{f}_2(\boldsymbol{x},t)$ may be known a priori on a part of the boundary. In this case it represents a data of the problem. The hydrostatic force exerted on a deformable body embedded in a fluid represent a canonical example; in this case $\boldsymbol{f}_2(\boldsymbol{x},t) = -p_0\boldsymbol{\nu}(\boldsymbol{x})$ where p_0 denotes the hydrostatic pressure and $\boldsymbol{\nu}(\boldsymbol{x})$ is the unit outward normal at $\boldsymbol{x}$ on Γ.

Next, we introduce the notion of internal forces. To this end, consider again a part $\mathcal{P}$ of the body which at the moment $t = 0$ occupies the open subset $\omega \subset \Omega$ and suppose that ω has a smooth boundary $\partial\omega$. Let $\boldsymbol{x} \in \partial\omega$ and denote by $\boldsymbol{\nu}$ the unit outward normal on $\partial\omega$ at $\boldsymbol{x}$. Let $t > 0$. We suppose that there exists a vector $\boldsymbol{s}(\boldsymbol{x},\boldsymbol{\nu},t)$ which depends on the point $\boldsymbol{x}$ and also on the direction of the normal $\boldsymbol{\nu}$ such that:

- the resultant of the action at time t of the rest of the medium on $\mathcal{P}$ is $\int_{\partial\omega} \boldsymbol{s}(\boldsymbol{x},\boldsymbol{\nu},t)\,da$, if $\boldsymbol{x} \in \Omega$;
- $\boldsymbol{s}(\boldsymbol{x},\boldsymbol{\nu},t)=\boldsymbol{f}_2(\boldsymbol{x},t)$ if $\boldsymbol{x} \in \Gamma$ and $\boldsymbol{\nu}$ is the unit outward normal on Γ.

The vector $\boldsymbol{s}(\boldsymbol{x}, \nu, t)$ is known as the *first Piola-Kirchhoff stress vector*. Since $\mathcal{P}$ is an arbitrary part of the body (i.e. ω is an arbitrary part of Ω), it follows that the stress vector defines a field $\boldsymbol{s}(\cdot,\cdot,t) : \overline{\Omega} \times S_d \to \mathbb{R}^d$ where S_d is the surface of the unit sphere in $\mathbb{R}^d$, that is $S_d = \{ \boldsymbol{\nu} \in \mathbb{R}^d \ : \ \|\boldsymbol{\nu}\| = 1 \}$.

A *system of forces* at time $t > 0$ is formed by the applied forces corresponding to the densities $\boldsymbol{f}_0(\cdot,t) : \Omega \to \mathbb{R}^d$, $\boldsymbol{f}_2(\cdot,t) : \Gamma \to \mathbb{R}^d$ and the surface forces corresponding to the vector field $\boldsymbol{s}(\cdot,\cdot,t) : \overline{\Omega} \times S_d \to \mathbb{R}^d$.

Balance laws of momentum, stress tensors. Two general principles hold for every continuous medium subjected to a system of forces. They are given by the following equalities:

$$\frac{d}{dt}\int_\omega \rho\dot{\boldsymbol{u}}\,dx = \int_\omega \boldsymbol{f}_0\,dx + \int_{\partial\omega} \boldsymbol{s}\,da, \tag{5.20}$$

$$\frac{d}{dt}\int_\omega \boldsymbol{x}\wedge\rho\dot{\boldsymbol{u}}\,dx = \int_\omega \boldsymbol{x}\wedge\boldsymbol{f}_0\,dx + \int_{\partial\omega} \boldsymbol{x}\wedge\boldsymbol{s}\,da \tag{5.21}$$

for any $\omega \subset \Omega$ at any moment $t > 0$. Equality (5.20) is called the *balance law of momentum* and it represents a version of Newton's second law for continuous media. Indeed, the domain integral $\int_\omega \boldsymbol{f}_0\,dx$ is the resultant of applied body forces acting on $\mathcal{P}$ and the surface integral $\int_{\partial\omega} \boldsymbol{s}\,da$ is the resultant of surface forces acting on $\mathcal{P}$ at time t. The sum of the two force resultants must equal to the time derivative of the kinetic resultant on $\mathcal{P}$, expressed by the left-hand side of (5.20).

Equality (5.21) is called the *balance law of angular momentum.* It states that the resultant momentum with respect to the origin O of the system of forces acting on $\mathcal{P}$ is equal to the time derivative of the kinetic momentum of $\mathcal{P}$ with respect to the origin. In (5.21) $\wedge$ represents the exterior product in $\mathbb{R}^d$.

One of the major consequences of the balance law (5.20) is the linearity of the first Piola-Kirchhoff stress vector $\boldsymbol{s}(\boldsymbol{x}, \boldsymbol{\nu}, t)$ with respect to its second argument $\boldsymbol{\nu} \in S_d$. More precisely, it can be shown that there exists a tensor field $\boldsymbol{\tau}(\cdot,t)$ defined in Ω such that

$$\boldsymbol{s}(\boldsymbol{x},\boldsymbol{\nu},t) = \boldsymbol{\tau}(\boldsymbol{x},t)\,\boldsymbol{\nu} \quad \forall \boldsymbol{x}\in\Omega,\ \boldsymbol{\nu}\in S_d,\ t>0. \tag{5.22}$$

The tensor $\boldsymbol{\tau}$ is called the *first Piola-Kirchhoff stress tensor.* Detailed proof of this result can be found, e.g., in [11, 94, 146].

We are now in a position to derive *local* forms of the balance laws. Using (5.22) and a variant of the Green-Gauss theorem we find

$$\int_{\partial\omega} \boldsymbol{s}\,da = \int_{\partial\omega} \boldsymbol{\tau\nu}\,da = \int_{\partial\omega} \operatorname{Div}\boldsymbol{\tau}\,dx, \tag{5.23}$$

where $\operatorname{Div}\boldsymbol{\tau}$ is the divergence of $\boldsymbol{\tau}$ with respect to the spatial variable of the actual configuration,

$$(\operatorname{Div}\boldsymbol{\tau})_i = \tau_{ij,j}, \quad 1 \le i \le d.$$

Since ω is chosen independent of time we have

$$\frac{d}{dt}\int_\omega \rho\,\dot{\boldsymbol{u}}\,dx = \int_\omega \rho\,\ddot{\boldsymbol{u}}\,dx. \tag{5.24}$$

Therefore, using (5.20), (5.23) and (5.24), we find

$$\int_\omega (\rho\,\ddot{\boldsymbol{u}} - \operatorname{Div}\boldsymbol{\tau} - \boldsymbol{f}_0)\,dx = \boldsymbol{0}.$$

Now the domain ω is arbitrary, the integrand in the previous identity must vanish,

$$\operatorname{Div}\boldsymbol{\tau} + \boldsymbol{f}_0 = \rho\,\ddot{\boldsymbol{u}}; \tag{5.25}$$

in component form,

$$\frac{\partial \tau_{ij}}{\partial x_j} + f_{0i} = \rho\,\ddot{u}_i, \quad 1 \le i \le d.$$

We have chosen to present the arguments leading to the first Piola-Kirchhoff stress tensor in the setting of the reference configuration, with $\boldsymbol{x}$ and t as the independent variables. Since the motion $\boldsymbol{y} = \boldsymbol{y}(\boldsymbol{x}, t)$ is invertible, it is also possible to treat $\boldsymbol{y}$ and t as the independent variables and to carry out the development in the *current* configuration Ω_t. Similar arguments as above lead to the existence of a tensor $\boldsymbol{\sigma}$, called the *Cauchy stress tensor*, which has the same relationship to the current configuration as does the first Piola-Kirchhoff stress to the reference configuration. It can be shown that the first Piola-Kirchhoff and Cauchy stresses are related:

$$\boldsymbol{\sigma} = J^{-1}\boldsymbol{\tau}(\boldsymbol{I} + \nabla\boldsymbol{u})^T. \tag{5.26}$$

We have not yet examined consequences of the equation (5.21) for the balance of angular momentum. By carrying out manipulations similar to those that lead to (5.25), it is possible to show that this balance law implies that $\boldsymbol{\tau}(\boldsymbol{I} + \nabla\boldsymbol{u})^T$ is a symmetric tensor. Equivalently, we have the classical result that the Cauchy stress is symmetric:

$$\boldsymbol{\sigma} = \boldsymbol{\sigma}^T. \tag{5.27}$$

The linearized theory. We now investigate the linearized theory. We assume $\eta = \|\nabla\boldsymbol{u}\| << 1$ that permits us to neglect all the quantities of order 2 or more in η. We also assume that the first Piola-Kirchhoff stress tensor (or, equivalently, the Cauchy stress tensor) is of order η. From the definitions (5.2) and (5.3) we have

$$J = \det(\boldsymbol{I} + \nabla\boldsymbol{u}) = 1 + \operatorname{div}\boldsymbol{u} + O(\eta^2).$$

Thus, $J \approx 1 + \operatorname{div}\boldsymbol{u}$ and $J^{-1} \approx 1 - \operatorname{div}\boldsymbol{u}$, that is $J = 1 + O(\eta)$ and $J^{-1} = 1 + O(\eta)$. Keeping in mind (5.26) we deduce that $\boldsymbol{\sigma} \approx \boldsymbol{\tau}$ so the distinction between the first

Piola-Kirchhoff stress and Cauchy stress tensors is negligible. For this reason, in the rest of the book we shall denote by $\boldsymbol{\sigma}$ the stress tensor given by $\boldsymbol{\sigma} \approx \boldsymbol{\tau}$ and we shall call $\boldsymbol{\sigma}$ the *stress tensor*. Similarly, the vector field $\boldsymbol{\sigma\nu} \approx \boldsymbol{\tau\nu}$ will be called the *stress vector*. From (5.27) we see that $\boldsymbol{\sigma}$ is a symmetric tensor. By (5.25) we obtain

$$\operatorname{Div} \boldsymbol{\sigma} + \boldsymbol{f}_0 = \rho \ddot{\boldsymbol{u}} \quad \text{in } \Omega, \ \text{for } t > 0. \tag{5.28}$$

In some cases, equation (5.28) can be reduced. For example, when the physical process is slow, we may neglect the inertial term $\rho\ddot{\boldsymbol{u}}$ to obtain

$$\operatorname{Div} \boldsymbol{\sigma} + \boldsymbol{f}_0 = \boldsymbol{0} \quad \text{in } \Omega, \ \text{for } t > 0. \tag{5.29}$$

Equation (5.28) is called the *equation of motion* whereas equation (5.29) is called the *equation of equilibrium*. Processes modeled by equation (5.28) are called *dynamic processes*. Processes modeled by equation (5.29) are called *quasistatic processes*.

In conclusion, in the linearized theory of deformation, the unknowns of the problem are the displacement field $\boldsymbol{u}$ and the stress field $\boldsymbol{\sigma}$ which satisfy (5.28) in the case of dynamic process and (5.29) in the case of quasistatic process. These unknowns are defined on the reference configuration Ω, for any moment $t > 0$. Moreover, since $\boldsymbol{\sigma}$ is symmetric, $\boldsymbol{\sigma}(\boldsymbol{x}, t) \in \mathbb{S}^d$ for all $\boldsymbol{x} \in \Omega$ and $t > 0$.

5.3 Physical Setting of Contact Problems

In this book we are mainly concerned with mathematical models describing the contact between a deformable body and an obstacle, the so-called foundation. Such kind of problems abound in industry and everyday life. Because of the importance of contact processes, a considerable effort has been made in its modelling, analysis, and numerical simulations. The literature concerning this topic is extensive.

The large variety of situations involving contact phenomena can be cast in the general physical setting shown in Figure 5.2 and described in the sequel. A deformable body occupies an open, bounded, connected set $\Omega \subset \mathbb{R}^d$ $(d = 1, 2, 3)$. The boundary $\Gamma = \partial\Omega$ is assumed to be composed of three sets $\overline{\Gamma}_1$, $\overline{\Gamma}_2$ and $\overline{\Gamma}_3$, with the mutually disjoint relatively open sets Γ_1, Γ_2 and Γ_3. We assume the boundary Γ is Lipschitz continuous and therefore the unit outward normal vector $\boldsymbol{\nu}$ exists a.e. on Γ. The body is clamped on Γ_1. Surface tractions of density $\boldsymbol{f}_2$ act on Γ_2 and volume forces of density $\boldsymbol{f}_0$ act in Ω. The body is or may become in contact on Γ_3 with an obstacle, the so-called foundation. The force densities $\boldsymbol{f}_2$ and $\boldsymbol{f}_0$ may depend on time. As a result, the mechanical state of the body evolves in the time interval of interest $\overline{I} = [0, T]$, $T > 0$.

We are interested in providing a mathematical model which describes the evolution of the body of the previous physical setting. To this end we denote by $\boldsymbol{\sigma} = \boldsymbol{\sigma}(\boldsymbol{x}, t)$ the stress field and by $\boldsymbol{u} = \boldsymbol{u}(\boldsymbol{x}, t)$ the displacement field. The functions $\boldsymbol{u} : \Omega \times \overline{I} \to \mathbb{R}^d$ and $\boldsymbol{\sigma} : \Omega \times \overline{I} \to \mathbb{S}^d$ will play the role as the unknowns

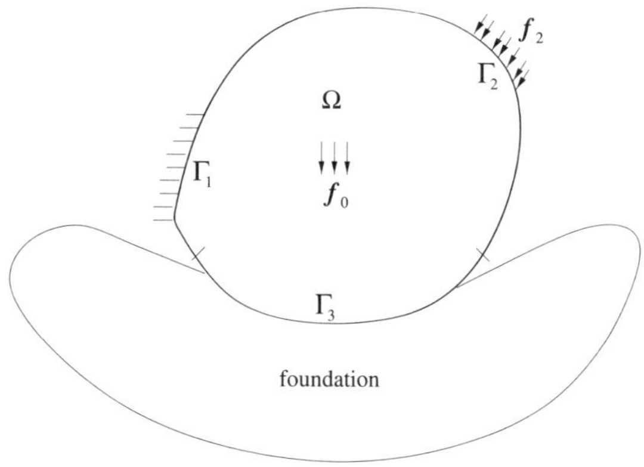

Figure 5.2: A deformable body in contact with a foundation

of the contact problem. Recall that the knowledge of the displacement field allows us to obtain the deformed configuration of the body by (5.3). From time to time, we suppress the explicit dependence of the quantities on the spatial variable $\boldsymbol{x}$, or both $\boldsymbol{x}$ and t; i.e. when it is convenient to do so, we write $\boldsymbol{\sigma}(t)$ and $\boldsymbol{u}(t)$, or even $\boldsymbol{\sigma}$ and $\boldsymbol{u}$.

We assume that the volume forces and surface tractions change slowly in time so that the inertia of the mechanical system is negligible. In other words, we neglect the acceleration term in the equation of motion and therefore we consider a quasistatic process of contact. Thus, the stress field satisfies the equation

$$\operatorname{Div} \boldsymbol{\sigma} + \boldsymbol{f}_0 = \boldsymbol{0} \quad \text{in } \Omega \times I. \tag{5.30}$$

Since the body is clamped on Γ_1, we impose the condition

$$\boldsymbol{u} = \boldsymbol{0} \quad \text{on } \Gamma_1 \times I. \tag{5.31}$$

Throughout the book, we assume meas $(\Gamma_1) > 0$. This condition is needed in order to prove existence and uniqueness of the displacement fields.

The traction boundary condition is

$$\boldsymbol{\sigma\nu} = \boldsymbol{f}_2 \quad \text{on } \Gamma_2 \times I. \tag{5.32}$$

It states that the stress vector $\boldsymbol{\sigma\nu}$ is given on part Γ_2 of the boundary during the contact process.

We will specify contact conditions on Γ_3. A survey of the various contact conditions we consider in this book will be provided in Section 5.4. The contact may be frictionless or frictional, with a rigid or with a deformable foundation and, in the case of frictional, a number of different friction conditions will be employed.

We now recall the strain-displacement relation

$$\boldsymbol{\varepsilon}(\boldsymbol{u}) = (\varepsilon_{ij}(\boldsymbol{u})), \quad \varepsilon_{ij}(\boldsymbol{u}) = \frac{1}{2}\Big(\frac{\partial u_i}{\partial x_j} + \frac{\partial u_j}{\partial x_i}\Big), \quad 1 \le i,j \le d \quad \text{in } \Omega \times I \tag{5.33}$$

which defines the infinitesimal strain tensor. Sometimes we omit the explicit dependence of ε on $\boldsymbol{u}$ by writing ε instead of $\varepsilon(\boldsymbol{u})$. Note that all the problems studied in this book are formulated in the framework of small strain theory.

At this stage description of our model is not completed: We have more unknown functions than the number of equations. Indeed, let us give a number count for the case $d = 3$. Equations (5.30) and (5.33) represent, when written out in their component forms, a total of nine equations: three from the equilibrium equation and six from strain-displacement relations (taking into account the symmetry of ε). The total number of unknowns is fifteen: three components of the displacement, six components of the strain and six components for the stress (again accounting for the symmetry of ε and $\boldsymbol{\sigma}$). Thus, it is clear that six additional equations are required if we are to have a problem that is at least in principle solvable.

Physical considerations also indicate that the description of the problem so far is incomplete: In addition to the kinematics and the balance laws which apply to all the materials, we also need description of the particular material behavior. This information, embodied in the *constitutive equation* or *constitutive law* of the material, will provide the remaining equations of the problem. A complete description of the constitutive laws considered in this book will be given in Chapter 6. Here we only remark that in this book we deal with elastic, viscoelastic and viscoplastic materials.

To conclude, the physical setting of the contact problems studied in this book will be modeled by the system of partial differential equations (5.30), (5.33) in $\Omega \times I$, with the boundary conditions (5.31) on $\Gamma_1 \times I$ and (5.32) on $\Gamma_2 \times I$, and contact boundary conditions on $\Gamma_3 \times I$, supplemented by the constitutive equations which describe the physical properties of the materials, and if necessary, by initial conditions for the displacement and stress field. Note that in the case of static problems for elastic materials, the data $\boldsymbol{f}_0$ and $\boldsymbol{f}_2$ do not depend on time and, since the time variable is not involved in the constitutive laws and boundary conditions, relations (5.30), (5.33) hold in Ω, while conditions (5.31) and (5.32) hold on Γ_1 and Γ_2, respectively.

5.4 Contact Boundary Conditions and Friction Laws

The science and technology of interacting surfaces in contact and relative motion is the *tribology*. The term tribology is derived from the word *tribos* which means rubbing in classical Greek. Friction and wear are two important subfields of tribology. Usually, friction is defined as the resisting force tangential to the common boundary between two bodies, that appears when one body moves or tends to move relative to the surface of the other under the action of an external force. Wear is defined as the progressive loss of substance from the operating surface of the body occuring as a result of relative motion of the surface.

Normally, the following guiding rules serve as good approximations for friction:

- the friction force is proportional to the normal force acting on the body;
- the friction force is independent of the apparent geometric area of contact;
- the friction force is independent of the amount of the sliding velocity, which implies that the force required to initiate sliding will be the same as the force required to maintain sliding at any specified velocity.

The first two rules were suggested by Amontons in 1699, and the third was stated by Coulomb in 1785. These rules have proven to be good approximations for modelling frictional phenomena.

However, the nature of friction forces developed between bodies in contact is extremely complex and is affected by a long list of factors: the constitution of the interface, the time scales and frequencies of the contact, the response of the interface to normal forces, inertia and thermal effects, roughness of the contacting surfaces, history of loadings, the presence or absence of lubricants and so on. Thus, friction is not a single phenomenon but is a collection of many complex mechanical and chemical phenomena entwined in a mosaic whose features cannot be grasped through isolated simple experiments. For this reason, the absence of a universal continuum model of friction is not surprising.

Generally, a mathematical theory of friction should be a generalization of experimental facts and it must be in agreement with the ideas of thermodynamics of irreversible processes. This subject may be studied and understood at different levels. Several approaches to such theory, employing different levels of mechanics and mathematics are shown in the works [60, 68, 157, 164, 223]. We agree that a useful theory should be set up as a continuum field theory but the development of a rational theory of contact with friction is not the purpose of the present book. Here, we limit our discussion to the description of a number of contact boundary conditions found in the literature and used in the rest of the book. The unifying element is that we can use them to formulate variational inequalities for contact problems.

We refer to the general physical setting described in Section 5.3. The equalities and inequalities below are assumed to be valid a.e. on $\Gamma_3 \times I$ and, in the case of static problems, a.e. on Γ_3. However, for simplicity, we do not exhibit explicitly the dependence of various functions on $\boldsymbol{x}$ and t. When we deal with the tangential components of the vector fields, we implicitly assume that $d \geq 2$.

We denote by u_ν and $\boldsymbol{u}_\tau$ the *normal* and *tangential* components of the displacement field $\boldsymbol{u}$ on the boundary, given by

$$u_\nu = \boldsymbol{u} \cdot \boldsymbol{\nu}, \quad \boldsymbol{u}_\tau = \boldsymbol{u} - u_\nu \boldsymbol{\nu}. \tag{5.34}$$

Similarly, the *normal* and *tangential* components of the velocity field $\dot{\boldsymbol{u}}$ on the boundary are defined by

$$\dot{u}_\nu = \dot{\boldsymbol{u}} \cdot \boldsymbol{\nu}, \quad \dot{\boldsymbol{u}}_\tau = \dot{\boldsymbol{u}} - \dot{u}_\nu \boldsymbol{\nu}. \tag{5.35}$$

We also denote by σ_ν and $\boldsymbol{\sigma}_\tau$ the *normal* and *tangential* components of the stress field $\boldsymbol{\sigma}$ on the boundary, that is

$$\sigma_\nu = (\boldsymbol{\sigma}\boldsymbol{\nu}) \cdot \boldsymbol{\nu}, \quad \boldsymbol{\sigma}_\tau = \boldsymbol{\sigma}\boldsymbol{\nu} - \sigma_\nu \boldsymbol{\nu}. \tag{5.36}$$

The component $\boldsymbol{\sigma}_\tau$ represents the *friction force* on the contact surface Γ_3.

To describe the contact of a body with a foundation we need to consider the normal approach and the tangential process. In the following, by a *contact condition* we understand a condition which describes the normal approach, that is a relation involving only the normal components of the displacement, velocity fields and stress field. By a *friction condition* we understand a condition which describe the tangential process, that is a relation involving the tangential stress $\boldsymbol{\sigma}_\tau$ and the tangential velocity $\dot{\boldsymbol{u}}_\tau$. In most of the cases, contact and friction conditions are formulated as nonlinear equalities or inequalities. As a result, highly nonlinear initial boundary value problems are obtained.

We start with the presentation of the contact conditions.

Bilateral contact. We assume that there is no loss of the contact during the process. Thus the normal displacement vanishes, that is

$$u_\nu = 0. \tag{5.37}$$

This condition is used when the body is somewhat glued to the foundation. From mathematical point of view this condition is very convenient since in general it leads to a linear subspace for the set of admissible displacements fields. Problems involving bilateral contact were considered by many authors, see for instance [70, 167].

Signorini contact condition. This condition models the contact with a perfectly rigid foundation, i.e. a foundation which is infinitely resistant to compression. We assume that there is a gap g between the body and the foundation, measured along the outward normal. Since the obstacle is rigid, we have the non-penetration condition

$$u_\nu \le g. \tag{5.38}$$

In the study of Signorini problem, the non-penetration condition (5.38) sets a constraint on the admissible displacements. Then the set of admissible displacement fields is not a linear subspace.

Now, if $u_\nu < g$, there is no contact between the body and the foundation and consequently the foundation does not produce a reaction towards the body, that is $\sigma_\nu = 0$. If $u_\nu = g$, there exists contact and the foundation exerts a normal compression force $\sigma_\nu \le 0$ on the body. We conclude that, in any case, the following relations hold:

$$\sigma_\nu \le 0, \quad \sigma_\nu(u_\nu - g) = 0. \tag{5.39}$$

Conditions (5.38)–(5.39) are called the *Signorini contact condition.* They model the contact with a rigid foundation. Note that the actual surface on which the body comes in contact with the foundation is not known in advance but it is contained in Γ_3.

In the case with a zero gap function (i.e. the body lays on the obstacle on the part Γ_3 of this boundary), the Signorini contact condition (5.38)–(5.39) reduces to

$$u_\nu \leq 0, \quad \sigma_\nu \leq 0, \quad \sigma_\nu u_\nu = 0. \tag{5.40}$$

The Signorini condition was proposed by Signorini [191], and have been used in a large number of papers, see e.g. [51, 53, 69, 85, 124, 125, 161].

Contact with normal compliance. In this condition the foundation is supposed to be deformable. Assume there is a gap g between the potential contact surface Γ_3 and the foundation, measured along the outward normal. We assume that the normal stress σ_ν satisfies the condition

$$-\sigma_\nu = p_\nu(u_\nu - g). \tag{5.41}$$

The equality (5.41) represents the *normal compliance* contact condition in which p_ν is a prescribed function such that $p_\nu(r) = 0$ if $r \leq 0$. Here, $u_\nu - g$, when positive, represents the penetration of the surface asperities of the body into those of the foundation. An example of the normal compliance function p_ν is

$$p_\nu(r) = c_\nu r_+, \tag{5.42}$$

or, more general,

$$p_\nu(r) = c_\nu (r_+)^m. \tag{5.43}$$

Here c_ν is a positive constant, $r_+ = \max\{0, r\}$ and $m > 0$. Formally, Signorini's nonpenetration condition is obtained in the limit $c_\nu \to \infty$. This leads to the idea of regarding the contact with a rigid support as a limiting case of contact with deformable supports whose resistance to compression subsequently increases. We can also consider the following normal compliance function

$$p_\nu(r) = \begin{cases} c_\nu r_+ & \text{if} \quad r \leq \alpha, \\ c_\nu \alpha & \text{if} \quad r > \alpha, \end{cases} \tag{5.44}$$

where α is a positive coefficient related to the wear and hardness of the surface. In this case the contact condition (5.41) means that when the penetration is too large, i.e. when it exceeds α, the obstacle disintegrates and offers no additional resistance to the penetration.

The normal compliance contact condition was proposed in [152], in the particular form of (5.43). Then it was used in a large number of papers, see e.g. [7, 8, 134, 137, 138, 178] and the references therein.

Contact with normal damped response. We assume that the normal stress on the contact surface is related to the normal velocity, that is

$$-\sigma_\nu = p_\nu(\dot{u}_\nu). \tag{5.45}$$

The equality (5.45) represents the *normal damped response* contact condition in which p_ν is a prescribed function. In the general form (5.45), the normal damped response contact condition was considered in [16, 40].

In [179] the following form of the function p_ν is employed

$$p_\nu(r) = \kappa\, r_+ + p_0 \tag{5.46}$$

in order to model the setting when the foundation is covered with a thin lubricant layer, say oil. Here $\kappa \geq 0$ is the damping resistance coefficient, assumed positive, and p_0 is the oil pressure, which is given and is nonnegative. In this case the lubricant layer presents resistance, or damping, only when the surface moves towards the foundation, but does nothing when it recedes.

Another choice of p_ν is

$$p_\nu(r) = \kappa\, |r|^{q-1} r. \tag{5.47}$$

Here $\kappa \geq 0$, $0 < q \leq 1$ and the normal contact stress depends on a power of the normal velocity, which mimics the behavior of a nonlinear viscous dashpot. For this reason we refer to the contact condition (5.45) associated to the function (5.47) as a *viscous contact*. This condition was considered for instance in [190, 196].

One more choice is

$$p_\nu(r) = S, \tag{5.48}$$

where S is a given function. This type of contact conditions in which the normal stress is prescribed arises in the study of some mechanisms and was considered by a number of authors (see, e.g. [70, 167]).

The bilateral contact, the Signorini contact, contact with normal compliance and normal damped response are the main types of contact conditions used in this book. They are associated with frictional contact conditions which will be described below. We note that additional contact boundary conditions will be presented in Sections 13.2 and 13.3. There the physical setting is different, since the contact arises with a rigid moving foundation, leading to the wear of the contacting surfaces.

We now present the main types of friction laws used in the book.

Frictionless condition. In frictionless contact the reaction of the foundation towards the body is in the direction of the normal only. Thus, the friction force on the contact surface vanishes, that is

$$\boldsymbol{\sigma}_\tau = \mathbf{0}. \tag{5.49}$$

This condition represents a first approximation of more realistic conditions involving friction and it is used when the contact surfaces are fully lubricated. Frictionless contact problems were considered, for instance, in [81, 82, 125].

Tresca friction law. This contact condition can be found in [70, 167], and more recently in [2, 5] (see references therein for further details). It is expressed in the following form:

$$\left.\begin{aligned} &\|\boldsymbol{\sigma}_\tau\| \le g, \\ &\qquad \|\boldsymbol{\sigma}_\tau\| < g \ \Rightarrow\ \dot{\boldsymbol{u}}_\tau = \boldsymbol{0}, \\ &\qquad \|\boldsymbol{\sigma}_\tau\| = g \ \Rightarrow\ \exists\,\lambda \ge 0 \text{ s.t. } \boldsymbol{\sigma}_\tau = -\lambda\,\dot{\boldsymbol{u}}_\tau. \end{aligned}\right\} \tag{5.50}$$

Here $g \ge 0$ represents the *friction bound*, i.e., the magnitude of the limiting friction traction at which slip begins. In (5.50) when the strict inequality holds, the material point is in the *stick* zone; when the equality holds, the material point is in the *slip* zone. The boundary of these zones is unknown a priori, and is part of the problems. Such a boundary is called a *free boundary.*

The main feature of the Tresca friction law (5.50) is the assumption that the friction bound g is given. However, when two surfaces come into contact, rearrangement of the surface asperities takes place. The surface asperities become more compliant and the slip becomes smoother. Clearly, these changes affect the friction process and we expect that the friction bound decreases. Therefore, there is a need, when we describe the physical setting, to allow for the resulting change or evolution of the contact and frictional boundary conditions. An important step in modeling the friction phenomenon was made in [171] where it was pointed out that the coefficient of friction varies with the tangential displacement. In this way a part of the elastic-plastic deformation of the interface is captured in the model. Other authors (see for instance [166] for a review) performed a series of experiments in which it was possible to measure directly the interdependence between the different local variables on the contact surface. These experiments show that the friction depends mostly on the slip. Following these experiments, slip-weakening type friction laws were considered by many authors, see for example [38, 119, 142]. The changes in the current friction with the slip speed reflect mainly the elastic deformation on the surface asperities and possible chemical bonding.

Slip dependent Tresca friction law. A variant of the Tresca friction law is obtained by assuming that the friction bound g depends on the accumulated slip of the surface. This represents changes in the contact surface structure that are resulted from sliding. Therefore, we take

$$g = g(S_t(\dot{\boldsymbol{u}})) \tag{5.51}$$

where $S_t(\dot{\boldsymbol{u}})(\boldsymbol{x})$ is the accumulated slip at the point $\boldsymbol{x}$ on Γ_3 over the time period $[0, t]$:

$$S_t(\dot{\boldsymbol{u}}) = \int_0^t \|\dot{\boldsymbol{u}}_\tau(s)\|\, ds, \quad t \in \overline{I}. \tag{5.52}$$

We use Tresca's friction law (5.50) with g given by (5.51) to obtain

$$\left.\begin{aligned} &\|\boldsymbol{\sigma}_\tau\| \le g(S_t(\dot{\boldsymbol{u}})), \\ &\qquad \|\boldsymbol{\sigma}_\tau\| < g(S_t(\dot{\boldsymbol{u}})) \ \Rightarrow\ \dot{\boldsymbol{u}}_\tau = \boldsymbol{0}, \\ &\qquad \|\boldsymbol{\sigma}_\tau\| = g(S_t(\dot{\boldsymbol{u}})) \ \Rightarrow\ \exists\,\lambda \ge 0 \text{ s.t. } \boldsymbol{\sigma}_\tau = -\lambda\,\dot{\boldsymbol{u}}_\tau. \end{aligned}\right\} \tag{5.53}$$

Inclusion of a time-dependent frictional bound makes the contact problem history dependent and leads to a new, interesting and nonstandard variational formulation, see [39].

Coulomb friction law. In the classical Coulomb law it is assumed that the friction bound is proportional to the magnitude of the normal stress, that is

$$\left.\begin{aligned} &\|\boldsymbol{\sigma}_\tau\| \le \mu\,|\sigma_\nu|, \\ &\quad \|\boldsymbol{\sigma}_\tau\| < \mu\,|\sigma_\nu| \;\Rightarrow\; \dot{\boldsymbol{u}}_\tau = \boldsymbol{0}, \\ &\quad \|\boldsymbol{\sigma}_\tau\| = \mu\,|\sigma_\nu| \;\Rightarrow\; \exists\,\lambda \ge 0 \text{ s.t. } \boldsymbol{\sigma}_\tau = -\lambda\,\dot{\boldsymbol{u}}_\tau. \end{aligned}\right\} \tag{5.54}$$

Here $\mu \ge 0$ represents the *coefficient of friction.* As in the case of the Tresca friction law, the contact boundary is divided, at any moment, into two zones: the stick zone and the slip zone. In the classical Coulomb law (5.54) the coefficient of friction μ is given. As it was already pointed out, one expects that μ can vary, depending on the roughness of the surfaces, on the slip velocity and on the temperature. The last one is very complicated to measure, since frictional sliding causes considerable heat generation which, in turn, leads to uneven temperature rises. Moreover, it is observed that as the melting temperature of one of the surfaces is approached, the friction coefficient first increases rapidly, and then decreases sharply, once melting commences (see, e.g. [172] for full details). Variational analysis of frictional problems with Coulomb law in which the coefficient of friction μ depends either on the total slip (5.52) or on the current slip $\|\dot{\boldsymbol{u}}_\tau\|$ was provided in [3].

A generalized version of Coulomb friction law. We consider now the following friction law:

$$\left.\begin{aligned} &\|\boldsymbol{\sigma}_\tau\| \le \mu\,p(|\sigma_\nu|), \\ &\quad \|\boldsymbol{\sigma}_\tau\| < \mu\,p(|\sigma_\nu|) \;\Rightarrow\; \dot{\boldsymbol{u}}_\tau = \boldsymbol{0}, \\ &\quad \|\boldsymbol{\sigma}_\tau\| = \mu\,p(|\sigma_\nu|) \;\Rightarrow\; \exists\,\lambda \ge 0 \text{ s.t. } \boldsymbol{\sigma}_\tau = -\lambda\,\dot{\boldsymbol{u}}_\tau. \end{aligned}\right\} \tag{5.55}$$

Here p is a nonnegative function, and $\mu \ge 0$ is the coefficient of friction.

In the case where p is a constant function, i.e.

$$p(r) = \kappa, \tag{5.56}$$

the friction law involved in (5.55) becomes Tresca's friction law (5.50). Here, $g = \mu\,\kappa$ is the friction bound, i.e. the magnitude of friction force at which slip begins. Taking

$$p(r) = r \tag{5.57}$$

in (5.55), we recover the Coulomb friction law (5.55). Recently, a new version for Coulomb's law of friction was derived in [201, 202] from thermodynamic consideration. It consists of using in (5.55) the function

$$p(r) = [r\,(1 - \delta\,r)]_+, \tag{5.58}$$

where δ is a small positive coefficient related to the wear and hardness of the surface. This friction law means that when the normal stress is large enough, i.e. when it exceeds $1/\delta$, the surface disintegrates and offers no resistance to the motion.

Nonlocal versions of Coulomb friction law. Consideration of the Coulomb friction law (5.55) together with Signorini or bilateral contact conditions leads to substantial mathematical difficulties, since the trace of the stress tensor on the boundary is too rough to be defined in the ordinary sense. For this reason, many authors (see, e.g. [51, 69]) replaced it by the nonlocal friction law

$$\left.\begin{array}{l} \|\boldsymbol{\sigma}_\tau\| \le \mu\, p(\mathcal{R}|\sigma_\nu|), \\ \qquad \|\boldsymbol{\sigma}_\tau\| < \mu\, p(\mathcal{R}|\sigma_\nu|) \;\Rightarrow\; \dot{\boldsymbol{u}}_\tau = \boldsymbol{0}, \\ \qquad \|\boldsymbol{\sigma}_\tau\| = \mu\, p(\mathcal{R}|\sigma_\nu|) \;\Rightarrow\; \exists\, \lambda \ge 0 \text{ s.t. } \boldsymbol{\sigma}_\tau = -\lambda\, \dot{\boldsymbol{u}}_\tau \end{array}\right\} \tag{5.59}$$

in which $\mathcal{R}$ is a smoothing operator. Examples of functions p are provided by (5.56)–(5.58). We shall use the friction law (5.59) in Section 13.1.

Viscous friction law. Note that the previous versions of the Coulomb law, including Tresca's friction law, are characterized by the existence of stick-slip zones on the contact boundary. All these laws model the phenomenon that slip occurs only when the friction force reaches a critical value, the friction bound. However, in the case when the contact surface is lubricated, slip appears even for small tangential shears. Such a phenomenon can be modeled by the so-called *viscous friction law* in which the friction force depends only on the tangential velocity, that is

$$-\boldsymbol{\sigma}_\tau = p_\tau(\dot{\boldsymbol{u}}_\tau). \tag{5.60}$$

Here p_τ is a prescribed vector-valued function. As an example we may consider the function

$$p_\tau(\boldsymbol{v}) = \mu\, \|\boldsymbol{v}\|^{m-1}\boldsymbol{v}, \tag{5.61}$$

where μ represents the coefficient of friction, assumed positive, and m is a positive exponent. This is the case when the contact surface is lubricated with a thin layer of non-Newtonian fluid, see e.g. [196]. Here, the tangential shear is proportional to a power of the tangential velocity. In the particular case $m = 1$, (5.60) and (5.61) imply that the tangential shear is proportional to the tangential velocity.

Tresca friction law, versions of Coulomb laws and the viscous friction law are the main types of frictional conditions used in this book. They are combined with various contact conditions. Additional frictional conditions will be presented in Chapters 13 and 18 in the study of contact problems with viscoelastic and viscoplastic materials, respectively.

Chapter 6

Constitutive Relations in Solid Mechanics

This chapter is concerned with constitutive laws in solid mechanics with an emphasis on those used in the rest of the book. We describe here some basic experiments on solids exhibiting phenomena typical to elastic, viscoelastic and plastic behavior. Then we review the constitutive laws for elastic, viscoelastic and viscoplastic materials and prescribe the assumptions on the constitutive functions involved therein. In the following, by a constitutive law we understand a relation between the stress tensor $\boldsymbol{\sigma}$, the small strain tensor $\boldsymbol{\varepsilon}$ and their temporal derivatives $\dot{\boldsymbol{\sigma}}$ and $\dot{\boldsymbol{\varepsilon}}$.

6.1 Physical Background and Experiments

Constitutive laws model the physical state of the medium under consideration. Though they must satisfy some basic axioms and invariance principles, they originate mostly from experiments. For this reason we give in this section a general description of several diagnostic experiments which are made with "standard" universal testing machines and we shall point out what sort of information they may reveal which could be useful for the formulation of the constitutive relations.

Stress-strain diagram. The simplest diagnostic test is the uniaxial tension of a cylindrical specimen. To describe it we consider a specimen in a form of rectilinear rod which is in the natural (stress-free) configuration at time $t = 0$. Time dependence tensile forces $F(t)$ are applied to the ends of the rod for $t > 0$, as shown in Figure 6.1.

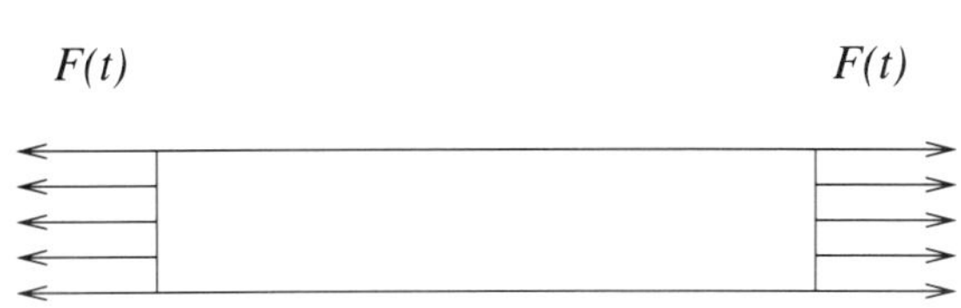

Figure 6.1: A specimen under axial load

As a measure of the stress we define the quantity $\sigma(t)$ by the equality

$$\sigma(t) = \frac{F(t)}{S_0}, \tag{6.1}$$

where S_0 is the cross-section area of the specimen in the reference configuration. The quantity $\sigma(t)$ represents the intensity of a force per unit area at the moment t. In the engineering literature, it is often called *the engineering stress.* Since we assume the idealized situation when the stress does not depend on the position and since in the case of one-dimensional bodies the stress is represented by real numbers, we can interpret the quantity $\sigma(t)$ given by (6.1) as the stress in the rod at the moment t.

Under the action of the force $F(t)$ the rod deforms. To characterize the rod's deformation we introduce the quantity

$$\varepsilon(t) = \frac{l(t) - l_0}{l_0}, \tag{6.2}$$

where l_0 and $l(t)$ denote the initial and the current lengths of the rod. In the engineering literature $\varepsilon(t)$ is called the *proof strain.* Assume the deformation of the rod is homogeneous. It follows that the quantity (6.2) plays the role of the small strain tensor which in the one-dimensional case reduces to a single component. As is common in practice, the proof strain ε is used when the corresponding elongations are rather small, i.e. less than 0.1. For deformations exceeding 0.1, either the *extension ratio*

$$\lambda(t) = \frac{l(t)}{l_0}$$

or the *logarithmic strain*

$$e(t) = \ln \frac{l(t)}{l_0}$$

is employed as a measure of stretches.

An important merit of the proof strains is the possibility to add them in order to calculate the total strain. Let us assume that the rod is stretched first from its initial length l_0 to a length l_1 and afterward, from the length l_1 to a length l_2. Then the total strain can be computed as

$$\varepsilon = \varepsilon_1 + \varepsilon_2, \tag{6.3}$$

where

$$\varepsilon_1 = \frac{l_1 - l_0}{l_0}, \quad \varepsilon_2 = \frac{l_2 - l_0}{l_0}$$

provided $|l_1 - l_0| << l_0$ and $|l_2 - l_0| << l_0$. The linear rule of summation (6.3) is valid in the framework of small deformations theory, i.e. for models which can bear only small strains.

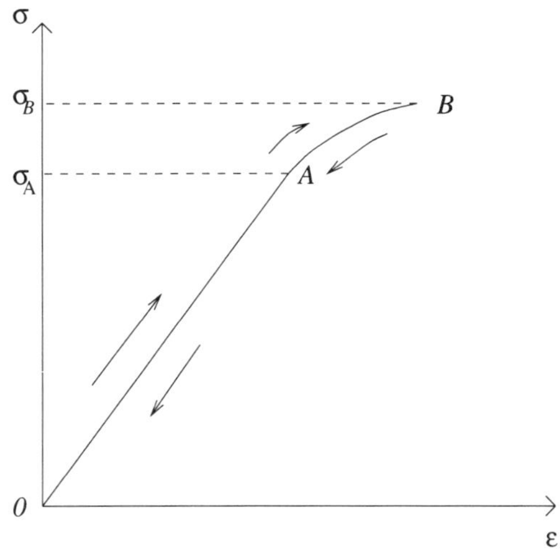

Figure 6.2: Stress-strain diagram in a loading process

Measuring the relative extension ε as a function of the stress σ we obtain a so-called *stress-strain* diagram. The feature of this diagram follows from some standard experiments on solids that we consider next.

Loading experiments. Suppose that the force $F(t)$ acting on the rod is gradually increased, i.e. we have a *loading process.* Then there is a related change in the length of the rod and hence a change in the strain as well. The corresponding dependence $\sigma \sim \varepsilon$ is plotted in Figure 6.2, which represents a typical stress-strain diagram. The part of the curve between the points O and A is straight indicating that for small stresses the relation between stress and strain is a linear one. The stress σ_A corresponding to A is called the *proportionality limit*, that may be determined experimentally (for example, when the angle between the tangent to the stress-strain curve and the stress axis is increased by a certain amount specified by convention, we say that the proportionality limit is reached). The part AB of the stress-strain diagram is curved; hence, for stresses higher than the proportionality limit the relation between σ and ε is no longer linear.

The ultimate strain ε_A (or the corresponding stress σ_A) characterizes a boundary between regions of linear and nonlinear elasticity. For a number of materials, the value of ε_A is rather small, usually of the order from 0.001 to 0.01. This is the standard ultimate strain for metals at room temperature, concrete, stones, semicrystalline plastics, etc. On the other hand, some media demonstrate a linear stress-strain dependence for deformations up to 0.05–0.1, e.g. several types of rubbers and biological tissues.

Loading and unloading experiments. For the purpose of establishing a mathematical model it is useful to perform *unloading experiments*, i.e. to reverse

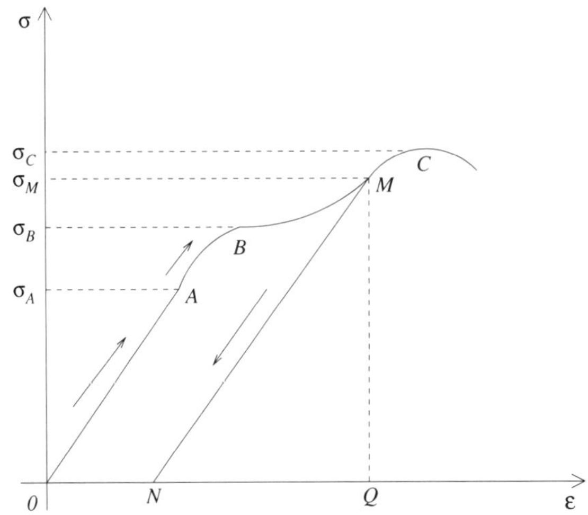

Figure 6.3: Stress-strain diagram for steel

the direction of the loading until the external force is completely removed. For some materials, if this reversal takes place at a sufficiently small stress, then the graph for the stress-strain relation is retraced to the origin along the same curve. In this case, i.e. when the stress-strain diagram is the same for loading and unloading processes, the deformation of the material is said to be *elastic.* Otherwise it is said to be *inelastic.* For an elastic specimen the strain vanishes completely if the load is entirely removed. Deformation of elastic materials is studied in the *theory of elasticity.*

When the unloading is performed after the strain reaches certain sufficient amount, the stress-strain diagram for a typical steel and some other metals corresponds to the graph presented in Figure 6.3.

It may be observed that at point B the strain increases dramatically without much change in the stress. The stress σ_B is called the *yield limit.* There is a point C on the curve such that once the point is reached, the strain increases without any increase in the stress. The stress σ_C corresponding to this point C is called the *strength limit.* Further stretching would finally cause the rod to break.

Let us suppose now that, after the loading reaches as far as a point M such that σ_M exceeds the yield limit σ_B, an unloading process is started. For most metals the unloading follows the path MN, which is a reasonable approximation to a straight line parallel to OA. We observe that after the unloading process, a part of the strain disappears (the portion NQ in Figure 6.3) although a part of it remains permanently (the portion ON). The portion of the strain that vanishes after unloading is called the *elastic* (*reversible*) strain, while the portion that remains permanently is called the *plastic* (*irreversible*) strain. We write

$$\varepsilon = \varepsilon^e + \varepsilon^p, \tag{6.4}$$

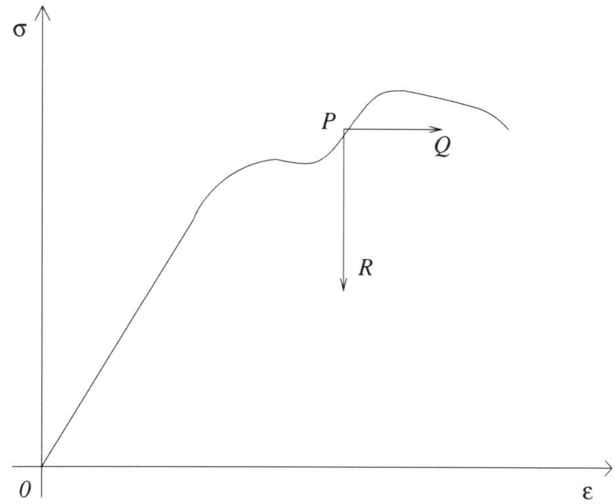

Figure 6.4: Creep and relaxation experiments

where ε^e is the elastic (reversible) part NQ of the strain and ε^p is the plastic (irreversible) part ON. Mathematical models where the strain can be decomposed according to (6.4) are called *elastoplastic.* In certain circumstances the elastic part of the strain can be neglected in the body as compared to the irreversible part; such a model is called *rigid-plastic.* The study of elastoplastic and rigid-plastic materials concerns the *theory of plasticity.*

Creep and relaxation experiments. We assume now that a loading experiment, which was performed to get the stress-strain curve of Figure 6.3 with a standard testing machine, is stopped at a certain stress and strain state (point P in Figure 6.4, say), at the moment $t = t_p$.

With a soft testing machine the stress can be kept constant for $t \geq t_p$. For a purely elastic material, the strain $\varepsilon(t)$ at any instant $t \geq t_p$ coincides with the strain $\varepsilon(t_p)$. For an inelastic material, an additional strain can be observed which increases in time: $\varepsilon(t) > \varepsilon(t_p)$, $\dot{\varepsilon}(t) > 0$ for $t > t_p$. Here the dot denotes the differentiation with respect to time. This phenomenon (growth of strain in time under a fixed load) is called *creep.* In a creep phenomenon the diagram in Figure 6.4 continues from P in the direction PQ.

The growth of strains in time is a characteristic feature of both inelastic liquids and inelastic solids. To distinguish these two classes of media we presume that for solid media $t \mapsto \varepsilon(t)$ tends to a finite limit $\varepsilon_\infty > 0$ and its derivative $t \mapsto \dot{\varepsilon}(t)$ tends to zero as $t \to \infty$, whereas for liquid media the rate of strains $t \mapsto \dot{\varepsilon}(t)$ tends to a finite limit $\dot{\varepsilon}_\infty > 0$ which corresponds to a Newtonian flow. Typical creep curves $t \mapsto \varepsilon(t)$ are plotted in Figure 6.5. The division of media into solids and liquids is not completely strict, and it depends essentially on temperature. For example, metals at room temperature are elastic solids, whereas at elevated temperatures they demonstrate the creep typical of liquids (see [207]). Note that the mechanical behavior of several polymers differs from the standard creep behavior in both solids and liquids: Strain $t \mapsto \varepsilon(t)$ increases monotonically and does not tend to any finite limit, while the rate of strain $t \mapsto \dot{\varepsilon}(t)$ tends to zero with the growth of time.

Another diagnostic test can be done with a hard testing machine if, after a continuously increasing loading to the state P (Figure 6.4), the strain is kept constant while the stress variation can be measured. Assume that the state P is reached at time $t = t_p$. For a purely elastic material, the stress $\sigma(t)$ at any instant $t \geq t_p$ coincides with the stress $\sigma(t_p)$. For an inelastic material, the stress $\sigma(t)$ decreases monotonically in time for $t \geq t_p$ tending to some limiting value σ_∞ as $t \to \infty$. This phenomenon (decreasing of the stress in time for a fixed strain) is called *relaxation*. In a relaxation phenomenon the diagram in Figure 6.4 continues from P in the direction PR.

Typical graphs of time dependence of the stress during a relaxation experiment is plotted in Figure 6.6. We distinguish two cases: the case when $\sigma_\infty > 0$, which corresponds to a solid medium, and the case when $\sigma_\infty = 0$ which corresponds to a liquid medium. Indeed, by stretching the specimen, we create a new shape for it. A medium demonstrates the liquid-type behavior if no additional stresses arise when we provide a new shape without changing its volume. For uniaxial loading, the condition $\sigma_\infty = 0$ means that asymptotically (i.e. for large times) the medium accepts its new shape without resistance, i.e. it behaves like a liquid.

A material which demonstrates both creep and relaxation phenomena is called *viscoelastic*. The study of viscoelastic materials concerns the *theory of viscoelasticity*.

Concluding remarks. In this section we described some standard experiments on solids which led us to characterize the features of elastic and inelastic behaviors. Briefly, we may say that elastic materials are those for which any reduction of the stress leads to a corresponding decrease in the strain, whereas the stress and the strain obey the same stress-strain relation both for loading and unloading processes. In an elastic material the stress is completely determined by the strain and vice versa. On the contrary, for an elastoplastic medium, a reduction of the stress leads to a decrease in the strain, but the stress-strain diagrams for loading and unloading differ from each other. As a consequence, elastoplastic materials are characterized by a decomposition of the strain into a reversible part and an irreversible part. Finally, the main features of the viscoelastic behavior are the creep and relaxation phenomena which arise in long-range experiments.

The analysis of the experimental data allows us to formulate mathematically the above physical features. In the following sections of this chapter we discuss mathematical formulas for constitutive relations of various materials. We start by presenting the constitutive laws in the one-dimensional case and discuss their capacity to model the elastic, viscoelastic and elastoplastic behaviors. Then, we provide the form of the constitutive laws in the general multi-dimensional case, by abstraction and generalization based on mathematical tools. In this way

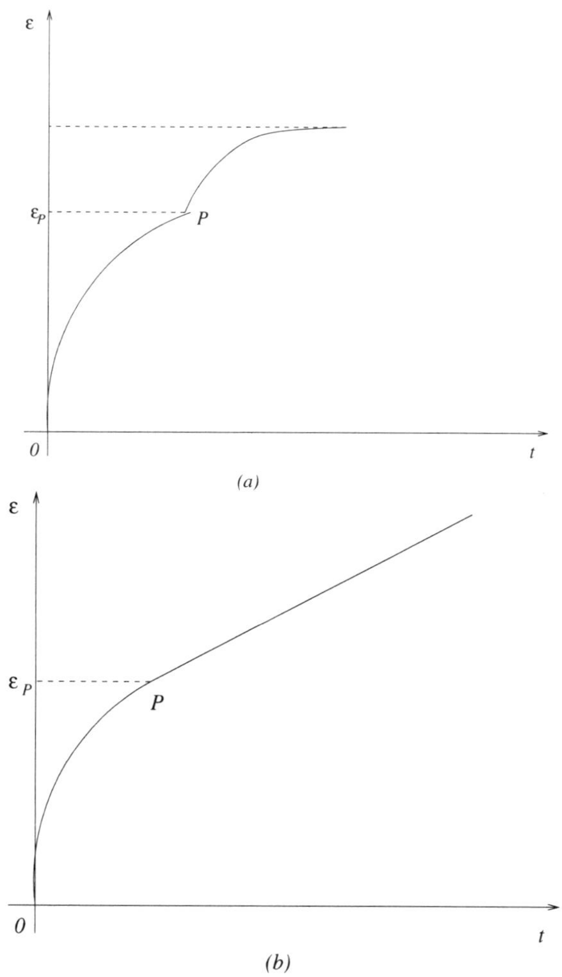

Figure 6.5: Typical creep curves for (a) inelastic solids and (b) inelastic liquids

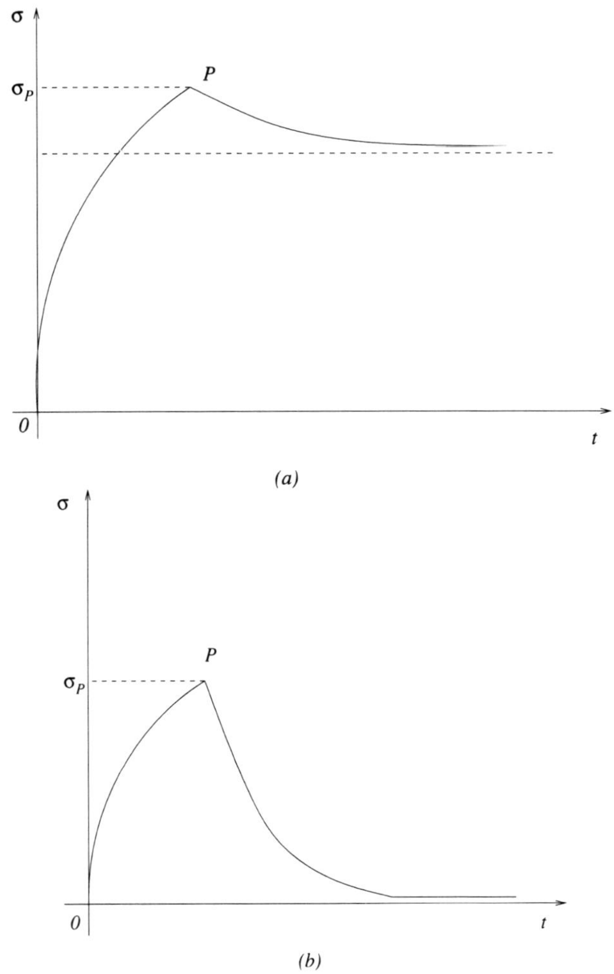

Figure 6.6: Typical relaxation curves for (a) inelastic solids and (b) inelastic liquids

we hope that the connection between a physical behavior and its mathematical idealization may become more apparent.

6.2 Constitutive Relations in Elasticity

The loading experiments allow us to divide the elastic materials into two groups: materials which demonstrate a purely linear response, i.e. for which the stress-strain diagram is straight, and materials which demonstrate both linear and nonlinear responses, i.e. for which the stress-strain diagram is not straight. The first group of materials is the subject of the linear theory of elasticity. The second group is considered in the (physically) nonlinear theory of elasticity.

In the case of linear theory of elasticity, the stress σ and the strain ε are assumed mutually proportional, which is expressed by *Hooke's law*

$$\sigma = E\varepsilon, \tag{6.5}$$

where E is the so-called *Young's modulus* of elasticity. This constitutive law models the behavior of the material corresponding to the segment OA in Figure 6.2. More generally, we can consider a nonlinear relation between the stress σ and the strain ε of the form

$$\sigma = f(\varepsilon) \tag{6.6}$$

where f is a nonlinear real-valued function. This constitutive law is able to describe the nonlinear behavior of the portion AB in Figure 6.2.

As a simple example we consider the function ([147, 167])

$$f(\varepsilon) = \begin{cases} a\,\varepsilon + b\,(\varepsilon + \varepsilon_0) & \text{if} \quad \varepsilon \le -\varepsilon_0, \\ a\,\varepsilon & \text{if} \quad -\varepsilon_0 < \varepsilon \le \varepsilon_0, \\ a\,\varepsilon + b\,(\varepsilon - \varepsilon_0) & \text{if} \quad \varepsilon \ge \varepsilon_0, \end{cases} \tag{6.7}$$

where a, b and ε_0 are positive constants. The stress-strain diagram of the elastic constitutive law (6.6) with f given by (6.7) is plotted in Figure 6.7. Various one-dimensional nonlinear elastic constitutive laws of the form (6.6) may be found in [167, pp. 92–97].

The elastic constitutive equations (6.5) and (6.6) are formulated in the homogeneous one-dimensional case. In order to consider the general case, we denote by $\Omega \subset \mathbb{R}^d$ $(d = 1, 2, 3)$ the reference configuration of the elastic body. To extend Hooke's law (6.5) we suppose that at every point $\boldsymbol{x} \in \Omega$ there exists a linear relation between the stress tensor $\boldsymbol{\varepsilon}$ and the small strain tensor $\boldsymbol{\varepsilon} = \boldsymbol{\varepsilon}(\boldsymbol{u})$, that is

$$\boldsymbol{\sigma} = \mathcal{E}\boldsymbol{\varepsilon}(\boldsymbol{u}). \tag{6.8}$$

Here and below, for simplicity, we do not specify explicitly the dependence of various functions on $\boldsymbol{x} \in \Omega$. The quantity $\mathcal{E} = (\mathcal{E}_{ijkl})$ is a fourth-order

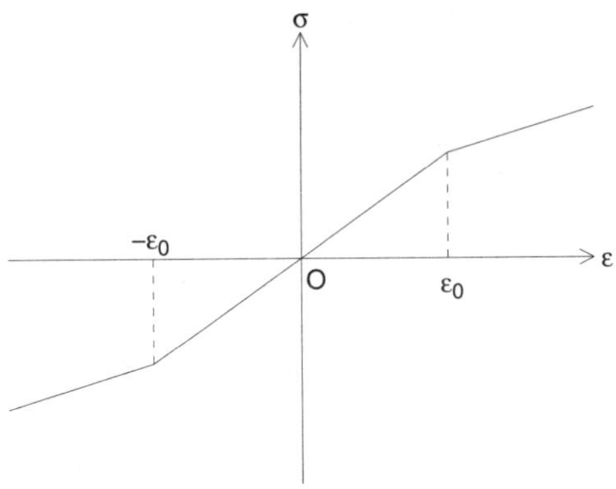

Figure 6.7: Stress-strain diagram for the function (6.7)

tensor, called the *elasticity tensor.* It is a linear map from the space $\mathbb{S}^d$ of symmetric second-order tensors on $\mathbb{R}^d$ into itself. In general the elasticity tensor is a function of the position in the body. We assume in this book that $\mathcal{E}$ is bounded, symmetric and positively definite in Ω:

$$\left.\begin{array}{l} \text{(a) } \mathcal{E} : \Omega \times \mathbb{S}^d \to \mathbb{S}^d. \\ \text{(b) } \mathcal{E}_{ijkl} \in L^\infty(\Omega),\ 1 \le i,j,k,l \le d. \\ \text{(c) } \mathcal{E}\boldsymbol{\sigma}\cdot\boldsymbol{\tau} = \boldsymbol{\sigma}\cdot\mathcal{E}\boldsymbol{\tau},\ \forall\boldsymbol{\sigma},\boldsymbol{\tau}\in\mathbb{S}^d,\ \text{ a.e. in } \Omega. \\ \text{(d) There exists } \alpha_0 > 0 \text{ such that} \\ \qquad \mathcal{E}\boldsymbol{\tau}\cdot\boldsymbol{\tau} \ge \alpha_0|\boldsymbol{\tau}|^2\ \ \forall\boldsymbol{\tau}\in\mathbb{S}^d,\ \text{ a.e. in } \Omega. \end{array}\right\} \tag{6.9}$$

In component form the constitutive equation (6.8) has the form

$$\sigma_{ij} = \mathcal{E}_{ijkl}\varepsilon_{kl}, \quad 1 \le i,j \le d, \tag{6.10}$$

and the symmetry condition (6.9) (b) is equivalent to the symmetry properties

$$\mathcal{E}_{ijkl} = \mathcal{E}_{klij} = \mathcal{E}_{ijlk}, \quad 1 \le i,j,k,l \le d.$$

These equalities indicate that the components of the elasticity tensor are not completely independent. In the case $d = 3$, there are at most 21 independent components that together characterize the elastic response of the material at any point $\boldsymbol{x} \in \Omega$.

Sometimes it is convenient to work not with the stress as a function of the strain, but rather the strain as a function of the stress. It follows from assumption (6.9) that $\mathcal{E}$ is an invertible tensor for a.e. $\boldsymbol{x} \in \Omega$. Therefore we may invert the equality (6.8) to obtain

$$\varepsilon(\boldsymbol{u}) = \mathcal{A}\boldsymbol{\sigma}. \tag{6.11}$$

Here $\mathcal{A} : \Omega \times \mathbb{S}^d \to \mathbb{S}^d$ is the inverse tensor of $\mathcal{E}$, called the *compliance tensor.* Notice that the compliance tensor enjoys the same properties as the elasticity tensor: It is bounded, symmetric, and positively definite.

If the components $\mathcal{E}_{ijkl}$ of the elasticity tensor do not depend on $\boldsymbol{x} \in \Omega$, the material is said to be *homogeneous*. Otherwise it is *nonhomogeneous*. For a fixed $\boldsymbol{x} \in \Omega$, if the tensor $\mathcal{E}(\boldsymbol{x})$ is invariant with respect to rotations of the coordinate system, the material is said to be *isotropic* at the point $\boldsymbol{x}$. Otherwise, the material is *anisotropic* at the point $\boldsymbol{x} \in \Omega$. For example, wood is anisotropic at any point since it possesses different properties along and across its fibers (such material is called *orthotropic*).

One can show that for homogeneous, isotropic, linearized elastic materials we have

$$\mathcal{E}_{ijkl} = \lambda\, \delta_{ij}\delta_{kl} + \mu\, (\delta_{ik}\delta_{jl} + \delta_{il}\delta_{jk}), \tag{6.12}$$

where the constants λ and μ are called the *Lamé coefficients* and δ_{ij} is the Kronecker delta. Combining (6.10) and (6.12) we obtain

$$\sigma_{ij} = \lambda\, (\varepsilon_{kk})\, \delta_{ij} + 2\, \mu\, \varepsilon_{ij}. \tag{6.13}$$

Inverting (6.13) we get

$$\varepsilon_{ij} = \frac{1+\nu}{E}\, \sigma_{ij} - \frac{\nu}{E}\, (\sigma_{kk})\, \delta_{ij} \tag{6.14}$$

where the constants E and ν are defined by the relations

$$E = \frac{\mu(3\lambda + 2\mu)}{\lambda + \mu}, \qquad \nu = \frac{\lambda}{2(\lambda + \mu)}\,.$$

Young's modulus (E) measures the stiffness in the tension (axial) direction, and *Poisson's ratio* (ν) measures the lateral contraction. Another quantity used often is the *bulk modulus* (K), defined by

$$K = \frac{E}{3\,(1 - 2\,\nu)} = \frac{3\lambda + 2\mu}{3}.$$

These quantities are used in the engineering literature more often than the Lamé coefficients λ and μ. Some experimental tests (see, e.g. [161]) lead to restrictions $\lambda > 0$, $\mu > 0$ for the Lamé coefficients and it follows that $K > 0$, $E > 0$, $0 < \nu < \frac{1}{2}$.

In a tensorial form the stress-strain relation (6.13) is given by

$$\boldsymbol{\sigma} = \lambda\, (\operatorname{tr} \boldsymbol{\varepsilon})\boldsymbol{I} + 2\, \mu\, \boldsymbol{\varepsilon}, \tag{6.15}$$

where $\operatorname{tr} \boldsymbol{\varepsilon} = \varepsilon_{kk}$ is the *trace* of the tensor $\boldsymbol{\varepsilon}$. We see that this is a particular case of the form (6.8) with the elasticity tensor

$$\mathcal{E}(\boldsymbol{\varepsilon}) = \lambda\, (\operatorname{tr} \boldsymbol{\varepsilon})\, \boldsymbol{I} + 2\, \mu\, \boldsymbol{\varepsilon}.$$

Evidently the conditions (6.9) are valid. Similarly, the tensorial form of the equation (6.14) is of the form (6.11) with

$$\mathcal{A}(\boldsymbol{\sigma}) = \frac{1+\nu}{E}\, \boldsymbol{\sigma} - \frac{\nu}{E}\, (\operatorname{tr} \boldsymbol{\sigma})\, \boldsymbol{I},$$

where $\operatorname{tr}\boldsymbol{\sigma} = \sigma_{kk}$.

We now describe constitutive laws for physically nonlinear elastic materials. The one-dimensional nonlinear constitutive law (6.6) may be extended to the d-dimensional case by taking

$$\boldsymbol{\sigma} = \mathcal{F}\boldsymbol{\varepsilon}(\boldsymbol{u}) \tag{6.16}$$

in which $\mathcal{F}$ is a given nonlinear function. We assume that $\mathcal{F}$ satisfies the following conditions:

$$\left.\begin{array}{l}
\text{(a) } \mathcal{F} : \Omega \times \mathbb{S}^d \to \mathbb{S}^d.\\
\text{(b) There exists } M > 0 \text{ such that}\\
\qquad \|\mathcal{F}(\boldsymbol{x}, \boldsymbol{\varepsilon}_1) - \mathcal{F}(\boldsymbol{x}, \boldsymbol{\varepsilon}_2)\| \le M\|\boldsymbol{\varepsilon}_1 - \boldsymbol{\varepsilon}_2\|\\
\qquad\quad \forall\, \boldsymbol{\varepsilon}_1, \boldsymbol{\varepsilon}_1 \in \mathbb{S}^d, \text{ a.e. } \boldsymbol{x} \in \Omega.\\
\text{(c) There exists } m > 0 \text{ such that}\\
\qquad (\mathcal{F}(\boldsymbol{x}, \boldsymbol{\varepsilon}_1) - \mathcal{F}(\boldsymbol{x}, \boldsymbol{\varepsilon}_2)) \cdot (\boldsymbol{\varepsilon}_1 - \boldsymbol{\varepsilon}_2) \ge m\,\|\boldsymbol{\varepsilon}_1 - \boldsymbol{\varepsilon}_2\|^2\\
\qquad\quad \forall\, \boldsymbol{\varepsilon}_1, \boldsymbol{\varepsilon}_1 \in \mathbb{S}^d, \text{ a.e. } \boldsymbol{x} \in \Omega.\\
\text{(d) For any } \boldsymbol{\varepsilon} \in \mathbb{S}^d,\ \boldsymbol{x} \mapsto \mathcal{F}(\boldsymbol{x}, \boldsymbol{\varepsilon}) \text{ is measurable in } \Omega.\\
\text{(e) The mapping } \boldsymbol{x} \mapsto \mathcal{F}(\boldsymbol{x}, \boldsymbol{0}) \in L^2(\Omega)^{d\times d}.
\end{array}\right\} \tag{6.17}$$

Clearly, a family of elasticity operators satisfying the conditions (6.17) is provided by the linearly elastic materials (6.8), if the elasticity tensor satisfies the conditions (6.9).

A second example is provided by the nonlinear constitutive law

$$\boldsymbol{\sigma} = \mathcal{E}\boldsymbol{\varepsilon} + \beta\,(\boldsymbol{\varepsilon} - \mathcal{P}_K\boldsymbol{\varepsilon}).$$

Here $\mathcal{E}$ is a fourth-order tensor, $\beta > 0$, K is a closed convex subset of $\mathbb{S}^d$ such that $\boldsymbol{0} \in K$ and $\mathcal{P}_K : \mathbb{S}^d \to K$ denotes the projection map. The corresponding elasticity operator is given by

$$\mathcal{F}\boldsymbol{\varepsilon} = \mathcal{E}\boldsymbol{\varepsilon} + \beta\,(\boldsymbol{\varepsilon} - \mathcal{P}_K\boldsymbol{\varepsilon}). \tag{6.18}$$

Assume the conditions (6.9). Recall the nonexpansivity of the projection map (see (1.5)). We can verify that the nonlinear operator (6.18) satisfies the conditions (6.17).

A family of elasticity operators satisfying the conditions (6.17) is provided by nonlinear *Hencky materials* (for detail, cf. e.g., [217]). For a Hencky material, the stress-strain relation is

$$\boldsymbol{\sigma} = K_0\,(\operatorname{tr}\boldsymbol{\varepsilon}(\boldsymbol{u}))\,\boldsymbol{I} + \psi(\|\boldsymbol{\varepsilon}^D(\boldsymbol{u})\|^2)\,\boldsymbol{\varepsilon}^D(\boldsymbol{u}),$$

so that the elasticity operator is

$$\mathcal{F}(\boldsymbol{\varepsilon}) = K_0\,(\operatorname{tr}\boldsymbol{\varepsilon})\,\boldsymbol{I} + \psi(\|\boldsymbol{\varepsilon}^D\|^2)\,\boldsymbol{\varepsilon}^D. \tag{6.19}$$

Here, $K_0 > 0$ is a material coefficient, $\boldsymbol{I}$ is the identity tensor of the second order, $\psi : \mathbb{R} \to \mathbb{R}$ is a constitutive function and $\boldsymbol{\varepsilon}^D = \boldsymbol{\varepsilon}^D(\boldsymbol{u})$ denotes the deviatoric part of $\boldsymbol{\varepsilon} = \boldsymbol{\varepsilon}(\boldsymbol{u})$:

$$\boldsymbol{\varepsilon}^D = \boldsymbol{\varepsilon} - \frac{1}{d}\,(\operatorname{tr}\boldsymbol{\varepsilon})\,\boldsymbol{I}.$$

The function ψ is assumed to be piecewise continuously differentiable, and there exist positive constants c_1, c_2, d_1 and d_2, such that for $\xi \geq 0$,

$$\begin{aligned}
&\psi(\xi) \leq d_1, \\
&-c_1 \leq \psi'(\xi) \leq 0, \\
&\;c_2 \leq \psi(\xi) + 2\,\psi'(\xi)\,\xi \leq d_2.
\end{aligned}$$

The conditions (6.17) are satisfied for the elasticity operator defined in (6.19). For instance, let us verify (6.17) (b) and (6.17) (c). For $\boldsymbol{\varepsilon}_1, \boldsymbol{\varepsilon}_2 \in \mathbb{S}^d$ and $t \in [0,1]$, we use the notation

$$\boldsymbol{\varepsilon}^D(t) = \boldsymbol{\varepsilon}_2^D + t\,(\boldsymbol{\varepsilon}_1^D - \boldsymbol{\varepsilon}_2^D).$$

We have

$$\mathcal{F}(\boldsymbol{\varepsilon}_1) - \mathcal{F}(\boldsymbol{\varepsilon}_2) = K_0 \mathrm{tr}\,(\boldsymbol{\varepsilon}_1 - \boldsymbol{\varepsilon}_2)\,\boldsymbol{I} + \psi(\|\boldsymbol{\varepsilon}_1^D\|^2)\,\boldsymbol{\varepsilon}_1^D - \psi(\|\boldsymbol{\varepsilon}_2^D\|^2)\,\boldsymbol{\varepsilon}_2^D,$$

and

$$\begin{aligned}
\psi(\|\boldsymbol{\varepsilon}_1^D\|^2)\,\boldsymbol{\varepsilon}_1^D - \psi(\|\boldsymbol{\varepsilon}_2^D\|^2)\,\boldsymbol{\varepsilon}_2^D &= \int_0^1 \frac{d}{dt}\left[\psi(\|\boldsymbol{\varepsilon}^D(t)\|^2)\,\boldsymbol{\varepsilon}^D(t)\right] dt \\
&= \int_0^1 \big[2\,\psi'(\|\boldsymbol{\varepsilon}^D(t)\|^2)\,(\boldsymbol{\varepsilon}^D(t)\cdot(\boldsymbol{\varepsilon}_1 - \boldsymbol{\varepsilon}_2))\,\boldsymbol{\varepsilon}^D(t) \\
&\qquad + \psi(\|\boldsymbol{\varepsilon}^D(t)\|^2)\,(\boldsymbol{\varepsilon}_1 - \boldsymbol{\varepsilon}_2)\big]\,dt.
\end{aligned}$$

Then the condition (6.17) (b) is satisfied for some constant M depending on K_0, d_1, d_2 and c_1. Similarly,

$$\begin{aligned}
&(\mathcal{F}(\boldsymbol{\varepsilon}_1) - \mathcal{F}(\boldsymbol{\varepsilon}_2))\cdot(\boldsymbol{\varepsilon}_1 - \boldsymbol{\varepsilon}_2) \\
&\quad = K_0|\mathrm{tr}\,(\boldsymbol{\varepsilon}_1 - \boldsymbol{\varepsilon}_2)|^2 + \int_0^1 \big[2\,\psi'(\|\boldsymbol{\varepsilon}^D(t)\|^2)\,|\boldsymbol{\varepsilon}^D(t)\cdot(\boldsymbol{\varepsilon}_1 - \boldsymbol{\varepsilon}_2)|^2 \\
&\qquad + \psi(\|\boldsymbol{\varepsilon}^D(t)\|^2)\,\|\boldsymbol{\varepsilon}_1 - \boldsymbol{\varepsilon}_2\|^2\big]\,dt \\
&\quad \geq K_0|\mathrm{tr}\,(\boldsymbol{\varepsilon}_1 - \boldsymbol{\varepsilon}_2)|^2 \\
&\qquad + \int_0^1 \big[2\,\psi'(\|\boldsymbol{\varepsilon}^D(t)\|^2)\,\|\boldsymbol{\varepsilon}^D(t)\|^2 + \psi(\|\boldsymbol{\varepsilon}^D(t)\|^2)\big]\,\|\boldsymbol{\varepsilon}_1 - \boldsymbol{\varepsilon}_2\|^2\,dt \\
&\quad \geq K_0|\mathrm{tr}\,(\boldsymbol{\varepsilon}_1 - \boldsymbol{\varepsilon}_2)|^2 + c_2\|\boldsymbol{\varepsilon}_1 - \boldsymbol{\varepsilon}_2\|^2.
\end{aligned}$$

Hence, the condition (6.17) (c) is satisfied with m depending on K_0 and c_2.

6.3 Constitutive Relations in Viscoelasticity

In this section we first provide a brief survey of one-dimensional differential models in viscoelasticity, i.e. constitutive models where the stress and the strain are connected by differential equations. Then we present the *Kelvin-Voigt model*

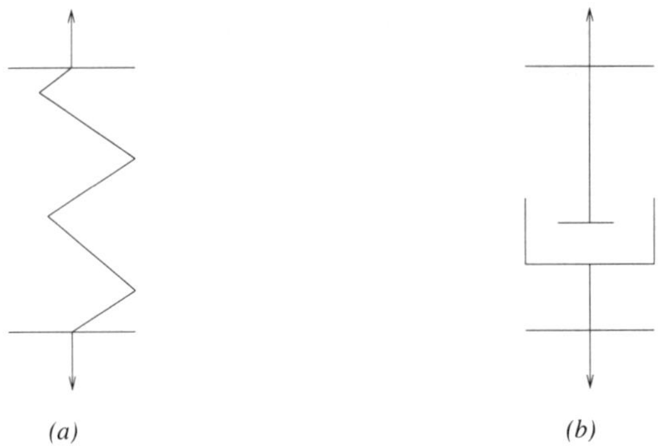

Figure 6.8: Basic rheological elements: a) spring; b) dashpot

used in Part III of this book and describe the conditions imposed on the constitutive functions. More details in the study of viscoelastic laws including integral models may be found in [67].

Viscoelasticity means that the material behavior presents both elastic and viscous features. In rheology, elastic properties are described by *springs*, i.e. elements which provide a one-to-one connection between the stress σ and the strain ε. A linearly elastic spring obeys Hooke's law (6.5), whereas a nonlinear elastic spring satisfies the relation (6.6). Viscous properties are modeled by *dashpots*, i.e. elements which provide a one-to-one connection between the stress σ and the strain rate $\dot{\varepsilon}$. The linear relation between the stress and the strain rate

$$\sigma = \eta\,\dot{\varepsilon} \tag{6.20}$$

is called *the Newton law*, where the coefficient $\eta > 0$ is called *Newtonian viscosity*. A more sophisticated relation may be developed by assuming a nonlinear dependence of the stress σ on the strain rate $\dot{\varepsilon}$,

$$\sigma = g(\dot{\varepsilon}), \tag{6.21}$$

where g is a given constitutive function. Basic rheological elements are plotted schematically in Figure 6.8.

To proceed with the construction of rheological models, it is natural to combine these elements in groups. Two possibilities arise in combining these elements: in serial and in parallel. The former leads to the *Maxwell model*, see Figure 6.9, whereas the latter leads to the *Kelvin-Voigt model*, see Figure 6.10.

To obtain more general models, we combine the basic elements (springs and dashpots) as well as the Maxwell and Kelvin-Voigt elements in parallel and serial. For example, taking the Maxwell model together with an elastic spring in parallel, we obtain the *standard viscoelastic model*, see Figure 6.11. By choosing an arbitrary number of Maxwell models connected in parallel we obtain the *generalized Maxwell model*, see Figure 6.12. Experimental data for this model

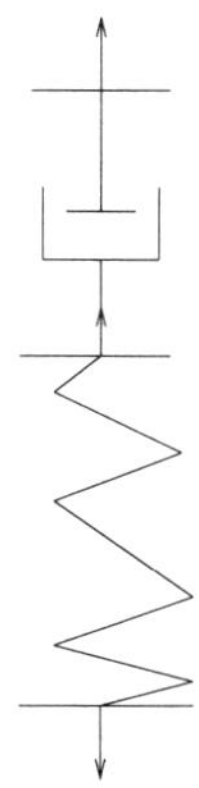

Figure 6.9: The Maxwell model

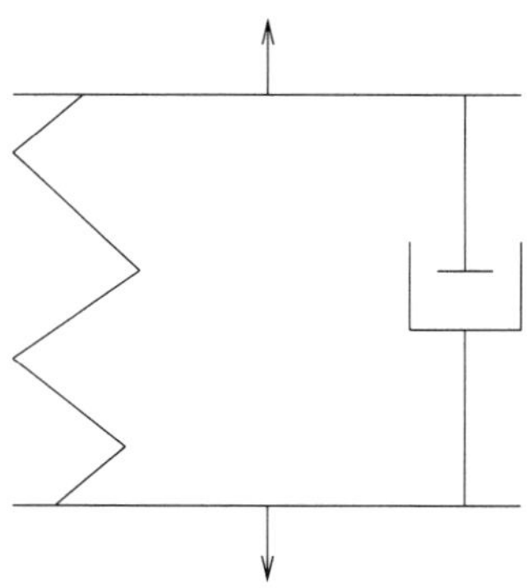

Figure 6.10: The Kelvin-Voigt model

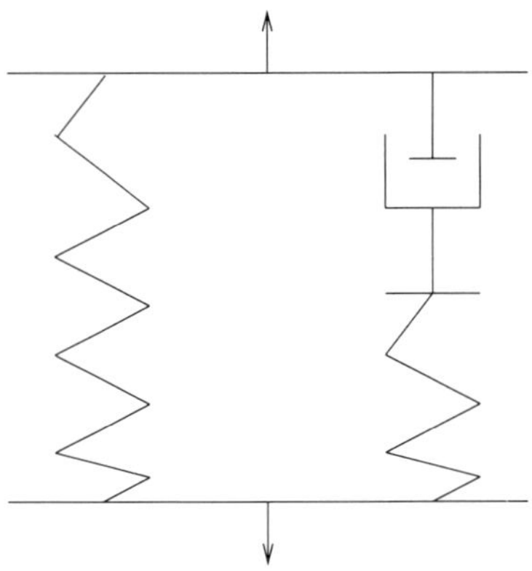

Figure 6.11: The standard viscoelastic model

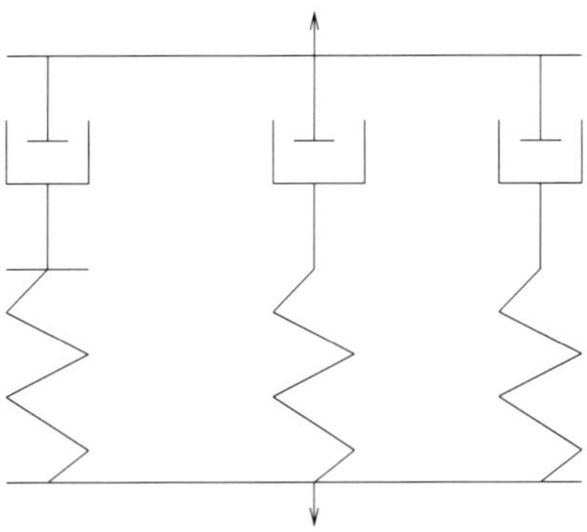

Figure 6.12: Generalized Maxwell model with 3 elements

are presented in [184]. By choosing an arbitrary number of Kelvin-Voigt models connected in serial we arrive at the *generalized Voigt model*, see Figure 6.13. For a discuss of this model and related experimental data see e.g. [89, 192]. Finally, by adding to the Kelvin-Voigt model a spring and a dashpot in serial, we obtain the *Burgers model*, see Figure 6.14.

By this approach, an arbitrary number of constitutive models may be constructed which differ from one another by number of elements and the way they are connected. In order to derive the differential equations corresponding to the rheological models introduced above, the following two rules are to be used:

- for a system of elements connected in serial, their stresses coincide and the total strain equals the sum of strains in separate elements;

- for a system of elements connected in parallel, their strains coincide and the total stress equals the sum of stresses in separate elements.

Let us apply these rules to the Maxwell model presented in Figure 6.9. We

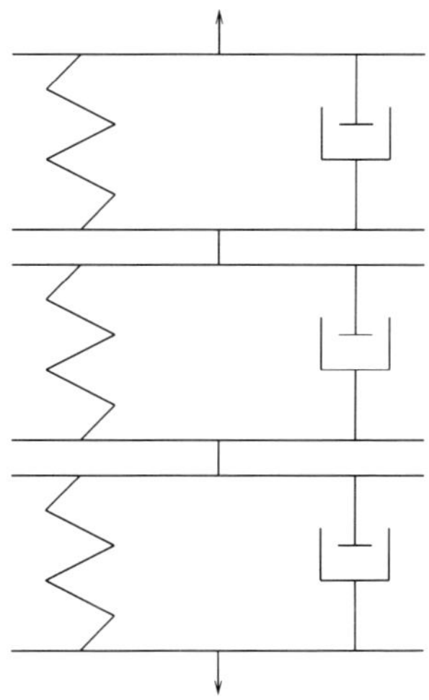

Figure 6.13: Generalized Voigt model with 3 elements

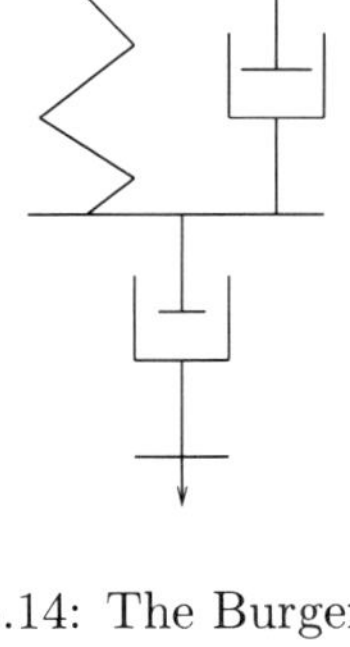

Figure 6.14: The Burgers model

find

$$\varepsilon = \varepsilon_e + \varepsilon_v \tag{6.22}$$

where ε_e and ε_v are the strains in the spring and in the dashpot. Assume the rheological elements are linear. Using (6.5), (6.20) we have

$$\dot{\varepsilon}_e = \frac{\dot{\sigma}}{E}, \qquad \dot{\varepsilon}_v = \frac{\sigma}{\eta}. \tag{6.23}$$

It follows now from (6.22) and (6.23) that

$$\dot{\varepsilon} = \frac{\dot{\sigma}}{E} + \frac{\sigma}{\eta}, \tag{6.24}$$

which represents the constitutive law of the linear Maxwell viscoelastic model.

We now apply the above procedure to the Kelvin-Voigt model described in Figure 6.10. We have

$$\sigma = \sigma_e + \sigma_v, \tag{6.25}$$

where σ_e and σ_v are the stresses in the spring and in the dashpot, respectively. Assume the rheological elements are linear. Using (6.5), (6.20) we have

$$\sigma_e = E\,\varepsilon, \qquad \sigma_v = \eta\,\dot{\varepsilon}. \tag{6.26}$$

Then by (6.25) and (6.26) we obtain

$$\sigma = E\,\varepsilon + \eta\,\dot{\varepsilon}, \tag{6.27}$$

which represents the constitutive law of the linear Kelvin-Voigt viscoelastic material.

We now derive the constitutive equation for the linear standard viscoelastic solid shown in Figure 6.11. We have

$$\sigma = \sigma_e + \sigma_m, \tag{6.28}$$

where σ_e is the stress in the spring and σ_m is the stress in the Maxwell element. Since the constitutive elements are assumed to be linear, using (6.5) and (6.24) we have

$$\sigma_e = E_1\varepsilon, \qquad \dot{\varepsilon} = \frac{\dot{\sigma}_m}{E} + \frac{\sigma_m}{\eta}, \tag{6.29}$$

where E_1 denotes the Young's modulus of the spring, whereas E and η represent Young's modulus and the Newtonian viscosity for the Maxwell element. We now eliminate σ_e and σ_m from (6.28) and (6.29) to obtain

$$\frac{\dot{\sigma}}{E} + \frac{\sigma}{\eta} = \left(1 + \frac{E_1}{E}\right)\dot{\varepsilon} + \frac{E_1}{\eta}\,\varepsilon, \tag{6.30}$$

which represents the constitutive equation of the linear standard viscoelastic material.

The previous rheological models can describe the creep and relaxation processes. For example, suppose $\sigma(0) = \sigma_0$, $\varepsilon(0) = \varepsilon_0$ and $\sigma(t) = \sigma_0$. Using (6.24) we find

$$\varepsilon(t) = \frac{\sigma_0}{\eta} t + \varepsilon_0 \quad \forall\, t \geq 0.$$

We conclude that the Maxwell model describes the creep phenomena of a Newtonian flow since

$$\dot{\varepsilon}(t) \to \frac{\sigma_0}{\eta} \quad \text{as } t \to \infty.$$

Similarly, supposing $\sigma(0) = \sigma_0$, $\varepsilon(0) = \varepsilon_0$ and $\varepsilon(t) = \varepsilon_0$, again by (6.24) we find

$$\sigma(t) = \sigma_0 e^{-\frac{E}{\eta} t} \quad \forall\, t \geq 0.$$

We conclude that the Maxwell model describes the relaxation phenomena of a liquid medium since

$$\sigma(t) \to 0 \quad \text{as } t \to \infty.$$

With the above procedure it can be shown that the Kelvin-Voigt model (6.27) and the linear viscoelastic model (6.30) describes the creep phenomena and behave like inelastic solids. Moreover, the linear viscoelastic law (6.30) models the relaxation phenomena in a solid medium.

It is easy to see that Maxwell constitutive law (6.24) and the standard viscoelastic constitutive law (6.30) may be written in a general form

$$\dot{\sigma} = \mathcal{E}\dot{\varepsilon} - k\,(\sigma - F\varepsilon) \tag{6.31}$$

where $\mathcal{E}$, k and F are appropriate constants such that $\mathcal{E} > 0$, $k > 0$, $F \geq 0$. Constitutive laws of the form (6.31) represent a particular case of a class of constitutive relations we shall discuss in the next section in the study of rate-type materials. For this reason we deal in the rest of the section only with Kelvin-Voigt constitutive law. Remember that spring and dashpot rheological models demonstrate linear or nonlinear behavior depending on assumptions regarding the mechanical response of the basic elements. Thus, using a nonlinear elastic spring $\sigma_e = b(\varepsilon)$ and a nonlinear dashpot $\sigma_v = a(\dot{\varepsilon})$ in (6.25) we obtain the nonlinear version of the one-dimensional Kelvin-Voigt constitutive law, that is

$$\sigma = a(\dot{\varepsilon}) + b(\varepsilon). \tag{6.32}$$

Here a and b are nonlinear real-valued functions. Various nonlinear viscoelastic constitutive laws which can be cast to the form (6.32) may be found in [168] and [67, p. 260].

We now extend (6.32) to the general d-dimensional case to obtain

$$\boldsymbol{\sigma} = \mathcal{A}\boldsymbol{\varepsilon}(\dot{\boldsymbol{u}}) + \mathcal{B}\boldsymbol{\varepsilon}(\boldsymbol{u}). \tag{6.33}$$

This is the *viscoelastic constitutive law* to be used in Part III. We allow the *viscosity operator* $\mathcal{A}$ and the *elasticity operator* $\mathcal{B}$ to depend on the location of the point. Thus here $\mathcal{A}\boldsymbol{\varepsilon}(\dot{\boldsymbol{u}})$ and $\mathcal{B}\boldsymbol{\varepsilon}(\boldsymbol{u})$ are short hand notations for $\mathcal{A}(\boldsymbol{x}, \boldsymbol{\varepsilon}(\dot{\boldsymbol{u}}))$ and $\mathcal{B}(\boldsymbol{x}, \boldsymbol{\varepsilon}(\boldsymbol{u}))$, respectively.

In the study of mechanical problems involving the viscoelastic material, we assume (6.34) and (6.35) below.

$$\left.\begin{array}{l}
\text{(a) } \mathcal{A} : \Omega \times \mathbb{S}^d \to \mathbb{S}^d. \\
\text{(b) There exists } L_{\mathcal{A}} > 0 \text{ such that} \\
\qquad \|\mathcal{A}(\boldsymbol{x}, \boldsymbol{\varepsilon}_1) - \mathcal{A}(\boldsymbol{x}, \boldsymbol{\varepsilon}_2)\| \le L_{\mathcal{A}} \|\boldsymbol{\varepsilon}_1 - \boldsymbol{\varepsilon}_2\| \\
\qquad \forall\, \boldsymbol{\varepsilon}_1, \boldsymbol{\varepsilon}_1 \in \mathbb{S}^d, \text{ a.e. } \boldsymbol{x} \in \Omega. \\
\text{(c) There exists } M > 0 \text{ such that} \\
\qquad (\mathcal{A}(\boldsymbol{x}, \boldsymbol{\varepsilon}_1) - \mathcal{A}(\boldsymbol{x}, \boldsymbol{\varepsilon}_2)) \cdot (\boldsymbol{\varepsilon}_1 - \boldsymbol{\varepsilon}_2) \ge M \, \|\boldsymbol{\varepsilon}_1 - \boldsymbol{\varepsilon}_2\|^2 \\
\qquad \forall\, \boldsymbol{\varepsilon}_1, \boldsymbol{\varepsilon}_1 \in \mathbb{S}^d, \text{ a.e. } \boldsymbol{x} \in \Omega. \\
\text{(d) For any } \boldsymbol{\varepsilon} \in \mathbb{S}^d, \ \boldsymbol{x} \mapsto \mathcal{A}(\boldsymbol{x}, \boldsymbol{\varepsilon}) \text{ is measurable on } \Omega. \\
\text{(e) The mapping } \boldsymbol{x} \mapsto \mathcal{A}(\boldsymbol{x}, \boldsymbol{0}) \in L^2(\Omega)^{d \times d}.
\end{array}\right\} \tag{6.34}$$

$$\left.\begin{array}{l}
\text{(a) } \mathcal{B} : \Omega \times \mathbb{S}^d \to \mathbb{S}^d. \\
\text{(b) There exists an } L_{\mathcal{B}} > 0 \text{ such that} \\
\qquad \|\mathcal{B}(\boldsymbol{x}, \boldsymbol{\varepsilon}_1) - \mathcal{B}(\boldsymbol{x}, \boldsymbol{\varepsilon}_2)\| \le L_{\mathcal{B}} \, \|\boldsymbol{\varepsilon}_1 - \boldsymbol{\varepsilon}_2\| \\
\qquad \forall\, \boldsymbol{\varepsilon}_1, \boldsymbol{\varepsilon}_2 \in \mathbb{S}^d, \text{ a.e. in } \Omega. \\
\text{(c) For any } \boldsymbol{\varepsilon} \in \mathbb{S}^d, \ \boldsymbol{x} \mapsto \mathcal{B}(\boldsymbol{x}, \boldsymbol{\varepsilon}) \text{ is measurable on } \Omega. \\
\text{(d) The mapping } \boldsymbol{x} \mapsto \mathcal{B}(\boldsymbol{x}, \boldsymbol{0}) \in L^2(\Omega)^{d \times d}.
\end{array}\right\} \tag{6.35}$$

We remark that in linear viscoelasticity, the stress tensor $\sigma = (\sigma_{ij})$ is given by

$$\sigma_{ij} = a_{ijkl}\varepsilon_{kl}(\dot{u}) + b_{ijkl}\varepsilon_{kl}(u), \tag{6.36}$$

where $\mathcal{A} = (a_{ijkl})$ is the viscosity tensor and $\mathcal{B} = (b_{ijkl})$ is the elasticity tensor, $1 \le i, j, k, l \le d$. Clearly, the assumption (6.34) is satisfied if the components a_{ijkl} belong to $L^\infty(\Omega)$ and satisfy the usual properties of symmetry and ellipticity. The assumption (6.35) is satisfied if b_{ijkl} belong to $L^\infty(\Omega)$ and satisfy the usual symmetry property.

A second example is provided by the nonlinear viscoelastic constitutive law

$$\boldsymbol{\sigma} = \mathcal{A}\dot{\boldsymbol{\varepsilon}} + \beta\,(\boldsymbol{\varepsilon} - \mathcal{P}_K\boldsymbol{\varepsilon}). \tag{6.37}$$

Here $\mathcal{A}$ is a fourth-oder tensor which satisfies (6.34), $\beta > 0$, K is a closed convex subset of $\mathbb{S}^d$ such that $\boldsymbol{0} \in K$, and $\mathcal{P}_K : \mathbb{S}^d \to K$ denotes the projection map. Since the projection map is non-expansive, we see that the elasticity operator $\mathcal{B}(\boldsymbol{x}, \boldsymbol{\varepsilon}) = \beta\,(\boldsymbol{\varepsilon} - \mathcal{P}_K\boldsymbol{\varepsilon})$ satisfies condition (6.35).

We conclude that the results in Part III of the book are valid for Kelvin-Voigt viscoelastic materials (6.36) and (6.37), under the above assumptions.

6.4 Constitutive Relations in Viscoplasticity

In this section we present the rate-type constitutive laws we use in Part IV of this book. These laws may model the fundamental phenomena described by loading and unloading experiments as well as the time-dependent phenomena (creep and relaxation). For this reason, they are called *elastic-visco-plastic* constitutive laws or, for simplicity, *viscoplastic constitutive laws*. Materials modeled by this kind of laws demonstrate elastic, viscous and plastic behaviors.

We start with the homogeneous one-dimensional case. A point $(\sigma, \varepsilon) \in \mathbb{R}^2$ will be called in what follows a *state*. We assume that the total strain ε can be decomposed into two parts, the elastic strain ε^e and the plastic strain ε^p such that (6.4) holds. The elastic strain satisfies Hooke's law (6.5), that is

$$\varepsilon^e = \frac{\sigma}{E}, \tag{6.38}$$

where $E > 0$ is Young's modulus. The rate of the plastic strain depends only on the state $(\sigma, \varepsilon) \in \mathbb{R}^2$, that is

$$\dot{\varepsilon}^p = \psi(\sigma, \varepsilon), \tag{6.39}$$

where $\psi : \mathbb{R}^2 \to \mathbb{R}$ is a constitutive function. Combining now (6.4), (6.38) and (6.39) we find

$$\dot{\varepsilon} = \frac{\dot{\sigma}}{E} + \psi(\sigma, \varepsilon).$$

With the notation $G(\sigma, \varepsilon) = -E\psi(\sigma, \varepsilon)$ for all $(\sigma, \varepsilon) \in \mathbb{R}^2$, we have

$$\dot{\sigma} = E\dot{\varepsilon} + G(\sigma, \varepsilon). \tag{6.40}$$

Any point $(\sigma_0, \varepsilon_0) \in \mathbb{R}^2$ with $G(\sigma_0, \varepsilon_0) = 0$ is called an *equilibrium point* or a *point of elastic behavior*. The set of equilibrium points form the *domain of elastic behavior* of the material. Following [59], we say that (6.40) represents a *viscoelastic* constitutive law if the set of equilibrium points may be described by an explicit equation $\sigma = F(\varepsilon)$ where F is a real-valued function; if the set of equilibrium points has a nonempty interior in $\mathbb{R}^2$, we say that (6.40) represents a *viscoplastic* constitutive law.

It is straightforward to see that the constitutive law (6.31) is of the form (6.40) and represents a viscoelastic constitutive law in the sense of the previous definition. We conclude that the Maxwell constitutive law (6.24) as well as the standard viscoelastic constitutive law (6.30) are example of viscoelastic constitutive laws of the form (6.40) and therefore such kind of models can describe the creep and relaxation phenomena.

We now present some common viscoplastic models of the form (6.40).

Example 6.1 *Take*

$$\mathcal{G}(\sigma, \varepsilon) = \begin{cases} 0 & \text{if} \quad |\sigma| < \sigma_Y, \\ -(\operatorname{sign} \sigma) F(|\sigma| - \sigma_Y) & \text{if} \quad |\sigma| \geq \sigma_Y, \end{cases} \tag{6.41}$$

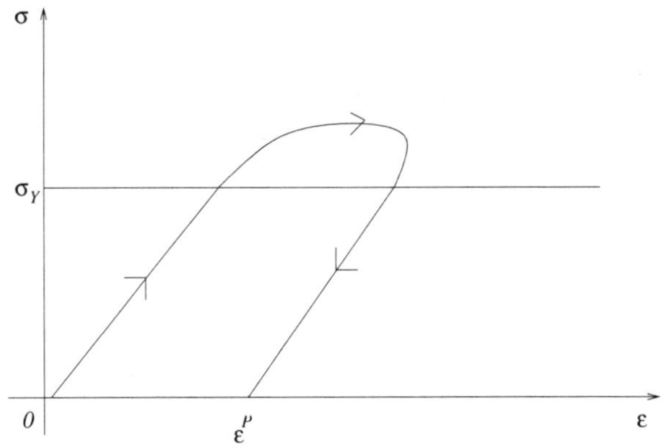

Figure 6.15: Loading and unloading process for Sokolowski's model

where σ_Y is the yield limit and $F : \mathbb{R}_+ \to \mathbb{R}$ is a smooth function with $F(0) = 0$, $F'(r) > 0$ for all $r \in \mathbb{R}_+$. The simplest example of such a function considered in the literature is $F(r) = kEr$ where $k > 0$ is a viscosity constant. Here the domain of elastic behavior is characterized by the inequality $|\sigma| \leq \sigma_Y$. For all stresses satisfying this inequality we have $\dot{\sigma} = E\dot{\varepsilon}$. Moreover, it follows that the constitutive law (6.40), (6.41), *known as the* Sokolowski model, *may predict the existence of plastic deformations observed in the unloading experiments, as is shown in Figure 6.15.*

One can also see that plastic deformations occur only for $\sigma = \sigma_Y$, i.e. only at a fixed value for the stress. For this reason, the model (6.40), (6.41) *is called a* perfectly plastic *model.*

A relatively simple one-dimensional example of constitutive law of the form (6.40) in the case when a full coupling of the stress and the strain is involved in $\mathcal{G}$ is the following one.

Example 6.2 *(see [59, p. 35]).*

$$\mathcal{G}(\sigma, \varepsilon) = \begin{cases} -k_1 F_1(\sigma - f(\varepsilon)) & \text{if} \quad \sigma > f(\varepsilon), \\ 0 & \text{if} \quad g(\varepsilon) \leq \sigma \leq f(\varepsilon), \\ k_2 F_2(g(\varepsilon) - \sigma) & \text{if} \quad \sigma < g(\varepsilon), \end{cases} \tag{6.42}$$

where k_1, $k_2 > 0$ are viscosity constants, f and g are Lipschitz continuous functions with $g(\varepsilon) \leq f(\varepsilon)$, and F_1, $F_2 : \mathbb{R}_+ \to \mathbb{R}$ are increasing functions with $F_1(0) = F_2(0) = 0$.

Note that the domain of elastic behavior is characterized by the inequalities $g(\varepsilon) \leq \sigma \leq f(\varepsilon)$ and therefore, if $f(\varepsilon) > g(\varepsilon)$ $\forall \varepsilon \in \mathbb{R}$, it follows that (6.40), (6.42) *represents a viscoplastic law. In this case plastic deformations occur only for $\sigma > f(\varepsilon)$ in extension or for $\sigma < g(\varepsilon)$ in compression. Therefore, since the yield limit (in extension and in compression) depends on the deformation, we say that the model* (6.40), (6.42) *represents a model with* hardening. *In the*

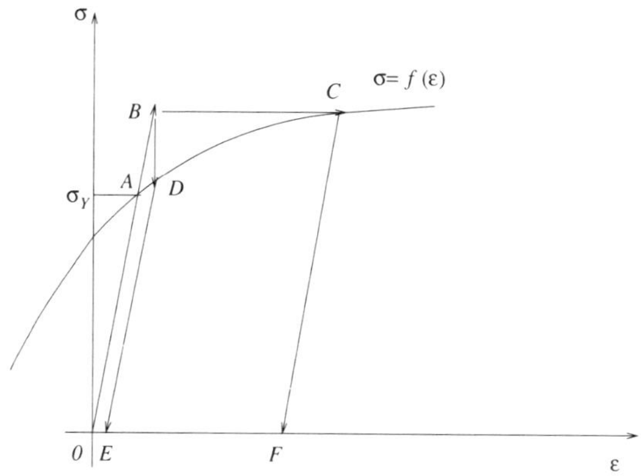

Figure 6.16: Loading and unloading process for materials obeying the constitutive relations (6.40), (6.42)

case $f(\varepsilon) = g(\varepsilon)\ \forall\, \varepsilon \in \mathbb{R}$, (6.40), (6.42) *represent a viscoelastic law. Note that the present example includes Sokolowski's constitutive equation as well as the viscoelastic models of the form* (6.31).

Now suppose $f(\varepsilon) > 0 > g(\varepsilon)$ *for all* $\varepsilon \in \mathbb{R}$. *With the help of Figure 6.16 let us explain the basic mechanical phenomena which can be described by the above model. Every loading that starts from* O *contains the elastic line* OA. *Let us imagine now that we have an instantaneous loading with* $\sigma > \sigma_Y$, *so that we have an elastic response and we reach point* B. *Further on, if we maintain constant stress, the strain will increase in time (the creep phenomenon) until point* C. *If we unload from* C, *then the behavior of the material is elastic, and we reach point* F. *In this way we obtain an irreversible strain* OF. *If we start again from* B *and we keep a constant strain, then the stress decreases in time (relaxation) until point* D. *If we unload from* D *the material behaves in an elastic way and we reach point* E. *Thus we obtain an irreversible strain* OE.

In the general d-dimensional setting ($d = 1, 2, 3$) the rate-type constitutive law (6.40) may be written as

$$\dot{\boldsymbol{\sigma}} = \mathcal{E}\boldsymbol{\varepsilon}(\dot{\boldsymbol{u}}) + \mathcal{G}(\boldsymbol{\sigma}, \boldsymbol{\varepsilon}(\boldsymbol{u})), \tag{6.43}$$

where $\mathcal{E}$ and $\mathcal{G}$ are material constitutive functions. This is the *viscoplastic constitutive law* to be used in Part IV. Here and below we do not specify explicitly the dependence of various functions on $\boldsymbol{x} \in \Omega$, Ω being the reference configuration of the body. The function $\mathcal{E}$ is assumed to be linear while $\mathcal{G}$ is in general nonlinear. We assume in this book that $\mathcal{E}$ and $\mathcal{G}$ satisfy the following conditions.

$$\left.\begin{array}{l}\text{(a) } \mathcal{E}:\Omega\times\mathbb{S}^d\to\mathbb{S}^d.\\ \text{(b) } \mathcal{E}_{ijkl}\in L^\infty(\Omega),\ 1\le i,j,k,l\le d.\\ \text{(c) } \mathcal{E}\boldsymbol{\sigma}\cdot\boldsymbol{\tau}=\boldsymbol{\sigma}\cdot\mathcal{E}\boldsymbol{\tau},\ \forall\,\boldsymbol{\sigma},\boldsymbol{\tau}\in\mathbb{S}^d,\ \text{a.e. in } \Omega.\\ \text{(d) There exists } \alpha_0>0 \text{ such that}\\ \qquad \mathcal{E}\boldsymbol{\tau}\cdot\boldsymbol{\tau}\ge\alpha_0\|\boldsymbol{\tau}\|^2\ \ \forall\,\boldsymbol{\tau}\in\mathbb{S}^d,\ \text{a.e. in }\Omega.\end{array}\right\}\qquad(6.44)$$

$$\left.\begin{array}{l}\text{(a) } \mathcal{G}:\Omega\times\mathbb{S}^d\times\mathbb{S}^d\to\mathbb{S}^d.\\ \text{(b) There exists } L_{\mathcal{G}}>0 \text{ such that}\\ \qquad \|\mathcal{G}(\boldsymbol{x},\boldsymbol{\sigma}_1,\boldsymbol{\varepsilon}_1)-\mathcal{G}(\boldsymbol{x},\boldsymbol{\sigma}_2,\boldsymbol{\varepsilon}_2)\|\\ \qquad\qquad \le L_{\mathcal{G}}\left(\|\boldsymbol{\sigma}_1-\boldsymbol{\sigma}_2\|+\|\boldsymbol{\varepsilon}_1-\boldsymbol{\varepsilon}_2\|\right)\\ \qquad \forall\,\boldsymbol{\sigma}_1,\boldsymbol{\sigma}_2,\boldsymbol{\varepsilon}_1,\boldsymbol{\varepsilon}_2\in\mathbb{S}^d,\ \text{a.e. in }\Omega.\\ \text{(c) For any } \boldsymbol{\sigma},\boldsymbol{\varepsilon}\in\mathbb{S}^d,\ \boldsymbol{x}\mapsto\mathcal{G}(\boldsymbol{x},\boldsymbol{\sigma},\boldsymbol{\varepsilon}) \text{ is measurable.}\\ \text{(d) The mapping } \boldsymbol{x}\mapsto\mathcal{G}(\boldsymbol{x},\mathbf{0},\mathbf{0})\in L^2(\Omega)^{d\times d}.\end{array}\right\}\qquad(6.45)$$

Rate-type viscoplastic models of the form (6.43) are considered in order to describe the behavior of real materials like rubbers, metals, pastes, rocks and so on. Various results and mechanical interpretation concerning models of this form may be found for instance in [59] (see also the references quoted there). Existence and uniqueness results for initial and boundary value problems for the deformation of a material following the constitutive law (6.43) can be found in [121] for the usual displacement-traction boundary conditions.

It follows now from assumptions (6.44) that $\mathcal{E}$ is an invertible tensor for almost any $\boldsymbol{x}\in\Omega$. Therefore the role of $\dot{\boldsymbol{\varepsilon}}$ can be interchanged with that of $\dot{\boldsymbol{\sigma}}$ in (6.43) to obtain

$$\boldsymbol{\varepsilon}(\dot{\boldsymbol{u}})=\mathcal{E}^{-1}\dot{\boldsymbol{\sigma}}+\tilde{\mathcal{G}}(\boldsymbol{\sigma},\boldsymbol{\varepsilon}(\boldsymbol{u})),\qquad(6.46)$$

where $\mathcal{E}^{-1}$ denotes the inverse of $\mathcal{E}$ and $\tilde{\mathcal{G}}(\boldsymbol{\sigma},\boldsymbol{\varepsilon}(\boldsymbol{u}))=-\mathcal{E}^{-1}\mathcal{G}(\boldsymbol{\sigma},\boldsymbol{\varepsilon}(\boldsymbol{u}))$. This formula expresses again the fact that the rate of deformation can be decomposed into two parts: the elastic component $\dot{\boldsymbol{\varepsilon}}^e=\mathcal{E}^{-1}\dot{\boldsymbol{\sigma}}$ and the plastic one given by $\dot{\boldsymbol{\varepsilon}}^p=\tilde{\mathcal{G}}(\boldsymbol{\sigma},\boldsymbol{\varepsilon}(\boldsymbol{u}))$.

The definitions presented above in the one-dimensional case can be extended straightforward to the general case: The set of points $(\boldsymbol{\sigma},\boldsymbol{\varepsilon})\in\mathbb{S}^d\times\mathbb{S}^d$ which satisfy $G(\boldsymbol{x},\boldsymbol{\sigma},\boldsymbol{\varepsilon})=\mathbf{0}$ is the *domain of elastic behavior* of the material at $\boldsymbol{x}\in\Omega$. If this set may be described by an explicit equation $\boldsymbol{\sigma}=F(\boldsymbol{x},\boldsymbol{\varepsilon})$ where F is a tensor-valued function, the material is viscoelastic at $\boldsymbol{x}\in\Omega$; if this set has a nonempty interior in $\mathbb{S}^d\times\mathbb{S}^d$, we say that (6.43) represents a *viscoplastic* constitutive law at $\boldsymbol{x}\in\Omega$.

Example 6.3 *A concrete example of an elastic-viscoplastic constitutive law of this form is* Perzyna's law *given by*

$$\dot{\boldsymbol{\varepsilon}}=\mathcal{E}^{-1}\dot{\boldsymbol{\sigma}}+\frac{1}{\mu}(\boldsymbol{\sigma}-\mathcal{P}_K\boldsymbol{\sigma}).\qquad(6.47)$$

Here $\mathcal{E}$ *is a fourth order tensor satisfying* (6.44), $\mu>0$ *is a viscosity constant,* $K\subset\mathbb{S}^d$ *is a nonempty, closed, convex set in the space* $\mathbb{S}^d$ *of symmetric tensors* *and $\mathcal{P}_K$ represents the projection mapping. Notice that the function $\mathcal{G}$ does not depend on ε and is given by*

$$\mathcal{G}(\boldsymbol{\sigma}, \boldsymbol{\varepsilon}) = -\frac{1}{\mu}\,\mathcal{E}(\boldsymbol{\sigma} - \mathcal{P}_K\boldsymbol{\sigma}).$$

The conditions (6.45) *are satisfied. Since $\boldsymbol{\sigma} = \mathcal{P}_K\boldsymbol{\sigma}$ iff $\boldsymbol{\sigma} \in K$, from* (6.47) *we see that viscoplastic deformations occur only for the stress tensors $\boldsymbol{\sigma}$ outside the set K. Thus, the set K represents the domain of elastic behavior of the material* (6.47)*. It is usually defined by*

$$K = \{\, \boldsymbol{\sigma} \in \mathbb{S}^d : \mathcal{F}(\boldsymbol{\sigma}) \le 0 \,\} \tag{6.48}$$

where $\mathcal{F} : \mathbb{S}^d \to \mathbb{R}^d$ is a convex function such that $\mathcal{F}(0) < 0$. The function $\mathcal{F}$ is called the yield function *and the equation $\mathcal{F}(\boldsymbol{\sigma}) = 0$ is called the* yield condition*. A well-known example of the yield function is the* von Mises function

$$\mathcal{F}(\boldsymbol{\sigma}) = \frac{1}{2}\,\|\boldsymbol{\sigma}^D\|^2 - k^2, \tag{6.49}$$

where $\boldsymbol{\sigma}^D$ is the deviator of $\boldsymbol{\sigma}$ and k is a strictly positive constant, called the yield limit.

The projection mapping on the von Mises convex set (6.48), (6.49) *can be easily computed. It follows that*

$$\mathcal{P}_K(\boldsymbol{\sigma}) = \begin{cases} \boldsymbol{\sigma} & \text{if} \quad \boldsymbol{\sigma} \in K, \\ \dfrac{1}{3}(\operatorname{tr}\boldsymbol{\sigma})\boldsymbol{I} + \dfrac{k\sqrt{2}}{\|\boldsymbol{\sigma}^D\|}\,\boldsymbol{\sigma}^D & \text{if} \quad \boldsymbol{\sigma} \notin K. \end{cases} \tag{6.50}$$

A derivation of this formula can be found, for instance, in [121, p. 46]. Thus, the Perzyna law (6.47) *associated with the von Mises convex set can be expressed as follows:*

$$\dot{\boldsymbol{\varepsilon}} = \begin{cases} \mathcal{E}\dot{\boldsymbol{\sigma}} & \text{if} \quad \|\boldsymbol{\sigma}^D\| \le k\sqrt{2}, \\ \mathcal{E}\dot{\boldsymbol{\sigma}} + \dfrac{1}{\mu}\Big(1 - \dfrac{k\sqrt{2}}{\|\boldsymbol{\sigma}^D\|}\Big)\boldsymbol{\sigma}^D & \text{if} \quad \|\boldsymbol{\sigma}^D\| \ge k\sqrt{2}. \end{cases} \tag{6.51}$$

Note that formally, when $\mu \to 0$ the Perzyna law (6.47) becomes the constitutive law

$$\mathcal{A}\dot{\boldsymbol{\sigma}} + \partial\psi_K(\boldsymbol{\sigma}) \ni \boldsymbol{\varepsilon}(\dot{\boldsymbol{u}}). \tag{6.52}$$

Here $\mathcal{A} = \mathcal{E}^{-1}$ is the tensor of elastic compliance, K is the set of elastic stresses and $\partial\psi_K$ denotes the subdifferential of the indicator function $\psi_K : \mathbb{S}^d \to (-\infty, +\infty]$. Recall that

$$\psi_K(\boldsymbol{\tau}) = \begin{cases} 0 & \text{if } \boldsymbol{\tau} \in K, \\ +\infty & \text{otherwise.} \end{cases}$$

Equation (6.52) shows that the total strain rate is the sum of the elastic part $\dot{\varepsilon}^e = \mathcal{A}\dot{\sigma}$ and the plastic part $\dot{\varepsilon}^p \in \partial\psi_K(\sigma)$. By the properties of the subdifferential it follows that

$$\sigma \in K, \quad \dot{\varepsilon}^p \cdot (\tau - \sigma) \leq 0 \quad \forall \tau \in K. \tag{6.53}$$

This is usually called the *Prandtl–Reuss flow rule.* From (6.53) it follows that if $\sigma \in \operatorname{int}(K)$ then $\partial\psi_K(\sigma) = \{\mathbf{0}\}$ and therefore $\dot{\varepsilon}^p = \mathbf{0}$. We conclude that for processes in which $\sigma \in \operatorname{int}(K)$, the elastoplastic material (6.52) behaves elastically. Plastic deformations occur only for $\sigma \in \partial K$. Since K is a fixed convex set, we deduce that plastic deformations occur only at *fixed values* of the stress tensor. For this reason, the model (6.52) is called an *elastic-perfectly plastic model* or, for simplicity, a *perfectly plastic model.* Note also that for the perfectly plastic model (6.52) the stress σ is not allowed to lie outside the set K.

We see that formally (6.52) may be recovered from (6.47) as $\mu \to 0$; this idea is used in the regularization technique for some existence results. For example, in Part IV we will show existence results for contact problems involving the Perzyna law (6.47). Then in Section 18.3 we will extend the existence results by the regularization method for frictional contact problems for perfectly plastic materials of the form (6.52).

In this book we also consider rate-type viscoplastic materials with constitutive relation

$$\dot{\sigma} = \mathcal{E}\varepsilon(\dot{u}) + \mathcal{G}(\sigma, \varepsilon(u), \beta) \tag{6.54}$$

in which β represents an *internal state variable.* Assume (6.44) for $\mathcal{E}$. An equivalent form of (6.54) is given by

$$\varepsilon(\dot{u}) = \mathcal{E}^{-1}\dot{\sigma} + \tilde{\mathcal{G}}(\sigma, \varepsilon(u), \beta), \tag{6.55}$$

where

$$\tilde{\mathcal{G}}(\sigma, \varepsilon(u), \beta) = -\mathcal{E}^{-1}\mathcal{G}(\sigma, \varepsilon(u), \beta). \tag{6.56}$$

Equation (6.55) shows that the internal state variable β is involved only in the plastic rate of the deformation tensor defined by

$$\dot{\varepsilon}^p = \tilde{\mathcal{G}}(\sigma, \varepsilon(u), \beta). \tag{6.57}$$

There is a variety of choices for the internal state variables; for reference in the field see [59, Ch. VI]. Some commonly used internal state variables are the plastic strain and a number of tensor variables that take into account the spatial display of dislocations and the work-hardening of the material. A major problem concerns the evolution laws for the internal state variables. Following [59, 121], we suppose that β is a vector-valued function whose evolution is governed by a differential equation of the form

$$\dot{\beta} = \varphi(\sigma, \varepsilon(u), \beta) \tag{6.58}$$

in which φ is a nonlinear constitutive function to be described below. Rate-type constitutive models of the form (6.54), (6.58) were used in the literature to describe the behavior of real materials like rubbers, pastes, metals, rocks and so on. Concrete examples, experimental background and mechanical interpretation concerning models of this form may be found in [59] and references therein.

Example 6.4 *A simple example of the internal state variable, for which an evolution equation of the form* (6.58) *holds, can be obtained by considering*

$$\varphi(\boldsymbol{\sigma}, \boldsymbol{\varepsilon}(\boldsymbol{u}), \boldsymbol{\beta}) = \Phi(\|\dot{\boldsymbol{\varepsilon}}^p\|)$$

where $\Phi : \mathbb{R}_+ \to \mathbb{R}$ *is given. From* (6.57) *it follows that*

$$\varphi(\boldsymbol{\sigma}, \boldsymbol{\varepsilon}(\boldsymbol{u}), \boldsymbol{\beta}) = \Phi(\|\dot{\boldsymbol{\varepsilon}}^p\|) = \Phi(\|\tilde{\mathcal{G}}(\boldsymbol{\sigma}, \boldsymbol{\varepsilon}(\boldsymbol{u}), \boldsymbol{\beta})\|). \tag{6.59}$$

The real-valued internal state variable defined in this way is called the strain-hardening parameter. *If we take* $\Phi(r) = \frac{2}{3}r$, *then* $\boldsymbol{\beta}$ *given by* (6.58) *and* (6.59) *represents the* irreversible equivalent strain.

Example 6.5 *Another example of the internal state variable can be obtained by considering in* (6.58) *the function* φ *given by*

$$\varphi(\boldsymbol{\sigma}, \boldsymbol{\varepsilon}(\boldsymbol{u}), \boldsymbol{\beta}) = \boldsymbol{\sigma} \cdot \dot{\boldsymbol{\varepsilon}}^p = \boldsymbol{\sigma} \cdot \tilde{\mathcal{G}}(\boldsymbol{\sigma}, \boldsymbol{\varepsilon}(\boldsymbol{u}), \boldsymbol{\beta}).$$

Since $\boldsymbol{\sigma} \cdot \dot{\boldsymbol{\varepsilon}}^p$ *represents the irreversible stress power, the parameter* $\boldsymbol{\beta}$ *defined in this way is called* the work hardening parameter.

Example 6.6 *A concrete example of an elastic-viscoplastic constitutive law of the form* (6.55) *is given by the following variant of Perzyna's law*

$$\dot{\boldsymbol{\varepsilon}} = \mathcal{E}^{-1}\dot{\boldsymbol{\sigma}} + \frac{1}{\mu}(\boldsymbol{\sigma} - \mathcal{P}_{K(\boldsymbol{\beta})}\boldsymbol{\sigma}). \tag{6.60}$$

Here $\mathcal{E}$ *is a fourth order tensor satisfying* (6.44), $\mu > 0$ *is a viscosity constant,* $\mathcal{P}_{K(\boldsymbol{\beta})}$ *represents the projection mapping on the von Mises convex set* $K(\boldsymbol{\beta})$ *defined by the yield function* $\boldsymbol{\beta} \mapsto k(\boldsymbol{\beta})$:

$$K(\boldsymbol{\beta}) = \{\, \boldsymbol{\sigma} \in \mathbb{S}^d : \|\boldsymbol{\sigma}^D\| \le k(\boldsymbol{\beta})\sqrt{2}\,\}. \tag{6.61}$$

Using (6.61), (6.50) *we can give the explicit form of* (6.60):

$$\dot{\boldsymbol{\varepsilon}} = \begin{cases} \mathcal{E}\dot{\boldsymbol{\sigma}} & \text{if } \|\boldsymbol{\sigma}^D\| \le k(\boldsymbol{\beta})\sqrt{2}, \\ \mathcal{E}\dot{\boldsymbol{\sigma}} + \dfrac{1}{\mu}\Big(1 - \dfrac{k(\boldsymbol{\beta})\sqrt{2}}{\|\boldsymbol{\sigma}^D\|}\Big)\boldsymbol{\sigma}^D & \text{if } \|\boldsymbol{\sigma}^D\| > k(\boldsymbol{\beta})\sqrt{2}. \end{cases} \tag{6.62}$$

The parameter $\boldsymbol{\beta}$ *may be chosen as the irreversible equivalent strain, that is*

$$\varphi(\boldsymbol{\sigma}, \boldsymbol{\varepsilon}(\boldsymbol{u}), \boldsymbol{\beta}) = \frac{2}{3\mu}\,\|\boldsymbol{\sigma} - P_{K(\boldsymbol{\beta})}\boldsymbol{\sigma}\|. \tag{6.63}$$

We assume in this book that the constitutive function $\mathcal{G}$ and φ have the properties:

$$\left.\begin{array}{l}\text{(a) } \mathcal{G} : \Omega \times \mathbb{S}^d \times \mathbb{S}^d \times \mathbb{R}^d \to \mathbb{S}^d.\\ \text{(b) There exists an } L_{\mathcal{G}} > 0 \text{ such that}\\ \qquad \|\mathcal{G}(\boldsymbol{x}, \boldsymbol{\sigma}_1, \boldsymbol{\varepsilon}_1, \boldsymbol{\beta}_1) - \mathcal{G}(\boldsymbol{x}, \boldsymbol{\sigma}_2, \boldsymbol{\varepsilon}_2, \boldsymbol{\beta}_2)\|\\ \qquad\qquad \le L_{\mathcal{G}}\,(\|\boldsymbol{\sigma}_1 - \boldsymbol{\sigma}_2\| + \|\boldsymbol{\varepsilon}_1 - \boldsymbol{\varepsilon}_2\| + \|\boldsymbol{\beta}_1 - \boldsymbol{\beta}_2\|)\\ \qquad \forall\, \boldsymbol{\sigma}_1, \boldsymbol{\sigma}_2, \boldsymbol{\varepsilon}_1, \boldsymbol{\varepsilon}_2 \in \mathbb{S}^d,\ \boldsymbol{\beta}_1, \boldsymbol{\beta}_2 \in \mathbb{R}^m, \text{ a.e. } \boldsymbol{x} \in \Omega.\\ \text{(c) The mapping } \boldsymbol{x} \mapsto \mathcal{G}(\boldsymbol{x}, \boldsymbol{\sigma}, \boldsymbol{\varepsilon}, \boldsymbol{\beta}) \text{ is measurable on } \Omega\\ \qquad \text{for any } \boldsymbol{\sigma}, \boldsymbol{\varepsilon} \in \mathbb{S}^d \text{ and } \boldsymbol{\beta} \in \mathbb{R}^m.\\ \text{(d) The mapping } \boldsymbol{x} \mapsto \mathcal{G}(\boldsymbol{x}, \boldsymbol{0}, \boldsymbol{0}, \boldsymbol{0}) \in L^2(\Omega)^{d\times d}.\end{array}\right\} \tag{6.64}$$

$$\left.\begin{array}{l}\text{(a) } \varphi : \Omega \times \mathbb{S}^d \times \mathbb{S}^d \times \mathbb{R}^m \to \mathbb{R}^m.\\ \text{(b) There exists an } L_{\varphi} > 0 \text{ such that}\\ \qquad \|\varphi(\boldsymbol{x}, \boldsymbol{\sigma}_1, \boldsymbol{\varepsilon}_1, \boldsymbol{\beta}_1) - \varphi(\boldsymbol{x}, \boldsymbol{\sigma}_2, \boldsymbol{\varepsilon}_2, \boldsymbol{\beta}_2)\|\\ \qquad\qquad \le L_{\varphi}(\|\boldsymbol{\sigma}_1 - \boldsymbol{\sigma}_2\| + \|\boldsymbol{\varepsilon}_1 - \boldsymbol{\varepsilon}_2\| + \|\boldsymbol{\beta}_1 - \boldsymbol{\beta}_2\|)\\ \qquad \forall\, \boldsymbol{\sigma}_1, \boldsymbol{\sigma}_2, \boldsymbol{\varepsilon}_1, \boldsymbol{\varepsilon}_2 \in \mathbb{S}^d,\ \boldsymbol{\beta}_1, \boldsymbol{\beta}_2 \in \mathbb{R}^m, \text{ a.e. } \boldsymbol{x} \in \Omega.\\ \text{(c) The mapping } \boldsymbol{x} \mapsto \varphi(\boldsymbol{x}, \boldsymbol{\sigma}, \boldsymbol{\varepsilon}, \boldsymbol{\beta}) \text{ is measurable on } \Omega\\ \qquad \text{for any } \boldsymbol{\sigma}, \boldsymbol{\varepsilon} \in \mathbb{S}^d \text{ and } \boldsymbol{\beta} \in \mathbb{R}^m.\\ \text{(d) The mapping } \boldsymbol{x} \mapsto \varphi(\boldsymbol{x}, \boldsymbol{0}, \boldsymbol{0}, \boldsymbol{0}) \in L^2(\Omega)^{m\times m}.\end{array}\right\} \tag{6.65}$$

It can be shown that conditions (6.64) and (6.65) are satisfied in the case of the constitutive equations (6.60) and (6.63) provided that $\boldsymbol{\beta} \mapsto k(\boldsymbol{\beta}) : \mathbb{R} \to \mathbb{R}_+$ is a Lipschitz continuous function (see [121, p. 92]). Moreover, it is straightforward to see that if the function $\mathcal{G}$ satisfies the conditions (6.64) and φ is given by (6.59), (6.56), then φ satisfies the conditions (6.65).

Chapter 7

Background on Variational and Numerical Analysis in Contact Mechanics

In this chapter we make preparations on variational and numerical analysis that will be used in the study of contact problems in all the following chapters of the book. We start with an introduction of function spaces and sets for the displacement variable and the stress field. We list some properties of the function spaces and sets needed in the analysis and approximations of contact problems. For contact problems in viscoelasticity and viscoplasticity, the mathematical models are evolutionary variational inequalities. For their numerical approximations, we will use finite differences to replace the time derivative and use finite elements to discretize the spatial domain. Estimates concerning the differences and finite elements that will be repeatedly used later are listed in Section 7.2. For numerical solutions of the contact problems, we will derive error estimates under appropriate solution regularity assumptions. However, since results on solution regularities of the variational inequalities arising from the contact problems are largely unavailable in the literature, it is desirable to know if the numerical solutions indeed converge to the exact solutions when no additional solution regularity assumptions are made. In Section 7.3 we provide several abstract frameworks for proving the convergence of numerical solutions without additional solution regularity assumptions. These abstract results will be applied later in the context of solving variational inequalities arising from the contact problems. In the last section of the chapter, we collect some inequalities useful for later use.

7.1 Function Spaces in Solid Mechanics

In the study of mechanical problems, we frequently use function spaces involving the deformation and divergence operators. From now on, we always assume $\Omega \subset \mathbb{R}^d$ is open, connected and bounded with a Lipschitz boundary Γ. Since the boundary is Lipschitz continuous, the outward unit normal exists a.e. on Γ and is denoted by $\boldsymbol{\nu}$.

Finite dimensional spaces and L^2 spaces. Recall that $\mathbb{S}^d$ is the space of second order symmetric tensors on $\mathbb{R}^d$ or, equivalently, the space of symmetric matrices of order d. Also recall that the canonical inner products and corresponding norms on $\mathbb{R}^d$ and $\mathbb{S}^d$ are

$$\boldsymbol{u}\cdot\boldsymbol{v} = u_i v_i, \quad \|\boldsymbol{v}\| = (\boldsymbol{v}\cdot\boldsymbol{v})^{1/2} \quad \forall\, \boldsymbol{u},\boldsymbol{v}\in\mathbb{R}^d,$$
$$\boldsymbol{\sigma}\cdot\boldsymbol{\tau} = \sigma_{ij}\tau_{ij}, \quad \|\boldsymbol{\tau}\| = (\boldsymbol{\tau}\cdot\boldsymbol{\tau})^{1/2} \quad \forall\, \boldsymbol{\sigma},\boldsymbol{\tau}\in\mathbb{S}^d.$$

Here and throughout this work, the indices i and j run between 1 and d; the summation convention over repeated indices is used, and the index following a comma indicates a partial derivative, in the sense of Definition 2.6.

We will frequently use the following spaces:

$$H = \big\{\, \boldsymbol{v} = (v_1,\dots,v_d)^T : v_i \in L^2(\Omega),\ 1\le i\le d \,\big\} = L^2(\Omega)^d, \tag{7.1}$$
$$Q = \big\{\, \boldsymbol{\tau} = (\tau_{ij}) : \tau_{ij} = \tau_{ji} \in L^2(\Omega),\ 1\le i,\, j\le d \,\big\} = L^2(\Omega)^{d\times d}_s. \tag{7.2}$$

These are Hilbert spaces with the canonical inner products

$$(\boldsymbol{u},\boldsymbol{v})_H = \int_\Omega u_i(\boldsymbol{x})\, v_i(\boldsymbol{x})\, dx,$$
$$(\boldsymbol{\sigma},\boldsymbol{\tau})_Q = \int_\Omega \sigma_{ij}(\boldsymbol{x})\, \tau_{ij}(\boldsymbol{x})\, dx.$$

The associated norms on these spaces will be denoted by $\|\cdot\|_H$ and $\|\cdot\|_Q$, respectively.

Function spaces for the displacement fields. Displacements will be sought in the space

$$H_1 = \big\{\, \boldsymbol{v} = (v_1,\dots,v_d)^T : v_i \in H^1(\Omega),\ 1\le i\le d \,\big\} = H^1(\Omega)^d$$

or its subspaces or subsets, depending on prescribed essential boundary conditions. The space H_1 is a Hilbert space with the canonical inner product

$$(\boldsymbol{u},\boldsymbol{v}) = \int_\Omega \left[u_i(\boldsymbol{x})\, v_i(\boldsymbol{x}) + u_{i,j}(\boldsymbol{x})\, v_{i,j}(\boldsymbol{x})\right] dx.$$

The corresponding norm is

$$\|\boldsymbol{v}\| = \left(\int_\Omega \left[v_i(\boldsymbol{x})\, v_i(\boldsymbol{x}) + v_{i,j}(\boldsymbol{x})\, v_{i,j}(\boldsymbol{x})\right] dx\right)^{1/2}. \tag{7.3}$$

Let $\boldsymbol{\varepsilon} : H_1 \to Q$ be the *deformation* operator defined by

$$\boldsymbol{\varepsilon}(\boldsymbol{u}) = (\varepsilon_{ij}(\boldsymbol{u})), \quad \varepsilon_{ij}(\boldsymbol{u}) = \frac{1}{2}\,(u_{i,j} + u_{j,i}).$$

The quantity $\boldsymbol{\varepsilon}(\boldsymbol{u})$ is the *linearized* (or *small*) *strain tensor* associated with the displacement $\boldsymbol{u}$. Define

$$(\boldsymbol{u},\boldsymbol{v})_{H_1} = (\boldsymbol{u},\boldsymbol{v})_H + (\boldsymbol{\varepsilon}(\boldsymbol{u}),\boldsymbol{\varepsilon}(\boldsymbol{v}))_Q \quad \forall\, \boldsymbol{u},\boldsymbol{v}\in H_1, \tag{7.4}$$

and the corresponding norm

$$\|\boldsymbol{v}\|_{H_1} = \sqrt{(\boldsymbol{v},\boldsymbol{v})_{H_1}} \quad \forall\, \boldsymbol{v} \in H_1. \tag{7.5}$$

Korn's first inequality states (cf. [134])

$$\int_\Omega (v_i v_i + v_{i,j} v_{i,j})\, dx \le c \int_\Omega (v_i v_i + \varepsilon_{ij}(\boldsymbol{v})\, \varepsilon_{ij}(\boldsymbol{v}))\, dx \quad \forall\, \boldsymbol{v} \in H_1.$$

It follows that $(\cdot,\cdot)_{H_1}$ defined in (7.4) is an inner product in H_1, and the corresponding norm $\|\cdot\|_{H_1}$ is equivalent to the canonical norm (7.3). Therefore, $(H_1, \|\cdot\|_{H_1})$ is a real Hilbert space. In this work we use (7.4) and (7.5) as the inner product and norm in the space H_1.

Theorem 2.15 on page 33 allows us to define the trace of a function in H_1 on the boundary Γ.

Theorem 7.1 *There exists a linear continuous operator $\gamma : H_1 \to L^2(\Gamma)^d$ such that $\gamma\boldsymbol{v} = \boldsymbol{v}|_\Gamma$ if $\boldsymbol{v} \in H^1 \cap C(\overline{\Omega})^d$. Moreover, the mapping $\gamma : H_1 \to L^2(\Gamma)^d$ is compact, i.e., for any bounded sequence $\{\boldsymbol{v}_n\}$ in H_1, there is a subsequence $\{\boldsymbol{v}_{n'}\} \subset \{\boldsymbol{v}_n\}$ such that $\{\gamma\boldsymbol{v}_{n'}\}$ is convergent in $L^2(\Gamma)^d$.*

The operator γ is called the *trace operator* and $\gamma\boldsymbol{v}$ is called the *trace* of $\boldsymbol{v} \in H_1$ on Γ. When no ambiguity occurs, we usually write $\boldsymbol{v}$ instead of $\gamma\boldsymbol{v}$. Note that the continuity of γ implies that there exists a constant $c > 0$, depending only on Ω, such that

$$\|\boldsymbol{v}\|_{L^2(\Gamma)^d} \le c\, \|\boldsymbol{v}\|_{H_1} \quad \forall\, \boldsymbol{v} \in H_1. \tag{7.6}$$

The trace operator is neither an injection nor a surjection from H_1 to $L^2(\Gamma)^d$. The range $\gamma(H_1)$ is a space smaller than $L^2(\Gamma)^d$, namely $H^{\frac{1}{2}}(\Gamma)^d$. We use the symbol H_Γ for this space, i.e.

$$H_\Gamma = \big\{\, \boldsymbol{\xi} = (\xi_1, \ldots, \xi_d)^T : \xi_i \in H^{\frac{1}{2}}(\Gamma),\ 1 \le i \le d \,\big\} = H^{\frac{1}{2}}(\Gamma)^d.$$

This is a Hilbert space with the canonical inner product

$$(\boldsymbol{\chi}, \boldsymbol{\xi})_{H_\Gamma} = (\chi_i, \xi_i)_{1/2}.$$

Recall that here $(\cdot,\cdot)_{1/2}$ denotes the inner product on the space $H^{\frac{1}{2}}(\Gamma)$, see Section 2.2.

The following properties of the trace operator are easily derived from Theorem 2.19 (page 36).

Theorem 7.2 *The operator $\gamma : H_1 \to H_\Gamma$ is linear, continuous and surjective. Moreover, there exists a linear continuous operator $z : H_\Gamma \to H_1$ such that*

$$\gamma(z(\boldsymbol{\xi})) = \boldsymbol{\xi} \quad \forall\, \boldsymbol{\xi} \in H_\Gamma.$$

Note that the continuity of z implies the existence of a constant $c > 0$, depending on Ω, such that

$$\|z(\boldsymbol{\xi})\|_{H_1} \le c\, \|\boldsymbol{\xi}\|_{H_\Gamma} \quad \forall\, \boldsymbol{\xi} \in H_\Gamma.$$

We denote by H'_Γ the dual space of H_Γ, i.e. $H'_\Gamma = H^{-\frac{1}{2}}(\Gamma)^d$. The duality pairing between these spaces will be denoted by $\langle \cdot, \cdot \rangle_\Gamma$. Recall that if $\boldsymbol{\xi}' = (\xi'_i) \in L^2(\Gamma)^d$ then $\boldsymbol{\xi}' \in H'_\Gamma$ and

$$\langle \boldsymbol{\xi}', \boldsymbol{\xi} \rangle_\Gamma = \int_\Gamma \xi'_i\, \xi_i\, da \quad \forall\, \boldsymbol{\xi} = (\xi_i) \in H_\Gamma.$$

Now, let Γ_1 be a measurable subset of the boundary Γ with meas$\,(\Gamma_1) > 0$. In the study of contact problems, we frequently use the following subspace of H_1:

$$V = \{\, \boldsymbol{v} \in H_1 \,:\, \boldsymbol{v} = \boldsymbol{0} \text{ on } \Gamma_1 \,\}. \tag{7.7}$$

Here, the equality $\boldsymbol{v} = \boldsymbol{0}$ on Γ_1 is understood in the sence of traces, i.e. almost everywhere on Γ_1. Since the trace operator is continuous, we see that V is a closed subspace of the space H_1. Moreover, since meas$\,(\Gamma_1) > 0$, Korn's inequality (see, e.g., [161, p. 79]) holds:

$$\|\boldsymbol{\varepsilon}(\boldsymbol{v})\|_Q \ge c\, \|\boldsymbol{v}\|_{H_1} \quad \forall\, \boldsymbol{v} \in V. \tag{7.8}$$

Here c denotes a positive constant depending only on Ω and Γ_1. We define the inner product $(\cdot\,,\cdot)_V$ on V by

$$(\boldsymbol{u}, \boldsymbol{v})_V = (\boldsymbol{\varepsilon}(\boldsymbol{u}), \boldsymbol{\varepsilon}(\boldsymbol{v}))_Q \tag{7.9}$$

which induces the norm

$$\|\boldsymbol{v}\|_V = \|\boldsymbol{\varepsilon}(\boldsymbol{v})\|_Q. \tag{7.10}$$

It follows from (7.8) and (7.10) that $\|\cdot\|_{H_1}$ and $\|\cdot\|_V$ are equivalent norms on V. Thus, $(V, \|\cdot\|_V)$ is a real Hilbert space.

In analysis of contact problems, the boundary Γ is decomposed into three parts $\overline{\Gamma}_1$, $\overline{\Gamma}_2$ and $\overline{\Gamma}_3$, with Γ_1, Γ_2 and Γ_3 being relatively open and mutually disjoint, meas$\,(\Gamma_1) > 0$. In addition to the space V defined in (7.7), we will also need its subspace

$$V_1 = \{\, \boldsymbol{v} \in V \,:\, v_\nu = 0 \text{ a.e. on } \Gamma_3 \,\} \tag{7.11}$$

and its subset

$$V_2 = \{\, \boldsymbol{v} \in V \,:\, v_\nu \le 0 \text{ a.e. on } \Gamma_3 \,\}. \tag{7.12}$$

Over V_1 and V_2, we use the inner product $(\cdot, \cdot)_V$ and norm $\|\cdot\|_V$ of the space V. We see that V_1 is itself a Hilbert space, while V_2 is a non-empty, closed convex set in V. By (7.6) and (7.8) we have a constant $C_0 > 0$ depending only on Ω, Γ_1, and Γ_3 such that

$$\|\boldsymbol{v}\|_{L^2(\Gamma_3)^d} \le C_0 \, \|\boldsymbol{v}\|_V \quad \forall\, v \in V. \tag{7.13}$$

Function spaces for the stress fields. To introduce function spaces for stress fields, we need to extend the definition of the divergence for a regular tensorial field. A general approach for this extension is to use the derivatives in the distributional sense, see e.g. [121]. Here we choose to introduce the concept of the weak divergence directly, which is sufficient for this work.

Definition 7.3 *Let $\boldsymbol{\sigma} = (\sigma_{ij})$ and $\boldsymbol{w} = (w_i)$ be such that $\sigma_{ij} = \sigma_{ji} \in L^1_{\text{loc}}(\Omega)$, $w_i \in L^1_{\text{loc}}(\Omega)$ for all $1 \le i, j \le d$. Then $\boldsymbol{w}$ is called a* weak divergence *of $\boldsymbol{\sigma}$ if*

$$\int_\Omega \sigma_{ij} \varphi_{i,j} \, dx = - \int_\Omega w_i \, \varphi_i \, dx \quad \forall\, \boldsymbol{\varphi} = (\varphi_i) \in C_0^\infty(\Omega)^d. \tag{7.14}$$

Applying Lemma 2.4 (page 29), we can show that when a weak divergence exists, it is uniquely determined up to a set of measure zero. Let $C^1(\Omega)^{d\times d}_s$ be the space of symmetric tensors with $C^1(\Omega)$ components. Similarly, let $C^1(\overline{\Omega})^{d\times d}_s$ be the space of symmetric tensors with $C^1(\overline{\Omega})$ components. From the definition of the weak divergence, we see that if $\boldsymbol{\sigma} \in C^1(\Omega)^{d\times d}_s$, then the classical divergence of $\boldsymbol{\sigma}$ is also the weak divergence of $\boldsymbol{\sigma}$. For this reason, we use the notation $\operatorname{Div} \boldsymbol{\sigma}$ or $(\sigma_{ij,j})$ for the weak divergence of $\boldsymbol{\sigma}$ and, for simplicity, we omit the adjective "weak".

We now define the space

$$Q_1 = \{\, \boldsymbol{\tau} \in Q : \operatorname{Div} \boldsymbol{\tau} \in H \,\}, \tag{7.15}$$

which is a Hilbert space endowed with the inner product

$$(\boldsymbol{\sigma}, \boldsymbol{\tau})_{Q_1} = (\boldsymbol{\sigma}, \boldsymbol{\tau})_Q + (\operatorname{Div} \boldsymbol{\sigma}, \operatorname{Div} \boldsymbol{\tau})_H$$

and the associated norm $\|\cdot\|_{Q_1}$.

If $\boldsymbol{\sigma}$ is a regular function, e.g. $\boldsymbol{\sigma} \in C^1(\overline{\Omega})^{d\times d}_s$, then the following Green's formula holds:

$$(\boldsymbol{\sigma}, \boldsymbol{\varepsilon}(\boldsymbol{v}))_Q + (\operatorname{Div} \boldsymbol{\sigma}, \boldsymbol{v})_H = \int_\Gamma \boldsymbol{\sigma\nu} \cdot \boldsymbol{v} \, da \quad \forall\, \boldsymbol{v} \in H_1. \tag{7.16}$$

A proof of the formula is based on a standard density argument: It follows from the classical Green-Gauss formula that (7.16) is valid for all $\boldsymbol{v} \in C^\infty(\overline{\Omega})^d$; then we use the density of the space $C^\infty(\overline{\Omega})^d$ in H_1 (see Theorem 2.11).

The Green formula (7.16) allows us to define the *trace* of any element $\boldsymbol{\sigma} \in Q_1$ on the boundary. More precisely, we have the following result.

Theorem 7.4 *There exists a linear continuous surjective operator $\overline{\gamma} : Q_1 \to H'_\Gamma$ such that*

$$\langle \overline{\gamma} \boldsymbol{\sigma}, \gamma \boldsymbol{v} \rangle_\Gamma = (\boldsymbol{\sigma}, \boldsymbol{\varepsilon}(\boldsymbol{v}))_Q + (\operatorname{Div} \boldsymbol{\sigma}, \boldsymbol{v})_H \quad \forall\, \boldsymbol{\sigma} \in Q_1, \ \boldsymbol{u} \in H_1, \tag{7.17}$$

$$\langle \overline{\gamma} \boldsymbol{\sigma}, \boldsymbol{\xi} \rangle_\Gamma = \int_\Gamma \boldsymbol{\sigma\nu} \cdot \boldsymbol{\xi} \, da \quad \forall\, \boldsymbol{\sigma} \in C^1(\overline{\Omega})^{d\times d}_s, \ \boldsymbol{\xi} \in H_\Gamma. \tag{7.18}$$

Moreover, there exists a linear continuous operator $\overline{z} : H'_\Gamma \to Q_1$ *such that*

$$\overline{\gamma}(\overline{z}(\boldsymbol{\xi}')) = \boldsymbol{\xi}' \quad \forall \boldsymbol{\xi}' \in H'_\Gamma.$$

The operator $\overline{\gamma}$ is called again the *trace operator*, and the element $\overline{\gamma}\boldsymbol{\sigma} \in H'_\Gamma$ will be called the *trace* of $\boldsymbol{\sigma} \in Q_1$ on Γ. Due to (7.18), for simplicity, when no ambiguity occurs, we will write $\boldsymbol{\sigma\nu}$ instead of $\overline{\gamma}\boldsymbol{\sigma}$.

Now assume the boundary Γ is decomposed into two parts $\overline{\Gamma}_1$ and $\overline{\Gamma}_2$, such that Γ_1 and Γ_2 are relatively open and mutually disjoint. Let $\boldsymbol{\sigma} \in Q_1$. We will use the definition

$$\boldsymbol{\sigma\nu} = \mathbf{0} \text{ on } \Gamma_2 \iff \langle \overline{\gamma}\boldsymbol{\sigma}, \gamma\boldsymbol{v}\rangle_\Gamma = 0 \quad \forall \boldsymbol{v} \in H_1 \text{ s.t. } \boldsymbol{v} = \mathbf{0} \text{ on } \Gamma_1. \tag{7.19}$$

We say in this case that $\boldsymbol{\sigma\nu}$ vanishes on Γ_2, in the sense of traces.

Given $\boldsymbol{f} \in H'_\Gamma$ we say that $\boldsymbol{\sigma\nu} = \boldsymbol{f}$ on Γ_2 if $\boldsymbol{\sigma\nu} - \boldsymbol{f} = \mathbf{0}$ on Γ_2. If in addition $\boldsymbol{f} \in L^2(\Gamma)^d$ and $\boldsymbol{\sigma\nu} \in L^2(\Gamma)^d$, then the equality $\boldsymbol{\sigma\nu} = \boldsymbol{f}$ is valid almost everywhere on Γ_2.

In the study of contact problems, we frequently use the following subspace of Q_1:

$$\mathcal{V} = \{\, \boldsymbol{\sigma} \in Q_1 : \operatorname{Div} \boldsymbol{\sigma} = \mathbf{0} \text{ in } \Omega,\ \boldsymbol{\sigma\nu} = \mathbf{0} \text{ on } \Gamma_2 \,\}.$$

Here, the equality $\operatorname{Div} \boldsymbol{\sigma} = \mathbf{0}$ is understood to be a relation almost everywhere in Ω, and the equality $\boldsymbol{\sigma\nu} = \mathbf{0}$ on Γ_2 is understood in the sense of traces, specified in (7.19). By the continuity of the divergence and trace operators, we deduce that $\mathcal{V}$ is a closed subspace of the space Q_1. Moreover, using (7.19) and (7.17), we obtain the orthogonality property

$$(\boldsymbol{\sigma}, \boldsymbol{\varepsilon}(\boldsymbol{v}))_Q = 0 \quad \forall \boldsymbol{\sigma} \in \mathcal{V},\ \boldsymbol{v} \in V, \tag{7.20}$$

where V denotes the space defined in (7.7).

We have the following result for an orthogonal decomposition of the space Q.

Theorem 7.5 *The set* $\boldsymbol{\varepsilon}(V)$ *is a closed subspace in* Q *and its orthogonal complement in* Q *is* $\mathcal{V}$, *i.e.* $Q = \boldsymbol{\varepsilon}(V) \oplus \mathcal{V}$.

Normal and tangential traces. We suppose here that Ω is a $C^{1,1}$ domain (cf. Definition 2.10 on page 31) or a polygonal domain. When Ω is a polygonal domain, the pointwise relations concerning normal and tangential traces are understood to be valid almost everywhere. For an element $\boldsymbol{\xi} \in H_\Gamma$, its *normal* and *tangential* components are given by

$$\xi_\nu = \boldsymbol{\xi} \cdot \boldsymbol{\nu}, \quad \boldsymbol{\xi}_\tau = \boldsymbol{\xi} - \xi_\nu \boldsymbol{\nu}, \tag{7.21}$$

respectively. It follows that $\xi_\nu \in H^{1/2}(\Gamma)$ and $\boldsymbol{\xi}_\tau \in H_\tau(\Gamma)$, where $H_\tau(\Gamma)$ is the closed subspace of H_Γ defined by

$$H_\tau(\Gamma) = \{\, \boldsymbol{\xi} \in H_\Gamma : \xi_\nu = 0 \text{ on } \Gamma \,\}.$$

Moreover, it can be shown that the mapping $\boldsymbol{\xi} \mapsto (\xi_\nu, \boldsymbol{\xi}_\tau)$ is an isomorphism from H_Γ onto $H^{1/2}(\Gamma) \times H_\tau(\Gamma)$.

We denote by $H'_\tau(\Gamma)$ the strong dual of $H_\tau(\Gamma)$ and let $\langle\cdot,\cdot\rangle_\tau$ be the duality pairing between $H'_\tau(\Gamma)$ and $H_\tau(\Gamma)$. Recall that $\langle\cdot,\cdot\rangle_{1/2}$ and $\langle\cdot,\cdot\rangle_\Gamma$ denote the duality pairing between $H^{-1/2}(\Gamma)$ and $H^{1/2}(\Gamma)$, H'_Γ and H_Γ, respectively.

For any $\boldsymbol{\xi}' \in H'_\Gamma$ its *normal* and *tangential* components are the elements $\xi'_\nu \in H^{-1/2}(\Gamma)$ and $\boldsymbol{\xi}'_\tau \in H'_\tau(\Gamma)$, defined as

$$\langle \xi'_\nu, \xi\rangle_{1/2} = \langle \boldsymbol{\xi}', \xi\boldsymbol{\nu}\rangle_\Gamma \quad \forall \xi \in H^{1/2}(\Gamma), \tag{7.22}$$

$$\langle \boldsymbol{\xi}'_\tau, \boldsymbol{\xi}\rangle_\tau = \langle \boldsymbol{\xi}', \boldsymbol{\xi}\rangle_\Gamma \quad \forall \boldsymbol{\xi} \in H_\tau(\Gamma), \tag{7.23}$$

respectively. The mapping $\boldsymbol{\xi}' \mapsto (\xi'_\nu, \boldsymbol{\xi}'_\tau)$ is an isomorphism from H'_Γ onto $H^{-1/2}(\Gamma) \times H'_\tau(\Gamma)$. From (7.21)–(7.23), it follows that

$$\langle \boldsymbol{\xi}', \boldsymbol{\xi}\rangle_\Gamma = \langle \xi'_\nu, \xi_\nu\rangle_{1/2} + \langle \boldsymbol{\xi}'_\tau, \boldsymbol{\xi}_\tau\rangle_\tau \quad \forall \boldsymbol{\xi}' \in H'_\Gamma,\ \boldsymbol{\xi} \in H_\Gamma. \tag{7.24}$$

We use these results to introduce the normal and tangential components of displacement and stress fields, in the sense of traces.

For an element $\boldsymbol{v} \in H_1$ we denote by $\boldsymbol{v}$ its trace on Γ. We have $\boldsymbol{v} \in H_\Gamma$ and therefore we may consider its *normal* and *tangential* components on the boundary, denoted by v_ν and $\boldsymbol{v}_\tau$, respectively. Using (7.21), we have

$$v_\nu = \boldsymbol{v}\cdot\boldsymbol{\nu}, \quad \boldsymbol{v}_\tau = \boldsymbol{v} - v_\nu\boldsymbol{\nu}. \tag{7.25}$$

For an element $\boldsymbol{\sigma} \in Q_1$ it follows from Theorem 7.4 that its trace $\boldsymbol{\sigma\nu}$ belongs to H'_Γ. Therefore, we may consider its *normal* and *tangential* components on the boundary, denoted by σ_ν and $\boldsymbol{\sigma}_\tau$, respectively. Then $\sigma_\nu \in H^{-1/2}(\Gamma)$ and $\boldsymbol{\sigma}_\tau \in H'_\tau(\Gamma)$. From (7.24) and (7.17) we obtain

$$\begin{aligned}\langle \boldsymbol{\sigma\nu}, \gamma\boldsymbol{v}\rangle_\Gamma &= \langle \sigma_\nu, v_\nu\rangle_{1/2} + \langle \boldsymbol{\sigma}_\tau, \boldsymbol{v}_\tau\rangle_\tau \\ &= (\boldsymbol{\sigma}, \boldsymbol{\varepsilon}(\boldsymbol{v}))_Q + (\operatorname{Div}\boldsymbol{\sigma}, \boldsymbol{v})_H \quad \forall \boldsymbol{\sigma} \in Q_1,\ \boldsymbol{v} \in H_1.\end{aligned} \tag{7.26}$$

Finally, we note that if $\boldsymbol{\sigma}$ is a regular function, e.g. $\boldsymbol{\sigma} \in C^1(\overline{\Omega})^{d\times d}_s$, then

$$\sigma_\nu = (\boldsymbol{\sigma\nu})\cdot\boldsymbol{\nu}, \quad \boldsymbol{\sigma}_\tau = \boldsymbol{\sigma\nu} - \sigma_\nu\boldsymbol{\nu}. \tag{7.27}$$

Now, we assume again that the boundary Γ is divided into two disjoint measurable parts Γ_1 and Γ_2. Let $\boldsymbol{\sigma}_1 \in Q_1$ be a stress field and let $f \in H^{1/2}(\Gamma)$, $\boldsymbol{f} \in H_\tau(\Gamma)$. The following definitions are justified by the formulas (7.25)–(7.27):

$$\begin{aligned}\sigma_\nu = f \text{ on } \Gamma_2 \iff &\langle \overline{\gamma}\boldsymbol{\sigma}, \gamma\boldsymbol{v}\rangle_\Gamma = \langle f, v_\nu\rangle_{1/2} \ \ \forall \boldsymbol{v} \in H_1 \\ &\text{s.t. } v_\nu = 0 \text{ on } \Gamma_1 \text{ and } \boldsymbol{v}_\tau = \boldsymbol{0} \text{ on } \Gamma.\end{aligned} \tag{7.28}$$

$$\begin{aligned}\boldsymbol{\sigma}_\tau = \boldsymbol{f} \text{ on } \Gamma_2 \iff &\langle \overline{\gamma}\boldsymbol{\sigma}, \gamma\boldsymbol{v}\rangle_\Gamma = \langle \boldsymbol{f}, \boldsymbol{v}_\tau\rangle_\tau \ \ \forall \boldsymbol{v} \in H_1 \\ &\text{s.t. } v_\nu = 0 \text{ on } \Gamma \text{ and } \boldsymbol{v}_\tau = \boldsymbol{0} \text{ on } \Gamma_1.\end{aligned} \tag{7.29}$$

We refer to the equalities $\sigma_\nu = f$ and $\boldsymbol{\sigma}_\tau = \boldsymbol{f}$ in (7.28) and (7.29) as equalities on Γ_2 in the sense of traces. If moreover $f \in L^2(\Gamma)$, $\boldsymbol{f} \in L^2(\Gamma)^d$, $\sigma_\nu \in L^2(\Gamma)$

and $\boldsymbol{\sigma}_\tau \in L^2(\Gamma)^d$, it follows that the previous equalities are valid almost everywhere on Γ_2.

Some results on density of smooth functions. In convergence analysis for numerical solutions of contact problems under minimal solution regularity, we need various density results. We mention some of them found in the literature. We already know from Section 2.2 that the space $C_0^\infty(\Omega)$ is dense in $H_0^1(\Omega)$; and if Ω is a Lipschitz domain, then $C^\infty(\overline{\Omega})$ is dense in $H^1(\Omega)$.

The following density result is proved in [64].

Theorem 7.6 *Let $\Omega \subset \mathbb{R}^d$, $d \geq 1$, be an open, bounded, Lipschitz domain, and let $\Gamma_1 \subset \Gamma$ be a relatively open set with a Lipschitz relative boundary. Then the space*

$$\{\, v \in C^\infty(\overline{\Omega}) : v = 0 \text{ in a neighborhood of } \Gamma_1 \,\}$$

is dense in the following subspace of $H^1(\Omega)$:

$$\{\, v \in H^1(\Omega) : v = 0 \text{ a.e. on } \Gamma_1 \,\}.$$

From this, we immediately obtain the next result.

Theorem 7.7 *Let $\Omega \subset \mathbb{R}^d$, $d \geq 1$, be an open, bounded, Lipschitz domain, and let $\Gamma_1 \subset \Gamma$ be a relatively open set with a Lipschitz relative boundary. Let V be the space defined in (7.7). Then*

$$V_0 = \{\, \boldsymbol{v} \in C^\infty(\overline{\Omega})^d : \boldsymbol{v} = \boldsymbol{0} \text{ in a neighborhood of } \Gamma_1 \,\} \tag{7.30}$$

is dense in V.

We turn to smooth density results concerning the subspace V_1 defined in (7.11) and the set V_2 defined in (7.12). We will assume Ω is a polygonal domain ($d = 2$) or a polyhedral domain ($d = 3$) and its boundary is decomposed as $\Gamma = \overline{\Gamma}_1 \cup \overline{\Gamma}_2 \cup \overline{\Gamma}_3$ such that Γ_1, Γ_2 and Γ_3 are relatively open on Γ and are mutually disjoint. The following result is proved in [113].

Lemma 7.8 *Let $d = 2$ and Ω be polygonal. Assume $\overline{\Gamma}_1$, $\overline{\Gamma}_2$ and $\overline{\Gamma}_3$ are finite unions of line segments. Then the set $V_2 \cap C^\infty(\overline{\Omega})^2$ is dense in V_2 with respect to the norm of the space V.*

Since the journal where [113] was published was not widely circulated, for the reader's convenience, we briefly sketch the main ideas in the proof of Lemma 7.8. A boundary point is called a singular point if it is a vertex of Γ or if it belongs to $\overline{\Gamma}_1 \cap \overline{\Gamma}_2$, $\overline{\Gamma}_2 \cap \overline{\Gamma}_3$ or $\overline{\Gamma}_1 \cap \overline{\Gamma}_3$. By the assumption of the lemma, the number of singular points is finite.

Let $\{\mathcal{O}_i\}_{i=0}^r$ be an open covering of Ω such that $\overline{\mathcal{O}}_0 \subset \Omega$ and for $i = 1, \ldots, r$, the set $\mathcal{O}_i$ covers part of the boundary and contains at most one singular point.

Let $\{\varphi_i\}_{i=0}^r$ be a smooth partition of unity subordinate to the open covering. Any $\boldsymbol{u} \in V_2$ can be written as

$$\boldsymbol{u} = \sum_{i=0}^{r} \boldsymbol{u}_i, \quad \boldsymbol{u}_i = \boldsymbol{u}\, \varphi_i.$$

We have $\boldsymbol{u}_i \in H^1(\Omega)^2$ and $\operatorname{supp} \boldsymbol{u}_i \subset \mathcal{O}_i$, $0 \le i \le r$. Since $\boldsymbol{u}_0 \in H_0^1(\Omega)^2$, it can be approximated by a sequence of $C_0^\infty(\Omega)^2$ functions. For $i = 1, \dots, r$, we consider eight cases:

Case 1. $\mathcal{O}_i \cap \Gamma \subset \Gamma_1$;

Case 2. $\mathcal{O}_i \cap \Gamma \subset \Gamma_3$ and does not contain any vertex of Γ;

Case 3. $\mathcal{O}_i \cap \Gamma \subset \Gamma_3$ and contains a vertex of Γ;

Case 4. $\mathcal{O}_i \cap \Gamma$ contains a singular point of $\overline{\Gamma}_2 \cap \overline{\Gamma}_3$;

Case 5. $\mathcal{O}_i \cap \Gamma$ contains a singular point of $\overline{\Gamma}_1 \cap \overline{\Gamma}_2$;

Case 6. $\mathcal{O}_i \cap \Gamma$ contains a non-vertex singular point of $\overline{\Gamma}_1 \cap \overline{\Gamma}_3$;

Case 7. $\mathcal{O}_i \cap \Gamma$ contains a vertex singular point of $\overline{\Gamma}_1 \cap \overline{\Gamma}_3$;

Case 8. $\mathcal{O}_i \cap \Gamma \subset \Gamma_2$.

For Case 1, $\boldsymbol{u}_i \in H_0^1(\mathcal{O}_i \cap \Omega)^2$ can be approximated by a sequence of $C_0^\infty(\mathcal{O}_i \cap \Omega)^2$ functions.

In Case 2, let $\boldsymbol{e}_\xi$ be a unit vector along the tangential direction of $\mathcal{O}_i \cap \Gamma$ and let $\boldsymbol{e}_\eta$ be the unit inward normal vector. In this way, a local Cartesian coordinate system (ξ, η) is established for $\mathcal{O}_i \cap \Omega$ and we can write $\boldsymbol{u} = u_\xi \boldsymbol{e}_\xi + u_\eta \boldsymbol{e}_\eta$ with $u_\eta = -u_\nu \ge 0$ for $\eta = 0$. Now u_ξ can be approximated by a sequence of smooth functions. By the trace theorem, we have a $v \in H^1(\mathcal{O}_i \cap \Omega)$ such that $v = u_\eta$ on Γ, $\operatorname{supp} v \subset \mathcal{O}_i$. Without loss of generality we may assume $v \ge 0$ in $\mathcal{O}_i \cap \Omega$, for otherwise, we can use $|v|$ to replace v. Note that the absolute value of an H^1 function is again an H^1 function. Then $u_\eta = v + z$ with $z \in H_0^1(\mathcal{O}_i \cap \Omega)$. The function z can be approximated by $C_0^\infty(\mathcal{O}_i \cap \Omega)$ functions. For v, one first extends it to $\mathcal{O}_i \backslash \Omega$ by a reflection: The value at $(\xi, -\eta)$ is defined to be that at (ξ, η). Then by the standard mollification procedure, it is seen that v can be approximated by a sequence of $C^\infty(\overline{\Omega})$ functions which are non-negative on Γ.

In Case 3, denote $\boldsymbol{e}_1$ and $\boldsymbol{e}_2$ the unit tangential vectors along the two legs of the boundary emanating from the vertex, the corresponding legs being denoted by $\Gamma^{(2)}$ and $\Gamma^{(1)}$ with the unit outward normal vectors $\boldsymbol{n}_2$ and $\boldsymbol{n}_1$. Then the function $\boldsymbol{u}_i$ can be decomposed to

$$\boldsymbol{u}_i = \alpha_1\, u^{(1} \boldsymbol{e}_1 + \alpha_2\, u^{(2} \boldsymbol{e}_2,$$

where $\alpha_1 = 1/(\boldsymbol{e}_1 \cdot \boldsymbol{n}_1)$, $\alpha_2 = 1/(\boldsymbol{e}_2 \cdot \boldsymbol{n}_2)$. We see that $u^{(1} = \boldsymbol{u}_i \cdot \boldsymbol{n}_1$, $u^{(2} = \boldsymbol{u}_i \cdot \boldsymbol{n}_2$. Consider $u^{(2}$. We have $u^{(2} = u_\nu \le 0$ on $\Gamma^{(2)}$. By mapping $\mathcal{O}_i \cap \Omega$ into the upper halfplane and mapping back, a function $v \in H^1(\mathcal{O}_i \cap \Omega)$ can be constructed such that $v \le 0$ in $\mathcal{O}_i \cap \Omega$, $\operatorname{supp} v \subset \mathcal{O}_i$ and $v = u^{(2}$ on $\Gamma^{(2)}$. Writing $u^{(2} = v + z$, we have $z \in H^1(\mathcal{O}_i \cap \Omega)$, $\operatorname{supp} z \subset \mathcal{O}_i$ and $z = 0$ on $\Gamma^{(2)}$. For the smooth approximation of v, define $v_\lambda(\boldsymbol{x}) = v(\boldsymbol{x} + \lambda\, \boldsymbol{e})$ where $\lambda > 0$ and $\boldsymbol{e}$ is a fixed unit vector in the open cone formed by $\Gamma^{(2)}$ and $\Gamma^{(1)}$. We have $v_\lambda \to v$ in $H^1(\mathcal{O}_i \cap \Omega)$ as $\lambda \to 0+$. Mollifying the function v_λ to get an approximating sequence of $C^\infty(\overline{\Omega})$ functions that are non-negative on $\Gamma^{(2)}$. Applying Theorem 7.6, we have a $C^\infty(\overline{\Omega})$ sequence approximating z in $H^1(\mathcal{O}_i \cap \Omega)$ such that each member in the sequence has a support contained in $\mathcal{O}_i$, vanishes in a neighborhood of $\Gamma^{(2)}$.

For the remaining cases, appropriate smooth approximating sequences can be constructed similarly.

A simplified version of the proof of Lemma 7.8 for the case of the subspace V_1 leads to the next result.

Lemma 7.9 *Let $d = 2$ and Ω be polygonal. Assume $\overline{\Gamma}_1$, $\overline{\Gamma}_2$ and $\overline{\Gamma}_3$ are finite unions of line segments. Then the set $V_1 \cap C^\infty(\overline{\Omega})^2$ is dense in V_1 with respect to the norm of the space V.*

The ideas of the proof of Lemma 7.8 can be extended for similar smooth density results over three-dimensional sets (with more cases to consider, though), and we have the following results.

Theorem 7.10 *Let $d = 2$ or 3, and let Ω be a polygon or a polyhedra. Assume $\overline{\Gamma}_1$, $\overline{\Gamma}_2$ and $\overline{\Gamma}_3$ are finite unions of line segments for $d = 2$, or finite unions of polygons for $d = 3$. Then the set $V_1 \cap C^\infty(\overline{\Omega})^d$ is dense in V_1 with respect to the norm of the space V.*

Theorem 7.11 *Let $d = 2$ or 3, and let Ω be a polygon or a polyhedra. Assume $\overline{\Gamma}_1$, $\overline{\Gamma}_2$ and $\overline{\Gamma}_3$ are finite unions of line segments for $d = 2$, or finite unions of polygons for $d = 3$. Then the set $V_2 \cap C^\infty(\overline{\Omega})^d$ is dense in V_2 with respect to the norm of the space V.*

Now we present some density results concerning vector-valued function spaces. The following density result is found in [218, p. 423].

Lemma 7.12 *Assume X is a Banach space or a closed set of a Banach space with the norm $\|\cdot\|_X$. Then $C^\infty(\overline{I}; X)$ is dense in $H^1(I; X)$; that is, given $w \in H^1(I; X)$, for any $\varepsilon > 0$ there exists $\bar{w} \in C^\infty(\overline{I}; X)$ such that*

$$\|w - \bar{w}\|_{H^1(I;X)} \le \varepsilon.$$

Lemma 7.13 *Assume X is a Banach space or a closed set of a Banach space with the norm $\|\cdot\|_X$. If $X_0 \subset X$ is dense in X, then $C(\overline{I}; X_0)$ is dense in $C(\overline{I}; X)$.*

Proof. Let $v \in C(\overline{I}; X)$. Then $v(t)$ is a uniformly continuous function on $\overline{I}$. Thus, for any $\varepsilon > 0$, we can find an integer $n > 0$ such that

$$\|v(t) - v(s)\|_X < \frac{\varepsilon}{2} \quad \text{if } |t - s| \le \frac{1}{n}.$$

Denote $t_j = j\,T/n$ and $v_j = v(t_j)$, $0 \le j \le n$. For any j, we choose $z_j \in V_0$ satisfying

$$\|z_j - v_j\|_X < \frac{\varepsilon}{2}.$$

Then we define a function $z : \overline{I} \to X_0$ by the formula

$$z(t) = n\,(t_j - t)\,z_{j-1} + n\,(t - t_{j-1})\,z_j, \quad t_{j-1} \le t \le t_j,\ 1 \le j \le n.$$

Obviously, $z \in C(\overline{I}; X_0)$. For $t \in [t_{j-1}, t_j]$, we have

$$z(t) - v(t) = n\,(t_j - t)\,(z_{j-1} - v(t)) + n\,(t - t_{j-1})\,(z_j - v(t)).$$

Thus

$$\begin{aligned}\|z(t) - v(t)\|_X &\le n\,(t_j - t)\,(\|z_{j-1} - v_{j-1}\|_X + \|v_{j-1} - v(t)\|_X)\\ &\quad + n\,(t - t_{j-1})\,(\|z_j - v_j\|_X + \|v_j - v(t)\|_X)\\ &< \varepsilon.\end{aligned}$$

Therefore

$$\|z - v\|_{C(\overline{I};X)} < \varepsilon,$$

and $C(\overline{I}; X_0)$ is dense in $C(\overline{I}; X)$. ■

From Lemma 7.13 we obtain the following result.

Lemma 7.14 *Assume X is a Banach space or a closed set of a Banach space with the norm $\|\cdot\|_X$. If $X_0 \subset X$ is dense in X, then $C^1(\overline{I}; X_0)$ is dense in $C^1(\overline{I}; X)$.*

Proof. Let $v \in C^1(\overline{I}; X)$. Then

$$v(t) = v(0) + \int_0^t \dot{v}(s)\,ds,$$

where $v(0) \in X$. Let $\varepsilon > 0$. Since $\dot{v} \in C(\overline{I}; X)$, by Lemma 7.13, there exists $z_1 \in C(\overline{I}; X_0)$ such that

$$\|\dot{v} - z_1\|_{C(\overline{I};X)} < \varepsilon.$$

Since X_0 is dense in X, there exists $z_0 \in X_0$ such that

$$\|v(0) - z_0\|_X < \varepsilon.$$

Now define

$$z(t) = z_0 + \int_0^t z_1(s)\,ds.$$

Then $z \in C^1(\overline{I}; X_0)$. From the equalities

$$\begin{aligned} v(t) - z(t) &= v(0) - z_0 + \int_0^t (\dot{v}(s) - z_1(s))\, ds, \\ \dot{v}(t) - \dot{z}(t) &= \dot{v}(t) - z_1(t), \end{aligned}$$

we obtain

$$\begin{aligned} \|v - z\|_{C^1(\overline{I};X)} &= \|v - z\|_{C(\overline{I};X)} + \|\dot{v} - \dot{z}\|_{C(\overline{I};X)} \\ &\le (T+2)\, \|\dot{v} - z_1\|_{C(\overline{I};X)} \\ &< (T+2)\, \varepsilon. \end{aligned}$$

Therefore, $C^1(\overline{I}; X_0)$ is dense in $C^1(\overline{I}; X)$. ■

By Lemma 2.27 (page 40), $W^{1,p}(I; X) \hookrightarrow C(\overline{I}; X)$ for any $p \in [1, \infty]$. A proof similar to those of Lemmas 7.13 and 7.14 yields the following result.

Lemma 7.15 *Assume X is a Banach space or a closed set of a Banach space with the norm $\|\cdot\|_X$. Let $p \in [1, \infty]$. If $X_0 \subset X$ is dense in X, then $W^{1,p}(I; X_0)$ is dense in $W^{1,p}(I; X)$.*

A sharper result is possible, although we do not need it in this book. Indeed, again let X be a Banach space or a closed set of a Banach space with the norm $\|\cdot\|_X$, and let X_0 be a dense subset in X. Then it can be shown that $C^\infty(\overline{I}; X_0)$ is dense in $C^m(\overline{I}; X)$ and $W^{m,p}(I; X)$.

7.2 Semi-discrete and Fully Discrete Approximations

For approximations of the contact problems, we use the finite element method to discretize the spatial domain. We briefly describe how to construct the finite dimensional space V^h via the finite element method. Some details have been reviewed in Chapter 3. For simplicity, we assume Ω is a polygon or polyhedron. We have $\overline{\Gamma}_3 = \cup_{i=1}^{i_0} \overline{\Gamma}_{3,i}$ with each piece $\overline{\Gamma}_{3,i}$ represented by an affine function. Let $\{\mathcal{T}^h\}$ be a regular family of finite element partitions of $\overline{\Omega}$ in such a way that if a side of an element lies on the boundary, the side belongs entirely to one of the subsets $\overline{\Gamma}_1$, $\overline{\Gamma}_2$ and $\overline{\Gamma}_3$. Let h be the maximal diameter of the elements in $\mathcal{T}^h$. Let $X^h \subset H^1(\Omega)$ be the finite element space of linear elements, corresponding to the partition $\mathcal{T}^h$. For the space V defined in (7.7), we use the finite element space

$$V^h = (X^h)^d \cap V.$$

Alternatively, we have the characterization

$$V^h = \{\, \boldsymbol{v}^h \in (X^h)^d : \boldsymbol{v}^h = \boldsymbol{0} \text{ at the nodes on } \overline{\Gamma}_1 \,\}.$$

For the subspace V_1 defined in (7.11), we use the finite element subspace

$$V_1^h = \{\, \boldsymbol{v}^h \in V^h : v_\nu^h = 0 \text{ at the nodes on } \overline{\Gamma_3} \,\},$$

and for the set V_2 defined in (7.12), we use the finite element set

$$V_2^h = \{\, \boldsymbol{v}^h \in V^h : v_\nu^h \le 0 \text{ at the nodes on } \overline{\Gamma_3} \,\}.$$

Since for any element K with one face on $\overline{\Gamma}_3$, $K \cap \overline{\Gamma}_3 \subset \overline{\Gamma}_{3,i}$ (for some i between 1 and i_0) is represented by an affine function and $\boldsymbol{v}^h \in V^h$ is a linear function on $K \cap \overline{\Gamma}_3$, we see that $V_1^h \subset V_1$ and $V_2^h \subset V_2$.

Now assume

$$\boldsymbol{u} \in V \cap H^2(\Omega)^d. \tag{7.31}$$

Since $d \le 3$, by Sobolev's embedding theorem, $\boldsymbol{u}$ is continuous: $\boldsymbol{u} \in C(\overline{\Omega})^d$. Denote its piecewise linear interpolant by $\Pi^h \boldsymbol{u} \in V^h$. Then we have the interpolation error estimates

$$\|\boldsymbol{u} - \Pi^h \boldsymbol{u}\|_{L^2(\Omega)^d} + h\, \|\nabla(\boldsymbol{u} - \Pi^h \boldsymbol{u})\|_{L^2(\Omega)^d} \le c\, h^2 |\boldsymbol{u}|_{H^2(\Omega)^d}. \tag{7.32}$$

If $\boldsymbol{u} \in V_1 \cap H^2(\Omega)^d$ (or $\boldsymbol{u} \in V_2 \cap H^2(\Omega)^d$), then we have $\Pi^h \boldsymbol{u} \in V_1^h$ ($\Pi^h \boldsymbol{u} \in V_2^h$, respectively) and the error estimates (7.32). Concerning the interpolation error on $\overline{\Gamma}_3$, assume

$$\boldsymbol{u}|_{\Gamma_{3,i}} \in H^2(\Gamma_{3,i}), \quad 1 \le i \le i_0. \tag{7.33}$$

The regular family of finite element partitions $\{\mathcal{T}_h\}$ induces a regular family of finite element partitions on $\overline{\Gamma}_3$. Then for the piecewise linear interpolant $\Pi^h \boldsymbol{u}$, restricted to $\overline{\Gamma}_3$, we have the error estimate

$$\|\boldsymbol{u} - \Pi^h \boldsymbol{u}\|_{L^2(\Gamma_3)^d} \le c\, h^2 \sum_{i=1}^{i_0} |\boldsymbol{u}|_{H^2(\Gamma_{3,i})^d}. \tag{7.34}$$

Similarly, if

$$\boldsymbol{u}_\tau|_{\Gamma_{3,i}} \in H^2(\Gamma_{3,i})^d, \quad 1 \le i \le i_0, \tag{7.35}$$

then

$$\|\boldsymbol{u}_\tau - \Pi^h \boldsymbol{u}_\tau\|_{L^2(\Gamma_3)^d} \le c\, h^2 \sum_{i=1}^{i_0} |\boldsymbol{u}_\tau|_{H^2(\Gamma_{3,i})^d}. \tag{7.36}$$

Also, if

$$u_\nu|_{\Gamma_{3,i}} \in H^2(\Gamma_{3,i}), \quad 1 \le i \le i_0, \tag{7.37}$$

then

$$\|u_\nu - \Pi^h u_\nu\|_{L^2(\Gamma_3)} \le c\, h^2 \sum_{i=1}^{i_0} |u_\nu|_{H^2(\Gamma_{3,i})}. \tag{7.38}$$

For the initial value $\boldsymbol{u}_0$, if we assume

$$\boldsymbol{u}_0 \in H^2(\Omega)^d, \tag{7.39}$$

then

$$\|\boldsymbol{u}_0 - \Pi^h \boldsymbol{u}_0\|_V \leq c\,h |\boldsymbol{u}_0|_{H^2(\Omega)^d}. \tag{7.40}$$

If the solution $\boldsymbol{u}$ has regularity higher than (7.31), (7.33), (7.35) and (7.39), we may use higher order elements, and the error estimates presented in this work can be easily extended to such a situation.

For the stress variable, we will usually assume $\boldsymbol{\sigma} \in H^1(\Omega)^{d\times d}$ (at any time, for time-dependent problems). This regularity assumption is not sufficient to ensure the continuity of $\boldsymbol{\sigma}$. Thus its finite element interpolants may not be well defined. We turn to finite element projections instead. Let $\{Q^h\}$ be the family of piecewise constants corresponding to the regular family of finite element partitions $\{\mathcal{T}^h\}$. For any $\boldsymbol{\tau} \in Q$, we define its finite element projection in Q^h to be

$$\mathcal{P}_{Q^h}\boldsymbol{\tau} \in Q^h, \quad (\mathcal{P}_{Q^h}\boldsymbol{\tau}, \boldsymbol{\tau}^h)_Q = (\boldsymbol{\tau}, \boldsymbol{\tau}^h)_Q \quad \forall\, \boldsymbol{\tau}^h \in Q^h.$$

This finite element projection is well defined and we have the error estimate (cf. [170])

$$\|\boldsymbol{\tau} - \mathcal{P}_{Q^h}\boldsymbol{\tau}\|_Q \leq c\,h\,|\boldsymbol{\tau}|_{1,\Omega} \quad \forall\, \boldsymbol{\tau} \in H^1(\Omega)^{d\times d} \cap Q. \tag{7.41}$$

For the time derivative occurring in the variational inequalities for quasistatic contact problems, we use finite difference approximations. The time interval of interest is $[0,T]$, and we use the notation $I = (0,T)$ and $\overline{I} = [0,T]$. For fully discrete approximations, we need a partition of the time interval $\overline{I}: 0 = t_0 < t_1 < \cdots < t_N = T$. We denote the time step-sizes $k_n = t_n - t_{n-1}$, $n = 1, \ldots, N$. We allow non-uniform partition of the time interval, and let $k = \max_n k_n$ be the maximal step-size.

At the abstract level, let X be a Hilbert space and $X^h \subset X$ be its finite-dimensional approximation. Consider a continuous function

$$u(t) = \int_0^t w(s)\,ds + u_0$$

for some $w \in C(\overline{I}; X)$ and $u_0 \in X$. Assume $u_0^h \in X^h$ is an approximation of u_0, and $w_j^{hk} \in X^h$ are approximations of $w_j \equiv w(t_j)$, $j = 0, 1, \ldots, N$. Let

$$u_n^{hk} = \sum_{j=1}^n w_j^{hk} k_j + u_0^h.$$

In several occasions in the rest of the book, we need to bound the term

$\|u_{n-1}^{hk} - u_n\|_X$. We have:

$$\begin{aligned}
u_{n-1}^{hk} - u_n &= \sum_{j=1}^{n-1} w_j^{hk} k_j + u_0^h - \int_0^{t_n} w(s)\,ds - u_0 \\
&= \sum_{j=1}^{n-1} (w_j^{hk} - w_j)\,k_j + u_0^h - u_0 \\
&\quad + \sum_{j=1}^{n-1} \left(w_j k_j - \int_{t_{j-1}}^{t_j} w(s)\,ds \right) - \int_{t_{n-1}}^{t_n} w(s)\,ds.
\end{aligned}$$

Now

$$\begin{aligned}
\left\| \sum_{j=1}^{n-1} \left(w_j k_j - \int_{t_{j-1}}^{t_j} w(s)\,ds \right) \right\|_X &= \left\| \sum_{j=1}^{n-1} \int_{t_{j-1}}^{t_j} (w_j - w(s))\,ds \right\|_X \\
&\le \sum_{j=1}^{n-1} \int_{t_{j-1}}^{t_j} \|w_j - w(s)\|_X\,ds,
\end{aligned}$$

and also

$$\left\| \int_{t_{n-1}}^{t_n} w(s)\,ds \right\|_X \le \int_{t_{n-1}}^{t_n} \|w(s)\|_X\,ds \le k\,\|w\|_{C(\overline{I};X)}.$$

Hence,

$$\begin{aligned}
\|u_{n-1}^{hk} - u_n\|_X \le \sum_{j=1}^{n-1} \|w_j^{hk} - w_j\|_X k_j + \|u_0^h - u_0\|_X \\
+ I_k(w) + k\,\|w\|_{C(\overline{I};X)},
\end{aligned} \tag{7.42}$$

where

$$I_k(w) = \sum_{j=1}^{N} \int_{t_{j-1}}^{t_j} \|w_j - w(s)\|_X ds. \tag{7.43}$$

Recall that $W^{1,1}(I;X) \hookrightarrow C(\overline{I};X)$ (Lemma 2.27 on page 40). Assume $w \in W^{1,1}(I;X)$. By writing

$$w_j - w(s) = \int_s^{t_j} \dot{w}(\tau)\,d\tau$$

we easily find the following bound for $I_k(w)$:

$$I_k(w) \le c\,k\,\|\dot{w}\|_{L^1(I;X)} \le c\,k\,\|f\|_{W^{1,1}(I;X)}. \tag{7.44}$$

Then we have

$$\|u_{n-1}^{hk} - u_n\|_X \le \sum_{j=1}^{n-1} \|w_j^{hk} - w_j\|_X k_j + \|u_0^h - u_0\|_X \\ + c\,k\,\|w\|_{W^{1,1}(I;X)}. \tag{7.45}$$

Similarly,

$$\|u_n^{hk} - u_n\|_X \le \sum_{j=1}^{n} \|w_j^{hk} - w_j\|_X k_j + \|u_0^h - u_0\|_X + I_k(w), \tag{7.46}$$

and if $w \in W^{1,1}(I;X)$, then we have

$$\|u_n^{hk} - u_n\|_X \le \sum_{j=1}^{n} \|w_j^{hk} - w_j\|_X k_j + \|u_0^h - u_0\|_X \\ + c\,k\,\|w\|_{W^{1,1}(I;X)}. \tag{7.47}$$

For later use, we record the next result.

Lemma 7.16 *For $w \in C(\overline{I};X)$, the quantity $I_k(w)$ defined in (7.43) converges to zero as $k \to 0$.*

Proof. Since $w \in C(\overline{I};X)$, $t \mapsto w(t)$ is uniformly continuous on $\overline{I}$. Thus for any $\varepsilon > 0$, there exists a $k_0 > 0$ such that if $k < k_0$, we have

$$\|w(t) - w(s)\|_X < \frac{\varepsilon}{T} \quad \forall\, s,t \in \overline{I},\ |t-s| < k.$$

Then by the definition (7.43), we have

$$I_k(w) \le \sum_{j=1}^{N} \int_{t_{j-1}}^{t_j} \frac{\varepsilon}{T}\, ds = \varepsilon.$$

Hence, $I_k(w) \to 0$ as $k \to 0$. ∎

7.3 Convergence under Basic Solution Regularity

In deriving error estimates for semi-discrete and fully discrete approximate solutions, we need to assume additional solution regularities. Since solution regularities of the variational inequalities studied in this work are largely unavailable from the literature, it is important to know if the numerical solutions converge under the basic solution regularity available from a well-posedness result.

A framework for convergence of semi-discrete solutions. This convergence framework is employed in Part III for semi-discrete solutions of quasistatic contact problems in viscoelasticity. Let X be a Banach space or a closed subset of a Banach space, $I = (0, T)$ be the time interval of interest, and $u \in C^1(\overline{I}; X)$ be the solution of a quasistatic variational inequality with initial value $u_0 \in X$. For a family of finite-dimensional subsets $X^h \subset X$, the corresponding semi-discrete solutions are denoted u^h with initial values $u_0^h \in X^h$, where $h \to 0+$ is a discretization parameter. We assume the following inequality

$$\|u - u^h\|_{C^1(\overline{I};X)} \le \inf_{v^h \in C(\overline{I};X^h)} g\left(\|\dot{u} - v^h\|_{C(\overline{I};X)}\right) + c_1 \|u_0 - u_0^h\|_X, \tag{7.48}$$

where $c_1 > 0$ is a constant, and

$$g : \mathbb{R}_+ \to \mathbb{R}_+ \text{ is nondecreasing, } g(e) \to 0 \text{ as } e \to 0+\,. \tag{7.49}$$

We see that by the assumption,

$$\forall \varepsilon > 0,\ \exists \delta > 0, \text{ such that } g(e) < \varepsilon \ \forall e \in (0, \delta). \tag{7.50}$$

We now introduce the following assumption.

Hypothesis 7.17 *There exists a subset $X_0 \subset X$ dense in X and a function $\alpha(h) > 0$ with the property $\lim_{h \to 0+} \alpha(h) = 0$ such that*

$$\|v - \Pi^h v\|_X \le \alpha(h)\, \|v\|_{X_0} \quad \forall v \in X_0,$$

where $\Pi^h v \in X^h$ is a finite-dimensional interpolant of $v \in X_0$.

In applications of the convergence framework for semi-discrete solutions of quasistatic variational inequalities, we usually have $X = V$, V_1 or V_2, and we can take $X_0 = H^2(\Omega)^d \cap X$ or even $X_0 = C^\infty(\overline{\Omega})^d \cap X$. Since X_0 is a set of smooth functions, finite-dimensional interpolants of the elements from X_0 are well-defined.

Theorem 7.18 *Assume* (7.48) *and* (7.49). *Suppose we choose the initial value $u_0^h \in X^h$ such that*

$$\|u_0 - u_0^h\|_X \to 0 \quad \text{as } h \to 0.$$

Then under Hypothesis 7.17, we have the convergence

$$\|u - u^h\|_{C^1(\overline{I};X)} \to 0 \quad \text{as } h \to 0.$$

Proof. Since $X_0 \subset X$ is dense in X, $C^1(\overline{I}; X_0)$ is dense in $C^1(\overline{I}; X)$ by Lemma 7.14. Recall that the solution $u \in C^1(\overline{I}; X)$. Let $\varepsilon > 0$. We can find $\tilde{u} \in C^1(\overline{I}; X_0)$ with

$$\|u - \tilde{u}\|_{C^1(\overline{I};X)} < \frac{\delta}{2},$$

where $\delta > 0$ satisfies the property (7.50). We then choose $h_0 > 0$ such that for $h \le h_0$,

$$c_1 \|u_0 - u_0^h\|_X < \varepsilon$$

and

$$\alpha(h)\, \|\tilde{u}\|_{C^1(\overline{I};X_0)} < \frac{\delta}{2}.$$

By Hypothesis 7.17,

$$\|\tilde{u} - \Pi^h \tilde{u}\|_{C^1(\overline{I};X)} \le \alpha(h)\, \|\tilde{u}\|_{C^1(\overline{I};X_0)} \le \frac{\delta}{2}.$$

Since

$$\|\dot{u} - \Pi^h \dot{\tilde{u}}\|_{C(\overline{I};X)} \le \|\dot{u} - \dot{\tilde{u}}\|_{C(\overline{I};X)} + \|\dot{\tilde{u}} - \Pi^h \dot{\tilde{u}}\|_{C(\overline{I};X)} < \delta,$$

we have

$$\inf_{v^h \in C(\overline{I};X^h)} g\left(\|\dot{u} - v^h\|_{C(\overline{I};X)}\right) \le g\left(\|\dot{u} - \Pi^h \dot{\tilde{u}}\|_{C(\overline{I};X)}\right) < \varepsilon.$$

Then using (7.48), we obtain

$$\|u - u^h\|_{C^1(\overline{I};X)} < 2\,\varepsilon \quad \text{for } h \le h_0.$$

Thus, the convergence theorem is proved. ■

Remark 7.19 *If in* (7.48), *the term*

$$\inf_{v^h \in C(\overline{I};X^h)} g\left(\|\dot{u} - v^h\|_{C(\overline{I};X)}\right)$$

is replaced by

$$\inf_{v^h \in C^1(\overline{I};X^h)} g\left(\|u - v^h\|_{C^1(\overline{I};X)}\right)$$

with g satisfying (7.49), *then the convergence result of Theorem 7.18 still holds.*

Another framework for convergence of semi-discrete solutions. The second convergence framework is employed in Part IV for semi-discrete solutions of quasistatic contact problems in viscoplasticity. Here, we need two spaces or sets. Let X be a Hilbert space or its closed subset with the inner product $(\cdot,\cdot)_X$, and let Y be another Hilbert space or its closed subset with the norm $(\cdot,\cdot)_Y$. The time interval of interest is still $I = (0,T)$. Let $(u,\sigma) \in W^{1,\infty}(I; X \times Y)$ be the solution of a quasistatic variational inequality with initial value $(u_0, \sigma_0) \in X \times Y$. Suppose $X^h \subset X$ and $Y^h \subset Y$ are finite-dimensional approximations of X and

Y. We denote $(u^h, \sigma^h) \in W^{1,\infty}(I; X^h \times Y^h)$ to be the corresponding semi-discrete solution with initial value $(u_0^h, \sigma_0^h) \in X^h \times Y^h$. We assume an estimate of the form

$$\begin{aligned}
&\|\sigma - \sigma^h\|_{L^\infty(I;Y)} + \|u - u^h\|_{L^\infty(I;X)} \\
&\quad \le c_1 \left(\|\sigma_0 - \sigma_0^h\|_Y + \|u_0 - u_0^h\|_X\right) \\
&\qquad + c_2 \|(I_Y - \mathcal{P}_{Y^h})(\sigma - \sigma_0)\|_{L^\infty(I;Y)} \\
&\qquad + \inf_{v^h \in L^\infty(I;X^h)} g\left(\|u - v^h\|_{L^\infty(I;X)}\right), \qquad (7.51)
\end{aligned}$$

where $c_1, c_2 > 0$ are constants, the function g has the property (7.50), $I_Y : Y \to Y$ is the identity mapping, and $\mathcal{P}_{Y^h} : Y \to Y^h$ is the projection operator defined by

$$(\mathcal{P}_{Y^h}\sigma, \sigma^h)_Y = (\sigma, \sigma^h)_Y \quad \forall\, \sigma \in Y,\ \sigma^h \in Y^h.$$

We introduce the following assumption.

Hypothesis 7.20 *There exist subsets $X_0 \subset X$, $Y_0 \subset Y$, and functions $\alpha_1(h) > 0$, $\alpha_2(h) > 0$ with the property*

$$\lim_{h \to 0+} [\alpha_1(h) + \alpha_2(h)] = 0$$

such that X_0 is dense in X, Y_0 is dense in Y, and

$$\begin{aligned}
\|v - \Pi^h v\|_X &\le \alpha_1(h)\, \|v\|_{X_0} \quad \forall\, v \in X_0, \\
\|\tau - \mathcal{P}_{Y^h}\tau\|_Y &\le \alpha_2(h)\, \|\tau\|_{Y_0} \quad \forall\, \tau \in Y_0,
\end{aligned}$$

where $\Pi^h v \in X^h$ is a finite-dimensional interpolant of $v \in X_0$ and $\mathcal{P}_{Y^h}\tau \in Y^h$ is the finite-dimensional projection of $\tau \in Y$.

In applications of the convergence framework for semi-discrete solutions of quasistatic variational inequalities in viscoplastic contact problems, we usually have $X = V$, V_1 or V_2, $Y = Q$, and we can take $X_0 = H^2(\Omega)^d \cap X$ or even $X_0 = C^\infty(\overline{\Omega})^d \cap X$, $Y_0 = H^1(\Omega)^{d \times d} \cap Y$. Since X_0 is a set of smooth functions, finite-dimensional interpolants of the elements from X_0 are well-defined.

Theorem 7.21 *Assume* (7.51) *and* (7.49). *Suppose the discrete initial values $u_0^h \in X^h$, $\sigma_0^h \in Y^h$ are chosen such that*

$$\|\sigma_0 - \sigma_0^h\|_Y + \|u_0 - u_0^h\|_X \to 0 \quad \text{as } h \to 0.$$

Then under the Hypothesis 7.20, we have the convergence

$$\|\sigma - \sigma^h\|_{L^\infty(I;Y)} + \|u - u^h\|_{L^\infty(I;X)} \to 0 \quad \text{as } h \to 0.$$

The proof of Theorem 7.21 is similar to that of Theorem 7.18, and is hence omitted. We only comment that the density result Lemma 7.15 is needed in the proof.

A framework for convergence of fully discrete solutions. This convergence framework will be used for quasistatic contact problems in viscoelasticity. Again let $u \in C^1(\overline{I}; X)$ be the solution of a quasistatic variational inequality with initial value $u_0 \in X$. In addition to the family of finite-dimensional subsets $X^h \subset X$, we also need the time interval partition described in Section 7.2. Let $\{u_n^{hk}\}_{n=0}^N$ be the corresponding fully discrete solution with the initial value $u_0^h \in X^h$. Here $h \to 0+$ is a spatial discretization parameter while $k \to 0+$ is a temporal discretization parameter. Usually the following inequality holds:

$$\max_{1\le n\le N} \left(\|u_n - u_n^{hk}\|_X + \|\dot{u}_n - \delta_n u_n^{hk}\|_X\right) \le \max_{1\le n\le N} \inf_{v^h\in X^h} g\left(\|\dot{u}_n - v^h\|_X\right) + c_1\|u_0 - u_0^h\|_X + b(k), \tag{7.52}$$

where $c_1 > 0$ is a constant, the function g satisfies the property (7.49), and

$$b(k) \to 0 \text{ as } k \to 0+\,. \tag{7.53}$$

Theorem 7.22 *Assume* (7.52) *with* (7.49) *and* (7.53). *Suppose we choose the initial value* $u_0^h \in X^h$ *such that*

$$\|u_0 - u_0^h\|_X \to 0 \quad \text{as } h \to 0.$$

Then under Hypothesis 7.17, we have the convergence

$$\max_{1\le n\le N} \left(\|u_n - u_n^{hk}\|_X + \|\dot{u}_n - \delta_n u_n^{hk}\|_X\right) \to 0 \quad \text{as } h, k \to 0.$$

Proof. Since $X_0 \subset X$ is dense in X, $C^1(\overline{I}; X_0)$ is dense in $C^1(\overline{I}; X)$ by Lemma 7.14. Let $\varepsilon > 0$. We can find $\tilde{u} \in C^1(\overline{I}; X_0)$ with

$$\|u - \tilde{u}\|_{C^1(\overline{I};X)} < \frac{\delta}{2}$$

which implies

$$\|\dot{u}_n - \dot{\tilde{u}}_n\|_X < \frac{\delta}{2}, \quad n = 1, \dots, N.$$

Here $\delta > 0$ satisfies the property (7.50). We then choose $h_0 > 0$ such that for $h \le h_0$,

$$c_1\|u_0 - u_0^h\|_X < \frac{\varepsilon}{2}$$

and

$$\alpha(h)\,\|\tilde{u}\|_{C^1(\overline{I};X_0)} \le \frac{\delta}{2}\,.$$

Since for any $n = 1, \ldots, N$,

$$\|\dot{u}_n - \Pi^h \dot{\tilde{u}}_n\|_X \le \|\dot{u}_n - \dot{\tilde{u}}_n\|_X + \|\dot{\tilde{u}}_n - \Pi^h \dot{\tilde{u}}_n\|_X < \delta,$$

we have

$$\inf_{v^h \in C^1(\bar{I}; X^h)} g\left(\|\dot{u}_n - v^h\|_X\right) \le g\left(\|\dot{u}_n - \Pi^h \dot{\tilde{u}}_n\|_X\right) < \varepsilon, \quad n = 1, \ldots, N.$$

We choose $k_0 > 0$ such that for $k \le k_0$,

$$b(k) < \frac{\varepsilon}{2}.$$

Then using (7.52), we obtain

$$\max_{1\le n\le N} \left(\|u_n - u_n^{hk}\|_X + \|\dot{u}_n - \delta_n u_n^{hk}\|_X\right) < 2\,\varepsilon \quad \text{for } h \le h_0,\ k \le k_0.$$

Thus, the convergence theorem is proved. ■

Another framework for convergence of fully discrete solutions. We finally present a convergence framework for fully discrete solutions of quasistatic contact problems in viscoplasticity. Again, let X be a Hilbert space or its closed subset with the inner product $(\cdot,\cdot)_X$, and let Y be another Hilbert space or its closed subset with the inner product $(\cdot,\cdot)_Y$. Denote $(u,\sigma) \in W^{1,\infty}(I; X\times Y)$ be the solution of a quasistatic variational inequality with initial value $(u_0, \sigma_0) \in X \times Y$. Let $X^h \subset X$ and $Y^h \subset Y$ be finite-dimensional approximations of X and Y and let $(u^h, \sigma^h) \in W^{1,\infty}(I; X^h \times Y^h)$ be the corresponding semi-discrete solution with initial value $(u_0^h, \sigma_0^h) \in X^h \times Y^h$. We assume an estimate of the form

$$\begin{aligned}
\max_{1\le n\le N} &\left(\|u_n - u_n^{hk}\|_X + \|\dot{u}_n - \delta_n u_n^{hk}\|_X\right) \\
&\le \max_{1\le n\le N} \inf_{v^h\in X^h} g\left(\|\dot{u}_n - v^h\|_X\right) \\
&\quad + c \max_{1\le n\le N} \|(I_Y - \mathcal{P}_{Y^h})(\sigma_n - \sigma_0)\|_Y \\
&\quad + c\,(\|\sigma_0 - \sigma_0^h\|_Y + \|u_0 - u_0^h\|_X) + b(k),
\end{aligned} \tag{7.54}$$

where the function g satisfies the property (7.49), and $b(k)$ satisfies (7.53).

Theorem 7.23 *Assume* (7.54) *with* (7.49) *and* (7.53). *Suppose we choose the initial value* $(u_0^h, \sigma_0^h) \in X^h \times Y^h$ *such that*

$$\|u_0 - u_0^h\|_X + \|\sigma_0 - \sigma_0^h\|_Y \to 0 \quad \text{as } h \to 0.$$

Then under Hypothesis 7.20, we have the convergence

$$\max_{1\le n\le N} \left(\|\sigma_n - \sigma_n^h\|_Y + \|u_n - u_n^{hk}\|_X\right) \to 0 \quad \text{as } h, k \to 0.$$

A proof of Theorem 7.23 can be made by combining the arguments in the proofs of Theorem 7.21 and Theorem 7.22. We omit the detailed argument here.

7.4 Some Inequalities

In this section we record some inequalities for later reference.

The following elementary inequality will be useful in several occasions, which has been used in the derivation of (4.88) before.

$$x, a, b \geq 0 \text{ and } x^2 \leq a\, x + b \Longrightarrow x^2 \leq a^2 + 2\, b. \tag{7.55}$$

Another elementary inequality is

$$\alpha \in (0,1],\ a, b \geq 0 \Longrightarrow |a^{\alpha+1} - b^{\alpha+1}| \leq (\alpha+1)\, |a-b|\, (|2a|^{\alpha} + |a-b|^{\alpha}). \tag{7.56}$$

The Gronwall inequality will be used frequently in variational analysis of the contact problems.

Lemma 7.24 (Gronwall's inequality) *Assume* $f, g \in C[a,b]$ *satisfying*

$$f(t) \leq g(t) + c \int_a^t f(s)\, ds, \quad t \in [a,b], \tag{7.57}$$

where $c > 0$ *is a constant. Then*

$$f(t) \leq g(t) + c \int_a^t g(s)\, e^{c\,(t-s)}\, ds, \quad t \in [a,b]. \tag{7.58}$$

Moreover if g *is nondecreasing, then*

$$f(t) \leq g(t)\, e^{c\,(t-a)}, \quad t \in [a,b]. \tag{7.59}$$

Proof. Define

$$F(s) = \int_a^s f(r)\, dr, \quad s \in [a,b].$$

Then $F'(s) = f(s)$. From the assumption (7.57), we get

$$F'(s) \leq g(s) + c\, F(s), \quad s \in [a,b].$$

Thus,

$$\left(e^{-c\,s} F(s)\right)' \leq g(s)\, e^{-c\,s}.$$

Integrate this inequality from a to $t \geq a$ and notice that $F(a) = 0$,

$$e^{-c\,t} F(t) \leq \int_a^t g(s)\, e^{-c\,s} ds, \quad t \in [a,b].$$

So,

$$F(t) \leq \int_a^t g(s)\, e^{c\,(t-s)} ds, \quad t \in [a,b]$$

and from (7.57), we obtain the inequality (7.58).

If g is nondecreasing, then

$$f(t) \le g(t) + g(t)\, c \int_a^t e^{c\,(t-s)}\, ds = g(t)\, e^{c\,(t-a)}.$$

The lemma is thus proved. ∎

Later in error analysis of numerical approximations of variational inequalities arising in contact problems, we will need to apply various discrete versions of Gronwall's inequality. Let T be fixed, and let $0 = t_0 < t_1 < \cdots < t_N = T$ be a partition of the interval $\overline{I}$. Define $k_j = t_j - t_{j-1}$, $j = 1, 2, \ldots, N$, and $k = \max_{1 \le j \le N} k_j$.

Lemma 7.25 *Assume $\{g_n\}_{n=1}^N$ and $\{e_n\}_{n=1}^N$ are two sequences of non-negative numbers satisfying*

$$e_n \le c\, g_n + c \sum_{j=1}^{n-1} k_j e_j, \quad n = 1, \ldots, N.$$

Then

$$e_n \le c \Big(g_n + \sum_{j=1}^{n-1} k_j g_j \Big), \quad n = 1, \ldots, N. \tag{7.60}$$

Therefore,

$$\max_{1 \le n \le N} e_n \le c \max_{1 \le n \le N} g_n. \tag{7.61}$$

Proof. Denote

$$E_n = \sum_{j=1}^{n} k_j e_j, \quad 1 \le n \le N,$$

and $E_0 = 0$. Then from the given condition,

$$e_n \le c\, g_n + c\, E_{n-1}, \quad n = 1, \ldots, N. \tag{7.62}$$

Now

$$E_n - E_{n-1} = k_n e_n \le c\, k_n g_n + c\, k_n E_{n-1},$$

which implies

$$E_n - (1 + c\, k_n)\, E_{n-1} \le c\, k_n g_n. \tag{7.63}$$

We introduce a sequence of numbers $\{z_n\}_{n=0}^N$ by $z_0 = 1$ and

$$z_n = \prod_{j=1}^{n} (1 + c\, k_j), \quad 1 \le n \le N.$$

Using the inequalities

$$1 \le 1 + c\,k_j \le e^{c\,k_j}, \quad j = 1, \dots, N,$$

we have the following bounds:

$$1 \le \prod_{i=j+1}^{N} (1 + ck_i) \le e^{c(T-t_j)}, \quad j = 1, \dots, N. \tag{7.64}$$

With the use of the sequence $\{z_n\}_{n=0}^N$, the inequality (7.63) can be rewritten as

$$\frac{E_n}{z_n} - \frac{E_{n-1}}{z_{n-1}} \le \frac{c\,k_n g_n}{z_n}.$$

A simple induction argument shows

$$E_n \le c\,z_n \sum_{j=1}^{n} k_j \frac{g_j}{z_j} = c \sum_{j=1}^{n} k_j \prod_{i=j+1}^{n} (1 + ck_i) g_j,$$

which can be combined with (7.62) and (7.64) to yield (7.60). The inequality (7.61) follows easily from (7.60). ■

Lemma 7.26 *Assume $\{g_n\}_{n=1}^N$ and $\{e_n\}_{n=1}^N$ are two sequences of non-negative numbers satisfying*

$$e_n \le c\,g_n + c \sum_{j=1}^{n} k_j e_j, \quad n = 1, \dots, N. \tag{7.65}$$

Then if $k = \max_{1 \le n \le N} k_n$ is sufficiently small,

$$e_n \le c\Big(g_n + \sum_{j=1}^{n} k_j g_j\Big), \quad n = 1, \dots, N. \tag{7.66}$$

Therefore,

$$\max_{1 \le n \le N} e_n \le c \max_{1 \le n \le N} g_n. \tag{7.67}$$

Proof. Denote

$$E_n = \sum_{j=1}^{n} k_j e_j.$$

Then (7.65) can be rewritten as

$$e_n \le c\,g_n + c\,E_n, \quad n = 1, \dots, N. \tag{7.68}$$

Now

$$E_n - E_{n-1} = k_n e_n \le c\,k_n g_n + c\,k_n E_n,$$

and hence,

$$(1 - c\,k_n)\,E_n - E_{n-1} \le c\,k_n g_n. \tag{7.69}$$

We introduce a sequence of numbers $\{z_n\}_{n=0}^N$ by $z_0 = 1$ and

$$z_n = (1 - c\,k_n)\,z_{n-1}, \quad 1 \le n \le N.$$

Obviously,

$$z_n = \prod_{j=1}^{n}(1 - c\,k_j).$$

Using the inequalities

$$e^{-2\,c\,k_j} \le 1 - c\,k_j \le e^{-c\,k_j} \quad \text{if } k_j \le k \text{ is sufficiently small,}$$

we have the following bounds on z_n:

$$e^{-2\,c\,t_n} \le z_n \le e^{-c\,t_n} \quad \text{if } k \text{ is sufficiently small.} \tag{7.70}$$

Now we rewrite (7.69),

$$z_n E_n - z_{n-1} E_{n-1} \le c\,k_n z_{n-1} g_n.$$

A simple induction argument yields

$$z_n E_n \le c \sum_{j=1}^{n} k_j z_{j-1} g_j.$$

Using the bounds (7.70), we obtain

$$E_n \le c \sum_{j=1}^{n} k_j g_j.$$

Then from (7.68) we get the bound (7.66). The inequality (7.67) follows easily from (7.66). ■

Chapter 8

Contact Problems in Elasticity

In this chapter we present the variational and numerical analysis of our first mathematical models describing the contact involving deformable bodies. We model the material's behavior with elastic constitutive laws. We start with a study of static frictionless contact problems, and then consider quasistatic frictional contact problems. For all these problems we prove existence and uniqueness results and derive error estimates for the corresponding discrete schemes. Similar studies will be presented for contact problems with viscoelastic materials in Part III and with viscoplastic materials in Part IV. For this reason, the contents of this chapter may be considered as a prelude to the variational and numerical analysis of the contact problems studied in the rest of the book.

8.1 Frictionless Contact Problems

In this section we consider two frictionless contact problems which model the contact between an elastic body and an obstacle, the so called foundation. We use a nonlinear elastic constitutive law of the form (6.16) (page 124). In the first problem, the contact is modelled with the well-known Signorini condition. Its variational formulation is an elliptic variational inequality of the first kind (4.1). In the second problem, the contact is modelled by a normal compliance condition, and we are led to an elliptic variational inequality of the second kind (4.2). In the variational formulations here, we use the displacement field as the unknown. The unique solvability of the two variational problems can be concluded by applying Corollary 4.3 (page 63) and Theorem 4.1 (page 60).

The Signorini frictionless contact problem. The physical setting is as follows. An elastic body occupies an open bounded connected domain Ω of $\mathbb{R}^d$ ($d = 1, 2, 3$ in applications) with a Lipschitz boundary Γ that is partitioned into three parts $\overline{\Gamma}_1$, $\overline{\Gamma}_2$ and $\overline{\Gamma}_3$ with Γ_1, Γ_2 and Γ_3 relatively open and mutually disjoint, such that $\operatorname{meas}(\Gamma_1) > 0$. The body is clamped on Γ_1 and so the displacement field vanishes there. Surface tractions of density $\boldsymbol{f}_2$ act on Γ_2 and volume forces of density $\boldsymbol{f}_0$ act in Ω. The body is in contact with a rigid foundation on Γ_3. The contact is frictionless and is modeled with the Signorini contact condition in a form with a zero gap function.

Under the previous assumptions, the equilibrium problem of the elastic body in contact with the obstacle can be formulated as follows:

Find a displacement field $\boldsymbol{u} : \Omega \to \mathbb{R}^d$ and a stress field $\boldsymbol{\sigma} : \Omega \to \mathbb{S}^d$ such that

$$\boldsymbol{\sigma} = \mathcal{F}(\boldsymbol{\varepsilon}(\boldsymbol{u})) \quad \text{in } \Omega, \tag{8.1}$$

$$\operatorname{Div} \boldsymbol{\sigma} + \boldsymbol{f}_0 = \boldsymbol{0} \quad \text{in } \Omega, \tag{8.2}$$

$$\boldsymbol{u} = \boldsymbol{0} \quad \text{on } \Gamma_1, \tag{8.3}$$

$$\boldsymbol{\sigma\nu} = \boldsymbol{f}_2 \quad \text{on } \Gamma_2, \tag{8.4}$$

$$u_\nu \le 0, \ \sigma_\nu \le 0, \ \sigma_\nu u_\nu = 0, \ \boldsymbol{\sigma}_\tau = \boldsymbol{0} \quad \text{on } \Gamma_3. \tag{8.5}$$

To obtain a variational formulation for the mechanical problem, we use the spaces V, Q and Q_1 defined in (7.7), (7.2) and (7.15):

$$V = \{\, \boldsymbol{v} \in H^1(\Omega)^d : \boldsymbol{v} = \boldsymbol{0} \text{ on } \Gamma_1 \,\},$$
$$Q = \{\, \boldsymbol{\tau} = (\tau_{ij}) \in L^2(\Omega)^{d \times d} : \tau_{ij} = \tau_{ji},\ 1 \le i, j \le d \,\},$$
$$Q_1 = \{\, \boldsymbol{\tau} \in Q : \operatorname{Div} \boldsymbol{\tau} \in L^2(\Omega)^d \,\}.$$

These are real Hilbert space with their canonical inner products. Recall that over the space V, we use the inner product

$$(\boldsymbol{u}, \boldsymbol{v})_V = (\boldsymbol{\varepsilon}(\boldsymbol{u}), \boldsymbol{\varepsilon}(\boldsymbol{v}))_Q \quad \forall\, \boldsymbol{u}, \boldsymbol{v} \in V.$$

We will use the set V_2 defined in (7.12) as *the set of admissible displacement fields.* Recall that

$$V_2 = \{\ \boldsymbol{v} \in V : v_\nu \le 0 \text{ a.e. on } \Gamma_3\ \}$$

is a nonempty closed convex set of V.

We assume that the elasticity operator $\mathcal{F}$ satisfies the condition

$$\left.\begin{array}{l}
\text{(a) } \mathcal{F} : \Omega \times \mathbb{S}^d \to \mathbb{S}^d. \\
\text{(b) There exists } M > 0 \text{ such that} \\
\qquad \|\mathcal{F}(\boldsymbol{x}, \boldsymbol{\varepsilon}_1) - \mathcal{F}(\boldsymbol{x}, \boldsymbol{\varepsilon}_2)\| \le M \|\boldsymbol{\varepsilon}_1 - \boldsymbol{\varepsilon}_2\| \\
\qquad \forall\, \boldsymbol{\varepsilon}_1, \boldsymbol{\varepsilon}_1 \in \mathbb{S}^d, \text{ a.e. } \boldsymbol{x} \in \Omega. \\
\text{(c) There exists } m > 0 \text{ such that} \\
\qquad (\mathcal{F}(\boldsymbol{x}, \boldsymbol{\varepsilon}_1) - \mathcal{F}(\boldsymbol{x}, \boldsymbol{\varepsilon}_2)) \cdot (\boldsymbol{\varepsilon}_1 - \boldsymbol{\varepsilon}_2) \ge m\, \|\boldsymbol{\varepsilon}_1 - \boldsymbol{\varepsilon}_2\|^2 \\
\qquad \forall\, \boldsymbol{\varepsilon}_1, \boldsymbol{\varepsilon}_1 \in \mathbb{S}^d, \text{ a.e. } \boldsymbol{x} \in \Omega. \\
\text{(d) For any } \boldsymbol{\varepsilon} \in \mathbb{S}^d,\ \boldsymbol{x} \mapsto \mathcal{F}(\boldsymbol{x}, \boldsymbol{\varepsilon}) \text{ is measurable in } \Omega. \\
\text{(e) The mapping } \boldsymbol{x} \mapsto \mathcal{F}(\boldsymbol{x}, \boldsymbol{0}) \in Q.
\end{array}\right\} \tag{8.6}$$

We also assume that the force and traction densities satisfy

$$\boldsymbol{f}_0 \in L^2(\Omega)^d, \quad \boldsymbol{f}_2 \in L^2(\Gamma_2)^d \tag{8.7}$$

and we denote by $\boldsymbol{f}$ the element of V given by

$$(\boldsymbol{f}, \boldsymbol{v})_V = \int_\Omega \boldsymbol{f}_0 \cdot \boldsymbol{v}\, dx + \int_{\Gamma_2} \boldsymbol{f}_2 \cdot \boldsymbol{v}\, da \quad \forall\, \boldsymbol{v} \in V. \tag{8.8}$$

We follow the standard procedure to derive the weak formulation of the mechanical problem (8.1)–(8.5). Assume the mechanical problem has a solution $(\boldsymbol{u}, \boldsymbol{\sigma})$, sufficiently smooth so that all the calculations below are meaningful. Let $\boldsymbol{v} \in V_2$ be arbitrary. We multiply the equation (8.2) by $(\boldsymbol{v} - \boldsymbol{u})$, integrate over Ω, and apply Green's formula (7.16) to obtain

$$(\boldsymbol{\sigma}, \boldsymbol{\varepsilon}(\boldsymbol{v}) - \boldsymbol{\varepsilon}(\boldsymbol{u}))_Q = \int_\Omega \boldsymbol{f}_0 \cdot (\boldsymbol{v} - \boldsymbol{u})\, dx + \int_\Gamma \boldsymbol{\sigma\nu} \cdot (\boldsymbol{v} - \boldsymbol{u})\, da.$$

Using the boundary conditions (8.3) and (8.4), we deduce the equality

$$(\boldsymbol{\sigma}, \boldsymbol{\varepsilon}(\boldsymbol{v}) - \boldsymbol{\varepsilon}(\boldsymbol{u}))_Q = (\boldsymbol{f}, \boldsymbol{v} - \boldsymbol{u})_V + \int_{\Gamma_3} \boldsymbol{\sigma\nu} \cdot (\boldsymbol{v} - \boldsymbol{u})\, da. \tag{8.9}$$

Over Γ_3,

$$\boldsymbol{\sigma\nu} \cdot (\boldsymbol{v} - \boldsymbol{u}) = \sigma_\nu (v_\nu - u_\nu) + \boldsymbol{\sigma}_\tau \cdot (\boldsymbol{v}_\tau - \boldsymbol{u}_\tau) = \sigma_\nu v_\nu,$$

where the boundary conditions (8.5) are used. Now over Γ_3, we also have $\sigma_\nu \le 0$ and $v_\nu \le 0$. Then

$$\boldsymbol{\sigma\nu} \cdot (\boldsymbol{v} - \boldsymbol{u}) \ge 0 \quad \text{on } \Gamma_3.$$

Therefore, we derive from (8.9) that

$$(\boldsymbol{\sigma}, \boldsymbol{\varepsilon}(\boldsymbol{v}) - \boldsymbol{\varepsilon}(\boldsymbol{u}))_Q \ge (\boldsymbol{f}, \boldsymbol{v} - \boldsymbol{u})_V. \tag{8.10}$$

Keeping the boundary conditions (8.3) and (8.5) in mind, we see that

$$\boldsymbol{u} \in V_2. \tag{8.11}$$

Summarizing, from (8.1), (8.10) and (8.11) we arrive at the following variational formulation of the contact problem (8.1)–(8.5).

Problem 8.1 *Find a displacement field $\boldsymbol{u}$ such that*

$$\boldsymbol{u} \in V_2, \quad (\mathcal{F}(\boldsymbol{\varepsilon}(\boldsymbol{u})), \boldsymbol{\varepsilon}(\boldsymbol{v}) - \boldsymbol{\varepsilon}(\boldsymbol{u}))_Q \ge (\boldsymbol{f}, \boldsymbol{v} - \boldsymbol{u})_V \quad \forall\, \boldsymbol{v} \in V_2. \tag{8.12}$$

We remark that Problem 8.1 is formally equivalent to the mechanical problem (8.1)–(8.5). Indeed, if $\boldsymbol{u}$ represents a regular solution of the variational problem 8.1 and $\boldsymbol{\sigma}$ is given by $\boldsymbol{\sigma} = \mathcal{F}(\boldsymbol{\varepsilon}(\boldsymbol{u}))$, by arguments similar to that used in [134], it follows that $(\boldsymbol{u}, \boldsymbol{\sigma})$ satisfy (8.1)–(8.5) in a generalized sense. See also the derivation of pointwise relations below, after Theorem 8.2.

Well-posedness of Problem 8.1 follows from the following result.

Theorem 8.2 *Assume* (8.6) *and* (8.7). *Then there exists a unique solution to Problem 8.1. Moreover, the mapping* $\boldsymbol{f} \mapsto \boldsymbol{u}$ *is Lipschitz continuous from* V *to* V.

Proof. Applying the Riesz representation theorem (page 15) we may define an operator $A : V \to V$ by

$$(A\boldsymbol{u}, \boldsymbol{v})_V = (\mathcal{F}(\boldsymbol{\varepsilon}(\boldsymbol{u})), \boldsymbol{\varepsilon}(\boldsymbol{v}))_Q \quad \forall\, \boldsymbol{u},\, \boldsymbol{v} \in V. \tag{8.13}$$

From (8.6) it follows that A is a Lipschitz continuous operator. Using Korn's inequality (7.8) (page 144) we find that A is strongly monotone. Notice that the set V_2 is a closed convex nonempty subset of V. Thus, Theorem 8.2 is deduced from Corollary 4.3 (page 63). ■

Let $\boldsymbol{u}$ be the solution of Problem 8.1 and define $\boldsymbol{\sigma}$ by (8.1). From (8.6) it follows that $\boldsymbol{\sigma} \in Q$. Taking $\boldsymbol{v} = \boldsymbol{u} \pm \boldsymbol{\varphi}$ with $\boldsymbol{\varphi} \in C_0^\infty(\Omega)^d$ in (8.12) we deduce that

$$\operatorname{Div} \boldsymbol{\sigma} = -\boldsymbol{f}_0 \quad \text{a.e. in } \Omega. \tag{8.14}$$

Using now assumption (8.7) we conclude that $\boldsymbol{\sigma} \in Q_1$.

A pair of functions $(\boldsymbol{u}, \boldsymbol{\sigma})$ which satisfies (8.12) and (8.1) is called a *weak solution* of the Signorini frictionless contact problem (8.1)–(8.5). Theorem 8.2 states that the problem (8.1)–(8.5) has a unique weak solution which depends Lipschitz continuously on the data.

The variational inequality (8.12) is derived from the classical formulation (8.1)–(8.5). We have seen that the weak solution $(\boldsymbol{u}, \boldsymbol{\sigma})$ satisfies the pointwise relation (8.14). Let us derive the other pointwise equality relations for the weak solution. Notice that $\boldsymbol{\sigma\nu} \in H^{-1/2}(\Gamma)^d$ is well-defined through the relation

$$\langle \boldsymbol{\sigma\nu}, \boldsymbol{w} \rangle_\Gamma = \int_\Omega [\operatorname{Div} \boldsymbol{\sigma} \cdot \boldsymbol{w} + \boldsymbol{\sigma} \cdot \boldsymbol{\varepsilon}(\boldsymbol{w})]\, dx \quad \forall\, \boldsymbol{w} \in H^1(\Omega)^d.$$

In (8.12) with $\mathcal{F}(\boldsymbol{\varepsilon}(\boldsymbol{u}))$ replaced by $\boldsymbol{\sigma}$ we perform an integration by parts on the left hand side and use (8.14) to obtain

$$\langle \boldsymbol{\sigma\nu}, \boldsymbol{v} - \boldsymbol{u} \rangle_\Gamma \geq \int_{\Gamma_2} \boldsymbol{f}_2 \cdot (\boldsymbol{v} - \boldsymbol{u})\, da \quad \forall\, \boldsymbol{v} \in V_2. \tag{8.15}$$

If we assume the regularity

$$\boldsymbol{\sigma\nu} \in L^2(\Gamma)^d, \tag{8.16}$$

then the relation (8.15) can be written as

$$\int_{\Gamma_2} \boldsymbol{\sigma\nu} \cdot (\boldsymbol{v} - \boldsymbol{u})\, da + \int_{\Gamma_3} \boldsymbol{\sigma\nu} \cdot (\boldsymbol{v} - \boldsymbol{u})\, da \geq \int_{\Gamma_2} \boldsymbol{f}_2 \cdot (\boldsymbol{v} - \boldsymbol{u})\, da \quad \forall\, \boldsymbol{v} \in V_2. \tag{8.17}$$

By taking $\boldsymbol{v} = \mathbf{0}$ and $2\,\boldsymbol{u}$ in (8.17), we have

$$\int_{\Gamma_2} \boldsymbol{\sigma\nu} \cdot \boldsymbol{u}\, da + \int_{\Gamma_3} \boldsymbol{\sigma\nu} \cdot \boldsymbol{u}\, da = \int_{\Gamma_2} \boldsymbol{f}_2 \cdot \boldsymbol{u}\, da. \tag{8.18}$$

Then from (8.17) again we have

$$\int_{\Gamma_2} \boldsymbol{\sigma\nu} \cdot \boldsymbol{v}\, da + \int_{\Gamma_3} \boldsymbol{\sigma\nu} \cdot \boldsymbol{v}\, da \ge \int_{\Gamma_2} \boldsymbol{f}_2 \cdot \boldsymbol{v}\, da \quad \forall\, \boldsymbol{v} \in V_2. \tag{8.19}$$

It is easy to verify that (8.17) is equivalent to (8.18) and (8.19).

Now since Γ_2 is relatively open in Γ, for any $\boldsymbol{x}_0 \in \Gamma_2$, we can find an $r_0 > 0$ such that $B(\boldsymbol{x}_0; r_0) \cap \Gamma_2 \subset \Gamma_2$. Fix an $r_1 \in (0, r_0)$. For any $r \in (0, r_1)$, we define

$$z(\boldsymbol{x}) = \begin{cases} 1 & \|\boldsymbol{x} - \boldsymbol{x}_0\| < r, \\ \dfrac{(r_1 - \|\boldsymbol{x} - \boldsymbol{x}_0\|)^2}{(r_1 - r)^3} [2\,(\|\boldsymbol{x} - \boldsymbol{x}_0\| - r) + r_1 - r] & r \le \|\boldsymbol{x} - \boldsymbol{x}_0\| < r_1, \\ 0 & \|\boldsymbol{x} - \boldsymbol{x}_0\| \ge r_1 \end{cases}$$

for $\boldsymbol{x} \in \Gamma$. Then $z \in H^1(\Gamma)$. Actually, we have $z \in W^{1,\infty}(\Gamma)$, and if $\Gamma \in C^1$, we further have $z \in C^1(\Gamma)$. Thus z can be extended to an $H^1(\Omega)$ function, still denoted by z, such that $z|_\Gamma$ is nonzero only on Γ_2. With $\boldsymbol{e}_1, \dots, \boldsymbol{e}_d$ being the canonical basis of $\mathbb{R}^d$, for any $i = 1, \dots, d$, we choose $\boldsymbol{v} = \pm z\, \boldsymbol{e}_i$ in (8.19) and let $r \to r_1-$ to obtain

$$\int_{B(\boldsymbol{x}_0; r_1) \cap \Gamma_2} (\boldsymbol{\sigma\nu})_i da = \int_{B(\boldsymbol{x}_0; r_1) \cap \Gamma_2} (\boldsymbol{f}_2)_i da, \quad i = 1, \dots, d.$$

Divide both sides by meas$\,(B(\boldsymbol{x}_0; r_1) \cap \Gamma_2)$ and let $r_1 \to 0+$. Recalling that $\boldsymbol{x}_0$ is arbitrary on Γ_2, we can apply Lebesgue's Theorem 2.5 (page 29) to conclude

$$(\boldsymbol{\sigma\nu})_i = (\boldsymbol{f}_2)_i \quad \text{a.e. on } \Gamma_2,\ i = 1, \dots, d,$$

i.e.,

$$\boldsymbol{\sigma\nu} = \boldsymbol{f}_2 \quad \text{a.e. on } \Gamma_2. \tag{8.20}$$

Using the relation (8.20) in the inequality (8.19), we have

$$\int_{\Gamma_3} \boldsymbol{\sigma\nu} \cdot \boldsymbol{v}\, da \ge 0 \quad \forall\, \boldsymbol{v} \in V_2,$$

or

$$\int_{\Gamma_3} (\sigma_\nu v_\nu + \boldsymbol{\sigma}_\tau \cdot \boldsymbol{v}_\tau)\, da \ge 0 \quad \forall\, \boldsymbol{v} \in V_2.$$

By the arbitrariness of $\boldsymbol{v}_\tau$ on Γ_3 for $\boldsymbol{v}$ in V_2, we have

$$\boldsymbol{\sigma}_\tau = \mathbf{0} \quad \text{a.e. on } \Gamma_3. \tag{8.21}$$

The relations (8.14), (8.20) and (8.21) will be needed in error estimates of numerical approximations.

It is possible to interpret pointwise inequality relations in (8.5) for the weak solution $(\boldsymbol{u}, \boldsymbol{\sigma})$. However, this requires some additional mathematical preparations (see [134]). In our error analysis of numerical schemes, we only need pointwise equality relations such as (8.14), (8.20) and (8.21).

The problem with deformable support. We now describe and analyze a similar problem in which the support is not rigid but elastic. We use a version of the normal compliance condition (5.41) (page 107), in a form with a zero gap function, to describe the reaction of the foundation when there is penetration, $u_\nu > 0$. We suppose that the normal stress on the contact surface is proportional to a power α of the penetration depth of the elastic body into the obstacle.

The classical formulation of the frictionless contact problem with a deformable obstacle can be stated as follows:

Find a displacement field $\boldsymbol{u} : \Omega \to \mathbb{R}^d$ and a stress field $\boldsymbol{\sigma} : \Omega \to \mathbb{S}^d$ such that

$$\boldsymbol{\sigma} = \mathcal{F}(\boldsymbol{\varepsilon}(\boldsymbol{u})) \quad \text{in } \Omega, \tag{8.22}$$

$$\operatorname{Div} \boldsymbol{\sigma} + \boldsymbol{f}_0 = \boldsymbol{0} \quad \text{in } \Omega, \tag{8.23}$$

$$\boldsymbol{u} = \boldsymbol{0} \quad \text{on } \Gamma_1, \tag{8.24}$$

$$\boldsymbol{\sigma\nu} = \boldsymbol{f}_2 \quad \text{on } \Gamma_2, \tag{8.25}$$

$$-\sigma_\nu = \mu\,(u_\nu)_+^\alpha,\ \boldsymbol{\sigma}_\tau = \boldsymbol{0} \quad \text{on } \Gamma_3. \tag{8.26}$$

Here $\mu > 0$ is a penalization parameter, which may be interpreted as a *deformability coefficient* of the foundation, $r_+ = \max\{0, r\}$ and $0 < \alpha \le 1$. Notice that formally (8.26) becomes (8.5) when $\mu \to 0+$.

In the study of this problem we use the same notation for the spaces V, Q and Q_1, and we assume the conditions (8.6) and (8.7). We define the element $\boldsymbol{f} \in V$ by (8.8) and denote by $j : V \to \mathbb{R}$ the functional

$$j(\boldsymbol{v}) = \frac{1}{\mu\,(\alpha+1)} \int_{\Gamma_3} (v_\nu)_+^{\alpha+1} da \quad \forall\, \boldsymbol{v} \in V. \tag{8.27}$$

It is easy to verify that if $(\boldsymbol{u}, \boldsymbol{\sigma})$ is a regular solution of the problem (8.22)–(8.26) then

$$\boldsymbol{\sigma\nu} \cdot (\boldsymbol{v} - \boldsymbol{u}) \ge \frac{1}{\mu\,(\alpha+1)} (u_\nu)_+^{\alpha+1} - \frac{1}{\mu\,(\alpha+1)} (v_\nu)_+^{\alpha+1} \quad \text{a.e. on } \Gamma_3,$$

for all $\boldsymbol{v} \in V$. Therefore, proceeding as in the case of the Signorini frictionless problem, we obtain the following variational formulation of the contact problem (8.22)–(8.26).

Problem 8.3 *Find a displacement field $\boldsymbol{u} \in V$ such that*

$$(\mathcal{F}(\boldsymbol{\varepsilon}(\boldsymbol{u})), \boldsymbol{\varepsilon}(\boldsymbol{v}) - \boldsymbol{\varepsilon}(\boldsymbol{u}))_Q + j(\boldsymbol{v}) - j(\boldsymbol{u}) \ge (\boldsymbol{f}, \boldsymbol{v} - \boldsymbol{u})_V \quad \forall\, \boldsymbol{v} \in V. \tag{8.28}$$

Well-posedness of Problem 8.3 is given in the following result.

Theorem 8.4 *Assume* (8.6) *and* (8.7). *Then there exists a unique solution to Problem 8.3. Moreover, the mapping* $\boldsymbol{f} \mapsto \boldsymbol{u}$ *is Lipschitz continuous from* V *to* V.

Proof. It follows from the proof of Theorem 8.2 that the operator A given by (8.13) is strongly monotone and Lipschitz continuous. The functional j defined by (8.27) is proper, convex and l.s.c. on V. Then Theorem 8.4 results from an application of Theorem 4.1. ■

Let now $\boldsymbol{u}$ be the solution of the problem 8.3 and define $\boldsymbol{\sigma}$ by (8.22). Similar to Problem 8.1, it can be shown that $\boldsymbol{\sigma} \in Q_1$.

A pair of functions $(\boldsymbol{u}, \boldsymbol{\sigma})$ which satisfies (8.28) and (8.22) is called a *weak solution* of the frictionless contact problem (8.22)–(8.26). We conclude from Theorem 8.2 that the problem (8.22)–(8.26) has a unique weak solution which depends Lipschitz continuously on the data. This solution depends on the parameter μ, and we denote it by $(\boldsymbol{u}_\mu, \boldsymbol{\sigma}_\mu)$. Let $(\boldsymbol{u}, \boldsymbol{\sigma})$ be the weak solution of the Signorini frictionless contact problem. It can be shown that ([66])

$$\boldsymbol{u}_\mu \to \boldsymbol{u} \quad \text{in } V, \qquad \boldsymbol{\sigma}_\mu \to \boldsymbol{\sigma} \quad \text{in } Q_1, \tag{8.29}$$

as $\mu \to 0$. A more general convergence result will be presented in Section 15.6 in the study of frictionless contact problems with viscoplastic materials.

Notice that (8.29) shows that the weak solution of the problem (8.1)–(8.5) modelling the frictionless contact between the elastic body and the rigid foundation may be obtained as the strong limit of the weak solution of the problem (8.22)–(8.26) modelling the frictionless contact with a deformable foundation, as the deformability coefficient of the foundation tends to zero.

8.2 Numerical Analysis of the Frictionless Contact Problems

In this section we study the finite element approximation of the frictionless contact problems for elastic materials presented in Section 8.1. Assume Ω is a polygon or polyhedron. Let $V^h \subset V$ be the finite element spaces of linear elements corresponding to a regular family of triangulations of $\overline{\Omega}$ that is compatible to the boundary decomposition $\Gamma = \overline{\Gamma}_1 \cup \overline{\Gamma}_2 \cup \overline{\Gamma}_3$.

The Signorini frictionless contact problem. Let $V_2^h = V^h \cap V_2$ be the finite element subset of V_2. An element from V_2^h is a continuous piecewise linear function with vanishing values at the finite element nodes lying on $\overline{\Gamma}_1$ such that its normal component at any node on $\overline{\Gamma}_3$ is non-positive. The discrete approximation of Problem 8.1 is:

Problem 8.5 *Find a displacement field* $\boldsymbol{u}^h$ *such that*

$$\boldsymbol{u}^h \in V_2^h, \quad (\mathcal{F}(\boldsymbol{\varepsilon}(\boldsymbol{u}^h)), \boldsymbol{\varepsilon}(\boldsymbol{v}^h) - \boldsymbol{\varepsilon}(\boldsymbol{u}^h))_Q \geq (\boldsymbol{f}, \boldsymbol{v}^h - \boldsymbol{u}^h)_V \quad \forall \boldsymbol{v}^h \in V_2^h. \tag{8.30}$$

With the assumptions made on the data, the discrete variational inequality (8.30) has a unique solution. Here we are interested in estimating the error $\boldsymbol{u} - \boldsymbol{u}^h$.

From the assumption (8.6) (c) it follows that for any $\boldsymbol{v}^h \in V_2^h$,

$$\begin{aligned} m\,\|\boldsymbol{u}-\boldsymbol{u}^h\|_V^2 &\le \big(\mathcal{F}(\varepsilon(\boldsymbol{u})) - \mathcal{F}(\varepsilon(\boldsymbol{u}^h)), \varepsilon(\boldsymbol{u}) - \varepsilon(\boldsymbol{u}^h)\big)_Q \\ &= \big(\mathcal{F}(\varepsilon(\boldsymbol{u})) - \mathcal{F}(\varepsilon(\boldsymbol{u}^h)), \varepsilon(\boldsymbol{u}) - \varepsilon(\boldsymbol{v}^h)\big)_Q \\ &\quad + \big(\mathcal{F}(\varepsilon(\boldsymbol{u})), \varepsilon(\boldsymbol{v}^h) - \varepsilon(\boldsymbol{u})\big)_Q - \big(\mathcal{F}(\varepsilon(\boldsymbol{u})), \varepsilon(\boldsymbol{u}^h) - \varepsilon(\boldsymbol{u})\big)_Q \\ &\quad - \big(\mathcal{F}(\varepsilon(\boldsymbol{u}^h)), \varepsilon(\boldsymbol{v}^h) - \varepsilon(\boldsymbol{u}^h)\big)_Q. \end{aligned}$$

We bound the last two terms by (8.12) and (8.30):

$$\begin{aligned} -\big(\mathcal{F}(\varepsilon(\boldsymbol{u})), \varepsilon(\boldsymbol{u}^h) - \varepsilon(\boldsymbol{u})\big)_Q &\le -(\boldsymbol{f}, \boldsymbol{u}^h - \boldsymbol{u})_V, \\ -\big(\mathcal{F}(\varepsilon(\boldsymbol{u}^h)), \varepsilon(\boldsymbol{v}^h) - \varepsilon(\boldsymbol{u}^h)\big)_Q &\le -(\boldsymbol{f}, \boldsymbol{v}^h - \boldsymbol{u}^h)_V. \end{aligned}$$

Then,

$$m\,\|\boldsymbol{u}-\boldsymbol{u}^h\|_V^2 \le \big(\mathcal{F}(\varepsilon(\boldsymbol{u})) - \mathcal{F}(\varepsilon(\boldsymbol{u}^h)), \varepsilon(\boldsymbol{u}) - \varepsilon(\boldsymbol{v}^h)\big)_Q + R(\boldsymbol{u}, \boldsymbol{v}^h), \qquad (8.31)$$

where

$$R(\boldsymbol{u}, \boldsymbol{v}^h) = \big(\mathcal{F}(\varepsilon(\boldsymbol{u})), \varepsilon(\boldsymbol{v}^h) - \varepsilon(\boldsymbol{u})\big)_Q - (\boldsymbol{f}, \boldsymbol{v}^h - \boldsymbol{u})_V. \qquad (8.32)$$

Using the assumption (8.6) (b) to bound the first term on the right hand side of (8.14), we obtain

$$m\,\|\boldsymbol{u}-\boldsymbol{u}^h\|_V^2 \le M\,\|\boldsymbol{u}-\boldsymbol{u}^h\|_V \|\boldsymbol{u}-\boldsymbol{v}^h\|_V + R(\boldsymbol{u}, \boldsymbol{v}^h).$$

Then, with a simple manipulation,

$$\|\boldsymbol{u}-\boldsymbol{u}^h\|_V \le c\,\big\{\|\boldsymbol{u}-\boldsymbol{v}^h\|_V + |R(\boldsymbol{u}, \boldsymbol{v}^h)|^{1/2}\big\}.$$

Since $\boldsymbol{v}^h \in V_2^h$ is arbitrary, we finally get the inequality

$$\|\boldsymbol{u}-\boldsymbol{u}^h\|_V \le c\,\inf_{\boldsymbol{v}^h \in V_2^h} \big\{\|\boldsymbol{u}-\boldsymbol{v}^h\|_V + |R(\boldsymbol{u}, \boldsymbol{v}^h)|^{1/2}\big\}. \qquad (8.33)$$

The inequality (8.33) is the basis for convergence analysis and error estimation. Let us first show the convergence of the numerical solution under the basic solution regularity $\boldsymbol{u} \in V_2$. From the definition

$$R(\boldsymbol{u}, \boldsymbol{v}^h) = \big(\boldsymbol{\sigma}, \varepsilon(\boldsymbol{v}^h) - \varepsilon(\boldsymbol{u})\big)_Q - (\boldsymbol{f}, \boldsymbol{v}^h - \boldsymbol{u})_V,$$

we obtain

$$|R(\boldsymbol{u}, \boldsymbol{v}^h)| \le c\,\|\boldsymbol{u}-\boldsymbol{v}^h\|_V,$$

where the constant c depends on $\|\boldsymbol{\sigma}\|_Q$ and $\|\boldsymbol{f}\|_V$. Thus by (8.33),

$$\|\boldsymbol{u}-\boldsymbol{u}^h\|_V \le c \inf_{\boldsymbol{v}^h\in V_2^h} \{\|\boldsymbol{u}-\boldsymbol{v}^h\|_V + \|\boldsymbol{u}-\boldsymbol{v}^h\|_V^{1/2}\}.$$

Let $\varepsilon > 0$. By Theorem 7.11 (page 150), there exists a $\tilde{\boldsymbol{u}} \in V_2 \cap C^\infty(\overline{\Omega})^d$ such that

$$\|\boldsymbol{u}-\tilde{\boldsymbol{u}}\|_V \le \varepsilon.$$

Let $\Pi^h\tilde{\boldsymbol{u}} \in V^h$ be the finite element interpolant of $\tilde{\boldsymbol{u}}$. We have $\Pi^h\tilde{\boldsymbol{u}} \in V_2^h$ and

$$\|\tilde{\boldsymbol{u}}-\Pi^h\tilde{\boldsymbol{u}}\|_V \le c\,h\,|\tilde{\boldsymbol{u}}|_{H^2(\Omega)}.$$

Thus,

$$\|\boldsymbol{u}-\Pi^h\tilde{\boldsymbol{u}}\|_V \le \|\boldsymbol{u}-\tilde{\boldsymbol{u}}\|_V + \|\tilde{\boldsymbol{u}}-\Pi^h\tilde{\boldsymbol{u}}\|_V \le \varepsilon + c\,h\,|\tilde{\boldsymbol{u}}|_{H^2(\Omega)}$$

and

$$\|\boldsymbol{u}-\boldsymbol{u}^h\|_V \le c\left\{\varepsilon + \varepsilon^{1/2} + h\,|\tilde{\boldsymbol{u}}|_{H^2(\Omega)} + h^{1/2}|\tilde{\boldsymbol{u}}|_{H^2(\Omega)}^{1/2}\right\}.$$

From this inequality we conclude

$$\|\boldsymbol{u}-\boldsymbol{u}^h\|_V \to 0 \quad \text{as } h \to 0.$$

We now derive an error estimate. We need to bound the residual term

$$R(\boldsymbol{u},\boldsymbol{v}^h) = \big(\boldsymbol{\sigma}, \boldsymbol{\varepsilon}(\boldsymbol{v}^h) - \boldsymbol{\varepsilon}(\boldsymbol{u})\big)_Q - (\boldsymbol{f}, \boldsymbol{v}^h - \boldsymbol{u})_V.$$

Integrate by parts on the first term and use the pointwise relation (8.14),

$$R(\boldsymbol{u},\boldsymbol{v}^h) = \langle\boldsymbol{\sigma}\boldsymbol{\nu}, \boldsymbol{v}^h - \boldsymbol{u}\rangle_\Gamma - \int_{\Gamma_2} \boldsymbol{f}_2\cdot(\boldsymbol{v}^h-\boldsymbol{u})\,da.$$

Under the regularity assumption (8.16), we can use the pointwise relations (8.20) and (8.21) to get

$$R(\boldsymbol{u},\boldsymbol{v}^h) = \int_{\Gamma_3} \sigma_\nu(v_\nu^h - u_\nu)\,da.$$

Then we obtain

$$|R(\boldsymbol{u},\boldsymbol{v}^h)| \le \|\sigma_\nu\|_{L^2(\Gamma_3)}\|v_\nu^h - u_\nu\|_{L^2(\Gamma_3)},$$

and (8.33) is reduced to

$$\|\boldsymbol{u}-\boldsymbol{u}^h\|_V \le c \inf_{\boldsymbol{v}^h\in V_2^h} \{\|\boldsymbol{u}-\boldsymbol{v}^h\|_V + \|v_\nu^h - u_\nu\|_{L^2(\Gamma_3)}^{1/2}\}. \tag{8.34}$$

Here we allow the constant c to depend on $\|\sigma_\nu\|_{L^2(\Gamma_3)}$.

Then under the additional regularity assumptions

$$\boldsymbol{u} \in H^2(\Omega), \quad u_\nu|_{\Gamma_{3,i}} \in H^2(\Gamma_{3,i}),\ 1 \le i \le i_0,$$

we can use the finite element interpolation error estimates (7.32) and (7.38) to obtain the optimal order error estimate from (8.34):

$$\|\boldsymbol{u} - \boldsymbol{u}^h\|_V = O(h).$$

The problem with deformable support. We now turn to a numerical analysis of Problem 8.3. Let $V^h \subset V$ be the finite element space of linear elements. Then the discrete approximation problem is

Problem 8.6 *Find a displacement field* $\boldsymbol{u}^h \in V^h$ *such that*

$$\begin{aligned}(\mathcal{F}(\boldsymbol{\varepsilon}(\boldsymbol{u}^h)), \boldsymbol{\varepsilon}(\boldsymbol{v}^h) - \boldsymbol{\varepsilon}(\boldsymbol{u}^h))_Q + j(\boldsymbol{v}^h) - j(\boldsymbol{u}^h)\\ \ge (\boldsymbol{f}, \boldsymbol{v}^h - \boldsymbol{u}^h)_V \quad \forall\, \boldsymbol{v}^h \in V^h.\end{aligned} \tag{8.35}$$

This problem has a unique solution by Theorem 4.1 (page 60). To estimate the error $\boldsymbol{u} - \boldsymbol{u}^h$, we take $\boldsymbol{v} = \boldsymbol{u}^h$ in (8.28) and add the resulting inequality to (8.35),

$$\begin{aligned}\big(\mathcal{F}(\boldsymbol{\varepsilon}(\boldsymbol{u})), \boldsymbol{\varepsilon}(\boldsymbol{u}) - \boldsymbol{\varepsilon}(\boldsymbol{u}^h)\big)_Q + \big(\mathcal{F}(\boldsymbol{\varepsilon}(\boldsymbol{u}^h)), \boldsymbol{\varepsilon}(\boldsymbol{u}^h) - \boldsymbol{\varepsilon}(\boldsymbol{v}^h)\big)_Q\\ \le j(\boldsymbol{v}^h) - j(\boldsymbol{u}) - (\boldsymbol{f}, \boldsymbol{v}^h - \boldsymbol{u})_V.\end{aligned} \tag{8.36}$$

Using the assumption (8.6) (c), we have

$$\begin{aligned}m\,\|\boldsymbol{u} - \boldsymbol{u}^h\|_V^2 &\le \big(\mathcal{F}(\boldsymbol{\varepsilon}(\boldsymbol{u})) - \mathcal{F}(\boldsymbol{\varepsilon}(\boldsymbol{u}^h)), \boldsymbol{\varepsilon}(\boldsymbol{u}) - \boldsymbol{\varepsilon}(\boldsymbol{u}^h)\big)_Q\\ &= \big(\mathcal{F}(\boldsymbol{\varepsilon}(\boldsymbol{u})), \boldsymbol{\varepsilon}(\boldsymbol{u}) - \boldsymbol{\varepsilon}(\boldsymbol{u}^h)\big)_Q + \big(\mathcal{F}(\boldsymbol{\varepsilon}(\boldsymbol{u}^h)), \boldsymbol{\varepsilon}(\boldsymbol{u}^h) - \boldsymbol{\varepsilon}(\boldsymbol{v}^h)\big)_Q\\ &\quad + \big(\mathcal{F}(\boldsymbol{\varepsilon}(\boldsymbol{u}^h)), \boldsymbol{\varepsilon}(\boldsymbol{v}^h) - \boldsymbol{\varepsilon}(\boldsymbol{u})\big)_Q.\end{aligned}$$

We then apply the inequality (8.36) to obtain

$$m\,\|\boldsymbol{u} - \boldsymbol{u}^h\|_V^2 \le \big(\mathcal{F}(\boldsymbol{\varepsilon}(\boldsymbol{u})) - \mathcal{F}(\boldsymbol{\varepsilon}(\boldsymbol{u}^h)), \boldsymbol{\varepsilon}(\boldsymbol{u}) - \boldsymbol{\varepsilon}(\boldsymbol{v}^h)\big)_Q + R(\boldsymbol{u}, \boldsymbol{v}^h),$$

where

$$R(\boldsymbol{u}, \boldsymbol{v}^h) = \big(\mathcal{F}(\boldsymbol{\varepsilon}(\boldsymbol{u})), \boldsymbol{\varepsilon}(\boldsymbol{v}^h) - \boldsymbol{\varepsilon}(\boldsymbol{u})\big)_Q + j(\boldsymbol{v}^h) - j(\boldsymbol{u}) - (\boldsymbol{f}, \boldsymbol{v}^h - \boldsymbol{u})_V. \tag{8.37}$$

Therefore,

$$\|\boldsymbol{u} - \boldsymbol{u}^h\|_V \le c\,\big\{\|\boldsymbol{u} - \boldsymbol{v}^h\|_V + |R(\boldsymbol{u}, \boldsymbol{v}^h)|^{1/2}\big\}.$$

Since $\boldsymbol{v}^h \in V^h$ is arbitrary, we finally get the inequality

$$\|\boldsymbol{u} - \boldsymbol{u}^h\|_V \le c \inf_{\boldsymbol{v}^h \in V^h} \big\{\|\boldsymbol{u} - \boldsymbol{v}^h\|_V + |R(\boldsymbol{u}, \boldsymbol{v}^h)|^{1/2}\big\}. \tag{8.38}$$

Again the inequality (8.38) is the basis for convergence analysis and error estimation. Like for Problem 8.5, we have the convergence

$$\|\boldsymbol{u} - \boldsymbol{u}^h\|_V \to 0 \quad \text{as } h \to 0$$

under the basic solution regularity $\boldsymbol{u} \in V$. To derive an error estimate, we need to bound the residual term $R(\boldsymbol{u}, \boldsymbol{v}^h)$ defined in (8.37). We first estimate

$$j(\boldsymbol{v}^h) - j(\boldsymbol{u}) = \frac{1}{\mu\,(\alpha+1)} \int_{\Gamma_3} \left[(v_\nu^h)_+^{\alpha+1} - (u_\nu)_+^{\alpha+1}\right] da.$$

Applying the elementary inequality (7.56) on page 162, we find

$$\begin{aligned} |j(\boldsymbol{v}^h) - j(\boldsymbol{u})| &\le c \int_{\Gamma_3} \left((u_\nu)_+^\alpha |(v_\nu^h)_+ - (u_\nu)_+| + |(v_\nu^h)_+ - (u_\nu)_+|^{1+\alpha}\right) da \\ &\le c\left[\|v_\nu^h - u_\nu\|_{L^2(\Gamma_3)} + \|v_\nu^h - u_\nu\|_{L^2(\Gamma_3)}^{1+\alpha}\right]. \end{aligned}$$

As in the previous example, under the regularity assumption (8.16) we finally have

$$\begin{aligned} \|\boldsymbol{u} - \boldsymbol{u}^h\|_V \le c \inf_{\boldsymbol{v}^h \in V^h} \Big\{ &\|\boldsymbol{u} - \boldsymbol{v}^h\|_V + \|v_\nu^h - u_\nu\|_{L^2(\Gamma_3)}^{1/2} \\ &+ \|v_\nu^h - u_\nu\|_{L^2(\Gamma_3)}^{(1+\alpha)/2} \Big\}. \end{aligned} \tag{8.39}$$

Here we allow the constant c to depend on the solution $(\boldsymbol{u}, \boldsymbol{\sigma})$.

Then with the additional regularity assumptions

$$\boldsymbol{u} \in H^2(\Omega), \quad u_\nu|_{\Gamma_{3,i}} \in H^2(\Gamma_{3,i}),\ 1 \le i \le i_0,$$

we obtain the optimal order error estimate from (8.39):

$$\|\boldsymbol{u} - \boldsymbol{u}^h\|_V = O(h).$$

8.3 Quasistatic Frictional Contact Problems

In this section we consider two frictional contact problems for linearly elastic materials which may be set in the variational formulation (4.31)–(4.32) (page 69). We apply Theorem 4.16 (page 69) to prove the existence, uniqueness and Lipschitz continuous dependence of the weak solution on the data.

Bilateral contact with Tresca's friction law. The physical setting is as follows. A linearly elastic body occupies an open bounded connected domain $\Omega \subset \mathbb{R}^d$ ($d = 1, 2, 3$ in applications) with a Lipschitz boundary Γ that is partitioned into three parts $\overline{\Gamma}_1$, $\overline{\Gamma}_2$ and $\overline{\Gamma}_3$ with Γ_1, Γ_2 and Γ_3 relatively open and mutually disjoiont, such that $\text{meas}\,(\Gamma_1) > 0$. Let $T > 0$ and let $I = (0, T)$ denote the time interval of interest. The body is clamped on $\Gamma_1 \times I$ and thus the displacement field vanishes there. A sentence such as "the body is clamped on $\Gamma_1 \times I$" means that the body is clamped on Γ_1 during the time interval I. We use this type of phrases throughout the book. A volume force of density $\boldsymbol{f}_0$ acts in $\Omega \times I$ and a surface traction of density $\boldsymbol{f}_2$ acts on $\Gamma_2 \times I$. The body is in frictional contact with an obstacle on $\Gamma_3 \times I$. We assume the contact is bilateral, i.e. there is no loss of contact during the process. Thus, the normal displacement u_ν vanishes on Γ_3 at any time. We model the friction with Tresca's friction law. The constitutive law of the material is (6.8) (page 121) and the process is quasistatic.

Under the previous assumptions, the classical formulation of the mechanical problem is the following:

Find a displacement field $\boldsymbol{u} : \Omega \times \overline{I} \to \mathbb{R}^d$ and a stress field $\boldsymbol{\sigma} : \Omega \times \overline{I} \to \mathbb{S}^d$ such that

$$\boldsymbol{\sigma} = \mathcal{E}\boldsymbol{\varepsilon}(\boldsymbol{u}) \qquad \text{in } \Omega \times I, \tag{8.40}$$

$$\operatorname{Div} \boldsymbol{\sigma} + \boldsymbol{f}_0 = \boldsymbol{0} \qquad \text{in } \Omega \times I, \tag{8.41}$$

$$\boldsymbol{u} = \boldsymbol{0} \qquad \text{on } \Gamma_1 \times I, \tag{8.42}$$

$$\boldsymbol{\sigma\nu} = \boldsymbol{f}_2 \qquad \text{on } \Gamma_2 \times I, \tag{8.43}$$

$$\left.\begin{array}{l} u_\nu = 0, \ \|\boldsymbol{\sigma}_\tau\| \le g, \\ \|\boldsymbol{\sigma}_\tau\| < g \ \Rightarrow \ \dot{\boldsymbol{u}}_\tau = 0, \\ \|\boldsymbol{\sigma}_\tau\| = g \ \Rightarrow \ \exists \lambda \ge 0 \text{ s.t. } \boldsymbol{\sigma}_\tau = -\lambda \dot{\boldsymbol{u}}_\tau \end{array}\right\} \qquad \text{on } \Gamma_3 \times I, \tag{8.44}$$

$$\boldsymbol{u}(0) = \boldsymbol{u}_0 \qquad \text{in } \Omega. \tag{8.45}$$

Here $\boldsymbol{u}_0$ denotes the given initial displacement field. In (8.44), $g \ge 0$ represents the friction bound function, i.e. the magnitude of the limiting friction traction at which slip begins. The strict inequality in (8.44) holds in the stick zone and the equality holds in the slip zone.

We now turn to a variational formulation for the mechanical problem (8.40)–(8.45). The displacement space is V_1 defined in (7.11) (page 144). We assume that the elasticity operator $\mathcal{E}$ satisfies the condition (6.9) (page 122). We also assume that the force and the traction densities satisfy

$$\boldsymbol{f}_0 \in W^{1,\infty}(I; L^2(\Omega)^d), \quad \boldsymbol{f}_2 \in W^{1,\infty}(I; L^2(\Gamma_2)^d), \tag{8.46}$$

and the friction bound satisfies

$$g \in L^\infty(\Gamma_3), \quad g \ge 0 \text{ a.e. on } \Gamma_3. \tag{8.47}$$

Next we define the bilinear form $a : V_1 \times V_1 \to \mathbb{R}$ by

$$a(\boldsymbol{u}, \boldsymbol{v}) = (\mathcal{E}\boldsymbol{\varepsilon}(\boldsymbol{u}), \boldsymbol{\varepsilon}(\boldsymbol{v}))_Q \tag{8.48}$$

and the functional $j : V_1 \to \mathbb{R}_+$ by

$$j(\boldsymbol{v}) = \int_{\Gamma_3} g \, \|\boldsymbol{v}_\tau\| \, da. \tag{8.49}$$

We denote by $\boldsymbol{f}(t)$ the element of V_1 given by

$$(\boldsymbol{f}(t), \boldsymbol{v})_V = \int_\Omega \boldsymbol{f}_0(t) \cdot \boldsymbol{v}\, dx + \int_{\Gamma_2} \boldsymbol{f}_2(t) \cdot \boldsymbol{v}\, da \tag{8.50}$$

for all $\boldsymbol{v} \in V_1$ and $t \in \overline{I}$. From (8.46) and (8.47), it follows that the integrals in (8.50) and (8.49) are well defined. Moreover,

$$\boldsymbol{f} \in W^{1,\infty}(I; V_1). \tag{8.51}$$

Finally we assume that the initial data satisfies

$$\boldsymbol{u}_0 \in V_1, \tag{8.52}$$

$$a(\boldsymbol{u}_0, \boldsymbol{v}) + j(\boldsymbol{v}) \geq (\boldsymbol{f}(0), \boldsymbol{v})_V \quad \forall\, \boldsymbol{v} \in V_1. \tag{8.53}$$

Proceeding as in the case of the static contact problems presented in Section 8.1, we obtain the following variational formulation of the contact problem (8.40)–(8.45).

Problem 8.7 *Find a displacement field $\boldsymbol{u} : \overline{I} \to V_1$ such that*

$$\begin{aligned} & a(\boldsymbol{u}(t), \boldsymbol{v} - \dot{\boldsymbol{u}}(t)) + j(\boldsymbol{v}) - j(\dot{\boldsymbol{u}}(t)) \\ & \qquad \geq (\boldsymbol{f}(t), \boldsymbol{v} - \dot{\boldsymbol{u}}(t))_V \quad \forall\, \boldsymbol{v} \in V_1, \text{ a.e. } t \in I, \end{aligned} \tag{8.54}$$

$$\boldsymbol{u}(0) = \boldsymbol{u}_0. \tag{8.55}$$

The assumption (8.47) implies that j is a continuous semi-norm on V_1. Assumption (6.44) implies that a is a continuous and coercive bilinear form on V_1. Therefore, recalling (8.51)–(8.53), by Theorem 4.16 (page 69) we obtain the following result.

Theorem 8.8 *Assume the conditions* (6.9), (8.46)–(8.47) *and* (8.52)–(8.53). *Then Problem 8.7 has a unique solution $\boldsymbol{u} \in W^{1,\infty}(I; V_1)$. Moreover, the mapping $(\boldsymbol{f}, \boldsymbol{u}_0) \mapsto \boldsymbol{u}$ is Lipschitz continuous from $W^{1,1}(I; V_1) \times V_1$ to $L^\infty(I; V_1)$.*

Let now $\boldsymbol{u} \in W^{1,\infty}(I; V_1)$ be the solution of Problem 8.7 and let $\boldsymbol{\sigma}$ be the stress field given by (8.40). It can be shown that $\boldsymbol{\sigma} \in W^{1,\infty}(I; Q_1)$. A pair of functions $(\boldsymbol{u}, \boldsymbol{\sigma})$ which satisfies (8.54), (8.55) and (8.40) is called a *weak solution* of the bilateral contact problem with Tresca's friction law. Thus problem (8.40)–(8.45) has a unique weak solution which depends Lipschitz continuously on the data.

Contact with simplified Coulomb's friction law. We will show that Problem 8.7 with a different choice of the functional j describes another frictional contact problem. Consider a contact problem modeled by a simplified version of Coulomb's law (see e.g. [70, 167]),

$$\left.\begin{array}{l} \sigma_\nu = S, \quad \|\boldsymbol{\sigma}_\tau\| \leq \mu\, |\sigma_\nu|, \\ \|\boldsymbol{\sigma}_\tau\| < \mu\, |\sigma_\nu| \;\Rightarrow\; \dot{\boldsymbol{u}}_\tau = \boldsymbol{0}, \\ \|\boldsymbol{\sigma}_\tau\| = \mu\, |\sigma_\nu| \;\Rightarrow \exists\, \lambda \geq 0 \text{ s.t. } \boldsymbol{\sigma}_\tau = -\lambda \dot{\boldsymbol{u}}_\tau \end{array}\right\} \quad \text{on } \Gamma_3 \times I. \tag{8.56}$$

Here S is a given function on Γ_3 and μ is the coefficient of friction.

The mechanical problem of frictional contact we consider is the following:

Find a displacement field $\boldsymbol{u} : \Omega \times \overline{I} \to \mathbb{R}^d$ and a stress field $\boldsymbol{\sigma} : \Omega \times \overline{I} \to \mathbb{S}^d$ which satisfy (8.40)–(8.43), (8.45) and (8.56).

We now turn to a variational formulation for this mechanical problem. The displacement space is V defined in (7.7) (page 144). We assume that the elasticity operator $\mathcal{E}$ satisfies the condition (6.9) (page 122) and we consider the bilinear form $a : V \times V \to \mathbb{R}$ given by (8.48).

We assume that the given normal stress satisfies

$$S \in L^\infty(\Gamma_3) \tag{8.57}$$

and the coefficient of friction is such that

$$\mu \in L^\infty(\Gamma_3), \quad \mu \geq 0 \text{ a.e. on } \Gamma_3. \tag{8.58}$$

We define the friction functional $j : V \to \mathbb{R}$ by

$$j(\boldsymbol{v}) = \int_{\Gamma_3} \mu \, |S| \, \|\boldsymbol{v}_\tau\| \, da \tag{8.59}$$

and let $\boldsymbol{f}(t)$ be the element of V given by

$$(\boldsymbol{f}(t), \boldsymbol{v})_V = \int_\Omega \boldsymbol{f}_0(t) \cdot \boldsymbol{v} \, dx + \int_{\Gamma_2} \boldsymbol{f}_2(t) \cdot \boldsymbol{v} \, da + \int_{\Gamma_3} S \, v_\nu \, da \tag{8.60}$$

for all $\boldsymbol{v} \in V$, $t \in \overline{I}$.

Finally we assume that the initial data satisfies

$$\boldsymbol{u}_0 \in V, \tag{8.61}$$

$$a(\boldsymbol{u}_0, \boldsymbol{v}) + j(\boldsymbol{v}) \geq (\boldsymbol{f}(0), \boldsymbol{v})_V \quad \forall \boldsymbol{v} \in V. \tag{8.62}$$

With these notations, the variational formulation of the mechanical problem (8.40)–(8.43), (8.45) and (8.56) is the following.

Problem 8.9 *Find a displacement field* $\boldsymbol{u} : \overline{I} \to V$ *such that*

$$\begin{aligned} &a(\boldsymbol{u}(t), \boldsymbol{v} - \dot{\boldsymbol{u}}(t)) + j(\boldsymbol{v}) - j(\dot{\boldsymbol{u}}(t)) \\ &\qquad \geq (\boldsymbol{f}(t), \boldsymbol{v} - \dot{\boldsymbol{u}}(t))_V \quad \forall \boldsymbol{v} \in V, \text{ a.e. } t \in I, \\ &\boldsymbol{u}(0) = \boldsymbol{u}_0. \end{aligned}$$

We observe that assumptions (8.57) and (8.58) imply that j is a continuous semi-norm on V. The condition (6.9) implies that a is a continuous and coercive bilinear form on V and conditions (8.46) and (8.57) imply the regularity $\boldsymbol{f} \in W^{1,\infty}(I;V)$. Therefore, using Theorem 4.16 we obtain the following result.

Theorem 8.10 *Assume* (6.9), (8.46), (8.57), (8.58), (8.61) *and* (8.62). *Then Problem 8.9 has a unique solution* $\boldsymbol{u} \in W^{1,\infty}(I;V)$. *Moreover, the mapping* $(\boldsymbol{f}, \boldsymbol{u}_0) \mapsto \boldsymbol{u}$ *is Lipschitz continuous from* $W^{1,1}(I;V) \times V$ *to* $L^\infty(I;V)$.

Again, we conclude that the mechanical problem (8.40)–(8.43), (8.45) and (8.56) has a unique *weak solution* which depends Lipschitz continuously on the data.

8.4 Numerical Analysis of Quasistatic Frictional Contact Problems

In this section we study the finite element approximation of the quasistatic frictionless contact problems for elastic materials presented in Section 8.3. Since the weak formulations of the two problems have the same structure, we will provide numerical analysis only for the contact problem of bilateral contact with Tresca's friction law. We will specialize the abstract results of Sections 4.4 and 4.5. An argument similar to that in Section 8.1 shows that the solution of Problem 8.7 satisfies

$$\operatorname{Div} \boldsymbol{\sigma} + \boldsymbol{f}_0 = \boldsymbol{0} \quad \text{a.e. in } \Omega \times I, \tag{8.63}$$

where $\boldsymbol{\sigma}$ is defined by (8.40). Under the assumption

$$\boldsymbol{\sigma\nu} \in L^\infty(I; L^2(\Gamma)^d) \tag{8.64}$$

we further have

$$\boldsymbol{\sigma\nu} = \boldsymbol{f}_2 \quad \text{a.e. on } \Gamma_2 \times I. \tag{8.65}$$

Semi-discrete approximations. We consider a semi-discrete approximation of Problem 8.7. Let $\{V_1^h\} \subset V_1$ be a family of piecewise linear finite element subspaces corresponding to a regular family of partitions of $\overline{\Omega}$ that are compatible to the boundary decomposition $\Gamma = \overline{\Gamma}_1 \cup \overline{\Gamma}_2 \cup \overline{\Gamma}_3$.

Problem 8.11 *Find a displacement field* $\boldsymbol{u}^h : \overline{I} \to V_1^h$ *such that*

$$\begin{aligned}
&a(\boldsymbol{u}^h(t), \boldsymbol{v}^h - \dot{\boldsymbol{u}}^h(t)) + j(\boldsymbol{v}^h) - j(\dot{\boldsymbol{u}}^h(t)) \\
&\qquad \geq (\boldsymbol{f}(t), \boldsymbol{v}^h - \dot{\boldsymbol{u}}^h(t))_V \quad \forall\, \boldsymbol{v}^h \in V_1^h, \text{ a.e. } t \in I, \\
&\boldsymbol{u}^h(0) = \mathcal{P}^h \boldsymbol{u}_0,
\end{aligned}$$

where $\mathcal{P}^h \boldsymbol{u}_0 \in V_1^h$ *is the finite element projection of* $\boldsymbol{u}_0$ *on* V_1^h *defined by*

$$a(\mathcal{P}^h \boldsymbol{u}_0, \boldsymbol{v}^h) = a(\boldsymbol{u}_0, \boldsymbol{v}^h) \quad \forall\, \boldsymbol{v}^h \in V_1^h. \tag{8.66}$$

From the discussion in Section 4.4, we know that Problem 8.11 has a unique solution. For error estimation, we have

$$\begin{aligned}
\|\boldsymbol{u} - \boldsymbol{u}^h\|_{L^\infty(I;V)} \leq c\, \|\boldsymbol{u}_0 - \mathcal{P}^h \boldsymbol{u}_0\|_V + c \inf_{\boldsymbol{v}^h \in L^2(I;V_1^h)} \big[\|\dot{\boldsymbol{u}} - \boldsymbol{v}^h\|_{L^2(I;V)} \\
+ \|R(\cdot; \dot{\boldsymbol{u}}(\cdot), \boldsymbol{v}^h(\cdot))\|_{L^1(I)}^{1/2}\big],
\end{aligned} \tag{8.67}$$

where

$$\begin{aligned}
R(t; \dot{\boldsymbol{u}}(t), \boldsymbol{v}^h) = a(\boldsymbol{u}(t), \boldsymbol{v}^h - \dot{\boldsymbol{u}}(t)) + j(\boldsymbol{v}^h) - j(\dot{\boldsymbol{u}}(t)) \\
- (\boldsymbol{f}(t), \boldsymbol{v}^h - \dot{\boldsymbol{u}}(t))_V
\end{aligned} \tag{8.68}$$

is a residual quantity.

The inequality (8.67) is the basis for convergence analysis and error estimation. For $j(\cdot)$ defined by (8.59), we have

$$\begin{aligned}|j(\boldsymbol{v}^h) - j(\dot{\boldsymbol{u}}(t))| &\le \|g\|_{L^\infty(\Gamma_3)} \|\boldsymbol{v}_\tau^h - \dot{\boldsymbol{u}}_\tau(t)\|_{L^1(\Gamma_3)^d} \\ &\le c\, \|g\|_{L^\infty(\Gamma_3)} \|\boldsymbol{v}^h - \dot{\boldsymbol{u}}(t)\|_V.\end{aligned}$$

Thus a direct estimate of the residual (8.68) is

$$|R(t; \dot{\boldsymbol{u}}(t), \boldsymbol{v}^h)| \le c\, \|\boldsymbol{v}^h - \dot{\boldsymbol{u}}(t)\|_V \tag{8.69}$$

where the constant c depends on $\|\boldsymbol{u}\|_{L^\infty(I;V)}$, $\|g\|_{L^\infty(\Gamma_3)}$, $\|\boldsymbol{f}_0\|_{L^\infty(I;L^2(\Omega)^d)}$ and $\|\boldsymbol{f}_2\|_{L^\infty(I;L^2(\Gamma_2)^d)}$. With the bound (8.69), we obtain from (8.67) that

$$\begin{aligned}\|\boldsymbol{u} - \boldsymbol{u}^h\|_{L^\infty(I;V)} \le c\, \|\boldsymbol{u}_0 - \mathcal{P}^h \boldsymbol{u}_0\|_V + c \inf_{\boldsymbol{v}^h \in L^2(I;V_1^h)} \big[\|\dot{\boldsymbol{u}} - \boldsymbol{v}^h\|_{L^2(I;V)} \\ + \|\dot{\boldsymbol{u}} - \boldsymbol{v}^h\|_{L^2(I;V)}^{1/2} \big].\end{aligned}$$

By Theorem 7.10 (page 150), $C^\infty(\overline{\Omega})^d \cap V_1$ is dense in V_1. We can use an argument similar to the proof of Theorem 7.18 (page 157) to show the convergence

$$\|\boldsymbol{u} - \boldsymbol{u}^h\|_{L^\infty(I;V)} \to 0 \quad \text{as } h \to 0.$$

For error estimation, we assume the regularity condition (8.64). Performing an integration by parts, we have for a.e. $t \in I$,

$$\begin{aligned}a(\boldsymbol{u}(t), \boldsymbol{v}) &= \int_\Gamma \boldsymbol{\sigma}(t)\boldsymbol{\nu} \cdot \boldsymbol{v}\, da - \int_\Omega \operatorname{Div} \boldsymbol{\sigma}(t) \cdot \boldsymbol{v}\, dx \\ &= \int_{\Gamma_3} (\sigma_\nu(t)\, v_\nu + \boldsymbol{\sigma}_\tau(t) \cdot \boldsymbol{v}_\tau)\, da \\ &\quad + \int_{\Gamma_2} \boldsymbol{f}_2(t) \cdot \boldsymbol{v}\, da + \int_\Omega \boldsymbol{f}_0(t) \cdot \boldsymbol{v}\, dx.\end{aligned}$$

Thus using the definition (8.68) and the pointwise relations (8.63), (8.65), we obtain

$$R(t; \dot{\boldsymbol{u}}(t), \boldsymbol{v}^h) = \int_{\Gamma_3} \big(\boldsymbol{\sigma}_\tau(t) \cdot (\boldsymbol{v}_\tau^h - \dot{\boldsymbol{u}}_\tau(t)) + g\,(\|\boldsymbol{v}_\tau^h\| - \|\dot{\boldsymbol{u}}_\tau(t)\|)\big)\, da.$$

Therefore, instead of (8.69), we have

$$|R(t; \dot{\boldsymbol{u}}(t), \boldsymbol{v}^h)| \le c\, \|\boldsymbol{v}^h - \dot{\boldsymbol{u}}(t)\|_{L^2(\Gamma_3)^d}. \tag{8.70}$$

Then from (8.67) we have

$$\begin{aligned}\|\boldsymbol{u} - \boldsymbol{u}^h\|_{L^\infty(I;V)} \le c\, \|\boldsymbol{u}_0 - \mathcal{P}^h \boldsymbol{u}_0\|_V + c \inf_{\boldsymbol{v}^h \in L^2(I;V_1^h)} \big[\|\dot{\boldsymbol{u}} - \boldsymbol{v}^h\|_{L^2(I;V)} \\ + \|\dot{\boldsymbol{u}} - \boldsymbol{v}^h\|_{L^2(I;L^2(\Gamma_3)^d)}^{1/2} \big].\end{aligned}$$

Under the additional regularity conditions

$$\begin{aligned}&\dot{\boldsymbol{u}} \in L^2(I; H^2(\Omega)^d),\ \dot{\boldsymbol{u}}|_{\Gamma_{3,i}} \in L^2(I; H^2(\Gamma_{3,i})^d),\ 1 \le i \le i_0,\\ &\boldsymbol{u}_0 \in H^2(\Omega)^d,\end{aligned}$$

we then obtain the optimal order error estimate

$$\|\boldsymbol{u} - \boldsymbol{u}^h\|_{L^\infty(I;V)} = O(h).$$

Fully discrete approximations. Now we consider fully discrete approximations, where the temporal and spatial variables are simultaneously discretized. In addition to the family of finite element subspaces $\{V_1^h\}$ of V_1, we need a partition of the time interval described in Section 4.5. Let $\theta \in [1/2, 1]$ be a parameter. The family of fully discrete approximation schemes is the following.

Problem 8.12 *Find* $\boldsymbol{u}^{hk} = \{\boldsymbol{u}_n^{hk}\}_{n=0}^N$, *where* $\boldsymbol{u}_n^{hk} \in V_1^h$, $0 \le n \le N$, *with* $\boldsymbol{u}_0^{hk} = \boldsymbol{u}_0^h$ *defined in* (8.66), *such that for* $n = 1, 2, \ldots, N$,

$$\begin{aligned}&a(\theta\, \boldsymbol{u}_n^{hk} + (1-\theta)\, \boldsymbol{u}_{n-1}^{hk}, \boldsymbol{v}^h - \delta_n \boldsymbol{u}_n^{hk}) + j(\boldsymbol{v}^h) - j(\delta_n \boldsymbol{u}_n^{hk})\\ &\quad \ge (\boldsymbol{f}_{n-1+\theta}, \boldsymbol{v}^h - \delta_n \boldsymbol{u}_n^{hk})_V \quad \forall\, \boldsymbol{v}^h \in V_1^h.\end{aligned}$$

From the discussion in Section 4.5, we know that Problem 8.12 has a unique solution. For error estimates, if $\boldsymbol{u} \in W^{3,1}(I;V)$, we have

$$\begin{aligned}\max_n \|\boldsymbol{u}_n - \boldsymbol{u}_n^{hk}\|_V &\le c\,k + c\,\|\boldsymbol{u}_0 - \mathcal{P}^h \boldsymbol{u}_0\|_V + c \sum_{j=1}^N k_j \|\dot{\boldsymbol{u}}_{j-1+\theta} - \boldsymbol{v}_j^h\|_V\\ &\quad + c\Big\{\sum_{j=1}^N k_j |R_j(\dot{\boldsymbol{u}}_{j-1+\theta}, \boldsymbol{v}_j^h)|\Big\}^{1/2} \text{ if } \theta \ne \frac{1}{2}\end{aligned} \tag{8.71}$$

and

$$\begin{aligned}\max_n \|\boldsymbol{u}_n - \boldsymbol{u}_n^{hk}\|_V &\le c\,k^2 + c\,\|\boldsymbol{u}_0 - \mathcal{P}^h \boldsymbol{u}_0\|_V + c \sum_{j=1}^N k_j \|\dot{\boldsymbol{u}}_{j-1/2} - \boldsymbol{v}_j^h\|_V\\ &\quad + c\Big\{\sum_{j=1}^N k_j |R_j(\dot{\boldsymbol{u}}_{j-1/2}, \boldsymbol{v}_j^h)|\Big\}^{1/2} \text{ if } \theta = \frac{1}{2},\end{aligned} \tag{8.72}$$

where for $1 \le j \le N$, $\boldsymbol{v}_j^h \in V^h$ are arbitrary, and

$$\begin{aligned}R_j(\dot{\boldsymbol{u}}_{j-1+\theta}, \boldsymbol{v}_j^h) &= a(\boldsymbol{u}_{j-1+\theta}, \boldsymbol{v}_j^h - \dot{\boldsymbol{u}}_{j-1+\theta}) + j(\boldsymbol{v}_j^h) - j(\dot{\boldsymbol{u}}_{j-1+\theta})\\ &\quad - (\boldsymbol{f}_{j-1+\theta}, \boldsymbol{v}_j^h - \dot{\boldsymbol{u}}_{j-1+\theta})_V\end{aligned} \tag{8.73}$$

is a residual quantity. Assume the regularity condition (8.64). Similar to (8.70), we have

$$|R_j(\dot{\boldsymbol{u}}_{j-1+\theta}, \boldsymbol{v}_j^h)| \le c\,\|\boldsymbol{v}_j^h - \dot{\boldsymbol{u}}_{j-1+\theta}\|_{L^2(\Gamma_3)^d}.$$

Then the estimates (8.71) and (8.72) imply

$$\max_n \|\boldsymbol{u}_n - \boldsymbol{u}_n^{hk}\|_V \le c\,k + c\,\|\boldsymbol{u}_0 - \mathcal{P}^h \boldsymbol{u}_0\|_V + c\sum_{j=1}^N k_j \|\dot{\boldsymbol{u}}_{j-1+\theta} - \boldsymbol{v}_j^h\|_V$$
$$+ c\Big\{\sum_{j=1}^N k_j \|\boldsymbol{v}_j^h - \dot{\boldsymbol{u}}_{j-1+\theta}\|_{L^2(\Gamma_3)^d}\Big\}^{1/2} \quad \text{if } \theta \ne \frac{1}{2}$$

and

$$\max_n \|\boldsymbol{u}_n - \boldsymbol{u}_n^{hk}\|_V \le c\,k^2 + c\,\|\boldsymbol{u}_0 - \mathcal{P}^h \boldsymbol{u}_0\|_V + c\sum_{j=1}^N k_j \|\dot{\boldsymbol{u}}_{j-1/2} - \boldsymbol{v}_j^h\|_V$$
$$+ c\Big\{\sum_{j=1}^N k_j \|\boldsymbol{v}_j^h - \dot{\boldsymbol{u}}_{j-1/2}\|_{L^2(\Gamma_3)^d}\Big\}^{1/2} \quad \text{if } \theta = \frac{1}{2},$$

for any $\boldsymbol{v}_j^h \in V_1^h$, $j = 1, \dots, N$.

Finally, under the additional regularity conditions

$$\dot{\boldsymbol{u}} \in L^2(I; H^2(\Omega)^d), \ \ \dot{\boldsymbol{u}}|_{\Gamma_{3,i}} \in L^2(I; H^2(\Gamma_{3,i})^d), \ \ 1 \le i \le i_0,$$
$$\boldsymbol{u}_0 \in H^2(\Omega)^d,$$

we have the optimal order error estimates

$$\max_n \|\boldsymbol{u}_n - \boldsymbol{u}_n^{hk}\|_V = O(k + h) \quad \text{if } \theta \ne \frac{1}{2}, \tag{8.74}$$

$$\max_n \|\boldsymbol{u}_n - \boldsymbol{u}_n^{hk}\|_V = O(k^2 + h) \quad \text{if } \theta = \frac{1}{2}. \tag{8.75}$$

Convergence proof of the fully discrete solutions under the basic solution regularity $\boldsymbol{u} \in W^{1,\infty}(I; V)$ is more involved, since the pointwise value of $\dot{\boldsymbol{u}}(t)$ is not well defined and we cannot use (8.71) or (8.72) for a convergence analysis. Details of such convergence analysis can be found in [24]. Convergence analysis for a similar problem in the context of plasticity theory is found in [96, 97]. See also Section 17.6 for such a convergence analysis in a more complicated setting of a bilateral contact with Tresca's friction law for a viscoplastic material.

8.5 Numerical Examples

We report here numerical results on some two dimensional examples for Problem 8.7 defined on page 179. We use the fully discrete method with $\theta = 1$, i.e., we use backward Euler difference for the time derivative.

As a first two-dimensional example of Problem 8.7 we consider the physical setting shown in Figure 8.1. There, the domain $\Omega = (0, 4) \times (0, 4)$ is the cross section of a three-dimensional linearly elastic body subjected to the action of tractions in such a way that a plane stress hypothesis is valid. On the part

$\Gamma_1 = \{4\} \times (0,4)$ the body is clamped and therefore the displacement field vanishes there. Oblique tractions act on the part $\{0\}\times(0,4)$ of the boundary and the part $(0,4)\times\{4\}$ is traction free. Thus, $\Gamma_2 = (\{0\}\times(0,4))\cup(\{4\}\times(0,4))$. No body forces are assumed to act on the elastic body during the process. The body is in bilateral frictional contact with a rigid obstacle on the part $\Gamma_3 = (0,4)\times\{0\}$ of the boundary. The friction is modeled with Tresca law.

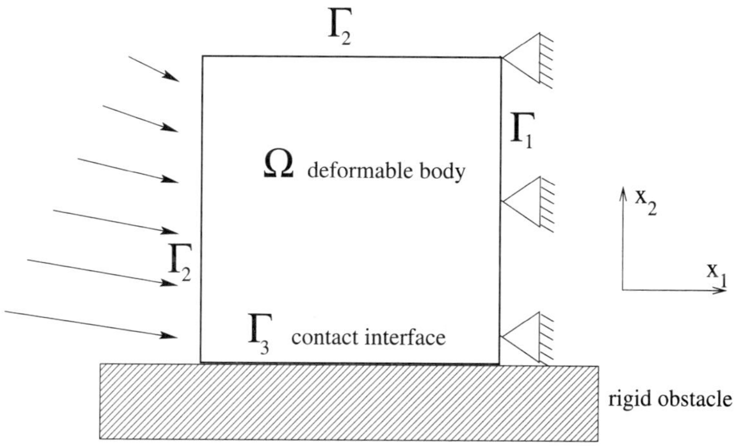

Figure 8.1: Initial configuration in the first two-dimensional example

The elasticity tensor $\mathcal{E}$ satisfies

$$(\mathcal{E}\boldsymbol{\tau})_{\alpha\beta} = \frac{E\nu}{1-\nu^2}(\tau_{11}+\tau_{22})\delta_{\alpha\beta} + \frac{E}{1+\nu}\tau_{\alpha\beta}, \quad 1 \le \alpha,\beta \le 2, \tag{8.76}$$

where E is the Young's modulus, ν the Poisson's ratio of the material and $\delta_{\alpha\beta}$ denotes the Kronecker symbol.

We consider discretization by linear elements for the space V and uniform partitions of the time interval $\overline{I}$. For computation we have used the following data (the unit daN/mm^2 stands for "decanewtons per square milimeter"):

$$E = 200\, daN/mm^2, \quad \nu = 0.3, \quad \boldsymbol{f}_0 = (0,0)\, daN/mm^2,$$
$$\boldsymbol{f}_2(x_1,x_2,t) = (0.02\,(5-x_2)\,t, -0.01\,t)\, daN/mm^2,$$
$$g = 0.0012\, daN/mm^2, \quad \boldsymbol{u}_0 = \boldsymbol{0}\, m, \quad T = 1\, sec.$$

In Figure 8.2 the deformed configuration as well as the contact interface forces on Γ_3 at moment $t = 1\, sec$ are plotted. Here, the time step considered is $k = 0.1\, sec$. It can be seen that the stick zone is located in a neighborhood of the corner $(4,0)$, while the rest of the contact boundary is in the slip zone.

To see the convergence behavior of the fully discrete scheme, we compute a sequence of numerical solutions based on uniform partitions of the time interval

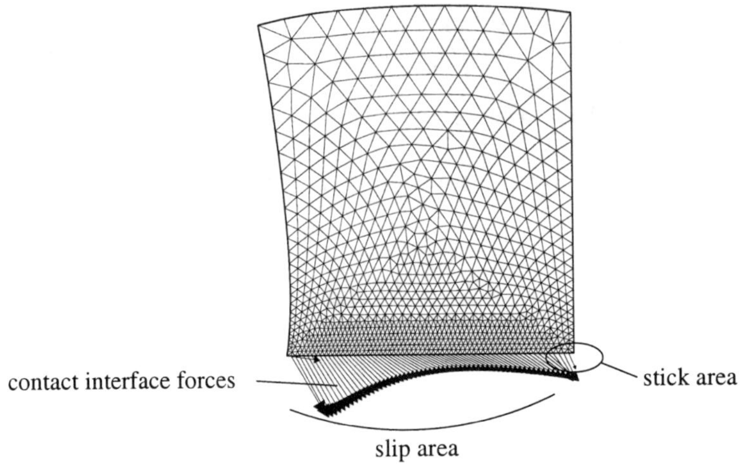

Figure 8.2: Deformed mesh (amplified by 100) and contact interface forces at $t = 1\,sec$ in the first two-dimensional example

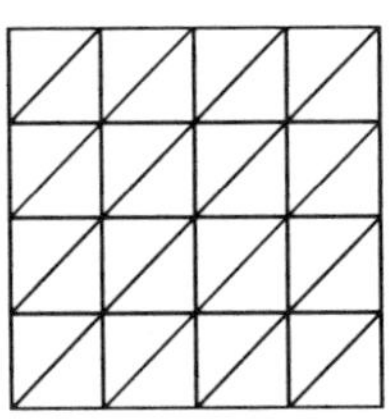

Figure 8.3: Uniform triangulation of the domain

$[0, 1]$ and uniform triangulations of the domain $[0, 4] \times [0, 4]$ of the type shown in Figure 8.3. Here the interval $[0, 4]$ is divided into $4/h$ equal parts.

We start with $h = 1/2$ and $k = 1/2$, which are successively halved. The numerical solution corresponding to $h = 1/512$ and $k = 1/512$ is taken as the "exact" solution that is used to compute the errors of the numerical solutions with larger values of h and k. Figure 8.4 provides the error values $\max_n \|\boldsymbol{u}_n - \boldsymbol{u}_n^{hk}\|_V$ for several discretization parameters of h and k. The linear asymptotic convergence behavior of the error, (8.74) with $\theta = 1$, is clearly observed.

As another two-dimensional example we consider the physical setting shown in Figure 8.5, which represents a dovetail assembly between a rigid and a deformable body, used as a locking system. Such kind of mechanical systems arise in some industrial processes, see for instance [173]. We choose a coordinate system so that the coordinates of the vertices of deformable body are: $A(-8, -8)$, $B(4, -8)$, $C(4, 0)$, $D(0, 0)$, $E(4, 4)$, $F(-8, 4)$, $G(-4, 0)$, and $H(-8, 0)$. The domain Ω represents again the cross section of a three-dimensional linearly elastic

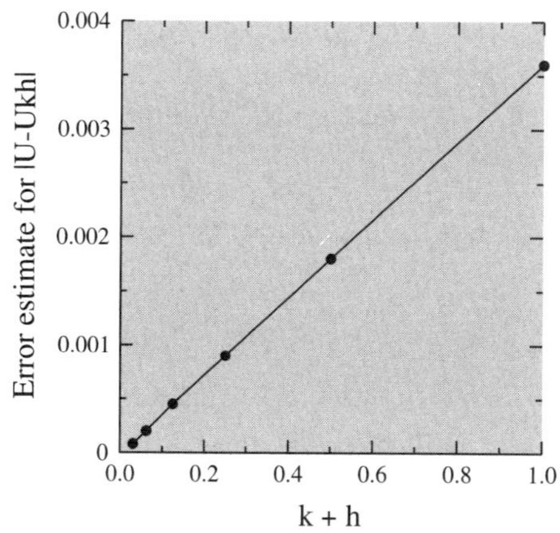

Figure 8.4: Asymptotic convergence in the first two-dimensional example

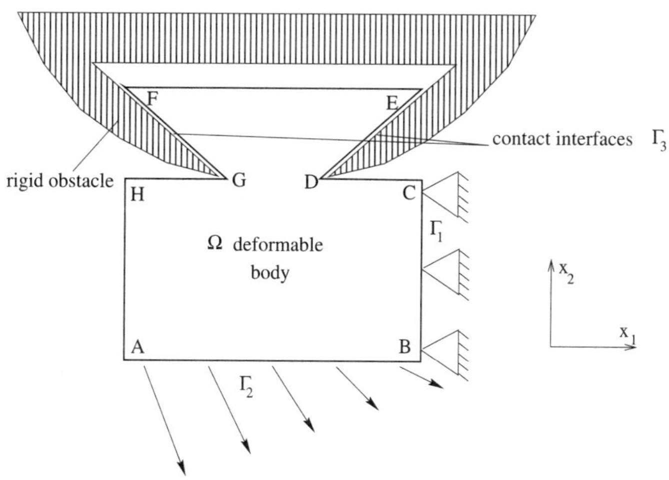

Figure 8.5: Initial configuration in the second two-dimensional example

body subjected to the action of oblique tractions on the segment AB. On the part $\Gamma_1 = BC$ the body is clamped and therefore the displacement field vanishes there. Here $\Gamma_2 = AB \cup CD \cup EF \cup GH \cup HA$ and the only nonzero traction $\boldsymbol{f}_2$ occurs on AB. No body forces are assumed to act on the elastic body during the process. The body is in bilateral frictional contact with a die on the boundary $\Gamma_3 = DE \cup FG$. The elasticity tensor is given by (8.76) and the friction is modeled again with Tresca's law.

We consider again a discretization by piecewise linear functions for the space V^h and a uniform partition of the time interval $\overline{I}$.

For computation we have used the following data:

$$E = 240\, daN/mm^2, \quad \nu = 0.3, \quad \boldsymbol{f}_0 = (0,0)\, daN/mm^2,$$
$$\boldsymbol{f}_2(x_1, x_2, t) = (0.024\, t, -0.024\,(5 - x_1)\, t)\, daN/mm^2,$$
$$g = 0.0038\, daN/mm^2, \quad \boldsymbol{u}_0 = \boldsymbol{0}\, m, \quad T = 1\, sec, \quad k = 0.1\, sec.$$

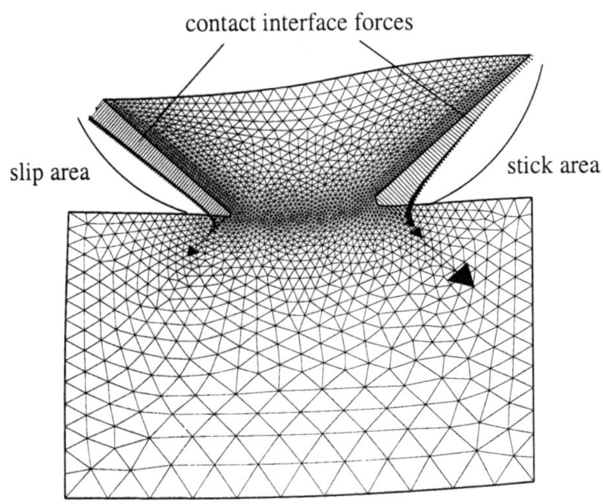

Figure 8.6: Deformed mesh (amplified by 100) and contact interface forces at $t = 1\, sec$ in the second two-dimensional example

In Figure 8.6 the deformed configuration as well as the contact interfaces forces on Γ_3 at moment $t = 1\, sec$ are plotted. The numerical computation show that on the part DE of Γ_3, the body is in the stick state during the process while on the line FG of Γ_3 the body is in the slip state.

Bibliographical Notes

The material presented in Sections 5.1 and 5.2 is standard and can be found in many books or surveys, see for instance [88, 94] and, more recently [132, 162, 211]. One is referred to [129, 130, 171, 172] for more details on the modeling aspects of contact mechanics treated in the rest of Chapter 5. Experimental background and elements of surface physics which justify some of the contact and frictional boundary conditions presented in Sections 5.4 can be found in [134, 151, 163].

More details concerning the description of the experiments presented in Section 6.1 may be found in [59, Ch. III] and the references therein. The reader is referred to [161] and [67] for more information concerning the elastic and viscoelastic constitutive laws presented in Sections 6.2 and 6.3, respectively, and to [48] for an in-depth mathematical treatment of three-dimensional elasticity. The constitutive law (6.47) was proposed in [169] in the three-dimensional case. Various results and mechanical interpretation concerning the constitutive relations for viscoplastic materials presented in Section 6.4 can be found in [58, 59] and, more recently, in [121].

The results presented in Section 7.1 on the function spaces in solid mechanics are well known and used by many authors. More details and proofs can be found in [70, 121, 134, 161, 167], for instance. The rest of Chapter 7 presents results that are common in the numerical approximation of many contact problems studied in the book; it was written based on papers [43] and [101]. Recently, application of non-conforming finite elements for an elastic plate contact problems has been analyzed in [105], where optimal order error estimates are derived. Notice that non-conforming finite elements are a natural choice in solving variational inequalities with fourth order differential operators that arise in plate contact problems.

Contact problems in elasticity have been studied by many authors. The famous Signorini problem was formulated in [191] to model the unilateral contact between an elastic material and a rigid foundation. Mathematical analysis of the Signorini problem was provided in [85] and its numerical analysis was performed in [134]. References concerning the Signorini contact problem for elastic and inelastic materials include [208, 209]. Results concerning the Signorini contact problem between two elastic bodies have been obtained in [106, 107, 108, 112] and, more recently, in [29, 30, 56, 110], where non-conforming finite element methods have been used. Various elastic contact problems were analyzed in [70] and [167] in the framework of variational inequalities.

The material in Sections 8.1 and 8.2 represents extensions of the standard results which may be found in the papers above to the case of the nonlinear elastic constitutive law (8.1). Existence of the solution to the static contact problem with nonlocal Coulomb law of friction for linearly elastic materials was obtained in [51, 61, 69]. Static problems with Coulomb's friction law was considered in [71] (the coercive case) and in [72] (the semi-coercive case).

The study of quasistatic frictional contact problems with linearly elastic materials is more recent. Dual variational formulations for quasistatic Signorini problem with friction has been considered in [210]. Existence and uniqueness results in the study of normal compliance and Coulomb friction law were obtained by [7, 8] for the case when the magnitude of the normal compliance term is restricted. This restriction was removed in [9]. A dynamic contact problem with normal compliance and regularized friction has been studied in [28] using an implicit time discretization method. The quasistatic Signorini frictional problem has been studied in [52, 53] in the case of nonlocal Coulomb law of friction and in [10, 176, 177] in the case of local Coulomb law.

Theorems 8.8 and 8.10 presented in Section 8.3 of this book were obtained in [158], as a corollary of an abstract result on evolutionary variational inequalities. The numerical analysis of the quasistatic frictional contact problem in Section 8.4 as well as the numerical examples in Section 8.5 are presented following [24].

Contact boundary problems with slip dependent coefficient of friction for elastic materials have been considered in [50, 119] in the static case and in [57] in the quasistatic case. Dynamic frictional problems for linearly elastic materials in which the coefficient of friction is assumed to depend on the slip velocity $\dot{\boldsymbol{u}}_\tau$ have been considered in [117, 118]. There it was shown that the solution to the model is not uniquely determined and presents shocks.

Recent references which provide the mathematical, mechanical and numerical state of the art for contact mechanics are the proceedings [92, 174], the surveys [93, 215] as well as the special issues [154, 188, 197]. A list of major directions in which mathematical contact mechanics will move in the near future is provided in the survey paper [189]. Optimal control of frictional contact problems, study of settings with large coefficient of friction, investigation of thermal effects, and inclusion of additional phenomena such as wear and adhesion into contact models are some of these directions.

Part III

Contact Problems in Viscoelasticity

Chapter 9

A Frictionless Contact Problem

For the problem studied in this chapter the contact is frictionless and is modeled with the well-known Signorini condition in a form with a zero gap function. We present the classical as well as the variational formulation of the problem in terms of displacement and establish the existence of a unique weak solution to the model. The proofs are based on a general result on evolution equations with maximal monotone operators. We then study a semi-discrete numerical scheme for the problem. The numerical scheme has a unique solution. We show the convergence of the scheme under the basic solution regularity. Under appropriate regularity assumptions on the solution, we also provide optimal order error estimates. A dual formulation of the contact problem in term of stress is analyzed; in particular, an existence and uniqueness result is presented.

9.1 Problem Statement

We assume the contact is frictionless and it is modeled with the Signorini contact condition, in the form with a zero gap function. The classical formulation of the frictionless contact problem of the viscoelastic body is the following:

Find a displacement field $\boldsymbol{u} : \Omega \times \overline{I} \to \mathbb{R}^d$ and a stress field $\boldsymbol{\sigma} : \Omega \times \overline{I} \to \mathbb{S}^d$ such that

$$\boldsymbol{\sigma} = \mathcal{A}\boldsymbol{\varepsilon}(\dot{\boldsymbol{u}}) + \mathcal{B}\boldsymbol{\varepsilon}(\boldsymbol{u}) \qquad \text{in } \Omega \times I, \tag{9.1}$$

$$\operatorname{Div} \boldsymbol{\sigma} + \boldsymbol{f}_0 = \boldsymbol{0} \qquad \text{in } \Omega \times I, \tag{9.2}$$

$$\boldsymbol{u} = \boldsymbol{0} \qquad \text{on } \Gamma_1 \times I, \tag{9.3}$$

$$\boldsymbol{\sigma\nu} = \boldsymbol{f}_2 \qquad \text{on } \Gamma_2 \times I, \tag{9.4}$$

$$u_\nu \le 0, \quad \sigma_\nu \le 0, \quad \sigma_\nu u_\nu = 0, \quad \boldsymbol{\sigma}_\tau = \boldsymbol{0} \qquad \text{on } \Gamma_3 \times I, \tag{9.5}$$

$$\boldsymbol{u}(0) = \boldsymbol{u}_0 \qquad \text{in } \Omega. \tag{9.6}$$

Here (9.1) represents the constitutive law of the viscoelastic material and $\boldsymbol{u}_0$ is a known initial displacement field.

To study the mechanical problem (9.1)–(9.6) we use the spaces V defined in (7.7) (page 144) and Q defined in (7.2) (page 142), and the set V_2 defined in (7.12) (page 144).

We assume that the viscosity operator $\mathcal{A} = (a_{ijkl})$ is linear, bounded, symmetric and positively definite; that is

$$\left.\begin{array}{l}\text{(a) } \mathcal{A} : \Omega \times \mathbb{S}^d \to \mathbb{S}^d.\\ \text{(b) } a_{ijkl} \in L^\infty(\Omega),\ 1 \le i,j,k,l \le d.\\ \text{(c) } \mathcal{A}\boldsymbol{\sigma} \cdot \boldsymbol{\tau} = \boldsymbol{\sigma} \cdot \mathcal{A}\boldsymbol{\tau} \quad \forall\, \boldsymbol{\sigma}, \boldsymbol{\tau} \in \mathbb{S}^d, \text{ a.e. in } \Omega.\\ \text{(d) There exists } m_{\mathcal{A}} > 0 \text{ such that}\\ \qquad \mathcal{A}\boldsymbol{\tau} \cdot \boldsymbol{\tau} \ge m_{\mathcal{A}} \|\boldsymbol{\tau}\|^2 \quad \forall\, \boldsymbol{\tau} \in \mathbb{S}^d, \text{ a.e. in } \Omega.\end{array}\right\} \tag{9.7}$$

The elasticity operator $\mathcal{B} : \Omega \times \mathbb{S}^d \to \mathbb{S}^d$ satisfies the assumption (6.35) (page 132). We also suppose that the body forces and surface tractions have the regularity

$$\boldsymbol{f}_0 \in W^{1,1}(I; H), \qquad \boldsymbol{f}_2 \in W^{1,1}(I; L^2(\Gamma_2)^d), \tag{9.8}$$

and the initial displacement satisfies

$$\boldsymbol{u}_0 \in V_2. \tag{9.9}$$

In this chapter, we use

$$(\boldsymbol{u}, \boldsymbol{v})_{\mathcal{A}} = (\mathcal{A}\boldsymbol{\varepsilon}(\boldsymbol{u}), \boldsymbol{\varepsilon}(\boldsymbol{v}))_Q, \quad \|\boldsymbol{u}\|_{\mathcal{A}} = (\boldsymbol{u}, \boldsymbol{u})_{\mathcal{A}}^{1/2}$$

as the inner product and norm on V. Notice that $(V, (\cdot,\cdot)_{\mathcal{A}})$ is a Hilbert space and $\|\cdot\|_{\mathcal{A}}$ is equivalent to the norm $\|\cdot\|_V$. The use of the inner product $(\cdot,\cdot)_{\mathcal{A}}$ simplifies some presentation of the chapter.

Next, we denote by $\boldsymbol{f}(t)$ the element of V given by

$$(\boldsymbol{f}(t), \boldsymbol{v})_{\mathcal{A}} = (\boldsymbol{f}_0(t), \boldsymbol{v})_H + (\boldsymbol{f}_2(t), \boldsymbol{v})_{L^2(\Gamma_2)^d} \quad \forall\, \boldsymbol{v} \in V,\ t \in \overline{I}, \tag{9.10}$$

and we note that conditions (9.8) imply

$$\boldsymbol{f} \in W^{1,1}(I; V). \tag{9.11}$$

It is straightforward to show that if $\boldsymbol{u}$ and $\boldsymbol{\sigma}$ are two regular functions satisfying (9.2)–(9.5) then $\boldsymbol{u}(t) \in V$, $\boldsymbol{\sigma}(t) \in Q_1$, and for any $t \in \overline{I}$,

$$\boldsymbol{u}(t) \in V_2, \quad (\boldsymbol{\sigma}(t), \boldsymbol{\varepsilon}(\boldsymbol{v}) - \boldsymbol{\varepsilon}(\boldsymbol{u}(t)))_Q \ge (\boldsymbol{f}(t), \boldsymbol{v} - \boldsymbol{u}(t))_{\mathcal{A}} \quad \forall\, \boldsymbol{v} \in V_2. \tag{9.12}$$

The inequality (9.12), together with the constitutive relation (9.1) and the initial condition (9.6), leads to the following variational problem in terms of the displacement.

Problem 9.1 *Find a displacement field $\boldsymbol{u} : \overline{I} \to V$ such that $\boldsymbol{u}(t) \in V_2$ for a.e. $t \in I$, and*

$$(\mathcal{A}\boldsymbol{\varepsilon}(\dot{\boldsymbol{u}}(t)), \boldsymbol{\varepsilon}(\boldsymbol{v}) - \boldsymbol{\varepsilon}(\boldsymbol{u}(t)))_Q + (\mathcal{B}\boldsymbol{\varepsilon}(\boldsymbol{u}(t)), \boldsymbol{\varepsilon}(\boldsymbol{v}) - \boldsymbol{\varepsilon}(\boldsymbol{u}(t)))_Q$$
$$\ge (\boldsymbol{f}(t), \boldsymbol{v} - \boldsymbol{u}(t))_{\mathcal{A}} \quad \forall\, \boldsymbol{v} \in V_2, \text{ a.e. } t \in I, \tag{9.13}$$
$$\boldsymbol{u}(0) = \boldsymbol{u}_0. \tag{9.14}$$

Note that Problem 9.1 is formally equivalent to the mechanical problem (9.1)–(9.6). Indeed, if $\boldsymbol{u}$ represents a regular solution of the variational problem 9.1 and $\boldsymbol{\sigma}$ is given by (9.1), using the arguments of [134], it follows that $(\boldsymbol{u}, \boldsymbol{\sigma})$ satisfies (9.1)–(9.6).

To prove an existence result for Problem 9.1, we will need an abstract result. Assume X is a real Hilbert space with the inner product $(\cdot,\cdot)_X$. Let $A : D(A) \subset X \to 2^X$ be a multivalued operator. Here, $D(A)$ denotes the domain of the multivalued operator A, and 2^X represents the set of the subsets of X. We say that the operator A is *monotone* if

$$(u_1 - u_2, w_1 - w_2)_X \geq 0 \quad \forall\, w_1 \in Au_1,\ w_2 \in Au_2,\ \forall\, u_1, u_2 \in D(A).$$

We see that this definition extends the definition of monotonicity of a (single-valued) operator. We say that the multivalued operator A is *maximal monotone* if there is no monotone multivalued operator $B : D(B) \subset X \to 2^X$ that is a proper extension of A. It can be shown that if $\phi : X \to \overline{\mathbb{R}}$ is a proper, convex, l.s.c. function, then its subdifferential $\partial\phi$ is a maximal monotone operator. It can also be shown that if $A_1 : D(A_1) \subset X \to 2^X$ is a maximal monotone operator and $A_2 : X \to X$ is a single-valued monotone continuous operator, then $A_1 + A_2$ is a maximal monotone operator. Proofs of these results as well as the next result can be found in [27, Chapter 1].

Theorem 9.2 *Let X be a real Hilbert space and denote $I_X : X \to X$ the identity operator. If $A : D(A) \subset X \to 2^X$ is a multivalued operator such that the operator $A + \omega I_X$ is maximal monotone for some real ω, then for any $f \in W^{1,1}(I;X)$ and $u_0 \in D(A)$, there exists a unique function $u \in W^{1,\infty}(I;X)$ which satisfies*

$$\dot{u}(t) + Au(t) \ni f(t) \qquad \text{a.e. } t \in I, \tag{9.15}$$

$$u(0) = u_0. \tag{9.16}$$

9.2 An Existence and Uniqueness Result

The main results of this section concerns the unique solvability of the variational problem 9.1.

Theorem 9.3 *Assume* (9.7)–(9.9) *and* (6.35). *Then there exists a unique solution $\boldsymbol{u}$ to Problem 9.1. Moreover, $\boldsymbol{u} \in W^{1,\infty}(I;V)$.*

Proof. By the Riesz representation theorem (page 15) we can define an operator $B : V \to V$ by

$$(B\boldsymbol{u}, \boldsymbol{v})_{\mathcal{A}} = (\mathcal{B}\boldsymbol{\varepsilon}(\boldsymbol{u}), \boldsymbol{\varepsilon}(\boldsymbol{v}))_Q \quad \forall\, \boldsymbol{u}, \boldsymbol{v} \in V.$$

It follows from the assumptions (6.35) on $\mathcal{B}$ and the assumptions (9.7) on $\mathcal{A}$ that

$$\|B\boldsymbol{u}_1 - B\boldsymbol{u}_2\|_{\mathcal{A}} \leq \frac{L_{\mathcal{B}}}{m_{\mathcal{A}}} \|\boldsymbol{u}_1 - \boldsymbol{u}_2\|_{\mathcal{A}} \quad \forall\, \boldsymbol{u}_1,\, \boldsymbol{u}_2 \in V,$$

i.e. B is a Lipschitz continuous operator. Moreover, the operator

$$B + \frac{L_{\mathcal{B}}}{m_{\mathcal{A}}} I : V \to V$$

is a monotone Lipschitz continuous operator on V. Let

$$\psi_{V_2} : V \to (-\infty, +\infty]$$

denote the indicator function of the set V_2 and let $\partial\psi_{V_2}$ be the subdifferential of ψ_{V_2}. Since V_2 is a nonempty convex closed set of V, it follows that $\partial\psi_{V_2}$ is a maximal monotone operator on V and $D(\partial\psi_{V_2}) = V_2$. Moreover, the sum

$$\partial\psi_{V_2} + B + \frac{L_{\mathcal{B}}}{m_{\mathcal{A}}} I : V_2 \subset V \to 2^V$$

is a maximal monotone operator. Thus, conditions (9.11) and (9.9) allow us to apply Theorem 9.2 with $X = V$ endowed with the inner product $(\cdot,\cdot)_{\mathcal{A}}$, $A = \partial\psi_{V_2} + B : D(A) = V_2 \subset V \to 2^V$ and $\omega = \mathcal{L}_{\mathcal{B}}/m_{\mathcal{A}}$. We deduce that there exists a unique element $\boldsymbol{u} \in W^{1,\infty}(I;V)$ such that

$$\dot{\boldsymbol{u}}(t) + \partial\psi_{V_2}(\boldsymbol{u}(t)) + B\boldsymbol{u}(t) \ni \boldsymbol{f}(t) \quad \text{a.e. } t \in I, \tag{9.17}$$

$$\boldsymbol{u}(0) = \boldsymbol{u}_0. \tag{9.18}$$

Observe that for any $\boldsymbol{u}, \boldsymbol{g} \in V$, the following equivalence holds:

$$\boldsymbol{g} \in \partial\psi_{V_2}(\boldsymbol{u}) \quad \Longleftrightarrow \quad \boldsymbol{u} \in V_2, \quad (\boldsymbol{g}, \boldsymbol{v} - \boldsymbol{u})_{\mathcal{A}} \le 0 \quad \forall\, \boldsymbol{v} \in V_2.$$

So the differential inclusion (9.17) is equivalent to $\boldsymbol{u}(t) \in V_2$ and the following variational inequality:

$$(\dot{\boldsymbol{u}}(t), \boldsymbol{v} - \boldsymbol{u}(t))_{\mathcal{A}} + (B\boldsymbol{u}(t), \boldsymbol{v} - \boldsymbol{u}(t))_{\mathcal{A}} \ge (\boldsymbol{f}(t), \boldsymbol{v} - \boldsymbol{u}(t))_{\mathcal{A}} \quad \forall\, \boldsymbol{v} \in V_2$$

for a.e. $t \in I$. It follows that $\boldsymbol{u}$ satisfies $\boldsymbol{u}(t) \in V_2$ and the inequality

$$\begin{aligned}(\mathcal{A}\boldsymbol{\varepsilon}(\dot{\boldsymbol{u}}(t)), \boldsymbol{\varepsilon}(\boldsymbol{v}) - \boldsymbol{\varepsilon}(\boldsymbol{u}(t)))_Q + (\mathcal{B}\boldsymbol{\varepsilon}(\boldsymbol{u}(t)), \boldsymbol{\varepsilon}(\boldsymbol{v}) - \boldsymbol{\varepsilon}(\boldsymbol{u}(t)))_Q \\ \ge (\boldsymbol{f}(t), \boldsymbol{v} - \boldsymbol{u}(t))_{\mathcal{A}} \quad \forall\, \boldsymbol{v} \in V_2\end{aligned} \tag{9.19}$$

for a.e. $t \in I$. Therefore from (9.19) and (9.18) we conclude the existence part in Theorem 9.3. The uniqueness part follows from the uniqueness of the element $\boldsymbol{u} \in W^{1,\infty}(I;V)$ which satisfies (9.17), (9.18), guaranteed by Theorem 9.2. ■

Let now $\boldsymbol{u}$ be the solution of Problem 9.1 and define $\boldsymbol{\sigma}$ by (9.1). Since $\boldsymbol{u} \in W^{1,\infty}(I;V)$ and assumptions (9.7) on $\mathcal{A}$ and (6.35) on $\mathcal{B}$ hold, we have $\boldsymbol{\sigma} \in L^\infty(I;Q)$. Taking $\boldsymbol{v} = \boldsymbol{u}(t) \pm \boldsymbol{\varphi}$ in (9.13), where $\boldsymbol{\varphi} \in C_0^\infty(\Omega)^d$ is arbitrary, and using the definition (9.10) for $\boldsymbol{f}(t)$ we find

$$\operatorname{Div} \boldsymbol{\sigma}(t) + \boldsymbol{f}_0(t) = \boldsymbol{0} \quad \text{a.e. } t \in I. \tag{9.20}$$

With the regularity assumption (9.8) on $\boldsymbol{f}_0$, we see that $\operatorname{Div} \boldsymbol{\sigma} \in L^\infty(I;H)$. Therefore, $\boldsymbol{\sigma} \in L^\infty(I;Q_1)$. A pair of functions $(\boldsymbol{u}, \boldsymbol{\sigma})$ which satisfies (9.1), (9.13) and (9.14) is called a *weak solution* of the problem (9.1)–(9.6). We conclude by Theorem 9.3 that the problem (9.1)–(9.6) has a unique weak solution.

By an argument similar to that used in Section 8.1, under the assumption

$$\boldsymbol{\sigma}\boldsymbol{\nu} \in L^\infty(I; L^2(\Gamma)^d), \tag{9.21}$$

we have

$$\boldsymbol{\sigma}\boldsymbol{\nu} = \boldsymbol{f}_2 \quad \text{a.e. on } \Gamma_2 \times I, \tag{9.22}$$

$$\boldsymbol{\sigma}_\tau = \boldsymbol{0} \quad \text{a.e. on } \Gamma_3 \times I. \tag{9.23}$$

These relations will be needed in error estimation of numerical solutions.

9.3 Numerical Approximations

We first consider a semi-discrtete approximation scheme for Problem 9.1 by discretizing only the spatial domain.

As in Section 7.2, we assume Ω is a polygon or polyhedron, and let $\{\mathcal{T}^h\}$ be a regular family of finite element partitions of $\overline{\Omega}$ compatible to the boundary subsets $\overline{\Gamma}_1$, $\overline{\Gamma}_2$ and $\overline{\Gamma}_3$. Let $V^h \subset V$ be the finite element space of linear elements corresponding to the partition $\mathcal{T}^h$. Then $V_2^h = V^h \cap V_2$ consists of continuous piecewise linear functions that vanish at the nodes on $\overline{\Gamma}_1$ and whose normal components are non-positive at the nodes on $\overline{\Gamma}_3$. Notice that V_2^h is a nonempty, closed, convex subset of V^h. Let $\boldsymbol{u}_0^h \in V_2^h$ be an approximation of $\boldsymbol{u}_0$. Then a spatially semi-discrete scheme of Problem 9.1 is:

Problem 9.4 *Find a displacement field* $\boldsymbol{u}^h : \overline{I} \to V^h$ *such that*

$$\boldsymbol{u}^h(0) = \boldsymbol{u}_0^h \tag{9.24}$$

and for a.e. $t \in I$, $\boldsymbol{u}^h(t) \in V_2^h$ *and*

$$\begin{aligned}(\mathcal{A}\varepsilon(\dot{\boldsymbol{u}}^h(t)), \varepsilon(\boldsymbol{v}^h) - \varepsilon(\boldsymbol{u}^h(t)))_Q + (\mathcal{B}\varepsilon(\boldsymbol{u}^h(t)), \varepsilon(\boldsymbol{v}^h) - \varepsilon(\boldsymbol{u}^h(t)))_Q \\ \geq (\boldsymbol{f}(t), \boldsymbol{v}^h - \boldsymbol{u}^h(t))_{\mathcal{A}} \quad \forall\, \boldsymbol{v}^h \in V_2^h.\end{aligned} \tag{9.25}$$

Let us first show the existence of a unique solution to Problem 9.4 by an argument similar to the proof of Theorem 9.3.

Theorem 9.5 *Assume* (9.7), (9.8) *and* (6.35). *Then there exists a unique solution* $\boldsymbol{u}^h \in W^{1,\infty}(I; V^h)$ *to Problem 9.4.*

Proof. We apply Theorem 9.2 for $X = V^h$ with the inner product $(\cdot,\cdot)_{\mathcal{A}}$. Denote $I^h : V^h \to V^h$ the identity operator on V^h. We define an operator $B^h : V^h \to V^h$ by

$$(B^h \boldsymbol{u}^h, \boldsymbol{v}^h)_{\mathcal{A}} = (\mathcal{B}\varepsilon(\boldsymbol{u}^h), \varepsilon(\boldsymbol{v}^h))_Q \quad \forall\, \boldsymbol{u}^h, \boldsymbol{v}^h \in V^h.$$

Then as in the proof of Theorem 9.3,

$$\|B^h\boldsymbol{u}_1^h - B^h\boldsymbol{u}_2^h\|_{\mathcal{A}} \le \frac{L_{\mathcal{B}}}{m_{\mathcal{A}}}\,\|\boldsymbol{u}_1^h - \boldsymbol{u}_2^h\|_{\mathcal{A}} \quad \forall\,\boldsymbol{u}_1^h, \boldsymbol{u}_2^h \in V^h.$$

So the operator

$$B^h + \frac{L_{\mathcal{B}}}{m_{\mathcal{A}}}\,I^h : V^h \to V^h$$

is monotone and Lipschitz continuous. For the indicator function $\psi^h_{V_2^h} : V^h \to (-\infty, \infty]$ defined by

$$\psi^h_{V_2^h}(\boldsymbol{v}^h) = \begin{cases} 0 & \text{if } \boldsymbol{v}^h \in V_2^h, \\ \infty & \text{otherwise,} \end{cases}$$

we denote its (discrete) subdifferential

$$\partial^h\psi^h_{V_2^h}(\boldsymbol{u}^h) = \left\{ \boldsymbol{g}^h \in V^h : (\boldsymbol{g}^h, \boldsymbol{v}^h - \boldsymbol{u}^h)_{\mathcal{A}} \le 0 \;\forall\,\boldsymbol{v}^h \in V_2^h \right\}$$

for any $\boldsymbol{u}^h \in V_2^h$. Like the subdifferential $\partial\psi_{V_2}$, $\partial^h\psi^h_{V_2^h}$ is a maximal monotone operator on V^h and the effective domain of $\partial^h\psi^h_{V_2^h}$ is V_2^h. We then define $\boldsymbol{f}^h : \overline{I} \to V^h$ by

$$(\boldsymbol{f}^h(t), \boldsymbol{v}^h)_{\mathcal{A}} = (\boldsymbol{f}(t), \boldsymbol{v}^h)_{\mathcal{A}} \quad \forall\,\boldsymbol{v}^h \in V^h. \tag{9.26}$$

We have the regularity $\boldsymbol{f}^h \in W^{1,1}(I; V^h)$. Applying Theorem 9.2 with $X = V^h$, $A^h = \partial^h\psi^h_{V_2^h} + B^h$, and $\boldsymbol{f}^h$ defined in (9.26), we obtain the existence of a unique $\boldsymbol{u}^h \in W^{1,\infty}(I; V^h)$ such that (9.24) and

$$\dot{\boldsymbol{u}}^h(t) + \partial^h\psi^h_{V_2^h}(\boldsymbol{u}^h(t)) + B^h\boldsymbol{u}^h(t) \ni \boldsymbol{f}^h(t) \quad \text{a.e. } t \in I \tag{9.27}$$

hold. It is easy to see that (9.27) is equivalent to $\boldsymbol{u}^h(t) \in V_2^h$ and the inequality (9.25) for a.e. $t \in I$. ■

Now we turn to an error analysis of the method. We take $\boldsymbol{v} = \boldsymbol{u}^h(s)$ in (9.13) and add it to (9.25) with $\boldsymbol{v}^h = \boldsymbol{v}^h(s) \in V_2^h$. After some rearrangement of the terms, we obtain

$$\begin{aligned}
&(\mathcal{A}\varepsilon(\dot{\boldsymbol{u}}(s)) - \mathcal{A}\varepsilon(\dot{\boldsymbol{u}}^h(s)), \varepsilon(\boldsymbol{u}(s) - \boldsymbol{u}^h(s)))_Q \\
&\quad \le (\mathcal{B}\varepsilon(\boldsymbol{u}(s)) - \mathcal{B}\varepsilon(\boldsymbol{u}^h(s)), \varepsilon(\boldsymbol{u}^h(s) - \boldsymbol{u}(s)))_Q \\
&\qquad + (\mathcal{A}\varepsilon(\dot{\boldsymbol{u}}^h(s)), \varepsilon(\boldsymbol{v}^h - \boldsymbol{u}(s)))_Q + (\mathcal{B}\varepsilon(\boldsymbol{u}^h(s)), \varepsilon(\boldsymbol{v}^h - \boldsymbol{u}(s)))_Q \\
&\qquad - (\boldsymbol{f}(s), \boldsymbol{v}^h - \boldsymbol{u}^h(s))_V \quad \text{a.e. } s \in I.
\end{aligned}$$

Then

$$\begin{aligned}
\frac{1}{2}\frac{d}{ds}\,\|\boldsymbol{u}(s) - \boldsymbol{u}^h(s)\|_{\mathcal{A}}^2 &\le (\mathcal{A}\varepsilon(\dot{\boldsymbol{u}}^h(s) - \dot{\boldsymbol{u}}(s)), \varepsilon(\boldsymbol{v}^h - \boldsymbol{u}(s)))_Q \\
&\quad + (\mathcal{B}\varepsilon(\boldsymbol{u}^h(s)) - \mathcal{B}\varepsilon(\boldsymbol{u}(s)), \varepsilon(\boldsymbol{v}^h - \boldsymbol{u}(s)))_Q \\
&\quad + (\mathcal{B}\varepsilon(\boldsymbol{u}(s)) - \mathcal{B}\varepsilon(\boldsymbol{u}^h(s)), \varepsilon(\boldsymbol{u}^h(s) - \boldsymbol{u}(s)))_Q \\
&\quad + R(s; \boldsymbol{v}^h - \boldsymbol{u}(s)) \quad \text{a.e. } s \in I,
\end{aligned} \tag{9.28}$$

where

$$R(s;\boldsymbol{v}) = (\mathcal{A}\boldsymbol{\varepsilon}(\dot{\boldsymbol{u}}(s)), \boldsymbol{\varepsilon}(\boldsymbol{v}))_Q + (\mathcal{B}\boldsymbol{\varepsilon}(\boldsymbol{u}(s)), \boldsymbol{\varepsilon}(\boldsymbol{v}))_Q - (\boldsymbol{f}(s), \boldsymbol{v})_V \tag{9.29}$$

for all $\boldsymbol{v} \in V$, a.e. $s \in I$.

Let $t \in \overline{I}$. Integrate the inequality (9.28) from 0 to t and use the initial conditions (9.14) and (9.24) to find

$$\begin{aligned}
&\frac{1}{2}\left[\|\boldsymbol{u}(t) - \boldsymbol{u}^h(t)\|_{\mathcal{A}}^2 - \|\boldsymbol{u}_0 - \boldsymbol{u}_0^h\|_{\mathcal{A}}^2\right] \\
&\qquad \le \int_0^t (\mathcal{A}\boldsymbol{\varepsilon}(\dot{\boldsymbol{u}}^h(s) - \dot{\boldsymbol{u}}(s)), \boldsymbol{\varepsilon}(\boldsymbol{v}^h - \boldsymbol{u}(s)))_Q ds \\
&\qquad\quad + \int_0^t L_{\mathcal{B}}\|\boldsymbol{u}(s) - \boldsymbol{u}^h(s)\|_V \|\boldsymbol{u}(s) - \boldsymbol{v}^h\|_V ds \\
&\qquad\quad + \int_0^t L_{\mathcal{B}}\|\boldsymbol{u}(s) - \boldsymbol{u}^h(s)\|_V^2 dt + \int_0^t \left|R(s;\boldsymbol{v}^h - \boldsymbol{u}(s))\right| ds.
\end{aligned}$$

We perform an integration by parts on the first term in the right-hand side:

$$\begin{aligned}
&\int_0^t (\mathcal{A}\boldsymbol{\varepsilon}(\dot{\boldsymbol{u}}^h(s) - \dot{\boldsymbol{u}}(s)), \boldsymbol{\varepsilon}(\boldsymbol{v}^h - \boldsymbol{u}(s)))_Q ds \\
&\qquad = (\mathcal{A}\boldsymbol{\varepsilon}(\boldsymbol{u}(t) - \boldsymbol{u}^h(t)), \boldsymbol{\varepsilon}(\boldsymbol{u}(t) - \boldsymbol{v}^h(t)))_Q \\
&\qquad\quad - (\mathcal{A}\boldsymbol{\varepsilon}(\boldsymbol{u}_0 - \boldsymbol{u}_0^h), \boldsymbol{\varepsilon}(\boldsymbol{u}_0 - \boldsymbol{v}_0^h))_Q \\
&\qquad\quad - \int_0^t (\mathcal{A}\boldsymbol{\varepsilon}(\boldsymbol{u}^h(s) - \boldsymbol{u}(s)), \boldsymbol{\varepsilon}(\dot{\boldsymbol{v}}^h - \dot{\boldsymbol{u}}(s)))_Q ds.
\end{aligned}$$

Then by the equivalence of the norms $\|\cdot\|_V$ and $\|\cdot\|_{\mathcal{A}}$, and assumptions on $\mathcal{A}$, we have

$$\begin{aligned}
\|\boldsymbol{u}(t) - \boldsymbol{u}^h(t)\|_V^2 &\le c\left(\|\boldsymbol{u}(t) - \boldsymbol{v}^h(t)\|_V^2 + \|\boldsymbol{u}_0 - \boldsymbol{u}_0^h\|_V^2 + \|\boldsymbol{u}_0 - \boldsymbol{v}_0^h\|_V^2\right) \\
&\quad + c\int_0^t \|\boldsymbol{u}(s) - \boldsymbol{u}^h(s)\|_V^2 ds + c\int_0^t \|\boldsymbol{u}(s) - \boldsymbol{v}^h(s)\|_V^2 ds \\
&\quad + c\int_0^t \|\dot{\boldsymbol{u}}(s) - \dot{\boldsymbol{v}}^h(s)\|_V^2 ds \\
&\quad + c\int_0^t \left|R(s;\boldsymbol{v}^h(s) - \boldsymbol{u}(s))\right| ds.
\end{aligned}$$

Since

$$\boldsymbol{u}(t) - \boldsymbol{v}^h(t) = \boldsymbol{u}_0 - \boldsymbol{v}_0^h + \int_0^t \left(\dot{\boldsymbol{u}}(s) - \dot{\boldsymbol{v}}^h(s)\right) ds,$$

we have

$$\|\boldsymbol{u}(t) - \boldsymbol{v}^h(t)\|_V^2 \le c\,\|\boldsymbol{u}_0 - \boldsymbol{v}_0^h\|_V^2 + c\int_0^t \|\dot{\boldsymbol{u}}(s) - \dot{\boldsymbol{v}}^h(s)\|_V^2 ds$$

and

$$\int_0^t \|\boldsymbol{u}(s) - \boldsymbol{v}^h(s)\|_V^2 ds \le c\,\|\boldsymbol{u}_0 - \boldsymbol{v}_0^h\|_V^2 + c\int_0^t \|\dot{\boldsymbol{u}}(s) - \dot{\boldsymbol{v}}^h(s)\|_V^2 ds.$$

Thus,

$$\begin{aligned}\|\boldsymbol{u}(t) - \boldsymbol{u}^h(t)\|_V^2 &\le c\left(\|\boldsymbol{u}_0 - \boldsymbol{u}_0^h\|_V^2 + \|\boldsymbol{u}_0 - \boldsymbol{v}_0^h\|_V^2\right)\\ &\quad + c\int_0^t \|\boldsymbol{u}(s) - \boldsymbol{u}^h(s)\|_V^2 ds + c\int_0^t \|\dot{\boldsymbol{u}}(s) - \dot{\boldsymbol{v}}^h(s)\|_V^2 ds\\ &\quad + c\int_0^t |R(s; \boldsymbol{v}^h(s) - \boldsymbol{u}(s))|\, ds.\end{aligned}$$

Applying Gronwall's inequality (Lemma 7.24), we obtain

$$\begin{aligned}\|\boldsymbol{u} - \boldsymbol{u}^h\|_{L^\infty(I;V)}^2 &\le c\Big(\|\boldsymbol{u}_0 - \boldsymbol{v}_0^h\|_V^2 + \|\dot{\boldsymbol{u}} - \dot{\boldsymbol{v}}^h\|_{L^2(I;V)}^2\\ &\quad + \|R(\cdot; \boldsymbol{v}^h - \boldsymbol{u})\|_{L^1(I;V)} + \|\boldsymbol{u}_0 - \boldsymbol{u}_0^h\|_V^2\Big).\end{aligned}$$

Since $\boldsymbol{v}^h \in H^1(I; V_2^h)$ is arbitrary, we then have

$$\begin{aligned}\|\boldsymbol{u} - \boldsymbol{u}^h\|_{L^\infty(I;V)} &\le c \inf_{\boldsymbol{v}^h \in H^1(I;V_2^h)} \Big(\|\boldsymbol{u}_0 - \boldsymbol{v}_0^h\|_V + \|\dot{\boldsymbol{u}} - \dot{\boldsymbol{v}}^h\|_{L^2(I;V)}\\ &\quad + \|R(\cdot; \boldsymbol{v}^h - \boldsymbol{u})\|_{L^1(I;V)}^{1/2}\Big) + c\,\|\boldsymbol{u}_0 - \boldsymbol{u}_0^h\|_V. \qquad (9.30)\end{aligned}$$

The inequality (9.30) is a basis for convergence analysis and error estimation. We have the following convergence result under the basic solution regularity.

Proposition 9.6 *Assume the conditions of Theorem 7.11 (page 150). If we choose $\boldsymbol{u}_0^h \in V_2^h$ such that*

$$\|\boldsymbol{u}_0 - \boldsymbol{u}_0^h\|_V \to 0 \quad \text{as } h \to 0, \qquad (9.31)$$

then the numerical method converges, that is

$$\|\boldsymbol{u} - \boldsymbol{u}^h\|_{L^\infty(I;V)} \to 0 \quad \text{as } h \to 0. \qquad (9.32)$$

Proof. From the definition (9.29), we immediately get

$$|R(s; \boldsymbol{v})| \le c\left(\|\boldsymbol{u}\|_{W^{1,\infty}(I;V)} + \|\boldsymbol{f}\|_{L^\infty(I;V)}\right) \|\boldsymbol{v}\|_V \quad \forall\, \boldsymbol{v} \in V, \text{ a.e. } s \in I.$$

Thus, we derive from (9.30) that

$$\begin{aligned}\|\boldsymbol{u} - \boldsymbol{u}^h\|_{L^\infty(I;V)} &\le c \inf_{\boldsymbol{v}^h \in H^1(I;V_2^h)} \Big(\|\boldsymbol{u}_0 - \boldsymbol{v}_0^h\|_V + \|\dot{\boldsymbol{u}} - \dot{\boldsymbol{v}}^h\|_{L^2(I;V)}\\ &\quad + \|\boldsymbol{u} - \boldsymbol{v}^h\|_{L^1(I;V)}^{1/2}\Big) + c\,\|\boldsymbol{u}_0 - \boldsymbol{u}_0^h\|_V. \qquad (9.33)\end{aligned}$$

By Theorem 7.11, $V_2 \cap C^\infty(\overline{\Omega})^d$ is dense in V_2 with respect to the norm of V. Then $H^1(I; V_2 \cap C^\infty(\overline{\Omega})^d)$ is dense in $H^1(I; V_2)$ with respect to the norm of $H^1(I; V)$. Based on the inequality (9.33), an argument similar to the proof of Theorem 7.18 (page 157) leads to the convergence (9.32). ∎

We remark that the assumption (9.31) is satisfied if $\boldsymbol{u}_0^h = \Pi^h \boldsymbol{u}_0$ when $\boldsymbol{u}_0 \in C(\overline{\Omega})^d$.

Now, we provide an optimal error estimates result under additional regularity on the solution.

Proposition 9.7 *Assume the additional solution regularities* (9.21) *and*

$$\boldsymbol{u} \in H^1(I; H^2(\Omega)^d), \quad u_\nu|_{\Gamma_{3,i}} \in L^1(I; H^2(\Gamma_{3,i})), \ 1 \le i \le i_0.$$

Then, choosing $\boldsymbol{u}_0^h = \Pi^h \boldsymbol{u}_0$, *we have the following optimal order error estimate:*

$$\|\boldsymbol{u} - \boldsymbol{u}^h\|_{L^\infty(I;V)} = O(h). \tag{9.34}$$

Proof. We first derive a sharper bound on the residual term under the assumption (9.21). Let $\boldsymbol{v} \in V$. The equalities and inequalities below involving the argument t are understood to be valid for almost any $t \in I$. From its definition (9.29) we obtain

$$R(t; \boldsymbol{v}) = (\boldsymbol{\sigma}(t), \boldsymbol{\varepsilon}(\boldsymbol{v}))_Q - (\boldsymbol{f}(t), \boldsymbol{v})_V \quad \forall\, \boldsymbol{v} \in V, \text{ a.e. } t \in I.$$

Performing an integration by parts on the first term and using the definition (9.10) of $\boldsymbol{f}(t)$, we obtain

$$\begin{aligned} R(t; \boldsymbol{v}) = & \int_\Gamma \boldsymbol{\sigma}(t)\boldsymbol{\nu} \cdot \boldsymbol{v}\, da - \int_\Omega \operatorname{Div} \boldsymbol{\sigma}(t) \cdot \boldsymbol{v}\, dx \\ & - \int_\Omega \boldsymbol{f}_0(t) \cdot \boldsymbol{v}\, dx - \int_{\Gamma_2} \boldsymbol{f}_2(t) \cdot \boldsymbol{v}\, da. \end{aligned}$$

Use now the pointwise relations (9.20), (9.22) and (9.23) of the weak solution,

$$R(t; \boldsymbol{v}) = \int_{\Gamma_3} \sigma_\nu(t)\, v_\nu\, da.$$

Then

$$|R(t; \boldsymbol{v})| \le c\, \|\sigma_\nu\|_{L^\infty(I;L^2(\Gamma))} \|v_\nu\|_{L^2(\Gamma_3)}.$$

By (9.30) we get

$$\begin{aligned} \|\boldsymbol{u} - \boldsymbol{u}^h\|_{L^\infty(I;V)} \le c & \inf_{\boldsymbol{v}^h \in H^1(I;V_2^h)} \Big(\|\boldsymbol{u}_0 - \boldsymbol{v}_0^h\|_V + \|\dot{\boldsymbol{u}} - \dot{\boldsymbol{v}}^h\|_{L^2(I;V)} \\ & + \|u_\nu - v_\nu^h\|_{L^1(I;L^2(\Gamma_3))}^{1/2} \Big) + c\, \|\boldsymbol{u}_0 - \Pi^h \boldsymbol{u}_0\|_V. \end{aligned} \tag{9.35}$$

For any $t \in \overline{I}$, let $\boldsymbol{v}^h(t) = \Pi^h \boldsymbol{u}(t) \in H^1(I; V_2^h)$ be the finite element interpolant of $\boldsymbol{u}(t)$. Then $\dot{\boldsymbol{v}}^h(t) = \Pi^h \dot{\boldsymbol{u}}(t)$ and $v_\nu^h(t) = \Pi^h u_\nu(t)$. Applying the finite element interpolation error estimates (7.34) and (7.38) in (9.35), we immediately get the error estimate (9.34). ∎

We now consider a fully discrete scheme for solving Problem 9.1. In addition to the finite element set V_2^h, we also need the time interval partition discussed in Section 7.2. Then the fully discrete scheme is

Problem 9.8 *Find a displacement field* $\boldsymbol{u}^{hk} = \{\boldsymbol{u}_n^{hk}\}_{n=0}^N \subset V_2^h$ *such that*

$$\boldsymbol{u}_0^{hk} = \boldsymbol{u}_0^h$$

and for $n = 1, \ldots, N$,

$$\begin{aligned}(\mathcal{A}\varepsilon(\delta_n \boldsymbol{u}_n^{hk}), \varepsilon(\boldsymbol{v}^h) - \varepsilon(\boldsymbol{u}_n^{hk}))_Q + (\mathcal{B}\varepsilon(\boldsymbol{u}_{n-1}^{hk}), \varepsilon(\boldsymbol{v}^h) - \varepsilon(\boldsymbol{u}_n^{hk}))_Q \\ \geq (\boldsymbol{f}(t), \boldsymbol{v}^h - \boldsymbol{u}_n^{hk})_A \quad \forall\, \boldsymbol{v}^h \in V_2^h.\end{aligned}$$

The fully discrete scheme has a unique solution. This can be proved directly without the need to apply Theorem 9.2. However, derivation of error estimates for the fully discrete solution remains an open problem.

9.4 Dual Formulation

Problem 9.1 on page 194 represents a variational formulation of the frictional contact problem (9.1)–(9.6) in terms of the displacement and may be referred to as the *primal formulation* of this mechanical problem. Our interest in this section lies in obtaining and analyzing a *dual formulation* in terms of the stress. In many engineering applications, knowledge of the contact stress is more important than that of the displacement.

We assume in this section that (9.7)–(9.9) and the assumptions (6.35) on $\mathcal{B}$ hold. To express the mechanical problem in terms of the stress we need the following result, which is one variant of the well-known Cauchy–Lipschitz theorem.

Theorem 9.9 *Assume* $(X, \|\cdot\|_X)$ *is a real Banach space. Let* $F(t,\cdot) : X \to X$ *be an operator defined a.e. on* I *satisfying the following conditions:*

- $\exists\, L_F > 0$ *such that* $\|F(t,x) - F(t,y)\|_X \leq L_F \|x - y\|_X \;\forall\, x, y \in X$, *a.e.* $t \in I$.
- $\exists\, p \geq 1$ *such that* $t \mapsto F(t,x) \in L^p(I;X)\;\forall\, x \in X$.

Then for any $x_0 \in X$ *there exists a unique function* $x \in W^{1,p}(I;X)$ *such that*

$$\begin{aligned}\dot{x}(t) &= F(t, x(t)) \quad \text{a.e. } t \in I, \\ x(0) &= x_0.\end{aligned}$$

A proof of Theorem 9.9 can be find in [205, p. 60]. We use this theorem with $X = Q$ and $p = \infty$ to deduce that for any $\boldsymbol{\sigma} \in L^\infty(I; Q)$ there exists a unique function $\boldsymbol{\tau} \in W^{1,\infty}(I; Q)$ such that

$$\mathcal{A}\dot{\boldsymbol{\tau}}(t) + \mathcal{B}\boldsymbol{\tau}(t) = \boldsymbol{\sigma}(t) \quad \text{a.e. } t \in I, \tag{9.36}$$

$$\boldsymbol{\tau}(0) = \boldsymbol{\varepsilon}(\boldsymbol{u}_0). \tag{9.37}$$

This result allows us to define an operator $\mathcal{T} : L^\infty(I; Q) \to W^{1,\infty}(I; Q)$ by the relation

$$\mathcal{T}\boldsymbol{\sigma} = \boldsymbol{\tau}, \tag{9.38}$$

where $\boldsymbol{\tau}$ is the unique solution of the Cauchy problem (9.36)–(9.37).

For any $t \in \overline{I}$, we denote the set of *admissible stress fields* by

$$\Sigma(t) = \{\, \boldsymbol{\tau} \in Q : (\boldsymbol{\tau}, \boldsymbol{\varepsilon}(\boldsymbol{v}))_Q \geq (\boldsymbol{f}(t), \boldsymbol{v})_{\mathcal{A}} \ \forall \boldsymbol{v} \in V_2 \,\}. \tag{9.39}$$

Consider the following variational problem.

Problem 9.10 *Find a stress field* $\boldsymbol{\sigma} : \overline{I} \to Q$ *such that*

$$\boldsymbol{\sigma}(t) \in \Sigma(t), \quad (\mathcal{T}\boldsymbol{\sigma}(t), \boldsymbol{\tau} - \boldsymbol{\sigma}(t))_Q \geq 0 \quad \forall \boldsymbol{\tau} \in \Sigma(t), \ \text{a.e. } t \in I. \tag{9.40}$$

The relation between Problem 9.1 and Problem 9.10 is given by the following result.

Theorem 9.11 *Assume* (9.7)–(9.9) *and the assumptions* (6.35) *on* $\mathcal{B}$.

i) *If* $\boldsymbol{u}$ *is the solution of Problem 9.1 and* $\boldsymbol{\sigma}$ *is defined by*

$$\boldsymbol{\sigma} = \mathcal{A}\boldsymbol{\varepsilon}(\dot{\boldsymbol{u}}) + \mathcal{B}\boldsymbol{\varepsilon}(\boldsymbol{u}), \tag{9.41}$$

then $\boldsymbol{\sigma}$ *is the solution of Problem 9.10 and* $\boldsymbol{\sigma} \in L^\infty(I; Q_1)$.

ii) *Conversely, let* $\boldsymbol{\sigma} \in L^\infty(I; Q_1)$ *be a solution of Problem 9.10. Then, there exists a unique function* $\boldsymbol{u} \in W^{1,\infty}(I; V)$ *such that* (9.41) *and* (9.14) *hold. Moreover,* $\boldsymbol{u}$ *is a solution of Problem 9.1.*

Proof. i) Using (9.13) and (9.41) we obtain

$$(\boldsymbol{\sigma}(t), \boldsymbol{\varepsilon}(\boldsymbol{v}) - \boldsymbol{\varepsilon}(\boldsymbol{u}(t)))_Q \geq (\boldsymbol{f}(t), \boldsymbol{v} - \boldsymbol{u}(t))_{\mathcal{A}} \quad \forall \boldsymbol{v} \in V_2. \tag{9.42}$$

This inequality as well as the equalities and inequalities below are valid a.e. $t \in I$. Choosing $\boldsymbol{v} = 2\boldsymbol{u}(t)$ and $\boldsymbol{v} = \boldsymbol{0}$ in (9.42) we find

$$(\boldsymbol{\sigma}(t), \boldsymbol{\varepsilon}(\boldsymbol{u}(t)))_Q = (\boldsymbol{f}(t), \boldsymbol{u})_{\mathcal{A}}. \tag{9.43}$$

Using now (9.42) and (9.43), we deduce that

$$(\boldsymbol{\sigma}(t), \boldsymbol{\varepsilon}(\boldsymbol{v}))_Q \geq (\boldsymbol{f}(t), \boldsymbol{v})_{\mathcal{A}} \quad \forall \boldsymbol{v} \in V_2.$$

Thus, $\boldsymbol{\sigma}(t) \in \Sigma(t)$. Moreover, from (9.39) and (9.43) we deduce that

$$(\boldsymbol{\varepsilon}(\boldsymbol{u}(t)), \boldsymbol{\tau} - \boldsymbol{\sigma}(t))_Q \geq 0 \quad \forall \boldsymbol{\tau} \in \Sigma(t). \tag{9.44}$$

Using (9.14) and (9.41) we have

$$\boldsymbol{\varepsilon}(\boldsymbol{u}) = \mathcal{T}\boldsymbol{\sigma}. \tag{9.45}$$

The inequality in (9.40) is now a consequence of (9.44) and (9.45). Moreover, since $\boldsymbol{\sigma}(t) \in \Sigma(t)$, it follows that $\operatorname{Div} \boldsymbol{\sigma}(t) + \boldsymbol{f}_0(t) = \boldsymbol{0}$ a.e. in Ω. Therefore, from (9.41) and (9.8) we find $\boldsymbol{\sigma} \in L^\infty(I; Q_1)$.

ii) Conversely, let $\boldsymbol{\sigma} \in L^\infty(I; Q_1)$ be a solution of Problem 9.10 and let $\boldsymbol{z}$ be given by

$$\boldsymbol{z} = \mathcal{T}\boldsymbol{\sigma} \in W^{1,\infty}(I; Q).$$

Using (9.40) we find

$$(\boldsymbol{z}(t), \boldsymbol{\tau} - \boldsymbol{\sigma}(t))_Q \geq 0 \quad \forall \boldsymbol{\tau} \in \Sigma(t). \tag{9.46}$$

Let $\tilde{\boldsymbol{z}} \in Q$ be such that $(\tilde{\boldsymbol{z}}, \boldsymbol{\varepsilon}(\boldsymbol{v}))_Q = 0$ for all $\boldsymbol{v} \in V$. Using (9.39) we see that $\boldsymbol{\sigma}(t) \pm \tilde{\boldsymbol{z}}$ belong to $\Sigma(t)$. By choosing $\boldsymbol{\tau} = \boldsymbol{\sigma}(t) \pm \tilde{\boldsymbol{z}}$ in (9.46) we have $(\boldsymbol{z}(t), \tilde{\boldsymbol{z}})_Q = 0$. Using Korn's inequality (7.8) (page 144) we know that the set

$$\boldsymbol{\varepsilon}(V) = \{\, \boldsymbol{\varepsilon}(\boldsymbol{v}) : \boldsymbol{v} \in V \,\}$$

is a closed subspace of Q. Therefore, $\boldsymbol{\varepsilon}(V^\perp)^\perp = \boldsymbol{\varepsilon}(V)$, where $\perp$ represents the orthogonal complement in Q. Then, $\boldsymbol{z}(t) \in \boldsymbol{\varepsilon}(V^\perp)^\perp$ and also $\boldsymbol{z}(t) \in \boldsymbol{\varepsilon}(V)$ for a.e. $t \in I$. Hence, there exists $\boldsymbol{u}(t) \in V$ such that

$$\boldsymbol{z}(t) = \boldsymbol{\varepsilon}(\boldsymbol{u}(t)) \quad \text{a.e. } t \in I. \tag{9.47}$$

Since $\boldsymbol{z} \in W^{1,\infty}(I; Q)$, using Korn's inequality again, we obtain that $\boldsymbol{u} \in W^{1,\infty}(I; V)$. Moreover, it follows from (9.36)–(9.38) and (9.47) that $\boldsymbol{u}$ satisfies (9.14) and (9.41). The uniqueness of $\boldsymbol{u}$ follows from (9.47) and Korn's inequality (7.8). We also note that by (9.46) and (9.47) we obtain (9.44).

We now prove that $\boldsymbol{u}(t) \in V_2$. We argue by contradiction. Assume $\boldsymbol{u}(t) \notin V_2$. Let us denote by $\mathcal{P}\boldsymbol{u}(t)$ the projection of $\boldsymbol{u}(t)$ on the closed convex subset $V_2 \subset V$. We have

$$(\mathcal{P}\boldsymbol{u}(t) - \boldsymbol{u}(t), \boldsymbol{v})_{\mathcal{A}} \geq (\mathcal{P}\boldsymbol{u}(t) - \boldsymbol{u}(t), \mathcal{P}\boldsymbol{u}(t))_{\mathcal{A}} > (\mathcal{P}\boldsymbol{u}(t) - \boldsymbol{u}(t), \boldsymbol{u}(t))_{\mathcal{A}}$$

for any $\boldsymbol{v} \in V_2$. From these inequalities we obtain that there exists $\alpha \in \mathbb{R}$ such that

$$(\mathcal{P}\boldsymbol{u}(t) - \boldsymbol{u}(t), \boldsymbol{v})_{\mathcal{A}} > \alpha > (\mathcal{P}\boldsymbol{u}(t) - \boldsymbol{u}(t), \boldsymbol{u}(t))_{\mathcal{A}} \quad \forall \boldsymbol{v} \in V_2. \tag{9.48}$$

Let $\tilde{\boldsymbol{\tau}}(t) \in Q$ be the element

$$\tilde{\boldsymbol{\tau}}(t) = \mathcal{A}\boldsymbol{\varepsilon}(\mathcal{P}\boldsymbol{u}(t) - \boldsymbol{u}(t)). \tag{9.49}$$

Using (9.48) and (9.49) we deduce that

$$(\tilde{\boldsymbol{\tau}}(t), \boldsymbol{\varepsilon}(\boldsymbol{v}))_Q > \alpha > (\tilde{\boldsymbol{\tau}}(t), \boldsymbol{\varepsilon}(\boldsymbol{u}(t)))_Q \qquad \forall\, \boldsymbol{v} \in V_2. \tag{9.50}$$

Taking $\boldsymbol{v} = \boldsymbol{0}$ in (9.50), we obtain

$$\alpha < 0. \tag{9.51}$$

Let us now suppose that there exists $\boldsymbol{v} \in V_2$ such that

$$(\tilde{\boldsymbol{\tau}}(t), \boldsymbol{\varepsilon}(\boldsymbol{v}))_Q < 0. \tag{9.52}$$

Since $\lambda\, \boldsymbol{v} \in V_2$ for $\lambda \geq 0$, it follows from (9.50) that

$$\lambda\, (\tilde{\boldsymbol{\tau}}(t), \boldsymbol{\varepsilon}(\boldsymbol{v}))_Q > \alpha \quad \forall\, \lambda \geq 0.$$

Passing to the limit $\lambda \to +\infty$, we obtain $\alpha \leq -\infty$ from (9.52). This contradicts to the fact that with $\alpha \in \mathbb{R}$. So we conclude that

$$(\tilde{\boldsymbol{\tau}}(t), \boldsymbol{\varepsilon}(\boldsymbol{v}))_Q \geq 0 \quad \forall\, \boldsymbol{v} \in V_2. \tag{9.53}$$

Let $\tilde{\boldsymbol{\sigma}}(t) = \mathcal{A}\boldsymbol{\varepsilon}(\boldsymbol{f}(t)) \in Q$. We find that

$$(\tilde{\boldsymbol{\sigma}}(t), \boldsymbol{\varepsilon}(\boldsymbol{v}))_Q = (\boldsymbol{f}(t), \boldsymbol{v})_{\mathcal{A}} \quad \forall\, \boldsymbol{v} \in V. \tag{9.54}$$

It follows from (9.53), (9.54) and (9.39) that $\tilde{\boldsymbol{\tau}}(t) + \tilde{\boldsymbol{\sigma}}(t) \in \Sigma(t)$. Taking $\boldsymbol{\tau} = \tilde{\boldsymbol{\tau}}(t) + \tilde{\boldsymbol{\sigma}}(t)$ in (9.44) we find

$$(\tilde{\boldsymbol{\tau}}(t), \boldsymbol{\varepsilon}(\boldsymbol{u}(t)))_Q \geq (\boldsymbol{\sigma}(t) - \tilde{\boldsymbol{\sigma}}(t), \boldsymbol{\varepsilon}(\boldsymbol{u}(t)))_Q.$$

Keeping in mind (9.50) and (9.51), from the previous inequality we deduce

$$(\boldsymbol{\sigma}(t) - \tilde{\boldsymbol{\sigma}}(t), \boldsymbol{\varepsilon}(\boldsymbol{u}(t)))_Q < 0. \tag{9.55}$$

On the other hand, (9.39) and (9.54) imply $2\boldsymbol{\sigma}(t) - \tilde{\boldsymbol{\sigma}}(t) \in \Sigma(t)$. Taking $\boldsymbol{\tau} = 2\boldsymbol{\sigma}(t) - \tilde{\boldsymbol{\sigma}}(t)$ in (9.44), we obtain

$$(\boldsymbol{\sigma}(t) - \tilde{\boldsymbol{\sigma}}(t), \boldsymbol{\varepsilon}(\boldsymbol{u}(t)))_Q \geq 0. \tag{9.56}$$

We observe that (9.55) and (9.56) cannot be both true. Therefore, $\boldsymbol{u}(t) \in V_2$. Taking now $\boldsymbol{\tau} = \tilde{\boldsymbol{\sigma}}(t)$ in (9.44) and using (9.54) we have

$$(\boldsymbol{f}(t), \boldsymbol{u}(t))_{\mathcal{A}} \geq (\boldsymbol{\sigma}(t), \boldsymbol{\varepsilon}(\boldsymbol{u}(t)))_Q. \tag{9.57}$$

As $\boldsymbol{\sigma}(t) \in \Sigma(t)$ and $\boldsymbol{u}(t) \in V_2$, it follows from (9.39) that

$$(\boldsymbol{\sigma}(t), \boldsymbol{\varepsilon}(\boldsymbol{u}(t)))_Q \geq (\boldsymbol{f}(t), \boldsymbol{u}(t))_{\mathcal{A}}. \tag{9.58}$$

So from (9.57) and (9.58) we obtain

$$(\boldsymbol{\sigma}(t), \boldsymbol{\varepsilon}(\boldsymbol{u}(t)))_Q = (\boldsymbol{f}(t), \boldsymbol{u}(t))_{\mathcal{A}}. \tag{9.59}$$

The inequality (9.13) follows now from (9.41), (9.59) and (9.39). ■

It is natural to call Problem 9.10 a *dual formulation* of the variational problem 9.1. Since Theorem 9.3 states that the primal problem 9.1 has a unique solution $\boldsymbol{u} \in W^{1,\infty}(I; V)$, it follows from Theorem 9.11 that the dual problem 9.10 has a unique solution $\boldsymbol{\sigma} \in L^\infty(I; Q_1)$. A careful analysis of the proof of Theorem 9.11 shows that in fact $\boldsymbol{\sigma} \in L^\infty(I; Q)$ and $\operatorname{Div} \boldsymbol{\sigma} \in W^{1,1}(I; H)$.

Chapter 10

Bilateral Contact with Slip Dependent Friction

For the problem studied in this chapter the contact is bilateral and the friction is modeled with Tresca's friction law with the friction bound depending on the total slip. We present the classical as well as variational formulations of the problem and show the existence and uniqueness of a weak solution. We then turn to numerical approximations of the problem. For both spatially semi-discrete and fully discrete schemes we show the existence of a unique solution and derive error estimates. We also study a dual formulation of the problem in the case of the classical Tresca's friction law, i.e. in the case when the friction bound is a given function.

10.1 Problem Statement

We assume that the contact is bilateral, i.e. there is no loss of contact between the body and the foundation. In other words, the normal displacement u_ν vanishes on Γ_3 at any time. We model the friction with Tresca's friction law, where the friction bound g is assumed to depend on the accumulated slip of the surface. In this model we try to incorporate changes in the contact surface structure resulted from sliding. Therefore,

$$g = g(S_t(\dot{\boldsymbol{u}})) \quad \text{on } \Gamma_3 \times I$$

with $S_t(\dot{\boldsymbol{u}})(\boldsymbol{x})$ being the accumulated slip at the point $\boldsymbol{x}$ on Γ_3 over the time period $[0, t]$:

$$S_t(\dot{\boldsymbol{u}}) = \int_0^t \|\dot{\boldsymbol{u}}_\tau(s)\| \, ds, \quad t \in \overline{I}. \tag{10.1}$$

Thus on Γ_3, we have $\|\boldsymbol{\sigma}_\tau\| \le g(S_t(\dot{\boldsymbol{u}}))$. When the strict inequality holds, $\|\boldsymbol{\sigma}_\tau\| < g(S_t(\dot{\boldsymbol{u}}))$, the material point is in the *stick* zone: $\dot{\boldsymbol{u}}_\tau = \boldsymbol{0}$, while when the equality holds, $\|\boldsymbol{\sigma}_\tau\| = g(S_t(\dot{\boldsymbol{u}}))$, the material point is in the *slip* zone: $\boldsymbol{\sigma}_\tau = -\lambda \dot{\boldsymbol{u}}_\tau$ for some $\lambda \ge 0$.

Under the previous assumptions, the classical formulation of the frictional contact problem with total slip dependent friction bound is the following:

Find a displacement field $\boldsymbol{u} : \Omega \times \overline{I} \to \mathbb{R}^d$ and a stress field $\boldsymbol{\sigma} : \Omega \times \overline{I} \to \mathbb{S}^d$ such that

$$\boldsymbol{\sigma} = \mathcal{A}\varepsilon(\dot{\boldsymbol{u}}) + \mathcal{B}\varepsilon(\boldsymbol{u}) \qquad \text{in } \Omega \times I, \tag{10.2}$$

$$\operatorname{Div} \boldsymbol{\sigma} + \boldsymbol{f}_0 = \boldsymbol{0} \qquad \text{in } \Omega \times I, \tag{10.3}$$

$$\boldsymbol{u} = \boldsymbol{0} \qquad \text{on } \Gamma_1 \times I, \tag{10.4}$$

$$\boldsymbol{\sigma\nu} = \boldsymbol{f}_2 \qquad \text{on } \Gamma_2 \times I, \tag{10.5}$$

$$\left.\begin{array}{l} u_\nu = 0, \quad \|\boldsymbol{\sigma}_\tau\| \le g(S_t(\dot{\boldsymbol{u}})), \\ \|\boldsymbol{\sigma}_\tau\| < g(S_t(\dot{\boldsymbol{u}})) \;\Rightarrow\; \dot{\boldsymbol{u}}_\tau = \boldsymbol{0}, \\ \|\boldsymbol{\sigma}_\tau\| = g(S_t(\dot{\boldsymbol{u}})) \;\Rightarrow\; \exists\, \lambda \ge 0 \text{ s.t. } \boldsymbol{\sigma}_\tau = -\lambda \dot{\boldsymbol{u}}_\tau \end{array}\right\} \qquad \text{on } \Gamma_3 \times I, \tag{10.6}$$

$$\boldsymbol{u}(0) = \boldsymbol{u}_0 \qquad \text{in } \Omega. \tag{10.7}$$

Here (10.2) represents the constitutive law of the viscoelastic material and $\boldsymbol{u}_0$ is a known initial displacement field.

To obtain a variational formulation of the mechanical problem, we use the space V_1 defined in (7.11) on page 144:

$$V_1 = \{\, \boldsymbol{v} \in V : \, v_\nu = 0 \text{ on } \Gamma_3 \,\}.$$

Recall that V_1 itself is a real Hilbert space under the inner product of the space V given by (7.9).

In the study of the mechanical problem (10.2)–(10.7) we assume that the viscosity operator $\mathcal{A}$ and the elasticity operator $\mathcal{B}$ satisfy (6.34) and (6.35) (page 132). In addition, we assume the friction bound g satisfies

$$\left.\begin{array}{l} \text{(a) } g : \Gamma_3 \times \mathbb{R} \to \mathbb{R}_+. \\ \text{(b) } \exists\, L_g > 0 \text{ s.t. } \forall\, u_1, u_2 \in \mathbb{R}, \text{ a.e. } \boldsymbol{x} \in \Omega, \\ \qquad |g(\boldsymbol{x}, u_1) - g(\boldsymbol{x}, u_2)| \le L_g\, |u_1 - u_2|. \\ \text{(c) For any } u \in \mathbb{R},\ \boldsymbol{x} \mapsto g(\boldsymbol{x}, u) \text{ is measurable.} \\ \text{(d) The mapping } \boldsymbol{x} \mapsto g(\boldsymbol{x}, 0) \in L^2(\Gamma_3). \end{array}\right\} \tag{10.8}$$

We also assume that the body force and surface traction satisfy

$$\boldsymbol{f}_0 \in C(\overline{I}; H), \qquad \boldsymbol{f}_2 \in C(\overline{I}; L^2(\Gamma_2)^d), \tag{10.9}$$

and the initial displacement

$$\boldsymbol{u}_0 \in V_1. \tag{10.10}$$

Next, we denote by $\boldsymbol{f}(t)$ the element of V_1 given by

$$(\boldsymbol{f}(t), \boldsymbol{v})_V = (\boldsymbol{f}_0(t), \boldsymbol{v})_H + (\boldsymbol{f}_2(t), \boldsymbol{v})_{L^2(\Gamma_2)^d} \quad \forall\, \boldsymbol{v} \in V_1, \tag{10.11}$$

for all $t \in \overline{I}$. Note that conditions (10.9) imply

$$\boldsymbol{f} \in C(\overline{I}; V_1). \tag{10.12}$$

Let $j : L^2(\Gamma_3) \times V_1 \to \mathbb{R}$ be the functional

$$j(v, \boldsymbol{w}) = \int_{\Gamma_3} g(v)\, \|\boldsymbol{w}_\tau\|\, da \quad \forall\, v \in L^2(\Gamma_3),\ \boldsymbol{w} \in V_1. \tag{10.13}$$

We notice that by the assumption (10.8) the integral in (10.13) is well defined.

With these notations, a standard calculation shows that if $(\boldsymbol{u}, \boldsymbol{\sigma})$ are sufficiently regular functions satisfying (10.3)–(10.6), then for all $t \in \overline{I}$, $\boldsymbol{u}(t) \in V_1$ and

$$\begin{aligned} &(\boldsymbol{\sigma}(t), \boldsymbol{\varepsilon}(\boldsymbol{v}) - \boldsymbol{\varepsilon}(\dot{\boldsymbol{u}}(t)))_Q + j(S_t(\dot{\boldsymbol{u}}), \boldsymbol{v}) - j(S_t(\dot{\boldsymbol{u}}), \dot{\boldsymbol{u}}(t)) \\ &\qquad \geq (\boldsymbol{f}(t), \boldsymbol{v} - \dot{\boldsymbol{u}}(t))_V \qquad \forall\, \boldsymbol{v} \in V_1. \end{aligned} \tag{10.14}$$

Thus, from (10.2), (10.14) and (10.7) we obtain the following variational formulation of the problem (10.2)–(10.7) in terms of the displacement.

Problem 10.1 *Find a displacement* $\boldsymbol{u} : \overline{I} \to V_1$ *such that*

$$\boldsymbol{u}(0) = \boldsymbol{u}_0 \tag{10.15}$$

and for all $t \in \overline{I}$,

$$\begin{aligned} &(\mathcal{A}\boldsymbol{\varepsilon}(\dot{\boldsymbol{u}}(t)), \boldsymbol{\varepsilon}(\boldsymbol{v} - \dot{\boldsymbol{u}}(t)))_Q + (\mathcal{B}\boldsymbol{\varepsilon}(\boldsymbol{u}(t)), \boldsymbol{\varepsilon}(\boldsymbol{v} - \dot{\boldsymbol{u}}(t)))_Q + j(S_t(\dot{\boldsymbol{u}}), \boldsymbol{v}) \\ &\quad - j(S_t(\dot{\boldsymbol{u}}), \dot{\boldsymbol{u}}(t)) \geq (\boldsymbol{f}(t), \boldsymbol{v} - \dot{\boldsymbol{u}}(t))_V \quad \forall\, \boldsymbol{v} \in V_1. \end{aligned} \tag{10.16}$$

10.2 An Existence and Uniqueness Result

Our main existence and uniqueness result is the following.

Theorem 10.2 *Assume* (6.34), (6.35) *and* (10.8)–(10.10). *Then Problem 10.1 has a unique solution* $\boldsymbol{u} \in C^1(\overline{I}; V_1)$.

The proof of Theorem 10.2 is based on fixed-point arguments, and is carried out in several steps. Let $\boldsymbol{\eta} \in C(\overline{I}; Q)$ and $\boldsymbol{\xi} \in C(\overline{I}; V_1)$ be arbitrarily given. We consider the following auxiliary variational problem.

Problem 10.3 *Find* $\boldsymbol{w}_{\eta\xi} : \overline{I} \to V_1$ *such that for any* $t \in \overline{I}$,

$$\begin{aligned} &(\mathcal{A}\boldsymbol{\varepsilon}(\boldsymbol{w}_{\eta\xi}(t)), \boldsymbol{\varepsilon}(\boldsymbol{v} - \boldsymbol{w}_{\eta\xi}(t)))_Q + (\boldsymbol{\eta}(t), \boldsymbol{\varepsilon}(\boldsymbol{v} - \boldsymbol{w}_{\eta\xi}(t)))_Q + j(S_t(\boldsymbol{\xi}), \boldsymbol{v}) \\ &\quad - j(S_t(\boldsymbol{\xi}), \boldsymbol{w}_{\eta\xi}(t)) \geq (\boldsymbol{f}(t), \boldsymbol{v} - \boldsymbol{w}_{\eta\xi}(t))_V \quad \forall\, \boldsymbol{v} \in V_1. \end{aligned} \tag{10.17}$$

Lemma 10.4 *There exists a unique solution* $\boldsymbol{w}_{\eta\xi}$ *to Problem 10.3. Moreover,* $\boldsymbol{w}_{\eta\xi} \in C(\overline{I}; V_1)$.

Proof. Let $t \in \overline{I}$. It follows from Theorem 4.1 that there exists a unique element $\boldsymbol{w}_{\eta\xi}(t) \in V_1$ that solves (10.17). Now, let $t_1, t_2 \in \overline{I}$. For simplicity we write $\boldsymbol{w}_{\eta\xi}(t_i) = \boldsymbol{w}_i$, $\boldsymbol{\eta}(t_i) = \boldsymbol{\eta}_i$, $\boldsymbol{f}(t_i) = \boldsymbol{f}_i$, $i = 1, 2$. Using (10.17) for $t = t_1$, t_2, after some manipulations we obtain

$$\begin{aligned}(\mathcal{A}\boldsymbol{\varepsilon}(\boldsymbol{w}_1) - \mathcal{A}\boldsymbol{\varepsilon}(\boldsymbol{w}_2), \boldsymbol{\varepsilon}(\boldsymbol{w}_1) - \boldsymbol{\varepsilon}(\boldsymbol{w}_2))_Q \\ \le (\boldsymbol{f}_1 - \boldsymbol{f}_2, \boldsymbol{w}_1 - \boldsymbol{w}_2)_V + (\boldsymbol{\eta}_1 - \boldsymbol{\eta}_2, \boldsymbol{\varepsilon}(\boldsymbol{w}_2) - \boldsymbol{\varepsilon}(\boldsymbol{w}_1))_Q \\ + j(S_{t_1}(\boldsymbol{\xi}), \boldsymbol{w}_2) - j(S_{t_1}(\boldsymbol{\xi}), \boldsymbol{w}_1) \\ + j(S_{t_2}(\boldsymbol{\xi}), \boldsymbol{w}_1) - j(S_{t_2}(\boldsymbol{\xi}), \boldsymbol{w}_2).\end{aligned} \tag{10.18}$$

By the assumptions (6.34) on $\mathcal{A}$, we find that

$$(\mathcal{A}\boldsymbol{\varepsilon}(\boldsymbol{w}_1) - \mathcal{A}\boldsymbol{\varepsilon}(\boldsymbol{w}_2), \boldsymbol{\varepsilon}(\boldsymbol{w}_1) - \boldsymbol{\varepsilon}(\boldsymbol{w}_2))_Q \ge M\, \|\boldsymbol{w}_1 - \boldsymbol{w}_2\|_V^2. \tag{10.19}$$

From (10.13), (10.8) and (10.1), we deduce

$$\begin{aligned}j(S_{t_1}(\boldsymbol{\xi}), \boldsymbol{w}_2) - j(S_{t_1}(\boldsymbol{\xi}), \boldsymbol{w}_1) + j(S_{t_2}(\boldsymbol{\xi}), \boldsymbol{w}_1) - j(S_{t_2}(\boldsymbol{\xi}), \boldsymbol{w}_2) \\ \le c\, |t_1 - t_2|\, \|\boldsymbol{\xi}\|_{C(\overline{I};V)}\, \|\boldsymbol{w}_1 - \boldsymbol{w}_2\|_V.\end{aligned} \tag{10.20}$$

Combining (10.18)–(10.20) we easily see that

$$\|\boldsymbol{w}_1 - \boldsymbol{w}_2\|_V \le c\Big(\|\boldsymbol{f}_1 - \boldsymbol{f}_2\|_V + \|\boldsymbol{\eta}_1 - \boldsymbol{\eta}_2\|_Q + |t_1 - t_2|\, \|\boldsymbol{\xi}\|_{C(\overline{I};V)}\Big), \tag{10.21}$$

which implies $\boldsymbol{w} \in C(\overline{I}; V_1)$. ■

Consider next the operator $\Lambda_\eta : C(\overline{I}; V_1) \to C(\overline{I}; V_1)$ defined by

$$\Lambda_\eta \boldsymbol{\xi} = \boldsymbol{w}_{\eta\xi} \quad \forall\, \boldsymbol{\xi} \in C(\overline{I}; V_1). \tag{10.22}$$

Lemma 10.5 *For any $\boldsymbol{\eta} \in C(\overline{I}; Q)$, the operator Λ_η has a unique fixed point $\boldsymbol{\xi}_\eta \in C(\overline{I}; V_1)$.*

Proof. Let $\boldsymbol{\eta} \in C(\overline{I}; Q)$ and $\boldsymbol{\xi}_1, \boldsymbol{\xi}_2 \in C(\overline{I}; V_1)$. We denote by $\boldsymbol{w}_i$ the solution of Problem 10.3 with $\boldsymbol{\xi} = \boldsymbol{\xi}_i$, $i = 1, 2$. By an argument similar to that used in obtaining (10.21) we deduce

$$\|\boldsymbol{w}_1(t) - \boldsymbol{w}_2(t)\|_V \le c \int_0^t \|\boldsymbol{\xi}_1(s) - \boldsymbol{\xi}_2(s)\|_V ds \quad \forall\, t \in \overline{I}.$$

Since $\boldsymbol{w}_i = \boldsymbol{w}_{\eta\xi_i} = \Lambda_\eta \boldsymbol{\xi}_i$, we rewrite the above inequality as

$$\|\Lambda_\eta \boldsymbol{\xi}_1(t) - \Lambda_\eta \boldsymbol{\xi}_2(t)\|_V \le c \int_0^t \|\boldsymbol{\xi}_1(s) - \boldsymbol{\xi}_2(s)\|_V ds \quad \forall\, t \in \overline{I}. \tag{10.23}$$

Let us denote

$$\|\boldsymbol{w}\|_\beta = \max_{t \in \overline{I}} e^{-\beta t}\, \|\boldsymbol{w}(t)\|_V, \quad \boldsymbol{w} \in C(\overline{I}; V_1)$$

with $\beta > 0$ to be chosen later. Clearly $\|\cdot\|_\beta$ defines a norm on the space $C(\overline{I}; V_1)$, that is equivalent to the standard norm $\|\cdot\|_{C(\overline{I};V)}$. Using (10.23) we have for any $t \in \overline{I}$,

$$\begin{aligned} e^{-\beta t} \|\Lambda_\eta \boldsymbol{\xi}_1(t) - \Lambda_\eta \boldsymbol{\xi}_2(t)\|_V &\le c\, e^{-\beta t} \int_0^t e^{\beta s} e^{-\beta s} \|\boldsymbol{\xi}_1(s) - \boldsymbol{\xi}_2(s)\|_V ds \\ &\le c\, e^{-\beta t} \int_0^t e^{\beta s} \|\boldsymbol{\xi}_1 - \boldsymbol{\xi}_2\|_\beta ds \\ &\le \frac{c}{\beta} \|\boldsymbol{\xi}_1 - \boldsymbol{\xi}_2\|_\beta. \end{aligned}$$

Therefore

$$\|\Lambda_\eta \boldsymbol{\xi}_1 - \Lambda_\eta \boldsymbol{\xi}_2\|_\beta \le \frac{c}{\beta} \|\boldsymbol{\xi}_1 - \boldsymbol{\xi}_2\|_\beta.$$

So if we choose β such that $\beta > c$, then the operator Λ_η is a contraction on the space $C(\overline{I}; V_1)$ endowed with the equivalent norm $\|\cdot\|_\beta$. By the Banach fixed-point theorem (Theorem 1.62, page 21), Λ_η has a unique fixed point $\boldsymbol{\xi}_\eta \in C(\overline{I}; V_1)$. ■

In what follows, for any $\boldsymbol{\eta} \in C(\overline{I}; Q)$ we write

$$\boldsymbol{w}_\eta = \boldsymbol{w}_{\eta \xi_\eta}. \tag{10.24}$$

We have $\Lambda_\eta \boldsymbol{\xi}_\eta = \boldsymbol{\xi}_\eta$ and using (10.22) and (10.24),

$$\boldsymbol{w}_\eta = \boldsymbol{\xi}_\eta. \tag{10.25}$$

Therefore, taking $\boldsymbol{\xi} = \boldsymbol{\xi}_\eta$ in (10.17) and using (10.24), (10.25) we deduce that for $t \in \overline{I}$,

$$\begin{aligned} &(\mathcal{A}\boldsymbol{\varepsilon}(\boldsymbol{w}_\eta(t)), \boldsymbol{\varepsilon}(\boldsymbol{v} - \boldsymbol{w}_\eta(t)))_Q + (\boldsymbol{\eta}, \boldsymbol{\varepsilon}(\boldsymbol{v} - \boldsymbol{w}_\eta(t)))_Q + j(S_t(\boldsymbol{w}_\eta), \boldsymbol{v}) \\ &\quad - j(S_t(\boldsymbol{w}_\eta), \boldsymbol{w}_\eta(t)) \ge (\boldsymbol{f}(t), \boldsymbol{v} - \boldsymbol{w}_\eta(t))_V \quad \forall \boldsymbol{v} \in V_1. \end{aligned} \tag{10.26}$$

Let $\boldsymbol{u}_\eta : \overline{I} \to V_1$ be the function given by

$$\boldsymbol{u}_\eta(t) = \int_0^t \boldsymbol{w}_\eta(s)\, ds + \boldsymbol{u}_0 \quad \forall t \in \overline{I}. \tag{10.27}$$

In addition, we define the operator $\Lambda : C(\overline{I}; Q) \to C(\overline{I}; Q)$ by the formula

$$\Lambda \boldsymbol{\eta} = \mathcal{B}(\boldsymbol{\varepsilon}(\boldsymbol{u}_\eta)) \quad \forall \boldsymbol{\eta} \in C(\overline{I}; Q). \tag{10.28}$$

Lemma 10.6 *The operator Λ has a unique fixed point $\boldsymbol{\eta}^* \in C(\overline{I}; Q)$.*

Proof. Let $\boldsymbol{\eta}_1, \boldsymbol{\eta}_2 \in C(\overline{I}; Q)$ and let $\boldsymbol{w}_i = \boldsymbol{w}_{\eta_i}$, $\boldsymbol{u}_i = \boldsymbol{u}_{\eta_i}$ for $i = 1, 2$. Using (10.26) and arguments similar to those used in the proof of Lemma 10.5 we obtain

$$\|\boldsymbol{w}_1(t) - \boldsymbol{w}_2(t)\|_V \le c\Big(\|\boldsymbol{\eta}_1(t) - \boldsymbol{\eta}_2(t)\|_Q + \int_0^t \|\boldsymbol{w}_1(s) - \boldsymbol{w}_2(s)\|_V\, ds\Big)$$

for all $t \in \overline{I}$. An application of the Gronwall inequality (Lemma 7.24) yields

$$\|\boldsymbol{w}_1(t) - \boldsymbol{w}_2(t)\|_V \leq c\Big(\|\boldsymbol{\eta}_1(t) - \boldsymbol{\eta}_2(t)\|_Q + \int_0^t \|\boldsymbol{\eta}_1(s) - \boldsymbol{\eta}_2(s)\|_Q\, ds\Big).$$

So

$$\int_0^t \|\boldsymbol{w}_1(s) - \boldsymbol{w}_2(s)\|_V\, ds \leq c \int_0^t \|\boldsymbol{\eta}_1(s) - \boldsymbol{\eta}_2(s)\|_Q ds \quad \forall\, t \in \overline{I}.$$

For the operator Λ defined in (10.28), we then obtain

$$\|\Lambda\boldsymbol{\eta}_1(t) - \Lambda\boldsymbol{\eta}_2(t)\|_Q \leq c \int_0^t \|\boldsymbol{\eta}_1(s) - \boldsymbol{\eta}_2(s)\|_V\, ds \quad \forall\, t \in \overline{I}. \tag{10.29}$$

Then the result of Lemma 10.6 follows from (10.29) after a fixed-point argument similar to that in the proof of Lemma 10.5. ■

Now we prove Theorem 10.2.

Proof of Theorem 10.2.
Existence. Let $\boldsymbol{\eta}^* \in C(\overline{I}; Q)$ be the fixed point of Λ and let $\boldsymbol{u}_{\eta^*} \in C^1(\overline{I}; V_1)$ be the function defined by (10.27) for $\boldsymbol{\eta} = \boldsymbol{\eta}^*$. We have $\dot{\boldsymbol{u}}_{\eta^*} = \boldsymbol{w}_{\eta^*}$ and use (10.26) for $\boldsymbol{\eta} = \boldsymbol{\eta}^*$ to obtain

$$\begin{aligned}(\mathcal{A}\boldsymbol{\varepsilon}(\dot{\boldsymbol{u}}_{\eta^*}(t)), \boldsymbol{\varepsilon}(\boldsymbol{v} - \dot{\boldsymbol{u}}_{\eta^*}(t)))_Q + (\boldsymbol{\eta}^*(t), \boldsymbol{\varepsilon}(\boldsymbol{v} - \dot{\boldsymbol{u}}_{\eta^*}(t)))_Q \\ + j(S_t(\dot{\boldsymbol{u}}_{\eta^*}), \boldsymbol{v}) - j(S_t(\dot{\boldsymbol{u}}_{\eta^*}), \dot{\boldsymbol{u}}_{\eta^*}(t)) \\ \geq (\boldsymbol{f}(t), \boldsymbol{v} - \dot{\boldsymbol{u}}_{\eta^*}(t))_V \quad \forall\, \boldsymbol{v} \in V_1\end{aligned} \tag{10.30}$$

for $t \in \overline{I}$. The inequality (10.16) follows now from (10.30) and (10.28) since $\boldsymbol{\eta}^* = \Lambda\boldsymbol{\eta}^* = \mathcal{B}\boldsymbol{\varepsilon}(\boldsymbol{u}_{\eta^*})$. Moreover, since (10.27) implies $\boldsymbol{u}_{\eta^*}(0) = \boldsymbol{u}_0$, we conclude that $\boldsymbol{u}_{\eta^*}$ is a solution of Problem 10.1.
Uniqueness. Let $\boldsymbol{u}_i \in C^1(\overline{I}; V_1)$, $i = 1, 2$, be two solutions to Problem 10.1 and let $\boldsymbol{w}_i = \dot{\boldsymbol{u}}_i$, $i = 1, 2$. We have

$$\boldsymbol{u}_i(t) = \int_0^t \boldsymbol{w}_i(s)\, ds + \boldsymbol{u}_0 \quad \forall\, t \in \overline{I}. \tag{10.31}$$

From (10.16) it follows that for $t \in \overline{I}$,

$$\begin{aligned}(\mathcal{A}\boldsymbol{\varepsilon}(\boldsymbol{w}_1(t)) - \mathcal{A}\boldsymbol{\varepsilon}(\boldsymbol{w}_2(t)), \boldsymbol{\varepsilon}(\boldsymbol{w}_1(t)) - \boldsymbol{\varepsilon}(\boldsymbol{w}_2(t)))_Q \\ \leq (\mathcal{B}\boldsymbol{\varepsilon}(\boldsymbol{u}_2(t)) - \mathcal{B}\boldsymbol{\varepsilon}(\boldsymbol{u}_1(t)), \boldsymbol{\varepsilon}(\boldsymbol{w}_1(t)) - \boldsymbol{\varepsilon}(\boldsymbol{w}_2(t)))_Q \\ + D(t; \boldsymbol{w}_1, \boldsymbol{w}_2),\end{aligned} \tag{10.32}$$

where

$$\begin{aligned}D(t; \boldsymbol{w}_1, \boldsymbol{w}_2) = j(S_t(\boldsymbol{w}_1), \boldsymbol{w}_2(t)) - j(S_t(\boldsymbol{w}_1), \boldsymbol{w}_1(t)) \\ + j(S_t(\boldsymbol{w}_2), \boldsymbol{w}_1(t)) - j(S_t(\boldsymbol{w}_2), \boldsymbol{w}_2(t)).\end{aligned}$$

Using now (10.13) and (10.8) we deduce that for $t \in \overline{I}$,

$$D(t; \boldsymbol{w}_1, \boldsymbol{w}_2) = \int_{\Gamma_3} g(S_t(\boldsymbol{w}_1)) - g(S_t(\boldsymbol{w}_2))(\|\boldsymbol{w}_{2\tau}(t)\| - \|\boldsymbol{w}_{1\tau}(t)\|)\, da$$
$$\leq c\, \|g(S_t(\boldsymbol{w}_1)) - g(S_t(\boldsymbol{w}_2))\|_{L^2(\Gamma_3)} \|\boldsymbol{w}_1(t) - \boldsymbol{w}_2(t)\|_{L^2(\Gamma_3)^d}$$

and then

$$D(t; \boldsymbol{w}_1, \boldsymbol{w}_2) \leq c \int_0^t \|\boldsymbol{w}_1(s) - \boldsymbol{w}_2(s)\|_V\, ds\, \|\boldsymbol{w}_1(t) - \boldsymbol{w}_2(t)\|_V \qquad (10.33)$$

for all $t \in \overline{I}$. Therefore, from the assumptions (6.34) on $\mathcal{A}$ and the relations (10.31)–(10.33) we obtain that for any $t \in \overline{I}$,

$$\|\boldsymbol{w}_1(t) - \boldsymbol{w}_2(t)\|_V \leq c \Big(\|\boldsymbol{u}_1(t) - \boldsymbol{u}_2(t)\|_V + \int_0^t \|\boldsymbol{w}_1(s) - \boldsymbol{w}_2(s)\|_V\, ds \Big).$$

An application of the Gronwall inequality yields

$$\|\boldsymbol{w}_1(t) - \boldsymbol{w}_2(t)\|_V \leq c \Big(\|\boldsymbol{u}_1(t) - \boldsymbol{u}_2(t)\|_V + \int_0^t \|\boldsymbol{u}_1(s) - \boldsymbol{u}_2(s)\|_V\, ds \Big).$$

Recalling the definition (10.31) of $\boldsymbol{u}_1$ and $\boldsymbol{u}_2$ we obtain

$$\|\boldsymbol{w}_1(t) - \boldsymbol{w}_2(t)\|_V \leq c \int_0^t \|\boldsymbol{w}_1(s) - \boldsymbol{w}_2(s)\|_V ds \quad \forall\, t \in \overline{I},$$

which implies $\boldsymbol{w}_1 = \boldsymbol{w}_2$ with another application of the Gronwall inequality. Then by definition (10.31) we see that $\boldsymbol{u}_1 = \boldsymbol{u}_2$. ∎

Let now $\boldsymbol{u} \in C^1(\overline{I}; V)$ be the solution of Problem 10.1 and let $\boldsymbol{\sigma}$ be the stress field defined by (10.2). It can be shown that $\boldsymbol{\sigma} \in C(\overline{I}; Q_1)$. A pair of functions $(\boldsymbol{u}, \boldsymbol{\sigma})$ which satisfies (10.2), (10.7) and (10.16) is called a *weak solution* of the problem (10.2)–(10.7). We conclude that, under the assumptions (6.34), (6.35) and (10.8)–(10.10), the mechanical problem (10.2)–(10.7) has a unique weak solution $\boldsymbol{u} \in C^1(\overline{I}; V_1)$, $\boldsymbol{\sigma} \in C(\overline{I}; Q_1)$.

Now we present some pointwise relations for the weak solution $(\boldsymbol{u}, \boldsymbol{\sigma})$. By an argument similar to that in Section 8.1, we have $\boldsymbol{\sigma} \in C(\overline{I}; Q_1)$ and

$$\operatorname{Div} \boldsymbol{\sigma}(t) = -\boldsymbol{f}_0(t) \quad \text{a.e. in } \Omega, \quad \forall\, t \in \overline{I}. \qquad (10.34)$$

Also if we make the assumption

$$\boldsymbol{\sigma\nu} \in C(\overline{I}; L^2(\Gamma)^d), \qquad (10.35)$$

then

$$\boldsymbol{\sigma}(t)\boldsymbol{\nu} = \boldsymbol{f}_2(t) \quad \text{a.e. on } \Gamma_2, \quad \forall\, t \in \overline{I}. \tag{10.36}$$

The relations (10.34) and (10.36) are needed in error estimation later.

Let us derive an equivalent form of Problem 10.1 in terms of the velocity field

$$\boldsymbol{w} = \dot{\boldsymbol{u}}.$$

By Theorem 10.2, we have $\boldsymbol{w} \in C(\overline{I}; V_1)$. Using the initial displacement condition (10.15), we can express the displacement $\boldsymbol{u}$ in terms of the velocity $\boldsymbol{w}$:

$$\boldsymbol{u}(t) = \mathcal{I}\boldsymbol{w}(t) \equiv \boldsymbol{u}_0 + \int_0^t \boldsymbol{w}(s)\, ds \quad \forall\, t \in \overline{I}, \tag{10.37}$$

we can rewrite (10.15)–(10.16) in the following equivalent form for all $t \in \overline{I}$:

$$\begin{aligned}(\mathcal{A}\boldsymbol{\varepsilon}(\boldsymbol{w}(t)), \boldsymbol{\varepsilon}(\boldsymbol{v} - \boldsymbol{w}(t)))_Q + (\mathcal{B}\boldsymbol{\varepsilon}(\mathcal{I}\boldsymbol{w}(t)), \boldsymbol{\varepsilon}(\boldsymbol{v} - \boldsymbol{w}(t)))_Q + j(S_t(\boldsymbol{w}), \boldsymbol{v})\\ - j(S_t(\boldsymbol{w}), \boldsymbol{w}(t)) \geq (\boldsymbol{f}(t), \boldsymbol{v} - \boldsymbol{w}(t))_V \quad \forall\, \boldsymbol{v} \in V_1.\end{aligned} \tag{10.38}$$

We will use the inequality (10.38) later to derive error estimates for numerical solutions of Problem 10.1.

10.3 Semi-discrete Approximation

We consider an approximation of Problem 10.1 by discretizing only the spatial space V_1. Equivalently, we consider the approximation of (10.38). Let $V_1^h \subset V_1$ be the finite element space discussed in Section 7.2. Then a semi-discrete scheme can be formed as following.

Problem 10.7 *Find a displacement field* $\boldsymbol{u}^h : \overline{I} \to V_1^h$ *such that*

$$\boldsymbol{u}^h(0) = \boldsymbol{u}_0^h \tag{10.39}$$

and for all $t \in \overline{I}$,

$$\begin{aligned}(\mathcal{A}\boldsymbol{\varepsilon}(\dot{\boldsymbol{u}}^h(t)), \boldsymbol{\varepsilon}(\boldsymbol{v}^h - \dot{\boldsymbol{u}}^h(t)))_Q + (\mathcal{B}\boldsymbol{\varepsilon}(\boldsymbol{u}^h(t)), \boldsymbol{\varepsilon}(\boldsymbol{v}^h - \dot{\boldsymbol{u}}^h(t)))_Q\\ + j(S_t(\dot{\boldsymbol{u}}^h), \boldsymbol{v}^h) - j(S_t(\dot{\boldsymbol{u}}^h), \dot{\boldsymbol{u}}^h(t))\\ \geq (\boldsymbol{f}(t), \boldsymbol{v}^h - \dot{\boldsymbol{u}}^h(t))_V \quad \forall\, \boldsymbol{v}^h \in V_1^h.\end{aligned} \tag{10.40}$$

Here, $\boldsymbol{u}_0^h \in V_1^h$ *is an appropriate approximation of* $\boldsymbol{u}_0$.

Using the arguments presented in the previous section, we see that under the conditions stated in Theorem 10.2, Problem 10.7 has a unique solution $\boldsymbol{u}^h \in C^1(\overline{I}; V_1^h)$. Our main purpose below is to derive estimates for the error $\boldsymbol{u} - \boldsymbol{u}^h$.

Let $t \in \overline{I}$ be arbitrary. We introduce the discrete velocity variable

$$\boldsymbol{w}^h(t) = \dot{\boldsymbol{u}}^h(t).$$

With the initial value condition (10.39), we have

$$\boldsymbol{u}^h(t) = \mathcal{I}^h \boldsymbol{w}^h(t) \equiv \boldsymbol{u}_0^h + \int_0^t \boldsymbol{w}^h(s)\, ds. \tag{10.41}$$

Subtracting (10.41) from (10.37)

$$\boldsymbol{u}(t) - \boldsymbol{u}^h(t) = \mathcal{I}\boldsymbol{w}(t) - \mathcal{I}^h \boldsymbol{w}^h(t),$$

we have

$$\|\mathcal{I}\boldsymbol{w}(t) - \mathcal{I}^h \boldsymbol{w}^h(t)\|_V \le \int_0^t \|\boldsymbol{w}(s) - \boldsymbol{w}^h(s)\|_V ds + \|\boldsymbol{u}_0 - \boldsymbol{u}_0^h\|_V. \tag{10.42}$$

The variational inequality (10.40) can be rewritten as

$$\begin{aligned}
&(\mathcal{A}\varepsilon(\boldsymbol{w}^h(t)), \varepsilon(\boldsymbol{v}^h - \boldsymbol{w}^h(t)))_Q + (\mathcal{B}\varepsilon(\mathcal{I}^h \boldsymbol{w}^h(t)), \varepsilon(\boldsymbol{v}^h - \boldsymbol{w}^h(t)))_Q \\
&\quad + j(S_t(\boldsymbol{w}^h), \boldsymbol{v}^h) - j(S_t(\boldsymbol{w}^h), \boldsymbol{w}^h(t)) \\
&\quad\quad \ge (\boldsymbol{f}(t), \boldsymbol{v}^h - \boldsymbol{w}^h(t))_V \quad \forall\, \boldsymbol{v}^h \in V_1^h.
\end{aligned} \tag{10.43}$$

We now take $\boldsymbol{v} = \boldsymbol{w}^h(t)$ in (10.38) and add the inequality to (10.43) with $\boldsymbol{v}^h = \boldsymbol{v}^h(t) \in V_1^h$. After some manipulations, we obtain

$$\begin{aligned}
&(\mathcal{A}\varepsilon(\boldsymbol{w}(t)) - \mathcal{A}\varepsilon(\boldsymbol{w}^h(t)), \varepsilon(\boldsymbol{w}(t)) - \varepsilon(\boldsymbol{w}^h(t)))_Q \\
&\quad \le (\mathcal{A}\varepsilon(\boldsymbol{w}^h(t)) - \mathcal{A}\varepsilon(\boldsymbol{w}(t)), \varepsilon(\boldsymbol{v}^h(t) - \boldsymbol{w}(t)))_Q \\
&\quad\quad + (\mathcal{B}\varepsilon(\mathcal{I}\boldsymbol{w}(t)) - \mathcal{B}\varepsilon(\mathcal{I}^h \boldsymbol{w}^h(t)), \varepsilon(\boldsymbol{w}^h(t) - \boldsymbol{v}^h(t)))_Q \\
&\quad\quad + R(t; \boldsymbol{w}(t), \boldsymbol{v}^h(t)) + D(t; \boldsymbol{w}^h(t), \boldsymbol{v}^h(t)),
\end{aligned} \tag{10.44}$$

where

$$\begin{aligned}
D(t; \boldsymbol{w}^h(t), \boldsymbol{v}^h(t)) &= j(S_t(\boldsymbol{w}), \boldsymbol{w}^h(t)) - j(S_t(\boldsymbol{w}), \boldsymbol{v}^h(t)) \\
&\quad + j(S_t(\boldsymbol{w}^h), \boldsymbol{v}^h(t)) - j(S_t(\boldsymbol{w}^h), \boldsymbol{w}^h(t)),
\end{aligned}$$

and

$$\begin{aligned}
R(t; \boldsymbol{w}(t), \boldsymbol{v}^h(t)) &= (\mathcal{A}\varepsilon(\boldsymbol{w}(t)), \varepsilon(\boldsymbol{v}^h(t) - \boldsymbol{w}(t)))_Q \\
&\quad + (\mathcal{B}\varepsilon(\mathcal{I}\boldsymbol{w}(t)), \varepsilon(\boldsymbol{v}^h(t) - \boldsymbol{w}(t)))_Q + j(S_t(\boldsymbol{w}), \boldsymbol{v}^h(t)) \\
&\quad - j(S_t(\boldsymbol{w}), \boldsymbol{w}(t)) - (\boldsymbol{f}(t), \boldsymbol{v}^h(t) - \boldsymbol{w}(t))_V
\end{aligned} \tag{10.45}$$

represents a residual quantity. Similar to the estimate (10.33), we can derive the following inequality

$$D(t; \boldsymbol{w}^h(t), \boldsymbol{v}^h(t)) \le c\, \|\boldsymbol{w}^h(t) - \boldsymbol{v}^h(t)\|_V \int_0^t \|\boldsymbol{w}(s) - \boldsymbol{w}^h(s)\|_V ds. \tag{10.46}$$

Using the assumptions (6.34), (6.35) and (10.8) on $\mathcal{A}$, $\mathcal{B}$ and g, from the inequalities (10.44) and (10.46) we have

$$\begin{aligned}\|\boldsymbol{w}(t)-\boldsymbol{w}^h(t)\|_V^2 \le c\Big\{ &\|\boldsymbol{w}(t)-\boldsymbol{w}^h(t)\|_V\,\|\boldsymbol{w}(t)-\boldsymbol{v}^h(t)\|_V \\ &+\|\mathcal{I}\boldsymbol{w}(t)-\mathcal{I}^h\boldsymbol{w}^h(t)\|_V\|\boldsymbol{w}^h(t)-\boldsymbol{v}^h(t)\|_V \\ &+|R(t;\boldsymbol{w}(t),\boldsymbol{v}^h(t))| \\ &+\|\boldsymbol{w}^h(t)-\boldsymbol{v}^h(t)\|_V\int_0^t\|\boldsymbol{w}(s)-\boldsymbol{w}^h(s)\|_V ds\Big\}.\end{aligned}$$

Use (10.42) to bound $\|\mathcal{I}\boldsymbol{w}(t)-\mathcal{I}^h\boldsymbol{w}^h(t)\|_V$,

$$\begin{aligned}\|\boldsymbol{w}(t)-\boldsymbol{w}^h(t)\|_V^2 \le c\Big\{ &\|\boldsymbol{w}(t)-\boldsymbol{w}^h(t)\|_V\,\|\boldsymbol{w}(t)-\boldsymbol{v}^h(t)\|_V \\ &+|R(t;\boldsymbol{w}(t),\boldsymbol{v}^h(t))| \\ &+\Big(\int_0^t\|\boldsymbol{w}(s)-\boldsymbol{w}^h(s)\|_V ds+\|\boldsymbol{u}_0-\boldsymbol{u}_0^h\|_V\Big) \\ &\times\big(\|\boldsymbol{w}(t)-\boldsymbol{w}^h(t)\|_V+\|\boldsymbol{w}(t)-\boldsymbol{v}^h(t)\|_V\big)\Big\}.\end{aligned}$$

After some elementary but lengthy manipulations, we obtain the inequality

$$\begin{aligned}\|\boldsymbol{w}(t)-\boldsymbol{w}^h(t)\|_V^2 \le c\Big(&\|\boldsymbol{u}_0-\boldsymbol{u}_0^h\|_V^2+\|\boldsymbol{w}(t)-\boldsymbol{v}^h(t)\|_V^2 \\ &+\int_0^t\|\boldsymbol{w}(s)-\boldsymbol{w}^h(s)\|_V^2\,ds+|R(t;\boldsymbol{w}(t),\boldsymbol{v}^h(t))|\Big).\end{aligned}$$

Applying the Gronwall inequality (Lemma 7.24) and recalling the arbitrariness of $\boldsymbol{v}^h\in C(\bar{I};V_1^h)$, we have

$$\begin{aligned}\|\boldsymbol{w}-\boldsymbol{w}^h\|_{C(\bar{I};V)} \le c &\inf_{\boldsymbol{v}^h\in C(\bar{I};V_1^h)}\Big(\|\boldsymbol{w}-\boldsymbol{v}^h\|_{C(\bar{I};V)}+\|R(\cdot;\boldsymbol{w},\boldsymbol{v}^h)\|_{C(\bar{I})}^{1/2}\Big) \\ &+c\,\|\boldsymbol{u}_0-\boldsymbol{u}_0^h\|_V.\end{aligned}\tag{10.47}$$

Combining (10.42) and (10.47) we obtain the following result.

Theorem 10.8 *Assume* (6.34), (6.35) *and* (10.8)–(10.10). *Then for the error of the semi-discrete solution of Problem 10.7, we have the estimate*

$$\begin{aligned}\|\boldsymbol{u}-\boldsymbol{u}^h\|_{C^1(\bar{I};V)} \le c &\inf_{\boldsymbol{v}^h\in C(\bar{I};V_1^h)}\Big(\|\dot{\boldsymbol{u}}-\boldsymbol{v}^h\|_{C(\bar{I};V)}+\|R(\cdot;\dot{\boldsymbol{u}},\boldsymbol{v}^h)\|_{C(\bar{I})}^{1/2}\Big) \\ &+c\,\|\boldsymbol{u}_0-\boldsymbol{u}_0^h\|_V,\end{aligned}\tag{10.48}$$

where $R(\cdot;\dot{\boldsymbol{u}}(\cdot),\boldsymbol{v}^h(\cdot))$ *is defined in* (10.45).

Theorem 10.8 is a basis for convergence analysis and for deriving error estimates for finite element solutions.

We first show the convergence of the semi-discrete scheme under the basic solution regularity $\boldsymbol{u} \in C^1(\overline{I}; V_1)$. From

$$\begin{aligned} &R(t; \dot{\boldsymbol{u}}(t), \boldsymbol{v}^h(t)) \\ &\quad = (\boldsymbol{\sigma}(t), \boldsymbol{\varepsilon}(\boldsymbol{v}^h(t) - \dot{\boldsymbol{u}}(t)))_Q + \int_{\Gamma_3} g(S_t(\dot{\boldsymbol{u}}))\,(\|\boldsymbol{v}_\tau^h(t)\| - \|\dot{\boldsymbol{u}}_\tau(t)\|)\,da \\ &\qquad - \int_\Omega \boldsymbol{f}_0(t)\cdot(\boldsymbol{v}^h(t) - \dot{\boldsymbol{u}}(t))\,dx - \int_{\Gamma_2} \boldsymbol{f}_2(t)\cdot(\boldsymbol{v}^h(t) - \dot{\boldsymbol{u}}(t))\,da, \end{aligned}$$

we have

$$|R(t; \dot{\boldsymbol{u}}(t), \boldsymbol{v}^h(t))| \le c\,\|\dot{\boldsymbol{u}}(t) - \boldsymbol{v}^h(t)\|_V,$$

where the constant c depends on $\|\boldsymbol{\sigma}(t)\|_Q$, $\|g(S_t(\dot{\boldsymbol{u}}))\|_{L^2(\Gamma_3)}$ and $\|\boldsymbol{f}(t)\|_V$. By the assumption (10.8)(b), we have

$$g(S_t(\dot{\boldsymbol{u}})) \le g(0) + L_g\|S_t(\dot{\boldsymbol{u}})\| = g(0) + L_g \int_0^t \|\dot{\boldsymbol{u}}_\tau(s)\|\,ds.$$

Then recalling the assumption (10.8) again, we see that

$$t \mapsto g(S_t(\dot{\boldsymbol{u}})) \in C(\overline{I}; L^2(\Gamma_3)).$$

Thus by (10.48),

$$\begin{aligned} \|\boldsymbol{u} - \boldsymbol{u}^h\|_{C^1(\overline{I};V)} \le c &\inf_{\boldsymbol{v}^h \in C(\overline{I};V_1^h)} \left(\|\dot{\boldsymbol{u}} - \boldsymbol{v}^h\|_{C(\overline{I};V)} + \|\dot{\boldsymbol{u}} - \boldsymbol{v}^h\|_{C(\overline{I};V)}^{1/2}\right) \\ &+ c\,\|\boldsymbol{u}_0 - \boldsymbol{u}_0^h\|_V. \end{aligned}$$

Let us apply Theorem 7.18 (page 157) with $g(e) = c\,(e + e^{1/2})$. Take $X = V_1$, $X_0 = V_1 \cap H^2(\Omega)^d$, and $X_1^h = V_1^h$. We see that Hypothesis 7.17 is satisfied by Theorem 7.10 and the finite element interpolation error estimates. So if we choose the initial value $\boldsymbol{u}_0^h \in V_1^h$ such that

$$\|\boldsymbol{u}_0 - \boldsymbol{u}_0^h\|_V \to 0 \quad \text{as } h \to 0,$$

then we have the convergence

$$\|\boldsymbol{u} - \boldsymbol{u}^h\|_{C^1(\overline{I};V)} \to 0 \quad \text{as } h \to 0.$$

Notice that we can choose the convergent initial value $\boldsymbol{u}_0^h$ by taking it to be the finite element projection of $\boldsymbol{u}_0$, or the finite element interpolant of $\boldsymbol{u}_0$ when it is continuous.

Now we derive an error estimate under solution regularity assumptions. Using the relation (10.34) we have the following expression

$$\begin{aligned} R(t;\dot{\boldsymbol{u}}(t),\boldsymbol{v}^h(t)) &= \langle \boldsymbol{\sigma}(t)\boldsymbol{\nu},\boldsymbol{v}^h(t)-\dot{\boldsymbol{u}}(t)\rangle_\Gamma \\ &\quad + \int_{\Gamma_3} g(S_t(\dot{\boldsymbol{u}}))\left(\|\boldsymbol{v}^h_\tau(t)\| - \|\dot{\boldsymbol{u}}_\tau(t)\|\right) da \\ &\quad - \int_{\Gamma_2} \boldsymbol{f}_2(t)\cdot(\boldsymbol{v}^h(t)-\dot{\boldsymbol{u}}(t))\, da. \end{aligned} \tag{10.49}$$

Assume the regularity condition (10.35). Then the relation (10.36) holds and we can simplify (10.49) to

$$\begin{aligned} R(t;\dot{\boldsymbol{u}}(t),\boldsymbol{v}^h(t)) = \int_{\Gamma_3} &\big(\boldsymbol{\sigma}_\tau(t)\cdot(\boldsymbol{v}^h_\tau(t)-\dot{\boldsymbol{u}}_\tau(t)) \\ &+ g(S_t(\dot{\boldsymbol{u}}))\left(\|\boldsymbol{v}^h_\tau(t)\| - \|\dot{\boldsymbol{u}}_\tau(t)\|\right)\big)\, da. \end{aligned} \tag{10.50}$$

Then

$$R(t;\dot{\boldsymbol{u}}(t),\boldsymbol{v}^h(t)) \le \int_{\Gamma_3} \left(\|\boldsymbol{\sigma}_\tau(t)\| + g(S_t(\dot{\boldsymbol{u}}))\right)\|\boldsymbol{v}^h_\tau(t)-\dot{\boldsymbol{u}}_\tau(t)\|\, da,$$

and we have the estimate

$$\begin{aligned} |R(t;\dot{\boldsymbol{u}}(t),\boldsymbol{v}^h(t))| &\le c\left(\|\boldsymbol{\sigma}_\tau\|_{C(\overline{I};L^2(\Gamma_3)^d)} + \|g(S_t(\dot{\boldsymbol{u}}))\|_{C(\overline{I};L^2(\Gamma_3))}\right) \\ &\quad \times \|\boldsymbol{v}^h_\tau(t)-\dot{\boldsymbol{u}}_\tau(t)\|_{L^2(\Gamma_3)^d}. \end{aligned} \tag{10.51}$$

Therefore, the estimate (10.48) reduces to

$$\begin{aligned} \|\boldsymbol{u}-\boldsymbol{u}^h\|_{C^1(\overline{I};V)} &\le c \inf_{\boldsymbol{v}^h\in C(\overline{I};V_1^h)} \left(\|\dot{\boldsymbol{u}}-\boldsymbol{v}^h\|_{C(\overline{I};V)} + \|\dot{\boldsymbol{u}}_\tau - \boldsymbol{v}^h_\tau\|^{1/2}_{C(\overline{I};L^2(\Gamma_3)^d)}\right) \\ &\quad + c\,\|\boldsymbol{u}_0-\boldsymbol{u}_0^h\|_V. \end{aligned} \tag{10.52}$$

Now assume the additional regularity conditions

$$\dot{\boldsymbol{u}} \in C(\overline{I};H^2(\Omega)^d),\ \ \dot{\boldsymbol{u}}_\tau|_{\Gamma_{3,i}} \in C(\overline{I};H^2(\Gamma_{3,i})^d),\ 1\le i\le i_0,\ \ \boldsymbol{u}_0\in H^2(\Omega)^d, \tag{10.53}$$

and take $\boldsymbol{u}_0^h = \Pi^h \boldsymbol{u}_0$. From the inequality (10.52) and the interpolation error estimates (7.32), (7.36) and (7.40) in Section 7.2, we obtain the optimal order error estimate

$$\|\boldsymbol{u}-\boldsymbol{u}^h\|_{C^1(\overline{I};V)} = O(h).$$

10.4 Fully Discrete Approximations

We consider a fully approximation of Problem 10.1 in this section. In addition to the finite element space V_1^h, we need a partition of the time interval $\overline{I}$ described in Section 7.2. Notice that no summation is implied over the repeated index n. For a sequence $\boldsymbol{u}^{hk} = \{\boldsymbol{u}_n^{hk}\}_{n=0}^N$, recall the notation $\delta_n \boldsymbol{u}_n^{hk} = (\boldsymbol{u}_n^{hk} - \boldsymbol{u}_{n-1}^{hk})/k_n$, $n = 1, \ldots, N$, for the backward divided differences. In this section, we will use the additional notation $\delta \boldsymbol{u}^{hk} = \{\delta_n \boldsymbol{u}_n^{hk}\}_{n=1}^N$.

Problem 10.9 *Find a displacement field* $\{\boldsymbol{u}_n^{hk}\}_{n=0}^N \subset V_1^h$ *such that*

$$\boldsymbol{u}_0^{hk} = \boldsymbol{u}_0^h \tag{10.54}$$

and for $n = 1, \ldots, N$,

$$\begin{aligned}(\mathcal{A}\boldsymbol{\varepsilon}(\delta_n \boldsymbol{u}_n^{hk}), \boldsymbol{\varepsilon}(\boldsymbol{v}^h - \delta_n \boldsymbol{u}_n^{hk}))_Q &+ (\mathcal{B}\boldsymbol{\varepsilon}(\boldsymbol{u}_{n-1}^{hk}), \boldsymbol{\varepsilon}(\boldsymbol{v}^h - \delta_n \boldsymbol{u}_n^{hk}))_Q \\ &+ j(S_{n-1}^k(\delta \boldsymbol{u}^{hk}), \boldsymbol{v}^h) - j(S_{n-1}^k(\delta \boldsymbol{u}^{hk}), \delta_n \boldsymbol{u}_n^{hk}) \\ &\geq (\boldsymbol{f}_n, \boldsymbol{v}^h - \delta_n \boldsymbol{u}_n^{hk})_V \quad \forall\, \boldsymbol{v}^h \in V_1^h.\end{aligned} \tag{10.55}$$

Here, $\boldsymbol{u}_0^h \in V_1^h$ is an appropriate approximation of $\boldsymbol{u}_0$ and

$$S_{n-1}^k(\delta \boldsymbol{u}^{hk}) = \sum_{j=1}^{n-1} \|(\delta_j \boldsymbol{u}_j^{hk})_\tau\|\, k_j.$$

Again to simplify the notation, we introduce the discrete velocity

$$\boldsymbol{w}^{hk} = \{\boldsymbol{w}_n^{hk}\}_{n=1}^N, \quad \boldsymbol{w}_n^{hk} = \delta_n \boldsymbol{u}_n^{hk}, \quad n = 1, \ldots, N. \tag{10.56}$$

Then using the initial value condition (10.54), we can express the discrete displacement in terms of the discrete velocity:

$$\boldsymbol{u}_n^{hk} = (\mathcal{I}^{hk} \boldsymbol{w}^{hk})_n \equiv \sum_{j=1}^n \boldsymbol{w}_j^{hk} k_j + \boldsymbol{u}_0^h. \tag{10.57}$$

Rewrite (10.55) in the form

$$\begin{aligned}(\mathcal{A}\boldsymbol{\varepsilon}(\boldsymbol{w}_n^{hk}), \boldsymbol{\varepsilon}(\boldsymbol{v}^h - \boldsymbol{w}_n^{hk}))_Q &+ (\mathcal{B}\boldsymbol{\varepsilon}((\mathcal{I}^{hk}\boldsymbol{w}^{hk})_{n-1}), \boldsymbol{\varepsilon}(\boldsymbol{v}^h - \boldsymbol{w}_n^{hk}))_Q \\ &+ j(S_{n-1}^k(\boldsymbol{w}^{hk}), \boldsymbol{v}^h) - j(S_{n-1}^k(\boldsymbol{w}^{hk}), \boldsymbol{w}_n^{hk}) \\ &\geq (\boldsymbol{f}_n, \boldsymbol{v}^h - \boldsymbol{w}_n^{hk})_V \quad \forall\, \boldsymbol{v}^h \in V_1^h.\end{aligned} \tag{10.58}$$

This is a discrete elliptic variational inequality and admits a unique solution $\boldsymbol{w}_n^{hk} \in V_1^h$. Note that $\boldsymbol{u}_0^{hk} = \boldsymbol{u}_0^h$ is given and we have the relations (10.56) and (10.57) between $\{\boldsymbol{u}_n^{hk}\}_{n=1}^N$ and $\{\boldsymbol{w}_n^{hk}\}_{n=1}^N$.

Next we give an error analysis for the fully discrete solution. Take $\boldsymbol{v} = \boldsymbol{w}_n^{hk}$ in (10.38) at $t = t_n$,

$$\begin{aligned}(\mathcal{A}\boldsymbol{\varepsilon}(\boldsymbol{w}_n), \boldsymbol{\varepsilon}(\boldsymbol{w}_n^{hk} - \boldsymbol{w}_n))_Q &+ (\mathcal{B}\boldsymbol{\varepsilon}(\boldsymbol{u}_n), \boldsymbol{\varepsilon}(\boldsymbol{w}_n^{hk} - \boldsymbol{w}_n))_Q + j(S_{t_n}(\boldsymbol{w}), \boldsymbol{w}_n^{hk}) \\ &- j(S_{t_n}(\boldsymbol{w}), \boldsymbol{w}_n) \geq (\boldsymbol{f}_n, \boldsymbol{w}_n^{hk} - \boldsymbol{w}_n)_V.\end{aligned} \tag{10.59}$$

We now add (10.58) with $\boldsymbol{v}^h = \boldsymbol{v}_n^h \in V_1^h$ and (10.59) to obtain an error relation

$$\begin{aligned}(\mathcal{A}\varepsilon(\boldsymbol{w}_n) - \mathcal{A}\varepsilon(\boldsymbol{w}_n^{hk}), &\varepsilon(\boldsymbol{w}_n) - \varepsilon(\boldsymbol{w}_n^{hk}))_Q \\ &\le (\mathcal{A}\varepsilon(\boldsymbol{w}_n^{hk}) - \mathcal{A}\varepsilon(\boldsymbol{w}_n), \varepsilon(\boldsymbol{v}_n^h) - \varepsilon(\boldsymbol{w}_n))_Q \\ &\quad + (\mathcal{B}\varepsilon(\boldsymbol{u}_{n-1}^{hk}) - \mathcal{B}\varepsilon(\boldsymbol{u}_n), \varepsilon(\boldsymbol{v}_n^h) - \varepsilon(\boldsymbol{w}_n^{hk}))_Q \\ &\quad + R_n(\boldsymbol{w}_n, \boldsymbol{v}_n^h) + D_n^k(\boldsymbol{w}^{hk}, \boldsymbol{v}_n^h), \end{aligned} \tag{10.60}$$

where

$$\begin{aligned} D_n^k(\boldsymbol{w}^{hk}, \boldsymbol{v}_n^h) = {} & j(S_{n-1}^k(\boldsymbol{w}^{hk}), \boldsymbol{v}_n^h) - j(S_{n-1}^k(\boldsymbol{w}^{hk}), \boldsymbol{w}_n^{hk}) \\ & + j(S_{t_n}(\boldsymbol{w}), \boldsymbol{w}_n^{hk}) - j(S_{t_n}(\boldsymbol{w}_n), \boldsymbol{v}_n^h), \end{aligned}$$

and

$$\begin{aligned} R_n(\boldsymbol{w}_n, \boldsymbol{v}_n^h) = {} & (\mathcal{A}\varepsilon(\boldsymbol{w}_n), \varepsilon(\boldsymbol{v}_n^h - \boldsymbol{w}_n))_Q + (\mathcal{B}\varepsilon(\boldsymbol{u}_n), \varepsilon(\boldsymbol{v}_n^h - \boldsymbol{w}_n))_Q \\ & + j(S_{t_n}(\boldsymbol{w}), \boldsymbol{v}_n^h) - j(S_{t_n}(\boldsymbol{w}), \boldsymbol{w}_n) \\ & - (\boldsymbol{f}_n, \boldsymbol{v}_n^h - \boldsymbol{w}_n)_V \end{aligned} \tag{10.61}$$

is the discrete residual. Let us estimate the terms on the right-hand side of (10.60). We have

$$\|\mathcal{B}\varepsilon(\boldsymbol{u}_{n-1}^{hk}) - \mathcal{B}\varepsilon(\boldsymbol{u}_n)\|_Q \le c\,\|\boldsymbol{u}_{n-1}^{hk} - \boldsymbol{u}_n\|_V.$$

We can apply (7.42)–(7.43) to bound the term $\|\boldsymbol{u}_{n-1}^{hk} - \boldsymbol{u}_n\|_V$. Then,

$$\begin{aligned} \|\mathcal{B}\varepsilon(\boldsymbol{u}_{n-1}^{hk}) - \mathcal{B}\varepsilon(\boldsymbol{u}_n)\|_Q \le {} & c\sum_{j=1}^{n-1} \|\boldsymbol{w}_j^{hk} - \boldsymbol{w}_j\|_V k_j + c\,\|\boldsymbol{u}_0^h - \boldsymbol{u}_0\|_V \\ & + c\,I_k(\boldsymbol{w}) + c\,k\,\|\boldsymbol{w}\|_{C(\overline{I};V)}, \end{aligned} \tag{10.62}$$

where

$$I_k(\boldsymbol{w}) = \sum_{j=1}^{N} \int_{t_{j-1}}^{t_j} \|\boldsymbol{w}_j - \boldsymbol{w}(s)\|_V ds.$$

Finally,

$$\begin{aligned} D_n^k(\boldsymbol{w}^{hk}, \boldsymbol{v}_n^h) &= \int_{\Gamma_3} (g(S_{t_n}(\boldsymbol{w})) - g(S_{n-1}^k \boldsymbol{w}^{hk}))\,(\|(\boldsymbol{w}_n^{hk})_\tau\| - \|(\boldsymbol{v}_n^h)_\tau\|)\,da \\ &\le L_g \int_{\Gamma_3} |S_{t_n}(\boldsymbol{w}) - S_{n-1}^k(\boldsymbol{w}^{hk})|\,\|(\boldsymbol{w}_n^{hk})_\tau - (\boldsymbol{v}_n^h)_\tau\|\,da. \end{aligned}$$

Here

$$S_{t_n}(\boldsymbol{w}) - S_{n-1}^k(\boldsymbol{w}^{hk}) = \int_0^{t_n} \|\boldsymbol{w}_\tau(s)\|\,ds - \sum_{j=1}^{n-1} \|(\boldsymbol{w}_j^{hk})_\tau\|\,k_j.$$

Similar to the inequality (10.62), we have

$$D_n^k(\boldsymbol{w}^{hk}, \boldsymbol{v}_n^h) \le c\Big(\sum_{j=1}^{n-1} \|\boldsymbol{w}_j^{hk} - \boldsymbol{w}_j\|_V k_j + I_k(\boldsymbol{w})\Big) \cdot \|(\boldsymbol{w}_n^{hk})_\tau - (\boldsymbol{v}_n^h)_\tau\|_{L^2(\Gamma_3)^d}. \tag{10.63}$$

From the inequalities (10.60), (10.62) and (10.63), and the assumptions (6.34), (6.35) and (10.8) on $\mathcal{A}$, $\mathcal{B}$ and g, we obtain

$$\begin{aligned}
&\|\boldsymbol{w}_n - \boldsymbol{w}_n^{hk}\|_V^2 \\
&\quad \le c\Big\{\|\boldsymbol{w}_n - \boldsymbol{w}_n^{hk}\|_V \, \|\boldsymbol{w}_n - \boldsymbol{v}_n^h\|_V + |R_n(\boldsymbol{w}_n, \boldsymbol{v}_n^h)| \\
&\qquad + \Big(\sum_{j=1}^{n-1} \|\boldsymbol{w}_j^{hk} - \boldsymbol{w}_j\|_V k_j + \|\boldsymbol{u}_0^h - \boldsymbol{u}_0\|_V + I_k(\boldsymbol{w}) + k\,\|\boldsymbol{w}\|_{C(\bar I;V)}\Big) \\
&\qquad \times \left(\|\boldsymbol{w}_n - \boldsymbol{w}_n^{hk}\|_V + \|\boldsymbol{w}_n - \boldsymbol{v}_n^h\|_V\right)\Big\}.
\end{aligned}$$

Hence, for $n = 1, 2, \dots, N$,

$$\begin{aligned}
\|\boldsymbol{w}_n - \boldsymbol{w}_n^{hk}\|_V \le c\Big\{&\|\boldsymbol{v}_n^h - \boldsymbol{w}_n\|_V + |R_n(\boldsymbol{w}_n, \boldsymbol{v}_n^h)|^{1/2} + \|\boldsymbol{u}_0^h - \boldsymbol{u}_0\|_V \\
&+ I_k(\boldsymbol{w}) + k\,\|\boldsymbol{w}\|_{C(\bar I;V)} + \sum_{j=1}^{n-1} \|\boldsymbol{w}_j^{hk} - \boldsymbol{w}_j\|_V k_j\Big\}.
\end{aligned} \tag{10.64}$$

To proceed further, we apply Lemma 7.25 (page 163) to obtain the following estimate

$$\begin{aligned}
\max_n \|\boldsymbol{w}_n - \boldsymbol{w}_n^{hk}\|_V \le c \max_n &\left\{\|\boldsymbol{v}_n^h - \boldsymbol{w}_n\|_V + |R_n(\boldsymbol{w}_n, \boldsymbol{v}_n^h)|^{1/2}\right\} \\
&+ c\left(\|\boldsymbol{u}_0^h - \boldsymbol{u}_0\|_V + I_k(\boldsymbol{w}) + k\,\|\boldsymbol{w}\|_{C(\bar I;V)}\right).
\end{aligned}$$

We can apply (7.46) to estimate the error in the displacement:

$$\|\boldsymbol{u}_n^{hk} - \boldsymbol{u}_n\|_V \le \sum_{j=1}^{n} \|\boldsymbol{w}_j^{hk} - \boldsymbol{w}_j\|_V k_j + \|\boldsymbol{u}_0^h - \boldsymbol{u}_0\|_V + c\, I_k(\boldsymbol{w}).$$

Combining the above inequalities, we arrive at the following result.

Theorem 10.10 *Assume* (6.34), (6.35) *and* (10.8)–(10.10). *Then for the error of the fully discrete solution of Problem 10.9, we have the estimate*

$$\begin{aligned}
&\max_{1\le n\le N} (\|\boldsymbol{u}_n^{hk} - \boldsymbol{u}_n\|_V + \|\dot{\boldsymbol{u}}_n - \delta_n \boldsymbol{u}_n^{hk}\|_V) \\
&\quad \le c \max_{1\le n\le N} \inf_{\boldsymbol{v}_n^h \in V_1^h} \left\{\|\boldsymbol{v}_n^h - \dot{\boldsymbol{u}}_n\|_V + |R_n(\dot{\boldsymbol{u}}_n, \boldsymbol{v}_n^h)|^{1/2}\right\} \\
&\qquad + c\left(\|\boldsymbol{u}_0^h - \boldsymbol{u}_0\|_V + I_k(\dot{\boldsymbol{u}}) + k\,\|\dot{\boldsymbol{u}}\|_{C(\bar I;V)}\right),
\end{aligned} \tag{10.65}$$

where $R_n(\dot{\boldsymbol{u}}_n, \boldsymbol{v}_n^h)$ *is defined by* (10.61).

The inequality (10.65) is the basis for convergence analysis and error estimation. First we have

$$|R_n(\boldsymbol{v}_n^h, \dot{\boldsymbol{u}}_n)| \le c\, \|\boldsymbol{v}_n^h - \dot{\boldsymbol{u}}_n\|_V$$

for some constant c depending on the solution. So from (10.65) we have

$$\begin{aligned} &\max_{1\le n\le N} (\|\boldsymbol{u}_n^{hk} - \boldsymbol{u}_n\|_V + \|\dot{\boldsymbol{u}}_n - \delta_n \boldsymbol{u}_n^{hk}\|_V) \\ &\quad \le c \max_{1\le n\le N} \inf_{\boldsymbol{v}_n^h \in V_1^h} \left\{ \|\boldsymbol{v}_n^h - \dot{\boldsymbol{u}}_n\|_V + \|\boldsymbol{v}_n^h - \dot{\boldsymbol{u}}_n\|_V^{1/2} \right\} \\ &\quad\quad + c\left(\|\boldsymbol{u}_0^h - \boldsymbol{u}_0\|_V + I_k(\dot{\boldsymbol{u}}) + k\, \|\dot{\boldsymbol{u}}\|_{C(\bar{I};V)}\right). \end{aligned}$$

Let us apply Theorem 7.22 (page 160). We have the inequality (7.52) with $X = V_1$, $X^h = V_1^h$, $g(e) = c\,(e + e^{1/2})$, $b(k) = c\,(I_k(\dot{\boldsymbol{u}}) + k\, \|\dot{\boldsymbol{u}}\|_{C(\bar{I};V)})$. The assumptions of Theorem 7.22 are satisfied. Therefore if we choose the initial value $\boldsymbol{u}_0^h$ such that

$$\|\boldsymbol{u}_0^h - \boldsymbol{u}_0\|_V \to 0 \quad \text{as } h \to 0,$$

we have the convergence

$$\max_{1\le n\le N} (\|\boldsymbol{u}_n^{hk} - \boldsymbol{u}_n\|_V + \|\dot{\boldsymbol{u}}_n - \delta_n \boldsymbol{u}_n^{hk}\|_V) \to 0 \quad \text{as } h, k \to 0.$$

Error estimates can be derived similar to that for the semi-discrete solutions. We conclude that under the additional regularity conditions (10.35) and (10.53), we have the optimal order error estimate

$$\max_{1\le n\le N} (\|\boldsymbol{u}_n^{hk} - \boldsymbol{u}_n\|_V + \|\dot{\boldsymbol{u}}_n - \delta_n \boldsymbol{u}_n^{hk}\|_V) = O(h + k).$$

10.5 Dual Formulation

Problem 10.1 represents a variational formulation of the frictional contact problem (10.2)–(10.7) in terms of the displacement and may be referred to as the *primal formulation* of this mechanical problem. As in Section 9.4, we can derive and study a *dual formulation* of the problem in terms of the stress. Dual variational formulations for various frictional contact problems involving Kelvin-Voigt viscoelastic materials of the form (10.2) are considered for example in [17, 18, 180]. There, existence and uniqueness results are shown using the equivalence between a primal problem and its dual.

We study in this section a dual formulation of the problem (10.2)–(10.7) in the case of the classical Tresca friction law, i.e. in the case when the friction bound g is given. Thus, we consider the following condition on the contact surface:

$$\left.\begin{array}{l} u_\nu = 0, \quad \|\boldsymbol{\sigma}_\tau\| \le g, \\ \|\boldsymbol{\sigma}_\tau\| < g \;\Rightarrow\; \dot{\boldsymbol{u}}_\tau = \boldsymbol{0}, \\ \|\boldsymbol{\sigma}_\tau\| = g \;\Rightarrow\; \exists\lambda \ge 0 \text{ s.t. } \boldsymbol{\sigma}_\tau = -\lambda \dot{\boldsymbol{u}}_\tau \end{array}\right\} \quad \text{on } \Gamma_3 \times I. \tag{10.66}$$

We assume that (6.34), (6.35), (10.9) and (10.10) hold and the friction bound g satisfies

$$g \in L^2(\Gamma_3), \quad g(\boldsymbol{x}) \geq 0 \text{ a.e. on } \Gamma_3. \tag{10.67}$$

Moreover, we denote by $j : V_1 \to \mathbb{R}$ the functional

$$j(\boldsymbol{v}) = \int_{\Gamma_3} g \, \|\boldsymbol{v}_\tau\| \, da \qquad \forall \boldsymbol{v} \in V_1. \tag{10.68}$$

Using the results in Section 10.1 we obtain the following variational formulation of problem (10.2)–(10.5), (10.7), (10.66), in terms of the displacement.

Problem 10.11 *Find a displacement* $\boldsymbol{u} : \overline{I} \to V_1$ *such that*

$$\boldsymbol{u}(0) = \boldsymbol{u}_0 \tag{10.69}$$

and for all $t \in \overline{I}$,

$$\begin{aligned}(\mathcal{A}\boldsymbol{\varepsilon}(\dot{\boldsymbol{u}}(t)), \boldsymbol{\varepsilon}(\boldsymbol{v} - \dot{\boldsymbol{u}}(t)))_Q + (\mathcal{B}\boldsymbol{\varepsilon}(\boldsymbol{u}(t)), \boldsymbol{\varepsilon}(\boldsymbol{v} - \dot{\boldsymbol{u}}(t)))_Q + j(\boldsymbol{v}) - j(\dot{\boldsymbol{u}}(t))\\ \geq (\boldsymbol{f}(t), \boldsymbol{v} - \dot{\boldsymbol{u}}(t))_V \quad \forall \boldsymbol{v} \in V_1.\end{aligned} \tag{10.70}$$

To express the mechanical problem in terms of the stress, we use similar arguments as in Section 9.4. We first notice that by the classical Cauchy-Lipschitz theorem (the C^1-version of Theorem 9.9, page 202), for any $\boldsymbol{\sigma} \in C(\overline{I}; Q)$ there exists a unique function $\boldsymbol{\tau} \in C^1(\overline{I}; Q)$ such that

$$\mathcal{A}\dot{\boldsymbol{\tau}}(t) + \mathcal{B}\boldsymbol{\tau}(t) = \boldsymbol{\sigma}(t) \quad \forall t \in \overline{I}, \tag{10.71}$$

$$\boldsymbol{\tau}(0) = \boldsymbol{\varepsilon}(\boldsymbol{u}_0). \tag{10.72}$$

This result allows us to define an operator $\mathcal{T} : C(\overline{I}; Q) \to C^1(\overline{I}; Q)$ by

$$\mathcal{T}\boldsymbol{\sigma} = \boldsymbol{\tau}, \tag{10.73}$$

where $\boldsymbol{\tau}$ is the unique solution of the Cauchy problem (10.71)–(10.72).

For all $t \in \overline{I}$ we define the set $\Sigma(t)$ by

$$\Sigma(t) = \{\, \boldsymbol{\tau} \in Q : (\boldsymbol{\tau}, \boldsymbol{\varepsilon}(\boldsymbol{v}))_Q + j(\boldsymbol{v}) \geq (\boldsymbol{f}(t), \boldsymbol{v})_V \ \forall \boldsymbol{v} \in V_1 \,\}. \tag{10.74}$$

Note that the set $\Sigma(t)$ is nonempty since by the definition of the norm of V, we trivially have

$$(\boldsymbol{\varepsilon}(\boldsymbol{f}(t)), \boldsymbol{\varepsilon}(\boldsymbol{v}))_Q = (\boldsymbol{f}(t), \boldsymbol{v})_V \quad \forall \boldsymbol{v} \in V_1.$$

Thus, $\boldsymbol{\varepsilon}(\boldsymbol{f}(t)) \in \Sigma(t)$.

We consider the following variational problem.

Problem 10.12 *Find a stress* $\boldsymbol{\sigma} : \overline{I} \to Q$ *such that for all* $t \in \overline{I}$,

$$\boldsymbol{\sigma}(t) \in \Sigma(t), \quad \Big(\frac{d}{dt}(\mathcal{T}\boldsymbol{\sigma})(t), \boldsymbol{\tau} - \boldsymbol{\sigma}(t)\Big)_Q \geq 0 \quad \forall \boldsymbol{\tau} \in \Sigma(t). \tag{10.75}$$

The relation between Problem 10.11 and Problem 10.12 is studied next.

Theorem 10.13 *Assume* (6.34), (6.35), (10.9), (10.10) *and* (10.67).
i) *If* $\boldsymbol{u}$ *is the solution of Problem 10.11 and* $\boldsymbol{\sigma}$ *is defined by*

$$\boldsymbol{\sigma} = \mathcal{A}\boldsymbol{\varepsilon}(\dot{\boldsymbol{u}}) + \mathcal{B}\boldsymbol{\varepsilon}(\boldsymbol{u}), \tag{10.76}$$

then $\boldsymbol{\sigma}$ *is the solution of Problem 10.12 and* $\boldsymbol{\sigma} \in C(\overline{I}; Q_1)$.
ii) *Conversely, let* $\boldsymbol{\sigma} \in C(\overline{I}; Q_1)$ *be a solution of Problem 10.12. Then, there exists a unique function* $\boldsymbol{u} \in C^1(\overline{I}; V_1)$ *such that* (10.76) *and* (10.69) *hold. Moreover,* $\boldsymbol{u}$ *is a solution of Problem 10.11.*

Proof. i) Let $t \in \overline{I}$. Using (10.76) we can rewrite inequality (10.70) as

$$(\boldsymbol{\sigma}(t), \boldsymbol{\varepsilon}(\boldsymbol{v}) - \boldsymbol{\varepsilon}(\dot{\boldsymbol{u}}(t)))_Q + j(\boldsymbol{v}) - j(\dot{\boldsymbol{u}}(t)) \geq (\boldsymbol{f}(t), \boldsymbol{v} - \dot{\boldsymbol{u}}(t))_V \quad \forall \boldsymbol{v} \in V_1. \tag{10.77}$$

Choosing in (10.77) $\boldsymbol{v} = 2\dot{\boldsymbol{u}}(t)$ and $\boldsymbol{v} = \boldsymbol{0}$, both in V_1, we find

$$(\boldsymbol{\sigma}(t), \boldsymbol{\varepsilon}(\dot{\boldsymbol{u}}(t)))_Q + j(\dot{\boldsymbol{u}}(t)) = (\boldsymbol{f}(t), \dot{\boldsymbol{u}}(t))_V. \tag{10.78}$$

Then from (10.77) again,

$$(\boldsymbol{\sigma}(t), \boldsymbol{\varepsilon}(\boldsymbol{v}))_Q + j(\boldsymbol{v}) \geq (\boldsymbol{f}(t), \boldsymbol{v})_V \quad \forall \boldsymbol{v} \in V_1. \tag{10.79}$$

It is easy to see that (10.77) is equivalent to (10.78) and (10.79). So we see that $\boldsymbol{\sigma}(t) \in \Sigma(t)$. Moreover, from (10.74) and (10.78) we deduce that

$$(\boldsymbol{\varepsilon}(\dot{\boldsymbol{u}}(t)), \boldsymbol{\tau} - \boldsymbol{\sigma}(t))_Q \geq 0 \qquad \forall \boldsymbol{\tau} \in \Sigma(t). \tag{10.80}$$

Using (10.69) and (10.76) we have

$$\boldsymbol{\varepsilon}(\boldsymbol{u}) = \mathcal{T}\boldsymbol{\sigma}. \tag{10.81}$$

The inequality (10.75) is now a consequence of (10.80) and (10.81). Moreover, since $\boldsymbol{\sigma}(t) \in \Sigma(t)$, it follows that $\operatorname{Div} \boldsymbol{\sigma}(t) + \boldsymbol{f}_0(t) = \boldsymbol{0}$ a.e. in Ω. Therefore, from (10.76) and (10.9) we find $\boldsymbol{\sigma} \in C(\overline{I}; Q_1)$.

ii) Conversely, let $\boldsymbol{\sigma} \in C(\overline{I}; Q_1)$ be a solution of Problem 10.12 and let $\boldsymbol{z}$ be given by

$$\boldsymbol{z} = \mathcal{T}\boldsymbol{\sigma} \in C^1(\overline{I}; Q).$$

Using (10.75) we find that for each $t \in \overline{I}$,

$$(\dot{\boldsymbol{z}}(t), \boldsymbol{\tau} - \boldsymbol{\sigma}(t)) \geq 0 \quad \forall \boldsymbol{\tau} \in \Sigma(t). \tag{10.82}$$

Let $\tilde{\boldsymbol{z}} \in Q$ be such that $(\tilde{\boldsymbol{z}}, \boldsymbol{\varepsilon}(\boldsymbol{v}))_Q = 0$ for all $\boldsymbol{v} \in V_1$. Using (10.74) we deduce that $\boldsymbol{\sigma}(t) \pm \tilde{\boldsymbol{z}}$ belong to $\Sigma(t)$. Choosing $\boldsymbol{\tau} = \boldsymbol{\sigma}(t) \pm \tilde{\boldsymbol{z}}$ in (10.82) we obtain $(\dot{\boldsymbol{z}}(t), \tilde{\boldsymbol{z}})_Q = 0$ for all $t \in \overline{I}$. Using Korn's inequality (7.8) (page 144) we know that the set $\boldsymbol{\varepsilon}(V_1) = \{\, \boldsymbol{\varepsilon}(\boldsymbol{v}) : \boldsymbol{v} \in V_1 \,\}$ is a closed subspace of Q. Therefore,

$\varepsilon(V_1^\perp)^\perp = \varepsilon(V_1)$, where $\perp$ represents the orthogonal complement in Q. Then, $\dot{\boldsymbol{z}}(t) \in \varepsilon(V_1^\perp)^\perp$ and also $\boldsymbol{z}(t) \in \varepsilon(V_1)$ for each $t \in \overline{I}$, since $\boldsymbol{z}(0) = \varepsilon(\boldsymbol{u}_0)$ and $\boldsymbol{u}_0 \in V_1$. Thus, there exists $\boldsymbol{u}(t) \in V_1$ such that

$$\boldsymbol{z}(t) = \varepsilon(\boldsymbol{u}(t)) \quad \forall\, t \in \overline{I}. \tag{10.83}$$

Since $\boldsymbol{z} \in C^1(\overline{I}; Q)$, using Korn's inequality again, we obtain that $\boldsymbol{u} \in C^1(\overline{I}; V_1)$. Moreover, it follows from (10.71)–(10.73) and (10.83) that $\boldsymbol{u}$ satisfies (10.69). The uniqueness of $\boldsymbol{u}$ follows from (10.83) and Korn's inequality.

We now prove that $\boldsymbol{u}$ is a solution of Problem 10.11. We note that, by the subdifferentiability of j on V_1, for any $t \in \overline{I}$, there exists $\bar{\boldsymbol{\tau}}(t) \in Q$ such that

$$(\bar{\boldsymbol{\tau}}(t), \varepsilon(\boldsymbol{v}) - \varepsilon(\dot{\boldsymbol{u}}(t)))_Q + j(\boldsymbol{v}) - j(\dot{\boldsymbol{u}}(t)) \geq (\boldsymbol{f}(t), \boldsymbol{v} - \dot{\boldsymbol{u}}(t))_V \quad \forall\, \boldsymbol{v} \in V_1. \tag{10.84}$$

We choose $\boldsymbol{v} = 2\dot{\boldsymbol{u}}(t)$ and then $\boldsymbol{v} = \boldsymbol{0}$ in (10.84) to find

$$(\bar{\boldsymbol{\tau}}(t), \varepsilon(\dot{\boldsymbol{u}}(t)))_Q + j(\dot{\boldsymbol{u}}(t)) = (\boldsymbol{f}(t), \dot{\boldsymbol{u}}(t))_V. \tag{10.85}$$

Using (10.84) and (10.85) again, we obtain $\bar{\boldsymbol{\tau}}(t) \in \Sigma(t)$. Therefore, since $\boldsymbol{z} = \mathcal{T}\boldsymbol{\sigma}$, by (10.83) and (10.75) we get

$$(\bar{\boldsymbol{\tau}}(t), \varepsilon(\dot{\boldsymbol{u}}(t)))_Q \geq (\boldsymbol{\sigma}(t), \varepsilon(\dot{\boldsymbol{u}}(t)))_Q. \tag{10.86}$$

From (10.85) and (10.86) it follows that

$$(\boldsymbol{f}(t), \dot{\boldsymbol{u}}(t))_V \geq (\boldsymbol{\sigma}(t), \varepsilon(\dot{\boldsymbol{u}}(t)))_Q + j(\dot{\boldsymbol{u}}(t)).$$

Since the reverse inequality follows from the fact that $\boldsymbol{\sigma}(t) \in \Sigma(t)$, for each $t \in \overline{I}$, we obtain (10.78) which, in turn, implies (10.77). The inequality (10.70) follows now from (10.76) and (10.77). This concludes the proof of Theorem 10.13. ■

Based on Theorem 10.13 it is natural to call Problem 10.12 a *dual formulation* of the variational problem 10.11. Moreover, since Theorem 10.2 states that the primal problem 10.11 has a unique solution $\boldsymbol{u} \in C^1(\overline{I}; V_1)$, it follows from Theorem 10.13 that the dual problem 10.12 has a unique solution $\boldsymbol{\sigma} \in C(\overline{I}; Q_1)$.

Chapter 11

Frictional Contact with Normal Compliance

For the problem studied in this chapter, the contact is modeled with normal compliance, i.e. the interpenetration of the body's surface into the foundation is allowed. The friction is modeled with a general version of Coulomb's law of dry friction. The weak formulation of the problem in terms of the displacement leads to an evolutionary variational inequality that we study in an abstract framework. We prove a general existence and uniqueness result, using standard arguments of time-dependent elliptic variational inequalities and Banach's fixed point theorem. We then consider numerical approximations of the abstract problem. We use the finite element method to discretize the spatial domain and we introduce spatially semi-discrete and fully discrete schemes. For both schemes, we show the existence of unique solutions and derive error estimates. We then apply the results on the abstract problem to the analysis and numerical approximations of the viscoelastic contact problem with normal compliance and friction. We also establish a continuous dependence result of the solution on the normal compliance functions. In the last section of the chapter, we present some numerical examples to show the performance of the fully discrete scheme.

11.1 Problem Statement

We assume there is a gap g between the potential contact surface Γ_3 and the foundation, measured along the outward normal direction. The classical formulation of the mechanical problem of frictional contact of the viscoelastic body is the following:

Find a displacement field $\boldsymbol{u} : \Omega \times \overline{I} \to \mathbb{R}^d$ and a stress field $\boldsymbol{\sigma} : \Omega \times \overline{I} \to \mathbb{S}^d$

such that

$$\boldsymbol{\sigma} = \mathcal{A}\varepsilon(\dot{\boldsymbol{u}}) + \mathcal{B}\varepsilon(\boldsymbol{u}) \quad \text{in } \Omega \times I, \tag{11.1}$$

$$\mathrm{Div}\,\boldsymbol{\sigma} + \boldsymbol{f}_0 = \boldsymbol{0} \quad \text{in } \Omega \times I, \tag{11.2}$$

$$\boldsymbol{u} = \boldsymbol{0} \quad \text{on } \Gamma_1 \times I, \tag{11.3}$$

$$\boldsymbol{\sigma\nu} = \boldsymbol{f}_2 \quad \text{on } \Gamma_2 \times I, \tag{11.4}$$

$$-\sigma_\nu = p_\nu(u_\nu - g) \quad \text{on } \Gamma_3 \times I, \tag{11.5}$$

$$\left.\begin{aligned} &\|\boldsymbol{\sigma}_\tau\| \le p_\tau(u_\nu - g), \\ &\|\boldsymbol{\sigma}_\tau\| < p_\tau(u_\nu - g) \;\Rightarrow\; \dot{\boldsymbol{u}}_\tau = \boldsymbol{0}, \\ &\|\boldsymbol{\sigma}_\tau\| = p_\tau(u_\nu - g) \;\Rightarrow\; \exists\, \lambda \ge 0 \text{ s.t. } \boldsymbol{\sigma}_\tau = -\lambda \dot{\boldsymbol{u}}_\tau \end{aligned}\right\} \quad \text{on } \Gamma_3 \times I, \tag{11.6}$$

$$\boldsymbol{u}(0) = \boldsymbol{u}_0 \quad \text{in } \Omega. \tag{11.7}$$

The equality (11.5) represents the *normal compliance* contact condition where p_ν is a prescribed function. The difference $u_\nu - g$, when positive, represents the penetration of the surface asperities into those of the foundation.

The relations (11.6) represent a version of Coulomb's law of dry friction. Here p_τ is a prescribed nonnegative function, called the *friction bound.* The law (11.6) states that the tangential shear cannot exceed the maximal frictional resistance $p_\tau(u_\nu - g)$. If the strict inequality holds, the surface adheres to the foundation and is in the so-called *stick* state, and when the equality holds there is relative sliding, the so-called *slip* state. Therefore, at any time instant the contact surface Γ_3 is divided into three zones: the stick zone, the slip zone, and the zone of separation in which $u_\nu < g$ and there is no contact. The boundaries of these zones are unknown a priori and are part of the problem. When

$$p_\tau = \mu\, p_\nu \tag{11.8}$$

we obtain the usual Coulomb law of dry friction, with $\mu \ge 0$ being the coefficient of friction (see, e.g., [70] or [167]). Recently a modified version of the Coulomb friction law has been derived in [201, 202] from thermodynamic considerations. It consists of the use of the friction law (11.6) with

$$p_\tau = \mu\, p_\nu (1 - \delta p_\nu)_+, \tag{11.9}$$

where δ is a small positive material constant related to the wear and hardness of the surface and $\mu \ge 0$ is the coefficient of friction.

In the study of the mechanical problem (11.1)–(11.7) we assume that the viscosity operator $\mathcal{A}$ and the elasticity operator $\mathcal{B}$ satisfy the conditions (6.34) and (6.35). The *normal compliance functions* p_r $(r = \nu, \tau)$ satisfy

$$\left.\begin{aligned} &\text{(a) } p_r : \Gamma_3 \times \mathbb{R} \to \mathbb{R}_+. \\ &\text{(b) } \exists\, L_r > 0 \text{ s.t. } \forall\, u_1, u_2 \in \mathbb{R},\ \text{a.e. } \boldsymbol{x} \in \Omega, \\ &\qquad |p_r(\boldsymbol{x}, u_1) - p_r(\boldsymbol{x}, u_2)| \le L_r\, |u_1 - u_2|. \\ &\text{(c) For any } u \in \mathbb{R},\ \boldsymbol{x} \mapsto p_r(\boldsymbol{x}, u) \text{ is measurable.} \\ &\text{(d) The mapping } \boldsymbol{x} \mapsto p_r(\boldsymbol{x}, 0) \in L^2(\Gamma_3). \end{aligned}\right\} \tag{11.10}$$

We observe that the assumptions (11.10) on the functions p_ν and p_τ are pretty general except the assumption (11.10)(b) which, roughly speaking, requires the functions to grow at most linearly. It is easily seen that the functions defined in (5.42) and (5.44) (page 107) satisfy the condition (11.10)(b). We also observe that if the functions p_ν and p_τ are related by (11.8) or (11.9) and p_ν satisfies the condition (11.10)(b), then p_τ also satisfies the condition (11.10)(b) with $L_\tau = \mu\, L_\nu$. So our results below are valid for the boundary value problems associated with these choices of the normal compliance functions.

We assume that the force and traction densities satisfy

$$\boldsymbol{f}_0 \in C(\overline{I}; L^2(\Omega)^d), \qquad \boldsymbol{f}_2 \in C(\overline{I}; L^2(\Gamma_2)^d), \tag{11.11}$$

and the gap function satisfies

$$g \in L^2(\Omega), \quad g \geq 0 \text{ a.e. on } \Gamma_3. \tag{11.12}$$

Finally, the initial displacement has the regularity

$$\boldsymbol{u}_0 \in V. \tag{11.13}$$

We denote by $\boldsymbol{f}(t)$ the element of V given by

$$(\boldsymbol{f}(t), \boldsymbol{v})_V = \int_\Omega \boldsymbol{f}_0(t) \cdot \boldsymbol{v}\, dx + \int_{\Gamma_2} \boldsymbol{f}_2(t) \cdot \boldsymbol{v}\, da \tag{11.14}$$

for all $\boldsymbol{v} \in V$ and $t \in \overline{I}$. Note that conditions (11.11) imply

$$\boldsymbol{f} \in C(\overline{I}; V). \tag{11.15}$$

Let $j : V \times V \to \mathbb{R}$ be the functional

$$j(\boldsymbol{v}, \boldsymbol{w}) = \int_{\Gamma_3} p_\nu(v_\nu - g)\, w_\nu\, da + \int_{\Gamma_3} p_\tau(v_\nu - g)\, \|\boldsymbol{w}_\tau\|\, da \quad \forall\, \boldsymbol{v}, \boldsymbol{w} \in V. \tag{11.16}$$

The integrals in (11.16) are well defined by the assumptions (11.10) and (11.12). With these notations, if $(\boldsymbol{u}, \boldsymbol{\sigma})$ are sufficiently regular functions satisfying (11.1)–(11.6), then for all $t \in \overline{I}$, $\boldsymbol{u}(t) \in V$ and

$$\begin{aligned}(\boldsymbol{\sigma}(t), \boldsymbol{\varepsilon}(\boldsymbol{v} - \dot{\boldsymbol{u}}(t)))_Q + j(\boldsymbol{u}(t), \boldsymbol{v}) - j(\boldsymbol{u}(t), \dot{\boldsymbol{u}}(t))& \\ \geq (\boldsymbol{f}(t), \boldsymbol{v} - \dot{\boldsymbol{u}}(t))_V \quad \forall\, \boldsymbol{v} \in V.&\end{aligned} \tag{11.17}$$

Thus, from (11.1), (11.7) and (11.17) we obtain the following variational formulation in terms of the displacement for the problem (11.1)–(11.7).

Problem 11.1 *Find a displacement $\boldsymbol{u} : \overline{I} \to V$ such that*

$$\boldsymbol{u}(0) = \boldsymbol{u}_0 \tag{11.18}$$

and for all $t \in \overline{I}$,

$$\begin{aligned}(\mathcal{A}\boldsymbol{\varepsilon}(\dot{\boldsymbol{u}}(t)), \boldsymbol{\varepsilon}(\boldsymbol{v}) - \boldsymbol{\varepsilon}(\dot{\boldsymbol{u}}(t)))_Q + (\mathcal{B}\boldsymbol{\varepsilon}(\boldsymbol{u}(t)), \boldsymbol{\varepsilon}(\boldsymbol{v}) - \boldsymbol{\varepsilon}(\dot{\boldsymbol{u}}(t)))_Q + j(\boldsymbol{u}(t), \boldsymbol{v})& \\ - j(\boldsymbol{u}(t), \dot{\boldsymbol{u}}(t)) \geq (\boldsymbol{f}(t), \boldsymbol{v} - \dot{\boldsymbol{u}}(t))_V \quad \forall\, \boldsymbol{v} \in V.&\end{aligned} \tag{11.19}$$

Well-posedness of Problem 11.1 is discussed in the framework of an abstract problem in the next section.

11.2 An Abstract Problem and its Well-posedness

The following abstract problem includes Problem 11.1 as a special case.

Problem 11.2 *Find* $u : \overline{I} \to X$ *such that*

$$u(0) = u_0 \tag{11.20}$$

and for $t \in \overline{I}$,

$$\begin{aligned}(A\dot{u}(t), v - \dot{u}(t))_X + (Bu(t), v - \dot{u}(t))_X + j(u(t), v)\\ - j(u(t), \dot{u}(t)) \geq (f(t), v - \dot{u}(t))_X \quad \forall\, v \in X.\end{aligned} \tag{11.21}$$

Here X is a real Hilbert space endowed with the inner product $(\cdot,\cdot)_X$ and the associated norm $\|\cdot\|_X$. The operators A and B are generally nonlinear on X, and u_0 is the initial datum.

We assume that $A : X \to X$ is a strongly monotone Lipschitz continuous operator, i.e.,

$$\left.\begin{array}{l}\text{(a) there exists } M > 0 \text{ such that}\\ \quad (Au_1 - Au_2, u_1 - u_2)_X \geq M\|u_1 - u_2\|_X^2\\ \quad \forall\, u_1, u_2 \in X.\\ \text{(b) there exists } L_A > 0 \text{ such that}\\ \quad \|Au_1 - Au_2\|_X \leq L_A\|u_1 - u_2\|_X \quad \forall\, u_1, u_2 \in X.\end{array}\right\} \tag{11.22}$$

The nonlinear operator $B : X \to X$ is Lipschitz continuous, i.e. there exists $L_B > 0$ such that

$$\|Bu_1 - Bu_2\|_X \leq L_B\,\|u_1 - u_2\|_X \quad \forall\, u_1, u_2 \in X. \tag{11.23}$$

The functional $j : X \times X \to \mathbb{R}$ satisfies

$$\left.\begin{array}{l}\text{(a) for all } g \in X,\ j(g,\cdot) \text{ is convex and l.s.c. on } X.\\ \text{(b) there exists } m > 0 \text{ such that}\\ \quad j(g_1, v_2) - j(g_1, v_1) + j(g_2, v_1) - j(g_2, v_2)\\ \quad \leq m\,\|g_1 - g_2\|_X\,\|v_1 - v_2\|_X \quad \forall\, g_1, g_2, v_1, v_2 \in X.\end{array}\right\} \tag{11.24}$$

Finally, we assume that

$$f \in C(\overline{I}; X) \tag{11.25}$$

and

$$u_0 \in X. \tag{11.26}$$

The main result of this section is the following.

Theorem 11.3 *Let* (11.22)–(11.26) *hold. Then there exists a unique solution* $u \in C^1(\overline{I}; X)$ *to Problem 11.2.*

The proof of Theorem 11.3 is based on fixed point arguments and is established in several steps. To simplify the notation, sometimes we will not indicate explicitly the dependence of various functions on the time variable t.

In the first step let $\eta \in C(\overline{I}; X)$ and $g \in C(\overline{I}; X)$ be given and we consider the following variational inequality of finding $v_{\eta g} : \overline{I} \to X$, such that for $t \in \overline{I}$,

$$\begin{aligned}(Av_{\eta g}(t), v - v_{\eta g}(t))_X + (\eta(t), v - v_{\eta g}(t))_X + j(g(t), v) \\ - j(g(t), v_{\eta g}(t)) \geq (f(t), v - v_{\eta g}(t))_X \quad \forall v \in X. \qquad (11.27)\end{aligned}$$

Lemma 11.4 *There exists a unique solution $v_{\eta g} \in C(\overline{I}; X)$ to the problem* (11.27).

Proof. For each $t \in \overline{I}$, it follows from classical results for elliptic variational inequalities (see Theorem 4.1, page 60) that there exists a unique element $v_{\eta g}(t) \in X$ that solves (11.27). It remains to prove the time continuity of the solution. Let $t_1, t_2 \in \overline{I}$. For simplicity in writing we denote $v_{\eta g}(t_i) = v_i$, $\eta(t_i) = \eta_i$, $g(t_i) = g_i$, $f(t_i) = f_i$, for $i = 1, 2$. Using (11.27) for $t = t_1$ and t_2, we easily derive the relation

$$\begin{aligned}(Av_1 - Av_2, v_1 - v_2)_X \leq (f_1 - f_2, v_1 - v_2)_X + (\eta_1 - \eta_2, v_2 - v_1)_X \\ + j(g_1, v_2) - j(g_1, v_1) + j(g_2, v_1) - j(g_2, v_2).\end{aligned}$$

Then we use the assumptions (11.22)(a) and (11.24)(b) to obtain

$$M\|v_1 - v_2\|_X \leq \|f_1 - f_2\|_X + \|\eta_1 - \eta_2\|_X + m\|g_1 - g_2\|_X. \qquad (11.28)$$

It follows that $t \mapsto v_{\eta g}(t) : \overline{I} \to X$ is a continuous function. ■

For each $\eta \in C(\overline{I}; X)$, we consider an operator $\Lambda_\eta : C(\overline{I}; X) \to C(\overline{I}; X)$ defined by

$$\Lambda_\eta g(t) = \int_0^t v_{\eta g}(s)\, ds + u_0 \quad \forall g \in C(\overline{I}; X),\ t \in \overline{I}. \qquad (11.29)$$

We have the following result.

Lemma 11.5 *The operator Λ_η has a unique fixed point $g_\eta \in C(\overline{I}; X)$.*

Proof. Let $g_1, g_2 \in C(\overline{I}; X)$, $\eta \in C(\overline{I}; X)$ and let v_i denote the solution of (11.27) for $g = g_i$, i.e. $v_i = v_{\eta g_i}$, $i = 1, 2$. From the definition (11.29) we derive the inequality

$$\|\Lambda_\eta g_1(t) - \Lambda_\eta g_2(t)\|_X \leq \int_0^t \|v_1(s) - v_2(s)\|_X ds \quad \forall t \in \overline{I}. \qquad (11.30)$$

An argument similar to that in the proof of Lemma 11.4 shows that

$$M\,\|v_1(s) - v_2(s)\|_X \leq m\,\|g_1(s) - g_2(s)\|_X \quad \forall s \in \overline{I}. \qquad (11.31)$$

Combining (11.30) and (11.31) yields

$$\|\Lambda_\eta g_1(t) - \Lambda_\eta g_2(t)\|_X \le c \int_0^t \|g_1(s) - g_2(s)\|_X ds \quad \forall\, t \in \overline{I}.$$

Reiterating the last inequality n times, we infer

$$\|\Lambda_\eta^n g_1(t) - \Lambda_\eta^n g_2(t)\|_{C(\overline{I};X)} \le \frac{c^n}{n!} \|g_1 - g_2\|_{C(\overline{I};X)}.$$

This inequality shows that for n large enough, the operator Λ_η^n is a contraction on $C(\overline{I};X)$. Thus, there exists a unique $g_\eta \in C(\overline{I};X)$ such that $\Lambda_\eta^n g_\eta = g_\eta$. Moreover g_η is the unique fixed point of Λ_η. ■

For each $\eta \in C(\overline{I};X)$, we denote by g_η the fixed point given in Lemma 11.5 and let $u_\eta \in C^1(\overline{I};X)$ be the function given by

$$u_\eta(t) = \int_0^t v_{\eta g_\eta}(s)\, ds + u_0, \quad t \in \overline{I}. \tag{11.32}$$

We have $\Lambda_\eta g_\eta = g_\eta$. From (11.29) and (11.32) it follows that

$$u_\eta = g_\eta. \tag{11.33}$$

Therefore, taking $g = g_\eta$ in (11.27) and using (11.32) and (11.33) we see that for all $t \in \overline{I}$, u_η satisfies

$$\begin{aligned}(A\dot{u}_\eta(t), v - \dot{u}_\eta(t))_X + (\eta(t), v - \dot{u}_\eta(t))_X + j(u_\eta(t), v)\\ - j(u_\eta(t), \dot{u}_\eta(t)) \ge (f(t), v - \dot{u}_{\eta g}(t))_X \quad \forall\, v \in X.\end{aligned} \tag{11.34}$$

Define an operator $\Lambda : C(\overline{I};X) \to C(\overline{I};X)$ by

$$\Lambda\eta = Bu_\eta \quad \forall\, \eta \in C(\overline{I};X). \tag{11.35}$$

We have the following result.

Lemma 11.6 *The operator Λ has a unique fixed point $\eta^* \in C(\overline{I};X)$.*

Proof. Let $\eta_1, \eta_2 \in C(\overline{I};X)$. We use the notation $u_i = u_{\eta_i}$ and $v_i = \dot{u}_{\eta_i}$ for $i = 1, 2$. Using the formula (11.32) we obtain

$$\|u_1(t) - u_2(t)\|_X \le \int_0^t \|v_1(s) - v_2(s)\|_X ds \quad \forall\, t \in \overline{I}. \tag{11.36}$$

Moreover, using (11.34) and estimates similar to those in the proof of Lemma 11.4 (see (11.28)) we find

$$\begin{aligned}M\,\|v_1(s) - v_2(s)\|_X \le \|\eta_1(s) - \eta_2(s)\|_X\\ + m\,\|u_1(s) - u_2(s)\|_X \quad \forall\, s \in \overline{I}.\end{aligned} \tag{11.37}$$

Combining (11.36) and (11.37) and applying the Gronwall inequality we obtain

$$\|u_1(t) - u_2(t)\|_X \leq c \int_0^t \|\eta_1(s) - \eta_2(s)\|_X ds \quad \forall\, t \in \overline{I}.$$

Taking (11.35) and (11.23) into account, we derive from the last inequality that

$$\|\Lambda\eta_1(t) - \Lambda\eta_2(t)\|_X \leq c \int_0^t \|\eta_1(s) - \eta_2(s)\|_X ds \quad \forall\, t \in \overline{I}. \tag{11.38}$$

The result of the lemma follows from (11.38) and an application of Banach's fixed point theorem as in the proof of Lemma 11.5. ■

We now prove Theorem 11.3.

Proof of Theorem 11.3.
Existence. Let $\eta^* \in C(\overline{I}; X)$ be the fixed point of Λ and let $u_{\eta^*} \in C^1(\overline{I}; X)$ be the function defined by (11.32) for $\eta = \eta^*$. From (11.34) for $\eta = \eta^*$ it follows that for any $t \in \overline{I}$,

$$\begin{aligned}(A\dot{u}_{\eta^*}(t), v - \dot{u}_{\eta^*}(t))_X + (\eta^*(t), v - \dot{u}_{\eta^*}(t))_X + j(u_{\eta^*}(t), v) \\ - j(u_{\eta^*}(t), \dot{u}_{\eta^*}(t)) \geq (f(t), v - \dot{u}_{\eta^*}(t))_X \quad \forall\, v \in X.\end{aligned} \tag{11.39}$$

The inequality (11.21) follows now from (11.39) and (11.35) since $\eta^* = \Lambda\eta^* = Bu_{\eta^*}$ and (11.20) results form (11.32). We conclude that u_{η^*} is a solution of (11.20)–(11.21).
Uniqueness. To prove the uniqueness of the solution assume $u_i \in C^1(\overline{I}; X)$ are two solutions of (11.20)–(11.21), $i = 1, 2$. We have

$$\begin{aligned}&(A\dot{u}_1(t), v - \dot{u}_1(t))_X + (Bu_1(t), v - \dot{u}_1(t))_X + j(u_1(t), v) \\ &\quad - j(u_1(t), \dot{u}_1(t)) \geq (f(t), v - \dot{u}_1(t))_X, \\ &(A\dot{u}_2(t), v - \dot{u}_2(t))_X + (Bu_2(t), v - \dot{u}_2(t))_X + j(u_2(t), v) \\ &\quad - j(u_2(t), \dot{u}_2(t)) \geq (f(t), v - \dot{u}_2(t))_X,\end{aligned}$$

for all $v \in X$ and $t \in \overline{I}$. We take $v = \dot{u}_2(t)$ in the first inequality, $v = \dot{u}_1(t)$ in the second inequality and add the two inequalities to obtain, for any $t \in \overline{I}$,

$$\begin{aligned}&(A\dot{u}_1(t) - A\dot{u}_2(t), \dot{u}_1(t) - \dot{u}_2(t))_X \\ &\quad \leq (Bu_1(t) - Bu_2(t), \dot{u}_2(t) - \dot{u}_1(t))_X + j(u_1(t), \dot{u}_2(t)) \\ &\qquad - j(u_1(t), \dot{u}_1(t)) + j(u_2(t), \dot{u}_1(t)) - j(u_2(t), \dot{u}_2(t)).\end{aligned}$$

Using (11.22), (11.23) and (11.24) we deduce

$$M\|\dot{u}_1(t) - \dot{u}_2(t)\|_X \leq (L_B + m)\, \|u_1(t) - u_2(t)\|_X \quad \forall\, t \in \overline{I}$$

which implies

$$\|\dot{u}_1(t) - \dot{u}_2(t)\|_X \leq c \int_0^t \|\dot{u}_1(s) - \dot{u}_2(s)\|_X ds \quad \forall\, t \in \overline{I}.$$

An application of the Gronwall inequality shows that $\dot{u}_1(t) = \dot{u}_2(t)$ for any $t \in \overline{I}$. Since u_1 and u_2 have the same initial condition (11.20), we conclude $u_1(t) = u_2(t)$ for any $t \in \overline{I}$. ■

We finish this section with a regularity result.

Proposition 11.7 *Under the conditions stated in Theorem 11.3, if $f \in W^{1,p}(I;X)$ for some $p \in [1,\infty]$, then $\dot{u} \in W^{1,p}(I;X)$, and*

$$\|\dot{u}\|_{W^{1,p}(I;X)} \le c\,(\|f\|_{W^{1,p}(I;X)} + \|u\|_{C^1(\overline{I};X)}).$$

Proof. For any $t_1, t_2 \in \overline{I}$, we apply the inequality (11.28) for the solution of Problem 11.2 to obtain

$$\begin{aligned} M\,\|\dot{u}(t_1) - \dot{u}(t_2)\|_X \le \|f(t_1) - f(t_2)\|_X + \|Bu(t_1) - Bu(t_2)\|_X \\ + m\,\|u(t_1) - u(t_2)\|_X. \end{aligned}$$

Using now the assumption (11.23), we find

$$\|\dot{u}(t_1) - \dot{u}(t_2)\|_X \le c\,(\|f(t_1) - f(t_2)\|_X + \|u(t_1) - u(t_2)\|_X),$$

from which the result of the proposition follows. ■

From the proof of Proposition 11.7, we observe that actually a more general result holds:

$$f \in W^{k,p}(I;X) \;\Rightarrow\; u \in W^{k+1,p}(I;X)$$

and

$$\|u\|_{W^{k+1,p}(I;X)} \le c\,(\|f\|_{W^{k,p}(I;X)} + \|u\|_{C^1(\overline{I};X)}).$$

In the next two sections, we will assume all the conditions stated in Theorem 11.3 are satisfied so that Problem 11.2 has a unique solution.

11.3 Semi-discrete Approximation

In this section we consider an approximation of Problem 11.2 by discretizing only the space X. Let $X^h \subset X$ be a finite-dimensional space which for example can be constructed by the finite element method. Then a semi-discrete scheme can be formed as following.

Problem 11.8 *Find $u^h : \overline{I} \to X^h$ such that*

$$u^h(0) = u_0^h \tag{11.40}$$

and for $t \in \overline{I}$,

$$\begin{aligned} (A\dot{u}^h(t), v^h - \dot{u}^h(t))_X + (Bu^h(t), v^h - \dot{u}^h(t))_X + j(u^h(t), v^h) \\ - j(u^h(t), \dot{u}^h(t)) \ge (f(t), v^h - \dot{u}^h(t))_X \quad \forall\, v^h \in X^h. \end{aligned} \tag{11.41}$$

Here, $u_0^h \in X^h$ is an appropriate approximation of u_0.

Using the arguments presented in the previous section, we see that under the conditions stated in Theorem 11.3, Problem 11.8 has a unique solution $u^h \in C^1(\overline{I}; X^h)$. Our main purpose here is to derive estimates for the errors $u - u^h$.

Let $t \in \overline{I}$. To simplify the writing, we introduce the velocity variable

$$w(t) = \dot{u}(t).$$

Then by using the initial value condition (11.20), we have the relation

$$u(t) = \int_0^t w(s)\, ds + u_0. \tag{11.42}$$

Similarly, we introduce the discrete velocity variable

$$w^h(t) = \dot{u}^h(t).$$

With the initial value condition (11.40), we have

$$u^h(t) = \int_0^t w^h(s)\, ds + u_0^h. \tag{11.43}$$

The variational inequalities (11.21) and (11.41) can be rewritten as

$$\begin{aligned}(Aw(t), v - w(t))_X + (Bu(t), v - w(t))_X + j(u(t), v)\\ - j(u(t), w(t)) \geq (f(t), v - w(t))_X \quad \forall\, v \in X,\end{aligned} \tag{11.44}$$

and

$$\begin{aligned}(Aw^h(t), v^h - w^h(t))_X + (Bu^h(t), v^h - w^h(t))_X + j(u^h(t), v^h)\\ - j(u^h(t), w^h(t)) \geq (f(t), v^h - w^h(t))_X \quad \forall\, v^h \in X^h.\end{aligned} \tag{11.45}$$

We now take $v = w^h(t)$ in (11.44) and add the inequality to (11.45) with $v^h = v^h(t) \in X^h$. After some manipulations, we have

$$\begin{aligned}&(Aw(t) - Aw^h(t), w(t) - w^h(t))_X\\ &\quad \leq (Aw^h(t) - Aw(t), v^h(t) - w(t))_X\\ &\qquad + (Bu(t) - Bu^h(t), w^h(t) - v^h(t))_X\\ &\qquad + R(t; w(t), v^h(t)) + j(u(t), w^h(t)) - j(u(t), v^h(t))\\ &\qquad + j(u^h(t), v^h(t)) - j(u^h(t), w^h(t)),\end{aligned}$$

where

$$\begin{aligned}R(t; w(t), v^h(t)) = {}&(Aw(t), v^h(t) - w(t))_X + (Bu(t), v^h(t) - w(t))_X\\ &+ j(u(t), v^h(t)) - j(u(t), w(t))\\ &- (f(t), v^h(t) - w(t))_X\end{aligned} \tag{11.46}$$

represents a residual quantity. Using the assumptions (11.22), (11.23) and (11.24), we have

$$\begin{aligned} M\,\|w(t)-w^h(t)\|_X^2 \le L_A\|w(t)-w^h(t)\|_X\,\|w(t)-v^h(t)\|_X \\ +|R(t;w(t),v^h(t))|+(L_B+m)\,\|u(t)-u^h(t)\|_X \\ \cdot\left(\|w(t)-w^h(t)\|_X+\|w(t)-v^h(t)\|_X\right). \end{aligned}$$

Thus, we have the inequality

$$\begin{aligned} \|w(t)-w^h(t)\|_X^2 \le c\left(\|w(t)-v^h(t)\|_X^2+\|u(t)-u^h(t)\|_X^2\right. \\ \left.+|R(t;w(t),v^h(t))|\right). \end{aligned} \tag{11.47}$$

Subtracting (11.43) from (11.42), we obtain

$$u(t)-u^h(t)=\int_0^t (w(s)-w^h(s))\,ds+u_0-u_0^h,$$

and so

$$\|u(t)-u^h(t)\|_X^2 \le c\left(\int_0^t \|w(s)-w^h(s)\|_X^2 ds+\|u_0-u_0^h\|_X^2\right). \tag{11.48}$$

Then the inequality (11.47) can be rewritten as

$$\begin{aligned} \|w(t)-w^h(t)\|_X^2 \le c\Bigg(\|w(t)-v^h(t)\|_X^2+\int_0^t \|w(s)-w^h(s)\|_X^2\,ds \\ +\|u_0-u_0^h\|_X^2+|R(t;w(t),v^h(t))|\Bigg). \end{aligned}$$

Applying the Gronwall inequality (Lemma 7.24, page 162), we have

$$\begin{aligned} \|w-w^h\|_{C(\overline{I};X)} \le c\,\|u_0-u_0^h\|_X+c\inf_{v^h\in C(\overline{I};X^h)}\left(\|w-v^h\|_{C(\overline{I};X)}\right. \\ \left.+\|R(\cdot;w(\cdot),v^h(\cdot))\|_{C(\overline{I})}^{1/2}\right). \end{aligned} \tag{11.49}$$

Summarizing, with (11.48) and (11.49), we have proved the following result.

Theorem 11.9 *Assume the conditions* (11.22)–(11.26). *Then for the error of the semi-discrete solution of Problem 11.8, we have the estimate*

$$\begin{aligned} \|u-u^h\|_{C^1(\overline{I};X)} \le c\,\|u_0-u_0^h\|_X+c\inf_{v^h\in C(\overline{I};X^h)}\left(\|\dot{u}-v^h\|_{C(\overline{I};X)}\right. \\ \left.+\|R(\cdot;\dot{u}(\cdot),v^h(\cdot))\|_{C(\overline{I})}^{1/2}\right), \end{aligned} \tag{11.50}$$

where $R(\cdot;\dot{u}(\cdot),v^h(\cdot))$ is defined in (11.46).

The inequality (11.50) is the basis for error analysis, as is shown later in Section 11.5 in the context of the viscoelastic frictional contact problem with normal compliance.

11.4 Fully Discrete Approximation

In this section we consider a fully discrete approximation of Problem 11.2. In addition to the finite dimensional space X^h, we need a partition of the time interval described in Section 7.2. The fully discrete approximation method we will analyze is the following.

Problem 11.10 *Find* $\{u_n^{hk}\}_{n=0}^N \subset X^h$ *such that*

$$u_0^{hk} = u_0^h \tag{11.51}$$

and for $n = 1, \dots, N$,

$$\begin{aligned}(A\delta_n u_n^{hk}, v^h - \delta_n u_n^{hk})_X + (Bu_{n-1}^{hk}, v^h - \delta_n u_n^{hk})_X + j(u_{n-1}^{hk}, v^h)\\ - j(u_{n-1}^{hk}, \delta_n u_n^{hk}) \ge (f_n, v^h - \delta_n u_n^{hk})_X \quad \forall\, v^h \in X^h.\end{aligned} \tag{11.52}$$

Here, $u_0^h \in X^h$ *is an appropriate approximation of* u_0.

Again to simplify the notation, we introduce the discrete velocity

$$w_n^{hk} = \delta_n u_n^{hk}, \quad n = 1, \dots, N.$$

Then using the initial value condition (11.51), we can determine the discrete displacement by the relation

$$u_n^{hk} = \sum_{j=1}^n w_j^{hk} k_j + u_0^h. \tag{11.53}$$

We can rewrite (11.52) in the form

$$\begin{aligned}(Aw_n^{hk}, v^h - w_n^{hk})_X + j(u_{n-1}^{hk}, v^h) - j(u_{n-1}^{hk}, w_n^{hk})\\ \ge (f_n, v^h - w_n^{hk})_X - (Bu_{n-1}^{hk}, v^h - w_n^{hk})_X \quad \forall\, v^h \in X^h.\end{aligned} \tag{11.54}$$

Given $u_{n-1}^{hk} \in X^h$, it follows from Theorem 4.1 that the inequality (11.54) has a unique solution $w_n^{hk} \in X^h$. Note that $u_0^{hk} = u_0^h$ is given and we can use the relation (11.53) to compute $\{u_n^{hk}\}_{n=1}^N$. An induction argument yields the existence and uniqueness of a solution of Problem 11.10. Our main objective of the section is to derive an error estimate for the fully discrete solution.

Take $v = w_n^{hk}$ in (11.44) at $t = t_n$,

$$\begin{aligned}(Aw_n, w_n^{hk} - w_n)_X + (Bu_n, w_n^{hk} - w_n) + j(u_n, w_n^{hk}) - j(u_n, w_n)\\ \ge (f_n, w_n^{hk} - w_n)_X.\end{aligned} \tag{11.55}$$

We now add (11.54) with $v^h = v_n^h \in X^h$ arbitrary and (11.55) to obtain an error relation

$$\begin{aligned}&(Aw_n - Aw_n^{hk}, w_n - w_n^{hk})_X\\ &\quad \le (Aw_n^{hk} - Aw_n, v_n^h - w_n)_X + (Bu_{n-1}^{hk} - Bu_n, v_n^h - w_n^{hk})_X\\ &\quad\quad + R_n(w_n, v_n^h) + j(u_{n-1}^{hk}, v_n^h) - j(u_n, v_n^h)\\ &\quad\quad + j(u_n, w_n^{hk}) - j(u_{n-1}^{hk}, w_n^{hk}),\end{aligned}$$

where

$$R_n(w_n, v_n^h) = (Aw_n, v_n^h - w_n)_X + (Bu_n, v_n^h - w_n)_X \\ + j(u_n, v_n^h) - j(u_n, w_n) - (f_n, v_n^h - w_n)_X. \qquad (11.56)$$

By the assumptions (11.22), (11.23) and (11.24) on A, B and j, we have

$$M \|w_n - w_n^{hk}\|_X^2 \le L_A \|w_n - w_n^{hk}\|_X \|w_n - v_n^h\|_X + |R_n(w_n, v_n^h)| \\ + (L_B + m) \|u_{n-1}^{hk} - u_n\|_X \|v_n^h - w_n^{hk}\|_X.$$

Here $\|v_n^h - w_n^{hk}\|_X$ will be bounded as follows:

$$\|v_n^h - w_n^{hk}\|_X \le \|v_n^h - w_n\|_X + \|w_n - w_n^{hk}\|_X.$$

Then we get the relation

$$\|w_n - w_n^{hk}\|_X^2 \le c\Big\{\|v_n^h - w_n\|_X^2 + \|u_{n-1}^{hk} - u_n\|_X^2 + |R_n(w_n, v_n^h)|\Big\},$$

or,

$$\|w_n - w_n^{hk}\|_X \le c\Big\{\|v_n^h - w_n\|_X + |R_n(w_n, v_n^h)|^{1/2} + \|u_{n-1}^{hk} - u_n\|_X\Big\}.$$

The term $\|u_{n-1}^{hk} - u_n\|_X$ is bounded by (7.42) (page 155). Then we have

$$\|w_n - w_n^{hk}\|_X \le c\Bigg\{\|v_n^h - w_n\|_X + |R_n(w_n, v_n^h)|^{1/2} + \|u_0^h - u_0\|_X + I_k(w) \\ + k\,\|w\|_{C(\bar{I};X)} + \sum_{j=1}^{n-1} \|w_j^{hk} - w_j\|_X k_j\Bigg\} \qquad (11.57)$$

where $I_k(w)$ is defined in (7.43).

Applying Lemma 7.25 (page 163) to the inequality (11.57), we obtain the following estimate

$$\max_n \|w_n - w_n^{hk}\|_X \le c \max_n \Big\{\|v_n^h - w_n\|_X + |R_n(w_n, v_n^h)|^{1/2}\Big\} \\ + c\,\big(\|u_0^h - u_0\|_X + I_k(w) + k\,\|w\|_{C(\bar{I};X)}\big).$$

By (7.46), we have

$$\|u_n^{hk} - u_n\|_X \le \sum_{j=1}^{n} \|w_j^{hk} - w_j\|_X k_j + \|u_0^h - u_0\|_X + I_k(w).$$

Then

$$\|u_n^{hk} - u_n\|_X \le T \max_j \|w_j^{hk} - w_j\|_X + \|u_0^h - u_0\|_X + I_k(w).$$

Therefore, we have proved the following result.

Theorem 11.11 *Assume* (11.22)–(11.26). *Then for the error of the fully discrete solution of Problem 11.10, we have the estimate*

$$\begin{aligned} \max_{1\le n\le N} &(\|u_n^{hk}-u_n\|_X+\|\dot u_n-\delta_n u_n^{hk}\|_X)\\ &\le c\max_{1\le n\le N}\left\{\|v_n^h-\dot u_n\|_X+|R_n(\dot u_n,v_n^h)|^{1/2}\right\}\\ &\quad +c\left(\|u_0^h-u_0\|_X+I_k(\dot u)+k\,\|\dot u\|_{C(\bar I;X)}\right), \end{aligned} \tag{11.58}$$

where $R_n(\dot u_n, v_n^h)$ *is defined by* (11.56).

When the data f is smoother, we can derive a more compact error bound from (11.58). Assume additionally $f\in W^{1,1}(I;X)$. Then by Proposition 11.7, we have $w\in W^{1,1}(I;X)$ and

$$\|w\|_{W^{1,1}(I;X)}\le c\,(\|f\|_{W^{1,1}(I;X)}+\|u\|_{C^1(\bar I;X)}).$$

We can use the inequality (7.44) (page 155) to bound the term $I_k(\dot u)$. Therefore, under the additional assumption $f\in W^{1,1}(I;X)$, the estimate (11.58) can be replaced by

$$\begin{aligned} \max_{1\le n\le N}&(\|u_n-u_n^{hk}\|_X+\|\dot u_n-\delta_n u_n^{hk}\|_X)\\ &\le c\max_{1\le n\le N}\inf_{v_n^h\in X^h}\left\{\|v_n^h-\dot u_n\|_X+|R_n(\dot u_n,v_n^h)|^{1/2}\right\}\\ &\quad +c\,\|u_0^h-u_0\|_X+c\,k\,(\|f\|_{W^{1,1}(I;X)}+\|u\|_{C^1(\bar I;X)}). \end{aligned}$$

11.5 Applications to the Contact Problem

In this section we apply the results of previous sections on the abstract problem to the study of the frictional problem with normal compliance (11.1)–(11.7). The well-posedness of Problem 11.1 follows from an application of Theorem 11.3.

Theorem 11.12 *Assume* (6.34), (6.35) *and* (11.10)–(11.13). *Then Problem 11.1 has a unique solution* $\boldsymbol{u}\in C^1(\bar I;V)$.

Proof. We take $X=V$. Let $A:V\to V$ and $B:V\to V$ be the operators defined by

$$(A\boldsymbol{v},\boldsymbol{w})_V=(\mathcal{A}\boldsymbol{\varepsilon}(\boldsymbol{v}),\boldsymbol{\varepsilon}(\boldsymbol{w}))_Q,\quad (B\boldsymbol{v},\boldsymbol{w})_V=(\mathcal{B}\boldsymbol{\varepsilon}(\boldsymbol{v}),\boldsymbol{\varepsilon}(\boldsymbol{w}))_Q \tag{11.59}$$

for all $\boldsymbol{v},\boldsymbol{w}\in V$. From the assumptions (6.34) and (6.35) on $\mathcal{A}$ and $\mathcal{B}$, it follows that A and B are Lipschitz continuous operators, and A is strongly monotone on V:

$$(A\boldsymbol{v}_1-A\boldsymbol{v}_2,\boldsymbol{v}_1-\boldsymbol{v}_2)_V\ge M\|\boldsymbol{v}_1-\boldsymbol{v}_2\|_V^2\quad\forall\,\boldsymbol{v}_1\,\boldsymbol{v}_2\in V. \tag{11.60}$$

Moreover, from the assumptions (11.10) on p_ν and p_τ it follows that the function j defined by (11.16) satisfies (11.24) and

$$\begin{aligned} &j(\boldsymbol{g}_1, \boldsymbol{v}_2) - j(\boldsymbol{g}_1, \boldsymbol{v}_1) + j(\boldsymbol{g}_2, \boldsymbol{v}_1) - j(\boldsymbol{g}_2, \boldsymbol{v}_2) \\ &\quad \le c_0(L_\nu + L_\tau)\, \|\boldsymbol{g}_1 - \boldsymbol{g}_2\|_V \, \|\boldsymbol{v}_1 - \boldsymbol{v}_2\|_V \quad \forall\, \boldsymbol{g}_1, \boldsymbol{g}_2, \boldsymbol{v}_1, \boldsymbol{v}_2 \in V, \end{aligned}$$

where $c_0 > 0$ depends only on Ω, Γ_1 and Γ_3. Applying Theorem 11.3, we conclude that Problem 11.1 has a unique solution $\boldsymbol{u} \in C^1(\overline{I}; V)$. ■

Let now $\boldsymbol{u} \in C^1(\overline{I}; V)$ be the solution of Problem 11.1 and let $\boldsymbol{\sigma}$ be the stress field defined by (11.1). A pair of functions $(\boldsymbol{u}, \boldsymbol{\sigma})$ which satisfies (11.1), (11.7) and (11.19) is called a *weak solution* of the problem (11.1)–(11.7). We conclude that, under the assumptions of Theorem 11.12, problem (11.1)–(11.7) has a unique weak solution.

By a procedure similar to that used in Section 8.1, we can show that for all $t \in \overline{I}$,

$$\operatorname{Div} \boldsymbol{\sigma} = -\boldsymbol{f}_0 \quad \text{a.e. in } \Omega. \tag{11.61}$$

In particular, by (11.11) we see that $\operatorname{Div} \boldsymbol{\sigma} \in C(\overline{I}; L^2(\Omega)^d)$; therefore $\boldsymbol{\sigma} \in C(\overline{I}; Q_1)$. Under the additional assumption

$$\boldsymbol{\sigma\nu} \in C(\overline{I}; L^2(\Gamma)^d), \tag{11.62}$$

for all $t \in \overline{I}$ we have

$$\boldsymbol{\sigma\nu} = \boldsymbol{f}_2 \quad \text{a.e. on } \Gamma_2, \tag{11.63}$$

$$-\sigma_\nu = p_\nu(u_\nu - g) \quad \text{a.e. on } \Gamma_3. \tag{11.64}$$

Now we state some sample results on error estimates for numerical approximations of Problem 11.1.

Let $\{V^h\} \subset V$ be the finite element spaces consisting of piecewise polynomials of degree less than or equal to 1, corresponding to a regular family of finite element partitions $\{\mathcal{T}^h\}$, compatible to the boundary decomposition $\Gamma = \overline{\Gamma}_1 \cup \overline{\Gamma}_2 \cup \overline{\Gamma}_3$. Then the semi-discrete approximation of Problem 11.1 is

Problem 11.13 *Find a displacement field* $\boldsymbol{u}^h : \overline{I} \to V^h$ *such that*

$$\boldsymbol{u}^h(0) = \boldsymbol{u}_0^h$$

and for any $t \in \overline{I}$,

$$\begin{aligned} &(\mathcal{A}\varepsilon(\dot{\boldsymbol{u}}^h(t)), \varepsilon(\boldsymbol{v}^h - \dot{\boldsymbol{u}}^h(t)))_Q + (\mathcal{B}\varepsilon(\boldsymbol{u}^h(t)), \varepsilon(\boldsymbol{v}^h - \dot{\boldsymbol{u}}^h(t)))_Q \\ &\quad + j(\boldsymbol{u}^h(t), \boldsymbol{v}^h) - j(\boldsymbol{u}^h(t), \dot{\boldsymbol{u}}^h(t)) \\ &\qquad \ge (\boldsymbol{f}(t), \boldsymbol{v}^h - \dot{\boldsymbol{u}}^h(t))_V \quad \forall\, \boldsymbol{v}^h \in V^h. \end{aligned}$$

Here $\boldsymbol{u}_0^h \in V^h$ *is a suitable approximation of* $\boldsymbol{u}_0$.

Problem 11.13 has a unique solution, and we have the following estimate from (11.50):

$$\|\boldsymbol{u} - \boldsymbol{u}^h\|_{C^1(\bar{I};V)} \le c \inf_{\boldsymbol{v}^h \in C(\bar{I};V^h)} \big(\|\dot{\boldsymbol{u}} - \boldsymbol{v}^h\|_{C(\bar{I};V)} + \|R(\cdot;\dot{\boldsymbol{u}}(\cdot),\boldsymbol{v}^h(\cdot))\|^{1/2}_{C(\bar{I})} \big) + c\,\|\boldsymbol{u}_0 - \boldsymbol{u}_0^h\|_V, \qquad (11.65)$$

where

$$R(t;\dot{\boldsymbol{u}}(t),\boldsymbol{v}^h(t)) = (\boldsymbol{\sigma}(t),\boldsymbol{\varepsilon}(\boldsymbol{v}^h(t)-\dot{\boldsymbol{u}}(t)))_Q + j(\boldsymbol{u}(t),\boldsymbol{v}^h(t)) - j(\boldsymbol{u}(t),\dot{\boldsymbol{u}}(t)) - (\boldsymbol{f}(t),\boldsymbol{v}^h(t)-\dot{\boldsymbol{u}}(t))_V.$$

For the first term, integrating by parts and using (11.61) we have

$$\begin{aligned}\int_\Omega \boldsymbol{\sigma}(t)\cdot\boldsymbol{\varepsilon}(\boldsymbol{v}^h(t)-\dot{\boldsymbol{u}}(t))\,dx &= \langle \boldsymbol{\sigma}(t)\boldsymbol{\nu},\boldsymbol{v}^h(t)-\dot{\boldsymbol{u}}(t)\rangle_\Gamma - \int_\Omega \operatorname{Div}\boldsymbol{\sigma}(t)\cdot(\boldsymbol{v}^h(t)-\dot{\boldsymbol{u}}(t))\,dx\\ &= \langle \boldsymbol{\sigma}(t)\boldsymbol{\nu},\boldsymbol{v}^h(t)-\dot{\boldsymbol{u}}(t)\rangle_\Gamma + \int_\Omega \boldsymbol{f}_0(t)\cdot(\boldsymbol{v}^h(t)-\dot{\boldsymbol{u}}(t))\,dx.\end{aligned}$$

Thus,

$$\begin{aligned}R(t;\dot{\boldsymbol{u}}(t),\boldsymbol{v}^h(t)) = {}&\langle \boldsymbol{\sigma}(t)\boldsymbol{\nu},\boldsymbol{v}^h(t)-\dot{\boldsymbol{u}}(t)\rangle_\Gamma\\ &- \int_{\Gamma_2} \boldsymbol{f}_2(t)\cdot(\boldsymbol{v}^h(t)-\dot{\boldsymbol{u}}(t))\,da\\ &+ \int_{\Gamma_3} p_\nu(u_\nu(t)-g)\,(v_\nu^h(t)-\dot{u}_\nu(t))\,da\\ &+ \int_{\Gamma_3} p_\tau(u_\nu(t)-g)\,(\|\boldsymbol{v}_\tau^h(t)\| - \|\dot{\boldsymbol{u}}_\tau(t)\|)\,da. \qquad (11.66)\end{aligned}$$

Then

$$|R(t;\dot{\boldsymbol{u}}(t),\boldsymbol{v}^h(t))| \le c(\boldsymbol{\sigma},\boldsymbol{u},\boldsymbol{f}_2)\,\|\boldsymbol{v}^h(t)-\dot{\boldsymbol{u}}(t)\|_V,$$

where

$$c(\boldsymbol{\sigma},\boldsymbol{u},\boldsymbol{f}_2) \le c\max_{t\in\bar{I}} \big(\|\boldsymbol{\sigma}(t)\boldsymbol{\nu}\|_{H^{-1/2}(\Gamma)^d} + \|\boldsymbol{f}_2(t)\|_{L^2(\Gamma_2)^d} + \|p_\nu(u_\nu(t)-g)\|_{L^2(\Gamma_3)^d} + \|p_\tau(u_\nu(t)-g)\|_{L^2(\Gamma_3)^d} \big).$$

Therefore, from (11.65),

$$\|\boldsymbol{u} - \boldsymbol{u}^h\|_{C^1(\bar{I};V)} \le c \inf_{\boldsymbol{v}^h \in C(\bar{I};V^h)} \big(\|\dot{\boldsymbol{u}} - \boldsymbol{v}^h\|_{C(\bar{I};V)} + \|\dot{\boldsymbol{u}} - \boldsymbol{v}^h\|^{1/2}_{C(\bar{I};V)} \big) + c\,\|\boldsymbol{u}_0 - \boldsymbol{u}_0^h\|_V.$$

We can now apply Theorem 7.18 (page 157) to conclude the convergence under the basic solution regularity $\boldsymbol{u} \in C^1(\overline{I}; V)$. The function g of (7.48) can be chosen as

$$g(e) = c\,(e + e^{1/2}),$$

which obviously satisfies the condition (7.49). By Theorm 7.7 (page 148), the subspace $V_0 = C^\infty(\overline{\Omega})^d \cap V$ is dense in V. The finite element interpolation operator Π^h satisfies

$$\|\boldsymbol{v} - \Pi^h \boldsymbol{v}\|_V \le c\,h\,|\boldsymbol{v}|_{H^2(\Omega)^d} \quad \forall\, \boldsymbol{v} \in C^\infty(\overline{\Omega})^d \cap V.$$

Thus Hypothesis 7.17 on page 157 is valid with $\alpha(h) = c\,h$. So if the initial value $\boldsymbol{u}_0^h \in V^h$ is chosen such that

$$\|\boldsymbol{u}_0 - \boldsymbol{u}_0^h\|_V \to 0 \quad \text{as } h \to 0,$$

then by Theorem 7.18 we have the convergence

$$\|\boldsymbol{u} - \boldsymbol{u}^h\|_{C^1(\overline{I};V)} \to 0 \quad \text{as } h \to 0.$$

Turning now to error estimation, we assume (11.62). Then

$$\sigma_\nu \in C(\overline{I}; L^2(\Gamma_3)), \quad \boldsymbol{\sigma}_\tau \in C(\overline{I}; L^2(\Gamma_3)^d).$$

Using (11.63) and (11.64), we derive from (11.66) that

$$R(t; \boldsymbol{v}^h(t), \dot{\boldsymbol{u}}(t)) = \int_{\Gamma_3} \big(\boldsymbol{\sigma}_\tau(t) \cdot (\boldsymbol{v}_\tau^h(t) - \dot{\boldsymbol{u}}_\tau(t)) \\ + p_\tau(u_\nu(t) - g)\,(\|\boldsymbol{v}_\tau^h(t)\| - \|\dot{\boldsymbol{u}}_\tau(t)\|)\big)\, da.$$

Thus,

$$|R(t; \dot{\boldsymbol{u}}(t), \boldsymbol{v}^h(t))| \le c\,\|\boldsymbol{v}_\tau^h(t) - \dot{\boldsymbol{u}}_\tau(t)\|_{L^2(\Gamma_3)^d}. \tag{11.67}$$

Here and in the remaining part of the section, the constant c may depend on the solution of the problem. Therefore, the estimate (11.65) in this case reduces to

$$\|\boldsymbol{u} - \boldsymbol{u}^h\|_{C^1(\overline{I};V)} \le c \inf_{\boldsymbol{v}^h \in C(\overline{I};V)} \Big(\|\dot{\boldsymbol{u}} - \boldsymbol{v}^h\|_{C(\overline{I};V)} + \|\dot{\boldsymbol{u}}_\tau - \boldsymbol{v}_\tau^h\|^{1/2}_{C(\overline{I};L^2(\Gamma_3)^d)}\Big) \\ + c\,\|\boldsymbol{u}_0 - \boldsymbol{u}_0^h\|_V.$$

Under the additional regularity conditions

$$\dot{\boldsymbol{u}} \in C(\overline{I}; H^2(\Omega)^d),\ \ \dot{\boldsymbol{u}}_\tau|_{\Gamma_{3,i}} \in C(\overline{I}; H^2(\Gamma_{3,i})^d),\ 1 \le i \le i_0,\ \ \boldsymbol{u}_0 \in H^2(\Omega)^d, \tag{11.68}$$

we have the interpolation error estimates (7.32), (7.36) and (7.40) in Section 7.2. Then if we take $\boldsymbol{u}_0^h = \Pi^h \boldsymbol{u}_0$, from (11.65) and (11.67) we have the optimal order error estimate

$$\|\boldsymbol{u} - \boldsymbol{u}^h\|_{C^1(\overline{I};V)} = O(h).$$

A fully discrete approximation for Problem 11.1 is

Problem 11.14 *Find a displacement field* $\boldsymbol{u}^{hk} = \{\boldsymbol{u}_n^{hk}\}_{n=0}^N \subset V^h$ *such that*

$$\boldsymbol{u}_0^{hk} = \boldsymbol{u}_0^h$$

and for $n = 1, \ldots, N$,

$$\begin{aligned}(\mathcal{A}\boldsymbol{\varepsilon}(\delta_n \boldsymbol{u}_n^{hk}), \boldsymbol{\varepsilon}(\boldsymbol{v}^h) - \boldsymbol{\varepsilon}(\delta_n \boldsymbol{u}_n^{hk}))_Q + j(\boldsymbol{u}_{n-1}^{hk}, \boldsymbol{v}^h) - j(\boldsymbol{u}_{n-1}^{hk}, \delta_n \boldsymbol{u}_n^{hk})& \\ \geq (\boldsymbol{f}(t), \boldsymbol{v}^h - \delta_n \boldsymbol{u}_n^{hk})_V - (\mathcal{B}\boldsymbol{\varepsilon}(\boldsymbol{u}_{n-1}^{hk}), \boldsymbol{\varepsilon}(\boldsymbol{v}^h - \delta_n \boldsymbol{u}_n^{hk}))_Q&\end{aligned}$$

for all $\boldsymbol{v}^h \in V^h$. *Here again* $\boldsymbol{u}_0^h \in V^h$ *is a suitable approximation of* $\boldsymbol{u}_0$.

Problem 11.14 admits a unique solution, and we have the following estimate from (11.58):

$$\begin{aligned}\max_{1\leq n\leq N} (\|\boldsymbol{u}_n^{hk} - \boldsymbol{u}_n\|_V + \|\dot{\boldsymbol{u}}_n - \delta_n \boldsymbol{u}_n^{hk}\|_V)& \\ \leq c \max_{1\leq n\leq N} \left\{ \|\boldsymbol{v}_n^h - \dot{\boldsymbol{u}}_n\|_V + |R_n(\dot{\boldsymbol{u}}_n, \boldsymbol{v}_n^h)|^{1/2} \right\}& \\ + c\left(\|\boldsymbol{u}_0^h - \boldsymbol{u}_0\|_V + I_k(\dot{\boldsymbol{u}}) + k\,\|\dot{\boldsymbol{u}}\|_{C(\bar{I};V)}\right).& \qquad (11.69)\end{aligned}$$

Applying Theorem 7.22 (page 160), we have the convergence

$$\max_{1\leq n\leq N} (\|\boldsymbol{u}_n^{hk} - \boldsymbol{u}_n\|_V + \|\dot{\boldsymbol{u}}_n - \delta_n \boldsymbol{u}_n^{hk}\|_V) \to 0 \quad \text{as } h, k \to 0$$

under the basic solution regularity $\boldsymbol{u} \in C^1(\bar{I}; V)$ and the condition

$$\|\boldsymbol{u}_0 - \boldsymbol{u}_0^h\|_V \to 0 \quad \text{as } h \to 0.$$

Under the additional regularity conditions (11.62) and (11.68), if we take $\boldsymbol{u}_0^h = \Pi^h \boldsymbol{u}_0$, then by a derivation similar to the error estimation of the semi-discrete scheme, we obtain the optimal order error estimate

$$\max_{1\leq n\leq N} (\|\boldsymbol{u}_n^{hk} - \boldsymbol{u}_n\|_V + \|\dot{\boldsymbol{u}}_n - \delta_n \boldsymbol{u}_n^{hk}\|_V) = O(h + k). \qquad (11.70)$$

We emphasize that the above error estimates are only sample results under the stated regularity conditions. If the regularity conditions are different, the error estimates need to be changed accordingly, but that follows easily from (11.65) and (11.69).

11.6 Continuous Dependence with Respect to Contact Conditions

In this section we study the dependence of the solution to the problem (11.1)–(11.7) with respect to perturbations of the normal compliance functions p_ν and p_τ. To this end we suppose in what follows that the assumptions (6.34), (6.35) on $\mathcal{A}$ and $\mathcal{B}$, and (11.10)–(11.13) hold. For every $\alpha \geq 0$, let p_r^α be a perturbation of p_r which satisfies (11.10) with the Lipschitz constant L_r^α ($r = \nu, \tau$). We introduce the functional j^α that is obtained from j by replacing p_ν and p_τ with p_ν^α and p_τ^α, respectively. We consider now the following variational problem.

Problem 11.15 *For $\alpha \geq 0$, find a displacement $\boldsymbol{u}^\alpha : \overline{I} \to V$ such that*

$$\boldsymbol{u}^\alpha(0) = \boldsymbol{u}_0 \tag{11.71}$$

and for all $t \in \overline{I}$,

$$\begin{aligned}&(\mathcal{A}\boldsymbol{\varepsilon}(\dot{\boldsymbol{u}}^\alpha(t)), \boldsymbol{\varepsilon}(\boldsymbol{v} - \dot{\boldsymbol{u}}^\alpha(t)))_Q + (\mathcal{B}\boldsymbol{\varepsilon}(\boldsymbol{u}^\alpha(t)), \boldsymbol{\varepsilon}(\boldsymbol{v} - \dot{\boldsymbol{u}}^\alpha(t)))_Q \\ &\quad + j^\alpha(\boldsymbol{u}^\alpha(t), \boldsymbol{v}) - j^\alpha(\boldsymbol{u}^\alpha(t), \dot{\boldsymbol{u}}^\alpha(t)) \\ &\quad \geq (\boldsymbol{f}(t), \boldsymbol{v} - \dot{\boldsymbol{u}}^\alpha(t))_V \quad \forall \boldsymbol{v} \in V.\end{aligned} \tag{11.72}$$

We deduce from Theorem 11.12 that for each $\alpha \geq 0$, Problem 11.15 has a unique solution $\boldsymbol{u}^\alpha \in C^1(\overline{I}; V)$.

Let us suppose now that the normal compliance functions satisfy the following assumptions for $r = \nu, \tau$:

$$\left.\begin{array}{l}\text{There exist } \beta_r \in \mathbb{R},\ \varphi_\nu : \mathbb{R}_+ \to \mathbb{R}_+ \text{ and } \mathcal{M} > 0 \text{ such that} \\ \text{(a) } |p_r^\alpha(\boldsymbol{x}, u) - p_r(\boldsymbol{x}, u)| \leq \varphi_r(\alpha)\,(|u| + \beta_r) \\ \qquad \forall\, u \in \mathbb{R},\ \text{a.e. } \boldsymbol{x} \in \Gamma_3; \\ \text{(b) } \lim\limits_{\alpha \to 0} \varphi_r(\alpha) = 0; \\ \text{(c) } L_r^\alpha \leq \mathcal{M} \quad \forall\, \alpha \geq 0.\end{array}\right\} \tag{11.73}$$

Under these assumptions, we have the following convergence result.

Theorem 11.16 *The solution $\boldsymbol{u}^\alpha$ of Problem 11.15 converges to the solution $\boldsymbol{u}$ of Problem 11.1, i.e.*

$$\boldsymbol{u}^\alpha \to \boldsymbol{u} \quad \text{in } C^1(\overline{I}; V) \text{ as } \alpha \to 0. \tag{11.74}$$

Proof. Let $\alpha \geq 0$. To simplify the notation we shall not indicate explicitly the dependence on $t \in \overline{I}$. Everywhere below c will represent a positive constant which depends on the data but is independent of α and whose value may change from place to place. From (11.19) and (11.72) we can find that

$$\begin{aligned}&(\mathcal{A}\boldsymbol{\varepsilon}(\dot{\boldsymbol{u}}^\alpha) - \mathcal{A}\boldsymbol{\varepsilon}(\dot{\boldsymbol{u}}), \boldsymbol{\varepsilon}(\dot{\boldsymbol{u}}^\alpha) - \boldsymbol{\varepsilon}(\dot{\boldsymbol{u}}))_Q \\ &\quad \leq (\mathcal{B}\boldsymbol{\varepsilon}(\boldsymbol{u}^\alpha) - \mathcal{B}\boldsymbol{\varepsilon}(\boldsymbol{u}), \boldsymbol{\varepsilon}(\dot{\boldsymbol{u}}) - \boldsymbol{\varepsilon}(\dot{\boldsymbol{u}}^\alpha))_Q \\ &\qquad + j(\boldsymbol{u}, \dot{\boldsymbol{u}}^\alpha) - j(\boldsymbol{u}, \dot{\boldsymbol{u}}) + j^\alpha(\boldsymbol{u}^\alpha, \dot{\boldsymbol{u}}) - j^\alpha(\boldsymbol{u}^\alpha, \dot{\boldsymbol{u}}^\alpha).\end{aligned} \tag{11.75}$$

Moreover, it follows from the assumptions (6.34), (6.35) on $\mathcal{A}$ and $\mathcal{B}$ that

$$(\mathcal{A}\boldsymbol{\varepsilon}(\dot{\boldsymbol{u}}^\alpha) - \mathcal{A}\boldsymbol{\varepsilon}(\dot{\boldsymbol{u}}), \boldsymbol{\varepsilon}(\dot{\boldsymbol{u}}^\alpha) - \boldsymbol{\varepsilon}(\dot{\boldsymbol{u}}))_Q \geq c\, \|\dot{\boldsymbol{u}}^\alpha - \dot{\boldsymbol{u}}\|_V^2, \tag{11.76}$$

$$(\mathcal{B}\boldsymbol{\varepsilon}(\boldsymbol{u}^\alpha) - \mathcal{B}\boldsymbol{\varepsilon}(\boldsymbol{u}), \boldsymbol{\varepsilon}(\dot{\boldsymbol{u}}) - \boldsymbol{\varepsilon}(\dot{\boldsymbol{u}}^\alpha))_Q \leq c\, \|\boldsymbol{u}^\alpha - \boldsymbol{u}\|_V \|\dot{\boldsymbol{u}}^\alpha - \dot{\boldsymbol{u}}\|_V. \tag{11.77}$$

Note that

$$\begin{aligned}&j(\boldsymbol{u}, \dot{\boldsymbol{u}}^\alpha) - j(\boldsymbol{u}, \dot{\boldsymbol{u}}) + j^\alpha(\boldsymbol{u}^\alpha, \dot{\boldsymbol{u}}) - j^\alpha(\boldsymbol{u}^\alpha, \dot{\boldsymbol{u}}^\alpha) \\ &\quad = \int_{\Gamma_3} (p_\nu(u_\nu - g) - p_\nu^\alpha(u_\nu^\alpha - g))(\dot{u}_\nu^\alpha - \dot{u}_\nu)\, da \\ &\qquad + \int_{\Gamma_3} (p_\tau(u_\nu - g) - p_\tau^\alpha(u_\nu^\alpha - g))\, (\|\dot{\boldsymbol{u}}_\tau^\alpha\| - \|\dot{\boldsymbol{u}}_\tau\|)\, da.\end{aligned}$$

We have

$$\begin{aligned} j(\boldsymbol{u}, \dot{\boldsymbol{u}}^\alpha) - j(\boldsymbol{u}, \dot{\boldsymbol{u}}) + j^\alpha(\boldsymbol{u}^\alpha, \dot{\boldsymbol{u}}) - j^\alpha(\boldsymbol{u}^\alpha, \dot{\boldsymbol{u}}^\alpha) \\ \le c\, \|p_\nu(u_\nu - g) - p_\nu^\alpha(u_\nu^\alpha - g)\|_{L^2(\Gamma_3)} \|\dot{\boldsymbol{u}}^\alpha - \dot{\boldsymbol{u}}\|_V \\ + c\, \|p_\tau(u_\nu - g) - p_\tau^\alpha(u_\nu^\alpha - g)\|_{L^2(\Gamma_3)} \|\dot{\boldsymbol{u}}^\alpha - \dot{\boldsymbol{u}}\|_V. \end{aligned} \tag{11.78}$$

For $r = \nu$ or τ,

$$\begin{aligned} \|p_r(u_\nu - g) - p_r^\alpha(u_\nu^\alpha - g)\|_{L^2(\Gamma_3)} \le c\, \|p_r(u_\nu - g) - p_r^\alpha(u_\nu - g)\|_{L^2(\Gamma_3)} \\ + c\, \|p_r^\alpha(u_\nu - g) - p_r^\alpha(u_\nu^\alpha - g)\|_{L^2(\Gamma_3)}. \end{aligned}$$

Taking into account (11.10) and (11.73) we obtain

$$\|p_r(u_\nu - g) - p_r^\alpha(u_\nu^\alpha - g)\|_{L^2(\Gamma_3)} \le c\, \|\boldsymbol{u}^\alpha - \boldsymbol{u}\|_V + c\, \varphi_r(\alpha). \tag{11.79}$$

Combining (11.75)–(11.79) we deduce that

$$\|\dot{\boldsymbol{u}}^\alpha - \dot{\boldsymbol{u}}\|_V \le c\, \|\boldsymbol{u}^\alpha - \boldsymbol{u}\|_V + c\, (\varphi_\nu(\alpha) + \varphi_\tau(\alpha)). \tag{11.80}$$

Now, with an integration over $[0, t]$, (11.7), (11.71) and (11.80) lead to

$$\begin{aligned} \|\boldsymbol{u}^\alpha(t) - \boldsymbol{u}(t)\|_V &\le c \int_0^t \|\dot{\boldsymbol{u}}^\alpha(s) - \dot{\boldsymbol{u}}(s)\|_V ds \\ &\le c \int_0^t \|\boldsymbol{u}^\alpha(s) - \boldsymbol{u}(s)\|_V ds + c\, (\varphi_\nu(\alpha) + \varphi_\tau(\alpha)) \end{aligned}$$

for all $t \in \overline{I}$. Applying the Gronwall inequality (Lemma 7.24, page 162), we have

$$\|\boldsymbol{u}^\alpha(t) - \boldsymbol{u}(t)\|_V \le c\, (\varphi_\nu(\alpha) + \varphi_\tau(\alpha)) \quad \forall\, t \in \overline{I}. \tag{11.81}$$

It follows from (11.80) and (11.81) that

$$\|\boldsymbol{u}^\alpha - \boldsymbol{u}\|_{C^1(\overline{I}; V)} \le c\, (\varphi_\nu(\alpha) + \varphi_\tau(\alpha)). \tag{11.82}$$

Theorem 11.16 is now a consequence of (11.82) and (11.73). ■

Let $\boldsymbol{\sigma}^\alpha \in C(\overline{I}; Q_1)$ and $\boldsymbol{\sigma} \in C(\overline{I}; Q_1)$ be the stress fields defined by

$$\boldsymbol{\sigma}^\alpha = \mathcal{A}\varepsilon(\dot{\boldsymbol{u}}^\alpha) + \mathcal{B}\varepsilon(\boldsymbol{u}^\alpha), \qquad \boldsymbol{\sigma} = \mathcal{A}\varepsilon(\dot{\boldsymbol{u}}) + \mathcal{B}\varepsilon(\boldsymbol{u}). \tag{11.83}$$

Using (11.19) and (11.72) we obtain

$$\operatorname{Div} \boldsymbol{\sigma}^\alpha = \operatorname{Div} \boldsymbol{\sigma} = -\boldsymbol{f}_0 \quad \text{a.e. in } \Omega \times I.$$

Therefore, from (11.83) and the assumptions (6.34), (6.35) on $\mathcal{A}$ and $\mathcal{B}$ we deduce that

$$\|\boldsymbol{\sigma}^\alpha - \boldsymbol{\sigma}\|_{C(\overline{I}; Q_1)} \le c\, \|\boldsymbol{u}^\alpha - \boldsymbol{u}\|_{C^1(\overline{I}; V)}. \tag{11.84}$$

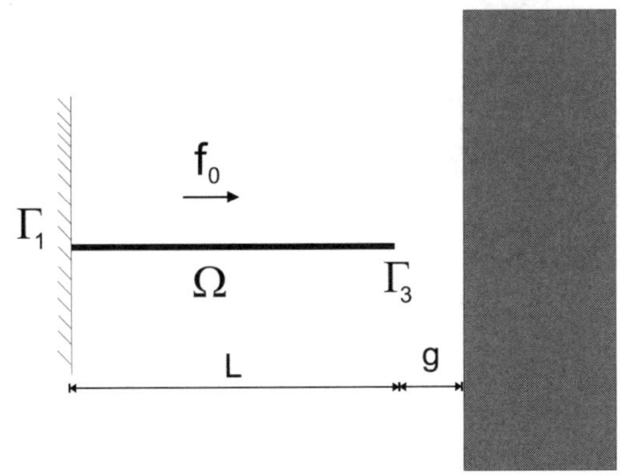

Figure 11.1: Contact of a rod with a deformable foundation

It follows now from (11.84) and (11.74) that

$$\boldsymbol{\sigma}^{\alpha} \to \boldsymbol{\sigma} \text{ in } C(\overline{I}; Q_1) \quad \text{as } \alpha \to 0. \tag{11.85}$$

In addition to the mathematical interest, the convergence results (11.74) and (11.85) are of importance in applications since they state that small errors in the contact and friction conditions lead to small errors in the solution. We conclude that the weak solution of the mechanical problem (11.1)–(11.7) depends continuously on the contact boundary conditions.

11.7 Numerical Examples

To see the performance of the fully discrete method described in Section 11.5, a number of numerical experiments have been done on test problems of the type (11.1)–(11.7) in one and two dimensions.

One-dimensional example. We consider a cantilever viscoelastic rod $\overline{\Omega} = [0, L]$ which is fixed at its left end $x = 0$ and is subjected to the action of a body force of density $f_0(x,t)$ in the x-direction (see Figure 11.1). An initial gap g is assumed between its right end $x = L$ and an obstacle with normal compliance. Contact occurs at certain time $t_c > 0$ and then compression starts to spread in from the contacting end. This problem corresponds to problem (11.1)–(11.7) with $\Omega = (0, L)$, $\Gamma_1 = \{0\}$, $\Gamma_2 = \emptyset$, $\Gamma_3 = \{L\}$. We use a linearly viscoelastic constitutive law, i.e.

$$\sigma = a\,\varepsilon(\dot{u}) + b\,\varepsilon(u).$$

Here $\varepsilon(u) = \frac{\partial u}{\partial x}$, while $a > 0$ and b are material constants, independent of x and t. We choose (5.42) (page 107) for the normal compliance contact function.

A complete description of this problem is the following one.

Problem 11.17 *Find a displacement field* $u : [0, L] \times [0, T] \to \mathbb{R}$ *and a stress field* $\sigma : [0, L] \times [0, T] \to \mathbb{R}$ *such that:*

$$\begin{aligned} \sigma(x,t) &= a\tfrac{\partial^2 u(x,t)}{\partial x \partial t} + b\tfrac{\partial u(x,t)}{\partial x} && \text{in } (0,L)\times(0,T),\\ \tfrac{\partial \sigma(x,t)}{\partial x} + f_0(x,t) &= 0 && \text{in } (0,L)\times(0,T),\\ u(0,t) &= 0 && \text{for } t \in (0,T),\\ -\sigma(L,t) &= c_\nu \max\{0, u(L,t) - g\} && \text{for } t \in (0,T),\\ u(x,0) &= u_0(x) && \text{in } (0,L). \end{aligned}$$

For computation we use the following data:

$$\begin{aligned} &L = 1\,m, \quad T = 0.1\,sec, \quad a = 1\,N sec/m, \quad b = 1\,N/m,\\ &f_0(x,t) = 10\,N/m \quad \forall\, x \in (0,1),\ t \in [0, 0.1],\\ &c_\nu = 1\,N/m, \quad g = 0.25\,m, \quad u_0(x) = 0\,m \quad \forall\, x \in (0,1). \end{aligned}$$

The exact solution of Problem 11.17 with the previous data can be obtained through an elementary but tedious calculation. It is given by the following:

- For $0 \le t \le t_c = \ln(\frac{20}{19})$, there is no contact and

$$\begin{aligned} u(x,t) &= 5\,(1 - e^{-t})\,(2x - x^2),\\ \sigma(x,t) &= 10 - 10\,x. \end{aligned}$$

- For $t_c = \ln(\frac{20}{19}) \le t \le 0.1$, there is contact and

$$\begin{aligned} u(x,t) &= -\frac{50}{19}\,e^{-2t}x + 5\,x\,(x-1)\,e^{-t} + \frac{61}{8}\,x - 5\,x^2,\\ \sigma(x,t) &= \frac{50}{19}\,e^{-2t} + \frac{61}{8} - 10\,x. \end{aligned}$$

We have implemented the fully discrete method described in the Section 11.5 on a typical workstation. We use a discretization by continuous piecewise linear functions for the space V^h and by piecewise constant functions for the space Q^h corresponding to a uniform partition of $[0, L]$ with meshsize h. For a time discretization, we use a uniform time step $k = h$. To solve the discrete systems, we employ a penalty-duality method (see e.g. [76]).

In Figure 11.2, we plot the error

$$E_{hk} = \max_{1 \le n \le N} \{\|u_n^{hk} - u_n\|_V + \|\delta u_n^{hk} - \dot{u}_n\|_V\}.$$

A linear convergence pattern is clearly observed, agreeing with the theoretical prediction (11.70).

Next, we report some numerical results with $h = k = 0.01$. In Figures 11.3 and 11.4 we show the displacement fields at times $t = 0.03,\ 0.06,\ 0.1$ and the corresponding errors. In Figures 11.5 and 11.6, we show the time evolution

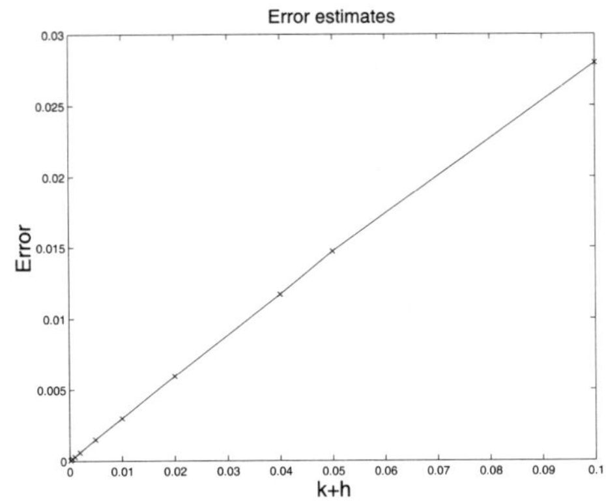

Figure 11.2: The 1D problem: linear convergence behavior

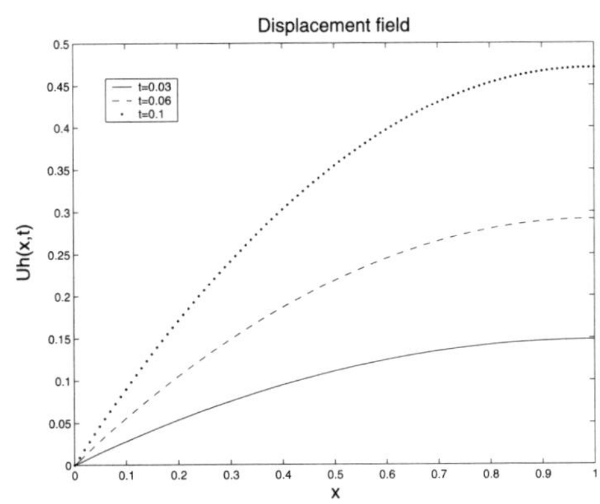

Figure 11.3: The 1D problem: displacement fields at several times

of the displacement at $x = 0.25,\ 0.5,\ 1$ for the approximate solutions and the corresponding errors. The time evolution of the approximate stress field at the contact node $x = 1$ and its difference with the exact solution are plotted in Figures 11.7 and 11.8. We observe that before contact the point $x = 1$ is stress free and, after the contact occurs, the reaction of the obstacle is towards the rod and increases when the penetration increases.

Two-dimensional example. As a two-dimensional example of the problem (11.1)–(11.7), we consider a plane stress viscoelasticity problem shown in Figure 11.9. The domain $\Omega = (0, 10) \times (0, 2)$ is a cross-section of a three-dimensional body subjected to the action of a vertical body force that is assumed to increase linearly in time. No tractions act on the part $\overline{\Gamma}_2 = \{[0, 10] \times \{2\}\} \cup \{[2, 8] \times \{0\}\}$ of the boundary. Both ends of the body are supposed to be clamped, i.e. $\overline{\Gamma}_1 = \{0, 10\} \times [0, 2]$. Finally, we assume that the body is in frictionless contact with an elastic foundation with normal compliance on $\overline{\Gamma}_3 = \{[0, 2] \times \{0\}\} \cup \{[8, 10] \times \{0\}\}$,

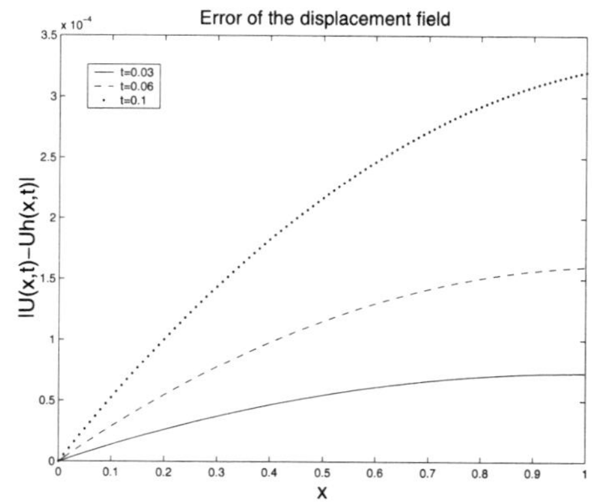

Figure 11.4: The 1D problem: error of displacement fields at several times

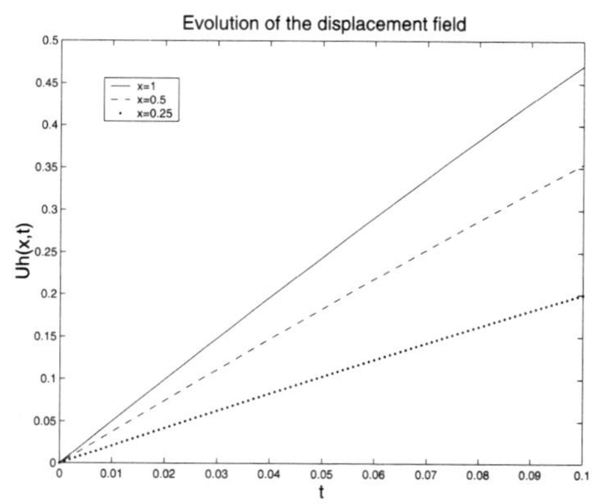

Figure 11.5: The 1D problem: time evolution of the displacement at $x = 0.25,\ 0.5,\ 1$

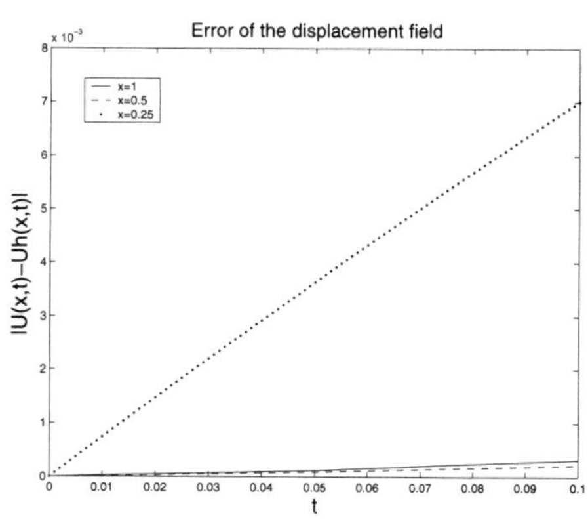

Figure 11.6: The 1D problem: errors of the displacement at $x = 0.25,\ 0.5,\ 1$

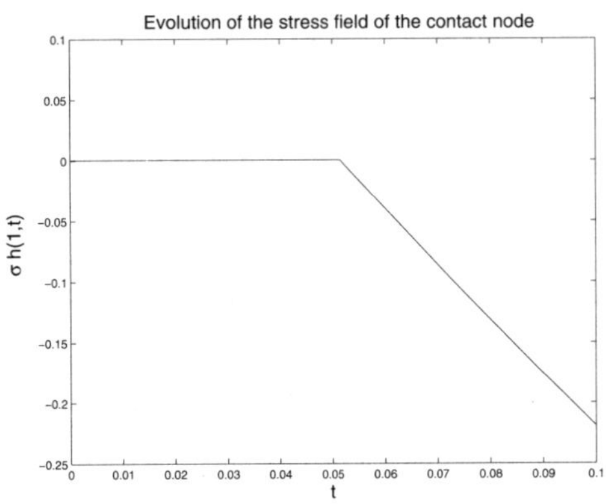

Figure 11.7: The 1D problem: time evolution of the stress at $x = 1$

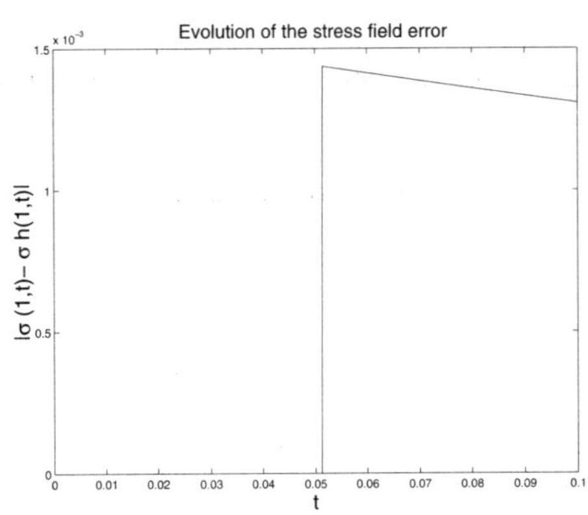

Figure 11.8: The 1D problem: error of the stress at $x = 1$

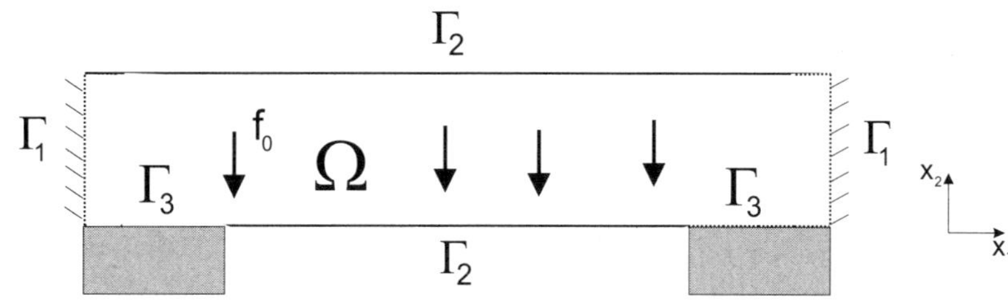

Figure 11.9: Contact of a 2D viscoelastic body

that is, we assume $p_\tau = 0$. We choose again (5.42) (page 107) as the normal compliance contact function.

The elasticity tensor $\mathcal{B}$ satisfies

$$(\mathcal{B}\boldsymbol{\tau})_{\alpha\beta} = \frac{E\nu}{1-\nu^2}\,(\tau_{11}+\tau_{22})\,\delta_{\alpha\beta} + \frac{E}{1+\nu}\,\tau_{\alpha\beta}, \quad 1 \le \alpha, \beta \le 2,$$

where E is Young's modulus, ν the Poisson ratio of the material and $\delta_{\alpha\beta}$ denotes the Kronecker symbol. The viscosity tensor $\mathcal{A}$ has a similar form, i.e.

$$(\mathcal{A}\boldsymbol{\tau})_{\alpha\beta} = \mu\,(\tau_{11}+\tau_{22})\,\delta_{\alpha\beta} + \eta\,\tau_{\alpha\beta}, \quad 1 \le \alpha, \beta \le 2,$$

where μ and η are viscosity constants. Recall also that the von Mises norm for a plane stress field $\boldsymbol{\tau} = (\tau_{\alpha\beta})$ is given by

$$\|\boldsymbol{\tau}\| = \left(\tau_{11}^2 + \tau_{22}^2 - \tau_{11}\tau_{22} + 3\tau_{12}^2\right)^{\frac{1}{2}}.$$

For computation we have used the following data:

$$\begin{aligned} &T = 1\,sec, \quad \boldsymbol{f}_0 = (0, -10t)\,N/m^2, \quad \boldsymbol{f}_2 = (0,0)\,N/m,\\ &g = 0\,m, \quad \boldsymbol{u}_0 = \mathbf{0}\,m, \quad E = 100\,N/m^2, \quad \nu = 0.3,\\ &\mu = 30\,N sec/m^2, \quad \eta = 20\,N sec/m^2. \end{aligned}$$

In Figures 11.10, 11.11, 11.12 and 11.13 the initial boundary and deformed mesh at final time $T = 1\,sec.$ are shown for values $c_\nu = 10\,N/m^2$, $100\,N/m^2$, $1000\,N/m^2$, and $10^8\,N/m^2$, respectively. Notice that the Signorini nonpenetration condition is progressively respected as c_ν gets bigger. In Figure 11.14 the von Mises norm for stress at final time $T = 1\,sec.$ is plotted in the deformed configuration for the value $c_\nu = 10^8\,N/m^2$. Finally, in Figure 11.15 the contact curves are shown for several values of c_ν.

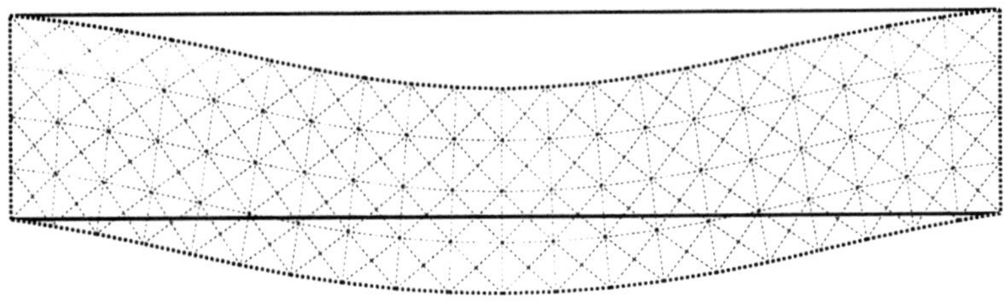

Figure 11.10: Initial boundary and deformed mesh at final time for $c_\nu = 10\,N/m^2$

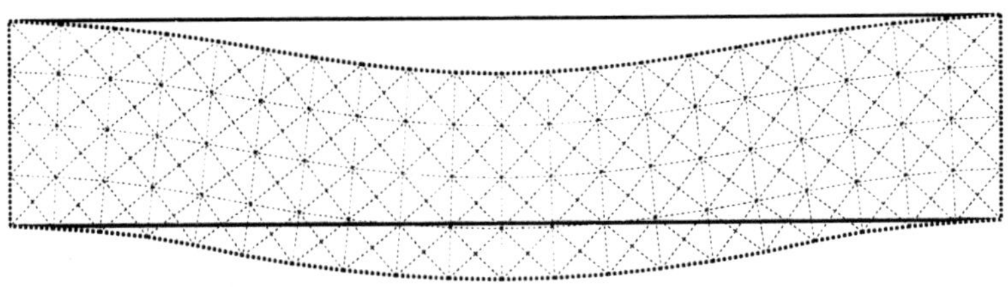

Figure 11.11: Initial boundary and deformed mesh at final time for $c_\nu = 100\,N/m^2$

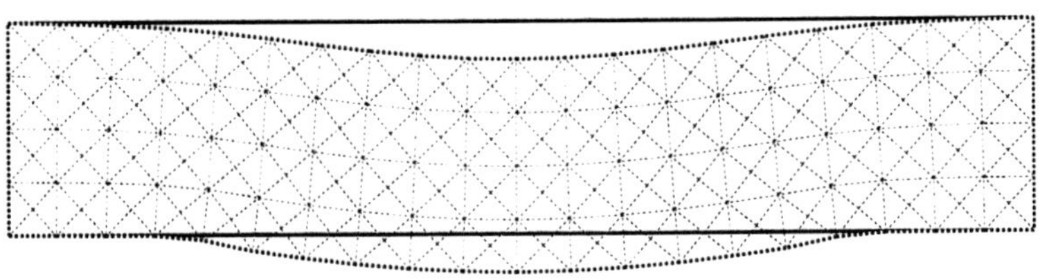

Figure 11.12: Initial boundary and deformed mesh at final time for $c_\nu = 1000\,N/m^2$

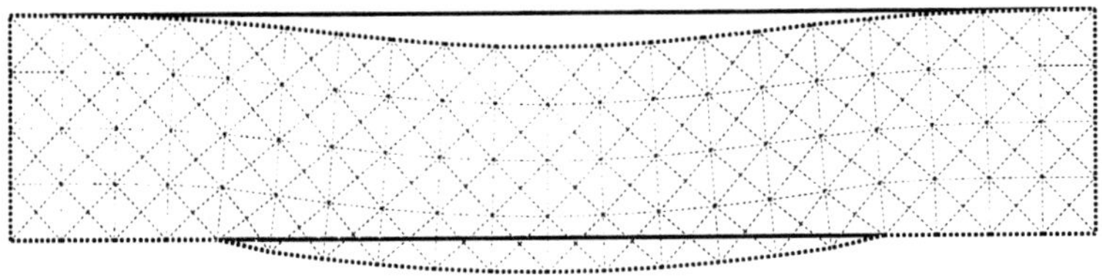

Figure 11.13: Initial boundary and deformed mesh at final time for $c_\nu = 10^8\,N/m^2$

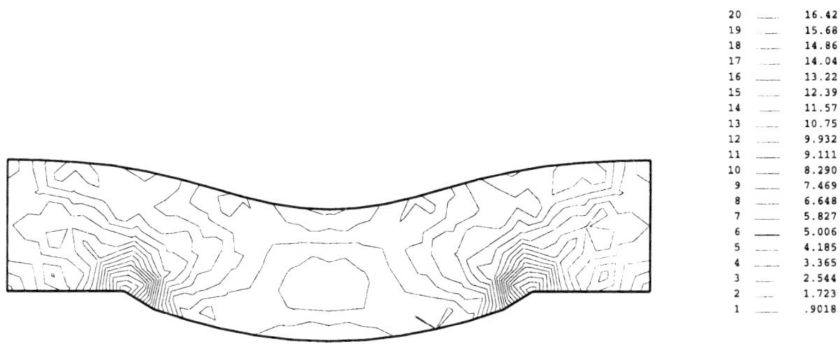

Figure 11.14: von Mises stress norm at final time in the deformed configuration

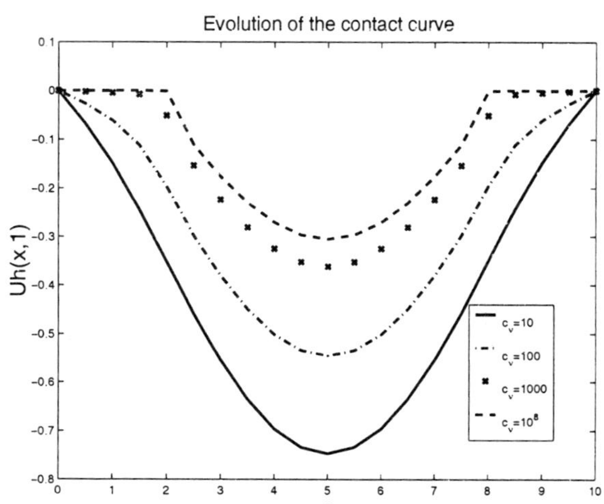

Figure 11.15: The contact curves for various values of c_ν

Chapter 12

Frictional Contact with Normal Damped Response

For the problem studied in this chapter the contact is described by a general normal damped response condition and a related version of the Coulomb law of dry friction. We derive a variational formulation for the problem in terms of the displacement, which leads to a new class of evolutionary variational inequalities. We study this class of variational inequalities in an abstract framework. Under the smallness assumption of certain coefficient, we prove an existence and uniqueness result using Banach's fixed point theorem. We then study two numerical approximation schemes of the problem: a semi-discrete scheme and a fully discrete scheme. For both schemes, we show the existence of a unique solution and derive error estimates. All these results on the abstract problem are applied to the analysis and numerical approximations of the viscoelastic frictional contact problem, with the finite element method used to discretize the spatial domain. We then introduce two variational formulations of the mechanical problem, both in terms of the stress and the displacement, and show their equivalence. Finally, we present some numerical examples to show the performance of the fully discrete scheme.

12.1 Problem Statement

The novelty of the mechanical problem we study in this chapter is the use of new boundary conditions formulated with a generalized version of the normal damped response and Coulomb's friction law. More precisely, the classical formulation of the mechanical problem we consider here is the following:

Find a displacement $\boldsymbol{u} : \Omega \times \overline{I} \to \mathbb{R}^d$ and a stress field $\boldsymbol{\sigma} : \Omega \times \overline{I} \to \mathbb{S}^d$ such that

$$\boldsymbol{\sigma} = \mathcal{A}\varepsilon(\dot{\boldsymbol{u}}) + \mathcal{B}\varepsilon(\boldsymbol{u}) \qquad \text{in } \Omega \times I, \tag{12.1}$$

$$\operatorname{Div} \boldsymbol{\sigma} + \boldsymbol{f}_0 = \boldsymbol{0} \qquad \text{in } \Omega \times I, \tag{12.2}$$

$$\boldsymbol{u} = \boldsymbol{0} \qquad \text{on } \Gamma_1 \times I, \tag{12.3}$$

$$\boldsymbol{\sigma\nu} = \boldsymbol{f}_2 \qquad \text{on } \Gamma_2 \times I, \tag{12.4}$$

$$\left.\begin{array}{l} -\sigma_\nu = p_\nu(\dot{u}_\nu), \ \|\boldsymbol{\sigma}_\tau\| \le p_\tau(\dot{u}_\nu), \\ \|\boldsymbol{\sigma}_\tau\| < p_\tau(\dot{u}_\nu) \ \Rightarrow \ \dot{\boldsymbol{u}}_\tau = \boldsymbol{0}, \\ \|\boldsymbol{\sigma}_\tau\| = p_\tau(\dot{u}_\nu) \ \Rightarrow \ \exists\, \lambda \ge 0 \text{ s.t. } \boldsymbol{\sigma}_\tau = -\lambda \dot{\boldsymbol{u}}_\tau \end{array}\right\} \qquad \text{on } \Gamma_3 \times I, \tag{12.5}$$

$$\boldsymbol{u}(0) = \boldsymbol{u}_0 \qquad \text{in } \Omega. \tag{12.6}$$

The contact condition (12.5) represents a normal damped response condition associated to a variant of Coulomb's law of friction. Examples of functions p_ν were provided in Section 5.4. Given p_ν, a choice of the friction bound function p_τ is provided by the equalities (11.8) or (11.9), page 228.

In the study of the mechanical problem (12.1)–(12.6) we recall the assumptions (6.34) and (6.35) on $\mathcal{A}$ and $\mathcal{B}$. We also assume the contact functions p_r ($r = \nu, \tau$) satisfy

$$\left.\begin{array}{l} \text{(a) } p_r : \Gamma_3 \times \mathbb{R} \to \mathbb{R}_+. \\ \text{(b) There exists an } L_r > 0 \text{ such that} \\ \qquad |p_r(\boldsymbol{x}, u_1) - p_r(\boldsymbol{x}, u_2)| \le L_r\, |u_1 - u_2| \\ \qquad \forall\, u_1, u_2 \in \mathbb{R}, \ \text{a.e. } \boldsymbol{x} \in \Omega. \\ \text{(c) For any } u \in \mathbb{R}, \ \boldsymbol{x} \mapsto p_r(\boldsymbol{x}, u) \text{ is measurable.} \\ \text{(d) The mapping } \boldsymbol{x} \mapsto p_r(\boldsymbol{x}, 0) \in L^2(\Gamma_3). \end{array}\right\} \tag{12.7}$$

We observe that the assumptions (12.7) on the functions p_ν and p_τ are pretty general. The only severe restriction comes from the condition (b), which, roughly speaking, requires the functions to grow at most linearly. Certainly the functions defined in (5.46) and (5.48) (page 108) satisfy the condition (12.7)(b). We also observe that if the functions p_ν and p_τ are related by (11.8) or (11.9) and p_ν satisfies the condition (12.7)(b), then p_τ also satisfies the condition (12.7)(b) with $L_\tau = \mu\, L_\nu$. We conclude that our results below are valid for the boundary value problems associated with these choices of the contact functions.

We assume that the force and traction densities have the regularities

$$\boldsymbol{f}_0 \in C(\overline{I}; L^2(\Omega)^d), \qquad \boldsymbol{f}_2 \in C(\overline{I}; L^2(\Gamma_2)^d), \tag{12.8}$$

and for the initial value,

$$\boldsymbol{u}_0 \in V. \tag{12.9}$$

Denote by $\boldsymbol{f}(t)$ the element of V given by

$$(\boldsymbol{f}(t), \boldsymbol{v})_V = \int_\Omega \boldsymbol{f}_0(t) \cdot \boldsymbol{v}\, dx + \int_{\Gamma_2} \boldsymbol{f}_2(t) \cdot \boldsymbol{v}\, da \tag{12.10}$$

for all $\boldsymbol{v} \in V$ and $t \in \overline{I}$. Note that conditions (12.8) imply

$$\boldsymbol{f} \in C(\overline{I}; V). \tag{12.11}$$

Let $j : V \times V \to \mathbb{R}$ be the functional defined by

$$j(\boldsymbol{v}, \boldsymbol{w}) = \int_{\Gamma_3} p_\nu(v_\nu)\, w_\nu \, da + \int_{\Gamma_3} p_\tau(v_\nu)\, \|\boldsymbol{w}_\tau\| \, da \quad \forall\, \boldsymbol{v}, \boldsymbol{w} \in V. \tag{12.12}$$

With these notations, a standard derivation shows that if $(\boldsymbol{u}, \boldsymbol{\sigma})$ is a pair of sufficiently regular functions satisfying (12.1)–(12.6), then $\boldsymbol{u}(t) \in V$ and for all $t \in \overline{I}$,

$$\begin{aligned} &(\boldsymbol{\sigma}(t), \boldsymbol{\varepsilon}(\boldsymbol{v}) - \boldsymbol{\varepsilon}(\dot{\boldsymbol{u}}(t)))_Q + j(\dot{\boldsymbol{u}}(t), \boldsymbol{v}) - j(\dot{\boldsymbol{u}}(t), \dot{\boldsymbol{u}}(t)) \\ &\qquad \geq (\boldsymbol{f}(t), \boldsymbol{v} - \dot{\boldsymbol{u}}(t))_V \quad \forall\, \boldsymbol{v} \in V. \end{aligned} \tag{12.13}$$

Thus, from (12.1), (12.6) and (12.13) we obtain the following variational formulation of the problem (12.1)–(12.6) in terms of the displacement.

Problem 12.1 *Find a displacement* $\boldsymbol{u} : \overline{I} \to V$ *such that*

$$\boldsymbol{u}(0) = \boldsymbol{u}_0 \tag{12.14}$$

and for all $t \in \overline{I}$,

$$\begin{aligned} &(\mathcal{A}\boldsymbol{\varepsilon}(\dot{\boldsymbol{u}}(t)), \boldsymbol{\varepsilon}(\boldsymbol{v}) - \boldsymbol{\varepsilon}(\dot{\boldsymbol{u}}(t)))_Q + (\mathcal{B}\boldsymbol{\varepsilon}(\boldsymbol{u}(t)), \boldsymbol{\varepsilon}(\boldsymbol{v}) - \boldsymbol{\varepsilon}(\dot{\boldsymbol{u}}(t)))_Q + \\ &\quad j(\dot{\boldsymbol{u}}(t), \boldsymbol{v}) - j(\dot{\boldsymbol{u}}(t), \dot{\boldsymbol{u}}(t)) \geq (\boldsymbol{f}(t), \boldsymbol{v} - \dot{\boldsymbol{u}}(t))_V \quad \forall\, \boldsymbol{v} \in V. \end{aligned} \tag{12.15}$$

Well-posedness of Problem 12.1 is discussed within the framework of an abstract problem in the next section.

12.2 An Abstract Problem and its Well-posedness

The following abstract problem includes Problem 12.1 as a special case.

Problem 12.2 *Find* $u : \overline{I} \to X$ *such that*

$$u(0) = u_0 \tag{12.16}$$

and for $t \in \overline{I}$,

$$\begin{aligned} &(A\dot{u}(t), v - \dot{u}(t))_X + (Bu(t), v - \dot{u}(t))_X + j(\dot{u}(t), v) - j(\dot{u}(t), \dot{u}(t)) \\ &\qquad \geq (f(t), v - \dot{u}(t))_X \qquad \forall\, v \in X. \end{aligned} \tag{12.17}$$

Here, X is a function space corresponding to admissible displacements, A and B are nonlinear operators corresponding to the viscoelastic constitutive law and the functional j is determined by the type of contact boundary conditions. The data f is related to the given body force and surface traction, and u_0 represents the initial displacement.

We assume in what follows X is a real Hilbert space endowed with the inner product $(\cdot,\cdot)_X$ and the associated norm $\|\cdot\|_X$. As in Section 11.2, we assume $A : X \to X$ satisfies (11.22), $B : X \to X$ satisfies (11.23), $j : X \times X \to \mathbb{R}$ satisfies (11.24), f satisfies (11.25), and u_0 satisfies (11.26).

The main result of this section is the following.

Theorem 12.3 *Let* (11.22)–(11.26) *hold. Then, if* $M > m$, *there exists a unique solution* $u \in C^1(\overline{I}; X)$ *to Problem 12.2.*

The proof of Theorem 12.3 is established in several steps. To simplify the notation, sometimes we will not indicate explicitly the dependence of various functions on the time variable t.

In the first step, let $\eta \in C(\overline{I}; X)$ and $g \in C(\overline{I}; X)$ be given and we consider the following auxiliary variational inequality of finding $v_{\eta g} : \overline{I} \to X$, such that for $t \in \overline{I}$,

$$\begin{aligned}(Av_{\eta g}(t), v - v_{\eta g}(t))_X + (\eta(t), v - v_{\eta g}(t))_X + j(g(t), v) - j(g(t), v_{\eta g}(t))\\ \geq (f(t), v - v_{\eta g}(t))_X \qquad \forall\, v \in X. \qquad (12.18)\end{aligned}$$

Lemma 12.4 *There exists a unique solution* $v_{\eta g} \in C(\overline{I}; X)$ *to the problem* (12.18).

Proof. It follows from Theorem 4.1 (page 60) that there exists a unique element $v_{\eta g}(t) \in X$ that solves (12.18) for each $t \in \overline{I}$. Let us show that $v_{\eta g} : \overline{I} \to X$ is continuous. Let $t_1, t_2 \in \overline{I}$. For simplicity we write $v_{\eta g}(t_i) = v_i$, $\eta(t_i) = \eta_i$, $g(t_i) = g_i$, for $i = 1, 2$. Using (12.18) we easily derive the relation

$$\begin{aligned}(Av_1 - Av_2, v_1 - v_2)_X \leq (f_1 - f_2, v_1 - v_2)_X + (\eta_1 - \eta_2, v_2 - v_1)_X\\ + j(g_1, v_2) - j(g_1, v_1) + j(g_2, v_1) - j(g_2, v_2).\end{aligned}$$

Then we use the conditions (11.22) (a) and (11.24) (b) to obtain

$$M\|v_1 - v_2\|_X \leq \|f_1 - f_2\|_X + \|\eta_1 - \eta_2\|_X + m\|g_1 - g_2\|_X. \qquad (12.19)$$

We deduce from (12.19) that $v_{\eta g}(t) : \overline{I} \to X$ is a continuous function. ∎

For each $\eta \in C(\overline{I}; X)$, we consider an operator $\Lambda_\eta : C(\overline{I}; X) \to C(\overline{I}; X)$ defined by

$$\Lambda_\eta g = v_{\eta g} \quad \forall\, g \in C(\overline{I}; X). \qquad (12.20)$$

We have the following result.

Lemma 12.5 *Let $M > m$. Then the operator Λ_η has a unique fixed-point $g_\eta \in C(\overline{I};X)$.*

Proof. Let $g_1, g_2 \in C(\overline{I};X)$, $\eta \in C(\overline{I};X)$ and let v_i denote the solution of (12.18) for $g = g_i$, i.e. $v_i = v_{\eta g_i}$, $i = 1, 2$. From the definition (12.20) we have

$$\|\Lambda_\eta g_1 - \Lambda_\eta g_2\|_X = \|v_1 - v_2\|_X.$$

An argument similar to that used in the proof of Lemma 12.4 shows

$$M\,\|v_1 - v_2\|_X \le m\,\|g_1 - g_2\|_X.$$

Then

$$\|\Lambda_\eta g_1 - \Lambda_\eta g_2\|_X \le \frac{m}{M}\,\|g_1 - g_2\|_X.$$

Thus if $M > m$, then the operator Λ_η is a contraction on the Banach space $C(\overline{I};X)$ and has a unique fixed-point $g_\eta \in C(\overline{I};X)$ by Banach's fixed point theorem (Theorem 1.62, page 21). ■

In what follows we assume $M > m$. Let $\eta \in C(\overline{I};X)$. Denote by g_η the fixed-point function stated in Lemma 12.5 and let $v_\eta \in C(\overline{I};X)$ be the function defined by

$$v_\eta = v_{\eta g_\eta}. \tag{12.21}$$

We have $\Lambda_\eta g_\eta = g_\eta$ and

$$v_\eta = g_\eta \tag{12.22}$$

by (12.20) and (12.21). Therefore, taking $g = g_\eta$ in (12.18) and using (12.21) and (12.22), we see that $v_\eta(t) \in X$ satisfies

$$\begin{aligned}(Av_\eta(t), v - v_\eta(t))_X + (\eta(t), v - v_\eta(t))_X + j(v_\eta(t), v) - j(v_\eta(t), v_\eta(t))\\ \ge (f(t), v - v_\eta(t))_X \quad \forall\, v \in X,\ t \in \overline{I}.\end{aligned} \tag{12.23}$$

We denote now by $u_\eta \in C^1(\overline{I};X)$ the function given by

$$u_\eta(t) = \int_0^t v_\eta(s)\,ds + u_0, \quad t \in \overline{I} \tag{12.24}$$

and define the operator $\Lambda : C(\overline{I};X) \to C(\overline{I};X)$ by

$$\Lambda\eta = Bu_\eta \quad \forall\, \eta \in C(\overline{I};X). \tag{12.25}$$

We have the next result.

Lemma 12.6 *Let $M > m$. Then the operator Λ has a unique fixed-point $\eta^* \in C(\overline{I};X)$.*

Proof. For a proof of this lemma, we use

$$\|v\|^*_{C(\overline{I};X)} = \max_{t\in\overline{I}} e^{-\beta\, t} \|v(t)\|_X$$

with $\beta > L_B/(M-m)$ as the norm in the space $C(\overline{I};X)$. This norm is equivalent to the standard norm $\|v\|_{C(\overline{I};X)}$. Let $\eta_1, \eta_2 \in C(\overline{I};X)$ and denote $v_i = v_{\eta_i}$, $u_i = u_{\eta_i}$ for $i = 1, 2$. Using (12.23) and the estimates in the proof of Lemma 12.4 (see (12.19)) we deduce that

$$M\,\|v_1 - v_2\|_X \le \|\eta_1 - \eta_2\|_X + m\,\|v_1 - v_2\|_X,$$

which implies

$$\|v_1 - v_2\|_X \le \frac{1}{M-m}\,\|\eta_1 - \eta_2\|_X. \tag{12.26}$$

Using now (12.25), (11.23) and (12.24) we obtain

$$\|\Lambda\eta_1(t) - \Lambda\eta_2(t)\|_X \le L_B \int_0^t \|v_1(s) - v_2(s)\|_X\,ds \quad \forall\, t \in \overline{I}.$$

By (12.26),

$$\|\Lambda\eta_1(t) - \Lambda\eta_2(t)\|_X \le \frac{L_B}{M-m} \int_0^t \|\eta_1(s) - \eta_2(s)\|_X\,ds \quad \forall\, t \in \overline{I}.$$

We then have

$$\begin{aligned} e^{-\beta\, t}\|\Lambda\eta_1(t) - \Lambda\eta_2(t)\|_X &\le \frac{L_B}{M-m}\, e^{-\beta\, t} \int_0^t e^{\beta\, s}\,ds\,\|\eta_1 - \eta_2\|^*_{C(\overline{I};X)} \\ &\le \frac{L_B}{M-m}\,\frac{1}{\beta}\,\|\eta_1 - \eta_2\|^*_{C(\overline{I};X)}, \end{aligned}$$

and so

$$\|\Lambda\eta_1 - \Lambda\eta_2\|^*_{C(\overline{I};X)} \le \frac{L_B}{M-m}\,\frac{1}{\beta}\,\|\eta_1 - \eta_2\|^*_{C(\overline{I};X)}.$$

Since $\beta > L_B/(M-m)$, the operator Λ is a contraction on the space $C(\overline{I};X)$ when the equivalent norm $\|\cdot\|^*_{C(\overline{I};X)}$ is used. By the Banach fixed-point theorem, Λ has a unique fixed point $\eta^* \in C(\overline{I};X)$. ■

We are now ready to prove Theorem 12.3.

Proof of Theorem 12.3.
Existence. Let $\eta^* \in C(\overline{I};X)$ be the fixed point of Λ and let $u_{\eta^*} \in C^1(\overline{I};X)$ be the function defined by (12.24) for $\eta = \eta^*$. We have $\dot{u}_{\eta^*} = v_{\eta^*}$. From (12.23) for $\eta = \eta^*$ it follows that for any $t \in \overline{I}$,

$$\begin{aligned} (A\dot{u}_{\eta^*}(t), v - \dot{u}_{\eta^*}(t))_X + (\eta^*(t), v - \dot{u}_{\eta^*}(t))_X + j(\dot{u}_{\eta^*}(t), v) \\ - j(\dot{u}_{\eta^*}(t), \dot{u}_{\eta^*}(t)) \ge (f(t), v - \dot{u}_{\eta^*}(t))_X \quad \forall\, v \in X. \end{aligned} \tag{12.27}$$

The inequality (12.17) follows now from (12.27) and (12.25) since $\eta^* = \Lambda\eta^* = Bu_{\eta^*}$, and (12.16) results from (12.24). We conclude that u_{η^*} is a solution of Problem 12.2.

Uniqueness. To prove the uniqueness of the solution let u_{η^*} be the solution of (12.17), (12.16) obtained above and let u be another solution such that $u \in C^1(\overline{I};X)$. We denote by $\eta \in C(\overline{I};X)$ the function given by

$$\eta = Bu \tag{12.28}$$

and let

$$\dot{u} = w. \tag{12.29}$$

Using (12.17) we obtain that w is a solution of the variational inequality (12.23) and since this problem has a unique solution $v_\eta \in C(\overline{I};X)$ (see, e.g. (12.26)) we conclude that

$$w = v_\eta. \tag{12.30}$$

Moreover, it follows from (12.29), (12.17), (12.30) and (12.24) that

$$u = u_\eta. \tag{12.31}$$

Using now (12.25), (12.28) and (12.31) we obtain that $\Lambda\eta = \eta$ and by the uniqueness of the fixed-point of Λ we have

$$\eta = \eta^*. \tag{12.32}$$

The uniqueness of the solution is a consequence of (12.31) and (12.32). The proof of Theorem 12.3 is now complete. ■

We end this section with the following regularity result.

Proposition 12.7 *Under the conditions stated in Theorem 12.3, if for some $p \in [1,\infty]$, $f \in W^{1,p}(I;X)$, then $\dot{u} \in W^{1,p}(I;X)$ and*

$$\|\dot{u}\|_{W^{1,p}(I;X)} \le c\,(\|f\|_{W^{1,p}(I;X)} + \|u\|_{C^1(\overline{I};X)}).$$

Proof. For any $t_1, t_2 \in \overline{I}$, we apply the inequality (12.19) to the inequality (12.17) to obtain

$$\begin{aligned} M\,\|\dot{u}(t_1) - \dot{u}(t_2)\|_X \le \|f(t_1) - f(t_2)\|_X &+ \|Bu(t_1) - Bu(t_2)\|_X \\ &+ m\,\|\dot{u}(t_1) - \dot{u}(t_2)\|_X. \end{aligned}$$

By the assumptions $M > m$ and (11.23), we find that

$$\|\dot{u}(t_1) - \dot{u}(t_2)\|_X \le c\,(\|f(t_1) - f(t_2)\|_X + \|u(t_1) - u(t_2)\|_X),$$

from which the result of the corollary follows. ■

In the next two sections, we assume all the conditions stated in Theorem 12.3 are satisfied.

12.3 Semi-discrete Approximation of the Abstract Problem

In this section we consider an approximation of Problem 12.2 by discretizing only the space X. Let $X^h \subset X$ be a finite-dimensional space which for example, is constructed by the finite element method. Then a semi-discrete scheme can be formed as following.

Problem 12.8 *Find $u^h : \overline{I} \to X^h$ such that*

$$u^h(0) = u_0^h \tag{12.33}$$

and for $t \in \overline{I}$,

$$\begin{aligned}(A\dot{u}^h(t), v^h - \dot{u}^h(t))_X + (Bu^h(t), v^h - \dot{u}^h(t))_X + j(\dot{u}^h(t), v^h) \\ - j(\dot{u}^h(t), \dot{u}^h(t)) \geq (f(t), v^h - \dot{u}^h(t))_X \quad \forall\, v^h \in X^h.\end{aligned} \tag{12.34}$$

Here, $u_0^h \in X^h$ is an appropriate approximation of u_0.

Using the arguments presented in the previous section, we see that under the conditions stated in Theorem 12.3, Problem 12.8 has a unique solution $u^h \in C^1(\overline{I}; X^h)$. Our main purpose here is to derive estimates for the error $u - u^h$.

Let $t \in \overline{I}$. To simplify the writing, we introduce the velocity variable

$$w(t) = \dot{u}(t).$$

Then by using the initial value condition (12.16), we express the displacement variable as

$$u(t) = \int_0^t w(s)\, ds + u_0. \tag{12.35}$$

Similarly, we introduce the discrete velocity variable

$$w^h(t) = \dot{u}^h(t).$$

With the initial value condition (12.33), we have

$$u^h(t) = \int_0^t w^h(s)\, ds + u_0^h. \tag{12.36}$$

Now the variational inequalities (12.17) and (12.34) can be rewritten as

$$\begin{aligned}(Aw(t), v - w(t))_X + (Bu(t), v - w(t))_X + j(w(t), v) \\ - j(w(t), w(t)) \geq (f(t), v - w(t))_X \quad \forall\, v \in X\end{aligned} \tag{12.37}$$

and

$$\begin{aligned}(Aw^h(t), v^h - w^h(t))_X + (Bu^h(t), v^h - w^h(t))_X + j(w^h(t), v^h) \\ - j(w^h(t), w^h(t)) \geq (f(t), v^h - w^h(t))_X \quad \forall\, v^h \in X^h,\end{aligned} \tag{12.38}$$

respectively. We now take $v = w^h(t)$ in (12.37) and add the inequality to (12.38) with $v^h = v^h(t) \in X^h$ arbitrary. After some manipulations, we have

$$\begin{aligned}
&(Aw(t) - Aw^h(t), w(t) - w^h(t))_X \\
&\quad \le (Aw^h(t) - Aw(t), v^h(t) - w(t))_X + R(w(t), v^h(t)) \\
&\qquad + (Bu(t) - Bu^h(t), w^h(t) - v^h(t))_X + j(w(t), w^h(t)) \\
&\qquad - j(w(t), v^h(t)) + j(w^h(t), v^h(t)) - j(w^h(t), w^h(t)),
\end{aligned}$$

where

$$\begin{aligned}
R(t; w(t), v^h(t)) &= (Aw(t), v^h(t) - w(t))_X + (Bu(t), v^h(t) - w(t))_X \\
&\quad + j(w(t), v^h(t)) - j(w(t), w(t)) \\
&\quad - (f(t), v^h(t) - w(t))_X
\end{aligned} \tag{12.39}$$

represents a residual quantity. Using the assumptions (11.22), (11.23) and (11.24) on the operators A, B and the functional j, we have

$$\begin{aligned}
&M\,\|w(t) - w^h(t)\|_X^2 \\
&\quad \le L_A\|w(t) - w^h(t)\|_X\,\|w(t) - v^h(t)\|_X + |R(t; w(t), v^h(t))| \\
&\qquad + (L_B\|u(t) - u^h(t)\|_X + m\,\|w(t) - w^h(t)\|_X)\,\|w^h(t) - v^h(t)\|_X \\
&\quad \le L_A\|w(t) - w^h(t)\|_X\,\|w(t) - v^h(t)\|_X + |R(t; w(t), v^h(t))| \\
&\qquad + (L_B\|u(t) - u^h(t)\|_X + m\,\|w(t) - w^h(t)\|_X) \\
&\qquad \times (\|w(t) - w^h(t)\|_X + \|w(t) - v^h(t)\|_X).
\end{aligned}$$

Thus, under the assumption $M > m$, we have the inequality

$$\begin{aligned}
\|w(t) - w^h(t)\|_X^2 \le c\,\big(&\|w(t) - v^h(t)\|_X^2 + \|u(t) - u^h(t)\|_X^2 \\
&+ |R(t; w(t), v^h(t))|\big).
\end{aligned} \tag{12.40}$$

Subtracting (12.36) from (12.35), we have

$$u(t) - u^h(t) = \int_0^t (w(s) - w^h(s))\,ds + u_0 - u_0^h,$$

and so

$$\|u(t) - u^h(t)\|_X^2 \le c\left(\int_0^t \|w(s) - w^h(s)\|_X^2 ds + \|u_0 - u_0^h\|_X^2\right). \tag{12.41}$$

Then the inequality (12.40) can be rewritten as

$$\begin{aligned}
\|w(t) - w^h(t)\|_X^2 \le c\Bigg(&\|w(t) - v^h(t)\|_X^2 + \|u_0 - u_0^h\|_X^2 \\
&+ \int_0^t \|w(s) - w^h(s)\|_X^2\,ds + |R(t; w(t), v^h(t))|\Bigg).
\end{aligned}$$

Applying the Gronwall inequality (Lemma 7.24), we have

$$\begin{aligned}\|w - w^h\|_{C(\overline{I};X)} \le c \inf_{v^h \in C(\overline{I};X^h)} \left(\|w - v^h\|_{C(\overline{I};X)} + \|R(\cdot; w, v^h)\|_{C(\overline{I})}^{1/2}\right) \\ + c\,\|u_0 - u_0^h\|_X. \qquad (12.42)\end{aligned}$$

Summarizing, with (12.41) and (12.42), we have proved the following result.

Theorem 12.9 *Assume the conditions* (11.22)–(11.26) *and* $M > m$. *Then for the error of the semi-discrete solution of Problem 12.8, we have the estimate*

$$\begin{aligned}\|u - u^h\|_{C^1(\overline{I};X)} \le c \inf_{v^h \in C(\overline{I};X^h)} \left(\|\dot{u} - v^h\|_{C(\overline{I};X)} + \|R(\cdot; \dot{u}, v^h)\|_{C(\overline{I})}^{1/2}\right) \\ + c\,\|u_0 - u_0^h\|_X, \qquad (12.43)\end{aligned}$$

where $R(\cdot; \dot{u}(\cdot), v^h(\cdot))$ *is defined in* (12.39).

The inequality (12.43) is the basis for convergence analysis and for deriving order error estimates, as is shown later in Section 12.5 in the context of the contact problem 12.1.

12.4 Fully Discrete Approximation of the Abstract Problem

In addition to the finite-dimensional subspace X^h, we use a partition of the time interval described in Section 7.2. The fully discrete approximation method we will analyze is the following.

Problem 12.10 *Find* $\{u_n^{hk}\}_{n=0}^N \subset X^h$ *such that*

$$u_0^{hk} = u_0^h \qquad (12.44)$$

and for $n = 1, \dots, N$,

$$\begin{aligned}(A\delta_n u_n^{hk}, v^h - \delta_n u_n^{hk})_X + (Bu_{n-1}^{hk}, v^h - \delta_n u_n^{hk})_X + j(\delta_n u_n^{hk}, v^h) \\ - j(\delta_n u_n^{hk}, \delta_n u_n^{hk}) \ge (f_n, v^h - \delta_n u_n^{hk})_X \quad \forall\, v^h \in X^h. \qquad (12.45)\end{aligned}$$

Here, $u_0^h \in X^h$ *is an appropriate approximation of* u_0.

Again to simplify the notation, we introduce the discrete velocity

$$w_n^{hk} = \delta_n u_n^{hk}, \quad n = 1, \dots, N.$$

Then using the initial value condition (12.44), we can express the discrete displacement as

$$u_n^{hk} = \sum_{j=1}^{n} w_j^{hk} k_j + u_0^h. \qquad (12.46)$$

We can rewrite (12.45) in the form

$$(Aw_n^{hk}, v^h - w_n^{hk})_X + (Bu_{n-1}^{hk}, v^h - w_n^{hk})_X + j(w_n^{hk}, v^h) \\ - j(w_n^{hk}, w_n^{hk}) \geq (f_n, v^h - w_n^{hk})_X \quad \forall\, v^h \in X^h. \tag{12.47}$$

By a discrete analogue of Theorem 12.3, we see that given $u_{n-1}^{hk} \in X^h$, the inequality (12.47) has a unique solution $w_n^{hk} \in X^h$. Note that $u_0^{hk} = u_0^h$ is given and we have the relation (12.46) between $\{u_n^{hk}\}_{n=1}^N$ and $\{w_n^{hk}\}_{n=1}^N$. A mathematical induction argument yields the existence and uniqueness of a solution of Problem 12.10. Our main objective here is to derive an estimate for the error of the fully discrete solution.

Take $v = w_n^{hk}$ in (12.37) at $t = t_n$,

$$(Aw_n, w_n^{hk} - w_n)_X + (Bu_n, w_n^{hk} - w_n)_X + j(w_n, w_n^{hk}) - j(w_n, w_n) \\ \geq (f_n, w_n^{hk} - w_n)_X. \tag{12.48}$$

We now add (12.47) with $v^h = v_n^h \in X^h$ arbitrary and (12.48) to obtain an error relation

$$\begin{aligned}(Aw_n - Aw_n^{hk}, w_n - w_n^{hk})_X &\\ \leq (Aw_n^{hk} - Aw_n, v_n^h - w_n)_X &\\ + (Bu_{n-1}^{hk} - Bu_n, v_n^h - w_n^{hk})_X + R_n(w_n, v_n^h) &\\ + j(w_n^{hk}, v_n^h) - j(w_n, v_n^h) + j(w_n, w_n^{hk}) - j(w_n^{hk}, w_n^{hk}),&\end{aligned}$$

where

$$R_n(w_n, v_n^h) = (Aw_n, v_n^h - w_n)_X + (Bu_n, v_n^h - w_n)_X \\ + j(w_n, v_n^h) - j(w_n, w_n) - (f_n, v_n^h - w_n)_X. \tag{12.49}$$

By the assumptions (11.22), (11.23) and (11.24) on the operators A, B and the functional j, we have

$$\begin{aligned}M\, \|w_n - w_n^{hk}\|_X^2 \leq L_A \|w_n - w_n^{hk}\|_X\, \|w_n - v_n^h\|_X &\\ + L_B \|u_{n-1}^{hk} - u_n\|_X\, \|v_n^h - w_n^{hk}\|_X + |R_n(w_n, v_n^h)| &\\ + m\, \|w_n - w_n^{hk}\|_X\, \|v_n^h - w_n^{hk}\|_X.&\end{aligned}$$

The term $\|v_n^h - w_n^{hk}\|_X$ is bounded as follows:

$$\|v_n^h - w_n^{hk}\|_X \leq \|v_n^h - w_n\|_X + \|w_n - w_n^{hk}\|_X.$$

Since $M > m$, we get the relation

$$\|w_n - w_n^{hk}\|_X^2 \leq c\, \big\{ \|v_n^h - w_n\|_X^2 + \|u_{n-1}^{hk} - u_n\|_X^2 + |R_n(w_n, v_n^h)| \big\},$$

or,

$$\|w_n - w_n^{hk}\|_X \leq c\, \big\{ \|v_n^h - w_n\|_X + |R_n(w_n, v_n^h)|^{1/2} + \|u_{n-1}^{hk} - u_n\|_X \big\}. \tag{12.50}$$

We can bound the term $\|u_{n-1}^{hk} - u_n\|_X$ by using (7.42) (page 155) with $I_k(w)$ defined in (7.43). Then from (12.50), we have

$$\begin{aligned}\|w_n - w_n^{hk}\|_X \le c \Big\{ &\|v_n^h - w_n\|_X + |R_n(w_n, v_n^h)|^{1/2} + \|u_0^h - u_0\|_X \\ &+ k\,\|u\|_{C^1(\bar{I};X)} + I_k(w) \\ &+ \sum_{j=1}^{n-1} \|w_j^{hk} - w_j\|_X k_j \Big\}. \end{aligned} \tag{12.51}$$

Applying Lemma 7.25 (page 163), we obtain the following estimate

$$\begin{aligned}\max_n \|w_n - w_n^{hk}\|_X \le c \max_n \Big\{ \|v_n^h - w_n\|_X + |R_n(w_n, v_n^h)|^{1/2} \Big\} \\ + c\,\big(\|u_0^h - u_0\|_X + I_k(w) + k\,\|u\|_{C^1(\bar{I};X)}\big).\end{aligned}$$

Similar to (7.42), we have

$$\begin{aligned}\|u_n^{hk} - u_n\|_X \le \sum_{j=1}^{n} \|w_j^{hk} - w_j\|_X k_j + \|u_0^h - u_0\|_X \\ + c\,\big(k\,\|u\|_{C^1(\bar{I};X)} + I_k(w)\big).\end{aligned}$$

Therefore, we have proved the following result.

Theorem 12.11 *Let the conditions* (11.22)–(11.26) *be satisfied and* $M > m$. *Then for the error of the fully discrete solution of Problem 12.10, we have the estimate*

$$\begin{aligned}\max_{1\le n\le N} &(\|u_n^{hk} - u_n\|_X + \|\dot{u}_n - \delta_n u_n^{hk}\|_X) \\ &\le c \max_{1\le n\le N} \inf_{v_n^h\in X^h} \Big\{ \|v_n^h - \dot{u}_n\|_X + |R_n(\dot{u}_n, v_n^h)|^{1/2} \Big\} \\ &\quad + c\,\|u_0^h - u_0\|_X + c\,(I_k(\dot{u}) + k\,\|u\|_{C^1(\bar{I};X)}),\end{aligned} \tag{12.52}$$

where $R_n(\dot{u}_n, v_n^h)$ *is defined by* (12.49), $I_k(\dot{u}) = I_k(w)$ *is defined in* (7.43).

When the data f is smoother, we can derive a more compact error bound from (12.52). For this purpose, we assume additionally $f \in W^{1,1}(I;X)$. Then by Proposition 12.7, we have $w \in W^{1,1}(I;X)$ and

$$\|w\|_{W^{1,1}(I;X)} \le c\,(\|f\|_{W^{1,1}(I;X)} + \|u\|_{C^1(\bar{I};X)}).$$

By using the bound (7.44) for $I_k(w)$ defined in (7.43), we can replace the estimate (12.52) by

$$\begin{aligned}\max_{1\le n\le N} &(\|u_n - u_n^{hk}\|_X + \|\dot{u}_n - \delta_n u_n^{hk}\|_X) \\ &\le c \max_{1\le n\le N} \inf_{v_n^h\in X^h} \Big\{ \|v_n^h - \dot{u}_n\|_X + |R_n(\dot{u}_n, v_n^h)|^{1/2} \Big\} \\ &\quad + c\,\|u_0^h - u_0\|_X + c\,k\,(\|f\|_{W^{1,1}(I;X)} + \|u\|_{C^1(\bar{I};X)}).\end{aligned} \tag{12.53}$$

12.5 Applications to the Contact Problem

Well-posedness of Problem 12.1 follows by an application of Theorem 12.3.

Theorem 12.12 *Assume* (6.34), (6.35) *and* (12.7)–(12.9). *Then there exists* $L_0 > 0$ *which depends only on* Ω, Γ_1 *and* $\mathcal{A}$ *such that Problem 12.1 has a unique solution* $\boldsymbol{u}$ *if* $L_\nu + L_\tau < L_0$. *Moreover,* $\boldsymbol{u} \in C^1(\overline{I}; V)$.

Proof. To apply Theorem 12.3, we take $X = V$, and let $A : V \to V$ and $B : V \to V$ be the operators defined by

$$(A\boldsymbol{v}, \boldsymbol{w})_V = (\mathcal{A}\boldsymbol{\varepsilon}(\boldsymbol{v}), \boldsymbol{\varepsilon}(\boldsymbol{w}))_Q, \quad (B\boldsymbol{v}, \boldsymbol{w})_V = (\mathcal{B}\boldsymbol{\varepsilon}(\boldsymbol{v}), \boldsymbol{\varepsilon}(\boldsymbol{w}))_Q$$

for all $\boldsymbol{v}, \boldsymbol{w} \in V$. From the assumptions (6.34) and (6.35) on $\mathcal{A}$ and $\mathcal{B}$, it follows that A and B are Lipschitz continuous operators, and A is a strongly monotone operator on V:

$$(A\boldsymbol{v}_1 - A\boldsymbol{v}_2, \boldsymbol{v}_1 - \boldsymbol{v}_2)_V \geq M\|\boldsymbol{v}_1 - \boldsymbol{v}_2\|_V^2 \quad \forall\, \boldsymbol{v}_1,\, \boldsymbol{v}_2 \in V.$$

By (12.7), the functional j defined in (12.12) satisfies (11.24) and for any $\boldsymbol{g}_1, \boldsymbol{g}_2, \boldsymbol{v}_1, \boldsymbol{v}_2 \in V$,

$$\begin{aligned} & j(\boldsymbol{g}_1, \boldsymbol{v}_2) - j(\boldsymbol{g}_1, \boldsymbol{v}_1) + j(\boldsymbol{g}_2, \boldsymbol{v}_1) - j(\boldsymbol{g}_2, \boldsymbol{v}_2) \\ & \quad \leq c_0(L_\nu + L_\tau)\, \|\boldsymbol{g}_1 - \boldsymbol{g}_2\|_V\, \|\boldsymbol{v}_1 - \boldsymbol{v}_2\|_V, \end{aligned}$$

where $c_0 > 0$ depends only on Ω, Γ_1 and Γ_3. Applying Theorem 12.3, we conclude that if

$$c_0(L_\nu + L_\tau) < M,$$

then Problem 12.1 has a unique solution $\boldsymbol{u} \in C^1(\overline{I}; V)$, and we may take $L_0 = M/c_0$. ∎

Let now $\boldsymbol{u} \in C^1(\overline{I}; V)$ be the solution of Problem 12.1 and let $\boldsymbol{\sigma}$ be the stress field given by (12.1). Using (12.15) and (12.8) it can be shown that $\operatorname{Div} \boldsymbol{\sigma} \in C(\overline{I}; L^2(\Omega)^d)$ and therefore $\boldsymbol{\sigma} \in C(\overline{I}; Q_1)$. A pair of functions $(\boldsymbol{u}, \boldsymbol{\sigma})$ which satisfies (12.1), (12.6) and (12.15) is called a *weak solution* of the problem (12.1)–(12.6). We conclude that the problem (12.1)–(12.6) has a unique weak solution provided $L_\nu + L_\tau$ is sufficiently small. The critical value L_0 depends only on the viscosity operator and on the geometry of the problem but does not depend on the elasticity operator, nor on the external forces, nor on the initial displacement.

We give some mechanical interpretation of the condition $L_\nu + L_\tau < L_0$ which guarantees the unique solvability of the problem. The verification of this condition as well as its interpretation depends on the specific mechanical problem. For example, consider the mechanical problem (12.1)–(12.6) in which the function p_ν is given by (5.46) (page 108) and the function p_τ is given by (11.8) or by (11.9). It follows that assumption (12.7)(b) is satisfied with $L_\nu = \beta$ and $L_\tau = \mu\,\beta$ and therefore the condition $L_\nu + L_\tau < L_0$ holds if $\beta \leq L_0/(\mu + 1)$, which may be interpreted as a smallness assumption on the damping resistance coefficient. We conclude that the corresponding mechanical problem has a unique weak solution if the damping resistance coefficient (of the oil layer, for example) is small enough. Consider now the mechanical problem (12.1)–(12.6) in the case when the function p_ν is given by (5.48) with $S \in L^\infty(\Gamma_3)$ and the function p_τ is given by (11.8) or by (11.9). In this case the assumption (12.7)(b) is satisfied with $L_\nu = L_\tau = 0$ and therefore the condition $L_\nu + L_\tau < L_0$ trivially holds. We conclude that the corresponding mechanical problem has a unique weak solution without any supplementary restriction on the coefficients μ and δ.

As in Section 8.1, it can be shown that

$$\mathrm{Div}\boldsymbol{\sigma}(t) + \boldsymbol{f}_0(t) = \boldsymbol{0} \quad \text{a.e. in } \Omega \tag{12.54}$$

for any $t \in \overline{I}$. Under the condition

$$\boldsymbol{\sigma\nu} \in C(\overline{I}; L^2(\Gamma)^d), \tag{12.55}$$

we further have the following equalities for any $t \in \overline{I}$:

$$\boldsymbol{\sigma}(t)\boldsymbol{\nu} = \boldsymbol{f}_2(t) \quad \text{a.e. on } \Gamma_2, \tag{12.56}$$

$$-\sigma_\nu(t) = p_\nu(\dot{u}_\nu(t)) \quad \text{a.e. on } \Gamma_3. \tag{12.57}$$

Now we state some sample results on error estimates for numerical approximations of Problem 12.1. In the rest of the section, we assume that the condition $L_\nu + L_\tau < L_0$ is valid.

We first consider the semi-discrete approximation of Problem 12.1.

Problem 12.13 *Find a displacement field* $\boldsymbol{u}^h : \overline{I} \to V^h$ *such that*

$$\boldsymbol{u}^h(0) = \boldsymbol{u}_0^h,$$

and for $t \in \overline{I}$,

$$\begin{aligned}(\mathcal{A}\boldsymbol{\varepsilon}(\dot{\boldsymbol{u}}^h(t)), \boldsymbol{\varepsilon}(\boldsymbol{v}^h - \dot{\boldsymbol{u}}^h(t)))_Q + (\mathcal{B}\boldsymbol{\varepsilon}(\boldsymbol{u}^h(t)), \boldsymbol{\varepsilon}(\boldsymbol{v}^h - \dot{\boldsymbol{u}}^h(t)))_Q \\ + j(\dot{\boldsymbol{u}}^h(t), \boldsymbol{v}^h) - j(\dot{\boldsymbol{u}}^h(t), \dot{\boldsymbol{u}}^h(t)) \\ \geq (\boldsymbol{f}(t), \boldsymbol{v}^h - \dot{\boldsymbol{u}}^h(t))_V \quad \forall \boldsymbol{v}^h \in V^h.\end{aligned}$$

Here, $\boldsymbol{u}_0^h \in V^h$ *is a suitable approximation of* $\boldsymbol{u}_0$.

Problem 12.13 has a unique solution. By Theorem 12.9, we have the estimate

$$\begin{aligned}\|\boldsymbol{u} - \boldsymbol{u}^h\|_{C^1(\overline{I};V)} \leq c \inf_{\boldsymbol{v}^h \in C(\overline{I};V^h)} \Big(\|\dot{\boldsymbol{u}} - \boldsymbol{v}^h\|_{C(\overline{I};V)} + \|R(\cdot; \dot{\boldsymbol{u}}(\cdot), \boldsymbol{v}^h(\cdot))\|_{C(\overline{I})}^{1/2}\Big) \\ + c\,\|\boldsymbol{u}_0 - \boldsymbol{u}_0^h\|_V,\end{aligned} \tag{12.58}$$

where

$$\begin{aligned}R(t; \dot{\boldsymbol{u}}(t), \boldsymbol{v}^h(t)) = (\boldsymbol{\sigma}(t), \boldsymbol{\varepsilon}(\boldsymbol{v}^h(t) - \dot{\boldsymbol{u}}(t)))_Q + j(\dot{\boldsymbol{u}}(t), \boldsymbol{v}^h(t)) \\ - j(\dot{\boldsymbol{u}}(t), \dot{\boldsymbol{u}}(t)) - (\boldsymbol{f}(t), \boldsymbol{v}^h(t) - \dot{\boldsymbol{u}}(t))_V.\end{aligned}$$

Perform an integration by parts on the first term and use the relation (12.54) and the definition (12.10),

$$R(t;\dot{\boldsymbol{u}}(t),\boldsymbol{v}^h(t)) = \langle \boldsymbol{\sigma}(t)\boldsymbol{\nu}, \boldsymbol{v}^h(t)-\dot{\boldsymbol{u}}(t)\rangle_\Gamma - \int_{\Gamma_2} \boldsymbol{f}_2(t)\cdot(\boldsymbol{v}^h(t)-\dot{\boldsymbol{u}}(t))\,da + j(\dot{\boldsymbol{u}}(t),\boldsymbol{v}^h(t)) - j(\dot{\boldsymbol{u}}(t),\dot{\boldsymbol{u}}(t)). \tag{12.59}$$

Now

$$j(\dot{\boldsymbol{u}}(t),\boldsymbol{v}^h(t)) - j(\dot{\boldsymbol{u}}(t),\dot{\boldsymbol{u}}(t)) = \int_{\Gamma_3} p_\nu(\dot{u}_\nu(t))\,\left(v_\nu^h(t)-\dot{u}_\nu(t)\right)\,da + \int_{\Gamma_3} p_\tau(\dot{u}_\nu(t))\,\left(\|\boldsymbol{v}_\tau^h(t)\| - \|\dot{\boldsymbol{u}}_\tau(t)\|\right)\,da.$$

By the assumptions (12.7) on p_r, we have

$$|p_r(\dot{u}_\nu(t))| \le |p_r(0)| + L_r|\dot{u}_\nu(t)|$$

for $r=\nu,\tau$. Hence,

$$\|p_r(\dot{u}_\nu(t))\|_{L^2(\Gamma_3)} \le c\,(1+\|\dot{u}_\nu(t)\|_{L^2(\Gamma_3)}) \le c\,(1+\|\dot{\boldsymbol{u}}(t)\|_V).$$

So

$$|j(\dot{\boldsymbol{u}}(t),\boldsymbol{v}^h(t)) - j(\dot{\boldsymbol{u}}(t),\dot{\boldsymbol{u}}(t))| \le c\,(1+\|\dot{\boldsymbol{u}}(t)\|_V)\,\|\boldsymbol{v}^h(t)-\dot{\boldsymbol{u}}(t)\|_{L^2(\Gamma_3)^d}$$

and then

$$|R(t;\dot{\boldsymbol{u}}(t),\boldsymbol{v}^h(t))| \le c\,\|\boldsymbol{v}^h(t)-\dot{\boldsymbol{u}}(t)\|_{H^{1/2}(\Gamma)^d} \le c\,\|\boldsymbol{v}^h(t)-\dot{\boldsymbol{u}}(t)\|_V,$$

where the constant c depends on

$$\max_{t\in\overline{I}}\left\{\|\dot{\boldsymbol{u}}(t)\|_V,\,\|\boldsymbol{\sigma}(t)\boldsymbol{\nu}\|_{H^{-1/2}(\Gamma)^d}\right\}.$$

Therefore, (12.58) implies

$$\|\boldsymbol{u}-\boldsymbol{u}^h\|_{C^1(\overline{I};V)} \le c \inf_{\boldsymbol{v}^h\in C(\overline{I};V^h)}\left(\|\dot{\boldsymbol{u}}-\boldsymbol{v}^h\|_{C(\overline{I};V)} + \|\dot{\boldsymbol{u}}-\boldsymbol{v}^h\|_{C(\overline{I};V)}^{1/2}\right) + c\,\|\boldsymbol{u}_0-\boldsymbol{u}_0^h\|_V.$$

So applying Theorem 7.18 (page 157), if we choose $\boldsymbol{u}_0^h$ such that $\|\boldsymbol{u}_0-\boldsymbol{u}_0^h\|_V\to 0$, then we have the convergence

$$\|\boldsymbol{u}-\boldsymbol{u}^h\|_{C^1(\overline{I};V)} \to 0 \quad \text{as } h\to 0.$$

For error estimation, we assume (12.55); therefore, (12.56) and (12.57) hold. We can further express $R(t;\dot{\boldsymbol{u}}(t),\boldsymbol{v}^h(t))$ as

$$R(t;\dot{\boldsymbol{u}}(t),\boldsymbol{v}^h(t)) = \int_{\Gamma_3}\left(\boldsymbol{\sigma}_\tau(t)\cdot(\boldsymbol{v}_\tau^h(t)-\dot{\boldsymbol{u}}_\tau(t)) + p_\tau(\dot{u}_\nu(t))\,(|\boldsymbol{v}_\tau^h(t)| - |\dot{\boldsymbol{u}}_\tau(t)|)\right)da.$$

We then have

$$|R(t;\boldsymbol{v}^h(t),\dot{\boldsymbol{u}}(t))| \le c\,\|\boldsymbol{v}^h_\tau(t)-\dot{\boldsymbol{u}}_\tau(t)\|_{L^2(\Gamma_3)^d}. \tag{12.60}$$

Therefore, with the assumption (12.55), the estimate (12.58) in this case reduces to

$$\begin{aligned}&\|\boldsymbol{u}-\boldsymbol{u}^h\|_{C^1(\overline{I};V)}\\ &\quad\le c\inf_{\boldsymbol{v}^h\in C(\overline{I};V^h)}\Big(\|\dot{\boldsymbol{u}}-\boldsymbol{v}^h\|_{C(\overline{I};V)}+\|\dot{\boldsymbol{u}}_\tau-\boldsymbol{v}^h_\tau\|^{1/2}_{C(\overline{I};L^2(\Gamma_3)^d)}\Big)\\ &\qquad+c\,\|\boldsymbol{u}_0-\boldsymbol{u}^h_0\|_V.\end{aligned}$$

Assume

$$\dot{\boldsymbol{u}}\in C(\overline{I};H^2(\Omega)^d),\ \ \dot{\boldsymbol{u}}_\tau|_{\Gamma_{3,i}}\in C(\overline{I};H^2(\Gamma_{3,i})^d),\ 1\le i\le i_0,\ \ \boldsymbol{u}_0\in H^2(\Omega)^d. \tag{12.61}$$

Then we have the interpolation error estimates (7.32), (7.36) and (7.40) from Section 7.2. If we take $\boldsymbol{u}^h_0=\Pi^h\boldsymbol{u}_0$, then from (12.58) and (12.60), we obtain the optimal order error estimate

$$\|\boldsymbol{u}-\boldsymbol{u}^h\|_{C^1(\overline{I};V)}=O(h).$$

We now turn to a study of a fully discrete approximation of the problem 12.1.

Problem 12.14 *Find a displacement field* $\boldsymbol{u}^{hk}=\{\boldsymbol{u}^{hk}_n\}_{n=0}^N\subset V^h$ *such that*

$$\boldsymbol{u}^{hk}_0=\boldsymbol{u}^h_0,$$

and for $n=1,\dots,N$,

$$\begin{aligned}&(\mathcal{A}\boldsymbol{\varepsilon}(\delta_n\boldsymbol{u}^{hk}_n),\boldsymbol{\varepsilon}(\boldsymbol{v}^h-\delta_n\boldsymbol{u}^{hk}_n))_Q+(\mathcal{B}\boldsymbol{\varepsilon}(\boldsymbol{u}^{hk}_{n-1}),\boldsymbol{\varepsilon}(\boldsymbol{v}^h-\delta_n\boldsymbol{u}^{hk}_n))_Q\\ &\quad+j(\delta_n\boldsymbol{u}^{hk}_n,\boldsymbol{v}^h)-j(\delta_n\boldsymbol{u}^{hk}_n,\delta_n\boldsymbol{u}^{hk}_n)\\ &\quad\ge(\boldsymbol{f}(t),\boldsymbol{v}^h-\delta_n\boldsymbol{u}^{hk}_n)_V\quad\forall\,\boldsymbol{v}^h\in V^h.\end{aligned}$$

Here again $\boldsymbol{u}^h_0\in V^h$ *is a suitable approximation of* $\boldsymbol{u}_0$.

Problem 12.14 has a unique solution. Convergence analysis and error estimation of the fully discrete scheme are similar to that of the semi-discrete scheme. Applying Theorem 7.22 (page 160), if we choose $\boldsymbol{u}^h_0$ such that $\|\boldsymbol{u}_0-\boldsymbol{u}^h_0\|_V\to0$, then we have the convergence

$$\max_{1\le n\le N}(\|\boldsymbol{u}_n-\boldsymbol{u}^{hk}_n\|_V+\|\dot{\boldsymbol{u}}_n-\delta_n\boldsymbol{u}^{hk}_n\|_V)\to0\quad\text{as } h,k\to0.$$

For convergence order error estimation, assuming $\boldsymbol{f}\in W^{1,1}(I;V)$, we have the following estimate from (12.52):

$$\begin{aligned}&\max_{1\le n\le N}(\|\boldsymbol{u}_n-\boldsymbol{u}^{hk}_n\|_V+\|\dot{\boldsymbol{u}}_n-\delta_n\boldsymbol{u}^{hk}_n\|_V)\\ &\quad\le c\max_{1\le n\le N}\inf_{\boldsymbol{v}^h_n\in V^h}\Big\{\|\boldsymbol{v}^h_n-\dot{\boldsymbol{u}}_n\|_V+|R_n(\dot{\boldsymbol{u}}_n,\boldsymbol{v}^h_n)|^{1/2}\Big\}\\ &\qquad+c\,\|\boldsymbol{u}^h_0-\boldsymbol{u}_0\|_V+c\,k\,(\|\boldsymbol{f}\|_{W^{1,1}(I;V)}+\|\boldsymbol{u}\|_{C^1(\overline{I};V)}),\end{aligned} \tag{12.62}$$

where

$$R_n(\dot{\boldsymbol{u}}_n, \boldsymbol{v}_n^h) = (\boldsymbol{\sigma}_n, \boldsymbol{v}_n^h - \dot{\boldsymbol{u}}_n)_V + j(\dot{\boldsymbol{u}}_n, \boldsymbol{v}_n^h) - j(\dot{\boldsymbol{u}}_n, \dot{\boldsymbol{u}}_n) - (\boldsymbol{f}_n, \boldsymbol{v}_n^h - \dot{\boldsymbol{u}}_n)_V.$$

Assume (12.55). Similar to (12.60), we have

$$|R_n(\dot{\boldsymbol{u}}_n, \boldsymbol{v}_n^h)| \le c\, \|(\boldsymbol{v}_n^h)_\tau - (\dot{\boldsymbol{u}}_n)_\tau\|_{L^2(\Gamma_3)^d}.$$

Then under the additional regularity conditions (12.61), we have the optimal order error estimate

$$\max_{1\le n\le N} (\|\boldsymbol{u}_n - \boldsymbol{u}_n^{hk}\|_V + \|\dot{\boldsymbol{u}}_n - \delta_n \boldsymbol{u}_n^{hk}\|_V) = O(h+k).$$

12.6 Two Field Variational Formulations

In this section we study two variational formulations of the mechanical problem (12.1)–(12.6) in which the unknowns are both the displacement and the stress field. The first variational formulation follows directly from (12.1), (12.6) and (12.13) and may be stated as follows.

Problem 12.15 *Find a displacement* $\boldsymbol{u} : \overline{I} \to V$ *and a stress field* $\boldsymbol{\sigma} : \overline{I} \to Q_1$ *such that*

$$\boldsymbol{u}(0) = \boldsymbol{u}_0 \tag{12.63}$$

and for all $t \in \overline{I}$,

$$\boldsymbol{\sigma} = \mathcal{A}\boldsymbol{\varepsilon}(\dot{\boldsymbol{u}}) + \mathcal{B}\boldsymbol{\varepsilon}(\boldsymbol{u}), \tag{12.64}$$

$$\begin{aligned}(\boldsymbol{\sigma}(t), \boldsymbol{\varepsilon}(\boldsymbol{v}) - \boldsymbol{\varepsilon}(\dot{\boldsymbol{u}}(t)))_Q + j(\dot{\boldsymbol{u}}(t), \boldsymbol{v}) - j(\dot{\boldsymbol{u}}(t), \dot{\boldsymbol{u}}(t))&\\ \ge (\boldsymbol{f}(t), \boldsymbol{v} - \dot{\boldsymbol{u}}(t))_V \quad \forall\, \boldsymbol{v} \in V.&\end{aligned} \tag{12.65}$$

By Theorem 12.12 we have the following existence and uniqueness result.

Theorem 12.16 *Assume* (6.34), (6.35) *and* (12.7)–(12.9). *Then there exists* $L_0 > 0$ *which depends only on* Ω, Γ_1 *and* $\mathcal{A}$ *such that Problem 12.15 has a unique solution* $(\boldsymbol{u}, \boldsymbol{\sigma})$ *if* $L_\nu + L_\tau < L_0$. *Moreover,* $\boldsymbol{u} \in C^1(\overline{I}; V)$ *and* $\boldsymbol{\sigma} \in C(\overline{I}; Q_1)$.

Clearly, the solution $(\boldsymbol{u}, \boldsymbol{\sigma})$ of Problem 12.15 is the weak solution of the mechanical problem (12.1)–(12.6).

We now establish a *dual formulation* of Problem 12.15. To this end we define the set $\Sigma(t, \boldsymbol{w})$ of admissible stress fields by

$$\Sigma(t, \boldsymbol{w}) = \{\, \boldsymbol{\sigma} \in Q : (\boldsymbol{\sigma}, \boldsymbol{\varepsilon}(\boldsymbol{v}))_Q + j(\boldsymbol{w}, \boldsymbol{v}) \ge (\boldsymbol{f}(t), \boldsymbol{v})_V \ \forall\, \boldsymbol{v} \in V \,\} \tag{12.66}$$

for all $t \in \overline{I}$ and all $\boldsymbol{w} \in V$. Choosing $\boldsymbol{v} = 2\dot{\boldsymbol{u}}(t)$ and $\boldsymbol{v} = \boldsymbol{0}$ in (12.65) we deduce

$$(\boldsymbol{\sigma}(t), \boldsymbol{\varepsilon}(\dot{\boldsymbol{u}}(t)))_Q + j(\dot{\boldsymbol{u}}(t), \dot{\boldsymbol{u}}(t)) = (\boldsymbol{f}(t), \dot{\boldsymbol{u}}(t))_V \tag{12.67}$$

for all $t \in \overline{I}$, which implies

$$\boldsymbol{\sigma}(t) \in \Sigma(t, \dot{\boldsymbol{u}}(t)), \quad (\boldsymbol{\tau} - \boldsymbol{\sigma}(t), \boldsymbol{\varepsilon}(\dot{\boldsymbol{u}}(t)))_Q \geq 0 \quad \forall \boldsymbol{\tau} \in \Sigma(t, \dot{\boldsymbol{u}}(t)),\ t \in \overline{I}. \tag{12.68}$$

Thus, by (12.1), (12.6) and (12.68) we obtain the following variational problem.

Problem 12.17 *Find a displacement $\boldsymbol{u} : \overline{I} \to V$ and a stress field $\boldsymbol{\sigma} : \overline{I} \to Q_1$ such that*

$$\boldsymbol{u}(0) = \boldsymbol{u}_0 \tag{12.69}$$

and for all $t \in \overline{I}$,

$$\boldsymbol{\sigma} = \mathcal{A}\boldsymbol{\varepsilon}(\dot{\boldsymbol{u}}) + \mathcal{B}\boldsymbol{\varepsilon}(\boldsymbol{u}), \tag{12.70}$$

$$\boldsymbol{\sigma}(t) \in \Sigma(t, \dot{\boldsymbol{u}}(t)), \quad (\boldsymbol{\tau} - \boldsymbol{\sigma}(t), \boldsymbol{\varepsilon}(\dot{\boldsymbol{u}}(t)))_Q \geq 0 \quad \forall \boldsymbol{\tau} \in \Sigma(t, \dot{\boldsymbol{u}}(t)). \tag{12.71}$$

We note that (12.71) is a so-called *quasivariational inequality* since the admissible set Σ depends on the solution.

The following result expresses the equivalence between Problem 12.15 and Problem 12.17.

Theorem 12.18 *Assume* (6.34), (6.35), (12.7)–(12.9). *Let $\boldsymbol{u} \in C^1(\overline{I}; V)$ and $\boldsymbol{\sigma} \in C(\overline{I}; Q_1)$. Then $(\boldsymbol{u}, \boldsymbol{\sigma})$ is a solution of Problem 12.15 if and only if $(\boldsymbol{u}, \boldsymbol{\sigma})$ is a solution of Problem 12.17.*

Proof. Let $(\boldsymbol{u}, \boldsymbol{\sigma}) \in C^1(\overline{I}; V \times Q_1)$. Let us establish the equivalence of the variational inequalities (12.65) and (12.71).

(i) (12.65) $\Rightarrow$ (12.71). Let $t \in \overline{I}$. We choose $\boldsymbol{v} = 2\dot{\boldsymbol{u}}(t)$ and $\boldsymbol{v} = \boldsymbol{0}$ in (12.65) and deduce (12.67). Using now (12.65) and (12.67) we obtain

$$(\boldsymbol{\sigma}(t), \boldsymbol{\varepsilon}(\boldsymbol{v}))_Q + j(\dot{\boldsymbol{u}}(t), \boldsymbol{v}) \geq (\boldsymbol{f}(t), \boldsymbol{v})_V \quad \forall \boldsymbol{v} \in V,$$

which implies that $\boldsymbol{\sigma}(t) \in \Sigma(t, \dot{\boldsymbol{u}}(t))$. On the other hand, we obtain from (12.66) and (12.67) that

$$(\boldsymbol{\tau} - \boldsymbol{\sigma}(t), \boldsymbol{\varepsilon}(\dot{\boldsymbol{u}}(t)))_Q \geq 0 \quad \forall \boldsymbol{\tau} \in \Sigma(t, \dot{\boldsymbol{u}}(t)),$$

which implies (12.71).

(ii) (12.71) $\Rightarrow$ (12.65). Let $t \in \overline{I}$. The subdifferentiability of the function $j(\dot{\boldsymbol{u}}(t), \cdot) : V \to \mathbb{R}$ implies that there exists an element $\tilde{\boldsymbol{f}}(t) \in V$ such that

$$j(\dot{\boldsymbol{u}}(t), \boldsymbol{v}) - j(\dot{\boldsymbol{u}}(t), \dot{\boldsymbol{u}}(t)) \geq (\tilde{\boldsymbol{f}}(t), \boldsymbol{v} - \dot{\boldsymbol{u}}(t))_V \quad \forall \boldsymbol{v} \in V.$$

Thus,

$$\begin{aligned} (\boldsymbol{f}(t) - \tilde{\boldsymbol{f}}(t), \boldsymbol{v} - \dot{\boldsymbol{u}}(t))_V + j(\dot{\boldsymbol{u}}(t), \boldsymbol{v}) - j(\dot{\boldsymbol{u}}(t), \dot{\boldsymbol{u}}(t)) \\ \geq (\boldsymbol{f}(t), \boldsymbol{v} - \dot{\boldsymbol{u}}(t))_V \quad \forall \boldsymbol{v} \in V. \end{aligned} \tag{12.72}$$

Choosing $\boldsymbol{v} = 2\dot{\boldsymbol{u}}(t)$ and $\boldsymbol{v} = \boldsymbol{0}$ in (12.72) we obtain that

$$(\boldsymbol{f}(t) - \tilde{\boldsymbol{f}}(t), \dot{\boldsymbol{u}}(t))_V + j(\dot{\boldsymbol{u}}(t), \dot{\boldsymbol{u}}(t)) = (\boldsymbol{f}(t), \dot{\boldsymbol{u}}(t))_V. \tag{12.73}$$

It follows from (12.72) and (12.73) that $\boldsymbol{\varepsilon}(\boldsymbol{f}(t) - \tilde{\boldsymbol{f}}(t)) \in \Sigma(t, \dot{\boldsymbol{u}}(t))$. Therefore, by using $\boldsymbol{\tau} = \boldsymbol{\varepsilon}(\boldsymbol{f}(t) - \tilde{\boldsymbol{f}}(t))$ as a test function in (12.71) we deduce

$$(\boldsymbol{f}(t) - \tilde{\boldsymbol{f}}(t), \dot{\boldsymbol{u}}(t))_V \geq (\boldsymbol{\sigma}(t), \boldsymbol{\varepsilon}(\dot{\boldsymbol{u}}(t)))_Q.$$

Adding $j(\dot{\boldsymbol{u}}(t), \dot{\boldsymbol{u}}(t))$ to this inequality and taking (12.73) into account we find that

$$(\boldsymbol{f}(t), \dot{\boldsymbol{u}}(t))_V \geq (\boldsymbol{\sigma}(t), \boldsymbol{\varepsilon}(\dot{\boldsymbol{u}}(t)))_Q + j(\dot{\boldsymbol{u}}(t), \dot{\boldsymbol{u}}(t)). \tag{12.74}$$

On the other hand, since $\boldsymbol{\sigma}(t) \in \Sigma(t, \dot{\boldsymbol{u}}(t))$ and $\dot{\boldsymbol{u}}(t) \in V$, we obtain

$$(\boldsymbol{\sigma}(t), \boldsymbol{\varepsilon}(\dot{\boldsymbol{u}}(t)))_Q + j(\dot{\boldsymbol{u}}(t), \dot{\boldsymbol{u}}(t)) \geq (\boldsymbol{f}(t), \dot{\boldsymbol{u}}(t))_V \tag{12.75}$$

and

$$(\boldsymbol{\sigma}(t), \boldsymbol{\varepsilon}(\boldsymbol{v}))_Q + j(\dot{\boldsymbol{u}}(t), \boldsymbol{v}) \geq (\boldsymbol{f}(t), \boldsymbol{v})_V \quad \forall\, \boldsymbol{v} \in V. \tag{12.76}$$

Inequality (12.65) is now a consequence of (12.74)–(12.76). ■

We conclude from Theorems 12.16 and 12.18 that, under assumptions (6.34), (6.35), (12.7)–(12.9), there exists $L_0 > 0$ depending only on Ω, Γ_1 and $\mathcal{A}$ such that Problem 12.17 has a unique solution $(\boldsymbol{u}, \boldsymbol{\sigma})$ if $L_\nu + L_\tau < L_0$. Moreover, the solution $(\boldsymbol{u}, \boldsymbol{\sigma})$ of Problem 12.17 is the weak solution of the mechanical problem (12.1)–(12.6).

12.7 Numerical Examples

A number of numerical experiments have been performed to show the performance of the fully discrete scheme described in Section 12.5. We describe in this sections numerical results for Problem (12.1)–(12.6) in one and two dimensions.

One dimensional example. We consider a viscoelastic rod $\overline{\Omega} = [0, L]$ which is fixed at its left end $x = 0$ and is subjected to the action of a body force of density $f_0(x,t)$ in the x-direction. Its right end $x = L$ is in contact with a reactive foundation and the contact is modeled with damped response (see Figure 11.1 on page 246 in which g is supposed to be zero). This problem corresponds to problem (12.1)–(12.6) with $\Omega = (0, L)$, $\Gamma_1 = \{0\}$, $\Gamma_2 = \emptyset$, $\Gamma_3 = \{L\}$. We use a linearly viscoelastic constitutive law, i.e.

$$\sigma = a\,\varepsilon(\dot{u}) + b\,\varepsilon(u).$$

Here $\varepsilon(u) = \frac{\partial u}{\partial x}$, while $a > 0$ and b are material constants, independent of x and t. We choose (5.46) (page 108) as normal damped response contact function.

A complete description of this problem is the following one.

Problem 12.19 *Find a displacement field* $u : [0, L] \times [0, T] \to \mathbb{R}$ *and a stress field* $\sigma : [0, L] \times [0, T] \to \mathbb{R}$ *such that*

$$
\begin{aligned}
\sigma(x,t) &= a\,\tfrac{\partial^2 u(x,t)}{\partial x \partial t} + b\,\tfrac{\partial u(x,t)}{\partial x} && \text{in } (0,L)\times(0,T),\\
\tfrac{\partial \sigma(x,t)}{\partial x} + f_0(x,t) &= 0 && \text{in } (0,L)\times(0,T),\\
u(0,t) &= 0 && \text{for } t \in (0,T),\\
-\sigma(L,t) &= \kappa\,\dot{u}(L,t)_+ + p_0 && \text{for } t \in (0,T),\\
u(x,0) &= u_0(x) && \text{in } (0,L).
\end{aligned}
$$

For computation we have used the following data:

$$
\begin{aligned}
&L = 1\,m, \quad T = 0.5\,sec, \quad a = 1\,N\,sec/m, \quad b = 1\,N/m,\\
&f_0(x,t) = 10\,N/m \quad \forall\, x \in (0,1),\ t \in [0, 0.1],\\
&\kappa = 1\,N\,sec/m, \quad u_0(x) = 0\,m \quad \forall\, x \in (0,1).
\end{aligned}
$$

After an elementary but tedious calculation, it can be proved that if $p_0 > 5$ then the velocity of the end $x = 1$ of the rod is negative at any $t \in [0,T]$ and therefore the rod is in compression during the process. It can also be proved that if $p_0 < 5$ then the velocity of the end $x = 1$ of the rod is positive for any $t \in [0,T]$ and therefore there is penetration into the obstacle. For this reason, we consider Problem 12.19 with the above data, for two different values of p_0. The corresponding exact solutions are the following:

$$
\begin{aligned}
u(x,t) &= (1 - e^{-t})\,(4x - 5x^2),\\
\sigma(x,t) &= 4 - 10\,x
\end{aligned}
$$

for $p_0 = 6$, and

$$
\begin{aligned}
u(x,t) &= 5\,x\,(x-1)\,e^{-t}x - 3\,x\,e^{-\frac{t}{2}} + 8\,x - 5\,x^2,\\
\sigma(x,t) &= -\frac{3}{2}\,e^{-\frac{t}{2}} + 8 - 10\,x
\end{aligned}
$$

for $p_0 = 2$.

We have implemented the fully discrete method described in Section 12.5 on a typical workstation. We use linear elements for the space V^h and piecewise constant functions for the space Q^h corresponding to a uniform partition of the domain $[0, L]$ with meshsize h. For time discretization we use a uniform time step $k = h$. To solve the discrete systems, we employ a penalty-duality method (see e.g. [76]).

We first present numerical results for the case $p_0 = 6$. In Figure 12.1, we plot the error

$$
E_{hk} = \max_{1 \le n \le N} \{ \|u_n^{hk} - u_n\|_V + \|\delta u_n^{hk} - \dot{u}_n\|_V \}.
$$

Linear convergence pattern is clearly observed, confirming the theoretical error estimate.

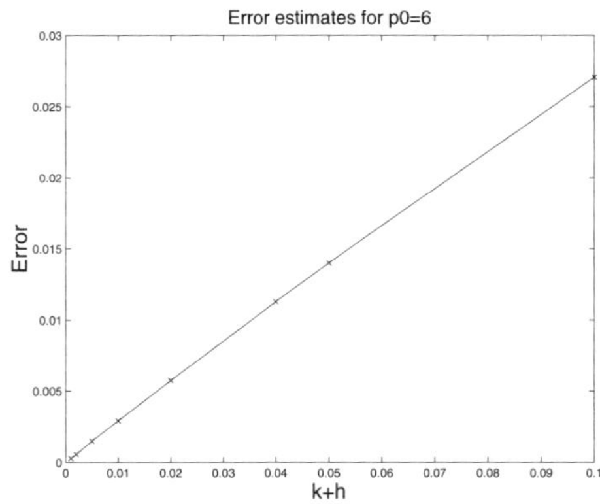

Figure 12.1: The 1D problem with $p_0 = 6$: linear convergence behavior

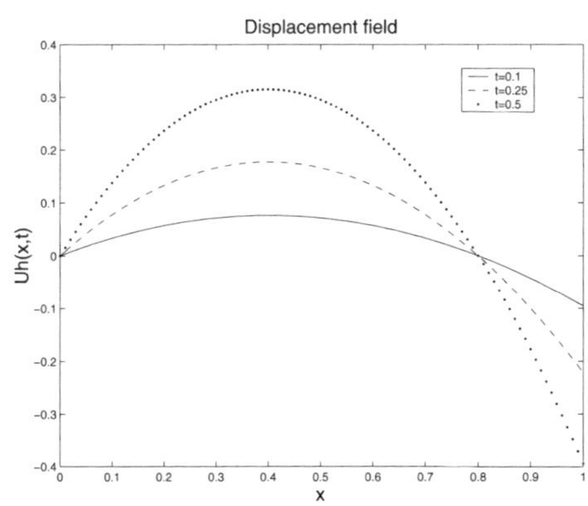

Figure 12.2: The 1D problem with $p_0 = 6$: displacement fields at several times

We then report some numerical results with $k = h = 0.01$. In Figures 12.2 and 12.3, the displacement field and the corresponding error are shown at several time values. Time evolution of the displacements at $x = 0.25$, 0.5, 1 and the corresponding errors are plotted in Figures 12.4 and 12.5. Time evolution of the stress field at the contact node $x = 1$ and the error are shown in Figures 12.6 and 12.7.

Numerical results for the case $p_0 = 2$ are reported in Figures 12.8 to 12.14.

Two-dimensional example. As a two-dimensional example of the problem (12.1)–(12.6), we consider a plane stress viscoelasticity problem shown in Figure 12.15. The physical setting is the following. We consider a three-dimensional linearly viscoelastic body of cross-section Ω under the plane stress hypothesis. The domain Ω is a square. We establish a Cartesian coordinate system in such a way that $\Omega = (0,6) \times (0,6)$. A linearly decreasing compression force is supposed to act on the part $\overline{\Gamma}_2 = [0,6] \times \{6\}$ of the boundary and no body forces act

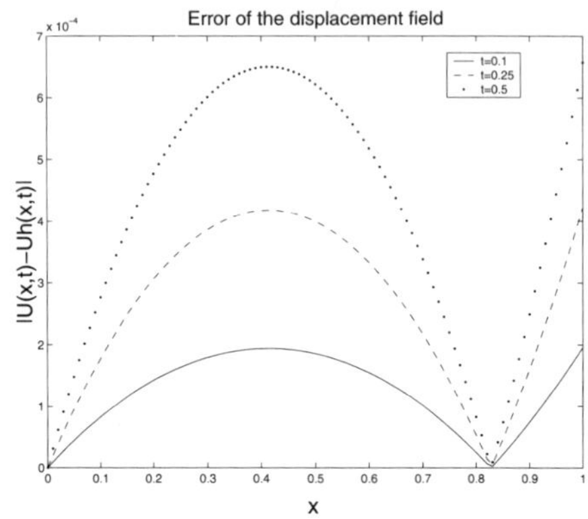

Figure 12.3: The 1D problem with $p_0 = 6$: error of displacement fields at several times

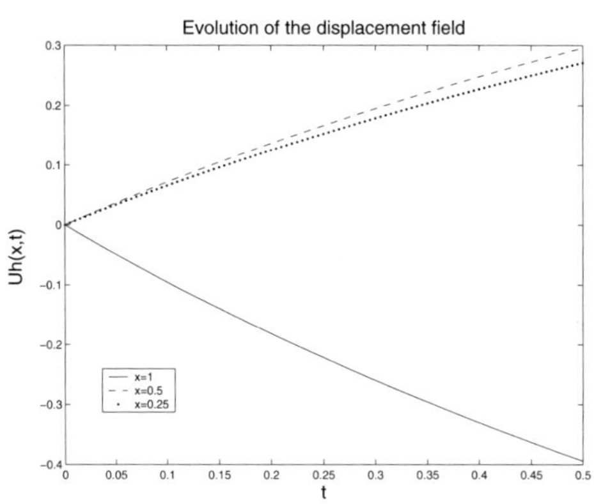

Figure 12.4: The 1D problem with $p_0 = 6$: time evolution of the displacement at several times

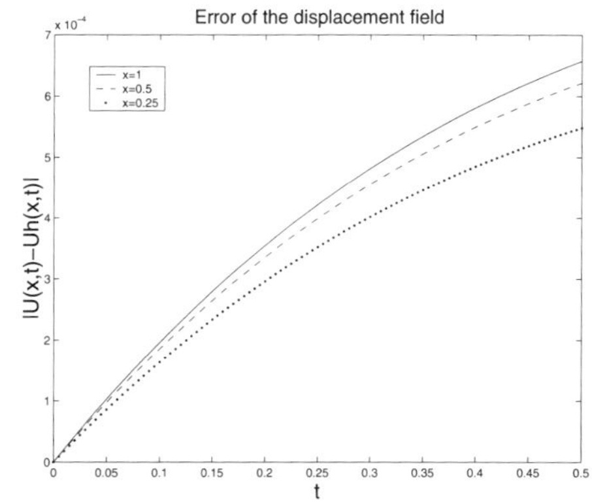

Figure 12.5: The 1D problem with $p_0 = 6$: error of the displacement fields at several times

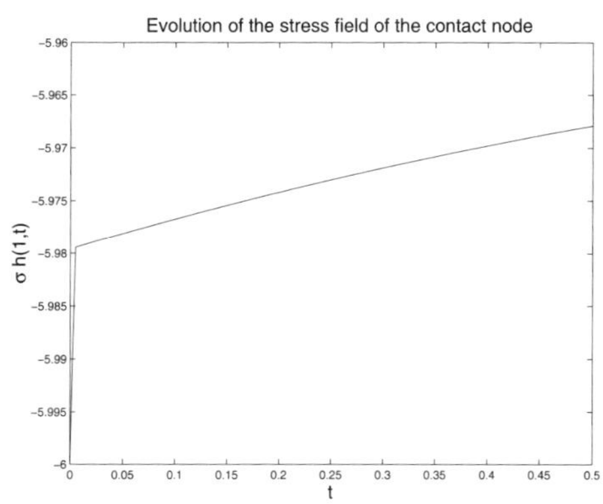

Figure 12.6: The 1D problem with $p_0 = 6$: time evolution of the stress field at $x = 1$

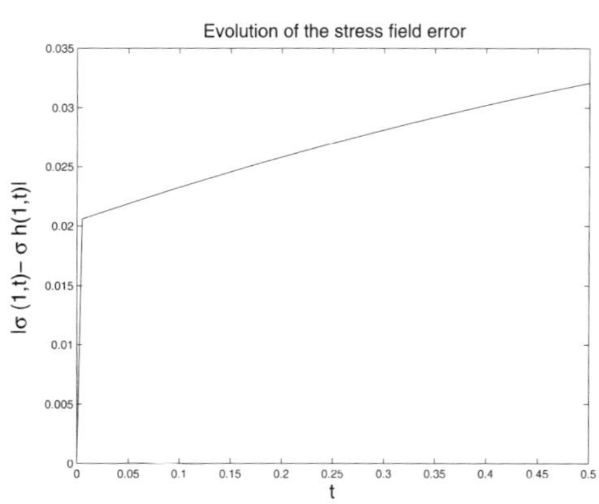

Figure 12.7: The 1D problem with $p_0 = 6$: error of the stress field at $x = 1$

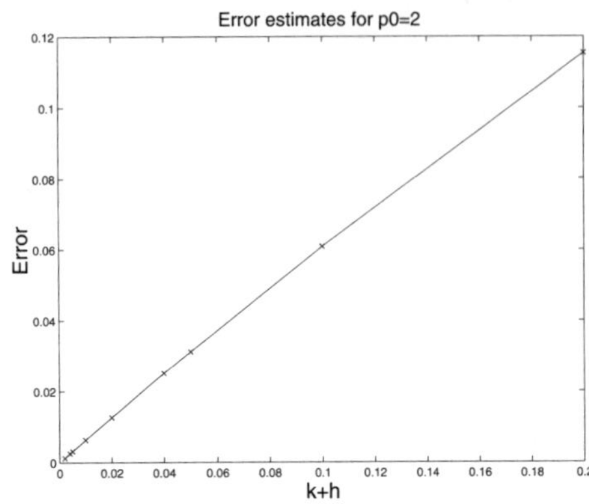

Figure 12.8: The 1D problem with $p_0 = 2$: linear convergence behavior

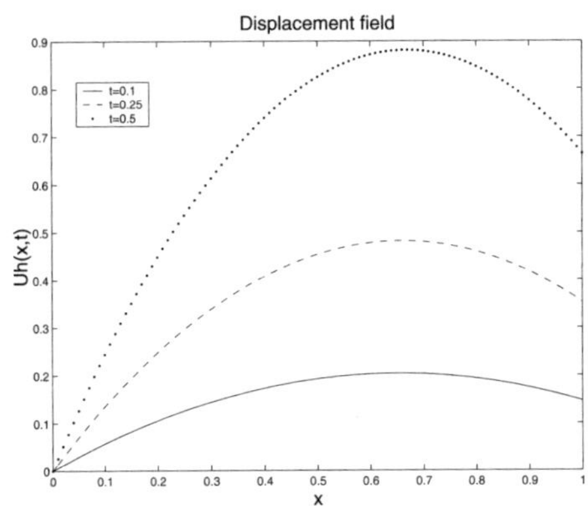

Figure 12.9: The 1D problem with $p_0 = 2$: displacement fields at several times

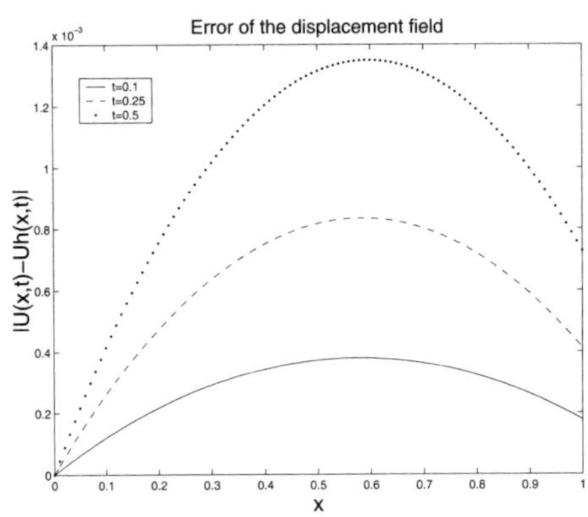

Figure 12.10: The 1D problem with $p_0 = 2$: error of displacement fields at several times

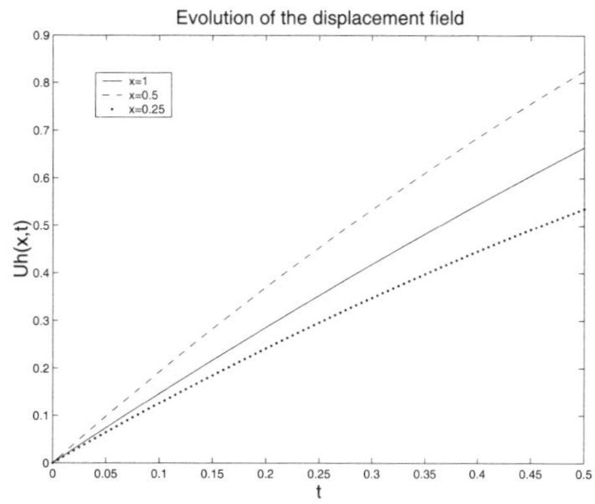

Figure 12.11: The 1D problem with $p_0 = 2$: time evolution of the displacement at several times

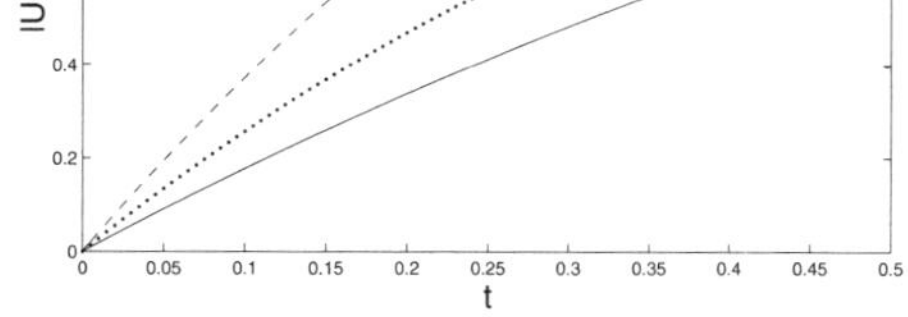

Figure 12.12: The 1D problem with $p_0 = 2$: error of the displacement fields at several times

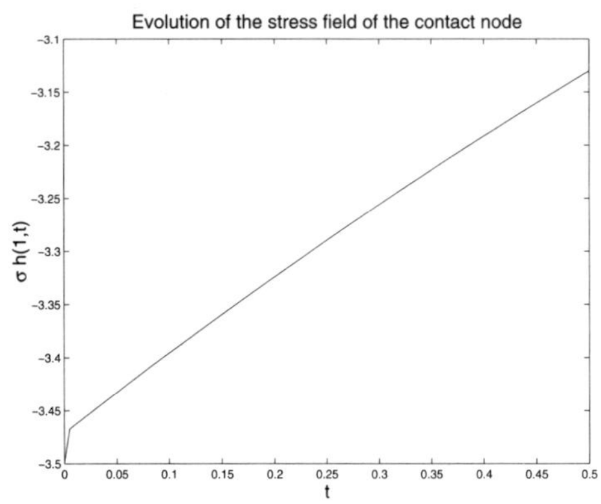

Figure 12.13: The 1D problem with $p_0 = 2$: time evolution of the stress field at $x = 1$

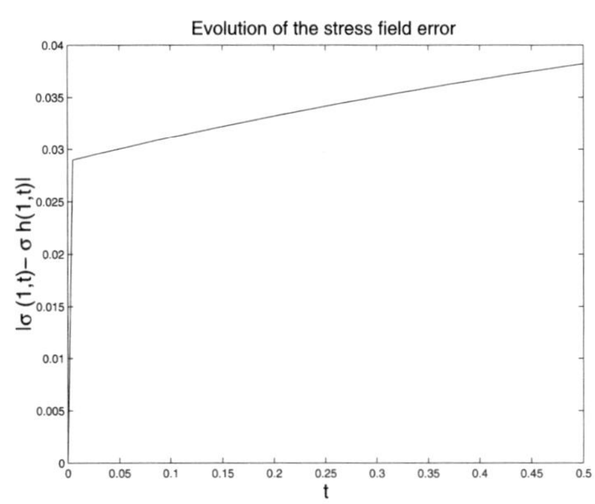

Figure 12.14: The 1D problem with $p_0 = 2$: error of the stress field at $x = 1$

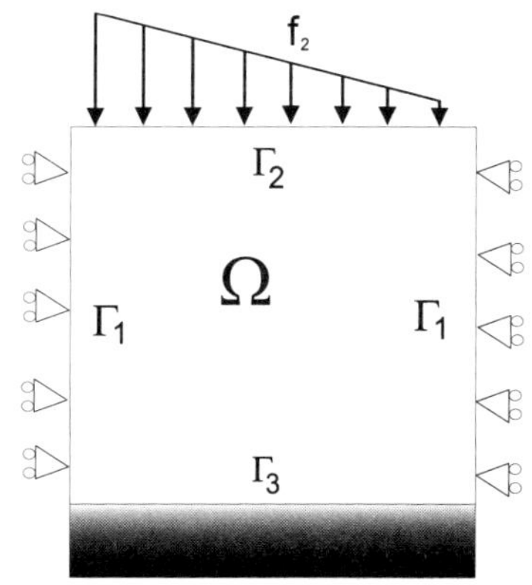

Figure 12.15: Contact of a 2D viscoelastic body

in Ω. The displacement on the lateral surface is supposed to homogeneous, i.e. $\overline{\Gamma}_1 = \{0,6\} \times [0,6]$. Finally, the body is in frictionless contact with normal damped response with a foundation on $\overline{\Gamma}_3 = [0,6] \times \{0\}$, that is, we assume $p_\tau = 0$. We choose (5.46) (page 108) as normal damped response contact function and we denote in the following $\kappa = c_\nu$.

The elasticity tensor $\mathcal{B}$ satisfies

$$(\mathcal{B}\boldsymbol{\tau})_{\alpha\beta} = \frac{E\nu}{1-\nu^2}\,(\tau_{11}+\tau_{22})\,\delta_{\alpha\beta} + \frac{E}{1+\nu}\,\tau_{\alpha\beta}, \quad 1 \le \alpha, \beta \le 2,$$

where E is Young's modulus and ν the Poisson ratio of the material. The viscosity tensor $\mathcal{A}$ has a similar form, i.e.

$$(\mathcal{A}\boldsymbol{\tau})_{\alpha\beta} = \mu\,(\tau_{11}+\tau_{22})\,\delta_{\alpha\beta} + \eta\,\tau_{\alpha\beta}, \quad 1 \le \alpha, \beta \le 2,$$

where μ and η are viscosity constants. The Von-Mises norm for a plane stress field $\boldsymbol{\tau} = (\tau_{\alpha\beta})$ is given by

$$\|\boldsymbol{\tau}\| = \left(\tau_{11}^2 + \tau_{22}^2 - \tau_{11}\tau_{22} + 3\,\tau_{12}^2\right)^{\frac{1}{2}}.$$

For computation we have used the following data:

$$T = 1\,sec, \quad \boldsymbol{f}_0 = \boldsymbol{0}\,N/m^2, \quad \boldsymbol{f}_2(x_1,x_2,t) = (0, -10\,(6-x_1)\,t)\,N/m,$$

$$p_0 = 2\,N/m, \quad \boldsymbol{u}_0 = \boldsymbol{0}\,m,$$

$$E = 100\,N/m^2, \quad \nu = 0.3, \quad \mu = 30\,N\,sec/m^2, \quad \eta = 20\,N\,sec/m^2.$$

Again, we use linear piecewise functions to approximate the space V^h and piecewise constant functions to approximate the space Q^h. For the time discretization, we use a uniform time step. The solution of the discrete problem at each time step is calculated by a penalty-duality method (see e.g. [213]) together with some ideas presented in [37] in solving frictionless contact problems for perfectly plastic materials.

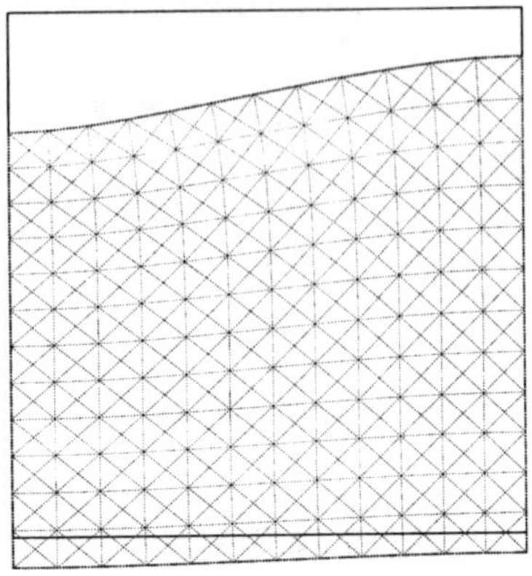

Figure 12.16: Initial boundary and deformed mesh at final time for $c_\nu = 10\,N\,sec/m^2$

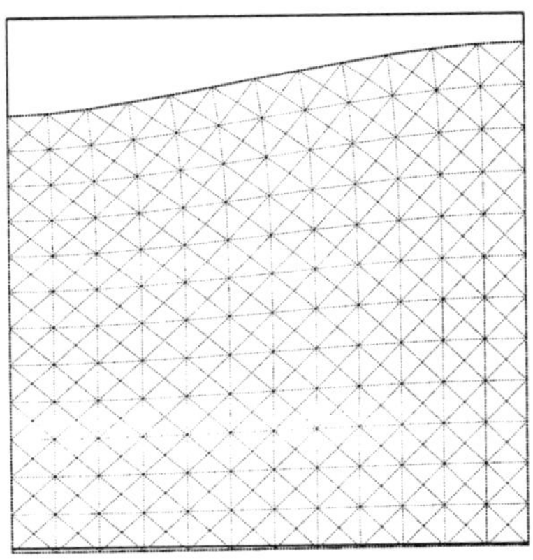

Figure 12.17: Initial boundary and deformed mesh at final time for $c_\nu = 100\,N\,sec/m^2$

In Figures 12.16, 12.17 and 12.18, the initial boundary and the deformed mesh at the final time $T = 1$ are shown for values $c_\nu = 10\,N\,sec/m^2$, 100 $N\,sec/m^2$, and $1000\,N\,sec/m^2$, respectively. The von Mises norm for the stress is plotted in Figure 12.19 at the final time. The contact boundary is depicted in Figure 12.20 for several values of c_ν.

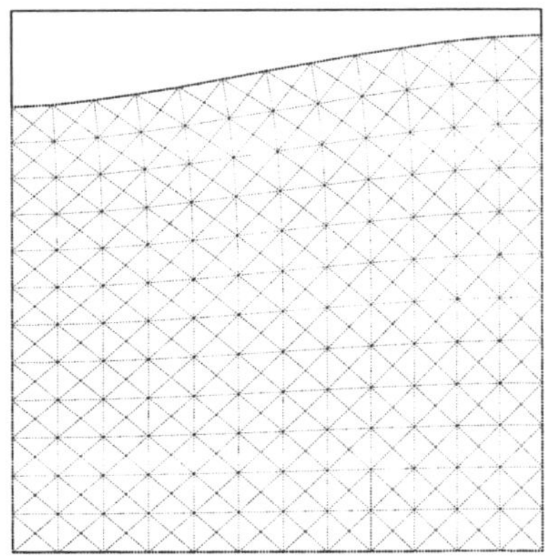

Figure 12.18: Initial boundary and deformed mesh at final time for $c_\nu = 1000\, N\, sec/m^2$

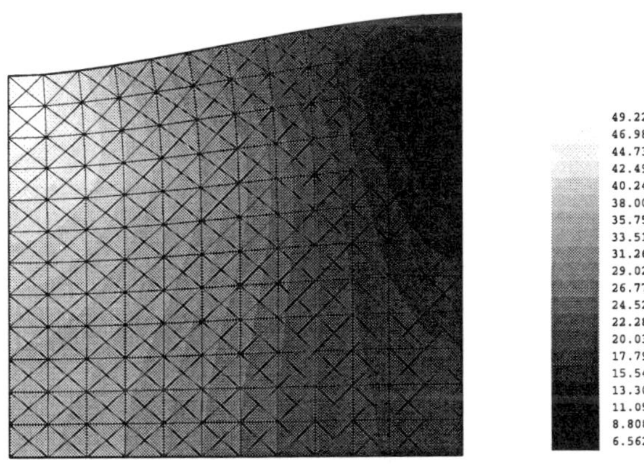

Figure 12.19: Von Mises stress norm at final time in the deformed configuration for $c_\nu = 1000\, N\, sec/m^2$

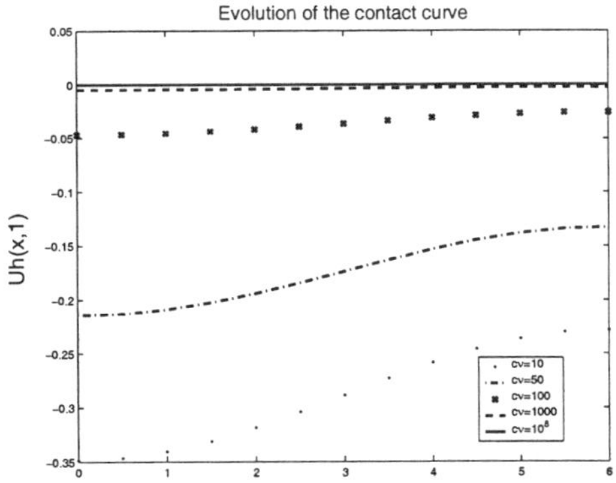

Figure 12.20: Contact boundary for several values of c_ν

Chapter 13

Other Viscoelastic Contact Problems

We have presented detailed variational and numerical analysis for several contact problems in viscoelasticity in the previous four chapters. In this chapter we will briefly discuss several more mathematical models for frictional contact of a viscoelastic body with an obstacle. Some of these models take into account the wear effect of the contacting surface due to the friction and therefore lead to new nonstandard contact boundary conditions. For all the frictional contact problems studied in this chapter we present weak formulations and establish existence and uniqueness results. The proofs are based on classical results for elliptic variational inequalities and fixed point arguments, and are similar to those in previous chapters. Since the modifications are straightforward, we omit sometimes the details. Numerical approximation schemes can be stated for the contact problems presented here similar to those provided in the previous chapters. Except for the contact problem discussed in Section 13.1, convergence analysis and error estimation can be given for the numerical schemes of the contact problems studied in this chapter.

13.1 Bilateral Contact with Nonlocal Coulomb Friction Law

We assume the contact is bilateral, i.e. there is no loss of contact between the body and the foundation. So the normal displacement u_ν vanishes on Γ_3 at any time. We model the friction with a nonlocal version of Coulomb's law involving a normal regularization operator $\mathcal{R} : H^{-\frac{1}{2}}(\Gamma) \to L^2(\Gamma)$. The classical formulation of the problem is the following:

Find a displacement field $\boldsymbol{u} : \Omega \times \overline{I} \to \mathbb{R}^d$ and a stress field $\boldsymbol{\sigma} : \Omega \times \overline{I} \to \mathbb{S}^d$

such that

$$\boldsymbol{\sigma} = \mathcal{A}\varepsilon(\dot{\boldsymbol{u}}) + \mathcal{B}\varepsilon(\boldsymbol{u}) \quad \text{in } \Omega \times I, \tag{13.1}$$

$$\operatorname{Div} \boldsymbol{\sigma} + \boldsymbol{f}_0 = \boldsymbol{0} \quad \text{in } \Omega \times I, \tag{13.2}$$

$$\boldsymbol{u} = \boldsymbol{0} \quad \text{on } \Gamma_1 \times I, \tag{13.3}$$

$$\boldsymbol{\sigma\nu} = \boldsymbol{f}_2 \quad \text{on } \Gamma_2 \times I, \tag{13.4}$$

$$\left.\begin{array}{l} u_\nu = 0, \quad \|\boldsymbol{\sigma}_\tau\| \le \mu\, p(|\mathcal{R}\sigma_\nu|), \\ \|\boldsymbol{\sigma}_\tau\| < \mu\, p(|\mathcal{R}\sigma_\nu|) \;\Rightarrow\; \dot{\boldsymbol{u}}_\tau = \boldsymbol{0}, \\ \|\boldsymbol{\sigma}_\tau\| = \mu\, p(|\mathcal{R}\sigma_\nu|) \;\Rightarrow\; \exists \lambda \ge 0 \text{ s.t. } \boldsymbol{\sigma}_\tau = -\lambda \dot{\boldsymbol{u}}_\tau \end{array}\right\} \quad \text{on } \Gamma_3 \times I, \tag{13.5}$$

$$\boldsymbol{u}(0) = \boldsymbol{u}_0 \quad \text{in } \Omega. \tag{13.6}$$

The conditions (13.5) describe a frictional bilateral contact process. Here p is a nonnegative valued function, and $\mu \ge 0$ is the coefficient of friction. The introduction of the nonlocal smoothing operator $\mathcal{R}$ is for technical reasons, since the trace of the stress tensor on the boundary is too rough to be defined in the ordinary sense (see, e.g. [51, 69]).

Using the continuity of the regularization operator $\mathcal{R} : H^{-\frac{1}{2}}(\Gamma) \to L^2(\Gamma)$ and the continuity of the normal trace mapping $\boldsymbol{\xi} \mapsto \xi_\nu : Q_1 \to H^{-\frac{1}{2}}(\Gamma)$ we deduce the existence of a constant $C_\mathcal{R} > 0$ depending only on Ω, Γ_1, Γ_3 and $\mathcal{R}$ such that

$$\|\mathcal{R}\xi_\nu\|_{L^2(\Gamma_3)} \le C_\mathcal{R}\, \|\boldsymbol{\xi}\|_{Q_1} \quad \forall\, \boldsymbol{\xi} \in Q_1. \tag{13.7}$$

In the study of the mechanical problem (13.1)–(13.6) we assume that the viscosity operator $\mathcal{A}$ and the elasticity operator $\mathcal{B}$ satisfy the conditions (6.34) and (6.35). Assumptions on the friction function p are:

$$\left.\begin{array}{l} \text{(a) } p : \Gamma_3 \times \mathbb{R} \to \mathbb{R}_+. \\ \text{(b) There exists an } L_p > 0 \text{ such that} \\ \qquad |p(\boldsymbol{x}, u_1) - p(\boldsymbol{x}, u_2)| \le L_p\, |u_1 - u_2| \\ \qquad \forall\, u_1, u_2 \in \mathbb{R}, \text{ a.e. in } \Omega. \\ \text{(c) For any } u \in \mathbb{R},\ \boldsymbol{x} \mapsto p(\boldsymbol{x}, u) \text{ is measurable.} \\ \text{(d) The mapping } \boldsymbol{x} \mapsto p(\boldsymbol{x}, 0) \in L^2(\Gamma_3). \end{array}\right\} \tag{13.8}$$

We observe that the assumptions (13.8) on the function p are pretty general except the assumption (13.8)(b) which, roughly speaking, requires the functions to grow at most linearly. Certainly the functions defined in (5.56)–(5.58) (page 110) satisfy this condition. So our results presented below are valid for the boundary value problems related to each of these examples.

We assume the force and traction densities have the smoothness

$$\boldsymbol{f}_0 \in C(\bar{I}; L^2(\Omega)^d), \qquad \boldsymbol{f}_2 \in C(\bar{I}; L^2(\Gamma_2)^d), \tag{13.9}$$

and the coefficient of friction μ has the properties

$$\mu \in L^\infty(\Omega), \quad \mu \ge 0 \text{ a.e. on } \Gamma_3. \tag{13.10}$$

Finally, we assume

$$\boldsymbol{u}_0 \in V_1. \tag{13.11}$$

Here and below we use the space V_1 defined in (7.11):

$$V_1 = \{\, \boldsymbol{v} \in V : v_\nu = 0 \text{ on } \Gamma_3 \,\}.$$

Recall that V_1 itself is a real Hilbert space under the inner product of the space V.

Denote by $\boldsymbol{f}(t)$ the element of V_1 given by

$$(\boldsymbol{f}(t), \boldsymbol{v})_V = \int_\Omega \boldsymbol{f}_0(t) \cdot \boldsymbol{v}\, dx + \int_{\Gamma_2} \boldsymbol{f}_2(t) \cdot \boldsymbol{v}\, da \tag{13.12}$$

for all $\boldsymbol{v} \in V_1$ and $t \in \overline{I}$ and let $j : Q_1 \times V_1 \to \mathbb{R}$ be the functional

$$j(\boldsymbol{\tau}, \boldsymbol{v}) = \int_{\Gamma_3} \mu\, p(|\mathcal{R}\tau_\nu|)\, \|\boldsymbol{v}_\tau\|\, da. \tag{13.13}$$

Since $\mathcal{R}\sigma_\nu$ lies in $L^2(\Gamma)$, from the assumptions (13.8) and (13.10), it follows that the integral in (13.13) is well defined on $Q_1 \times V_1$. For all $t \in \overline{I}$ and $\boldsymbol{\tau} \in Q_1$, we introduce the set

$$\Sigma(t, \boldsymbol{\tau}) = \{\, \boldsymbol{\xi} \in Q : (\boldsymbol{\xi}, \boldsymbol{\varepsilon}(\boldsymbol{v}))_Q + j(\boldsymbol{\tau}, \boldsymbol{v}) \ge (\boldsymbol{f}(t), \boldsymbol{v})_V \ \forall \boldsymbol{v} \in V_1 \,\}. \tag{13.14}$$

It is straightforward to show that if $\boldsymbol{u}$ and $\boldsymbol{\sigma}$ are sufficiently regular functions satisfying (13.2)–(13.5), then for all $t \in \overline{I}$,

$$\boldsymbol{u}(t) \in V_1, \quad \boldsymbol{\sigma}(t) \in \Sigma(t, \boldsymbol{\sigma}(t)), \tag{13.15}$$

$$\begin{aligned}(\boldsymbol{\sigma}(t), \boldsymbol{\varepsilon}(\boldsymbol{v}) - \boldsymbol{\varepsilon}(\dot{\boldsymbol{u}}(t)))_Q + j(\boldsymbol{\sigma}(t), \boldsymbol{v}) - j(\boldsymbol{\sigma}(t), \dot{\boldsymbol{u}}(t))&\\ \ge (\boldsymbol{f}(t), \boldsymbol{v} - \dot{\boldsymbol{u}}(t))_V \quad \forall \boldsymbol{v} \in V_1,&\end{aligned} \tag{13.16}$$

$$(\boldsymbol{\tau} - \boldsymbol{\sigma}(t), \boldsymbol{\varepsilon}(\boldsymbol{u}(t)))_Q \ge 0 \quad \forall \boldsymbol{\tau} \in \Sigma(t, \boldsymbol{\sigma}(t)). \tag{13.17}$$

Thus, from (13.1), (13.6), (13.15)–(13.17) we obtain the following two variational formulations of the problem (13.1)–(13.6).

Problem 13.1 *Find a displacement field* $\boldsymbol{u} : \overline{I} \to V_1$ *and a stress field* $\boldsymbol{\sigma} : \overline{I} \to Q_1$ *such that*

$$\boldsymbol{u}(0) = \boldsymbol{u}_0 \tag{13.18}$$

and for all $t \in \overline{I}$,

$$\boldsymbol{\sigma}(t) = \mathcal{A}\boldsymbol{\varepsilon}(\dot{\boldsymbol{u}}(t)) + \mathcal{B}\boldsymbol{\varepsilon}(\boldsymbol{u}(t)), \tag{13.19}$$

$$\begin{aligned}(\boldsymbol{\sigma}(t), \boldsymbol{\varepsilon}(\boldsymbol{v}) - \boldsymbol{\varepsilon}(\dot{\boldsymbol{u}}(t)))_Q + j(\boldsymbol{\sigma}(t), \boldsymbol{v}) - j(\boldsymbol{\sigma}(t), \dot{\boldsymbol{u}}(t))&\\ \ge (\boldsymbol{f}(t), \boldsymbol{v} - \dot{\boldsymbol{u}}(t))_V \quad \forall \boldsymbol{v} \in V_1.&\end{aligned} \tag{13.20}$$

Problem 13.2 *Find a displacement field* $\boldsymbol{u}:\overline{I}\to V_1$ *and a stress field* $\boldsymbol{\sigma}:\overline{I}\to Q_1$ *such that*

$$\boldsymbol{u}(0)=\boldsymbol{u}_0 \tag{13.21}$$

and for all $t\in\overline{I}$,

$$\boldsymbol{\sigma}(t)=\mathcal{A}\boldsymbol{\varepsilon}(\dot{\boldsymbol{u}}(t))+\mathcal{B}\boldsymbol{\varepsilon}(\boldsymbol{u}(t)), \tag{13.22}$$

$$\boldsymbol{\sigma}(t)\in\Sigma(t,\boldsymbol{\sigma}(t)),\quad (\boldsymbol{\tau}-\boldsymbol{\sigma}(t),\boldsymbol{\varepsilon}(\dot{\boldsymbol{u}}(t)))_Q\geq 0\quad\forall\,\boldsymbol{\tau}\in\Sigma(t,\boldsymbol{\sigma}(t)). \tag{13.23}$$

In the study of Problem 13.1 we have the following result.

Theorem 13.3 *Assume* (6.34), (6.35) *and* (13.8)–(13.11). *Then there exists* $\mu_0>0$ *which depends only on* Ω, Γ_1, Γ_3, $\mathcal{A}$, p *and* $\mathcal{R}$ *such that Problem 13.1 has a unique solution* $(\boldsymbol{u},\boldsymbol{\sigma})$ *if* $\|\mu\|_{L^\infty(\Gamma_3)}<\mu_0$. *Moreover, the solution satisfies*

$$\boldsymbol{u}\in C^1(\overline{I};V_1),\qquad \boldsymbol{\sigma}\in C(\overline{I};Q_1). \tag{13.24}$$

Proof. The proof is carried out in four steps sketched below.

i) Let $\boldsymbol{\eta}\in C(\overline{I};Q)$, and $\boldsymbol{\xi}\in C(\overline{I};Q_1)$. Then there exists a unique solution

$$\boldsymbol{v}_{\eta\xi}\in C(\overline{I};V_1),\quad \boldsymbol{\sigma}_{\eta\xi}\in C(\overline{I};Q_1)$$

to the following variational problem:

$$\begin{aligned}&\boldsymbol{\sigma}_{\eta\xi}(t)=\mathcal{A}\boldsymbol{\varepsilon}(\boldsymbol{v}_{\eta\xi}(t))+\boldsymbol{\eta}(t),\\ &(\boldsymbol{\sigma}_{\eta\xi}(t),\boldsymbol{\varepsilon}(\boldsymbol{v})-\boldsymbol{\varepsilon}(\boldsymbol{v}_{\eta\xi}(t)))_Q+j(\boldsymbol{\xi}(t),\boldsymbol{v})-j(\boldsymbol{\xi}(t),\boldsymbol{v}_{\eta\xi}(t))\\ &\qquad\geq(\boldsymbol{f}(t),\boldsymbol{v}-\boldsymbol{v}_{\eta\xi}(t))_V\quad\forall\,\boldsymbol{v}\in V_1\end{aligned}$$

for all $t\in\overline{I}$.

ii) Let

$$\mu_0=\frac{M}{L_{\mathcal{A}}L_pC_{\mathcal{R}}C_0},$$

where M, $L_{\mathcal{A}}$, L_p, $C_{\mathcal{R}}$, C_0 are the constants found in the assumptions (6.34), (13.8), (13.7) and (7.13), respectively. Then, the operator $\Lambda_\eta : C(\overline{I};Q_1)\to C(\overline{I};Q_1)$ defined by

$$\Lambda_\eta\boldsymbol{\xi}=\boldsymbol{\sigma}_{\eta\xi}\quad\forall\,\boldsymbol{\xi}\in C(\overline{I};Q_1)$$

has a unique fixed point $\boldsymbol{\xi}_\eta$, if $\|\mu\|_{L^\infty(\Gamma_3)}<\mu_0$.

iii) If $\|\mu\|_{L^\infty(\Gamma_3)}<\mu_0$, then the operator $\Lambda: C(\overline{I};Q)\to C(\overline{I};Q)$ defined by

$$\Lambda\boldsymbol{\eta}(t)=\mathcal{B}\boldsymbol{\varepsilon}(\boldsymbol{u}_\eta(t)),\quad \boldsymbol{u}_\eta(t)=\int_0^t\boldsymbol{v}_{\eta\xi_\eta}(s)\,ds+\boldsymbol{u}_0,$$

for all $\boldsymbol{\eta}\in C(\overline{I};Q)$ and $t\in\overline{I}$, has a unique fixed point $\boldsymbol{\eta}^*\in C(\overline{I};Q)$.

iv) Let $\|\mu\|_{L^\infty(\Gamma_3)}<\mu_0$. Then, $(\boldsymbol{u},\boldsymbol{\sigma})$ with $\boldsymbol{u}=\boldsymbol{u}_{\eta^*}$ and $\boldsymbol{\sigma}=\boldsymbol{\sigma}_{\eta^*\xi_{\eta^*}}$ is the unique solution of Problem 13.1 satisfying the regularity condition (13.24). ■

The following result shows that Problems 13.1 and 13.2 are equivalent.

Theorem 13.4 *Let conditions* (6.34), (6.35) *and* (13.8)–(13.11) *hold and assume* $(\boldsymbol{u}, \boldsymbol{\sigma})$ *satisfies* (13.24). *Then,* $(\boldsymbol{u}, \boldsymbol{\sigma})$ *is a solution of Problem 13.1 if and only if it is a solution of Problem 13.2.*

Proof. The proof of Theorem 13.4 is based on the equivalence between the variational inequalities (13.20) and (13.23) and is similar to the proof of Theorem 12.18. ∎

As a consequence of Theorems 13.3 and 13.4 we have the next result.

Theorem 13.5 *Let conditions* (6.34), (6.35) *and* (13.8)–(13.11) *hold and let* $\mu_0 > 0$ *be defined as in Theorem 13.3. Then if* $\|\mu\|_{L^\infty(\Gamma_3)} < \mu_0$, *Problem 13.2 has a unique solution* $(\boldsymbol{u}, \boldsymbol{\sigma})$ *with the regularity* (13.24).

Theorems 13.3 and 13.5 provide existence and uniqueness results for Problems 13.1 and 13.2. Moreover, these problems are equivalent by Theorem 13.4. From Theorems 13.3 and 13.5 we conclude that if the coefficient of friction is small enough, then the mechanical problem has a unique weak solution which satisfies both Problems 13.1 and 13.2.

More detailed study of the mechanical problem (13.1)–(13.6) including the continuous dependence of the weak solution with respect to the data as well as with respect the coefficient of friction may be found in [15].

13.2 Bilateral Contact with Friction and Wear

In this section we consider a contact problem with wear of the contacting surface taken into account. The contact is frictional and bilateral with a moving rigid foundation, which results in the wear of the contacting surface. We assume that there is only sliding contact which is maintained all the time.

We now briefly describe the boundary conditions on the contact surface Γ_3, based on the model derived in [201, 202]. We introduce the wear function $w : \Gamma_3 \times \overline{I} \to \mathbb{R}$ which measures the wear of the surface. The wear is identified as the normal depth of the material that is lost. Since the body is in bilateral contact with the foundation it follows that

$$u_\nu = -w \quad \text{on } \Gamma_3. \tag{13.25}$$

Thus the location of the contact evolves with the wear. We remark that the effect of the wear is the recession on Γ_3 and therefore it is natural to expect that $u_\nu \leq 0$ on Γ_3, which implies $w \geq 0$ on Γ_3.

The evolution of the wear of the contacting surface is governed by a simplified version of Archard's law (cf. [201, 202]) which we now describe. The rate form of Archard's law is

$$\dot{w} = -\kappa\sigma_\nu\|\dot{\boldsymbol{u}}_\tau - \boldsymbol{v}^*\|,$$

where $\kappa > 0$ is a wear coefficient, $\boldsymbol{v}^*$ is the tangential velocity of the foundation, and $\|\dot{\boldsymbol{u}}_\tau - \boldsymbol{v}^*\|$ represents the slip speed between the contact surface and the foundation. We see that the rate of wear is assumed to be proportional to the contact stress and the slip speed. For the sake of simplicity we assume in the rest of the section that the motion of the foundation is uniform, i.e. $\boldsymbol{v}^*$ does not vary in time. Denote $v^* = \|\boldsymbol{v}^*\| > 0$. We assume that v^* is large so that we can neglect in the sequel $\dot{\boldsymbol{u}}_\tau$ as compared with $\boldsymbol{v}^*$ to obtain the following version of Archard law

$$\dot{w} = -\kappa v^* \sigma_\nu. \tag{13.26}$$

Use of the simplified law (13.26) for the evolution of the wear avoids some mathematical difficulties in the study of the quasistatic viscoelastic contact problem.

We can now eliminate the unknown function w from the problem. Let $\alpha = \kappa v^*$ and $\beta = 1/\alpha$. Using (13.25) and (13.26) we have

$$\sigma_\nu = \beta\, \dot{u}_\nu. \tag{13.27}$$

We model the frictional contact between the viscoelastic body and the foundation with Coulomb's law of dry friction. Since there is only sliding contact it follows that

$$\|\boldsymbol{\sigma}_\tau\| = \mu\, |\sigma_\nu|, \quad \boldsymbol{\sigma}_\tau = -\lambda\, (\dot{\boldsymbol{u}}_\tau - \boldsymbol{v}^*), \quad \lambda \geq 0, \tag{13.28}$$

where $\mu > 0$ is the coefficient of friction. These relations set constraints on the evolution of the tangential stress; in particular, the tangential stress is in the direction opposite to the relative sliding velocity $\dot{\boldsymbol{u}}_\tau - \boldsymbol{v}^*$.

Naturally, the wear increases in time, i.e. $\dot{w} \geq 0$. Hence, it follows from (13.25) and (13.26) that $\dot{u}_\nu \leq 0$ and $\sigma_\nu \leq 0$ on Γ_3. Thus, the conditions (13.27) and (13.28) imply

$$-\sigma_\nu = \beta\, |\dot{u}_\nu|, \quad \|\boldsymbol{\sigma}_\tau\| = -\mu\, \sigma_\nu, \quad \boldsymbol{\sigma}_\tau = -\lambda\, (\dot{\boldsymbol{u}}_\tau - \boldsymbol{v}^*), \quad \lambda \geq 0. \tag{13.29}$$

With (13.1) as the constitutive law and (13.29) as the contact conditions, the classical formulation of the mechanical problem of sliding frictional contact with wear is the following:

Find a displacement field $\boldsymbol{u} : \Omega \times \overline{I} \to \mathbb{R}^d$ and a stress field $\boldsymbol{\sigma} : \Omega \times \overline{I} \to \mathbb{S}^d$ such that

$$\boldsymbol{\sigma} = \mathcal{A}\varepsilon(\dot{\boldsymbol{u}}) + \mathcal{B}\varepsilon(\boldsymbol{u}) \qquad \text{in } \Omega \times I, \tag{13.30}$$

$$\operatorname{Div} \boldsymbol{\sigma} + \boldsymbol{f}_0 = \boldsymbol{0} \qquad \text{in } \Omega \times I, \tag{13.31}$$

$$\boldsymbol{u} = \boldsymbol{0} \qquad \text{on } \Gamma_1 \times I, \tag{13.32}$$

$$\boldsymbol{\sigma\nu} = \boldsymbol{f}_2 \qquad \text{on } \Gamma_2 \times I, \tag{13.33}$$

$$\left.\begin{array}{l} -\sigma_\nu = \beta\, |\dot{u}_\nu|, \quad \|\boldsymbol{\sigma}_\tau\| = -\mu\, \sigma_\nu, \\ \exists\, \lambda \geq 0 \text{ s.t. } \boldsymbol{\sigma}_\tau = -\lambda\, (\dot{\boldsymbol{u}}_\tau - \boldsymbol{v}^*) \end{array}\right\} \qquad \text{on } \Gamma_3 \times I, \tag{13.34}$$

$$\boldsymbol{u}(0) = \boldsymbol{u}_0 \qquad \text{in } \Omega. \tag{13.35}$$

We assume in what follows that (6.34), (6.35), (13.9)–(13.11) hold and

$$\beta \in L^\infty(\Gamma_3), \quad \beta(\boldsymbol{x}) \geq \beta_* > 0 \text{ a.e. on } \Gamma_3. \tag{13.36}$$

Denote by $\boldsymbol{f}$ the element given by (13.12) and let $j : V \times V \to \mathbb{R}$ be the functional

$$j(\boldsymbol{u}, \boldsymbol{v}) = \int_{\Gamma_3} \beta \, |u_\nu| \, (\mu \, \|\boldsymbol{v}_\tau - \boldsymbol{v}^*\| + v_\nu) \, da \quad \forall \, \boldsymbol{u}, \boldsymbol{v} \in V. \tag{13.37}$$

The variational formulation of the mechanical problem (13.30)–(13.35) can be stated as follows.

Problem 13.6 *Find a displacement field* $\boldsymbol{u} : \overline{I} \to V$ *and a stress field* $\boldsymbol{\sigma} : \overline{I} \to Q_1$ *such that*

$$\boldsymbol{u}(0) = \boldsymbol{u}_0 \tag{13.38}$$

and for all $t \in \overline{I}$,

$$\boldsymbol{\sigma}(t) = \mathcal{A}\varepsilon(\dot{\boldsymbol{u}}(t)) + \mathcal{B}\varepsilon(\boldsymbol{u})(t), \tag{13.39}$$

$$(\boldsymbol{\sigma}(t), \varepsilon(\boldsymbol{v}) - \varepsilon(\dot{\boldsymbol{u}}(t)))_Q + j(\dot{\boldsymbol{u}}(t), \boldsymbol{v}) - j(\dot{\boldsymbol{u}}(t), \dot{\boldsymbol{u}}(t)) \geq (\boldsymbol{f}(t), \boldsymbol{v} - \dot{\boldsymbol{u}}(t))_V \quad \forall \, \boldsymbol{v} \in V. \tag{13.40}$$

Well-posedness of this problem is stated in the following result.

Theorem 13.7 *Let conditions* (6.34), (6.35), (13.9)–(13.11) *and* (13.36) *hold. Then there exists a constant* $\alpha_0 > 0$ *which depends only on* Ω, Γ_1, Γ_3 *and* $\mathcal{A}$ *such that if*

$$\|\beta\|_{L^\infty(\Gamma_3)} \Big(\|\mu\|_{L^\infty(\Gamma_3)} + 1 \Big) < \alpha_0, \tag{13.41}$$

then Problem 13.6 has a unique solution $(\boldsymbol{u}, \boldsymbol{\sigma})$ *with the regularity*

$$\boldsymbol{u} \in C^1(\overline{I}; V_1), \qquad \boldsymbol{\sigma} \in C(\overline{I}; Q_1).$$

The proof of Theorem 13.7 may be carried out in several steps, using the same arguments as in the proof of Theorem 13.3. The result may also be deduced from Theorem 12.3. Indeed, replacing (13.39) in (13.40) we obtain a variational formulation of Problem 13.6 in terms of the displacement which is a special case of the abstract Problem 12.2.

We remark that if v^* is large enough then $\beta = 1/(k_w v^*)$ is sufficiently small and therefore the condition (13.41) for the unique solvability of Problem 13.6 is satisfied. We conclude that the mechanical problem (13.30)–(13.35) has a unique weak solution if the tangential velocity of the foundation is large enough. Moreover, having solved the problem (13.30)–(13.35), we can find the wear function by integrating (13.26) and use the initial condition $w(0) = 0$ which represents that at the initial moment the body is not subject to any prior wear.

Details on the variational analysis of Problem 13.6 can be found in [190]. Numerical analysis of the problem, including error estimates for semi-discrete and fully discrete schemes is provided in [45].

13.3 Contact with Normal Compliance, Friction and Wear

In this section we study the contact problem with normal compliance and friction contact conditions when the wear of the contact surface due to friction is taken in account. As in Section 13.2, the foundation is assumed to move steadily and only sliding contact takes place. Let g be the initial gap between the body and the foundation and let p_ν, p_τ denote the normal compliance functions. As in the previous section, we introduce the wear function $w : \Gamma_3 \times \overline{I} \to \mathbb{R}$ which measures the accumulated wear of the surface. Denote by $\boldsymbol{v}^*$ and $v^* = \|\boldsymbol{v}^*\|$ the tangential velocity and the tangential speed of the foundation, respectively. We use the modified version of Archard law (13.26) to describe the evolution of the wear. We modify the contact conditions (11.5) and (11.6) (page 228) in order to take into account the material removal that takes place on the contact surface. More precisely, the classical formulation of the problem of sliding frictional contact with wear we consider is the following:

Find a displacement field $\boldsymbol{u} : \Omega \times \overline{I} \to \mathbb{R}^d$, a stress field $\boldsymbol{\sigma} : \Omega \times \overline{I} \to \mathbb{S}^d$ and a wear function $w : \Gamma_3 \times \overline{I} \to \mathbb{R}$ such that

$$\boldsymbol{\sigma} = \mathcal{A}\boldsymbol{\varepsilon}(\dot{\boldsymbol{u}}) + \mathcal{B}\boldsymbol{\varepsilon}(\boldsymbol{u}) \qquad \text{in } \Omega \times I, \tag{13.42}$$

$$\operatorname{Div} \boldsymbol{\sigma} + \boldsymbol{f}_0 = \boldsymbol{0} \qquad \text{in } \Omega \times I, \tag{13.43}$$

$$\boldsymbol{u} = \boldsymbol{0} \qquad \text{on } \Gamma_1 \times I, \tag{13.44}$$

$$\boldsymbol{\sigma\nu} = \boldsymbol{f}_2 \qquad \text{on } \Gamma_2 \times I, \tag{13.45}$$

$$-\sigma_\nu = p_\nu(u_\nu - w - g) \qquad \text{on } \Gamma_3 \times I, \tag{13.46}$$

$$\left.\begin{array}{l} |\boldsymbol{\sigma}_\tau| = p_\tau(u_\nu - w - g), \\ \exists\, \lambda \geq 0 \text{ s.t. } \boldsymbol{\sigma}_\tau = -\lambda\,(\boldsymbol{v}^* - \boldsymbol{u}_\tau) \end{array}\right\} \qquad \text{on } \Gamma_3 \times I, \tag{13.47}$$

$$\dot{w} = -\kappa\, v^* \sigma_\nu \qquad \text{on } \Gamma_3 \times I, \tag{13.48}$$

$$\boldsymbol{u}(0) = \boldsymbol{u}_0, \quad w(0) = 0 \qquad \text{in } \Omega. \tag{13.49}$$

Note that the wear appears in the normal compliance condition. Therefore, in comparison with the bilateral contact problem with friction and wear studied in the previous section, the mechanical problem (13.42)–(13.49) is coupled and therefore more complicated.

In the study of the mechanical problem (13.42)–(13.49) we assume that the viscosity operator $\mathcal{A}$ and the elasticity operator $\mathcal{B}$ satisfy the conditions (6.34) and (6.35) and the normal compliance functions p_r $(r = \nu, \tau)$ satisfy the condition (11.10). We also assume that the force and traction densities satisfy (11.11), the gap function satisfies (11.12) and the initial displacement has the regularity (11.13).

We denote by $\boldsymbol{f}(t)$ the element of V given by (11.14) for all $\boldsymbol{v} \in V$ and $t \in \overline{I}$, and we use the functional $j : V \times V \times L^2(\Gamma_3) \to \mathbb{R}$ given by

$$j(\boldsymbol{u}, \boldsymbol{v}, w) = \int_{\Gamma_3} p_\nu(u_\nu - w - g)\, v_\nu\, da + \int_{\Gamma_3} p_\tau(u_\nu - w - g)\, \|\boldsymbol{v}_\tau - \boldsymbol{v}^*\|\, da \tag{13.50}$$

for $\boldsymbol{u}, \boldsymbol{v} \in V$, $w \in L^2(\Gamma_3)$.

The variational formulation of the mechanical problem (13.42)–(13.49) can be stated as follows.

Problem 13.8 *Find a displacement field* $\boldsymbol{u} : \overline{I} \to V$, *a stress field* $\boldsymbol{\sigma} : \overline{I} \to Q_1$ *and a wear function* $w : \overline{I} \to L^2(\Gamma_3)$ *such that*

$$\boldsymbol{u}(0) = \boldsymbol{u}_0, \quad w(0) = 0 \tag{13.51}$$

and for all $t \in \overline{I}$,

$$\boldsymbol{\sigma}(t) = \mathcal{A}\boldsymbol{\varepsilon}(\dot{\boldsymbol{u}}(t)) + \mathcal{B}\boldsymbol{\varepsilon}(\boldsymbol{u}(t)), \tag{13.52}$$

$$\dot{w} = -\kappa\, v^* \sigma_\nu, \tag{13.53}$$

$$\begin{aligned}(\boldsymbol{\sigma}(t), \boldsymbol{\varepsilon}(\boldsymbol{v}) - \boldsymbol{\varepsilon}(\dot{\boldsymbol{u}}(t)))_Q + j(\boldsymbol{u}(t), \boldsymbol{v}, w(t)) - j(\boldsymbol{u}(t), \dot{\boldsymbol{u}}(t), w(t))&\\ \geq (\boldsymbol{f}(t), \boldsymbol{v} - \dot{\boldsymbol{u}}(t))_V \quad \forall \boldsymbol{v} \in V.&\end{aligned} \tag{13.54}$$

A triplet $(\boldsymbol{u}, \boldsymbol{\sigma}, w)$ which satisfies (13.51)–(13.54) is called a *weak solution* of the mechanical problem of sliding frictional contact with normal compliance and wear.

The main result of this section is the following.

Theorem 13.9 *Assume* (6.34), (6.35) *and* (11.10)–(11.13). *Then there exists a unique solution* $(\boldsymbol{u}, \boldsymbol{\sigma}, w)$ *to Problem 13.8. Moreover, the solution satisfies*

$$\boldsymbol{u} \in C^1(\overline{I}; V), \quad \boldsymbol{\sigma} \in C(\overline{I}; Q_1), \quad w \in C^1(\overline{I}; L^2(\Gamma_3)). \tag{13.55}$$

The proof of Theorem 13.9 is carried out in several steps. In the first step we assume the wear is given and we try to find the corresponding displacement and stress. Thus, let $w \in C(\overline{I}; L^2(\Gamma_3))$ and we consider the following auxiliary variational problem.

Problem 13.10 *Find* $\boldsymbol{u}_w : \overline{I} \to V$ *and* $\boldsymbol{\sigma}_w : \overline{I} \to Q_1$ *such that*

$$\boldsymbol{u}_w(0) = \boldsymbol{u}_0 \tag{13.56}$$

and for all $t \in \overline{I}$,

$$\boldsymbol{\sigma}_w(t) = \mathcal{A}\boldsymbol{\varepsilon}(\dot{\boldsymbol{u}}_w(t)) + \mathcal{B}\boldsymbol{\varepsilon}(\boldsymbol{u}_w(t)), \tag{13.57}$$

$$\begin{aligned}(\boldsymbol{\sigma}_w, \boldsymbol{\varepsilon}(\boldsymbol{v}) - \boldsymbol{\varepsilon}(\dot{\boldsymbol{u}}(t)))_Q + j(\boldsymbol{u}_w(t), \boldsymbol{v}, w(t)) - j(\boldsymbol{u}_w(t), \dot{\boldsymbol{u}}_w(t), w(t))&\\ \geq (\boldsymbol{f}(t), \boldsymbol{v} - \dot{\boldsymbol{u}}_w(t))_V \quad \forall \boldsymbol{v} \in V.&\end{aligned} \tag{13.58}$$

Arguing as in the proof of Theorem 11.12 we have

Lemma 13.11 *There exists a unique solution* $(\boldsymbol{u}_w, \boldsymbol{\sigma}_w)$ *to Problem 13.10 such that* $\boldsymbol{u}_w \in C^1(\overline{I}; V)$, $\boldsymbol{\sigma}_w \in C(\overline{I}; Q_1)$.

Let us now consider the operator $\mathcal{L} : C(\overline{I}; L^2(\Gamma_3)) \to C(\overline{I}; L^2(\Gamma_3))$ defined by

$$\mathcal{L}w(t) = -\kappa\, v^* \int_0^t (\sigma_w)_\nu(s)\, ds \quad \forall\, t \in \overline{I}. \tag{13.59}$$

The second step in the proof of Theorem 13.9 is the next result.

Lemma 13.12 *The operator $\mathcal{L}$ has a unique fixed point*

$$w^* \in C(\overline{I}; L^2(\Gamma_3)).$$

Proof. Let $w_1, w_2 \in C(\overline{I}; L^2(\Gamma_3))$ and denote by $(\boldsymbol{u}_i, \boldsymbol{\sigma}_i)$, $i = 1, 2$, the solutions of Problems 13.10 for $w = w_i$, i.e. $\boldsymbol{u}_i = \boldsymbol{u}_{w_i}$, $\boldsymbol{\sigma}_i = \boldsymbol{\sigma}_{w_i}$. Moreover, we denote in the sequel by c various positive constants which may depend on κ and v^*. Then we can show that

$$\|\dot{\boldsymbol{u}}_1(s) - \dot{\boldsymbol{u}}_2(s)\|_V \le c\Big(\|\boldsymbol{u}_1(s) - \boldsymbol{u}_2(s)\|_V + \|w_1(s) - w_2(s)\|_{L^2(\Gamma_3)}\Big) \tag{13.60}$$

for all $s \in \overline{I}$. Let $t \in \overline{I}$. Since $\boldsymbol{u}_1$ and $\boldsymbol{u}_2$ have the same initial value $\boldsymbol{u}_0$, we have

$$\boldsymbol{u}_1(t) - \boldsymbol{u}_2(t) = \int_0^t (\dot{\boldsymbol{u}}_1(s) - \dot{\boldsymbol{u}}_2(s))\, ds.$$

Using (13.60) we obtain

$$\begin{aligned}\|\boldsymbol{u}_1(t) - \boldsymbol{u}_2(t)\|_V &\le \int_0^t \|\dot{\boldsymbol{u}}_1(s) - \dot{\boldsymbol{u}}_2(s)\|_V ds \\ &\le c\int_0^t \|\boldsymbol{u}_1(s) - \boldsymbol{u}_2(s)\|_V ds \\ &\quad + c\int_0^t \|w_1(s) - w_2(s)\|_{L^2(\Gamma_3)} ds.\end{aligned}$$

Applying the Gronwall inequality, we deduce that

$$\|\boldsymbol{u}_1(t) - \boldsymbol{u}_2(t)\|_V \le c\int_0^t \|w_1(s) - w_2(s)\|_{L^2(\Gamma_3)} ds. \tag{13.61}$$

It follows now from (13.60) and (13.61) that

$$\begin{aligned}\|\dot{\boldsymbol{u}}_1(s) - \dot{\boldsymbol{u}}_2(s)\|_V &\le c\,\|w_1(s) - w_2(s)\|_{L^2(\Gamma_3)} \\ &\quad + c\int_0^s \|w_1(s) - w_2(s)\|_{L^2(\Gamma_3)} ds.\end{aligned} \tag{13.62}$$

On the other hand, by the assumptions (6.34) and (6.35) on the operators $\mathcal{A}$, $\mathcal{B}$, we can derive the following inequalities from (13.57) and (13.58):

$$\|\boldsymbol{\sigma}_1(t) - \boldsymbol{\sigma}_1(t)\|_{Q_1} \le c\Big(\|\dot{\boldsymbol{u}}_1(t) - \dot{\boldsymbol{u}}_2(t)\|_V + \|\boldsymbol{u}_1(t) - \boldsymbol{u}_2(t)\|_V\Big).$$

Using the bounds (13.61) and (13.62), we are led to

$$\begin{aligned}\|\boldsymbol{\sigma}_1(t)-\boldsymbol{\sigma}_2(t)\|_{Q_1} &\le c\,\|w_1(t)-w_2(t)\|_{L^2(\Gamma_3)}\\ &\quad + c\int_0^t \|w_1(s)-w_2(s)\|_{L^2(\Gamma_3)}ds.\end{aligned}$$

Now from the definition of the operator $\mathcal{L}$, (13.59), we have

$$\|\mathcal{L}w_1(t)-\mathcal{L}w_2(t)\|_{L^2(\Gamma_3)} \le c\int_0^t \|\boldsymbol{\sigma}_1(s)-\boldsymbol{\sigma}_2(s)\|_{Q_1}ds.$$

Thus, for all $t\in\overline{I}$,

$$\begin{aligned}\|\mathcal{L}w_1(t)-\mathcal{L}w_2(t)\|_{L^2(\Gamma_3)} &\le c\int_0^t \|w_1(s)-w_2(s)\|_{L^2(\Gamma_3)}ds\\ &\quad + c\int_0^t\int_0^s \|w_1(r)-w_2(r)\|_{L^2(\Gamma_3)}dr\,ds.\end{aligned}$$

Since

$$\begin{aligned}\int_0^t\int_0^s \|w_1(r)-w_2(r)\|_{L^2(\Gamma_3)}dr\,ds &\le \int_0^t\int_0^t \|w_1(r)-w_2(r)\|_{L^2(\Gamma_3)}dr\,ds\\ &\le T\int_0^t \|w_1(r)-w_2(r)\|_{L^2(\Gamma_3)}dr,\end{aligned}$$

we have

$$\|\mathcal{L}w_1(t)-\mathcal{L}w_2(t)\|_{L^2(\Gamma_3)} \le c\int_0^t \|w_1(s)-w_2(s)\|_{L^2(\Gamma_3)}ds. \tag{13.63}$$

Thus,

$$\|\mathcal{L}w_1(t)-\mathcal{L}w_2(t)\|_{L^2(\Gamma_3)} \le c\,t\,\|w_1-w_2\|_{C(\overline{I};L^2(\Gamma_3))},\quad t\in\overline{I}.$$

Using this inequality in (13.63) with w_1 and w_2 replaced by $\mathcal{L}w_1$ and $\mathcal{L}w_2$, respectively, we obtain

$$\begin{aligned}\|\mathcal{L}^2w_1(t)-\mathcal{L}^2w_2(t)\|_{L^2(\Gamma_3)} &\le c\int_0^t \|\mathcal{L}w_1(s)-\mathcal{L}w_2(s)\|_{L^2(\Gamma_3)}ds\\ &\le c^2\int_0^t s\,\|w_1-w_2\|_{C(\overline{I};L^2(\Gamma_3))}ds\\ &= \frac{c^2}{2!}\,t^2\|w_1-w_2\|_{C(\overline{I};L^2(\Gamma_3))}.\end{aligned}$$

Repeat this procedure, we obtain

$$\|\mathcal{L}^nw_1(t)-\mathcal{L}^nw_2(t)\|_{L^2(\Gamma_3)} \le \frac{c^n}{n!}\,t^n\|w_1-w_2\|_{C(\overline{I};L^2(\Gamma_3))},\quad t\in\overline{I}.$$

Hence,

$$\|\mathcal{L}^n w_1 - \mathcal{L}^n w_2\|_{C(\overline{I};L^2(\Gamma_3))} \le \frac{(cT)^n}{n!} \|w_1 - w_2\|_{C(\overline{I};L^2(\Gamma_3))}.$$

Therefore, for n large enough, $\mathcal{L}^n$ is a contractive operator on the Banach space $C(\overline{I};L^2(\Gamma_3))$. By Corollary 1.63, the operator $\mathcal{L}$ has a unique fixed-point $w^* \in C(\overline{I};L^2(\Gamma_3))$. ■

Proof of Theorem 13.9.
Let w^* be the fixed point of the operator $\mathcal{L}$. With (13.56)–(13.59) it is easy to verify that $(\boldsymbol{u}_{w^*}, \boldsymbol{\sigma}_{w^*}, w^*)$ is the unique solution of Problem 13.8 satisfying the regularity (13.55). ■

More details concerning the variational analysis of the problem (13.42)–(13.49) can be found in [178]. It is also possible to introduce semi-discrete and fully discrete schemes of the contact problem and derive error estimates. The presentation is similar to that in the previous section, although in a slightly more complicated setting.

13.4 Contact with Dissipative Frictional Potential

In the frictional contact problem considered in this section, the friction is modelled by a general velocity dependent dissipation functional. Existence and uniqueness of a weak solution to the problem is shown by standard arguments, similar to those presented in the previous sections and chapters. Here we provide a detailed analysis for the behavior of the solution as the viscosity tensor converges to zero. The results are applicable to a variety of concrete frictional contact problems.

The classical formulation of the problem is the following:

Find a displacement field $\boldsymbol{u} : \Omega \times \overline{I} \to \mathbb{R}^d$ and a stress field $\boldsymbol{\sigma} : \Omega \times \overline{I} \to \mathbb{S}^d$ such that

$$\boldsymbol{\sigma} = \mathcal{A}\varepsilon(\dot{\boldsymbol{u}}) + \mathcal{B}\varepsilon(\boldsymbol{u}) \quad \text{in } \Omega \times I, \qquad (13.64)$$

$$\operatorname{Div} \boldsymbol{\sigma} + \boldsymbol{f}_0 = \boldsymbol{0} \quad \text{in } \Omega \times I, \qquad (13.65)$$

$$\boldsymbol{u} = \boldsymbol{0} \quad \text{on } \Gamma_1 \times I, \qquad (13.66)$$

$$\boldsymbol{\sigma\nu} = \boldsymbol{f}_2 \quad \text{on } \Gamma_2 \times I, \qquad (13.67)$$

$$\boldsymbol{u} \in U,\ -\boldsymbol{\sigma}\boldsymbol{\nu}(\boldsymbol{v} - \dot{\boldsymbol{u}}) \le \varphi(\boldsymbol{v}) - \varphi(\dot{\boldsymbol{u}}) \quad \forall \boldsymbol{v} \in U \quad \text{on } \Gamma_3 \times I, \qquad (13.68)$$

$$\boldsymbol{u}(0) = \boldsymbol{u}_0 \quad \text{in } \Omega. \qquad (13.69)$$

Here (13.64) represents the linear version of Kelvin-Voigt constitutive law, i.e.

$$\sigma_{ij} = a_{ijkl}\, \varepsilon_{kl}(\dot{\boldsymbol{u}}) + b_{ijkl}\, \varepsilon_{kl}(\boldsymbol{u}),$$

with $\mathcal{A} = (a_{ijkl})$ being the viscosity tensor and $\mathcal{B} = (b_{ijkl})$ the elasticity tensor. The frictional contact is modelled with the subdifferential boundary condition

(13.68) in which U represents the set of contact admissible functions, and φ is a given function. Examples and detailed explanations of inequality problems in contact mechanics which lead to boundary conditions of this form will be presented at the end of this section.

We suppose that $U \subset H^1(\Omega)^d$, $\varphi : \Gamma_3 \times \mathbb{R}^d \to \mathbb{R}$ and, to accommodate the homogeneous displacement boundary condition (13.66) and the contact condition (13.68), we define

$$U_1 = V \cap U. \tag{13.70}$$

Let $j : U_1 \to (-\infty, +\infty]$ be the functional

$$j(\boldsymbol{v}) = \begin{cases} \displaystyle\int_{\Gamma_3} \varphi(\boldsymbol{v})\, da & \text{if } \varphi(\boldsymbol{v}) \in L^1(\Gamma_3), \\ +\infty & \text{otherwise.} \end{cases} \tag{13.71}$$

We assume in what follows that

$$U_1 \text{ is a closed subspace of } H_1 \text{ such that } C_0^\infty(\Omega)^d \subset U_1. \tag{13.72}$$

$$j \text{ is proper, convex and lower semicontinuous on } U_1. \tag{13.73}$$

We assume that the viscosity and elasticity tensors satisfy the usual properties of ellipticity and symmetry, i.e.

$$\left.\begin{array}{l} \text{(a) } \mathcal{A} : \Omega \times \mathbb{S}^d \to \mathbb{S}^d. \\ \text{(b) } a_{ijkl} \in L^\infty(\Omega). \\ \text{(c) } \mathcal{A}\boldsymbol{\sigma} \cdot \boldsymbol{\tau} = \boldsymbol{\sigma} \cdot \mathcal{A}\boldsymbol{\tau} \quad \forall\, \boldsymbol{\sigma}, \boldsymbol{\tau} \in \mathbb{S}^d, \text{ a.e. in } \Omega. \\ \text{(d) There exists } m_{\mathcal{A}} > 0 \text{ such that} \\ \qquad \mathcal{A}\boldsymbol{\tau} \cdot \boldsymbol{\tau} \geq m_{\mathcal{A}} \|\boldsymbol{\tau}\|^2 \quad \forall\, \boldsymbol{\tau} \in \mathbb{S}^d, \text{ a.e. in } \Omega. \end{array}\right\} \tag{13.74}$$

$$\left.\begin{array}{l} \text{(a) } \mathcal{B} : \Omega \times \mathbb{S}^d \to \mathbb{S}^d. \\ \text{(b) } b_{ijkl} \in L^\infty(\Omega). \\ \text{(c) } \mathcal{B}\boldsymbol{\sigma} \cdot \boldsymbol{\tau} = \boldsymbol{\sigma} \cdot \mathcal{B}\boldsymbol{\tau} \quad \forall\, \boldsymbol{\sigma}, \boldsymbol{\tau} \in \mathbb{S}^d, \text{ a.e. in } \Omega. \\ \text{(d) There exists } m_{\mathcal{B}} > 0 \text{ such that} \\ \qquad \mathcal{B}\boldsymbol{\tau} \cdot \boldsymbol{\tau} \geq m_{\mathcal{B}} \|\boldsymbol{\tau}\|^2 \quad \forall\, \boldsymbol{\tau} \in \mathbb{S}^d, \text{ a.e. in } \Omega. \end{array}\right\} \tag{13.75}$$

For the body force, surface tractions, and the initial displacement, we assume

$$\boldsymbol{f}_0 \in H^1(I; L^2(\Omega)^d), \quad \boldsymbol{f}_2 \in H^1(I; L^2(\Gamma_2)^d), \tag{13.76}$$

$$\boldsymbol{u}_0 \in U_1. \tag{13.77}$$

For $\boldsymbol{u}, \boldsymbol{v} \in U_1$ let

$$(\boldsymbol{u}, \boldsymbol{v})_{U_1} = (\mathcal{B}\boldsymbol{\varepsilon}(\boldsymbol{u}), \boldsymbol{\varepsilon}(\boldsymbol{v}))_Q, \quad \|\boldsymbol{u}\|_{U_1} = (\boldsymbol{u}, \boldsymbol{u})_{U_1}^{1/2}.$$

By the assumptions (13.75) on $\mathcal{B}$ and Korn's inequality (7.8) (page 144) we see that $(\cdot,\cdot)_{U_1}$ is an inner product on U_1, and $\|\cdot\|_{U_1}$ and $\|\cdot\|_{H^1(\Omega)^d}$ are equivalent norms on U_1. Therefore, $(U_1, \|\cdot\|_{U_1})$ is a real Hilbert space.

For any $t \in \overline{I}$, we denote by $\boldsymbol{f}(t)$ the element of U_1 given by

$$(\boldsymbol{f}(t), \boldsymbol{v})_{U_1} = \int_\Omega \boldsymbol{f}_0(t) \cdot \boldsymbol{v}\, dx + \int_{\Gamma_2} \boldsymbol{f}_2(t) \cdot \boldsymbol{v}\, da \quad \forall\, \boldsymbol{v} \in U_1. \tag{13.78}$$

Conditions (13.76) imply

$$\boldsymbol{f} \in H^1(I; U_1). \tag{13.79}$$

It is straightforward to show that if $(\boldsymbol{u}, \boldsymbol{\sigma})$ is a pair of sufficiently regular functions satisfying the relations (13.65)–(13.68), then for all $t \in \overline{I}$, $\boldsymbol{u}(t) \in U_1$ and

$$\begin{aligned}(\boldsymbol{\sigma}(t), \boldsymbol{\varepsilon}(\boldsymbol{v}) - \boldsymbol{\varepsilon}(\dot{\boldsymbol{u}}(t)))_Q + j(\boldsymbol{v}) - j(\dot{\boldsymbol{u}}(t))&\\ \geq (\boldsymbol{f}(t), \boldsymbol{v} - \dot{\boldsymbol{u}}(t))_V \quad \forall\, \boldsymbol{v} \in U_1.&\end{aligned}$$

Combining this inequality with the constitutive relation (13.64) and the initial condition (13.69) we obtain the following variational formulation of the problem (13.64)–(13.69) with the displacement as the unknown.

Problem 13.13 *Find a displacement field $\boldsymbol{u} : \overline{I} \to U_1$ such that*

$$\boldsymbol{u}(0) = \boldsymbol{u}_0 \tag{13.80}$$

and for all $t \in \overline{I}$,

$$\begin{aligned}(\mathcal{A}\boldsymbol{\varepsilon}(\dot{\boldsymbol{u}}(t)), \boldsymbol{\varepsilon}(\boldsymbol{v}) - \boldsymbol{\varepsilon}(\dot{\boldsymbol{u}}(t)))_Q + (\mathcal{B}\boldsymbol{\varepsilon}(\boldsymbol{u}(t)), \boldsymbol{\varepsilon}(\boldsymbol{v}) - \boldsymbol{\varepsilon}(\dot{\boldsymbol{u}}(t)))_Q&\\ + j(\boldsymbol{v}) - j(\dot{\boldsymbol{u}}(t)) \geq (\boldsymbol{f}(t), \boldsymbol{v} - \dot{\boldsymbol{u}}(t))_{U_1} \quad \forall\, \boldsymbol{v} \in U_1.&\end{aligned} \tag{13.81}$$

We have the following existence and uniqueness result.

Theorem 13.14 *Assume* (13.72)–(13.77). *Then there exists a unique solution $\boldsymbol{u} \in H^2(I; U_1)$ to Problem 13.13.*

Proof. Theorem 13.14 follows from an application of Theorem 11.3. The regularity $\boldsymbol{u} \in H^2(I; U_1)$ is a consequence of the data regularity (13.79) (see Proposition 11.7 on page 234). ■

Let $\boldsymbol{u} \in H^2(I; U_1)$ be the solution of Problem 13.13 and let $\boldsymbol{\sigma}$ be the stress field defined by the constitutive relation (13.64). Using the assumptions (13.74) and (13.75) on the constitutive operators $\mathcal{A}$ and $\mathcal{B}$, we have $\boldsymbol{\sigma} \in H^1(I; Q)$. Letting $\boldsymbol{v} \in C_0^\infty(\Omega)^d$ arbitrary in (13.81), we get

$$\operatorname{Div} \boldsymbol{\sigma}(t) = -\boldsymbol{f}_0(t) \quad \text{a.e. in } \Omega,\ t \in \overline{I}.$$

Thus, by the regularity assumption (13.76) on $\boldsymbol{f}_0$, we further have $\boldsymbol{\sigma} \in W^{1,2}(I; Q_1)$. This pair of functions $(\boldsymbol{u}, \boldsymbol{\sigma})$ is called a *weak solution* of the problem (13.64)–(13.69). We conclude that the viscoelastic contact problem (13.64)–(13.69) has a unique weak solution.

Now we investigate the behavior of the solution to Problem 13.13 when the viscosity operator converges to zero. To this end, we assume in the sequel that (13.72)–(13.77) hold. Introduce a parameter $\theta > 0$. We replace in (13.81) the operator $\mathcal{A}$ with $\theta\mathcal{A}$ to obtain the following viscoelastic problem.

Problem 13.15 *Find a displacement field* $\boldsymbol{u}_\theta : \overline{I} \to U_1$ *such that*

$$\boldsymbol{u}_\theta(0) = \boldsymbol{u}_0 \tag{13.82}$$

and for all $t \in \overline{I}$,

$$\begin{aligned} &\theta(\mathcal{A}\boldsymbol{\varepsilon}(\dot{\boldsymbol{u}}_\theta(t)), \boldsymbol{\varepsilon}(\boldsymbol{v}) - \boldsymbol{\varepsilon}(\dot{\boldsymbol{u}}_\theta(t)))_Q + (\mathcal{B}\boldsymbol{\varepsilon}(\boldsymbol{u}_\theta(t)), \boldsymbol{\varepsilon}(\boldsymbol{v}) - \boldsymbol{\varepsilon}(\dot{\boldsymbol{u}}_\theta(t)))_Q \\ &\quad + j(\boldsymbol{v}) - j(\dot{\boldsymbol{u}}_\theta(t)) \geq (\boldsymbol{f}(t), \boldsymbol{v} - \dot{\boldsymbol{u}}_\theta(t))_{U_1} \quad \forall\, \boldsymbol{v} \in U_1. \end{aligned} \tag{13.83}$$

By Theorem 13.14, Problem 13.15 has a unique solution $\boldsymbol{u}_\theta \in H^2(I; U_1)$.
Let us introduce the following "limiting" variational problem.

Problem 13.16 *Find a displacement field* $\boldsymbol{u} : \overline{I} \to U_1$ *such that*

$$\boldsymbol{u}(0) = \boldsymbol{u}_0 \tag{13.84}$$

and for almost any $t \in \overline{I}$,

$$\begin{aligned} &(\mathcal{B}\boldsymbol{\varepsilon}(\boldsymbol{u}(t)), \boldsymbol{\varepsilon}(\boldsymbol{v}) - \boldsymbol{\varepsilon}(\dot{\boldsymbol{u}}(t)))_Q + j(\boldsymbol{v}) - j(\dot{\boldsymbol{u}}(t)) \\ &\qquad \geq (\boldsymbol{f}(t), \boldsymbol{v} - \dot{\boldsymbol{u}}(t))_{U_1} \quad \forall\, \boldsymbol{v} \in U_1. \end{aligned} \tag{13.85}$$

Clearly, Problem 13.16 represents the variational formulation of the contact problem when the viscoelastic constitutive law (13.64) is replaced with the linearly elastic constitutive law

$$\boldsymbol{\sigma} = \mathcal{B}\boldsymbol{\varepsilon}(\boldsymbol{u}) \quad \text{in } \Omega \times I. \tag{13.86}$$

To study the elastic problem 13.16, we need an additional assumption on the initial displacement $\boldsymbol{u}_0 \in V$:

$$(\boldsymbol{u}_0, \boldsymbol{v})_{U_1} + j(\boldsymbol{v}) \geq (\boldsymbol{f}(0), \boldsymbol{v})_{U_1} \quad \forall\, \boldsymbol{v} \in U_1. \tag{13.87}$$

Note that Problem 13.16 resembles Problem 4.15 on page 69. However, here the assumption on the functional j is more general. For the well-posedness of Problem 13.16, we will apply the following existence and uniqueness result.

Theorem 13.17 *Let* $(X, (\cdot,\cdot)_X)$ *be a real Hilbert space and let* $j : X \to (-\infty, +\infty]$ *be proper, convex and lower semicontinuous. Let* $f \in H^1(I; X)$ *and* $u_0 \in X$ *be such that*

$$\sup_{v \in D(j)} \{(f(0), v)_X - (u_0, v)_X - j(v)\} < +\infty, \tag{13.88}$$

where $D(j)$ *represents the effective domain of* j,

$$D(j) = \{v \in X : j(v) < +\infty\}.$$

Then there exists a unique element $u \in H^1(I; X)$ *satisfying*

$$(u(t), v - \dot{u}(t))_X + j(v) - j(\dot{u}(t)) \geq (f(t), v - \dot{u}(t))_X$$

for all $v \in X$, *a.e.* $t \in I$, *and*

$$u(0) = u_0.$$

Theorem 13.17 is proved in [34, p. 117] by arguments of evolution equations with maximal monotone operators. A variant of this theorem is given in [96], where the proof is based on a time-discretization method. A version of Theorem 13.17 in the case when j depends on the solution, i.e. in the case when $j = j(\boldsymbol{u}, \boldsymbol{v})$, is recently established in [158].

Theorem 13.18 *Assume the conditions* (13.72)–(13.77) *and* (13.87). *Then Problem 13.16 has a unique solution* $\boldsymbol{u} \in H^1(I; U_1)$.

Proof. We take $X = U_1$. Note that (13.87) implies (13.88), and assumptions (13.72)–(13.77) allow us to use Theorem 13.17 to conclude the existence and uniqueness of an element $\boldsymbol{u} \in H^1(I; U_1)$ which solves Problem 13.16. ■

A pair of functions $(\boldsymbol{u}, \boldsymbol{\sigma})$ which satisfies (13.84)–(13.86) is called a *weak solution* of the elastic problem (13.65)–(13.69), (13.86). We conclude that the quasistatic elastic problem has a unique weak solution.

We have the following result concerning a relation between Problem 13.15 and Problem 13.16. Below c denotes a generic positive constant, independent of θ.

Theorem 13.19 *Assume* (13.72)–(13.77) *and* (13.87). *Then there exists* $c > 0$ *which is independent of* θ *such that*

$$\|\boldsymbol{u}_\theta(t) - \boldsymbol{u}(t)\|_{U_1} \leq c\sqrt{\theta}\,\|\dot{\boldsymbol{u}}\|_{L^2(I;U_1)}, \tag{13.89}$$

for all $t \in \overline{I}$.

Proof. Let $t \in \overline{I}$. To simplify the notation, we will not indicate explicitly the dependence of the various functions on time. It follows from the variational inequalities (13.83) and (13.85) that

$$\theta\,(\mathcal{A}\boldsymbol{\varepsilon}(\dot{\boldsymbol{u}}_\theta), \boldsymbol{\varepsilon}(\dot{\boldsymbol{u}}) - \boldsymbol{\varepsilon}(\dot{\boldsymbol{u}}_\theta))_Q + (\boldsymbol{u}_\theta - \boldsymbol{u}, \dot{\boldsymbol{u}} - \dot{\boldsymbol{u}}_\theta)_{U_1} \geq 0 \quad \text{a.e. on } I,$$

which implies

$$\begin{aligned}\theta\,(\mathcal{A}\boldsymbol{\varepsilon}(\dot{\boldsymbol{u}}_\theta) - \mathcal{A}\boldsymbol{\varepsilon}(\dot{\boldsymbol{u}}), \boldsymbol{\varepsilon}(\dot{\boldsymbol{u}}_\theta) - \boldsymbol{\varepsilon}(\dot{\boldsymbol{u}}))_Q + (\boldsymbol{u}_\theta - \boldsymbol{u}, \dot{\boldsymbol{u}}_\theta - \dot{\boldsymbol{u}})_{U_1}&\\ \leq \theta\,(\mathcal{A}\boldsymbol{\varepsilon}(\dot{\boldsymbol{u}}), \boldsymbol{\varepsilon}(\dot{\boldsymbol{u}}) - \boldsymbol{\varepsilon}(\dot{\boldsymbol{u}}_\theta))_Q \quad \text{a.e. on } I.&\end{aligned} \tag{13.90}$$

Integrate this inequality over $[0, t]$ and use the initial conditions (13.82), (13.84) and the assumptions (13.74) on $\mathcal{A}$ to obtain

$$\begin{aligned}\theta\, c \int_0^t \|\boldsymbol{\varepsilon}(\dot{\boldsymbol{u}}_\theta) - \boldsymbol{\varepsilon}(\dot{\boldsymbol{u}})\|_Q^2\, ds + \frac{1}{2}\,\|\boldsymbol{u}_\theta(t) - \boldsymbol{u}(t)\|_{U_1}^2&\\ \leq \theta \int_0^t \|\mathcal{A}\boldsymbol{\varepsilon}(\dot{\boldsymbol{u}})\|_Q \|\boldsymbol{\varepsilon}(\dot{\boldsymbol{u}}_\theta) - \boldsymbol{\varepsilon}(\dot{\boldsymbol{u}})\|_Q\, ds.&\end{aligned}$$

Applying the inequality

$$a\,b \leq \frac{a^2}{2\,c} + \frac{c\,b^2}{2}$$

to the integrand on the right hand side, we find

$$c\,\theta \int_0^t \|\varepsilon(\dot{\boldsymbol{u}}_\theta) - \varepsilon(\dot{\boldsymbol{u}})\|_Q^2\, ds + \|\boldsymbol{u}_\theta(t) - \boldsymbol{u}(t)\|_{U_1}^2 \le c\,\theta \int_0^t \|\mathcal{A}\varepsilon(\dot{\boldsymbol{u}})\|_Q^2\, ds.$$

Therefore,

$$\|\boldsymbol{u}_\theta(t) - \boldsymbol{u}(t)\|_{U_1} \le c\,\sqrt{\theta}\,\|\mathcal{A}\varepsilon(\dot{\boldsymbol{u}})\|_{L^2(I;Q)}.$$

Then the inequality (13.89) holds. ■

Let us define

$$\boldsymbol{\sigma}_\theta = \theta\,\mathcal{A}\varepsilon(\dot{\boldsymbol{u}}_\theta) + \mathcal{B}\varepsilon(\boldsymbol{u}_\theta) \quad \text{in } \Omega \times I,$$

where $\boldsymbol{u}_\theta$ is the solution of Problem 13.15 and let $\boldsymbol{\sigma}$ denote the function given by (13.86) with $\boldsymbol{u}$ being the solution of Problem 13.16. We have

$$\operatorname{Div} \boldsymbol{\sigma}_\theta = \operatorname{Div} \boldsymbol{\sigma} = -\boldsymbol{f}_0 \quad \text{a.e. in } \Omega,$$

for any $t \in \overline{I}$. Thus, $\boldsymbol{\sigma}_\theta \in H^1(I;Q_1)$, $\boldsymbol{\sigma} \in H^1(I;Q_1)$. Now

$$\boldsymbol{\sigma}_\theta(t) - \boldsymbol{\sigma}(t) = \theta\,\mathcal{A}\varepsilon(\dot{\boldsymbol{u}}_\theta(t)) + \mathcal{B}\varepsilon(\boldsymbol{u}_\theta(t) - \boldsymbol{u}(t)).$$

Using the assumptions (13.74) and (13.75) on the operators $\mathcal{A}$ and $\mathcal{B}$, we obtain

$$\|\boldsymbol{\sigma}_\theta(t) - \boldsymbol{\sigma}(t)\|_{Q_1} \le c\,\theta\,\|\dot{\boldsymbol{u}}_\theta(t)\|_{U_1} + c\,\|\boldsymbol{u}_\theta(t) - \boldsymbol{u}(t)\|_{U_1} \quad \forall\, t \in \overline{I}. \tag{13.91}$$

We note that (13.90) yields

$$\theta\,\|\dot{\boldsymbol{u}}_\theta(t) - \dot{\boldsymbol{u}}(t)\|_{U_1} \le \|\boldsymbol{u}_\theta(t) - \boldsymbol{u}(t)\|_{U_1} + c\,\theta\,\|\dot{\boldsymbol{u}}(t)\|_{U_1} \quad \text{a.e. } t \in I.$$

Therefore,

$$\theta\,\|\dot{\boldsymbol{u}}_\theta(t)\|_{U_1} \le \|\boldsymbol{u}_\theta(t) - \boldsymbol{u}(t)\|_{U_1} + c\,\theta\,\|\dot{\boldsymbol{u}}(t)\|_{U_1} \quad \text{a.e. } t \in I.$$

Using this bound in (13.91), we infer that

$$\|\boldsymbol{\sigma}_\theta(t) - \boldsymbol{\sigma}(t)\|_{Q_1} \le c\,\|\boldsymbol{u}_\theta(t) - \boldsymbol{u}(t)\|_{U_1} + c\,\theta\,\|\dot{\boldsymbol{u}}(t)\|_{U_1} \quad \text{a.e. } t \in I.$$

This inequality is combined with (13.89) to yield

$$\|\boldsymbol{\sigma}_\theta(t) - \boldsymbol{\sigma}(t)\|_{Q_1} \le c\,\sqrt{\theta}\,(\|\dot{\boldsymbol{u}}\|_{L^2(I;U_1)} + \sqrt{\theta}\,\|\dot{\boldsymbol{u}}(t)\|_{U_1}) \quad \text{a.e. } t \in I. \tag{13.92}$$

Therefore, as a consequence of the estimates (13.89) and (13.92), we obtain the convergence of the viscoelastic solution $(\boldsymbol{u}_\theta, \boldsymbol{\sigma}_\theta)$ to the elastic solution $(\boldsymbol{u}, \boldsymbol{\sigma})$:

$$\max_{t \in \overline{I}} \|\boldsymbol{u}_\theta(t) - \boldsymbol{u}(t)\|_{U_1} \to 0, \quad \|\boldsymbol{\sigma}_\theta - \boldsymbol{\sigma}\|_{L^2(I;Q_1)} \to 0 \quad \text{as } \theta \to 0. \tag{13.93}$$

We conclude that the weak solution of the frictional elastic problem (13.65)–(13.69), (13.86) may be approximated by the weak solution of the frictional viscoelastic problem (13.64)–(13.69), when the viscosity tensor is “small enough”.

In other words, the elasticity problem with friction may be viewed as a limiting case of viscoelasticity problem with friction.

Examples of subdifferential conditions with friction. We end this section by presenting examples of contact and friction laws which lead to an inequality of the form (13.68) such that (13.72) and (13.73) hold. Then Theorem 13.14 can be applied to conclude that the relevant viscoelastic boundary value problem for each of the examples has a unique weak solution. Moreover, we deduce from Theorem 13.18 and (13.93) that the corresponding quasistatic elastic problem has a unique weak solution which can be obtained as the limit of the weak solution of the viscoelastic problem as the viscosity converges to zero.

Example 13.20 *Bilateral contact with Tresca's friction law.*

This contact condition is of the form:

$$\left.\begin{array}{l} u_\nu = 0, \quad \|\boldsymbol{\sigma}_\tau\| \le g, \\ \|\boldsymbol{\sigma}_\tau\| < g \;\Rightarrow\; \dot{\boldsymbol{u}}_\tau = \boldsymbol{0}, \\ \|\boldsymbol{\sigma}_\tau\| = g \;\Rightarrow\; \exists\lambda \ge 0 \text{ s.t. } \boldsymbol{\sigma}_\tau = -\lambda\dot{\boldsymbol{u}}_\tau \end{array}\right\} \quad \text{on } \Gamma_3 \times I. \tag{13.94}$$

Here $g \in L^\infty(\Gamma_3)$ is a positive function which represents the friction bound, i.e., the magnitude of the limiting friction traction at which slip begins. The contact is assumed to be bilateral, i.e., there is no loss of contact during the process.

The set of admissible test functions U consists of those elements of H_1 whose normal component vanishes on Γ_3. Thus by definition (13.70), $U_1 = V_1$, where V_1 is the space given by (7.11) (page 144). It is straightforward to show that if $(\boldsymbol{u}, \boldsymbol{\sigma})$ is a pair of regular functions satisfying (13.94) then

$$\boldsymbol{\sigma}\boldsymbol{\nu} \cdot (\boldsymbol{v} - \dot{\boldsymbol{u}}) \ge g\,\|\dot{\boldsymbol{u}}_\tau\| - g\,\|\boldsymbol{v}_\tau\| \quad \forall\,\boldsymbol{v} \in U,$$

a.e. on $\Gamma_3 \times I$. So the contact condition (13.68) holds with $\varphi(\boldsymbol{v}) = g\,\|\boldsymbol{v}_\tau\|$ and the functional j of (13.71) is

$$j(\boldsymbol{v}) = \int_{\Gamma_3} g\,\|\boldsymbol{v}_\tau\|\,da \quad \forall\,\boldsymbol{v} \in V_1.$$

In this case conditions (13.72) and (13.73) hold. We conclude that the viscoelastic problem (13.64)–(13.67), (13.69), (13.94) has a unique weak solution, and so does the elastic problem (13.65)–(13.67), (13.69), (13.86), (13.94). Moreover, the solution of the viscoelastic problem converges to the solution of the elastic problem when the viscosity converges to zero.

Example 13.21 *Bilateral contact with viscous friction.*

We consider problems with the boundary conditions

$$u_\nu = 0, \quad \boldsymbol{\sigma}_\tau = -\mu\,\|\dot{\boldsymbol{u}}_\tau\|^{p-1}\dot{\boldsymbol{u}}_\tau \quad \text{on } \Gamma_3 \times I, \tag{13.95}$$

where $\mu \ge 0$ is the coefficient of friction and $0 < p \le 1$. Here, the tangential shear is proportional to the power p of the tangential speed. Such a boundary condition arises when the contact surface is lubricated with a thin layer of non-Newtonian fluid.

It is straightforward to show that if $(\boldsymbol{u}, \boldsymbol{\sigma})$ is a pair of regular functions satisfying (13.95), then the contact condition (13.68) holds with

$$U = \{\, \boldsymbol{v} \in H_1 \,:\, v_\nu = 0 \text{ on } \Gamma_3 \,\}$$

and

$$\varphi(\boldsymbol{v}) = \frac{\mu}{p+1}\, \|\boldsymbol{v}_\tau\|^{p+1}.$$

By the definition (13.70), $U_1 = V_1$. Assuming $\mu \in L^\infty(\Gamma_3)$, we deduce from the definition (13.71) of the functional j that

$$j(\boldsymbol{v}) = \frac{1}{p+1} \int_{\Gamma_3} \mu\, \|\boldsymbol{v}_\tau\|^{p+1} da \quad \forall\, \boldsymbol{v} \in V_1.$$

Since in this case the assumptions (13.72) and (13.73) are satisfied, we may apply our results to the mechanical problem (13.64)–(13.67), (13.69), (13.95).

Example 13.22 *Viscous contact with Tresca's friction law.*

We consider the contact problem with the boundary conditions

$$\left.\begin{array}{l} -\sigma_\nu = \kappa\, |\dot{u}_\nu|^{q-1} \dot{u}_\nu, \quad \|\boldsymbol{\sigma}_\tau\| \le g, \\ \|\boldsymbol{\sigma}_\tau\| < g \;\Rightarrow\; \dot{\boldsymbol{u}}_\tau = \boldsymbol{0}, \\ \|\boldsymbol{\sigma}_\tau\| = g \;\Rightarrow\; \exists \lambda \ge 0 \text{ s.t. } \boldsymbol{\sigma}_\tau = -\lambda \dot{\boldsymbol{u}}_\tau \end{array}\right\} \quad \text{on } \Gamma_3 \times I. \tag{13.96}$$

Here $g \in L^\infty(\Gamma_3)$ and $\kappa \in L^\infty(\Gamma_3)$ are positive functions, $0 < q \le 1$. The normal contact stress depends on a power of the normal speed. Such a condition has been considered for instance in [190, 196]. We have $U = H_1$, $U_1 = V$ and

$$\varphi(\boldsymbol{v}) = \frac{\kappa}{q+1} |v_\nu|^{q+1} + g\, \|\boldsymbol{v}_\tau\|.$$

We deduce from the definition (13.71) of j that

$$j(\boldsymbol{v}) = \frac{1}{q+1} \int_{\Gamma_3} (\kappa\, |v_\nu|^{q+1} + g\, \|\boldsymbol{v}_\tau\|)\, da \quad \forall\, \boldsymbol{v} \in V.$$

Assumptions (13.72) and (13.73) hold and we conclude that our abstract results apply in the study of the frictional contact problem (13.64)–(13.67), (13.69), (13.96).

Example 13.23 *Viscous contact with power-law friction.*

In this example the normal stress is proportional to a power of the normal speed, while the tangential shear is proportional to another power of the tangential speed. We choose the following contact boundary conditions:

$$-\sigma_\nu = \kappa\, |\dot{u}_\nu|^{q-1} \dot{u}_\nu, \quad \boldsymbol{\sigma}_\tau = -\mu\, \|\dot{\boldsymbol{u}}_\tau\|^{p-1} \dot{\boldsymbol{u}}_\tau \quad \text{on } \Gamma_3 \times I. \tag{13.97}$$

Here $\mu \in L^\infty(\Gamma_3)$ and $\kappa \in L^\infty(\Gamma_3)$ are positive functions, $0 < p, q \leq 1$. We choose $U = H_1$, $U_1 = V$ and

$$\varphi(\boldsymbol{v}) = \frac{\kappa}{q+1}|v_\nu|^{q+1} + \frac{\mu}{p+1}\,\|\boldsymbol{v}_\tau\|^{p+1}.$$

We may apply Theorems 13.14 and 13.19 to Problem (13.64)–(13.67), (13.69), (13.97), since assumptions (13.72) and (13.73) hold.

Example 13.24 *Normal damped response with Tresca's friction law.*

In this problem the contact condition describes the normal damped response of a thin lubricant layer. The contact pressure depends on the velocity, but only under compression, and the contact conditions are

$$\left.\begin{array}{l} -\sigma_\nu = \kappa\,(\dot{u}_\nu)_+ + p_0, \quad \|\boldsymbol{\sigma}_\tau\| \leq g, \\ \|\boldsymbol{\sigma}_\tau\| < g \;\Rightarrow\; \dot{\boldsymbol{u}}_\tau = \boldsymbol{0}, \\ \|\boldsymbol{\sigma}_\tau\| = g \;\Rightarrow\; \exists\lambda \geq 0 \text{ s.t. } \boldsymbol{\sigma}_\tau = -\lambda\dot{\boldsymbol{u}}_\tau \end{array}\right\} \quad \text{on } \Gamma_3 \times I. \tag{13.98}$$

Here, $g \in L^\infty(\Gamma_3)$, $\kappa \in L^\infty(\Gamma_3)$ and $p_0 \in L^\infty(\Gamma_3)$ are positive functions. We choose $U = H_1$, $U_1 = V$, and

$$\varphi(\boldsymbol{v}) = \frac{\kappa}{2}\left((v_\nu)_+\right)^2 + p_0 v_\nu + g\,\|\boldsymbol{v}_\tau\|.$$

Our results apply in the study of the viscoelastic problem (13.64)–(13.67), (13.69), (13.98), since (13.72) and (13.73) are satisfied.

Example 13.25 *Normal damped response with power-law friction.*

This is a variant of Examples 13.21 and 13.24 above and the contact conditions are

$$-\sigma_\nu = \kappa\,(\dot{u}_\nu)_+ + p_0, \quad \boldsymbol{\sigma}_\tau = -\mu\,\|\dot{\boldsymbol{u}}_\tau\|^{p-1}\dot{\boldsymbol{u}}_\tau \quad \text{on } \Gamma_3 \times I. \tag{13.99}$$

Here, $\kappa \in L^\infty(\Gamma_3)$, $\mu \in L^\infty(\Gamma_3)$ and $p_0 \in L^\infty(\Gamma_3)$ are positive functions, $0 < p \leq 1$. We choose $U = H_1$, $U_1 = V$, and

$$\varphi(\boldsymbol{v}) = \frac{\kappa}{2}\left((v_\nu)_+\right)^2 + p_0 v_\nu + \frac{\mu}{p+1}\,\|\boldsymbol{v}_\tau\|^{p+1}.$$

Since assumptions (13.72) and (13.73) hold, our results apply for the study of the problem (13.64)–(13.67), (13.69), (13.99).

Remark 13.26 *In the examples above the normal pressure and tangential stress are related to the normal and tangential velocities, and the set of admissible displacement fields is a linear subspace. This is dictated by the structure of the functional φ (which depends only on the surface velocity) and by condition* (13.72)*, respectively. Important extensions of the results presented in this section would allow additional dependence of φ on the displacement (such as that in the normal compliance contact condition) or unilateral conditions for the admissible displacement fields (such as those in the Signorini contact condition).*

Bibliographical Notes

The results presented in Chapter 9 in the study of Signorini frictionless contact problems for viscoelastic materials were obtained in [25]. A different proof of Theorem 9.3 in the case of linearly Kelvin-Voigt viscoelastic materials was recently obtained in [153]. Existence and uniqueness results in the study of dynamic Signorini problems with viscoelastic materials can be found in [62, 135] for the one-dimensional case and in [123, 124, 125] for general domains. Dynamic frictional contact problems for viscoelastic materials with unilateral conditions formulated in velocities have been considered in [122, 126]; there, the coefficient of friction was assumed to depend on the solution and the solvability of the problem was proved by arguments of penalization and regularization.

Sections 10.1–10.4 of Chapter 10 were written following [39]. The results concerning the dual formulation of the quasistatic problem presented in Section 10.5 were obtained in [18]. There, a continuous dependence result of the solution with respect the friction yield limit g is also presented.

The frictional contact problem with normal compliance (11.1)–(11.7) was first studied in [178]. There, the existence and uniqueness of a weak solution (Theorem 11.12) was proved and the continuous dependence of the solution with respect to the contact conditions (Theorem 11.16) was derived. The study of the abstract problem 11.2 including Theorems 11.3, 11.9 and 11.11 as well as their applications in the study of the frictional contact problem with normal compliance (11.1)–(11.7) was carried out in [104]. The numerical results in Section 11.7 were presented following [79]. An extension of the results of this chapter in the study of viscoelastic materials with damage may be found in [98].

Dynamic contact problems with normal compliance and Coulomb friction for linearly Kelvin-Voigt viscoelastic materials have been considered in [116, 141, 142]. In [116] the coefficient of friction is assumed to depend on the slip $\boldsymbol{u}_\tau$ and the existence of a weak solution is proved by using the Galerkin method. In [142] the coefficient of friction is assumed to depend on the slip velocity $\dot{\boldsymbol{u}}_\tau$ and the weak solution of the model is obtained using a theory of evolutionary inclusions for set-valued pseudomonotone maps. A similar method is used in [143] in the study of a bilateral frictional contact problem with discontinuous coefficient of friction. The analysis of a class of implicit evolutionary variational inequalities with emphasis in the study of dynamic contact problems for linearly viscoelastic materials may be found in [54, 55].

The results presented in Sections 12.1–12.6 were obtained in [101]. The frictional contact problem with normal damped response (12.1)–(12.6) was also considered in [180]. There the variational formulations (12.63)–(12.65) and (12.69)–(12.71) for the mechanical problem were derived and the equivalence result stated in Theorem 12.18 was obtained. A direct proof of Theorem 12.16 was also presented and an extension of the results to the study of viscoelastic materials with damage was provided. The numerical results presented in Section 12.7 were obtained in [80].

The results presented in Section 13.1 and 13.2 were obtained in [15] and [190], respectively. The numerical approximation of the sliding frictional contact problem (13.30)–(13.35) was carried out in [45], using semi-discrete and fully discrete schemes. Details on surface physics involving models for wear can be found in [13, 171, 172]. The first extensive thermodynamical model for contact friction and wear has been derived in [222] and detailed in [223]. Variational formulations of contact problems involving friction and wear can be found in [224, 225]. Mathematical models from problems involving friction, adhesion and wear with application in the medical field can be found in [186, 187].

Theorem 13.9 on the existence and uniqueness of a weak solution of the frictional contact problem with normal compliance and wear was obtained in [178]. A mathematical model for a dynamic frictional contact problem with normal compliance and wear was recently considered in [144]; there, the material is linearly viscoelastic, the coefficient of friction is assumed to be discontinuous, and the existence of a weak solution to the model is obtained by using arguments of set-valued pseudomonotone operators. Finally, the results in Section 13.4 were obtained in [41].

Part IV

Contact Problems in Viscoplasticity

Chapter 14

A Signorini Contact Problem

In this chapter we study a mathematical model for the unilateral contact of an elastic-viscoplastic body with a rigid foundation. The contact is frictionless and it is modeled by the classical Signorini boundary condition. We provide two different, yet equivalent, variational formulations of the model for which we prove existence and uniqueness results. We also analyze the continuous dependence of the solution with respect to the data. Numerical analysis of the contact problem is omitted since in Chapter 16 we will provide numerical analysis of a similar yet more complicated contact problem.

14.1 Problem Statement

We assume the contact is without friction and is modeled by the well known Signorini's contact condition with a zero gap function. The viscoplastic constitutive law of the material is (6.43) (page 135). Under the previous assumptions, the classical formulation of the mechanical problem is the following:

Find a displacement field $\boldsymbol{u} : \Omega \times \overline{I} \to \mathbb{R}^d$ and a stress field $\boldsymbol{\sigma} : \Omega \times \overline{I} \to \mathbb{S}^d$ such that

$$\dot{\boldsymbol{\sigma}} = \mathcal{E}\boldsymbol{\varepsilon}(\dot{\boldsymbol{u}}) + \mathcal{G}(\boldsymbol{\sigma}, \boldsymbol{\varepsilon}(\boldsymbol{u})) \quad \text{in } \Omega \times I, \tag{14.1}$$

$$\operatorname{Div} \boldsymbol{\sigma} + \boldsymbol{f}_0 = \boldsymbol{0} \quad \text{in } \Omega \times I, \tag{14.2}$$

$$\boldsymbol{u} = \boldsymbol{0} \quad \text{on } \Gamma_1 \times I, \tag{14.3}$$

$$\boldsymbol{\sigma\nu} = \boldsymbol{f}_2 \quad \text{on } \Gamma_2 \times I, \tag{14.4}$$

$$u_\nu \le 0,\ \sigma_\nu \le 0,\ \sigma_\nu u_\nu = 0,\ \boldsymbol{\sigma}_\tau = \boldsymbol{0} \quad \text{on } \Gamma_3 \times I, \tag{14.5}$$

$$\boldsymbol{u}(0) = \boldsymbol{u}_0, \quad \boldsymbol{\sigma}(0) = \boldsymbol{\sigma}_0 \quad \text{in } \Omega. \tag{14.6}$$

Here $\boldsymbol{u}_0$ and $\boldsymbol{\sigma}_0$ represent the given initial displacement and initial stress fields, respectively.

We assume that the elasticity operator $\mathcal{E}$ and the viscoplasticity operator $\mathcal{G}$ satisfy the conditions (6.44) and (6.45). We also assume that the force and traction densities satisfy

$$\boldsymbol{f}_0 \in W^{1,\infty}(I; L^2(\Omega)^d), \quad \boldsymbol{f}_2 \in W^{1,\infty}(I; L^2(\Gamma_2)^d). \tag{14.7}$$

Denote by $\boldsymbol{f}(t)$ the element of V given by

$$(\boldsymbol{f}(t), \boldsymbol{v})_V = \int_\Omega \boldsymbol{f}_0(t) \cdot \boldsymbol{v}\, dx + \int_{\Gamma_2} \boldsymbol{f}_2(t) \cdot \boldsymbol{v}\, da \quad \forall\, \boldsymbol{v} \in V \tag{14.8}$$

for all $t \in \overline{I}$. Note that conditions (14.7) imply

$$\boldsymbol{f} \in W^{1,\infty}(I; V). \tag{14.9}$$

Due to the displacement constraints (14.3) and (14.5), we will use V_2 defined in (7.12) as the set of admissible displacement fields. We also introduce a time dependent set $\Sigma(t)$ of admissible stress fields by

$$\Sigma(t) = \{\, \boldsymbol{\tau} \in Q : (\boldsymbol{\tau}, \boldsymbol{\varepsilon}(\boldsymbol{v}))_Q \geq (\boldsymbol{f}(t), \boldsymbol{v})_V \ \forall\, \boldsymbol{v} \in V_2 \,\}, \quad t \in \overline{I}. \tag{14.10}$$

Finally, we assume that the initial data satisfy

$$\boldsymbol{u}_0 \in V_2, \quad \boldsymbol{\sigma}_0 \in \Sigma(0), \quad (\boldsymbol{\sigma}_0, \boldsymbol{\varepsilon}(\boldsymbol{u}_0))_Q = (\boldsymbol{f}(0), \boldsymbol{u}_0)_V. \tag{14.11}$$

We have the following result.

Lemma 14.1 *If $\boldsymbol{u}$ and $\boldsymbol{\sigma}$ are regular functions satisfying* (14.2)–(14.5)*, then for all $t \in \overline{I}$, $\boldsymbol{u}(t) \in V_2$, $\boldsymbol{\sigma}(t) \in \Sigma(t)$, and*

$$(\boldsymbol{\sigma}(t), \boldsymbol{\varepsilon}(\boldsymbol{v}) - \boldsymbol{\varepsilon}(\boldsymbol{u}(t)))_Q \geq (\boldsymbol{f}(t), \boldsymbol{v} - \boldsymbol{u}(t))_V \quad \forall\, \boldsymbol{v} \in V_2, \tag{14.12}$$

$$(\boldsymbol{\tau} - \boldsymbol{\sigma}(t), \boldsymbol{\varepsilon}(\boldsymbol{u}(t)))_Q \geq 0 \quad \forall\, \boldsymbol{\tau} \in \Sigma(t). \tag{14.13}$$

Proof. The boundary conditions (14.3) and (14.5) imply that $\boldsymbol{u}(t) \in V_2$.

Let $\boldsymbol{v} \in V_2$ be arbitrary and $t \in \overline{I}$. Multiplying the equilibrium equation (14.2) by $\boldsymbol{v} - \boldsymbol{u}(t)$ and integrating by parts, we have

$$(\boldsymbol{\sigma}(t), \boldsymbol{\varepsilon}(\boldsymbol{v}) - \boldsymbol{\varepsilon}(\boldsymbol{u}(t)))_Q = \int_\Omega \boldsymbol{f}_0(t) \cdot \boldsymbol{v}\, da + \int_\Gamma \boldsymbol{\sigma}(t)\boldsymbol{\nu} \cdot (\boldsymbol{v} - \boldsymbol{u}(t))\, da.$$

We split the boundary integral term into subintegrals over Γ_1, Γ_2 and Γ_3, and use the boundary conditions (14.3) and (14.4) to obtain

$$(\boldsymbol{\sigma}(t), \boldsymbol{\varepsilon}(\boldsymbol{v}) - \boldsymbol{\varepsilon}(\boldsymbol{u}(t)))_Q = (\boldsymbol{f}(t), \boldsymbol{v} - \boldsymbol{u}(t))_V + \int_{\Gamma_3} \boldsymbol{\sigma}(t)\boldsymbol{\nu} \cdot (\boldsymbol{v} - \boldsymbol{u}(t))\, da,$$

where the definition (14.8) of $\boldsymbol{f}(t)$ is used. Using the contact boundary condition (14.5) and the assumption $\boldsymbol{v} \in V_2$, we have the following relation on Γ_3:

$$\boldsymbol{\sigma}(t)\boldsymbol{\nu} \cdot (\boldsymbol{v} - \boldsymbol{u}(t)) = \sigma_\nu(t)\,(v_\nu - u_\nu(t)) = \sigma_\nu(t)\, v_\nu \geq 0.$$

Thus the inequality (14.12) holds.

Taking $\boldsymbol{v} = 2\boldsymbol{u}(t)$ and $\boldsymbol{v} = \boldsymbol{0}$ in (14.12) we obtain

$$(\boldsymbol{\sigma}(t), \boldsymbol{\varepsilon}(\boldsymbol{u}(t)))_Q = (\boldsymbol{f}(t), \boldsymbol{u}(t))_V. \tag{14.14}$$

Then from (14.12) again, we have

$$(\boldsymbol{\sigma}(t), \boldsymbol{\varepsilon}(\boldsymbol{v}))_Q \geq (\boldsymbol{f}(t), \boldsymbol{v})_V \quad \forall\, \boldsymbol{v} \in V_2.$$

Thus, $\boldsymbol{\sigma}(t) \in \Sigma(t)$. The inequality (14.13) follows easily from the definition (14.10) and the relation (14.14). ■

The previous lemma leads us to consider two weak formulation of the Signorini problem (14.1)–(14.6).

Problem 14.2 *Find a displacement field* $\boldsymbol{u} : \overline{I} \to V$ *and a stress field* $\boldsymbol{\sigma} : \overline{I} \to Q_1$ *such that*

$$\dot{\boldsymbol{\sigma}}(t) = \mathcal{E}\boldsymbol{\varepsilon}(\dot{\boldsymbol{u}}(t)) + \mathcal{G}(\boldsymbol{\sigma}(t), \boldsymbol{\varepsilon}(\boldsymbol{u}(t))) \quad \text{a.e. } t \in I, \tag{14.15}$$

$$\boldsymbol{u}(t) \in V_2, \quad (\boldsymbol{\sigma}(t), \boldsymbol{\varepsilon}(\boldsymbol{v}) - \boldsymbol{\varepsilon}(\boldsymbol{u}(t)))_Q \geq (\boldsymbol{f}(t), \boldsymbol{v} - \boldsymbol{u}(t))_V \quad \forall\, \boldsymbol{v} \in V_2,\ t \in \overline{I}, \tag{14.16}$$

$$\boldsymbol{u}(0) = \boldsymbol{u}_0, \quad \boldsymbol{\sigma}(0) = \boldsymbol{\sigma}_0. \tag{14.17}$$

Problem 14.3 *Find a displacement field* $\boldsymbol{u} : \overline{I} \to V$ *and a stress field* $\boldsymbol{\sigma} : \overline{I} \to Q_1$ *such that*

$$\dot{\boldsymbol{\sigma}}(t) = \mathcal{E}\boldsymbol{\varepsilon}(\dot{\boldsymbol{u}}(t)) + \mathcal{G}(\boldsymbol{\sigma}(t), \boldsymbol{\varepsilon}(\boldsymbol{u}(t))) \quad \text{a.e. } t \in I, \tag{14.18}$$

$$\boldsymbol{\sigma}(t) \in \Sigma(t), \quad (\boldsymbol{\tau} - \boldsymbol{\sigma}(t), \boldsymbol{\varepsilon}(\boldsymbol{u}(t)))_Q \geq 0 \quad \forall\, \boldsymbol{\tau} \in \Sigma(t),\ t \in \overline{I}, \tag{14.19}$$

$$\boldsymbol{u}(0) = \boldsymbol{u}_0, \quad \boldsymbol{\sigma}(0) = \boldsymbol{\sigma}_0. \tag{14.20}$$

We remark that Problems 14.2 and 14.3 are formally equivalent to the mechanical problem (14.1)–(14.6). Indeed, if $(\boldsymbol{u}, \boldsymbol{\sigma})$ represents a regular solution of the variational problem 14.2 or 14.3, then it can be shown that $(\boldsymbol{u}, \boldsymbol{\sigma})$ satisfies (14.1)–(14.6).

Well-posedness of Problems 14.2 and 14.3 is the topic of the next section.

14.2 Existence and Uniqueness Results

We first consider the existence and uniqueness for Problem 14.2.

Theorem 14.4 *Assume* (6.44), (6.45), (14.7) *and* (14.11). *Then there exists a unique solution* $(\boldsymbol{u}, \boldsymbol{\sigma})$ *of Problem 14.2. Moreover, the solution satisfies*

$$\boldsymbol{u} \in W^{1,\infty}(I; V), \quad \boldsymbol{\sigma} \in W^{1,\infty}(I; Q_1). \tag{14.21}$$

The proof of Theorem 14.4 is carried out in several steps. In the first step, we assume the viscoplastic part of the stress field is given as $\boldsymbol{\eta} \in L^\infty(I; Q)$. We assume in what follows that (6.44), (6.45), (14.7) and (14.11) hold and define a function $\boldsymbol{z}_\eta \in W^{1,\infty}(I; Q)$ by

$$\boldsymbol{z}_\eta(t) = \int_0^t \boldsymbol{\eta}(s)\, ds + \boldsymbol{z}_0, \tag{14.22}$$

$$\boldsymbol{z}_0 = \boldsymbol{\sigma}_0 - \mathcal{E}\boldsymbol{\varepsilon}(\boldsymbol{u}_0). \tag{14.23}$$

We consider the following auxiliary variational problem.

Problem 14.5 *Find* $\boldsymbol{u}_\eta : \overline{I} \to V_2$ *and* $\boldsymbol{\sigma}_\eta : \overline{I} \to Q_1$ *such that*

$$\boldsymbol{\sigma}_\eta(t) = \mathcal{E}\boldsymbol{\varepsilon}(\boldsymbol{u}_\eta(t)) + \boldsymbol{z}_\eta(t), \tag{14.24}$$

$$(\boldsymbol{\sigma}_\eta(t), \boldsymbol{\varepsilon}(\boldsymbol{v}) - \boldsymbol{\varepsilon}(\boldsymbol{u}_\eta(t)))_Q \geq (\boldsymbol{f}(t), \boldsymbol{v} - \boldsymbol{u}_\eta(t))_V \quad \forall \boldsymbol{v} \in V_2 \tag{14.25}$$

for all $t \in \overline{I}$.

This problem is well-posed.

Lemma 14.6 *The variational problem* 14.5 *has a unique solution* $(\boldsymbol{u}_\eta, \boldsymbol{\sigma}_\eta)$ *such that* $\boldsymbol{u}_\eta \in W^{1,\infty}(I;V)$ *and* $\boldsymbol{\sigma}_\eta \in H^1(I;Q_1)$. *Moreover, the solution satisfies*

$$\boldsymbol{u}_\eta(0) = \boldsymbol{u}_0, \quad \boldsymbol{\sigma}_\eta(0) = \boldsymbol{\sigma}_0. \tag{14.26}$$

Proof. Let $t \in \overline{I}$. It follows from Corollary 4.3 that there exists a unique element $\boldsymbol{u}_\eta(t)$ such that

$$\boldsymbol{u}_\eta(t) \in V_2, \quad (\mathcal{E}\boldsymbol{\varepsilon}(\boldsymbol{u}_\eta(t)), \boldsymbol{\varepsilon}(\boldsymbol{v}) - \boldsymbol{\varepsilon}(\boldsymbol{u}_\eta(t)))_Q + (\boldsymbol{z}_\eta(t), \boldsymbol{\varepsilon}(\boldsymbol{v}) - \boldsymbol{\varepsilon}(\boldsymbol{u}_\eta(t)))_Q \geq (\boldsymbol{f}(t), \boldsymbol{v} - \boldsymbol{u}_\eta(t))_V \quad \forall \boldsymbol{v} \in V_2. \tag{14.27}$$

Defining $\boldsymbol{\sigma}_\eta(t)$ by (14.24), we deduce that $(\boldsymbol{u}_\eta, \boldsymbol{\sigma}_\eta)$ solves Problem 14.5. By taking $\boldsymbol{v} = \boldsymbol{u}_\eta(t) \pm \boldsymbol{\varphi}$ with $\boldsymbol{\varphi} \in C_0^\infty(\Omega)^d$ arbitrary, from the inequality (14.25) we obtain the following:

$$\operatorname{Div} \boldsymbol{\sigma}_\eta(t) + \boldsymbol{f}_0(t) = \boldsymbol{0} \quad \text{a.e. in } \Omega. \tag{14.28}$$

Hence $\boldsymbol{\sigma}_\eta(t) \in Q_1$. The initial conditions (14.26) follows from (14.11), (14.22), (14.23) and the uniqueness of the solution of Problem 14.5 for $t = 0$.

Let $t_1, t_2 \in \overline{I}$. Using the inequality (14.27) at $t = t_1$, t_2, and the assumptions (6.44) on the operator $\mathcal{E}$, we obtain

$$\|\boldsymbol{u}_\eta(t_1) - \boldsymbol{u}_\eta(t_2)\|_V \leq c\,(\|\boldsymbol{f}(t_1) - \boldsymbol{f}(t_2)\|_V + \|\boldsymbol{z}_\eta(t_1) - \boldsymbol{z}_\eta(t_2)\|_Q). \tag{14.29}$$

By the definition (14.24) of $\boldsymbol{\sigma}_\eta$, from (14.28) and (14.29) it follows that

$$\|\boldsymbol{\sigma}_\eta(t_1) - \boldsymbol{\sigma}_\eta(t_2)\|_{Q_1} \leq c\,(\|\boldsymbol{f}(t_1) - \boldsymbol{f}(t_2)\|_V + \|\boldsymbol{z}_\eta(t_1) - \boldsymbol{z}_\eta(t_2)\|_Q). \tag{14.30}$$

Since $\boldsymbol{f} \in W^{1,\infty}(I;V)$ and $\boldsymbol{z}_\eta \in W^{1,\infty}(I;Q)$, we have $\boldsymbol{u}_\eta \in W^{1,\infty}(I;V)$ and $\boldsymbol{\sigma}_\eta \in W^{1,\infty}(I;Q_1)$ from (14.29) and (14.30). This proves the existence part of the lemma. The uniqueness part follows from the unique solvability of the variational inequality (14.27). ■

We consider next an operator $\Lambda : L^\infty(I;Q) \to L^\infty(I;Q)$ defined by

$$\Lambda\boldsymbol{\eta} = \mathcal{G}(\boldsymbol{\sigma}_\eta, \boldsymbol{\varepsilon}(\boldsymbol{u}_\eta)) \quad \forall \boldsymbol{\eta} \in L^\infty(I;Q), \tag{14.31}$$

where $(\boldsymbol{u}_\eta, \boldsymbol{\sigma}_\eta)$ is the solution of Problem 14.5.

Lemma 14.7 *The operator* Λ *has a unique fixed point* $\boldsymbol{\eta}^* \in L^\infty(I;Q)$.

Proof. Let $t \in \overline{I}$, $\boldsymbol{\eta}_1, \boldsymbol{\eta}_2 \in L^\infty(I;Q)$ and let $\boldsymbol{u}_i = \boldsymbol{u}_{\eta_i}$, $\boldsymbol{\sigma}_i = \boldsymbol{\sigma}_{\eta_i}$ for $i = 1, 2$. Using the assumptions (6.45) on the operator $\mathcal{G}$, we have

$$\|\Lambda\boldsymbol{\eta}_1(t) - \Lambda\boldsymbol{\eta}_2(t)\|_Q \leq c\left(\|\boldsymbol{\sigma}_1(t) - \boldsymbol{\sigma}_2(t)\|_Q + \|\boldsymbol{u}_1(t) - \boldsymbol{u}_2(t)\|_V\right).$$

By arguments similar to those leading to the inequalities (14.29) and (14.30) we find that

$$\|\boldsymbol{u}_1(t) - \boldsymbol{u}_2(t)\|_V \leq c\,\|\boldsymbol{z}_1(t) - \boldsymbol{z}_2(t)\|_Q,$$
$$\|\boldsymbol{\sigma}_1(t) - \boldsymbol{\sigma}_2(t)\|_{Q_1} \leq c\,\|\boldsymbol{z}_1(t) - \boldsymbol{z}_2(t)\|_Q.$$

Thus,

$$\|\Lambda\boldsymbol{\eta}_1(t) - \Lambda\boldsymbol{\eta}_2(t)\|_Q \leq c\,\|\boldsymbol{z}_1(t) - \boldsymbol{z}_2(t)\|_Q.$$

Recalling the definitions (14.22), (14.23), we can rewrite the inequality as

$$\|\Lambda\boldsymbol{\eta}_1(t) - \Lambda\boldsymbol{\eta}_2(t)\|_Q \leq c\int_0^t \|\boldsymbol{\eta}_1(s) - \boldsymbol{\eta}_2(s)\|_Q ds.$$

Similar to the proof of Lemma 13.12, we know that for some integer n large enough, Λ^n is a contraction on the Banach space $L^\infty(I;Q)$. Applying Corollary 1.63 (page 23), we know that Λ has a unique fixed point $\boldsymbol{\eta}^* \in L^\infty(I;Q)$. ■

Now we are ready to prove Theorem 14.4.

Proof of Theorem 14.4.
Existence. Let $\boldsymbol{\eta}^* \in L^\infty(I;Q)$ be the fixed point of Λ and let $(\boldsymbol{u}_{\eta^*}, \boldsymbol{\sigma}_{\eta^*})$ be the solution of Problem 14.5 for $\boldsymbol{\eta} = \boldsymbol{\eta}^*$. We differentiate (14.24) with respect to t and use the definition (14.22) to obtain

$$\dot{\boldsymbol{\sigma}}_{\eta^*}(t) = \mathcal{E}\boldsymbol{\varepsilon}(\dot{\boldsymbol{u}}_{\eta^*}(t)) + \boldsymbol{\eta}^*(t) \quad \text{a.e. } t \in I.$$

Since

$$\boldsymbol{\eta}^*(t) = \Lambda\boldsymbol{\eta}^*(t) = \mathcal{G}(\boldsymbol{\sigma}_{\eta^*}(t), \boldsymbol{\varepsilon}(\boldsymbol{u}_{\eta^*}(t))) \quad \forall\, t \in \overline{I},$$

it follows that $(\boldsymbol{u}_{\eta^*}, \boldsymbol{\sigma}_{\eta^*})$ satisfies (14.15). This, together with (14.25) and (14.26), implies that $(\boldsymbol{u}_{\eta^*}, \boldsymbol{\sigma}_{\eta^*})$ is a solution of Problem 14.2 with regularity (14.21).
Uniqueness. Let $(\boldsymbol{u}, \boldsymbol{\sigma})$ be a solution of Problem 14.2 with regularity (14.21). Define $\boldsymbol{\eta} \in L^\infty(I;Q)$ by

$$\boldsymbol{\eta} = \mathcal{G}(\boldsymbol{\sigma}, \boldsymbol{\varepsilon}(\boldsymbol{u})).$$

We denote by $\boldsymbol{z}_\eta$ the function given by (14.22), (14.23). Then $(\boldsymbol{u}, \boldsymbol{\sigma})$ is a solution of Problem 14.5. By the uniqueness of a solution of Problem 14.5, we have

$$\boldsymbol{u} = \boldsymbol{u}_\eta, \quad \boldsymbol{\sigma} = \boldsymbol{\sigma}_\eta.$$

Then we conclude that $\Lambda\boldsymbol{\eta} = \boldsymbol{\eta}$ and by the uniqueness of the fixed point of Λ, it follows that $\boldsymbol{\eta} = \boldsymbol{\eta}^*$. Thus, a solution of Problem 14.2 is unique. ■

Remark 14.8 *A careful examination of the previous arguments shows that if the space $W^{1,\infty}$ in the condition* (14.7) *is replaced by $W^{1,p}$, $1 \le p < \infty$, then Theorem 14.4 is still valid with the conclusion that Problem 14.2 has a unique solution $(\boldsymbol{u}, \boldsymbol{\sigma}) \in W^{1,p}(I; V_2 \times Q_1)$.*

We now turn to an existence and uniqueness result for Problem 14.3.

Theorem 14.9 *Assume* (6.44), (6.45), (14.7) *and* (14.11). *Then there exists a unique solution $(\boldsymbol{u}, \boldsymbol{\sigma})$ to Problem 14.3. Moreover, the solution has the regularity* (14.21).

To prove Theorem 14.9 we suppose in what follows that the assumptions (6.44), (6.45), (14.7) and (14.11) hold. We first observe that the variational inequality (14.19) is defined over the time-dependent convex set $\Sigma(t)$. Let us use a change of variable to convert the variational inequality into one associated with a fixed convex set. To this end, let us define $\boldsymbol{\sigma_f} : \overline{I} \to Q$ by $\boldsymbol{\sigma_f}(t) = \boldsymbol{\varepsilon}(\boldsymbol{f}(t))$. Then, using the regularity assumption (14.7), we can show that $\boldsymbol{\sigma_f} \in W^{1,\infty}(\overline{I}; Q_1)$. We can now express $\Sigma(t)$ as

$$\Sigma(t) = \Sigma_0 + \{\boldsymbol{\sigma_f}(t)\} \quad \forall\, t \in \overline{I},$$

where

$$\Sigma_0 = \{\, \boldsymbol{\tau} \in Q : (\boldsymbol{\tau}, \boldsymbol{\varepsilon}(\boldsymbol{v}))_Q \ge 0 \;\forall\, \boldsymbol{v} \in V_2 \,\} \tag{14.32}$$

is time independent. Introduce

$$\bar{\boldsymbol{\sigma}} = \boldsymbol{\sigma} - \boldsymbol{\sigma_f}, \quad \bar{\boldsymbol{\sigma}}_0 = \boldsymbol{\sigma}_0 - \boldsymbol{\sigma_f}(0), \tag{14.33}$$

and consider the following variational problem.

Problem 14.10 *Find $\boldsymbol{u} : \overline{I} \to V$ and $\bar{\boldsymbol{\sigma}} : \overline{I} \to Q_1$ such that*

$$\begin{aligned} \boldsymbol{\varepsilon}(\dot{\boldsymbol{u}}(t)) = \mathcal{E}^{-1}\dot{\bar{\boldsymbol{\sigma}}}(t) - \mathcal{E}^{-1}\mathcal{G}(\bar{\boldsymbol{\sigma}}(t) + \boldsymbol{\sigma_f}(t), \boldsymbol{\varepsilon}(\boldsymbol{u}(t)))& \\ + \mathcal{E}^{-1}\dot{\boldsymbol{\sigma}}_{\boldsymbol{f}}(t) \quad \text{a.e. } t \in I,& \qquad (14.34) \\ \bar{\boldsymbol{\sigma}}(t) \in \Sigma_0, \quad (\boldsymbol{\tau} - \bar{\boldsymbol{\sigma}}(t), \boldsymbol{\varepsilon}(\boldsymbol{u}(t)))_Q \ge 0 \quad \forall\, \boldsymbol{\tau} \in \Sigma_0,\ t \in \overline{I},& \qquad (14.35) \\ \boldsymbol{u}(0) = \boldsymbol{u}_0 \quad \bar{\boldsymbol{\sigma}}(0) = \bar{\boldsymbol{\sigma}}_0.& \qquad (14.36) \end{aligned}$$

It is straightforward to show the following result.

Lemma 14.11 *The couple $(\boldsymbol{u}, \boldsymbol{\sigma})$ is a solution to Problem 14.3 with regularity $\boldsymbol{u} \in W^{1,\infty}(I; V)$, $\boldsymbol{\sigma} \in W^{1,\infty}(I; Q_1)$ iff the couple $(\boldsymbol{u}, \bar{\boldsymbol{\sigma}})$ is a solution to Problem 14.10 with regularity $\boldsymbol{u} \in W^{1,\infty}(I; V)$, $\bar{\boldsymbol{\sigma}} \in W^{1,\infty}(I; Q_1)$.*

We turn now to Problem 14.10 and show its well-posedness by using again a fixed point argument. Let $\boldsymbol{\eta} \in L^\infty(I; Q)$ and define $\boldsymbol{z}_\eta \in W^{1,\infty}(I; Q)$ by

$$\boldsymbol{z}_\eta(t) = \int_0^t \boldsymbol{\eta}(s)\, ds + \boldsymbol{\varepsilon}(\boldsymbol{u}_0) - \mathcal{E}^{-1}\boldsymbol{\sigma}_0 \quad \forall\, t \in \overline{I}. \tag{14.37}$$

We consider the following auxiliary problem.

Problem 14.12 *Find* $\boldsymbol{u}_\eta : \overline{I} \to V$ *and* $\boldsymbol{\sigma}_\eta : \overline{I} \to Q_1$ *such that*

$$\boldsymbol{\varepsilon}(\boldsymbol{u}_\eta(t)) = \mathcal{E}^{-1}\boldsymbol{\sigma}_\eta(t) + \boldsymbol{z}_\eta(t) + \mathcal{E}^{-1}\boldsymbol{\sigma}_{\boldsymbol{f}}(t), \tag{14.38}$$

$$\boldsymbol{\sigma}_\eta(t) \in \Sigma_0, \quad (\boldsymbol{\tau} - \boldsymbol{\sigma}_\eta(t), \boldsymbol{\varepsilon}(\boldsymbol{u}_\eta(t)))_Q \geq 0 \quad \forall \boldsymbol{\tau} \in \Sigma_0, \tag{14.39}$$

for all $t \in \overline{I}$.

We have the following result.

Lemma 14.13 *Problem 14.12 has a unique solution* $(\boldsymbol{u}_\eta, \boldsymbol{\sigma}_\eta)$ *such that* $\boldsymbol{u}_\eta \in W^{1,\infty}(I;V)$ *and* $\boldsymbol{\sigma}_\eta \in W^{1,\infty}(I;Q_1)$. *Moreover, the solution satisfies*

$$\boldsymbol{u}_\eta(0) = \boldsymbol{u}_0, \quad \boldsymbol{\sigma}_\eta(0) = \bar{\boldsymbol{\sigma}}_0. \tag{14.40}$$

Proof. Let $t \in \overline{I}$. Since Σ_0 is a nonempty closed convex set in Q, using Corollary 4.3 we obtain the existence and uniqueness of an element $\boldsymbol{\sigma}_\eta(t)$ which satisfies

$$\boldsymbol{\sigma}_\eta(t) \in \Sigma_0, \ (\boldsymbol{\tau} - \boldsymbol{\sigma}_\eta(t), \mathcal{E}^{-1}\boldsymbol{\sigma}_\eta(t) + \boldsymbol{z}_\eta(t) + \mathcal{E}^{-1}\boldsymbol{\sigma}_{\boldsymbol{f}}(t))_Q \geq 0 \ \ \forall \boldsymbol{\tau} \in \Sigma_0. \tag{14.41}$$

This inequality implies

$$(\boldsymbol{\tau} - \boldsymbol{\sigma}_\eta(t), \boldsymbol{\varepsilon}_\eta(t))_Q \geq 0 \quad \forall \boldsymbol{\tau} \in \Sigma_0, \tag{14.42}$$

where $\boldsymbol{\varepsilon}_\eta(t) \in Q$ is defined by

$$\boldsymbol{\varepsilon}_\eta(t) = \mathcal{E}^{-1}\boldsymbol{\sigma}_\eta(t) + \boldsymbol{z}_\eta(t) + \mathcal{E}^{-1}\boldsymbol{\sigma}_{\boldsymbol{f}}(t). \tag{14.43}$$

Let us introduce the space

$$\mathcal{V} = \{\, \boldsymbol{\tau} \in Q : \operatorname{Div} \boldsymbol{\tau} = \boldsymbol{0} \text{ in } \Omega, \ \boldsymbol{\tau}\boldsymbol{\nu} = \boldsymbol{0} \text{ on } \Gamma_2 \cup \Gamma_3 \,\}.$$

Since $\boldsymbol{\sigma}_\eta(t) \in \Sigma_0$ it follows that $\boldsymbol{\sigma}_\eta(t) \pm \boldsymbol{z} \in \Sigma_0$ for all $\boldsymbol{z} \in \mathcal{V}$. Thus, taking $\boldsymbol{\tau} = \boldsymbol{\sigma}_\eta(t) \pm \boldsymbol{z}$ in (14.42) we obtain that $(\boldsymbol{z}, \boldsymbol{\varepsilon}_\eta(t))_Q = 0$ for all $\boldsymbol{z} \in \mathcal{V}$. Since the orthogonal complement of $\mathcal{V}$ in Q is the space $\boldsymbol{\varepsilon}(V)$ (see Theorem 7.5 on page 146), we deduce the existence of an element $\boldsymbol{u}_\eta(t) \in V$ such that

$$\boldsymbol{\varepsilon}_\eta(t) = \boldsymbol{\varepsilon}(\boldsymbol{u}_\eta(t)). \tag{14.44}$$

Turn now to the regularity of $\boldsymbol{\sigma}_\eta$ and $\boldsymbol{u}_\eta$. From the inequality (14.41) and the assumptions (6.44) on $\mathcal{E}$ it follows that for $t_1, t_2 \in \overline{I}$,

$$\|\boldsymbol{\sigma}_\eta(t_1) - \boldsymbol{\sigma}_\eta(t_2)\|_Q \leq c\,(\|\boldsymbol{z}_\eta(t_1) - \boldsymbol{z}_\eta(t_2)\|_Q + \|\boldsymbol{\sigma}_{\boldsymbol{f}}(t_1) - \boldsymbol{\sigma}_{\boldsymbol{f}}(t_2)\|_Q).$$

The condition $\boldsymbol{\sigma}_\eta(t) \in \Sigma_0$ yields $\operatorname{Div} \boldsymbol{\sigma}_\eta(t) = \boldsymbol{0}$ a.e. in Ω, for all $t \in \overline{I}$. Thus,

$$\|\boldsymbol{\sigma}_\eta(t_1) - \boldsymbol{\sigma}_\eta(t_2)\|_{Q_1} \leq c\,(\|\boldsymbol{z}_\eta(t_1) - \boldsymbol{z}_\eta(t_2)\|_Q + \|\boldsymbol{\sigma}_{\boldsymbol{f}}(t_1) - \boldsymbol{\sigma}_{\boldsymbol{f}}(t_2)\|_Q).$$

Since $\boldsymbol{z}_\eta \in W^{1,\infty}(I;Q)$ and $\boldsymbol{\sigma_f} \in W^{1,\infty}(I;Q)$, we have $\boldsymbol{\sigma}_\eta \in W^{1,\infty}(I;Q_1)$. Then from the definition (14.43), we have $\boldsymbol{\varepsilon}_\eta \in W^{1,\infty}(I;Q)$. Applying Korn's inequality (7.8) we deduce that $\boldsymbol{u}_\eta \in W^{1,\infty}(I;V)$.

Finally, using (14.41), (14.43) and (14.44) we see that the couple $(\boldsymbol{u}_\eta, \boldsymbol{\sigma}_\eta)$ is a solution to Problem 14.12. The uniqueness of this solution follows from the uniqueness of the solution of the variational inequality (14.41) and Korn's inequality (7.8) (page 144). The initial conditions (14.40) follow from (14.11), (14.33) and the uniqueness of the solution to Problem 14.12 at $t = 0$. ■

We consider next an operator $\Lambda : L^\infty(I;Q) \to L^\infty(I;Q)$ defined by

$$\Lambda\boldsymbol{\eta} = -\mathcal{E}^{-1}\mathcal{G}(\boldsymbol{\sigma}_\eta + \boldsymbol{\sigma_f}, \boldsymbol{\varepsilon}(\boldsymbol{u}_\eta)) \tag{14.45}$$

where, for any $\boldsymbol{\eta} \in L^\infty(I;Q)$, $(\boldsymbol{u}_\eta, \boldsymbol{\sigma}_\eta)$ is the solution of Problem 14.12.

Using arguments similar to those used in the proof of Lemma 14.7 we obtain the following result.

Lemma 14.14 *The operator Λ has a unique fixed point $\boldsymbol{\eta}^* \in L^\infty(I;Q)$.*

Now we are ready to prove Theorem 14.9.

Proof of Theorem 14.9.
Existence. Let $\boldsymbol{\eta}^* \in L^\infty(I;Q)$ be the fixed point of Λ and let $(\boldsymbol{u}_{\eta^*}, \boldsymbol{\sigma}_{\eta^*})$ be the solution of Problem 14.12 for $\boldsymbol{\eta} = \boldsymbol{\eta}^*$. From (14.45) and Lemma 14.13 it follows that $(\boldsymbol{u}_{\eta^*}, \boldsymbol{\sigma}_{\eta^*})$ is a solution of Problem 14.10 with regularity $\boldsymbol{u}_{\eta^*} \in W^{1,\infty}(I;V)$, $\boldsymbol{\sigma}_{\eta^*} \in W^{1,\infty}(I;Q_1)$. It follows now from Lemma 14.11 that the couple $(\boldsymbol{u}_{\eta^*}, \boldsymbol{\sigma}_{\eta^*} + \boldsymbol{\sigma_f})$ is a solution to Problem 14.3 with regularity (14.21).
Uniqueness. The uniqueness part follows from the uniqueness of the fixed point of the operator Λ defined by (14.45) and Lemma 14.11. It can also be deduced directly from (14.18)–(14.20) using the assumptions (6.44) and (6.45) on the operators $\mathcal{E}$ and $\mathcal{G}$, and a Gronwall-type argument. ■

14.3 Some Properties of the Solution

We start this section with the study of the link between Problems 14.2 and 14.3 which represent two different variational formulations of the mechanical problem (14.1)–(14.6).

We have the following equivalence result.

Theorem 14.15 *Assume the conditions* (6.44), (6.45), (14.7) *and* (14.11). *Let $\boldsymbol{u} \in W^{1,\infty}(I;V)$, $\boldsymbol{\sigma} \in W^{1,\infty}(I;Q_1)$. Then the couple $(\boldsymbol{u}, \boldsymbol{\sigma})$ is a solution of Problem 14.2 iff it is a solution of Problem 14.3.*

Proof. First we suppose that $(\boldsymbol{u}, \boldsymbol{\sigma})$ is a solution of Problem 14.2. Let $t \in \overline{I}$. Taking $\boldsymbol{v} = 2\boldsymbol{u}(t) \in V_2$ and $\boldsymbol{v} = \boldsymbol{0} \in V_2$ in the variational inequality (14.16) we obtain

$$(\boldsymbol{\sigma}(t), \boldsymbol{\varepsilon}(\boldsymbol{u}(t)))_Q = (\boldsymbol{f}(t), \boldsymbol{u}(t))_V. \tag{14.46}$$

It is easily seen that the variational inequality (14.16) is equivalent to (14.46) and

$$(\boldsymbol{\sigma}(t), \boldsymbol{\varepsilon}(\boldsymbol{v}))_Q \geq (\boldsymbol{f}(t), \boldsymbol{v})_V \quad \forall \boldsymbol{v} \in V_2.$$

So $\boldsymbol{\sigma} \in \Sigma(t)$. The inequality in (14.19) follows now from the definition (14.10) of the set $\Sigma(t)$ and (14.46). We conclude that $(\boldsymbol{u}, \boldsymbol{\sigma})$ is a solution of Problem 14.3.

Conversely, let us suppose that $(\boldsymbol{u}, \boldsymbol{\sigma})$ is a solution of Problem 14.3. We shall first prove that $\boldsymbol{u}(t) \in V_2$ for all $t \in \overline{I}$. The argument below is similar to that used in the proof of Theorem 9.11 on page 203. For reader's convenience, we provide the detailed argument below. Arguing by contradiction, let us assume that there exists $t \in \overline{I}$ such that $\boldsymbol{u}(t) \notin V_2$. Denote by $\mathcal{P}\boldsymbol{u}(t)$ the projection of $\boldsymbol{u}(t)$ on the nonempty closed convex set $V_2 \subset V$. We have

$$\begin{aligned}(\mathcal{P}\boldsymbol{u}(t) - \boldsymbol{u}(t), \boldsymbol{v})_V &\geq (\mathcal{P}\boldsymbol{u}(t) - \boldsymbol{u}(t), \mathcal{P}\boldsymbol{u}(t))_V \\ &> (\mathcal{P}\boldsymbol{u}(t) - \boldsymbol{u}(t), \boldsymbol{u}(t))_V \quad \forall \boldsymbol{v} \in V_2.\end{aligned}$$

Therefore, there exists $\alpha \in \mathbb{R}$ such that

$$(\mathcal{P}\boldsymbol{u}(t) - \boldsymbol{u}(t), \boldsymbol{v})_V > \alpha > (\mathcal{P}\boldsymbol{u}(t) - \boldsymbol{u}(t), \mathcal{P}\boldsymbol{u}(t))_V \quad \forall \boldsymbol{v} \in V_2. \tag{14.47}$$

Denote $\tilde{\boldsymbol{\tau}}(t) = \boldsymbol{\varepsilon}(\mathcal{P}\boldsymbol{u}(t) - \boldsymbol{u}(t)) \in Q$. Realling the definition of the inner product in V, we can rewrite (14.47) as

$$(\tilde{\boldsymbol{\tau}}(t), \boldsymbol{\varepsilon}(\boldsymbol{v}))_Q > \alpha > (\tilde{\boldsymbol{\tau}}(t), \boldsymbol{\varepsilon}(\boldsymbol{u}(t)))_Q \quad \forall \boldsymbol{v} \in V_2. \tag{14.48}$$

Taking $\boldsymbol{v} = \boldsymbol{0}$ in (14.48) we deduce that

$$\alpha < 0. \tag{14.49}$$

Assume there exists $\boldsymbol{w} \in V_2$ such that

$$(\tilde{\boldsymbol{\tau}}(t), \boldsymbol{\varepsilon}(\boldsymbol{w}))_Q < 0.$$

Since $\lambda\, \boldsymbol{w} \in V_2$ for $\lambda \geq 0$, it follows from (14.48) that

$$\lambda\, (\tilde{\boldsymbol{\tau}}(t), \boldsymbol{\varepsilon}(\boldsymbol{w}))_Q > \alpha \quad \forall \lambda \geq 0.$$

Passing to the limit $\lambda \to +\infty$, we obtain $\alpha \leq -\infty$, which contradicts to the assumption $\alpha \in \mathbb{R}$. We conclude that

$$(\tilde{\boldsymbol{\tau}}(t), \boldsymbol{\varepsilon}(\boldsymbol{w}))_Q \geq 0 \quad \forall \boldsymbol{w} \in V_2,$$

i.e. $\tilde{\boldsymbol{\tau}}(t) \in \Sigma_0$ (see (14.32) for the definition of Σ_0). Then $\tilde{\boldsymbol{\tau}}(t) + \boldsymbol{\sigma_f}(t) \in \Sigma(t)$. Use the inequality (14.19) to obtain

$$(\tilde{\boldsymbol{\tau}}(t), \boldsymbol{\varepsilon}(\boldsymbol{u}(t)))_Q \geq (\boldsymbol{\sigma}(t) - \boldsymbol{\sigma_f}(t), \boldsymbol{\varepsilon}(\boldsymbol{u}(t)))_Q.$$

By (14.49), (14.48), we have

$$(\tilde{\boldsymbol{\tau}}(t), \boldsymbol{\varepsilon}(\boldsymbol{u}(t)))_Q < 0.$$

Hence,

$$(\boldsymbol{\sigma}(t) - \boldsymbol{\sigma}_{\boldsymbol{f}}(t), \boldsymbol{\varepsilon}(\boldsymbol{u}(t)))_Q < 0. \tag{14.50}$$

Moreover, letting $\boldsymbol{\tau} = 2(\boldsymbol{\sigma}(t) - \boldsymbol{\sigma}_{\boldsymbol{f}}(t)) \in \Sigma_0$ in (14.35) we obtain

$$(\boldsymbol{\sigma}(t) - \boldsymbol{\sigma}_{\boldsymbol{f}}(t), \boldsymbol{\varepsilon}(\boldsymbol{u}(t)))_Q \geq 0. \tag{14.51}$$

We note that inequalities (14.50) and (14.51) contradict to each other. Therefore, $\boldsymbol{u}(t) \in V_2$.

Since $\tilde{\boldsymbol{\sigma}}(t) \in \Sigma(t)$, we can take $\boldsymbol{\tau} = \tilde{\boldsymbol{\sigma}}_{\boldsymbol{f}}(t)$ in the variational inequality (14.19) to find

$$(\boldsymbol{f}(t), \boldsymbol{u}(t))_V \geq (\boldsymbol{\sigma}(t), \boldsymbol{\varepsilon}(\boldsymbol{u}(t)))_Q.$$

Since $\boldsymbol{\sigma}(t) \in \Sigma(t)$ and $\boldsymbol{u}(t) \in V_2$, by the definition of the set $\Sigma(t)$ we have

$$(\boldsymbol{\sigma}(t), \boldsymbol{\varepsilon}(\boldsymbol{u}(t)))_Q \geq (\boldsymbol{f}(t), \boldsymbol{u}(t))_V.$$

Combining these inequalities, we obtain

$$(\boldsymbol{\sigma}(t), \boldsymbol{\varepsilon}(\boldsymbol{u}(t)))_Q = (\boldsymbol{f}(t), \boldsymbol{u}(t))_V. \tag{14.52}$$

The inequality in (14.16) follows now from (14.10) and (14.52). We conclude that the couple $(\boldsymbol{u}, \boldsymbol{\sigma})$ is a solution of Problem 14.2. ∎

Theorems 14.4 and 14.9 state the unique solvability of Problems 14.2 and 14.3, respectively, while Theorem 14.15 expresses the equivalence of these variational problems. From these theorems we conclude that the mechanical problem (14.1)–(14.6) has a unique *weak solution* which solves both Problems 14.2 and 14.3.

We now provide a continuous dependence result of the solution on the input data.

Theorem 14.16 *Assume* (6.44) *and* (6.45). *For* $i = 1, 2$, *let* $(\boldsymbol{u}_i, \boldsymbol{\sigma}_i)$ *be the solution of Problems 14.2 or 14.3 for the data* $\boldsymbol{f}_{0i}$, $\boldsymbol{f}_{2i}$, $\boldsymbol{u}_{0i}$, $\boldsymbol{\sigma}_{0i}$ *which satisfy* (14.7), (14.11). *Then there exists* $c > 0$ *depending only on* Ω, Γ_1, $\mathcal{E}$, $\mathcal{G}$ *and* T *such that*

$$\begin{aligned}
&\|\boldsymbol{u}_1 - \boldsymbol{u}_2\|_{L^\infty(I;V)} + \|\boldsymbol{\sigma}_1 - \boldsymbol{\sigma}_2\|_{L^\infty(I;Q_1)} \\
&\quad \leq c \Big(\|\boldsymbol{u}_{01} - \boldsymbol{u}_{02}\|_V + \|\boldsymbol{\sigma}_{01} - \boldsymbol{\sigma}_{02}\|_{Q_1} \\
&\qquad + \|\boldsymbol{f}_{01} - \boldsymbol{f}_{02}\|_{L^\infty(I;H)} + \|\boldsymbol{f}_{21} - \boldsymbol{f}_{22}\|_{L^\infty(I;L^2(\Gamma_2)^d)} \Big).
\end{aligned} \tag{14.53}$$

Proof. Let $t \in \overline{I}$ and $i = 1, 2$. Using (14.16) we have

$$\boldsymbol{u}_i(t) \in V_2, \quad (\boldsymbol{\sigma}_i(t), \boldsymbol{\varepsilon}(\boldsymbol{v}) - \boldsymbol{\varepsilon}(\boldsymbol{u}_i(t)))_Q \geq (\boldsymbol{f}_i(t), \boldsymbol{v} - \boldsymbol{u}_i(t))_V \quad \forall \boldsymbol{v} \in V_2. \tag{14.54}$$

Using (14.15) and (14.17) we deduce that

$$\boldsymbol{\sigma}_i(t) = \mathcal{E}\boldsymbol{\varepsilon}(\boldsymbol{u}_i(t)) + \boldsymbol{z}_i(t), \tag{14.55}$$

where

$$\boldsymbol{z}_i(t) = \int_0^t \mathcal{G}(\boldsymbol{\sigma}_i(s), \boldsymbol{\varepsilon}(\boldsymbol{u}_i(s)))\, ds + \boldsymbol{\sigma}_{0i} - \mathcal{E}\boldsymbol{\varepsilon}(\boldsymbol{u}_{0i}). \tag{14.56}$$

After some algebraic manipulations, it follows from (14.54), (14.55) that

$$\|\boldsymbol{u}_1(t) - \boldsymbol{u}_2(t)\|_V \leq c\,(\|\boldsymbol{f}_1(t) - \boldsymbol{f}_2(t)\|_V + \|\boldsymbol{z}_1(t) - \boldsymbol{z}_2(t)\|_Q).$$

Using (14.55) and the previous inequality we deduce that

$$\begin{aligned}&\|\boldsymbol{u}_1(t) - \boldsymbol{u}_2(t)\|_V + \|\boldsymbol{\sigma}_1(t) - \boldsymbol{\sigma}_2(t)\|_Q \\ &\qquad \leq c\,(\|\boldsymbol{f}_1(t) - \boldsymbol{f}_2(t)\|_V + \|\boldsymbol{z}_1(t) - \boldsymbol{z}_2(t)\|_Q).\end{aligned}$$

Since $\operatorname{Div} \boldsymbol{\sigma}_i(t) + \boldsymbol{f}_{0i}(t) = \boldsymbol{0}$ a.e. in Ω we have

$$\begin{aligned}&\|\boldsymbol{u}_1(t) - \boldsymbol{u}_2(t)\|_V + \|\boldsymbol{\sigma}_1(t) - \boldsymbol{\sigma}_2(t)\|_{Q_1} \\ &\qquad \leq c\,(\|\boldsymbol{f}_1(t) - \boldsymbol{f}_2(t)\|_V + \|\boldsymbol{z}_1(t) - \boldsymbol{z}_2(t)\|_Q).\end{aligned} \tag{14.57}$$

Using now (14.56) we find

$$\begin{aligned}\|\boldsymbol{z}_1(t) - \boldsymbol{z}_2(t)\|_Q \leq c\Big(&\int_0^t (\|\boldsymbol{u}_1(s) - \boldsymbol{u}_2(s)\|_V + \|\boldsymbol{\sigma}_1(s) - \boldsymbol{\sigma}_2(s)\|_{Q_1})\, ds \\ &+ \|\boldsymbol{u}_{01} - \boldsymbol{u}_{02}\|_V + \|\boldsymbol{\sigma}_{01} - \boldsymbol{\sigma}_{02}\|_{Q_1}\Big).\end{aligned} \tag{14.58}$$

Combining (14.57) with (14.58) and using Gronwall's inequality we obtain

$$\begin{aligned}\|\boldsymbol{u}_1(t) - \boldsymbol{u}_2(t)\|_V + \|\boldsymbol{\sigma}_1(t) - \boldsymbol{\sigma}_2(t)\|_{Q_1} \leq c\,(&\|\boldsymbol{u}_{01} - \boldsymbol{u}_{02}\|_V + \|\boldsymbol{\sigma}_{01} - \boldsymbol{\sigma}_{02}\|_{Q_1} \\ &+ \|\boldsymbol{f}_1 - \boldsymbol{f}_2\|_{L^\infty(I;V)}).\end{aligned}$$

This inequality and (14.8) imply the inequality (14.53). ∎

Remark 14.17 *The estimate* (14.53) *provides a* continuous dependence result *for the solution of Problem 14.2 or 14.3 with respect to the input data. Moreover, under the assumption of Theorem 14.16, if* $\boldsymbol{f}_{01} = \boldsymbol{f}_{02}$ *and* $\boldsymbol{f}_{21} = \boldsymbol{f}_{22}$, *from* (14.53) *we obtain*

$$\begin{aligned}&\|\boldsymbol{u}_1 - \boldsymbol{u}_2\|_{L^\infty(I;V)} + \|\boldsymbol{\sigma}_1 - \boldsymbol{\sigma}_2\|_{L^\infty(I;Q_1)} \\ &\qquad \leq c\,\big(\|\boldsymbol{u}_{01} - \boldsymbol{u}_{02}\|_V + \|\boldsymbol{\sigma}_{01} - \boldsymbol{\sigma}_{02}\|_{Q_1}\big),\end{aligned}$$

which represents a finite-time stability result *for every solution of the variational problem 14.2 or 14.3.*

Chapter 15

Frictionless Contact with Dissipative Potential

In this chapter we analyze a class of viscoplastic frictionless contact problems for which the contact is modelled by a general displacement dependent dissipation functional. We provide a variational formulation of the model which consists of an evolutionary equation coupled with a time-dependent variational inequality and prove an existence and uniqueness result. We also provide numerical analysis of the problem, involving semi-discrete and fully discrete approximation schemes, for which we show existence of unique solutions and derive error estimates. We then specify the dissipation potential to obtain as special cases the Signorini contact problem studied in the previous chapter and a frictionless contact problem with normal compliance, for which we provide some additional results. Finally, we present some numerical examples for one, two and three dimensional contact problems to show the performance of the numerical schemes.

15.1 Problem Statement and Variational Analysis

We assume that the solid may be in frictionless contact with a foundation on Γ_3 and use (6.43) (page 135) as the viscoplastic constitutive law. The classical formulation of the mechanical problem is the following:

Find a displacement field $\boldsymbol{u} : \Omega \times \overline{I} \to \mathbb{R}^d$ and a stress field $\boldsymbol{\sigma} : \Omega \times \overline{I} \to \mathbb{S}^d$ such that

$$\dot{\boldsymbol{\sigma}} = \mathcal{E}\boldsymbol{\varepsilon}(\dot{\boldsymbol{u}}) + \mathcal{G}(\boldsymbol{\sigma}, \boldsymbol{\varepsilon}(\boldsymbol{u})) \qquad \text{in } \Omega \times I, \tag{15.1}$$

$$\operatorname{Div} \boldsymbol{\sigma} + \boldsymbol{f}_0 = \boldsymbol{0} \qquad \text{in } \Omega \times I, \tag{15.2}$$

$$\boldsymbol{u} = \boldsymbol{0} \qquad \text{on } \Gamma_1 \times I, \tag{15.3}$$

$$\boldsymbol{\sigma\nu} = \boldsymbol{f}_2 \qquad \text{on } \Gamma_2 \times I, \tag{15.4}$$

$$\boldsymbol{u} \in U, \quad -\boldsymbol{\sigma\nu} \cdot (\boldsymbol{v} - \boldsymbol{u}) \le \varphi(\boldsymbol{v}) - \varphi(\boldsymbol{u}) \ \forall\, \boldsymbol{v} \in U \quad \text{on } \Gamma_3 \times I, \tag{15.5}$$

$$\boldsymbol{u}(0) = \boldsymbol{u}_0, \ \boldsymbol{\sigma}(0) = \boldsymbol{\sigma}_0 \qquad \text{in } \Omega. \tag{15.6}$$

Here $\boldsymbol{u}_0$ and $\boldsymbol{\sigma}_0$ represent the given initial displacement and initial stress field, respectively. The frictionless contact with the foundation is modelled with the subdifferential boundary condition (15.5) in which U represents the set of admissible contact displacement functions and φ is a given function. Examples of inequality problems which lead to boundary conditions of this form are the Signorini's contact condition and the normal compliance contact condition which will be described later in this chapter.

We suppose that $U \subset H^1(\Omega)^d$, $\varphi : \Gamma_3 \times \mathbb{R}^d \to (-\infty, +\infty]$ and we define the functional $j : V \to (-\infty, +\infty]$,

$$j(\boldsymbol{v}) = \begin{cases} \displaystyle\int_{\Gamma_3} \varphi(\boldsymbol{v})\, da & \text{if } \boldsymbol{v} \in U \text{ and } \varphi(\boldsymbol{v}) \in L^1(\Gamma_3), \\ +\infty & \text{otherwise.} \end{cases} \tag{15.7}$$

We assume

$$j \text{ is proper, convex and lower semicontinuous on } V, \tag{15.8}$$

and for the given force densities,

$$\boldsymbol{f}_0 \in W^{1,\infty}(I; L^2(\Omega)^d), \quad \boldsymbol{f}_2 \in W^{1,\infty}(I; L^2(\Gamma_2)^d). \tag{15.9}$$

The initial data satisfy

$$\boldsymbol{u}_0 \in V, \quad \boldsymbol{\sigma}_0 \in Q_1, \tag{15.10}$$

$$(\boldsymbol{\sigma}_0, \boldsymbol{\varepsilon}(\boldsymbol{v}) - \boldsymbol{\varepsilon}(\boldsymbol{u}_0))_Q + j(\boldsymbol{v}) - j(\boldsymbol{u}_0) \geq (\boldsymbol{f}(0), \boldsymbol{v} - \boldsymbol{u}_0)_V \ \forall \boldsymbol{v} \in V, \tag{15.11}$$

where $\boldsymbol{f}(t)$ is the element of V given by

$$(\boldsymbol{f}(t), \boldsymbol{v})_V = \int_\Omega \boldsymbol{f}_0(t) \cdot \boldsymbol{v}\, dx + \int_{\Gamma_2} \boldsymbol{f}_2(t) \cdot \boldsymbol{v}\, da, \tag{15.12}$$

for all $\boldsymbol{v} \in V$ and $t \in \overline{I}$. Here the relation (15.11) represents a compatibility condition at the initial time, which in particular implies $\boldsymbol{u}_0 \in U$.

A standard procedure yields the following weak formulation of the contact problem (15.1)–(15.6).

Problem 15.1 *Find a displacement field $\boldsymbol{u} : \overline{I} \to V$ and a stress field $\boldsymbol{\sigma} : \overline{I} \to Q_1$ such that*

$$\boldsymbol{u}(0) = \boldsymbol{u}_0, \quad \boldsymbol{\sigma}(0) = \boldsymbol{\sigma}_0, \tag{15.13}$$

$$\dot{\boldsymbol{\sigma}}(t) = \mathcal{E}\boldsymbol{\varepsilon}(\dot{\boldsymbol{u}}(t)) + \mathcal{G}(\boldsymbol{\sigma}(t), \boldsymbol{\varepsilon}(\boldsymbol{u}(t))), \quad \text{a.e. } t \in I, \tag{15.14}$$

$$\begin{aligned} &(\boldsymbol{\sigma}(t), \boldsymbol{\varepsilon}(\boldsymbol{v}) - \boldsymbol{\varepsilon}(\boldsymbol{u}(t)))_Q + j(\boldsymbol{v}) - j(\boldsymbol{u}(t)) \\ &\qquad \geq (\boldsymbol{f}(t), \boldsymbol{v} - \boldsymbol{u}(t))_V \quad \forall \boldsymbol{v} \in V,\ t \in \overline{I}. \end{aligned} \tag{15.15}$$

Well-posedness of Problem 15.1 is given in the following result.

Theorem 15.2 *Under the assumptions* (6.44), (6.45) *and* (15.8)–(15.11), *there exists a unique solution $(\boldsymbol{u}, \boldsymbol{\sigma})$ of Problem 15.1. Moreover, the solution satisfies*

$$\boldsymbol{u} \in W^{1,\infty}(I; V), \quad \boldsymbol{\sigma} \in W^{1,\infty}(I; Q_1). \tag{15.16}$$

Proof. The proof of Theorem 15.2 is carried out in several steps, and is similar to the proof of Theorem 14.4. Here we only give a sketch of the main steps.

i) For any $\boldsymbol{\eta} \in L^\infty(I;Q)$ there exists a unique pair of functions $(\boldsymbol{u}_\eta, \boldsymbol{\sigma}_\eta) \in W^{1,\infty}(I;V \times Q_1)$ such that

$$\boldsymbol{\sigma}_\eta(t) = \mathcal{E}\boldsymbol{\varepsilon}(\boldsymbol{u}_\eta(t)) + \boldsymbol{z}_\eta(t),$$
$$(\boldsymbol{\sigma}_\eta(t), \boldsymbol{\varepsilon}(\boldsymbol{v}) - \boldsymbol{\varepsilon}(\boldsymbol{u}_\eta(t)))_Q + j(\boldsymbol{v}) - j(\boldsymbol{u}_\eta(t)) \geq (\boldsymbol{f}(t), \boldsymbol{v} - \boldsymbol{u}_\eta(t))_V \quad \forall\, \boldsymbol{v} \in U_1$$

for all $t \in \overline{I}$, where

$$\boldsymbol{z}_\eta(t) = \int_0^t \boldsymbol{\eta}(s)\, ds + \boldsymbol{\sigma}_0 - \mathcal{E}\boldsymbol{\varepsilon}(\boldsymbol{u}_0).$$

Moreover, the solution satisfies

$$\boldsymbol{u}_\eta(0) = \boldsymbol{u}_0, \quad \boldsymbol{\sigma}_\eta(0) = \boldsymbol{\sigma}_0.$$

ii) The operator $\Lambda : L^\infty(I;Q) \to L^\infty(I;Q)$ defined by

$$\Lambda\boldsymbol{\eta} = \mathcal{G}(\boldsymbol{\sigma}_\eta, \boldsymbol{\varepsilon}(\boldsymbol{u}_\eta)) \quad \forall\, \boldsymbol{\eta} \in L^\infty(I;Q)$$

has a unique fixed point $\boldsymbol{\eta}^* \in L^\infty(I;Q)$.

iii) The pair $(\boldsymbol{u}, \boldsymbol{\sigma})$ with $\boldsymbol{u} = \boldsymbol{u}_{\eta^*}$ and $\boldsymbol{\sigma} = \boldsymbol{\sigma}_{\eta^*}$, is the solution of Problem 15.1 and satisfies (15.16). ■

Since j is a proper functional, from the variational inequality (15.15) we see that $\boldsymbol{u}(t) \in U$ for all $t \in \overline{I}$.

A couple $(\boldsymbol{u}, \boldsymbol{\sigma})$ which solves Problem 15.1 is called a *weak solution* to the mechanical problem (15.1)–(15.6). We conclude from Theorem 15.2 that under the assumptions (6.44), (6.45) and (15.8)–(15.11), the mechanical problem (15.1)–(15.6) has a unique weak solution with regularity (15.16).

As in Section 10.2, we can show that

$$\operatorname{Div} \boldsymbol{\sigma} + \boldsymbol{f}_0 = \boldsymbol{0} \quad \text{a.e. in } \Omega \times I. \tag{15.17}$$

Under the regularity assumption

$$\boldsymbol{\sigma\nu} \in L^\infty(\overline{I}; L^2(\Gamma)^d), \tag{15.18}$$

we further have

$$\boldsymbol{\sigma\nu} = \boldsymbol{f}_2 \quad \text{a.e. on } \Gamma_2 \times I. \tag{15.19}$$

In the next two sections we suppose that assumptions of Theorem 15.2 hold and we consider numerical approximations of Problem 15.1.

15.2 Semi-discrete Approximation

In this section and the next, we present and analyze approximation schemes for solving the contact problem 15.1. We will give some preliminary results that will be used for error estimation in Sections 15.4 in the context of the particular problem 15.1 with Signorini contact condition, and in Section 15.5 for normal compliance contact condition.

In this section we consider an approximation of Problem 15.1 by discretizing only the spatial domain. A fully discrete scheme will be studied in the next section. Let $V^h \subset V$ and $Q^h \subset Q$ be finite-dimensional spaces which, for example, can be constructed by the finite element method. We assume that these spaces satisfy

$$\varepsilon(V^h) \subset Q^h. \tag{15.20}$$

This assumption is very natural and is valid when the polynomial degree for the space V^h is at most one higher than that for the space Q^h.

Let $\mathcal{P}_{Q^h} : Q \to Q^h$ be the orthogonal projection defined through the relation

$$(\mathcal{P}_{Q^h}\boldsymbol{q}, \boldsymbol{q}^h)_Q = (\boldsymbol{q}, \boldsymbol{q}^h)_Q \quad \forall\, \boldsymbol{q} \in Q,\ \boldsymbol{q}^h \in Q^h.$$

Taking $\boldsymbol{q}^h = \mathcal{P}_{Q^h}\boldsymbol{q}$ in this relation, we deduce that

$$\|\mathcal{P}_{Q^h}\boldsymbol{q}\|_Q \le \|\boldsymbol{q}\|_Q \quad \forall\, \boldsymbol{q} \in Q. \tag{15.21}$$

This property will be used on various occasions.

We now study a semi-discrete scheme.

Problem 15.3 *Find a displacement field $\boldsymbol{u}^h : \overline{I} \to V^h$ and a stress field $\boldsymbol{\sigma}^h : \overline{I} \to Q^h$ such that*

$$\boldsymbol{u}^h(0) = \boldsymbol{u}_0^h, \quad \boldsymbol{\sigma}^h(0) = \boldsymbol{\sigma}_0^h, \tag{15.22}$$

$$\dot{\boldsymbol{\sigma}}^h(t) = \mathcal{P}_{Q^h}\mathcal{E}\varepsilon(\dot{\boldsymbol{u}}^h(t)) + \mathcal{P}_{Q^h}\mathcal{G}(\boldsymbol{\sigma}^h(t), \varepsilon(\boldsymbol{u}^h(t))) \text{ a.e. } t \in I, \tag{15.23}$$

$$\begin{aligned}&(\boldsymbol{\sigma}^h(t), \varepsilon(\boldsymbol{v}^h - \boldsymbol{u}^h(t)))_Q + j(\boldsymbol{v}^h) - j(\boldsymbol{u}^h(t))\\ &\qquad \ge (\boldsymbol{f}(t), \boldsymbol{v}^h - \boldsymbol{u}^h(t))_V \quad \forall\, \boldsymbol{v}^h \in V^h,\ t \in I.\end{aligned} \tag{15.24}$$

Here, $\boldsymbol{u}_0^h \in V^h$, $\boldsymbol{\sigma}_0^h \in Q^h$ are appropriate approximations of $\boldsymbol{u}_0$ and $\boldsymbol{\sigma}_0$.

We remark that the projection operator $\mathcal{P}_{Q^h}$ is introduced to ensure that the relation (15.23) is well-defined on the space Q^h. When $\mathcal{E}$ is constant and $\varepsilon(V^h) \subset Q^h$ (a natural assumption valid when the polynomial degree for the space V^h is at most one higher than that for the space Q^h), $\mathcal{P}_{Q^h}\mathcal{E}\varepsilon(\dot{\boldsymbol{u}}^h(t)) = \mathcal{E}\varepsilon(\dot{\boldsymbol{u}}^h(t))$, and the equation (15.23) reduces to

$$\dot{\boldsymbol{\sigma}}^h(t) = \mathcal{E}\varepsilon(\dot{\boldsymbol{u}}^h(t)) + \mathcal{P}_{Q^h}\mathcal{G}(\boldsymbol{\sigma}^h(t), \varepsilon(\boldsymbol{u}^h(t))) \quad \text{a.e. } t \in I.$$

Using the arguments in the previous section, it can be shown that Problem 15.3 has a unique solution $(\boldsymbol{u}^h, \boldsymbol{\sigma}^h) \in W^{1,\infty}(I; V^h \times Q^h)$. Our main purpose here is to derive estimates for the errors $\boldsymbol{u} - \boldsymbol{u}^h$ and $\boldsymbol{\sigma} - \boldsymbol{\sigma}^h$.

Let $t \in \overline{I}$. First we integrate the constitutive relation (15.14) and its discrete analogue (15.23), and use the initial conditions (15.13) and (15.22) to obtain

$$\boldsymbol{\sigma}(t) = \mathcal{E}\varepsilon(\boldsymbol{u}(t)) + \int_0^t \mathcal{G}(\boldsymbol{\sigma}(s), \varepsilon(\boldsymbol{u}(s)))\, ds + \boldsymbol{\sigma}_0 - \mathcal{E}\varepsilon(\boldsymbol{u}_0), \tag{15.25}$$

$$\begin{aligned}\boldsymbol{\sigma}^h(t) = \mathcal{P}_{Q^h}\mathcal{E}\varepsilon(\boldsymbol{u}^h(t)) + \mathcal{P}_{Q^h}\int_0^t \mathcal{G}(\boldsymbol{\sigma}^h(s), \varepsilon(\boldsymbol{u}^h(s)))\, ds \\ + \boldsymbol{\sigma}_0^h - \mathcal{P}_{Q^h}\mathcal{E}\varepsilon(\boldsymbol{u}_0^h).\end{aligned} \tag{15.26}$$

We then subtract (15.26) from (15.25) to get the first error relation,

$$\begin{aligned}\boldsymbol{\sigma}(t) - \boldsymbol{\sigma}^h(t) &= \mathcal{P}_{Q^h}\mathcal{E}\varepsilon(\boldsymbol{u}(t) - \boldsymbol{u}^h(t)) + (I_Q - \mathcal{P}_{Q^h})(\boldsymbol{\sigma}(t) - \boldsymbol{\sigma}_0) \\ &\quad + \mathcal{P}_{Q^h}\int_0^t \left[\mathcal{G}(\boldsymbol{\sigma}(s), \varepsilon(\boldsymbol{u}(s))) - \mathcal{G}(\boldsymbol{\sigma}^h(s), \varepsilon(\boldsymbol{u}^h(s)))\right] ds \\ &\quad + \boldsymbol{\sigma}_0 - \boldsymbol{\sigma}_0^h - \mathcal{P}_{Q^h}\mathcal{E}\varepsilon(\boldsymbol{u}_0 - \boldsymbol{u}_0^h),\end{aligned} \tag{15.27}$$

where $I_Q : Q \to Q$ is the identity operator.

Denote

$$e_0 = \|\boldsymbol{\sigma}_0 - \boldsymbol{\sigma}_0^h\|_Q + \|\boldsymbol{u}_0 - \boldsymbol{u}_0^h\|_V. \tag{15.28}$$

We then derive the following inequality from (15.27), using the assumptions (6.44), (6.45) on $\mathcal{E}$ and $\mathcal{G}$, and the property (15.21),

$$\begin{aligned}\|\boldsymbol{\sigma}(t) - \boldsymbol{\sigma}^h(t)\|_Q &\le c\,\|\boldsymbol{u}(t) - \boldsymbol{u}^h(t)\|_V + \|(I_Q - \mathcal{P}_{Q^h})(\boldsymbol{\sigma}(t) - \boldsymbol{\sigma}_0)\|_Q \\ &\quad + c\int_0^t \left(\|\boldsymbol{\sigma}(s) - \boldsymbol{\sigma}^h(s)\|_Q + \|\boldsymbol{u}(s) - \boldsymbol{u}^h(s)\|_V\right) ds \\ &\quad + c\, e_0.\end{aligned} \tag{15.29}$$

Now using the expression (15.25) in the inequality (15.15) with $\boldsymbol{v} = \boldsymbol{u}^h(t)$, we obtain

$$\begin{aligned}&(\mathcal{E}\varepsilon(\boldsymbol{u}(t)), \varepsilon(\boldsymbol{u}(t) - \boldsymbol{u}^h(t)))_Q \\ &\quad \le (\boldsymbol{f}(t), \boldsymbol{u}(t) - \boldsymbol{u}^h(t))_V + j(\boldsymbol{u}^h(t)) - j(\boldsymbol{u}(t)) \\ &\qquad + \Big(\int_0^t \mathcal{G}(\boldsymbol{\sigma}(s), \varepsilon(\boldsymbol{u}(s)))\, ds, \varepsilon(\boldsymbol{u}^h(t) - \boldsymbol{u}(t))\Big)_Q \\ &\qquad + (\boldsymbol{\sigma}_0 - \mathcal{E}\varepsilon(\boldsymbol{u}_0), \varepsilon(\boldsymbol{u}^h(t) - \boldsymbol{u}(t)))_Q.\end{aligned}$$

Also using the expression (15.26) in the inequality (15.24), we obtain

$$\begin{aligned}&-(\mathcal{E}\varepsilon(\boldsymbol{u}^h(t)), \varepsilon(\boldsymbol{u}(t) - \boldsymbol{u}^h(t)))_Q \\ &\quad \le (\boldsymbol{f}(t), \boldsymbol{u}^h(t) - \boldsymbol{v}^h)_V + j(\boldsymbol{v}^h) - j(\boldsymbol{u}^h(t)) \\ &\qquad + \Big(\int_0^t \mathcal{G}(\boldsymbol{\sigma}^h(s), \varepsilon(\boldsymbol{u}^h(s)))\, ds, \varepsilon(\boldsymbol{v}^h - \boldsymbol{u}^h(t))\Big)_Q \\ &\qquad + (\boldsymbol{\sigma}_0^h - \mathcal{E}\varepsilon(\boldsymbol{u}_0^h), \varepsilon(\boldsymbol{v}^h - \boldsymbol{u}^h(t)))_Q \\ &\qquad + (\mathcal{E}\varepsilon(\boldsymbol{u}^h(t)), \varepsilon(\boldsymbol{v}^h - \boldsymbol{u}(t)))_Q.\end{aligned}$$

Adding the above two inequalities, we obtain the second error relation,

$$\begin{aligned}
&(\mathcal{E}\varepsilon(\boldsymbol{u}(t)-\boldsymbol{u}^h(t)),\varepsilon(\boldsymbol{u}(t)-\boldsymbol{u}^h(t)))_Q \\
&\quad\le (\mathcal{E}\varepsilon(\boldsymbol{u}(t)-\boldsymbol{u}^h(t)),\varepsilon(\boldsymbol{u}(t)-\boldsymbol{v}^h))_Q + R(t;\boldsymbol{u}(t),\boldsymbol{v}^h) \\
&\quad\quad + \Big(\int_0^t \big[\mathcal{G}(\boldsymbol{\sigma}(s),\varepsilon(\boldsymbol{u}(s))) - \mathcal{G}(\boldsymbol{\sigma}^h(s),\varepsilon(\boldsymbol{u}^h(s)))\big]\,ds, \varepsilon(\boldsymbol{u}^h(t)-\boldsymbol{v}^h)\Big)_Q \\
&\quad\quad + (\boldsymbol{\sigma}_0-\boldsymbol{\sigma}_0^h-\mathcal{E}\varepsilon(\boldsymbol{u}_0-\boldsymbol{u}_0^h),\varepsilon(\boldsymbol{u}^h(t)-\boldsymbol{v}^h))_Q,
\end{aligned}$$

where

$$\begin{aligned}
R(t;\boldsymbol{u}(t),\boldsymbol{v}^h) &= (\boldsymbol{f}(t),\boldsymbol{u}(t)-\boldsymbol{v}^h)_V + j(\boldsymbol{v}^h) - j(\boldsymbol{u}(t)) \\
&\quad - (\boldsymbol{\sigma}(t),\varepsilon(\boldsymbol{u}(t)-\boldsymbol{v}^h))_Q. \qquad (15.30)
\end{aligned}$$

After some elementary but tedious manipulations with the use of the assumptions (6.44) and (6.45) on $\mathcal{E}$ and $\mathcal{G}$, we derive from the above inequality that

$$\begin{aligned}
\|\boldsymbol{u}(t)-\boldsymbol{u}^h(t)\|_V &\le c\,\|\varepsilon(\boldsymbol{u}(t)-\boldsymbol{v}^h)\|_Q + c\,e_0 + c\,|R(t;\boldsymbol{u}(t),\boldsymbol{v}^h)|^{1/2} \\
&\quad + c\int_0^t \big(\|\boldsymbol{\sigma}(s)-\boldsymbol{\sigma}^h(s)\|_Q + \|\boldsymbol{u}(s)-\boldsymbol{u}^h(s)\|_V\big)\,ds,
\end{aligned}$$

which can be combined with (15.29) to yield that, for any $\boldsymbol{v}^h=\boldsymbol{v}^h(t)\in V^h$,

$$\begin{aligned}
&\|\boldsymbol{\sigma}(t)-\boldsymbol{\sigma}^h(t)\|_Q + \|\boldsymbol{u}(t)-\boldsymbol{u}^h(t)\|_V \\
&\quad\le c\,\|\boldsymbol{u}(t)-\boldsymbol{v}^h(t)\|_V + c\,e_0 + c\,\|(I_Q-\mathcal{P}_{Q^h})(\boldsymbol{\sigma}(t)-\boldsymbol{\sigma}_0)\|_Q \\
&\quad\quad + c\,|R(t;\boldsymbol{u}(t),\boldsymbol{v}^h(t))|^{1/2} \\
&\quad\quad + c\int_0^t \big(\|\boldsymbol{\sigma}(s)-\boldsymbol{\sigma}^h(s)\|_Q + \|\boldsymbol{u}(s)-\boldsymbol{u}^h(s)\|_V\big)\,ds.
\end{aligned}$$

Since $\boldsymbol{v}^h(t)\in V^h$ is arbitrary, applying the Gronwall's inequality (Lemma 7.24) we obtain the following result.

Theorem 15.4 *Let $(\boldsymbol{u},\boldsymbol{\sigma})\in W^{1,\infty}(I;V\times Q_1)$ be the solution of Problem 15.1, and $(\boldsymbol{u}^h,\boldsymbol{\sigma}^h)\in W^{1,\infty}(I;V^h\times Q^h)$ be the solution of the semi-discrete problem 15.3. Then for any $p\in[1,\infty]$, we have the error estimate*

$$\begin{aligned}
&\|\boldsymbol{\sigma}-\boldsymbol{\sigma}^h\|_{L^p(I;Q)} + \|\boldsymbol{u}-\boldsymbol{u}^h\|_{L^p(I;V)} \\
&\quad\le c\Big(\|\boldsymbol{\sigma}_0-\boldsymbol{\sigma}_0^h\|_Q + \|\boldsymbol{u}_0-\boldsymbol{u}_0^h\|_V \\
&\quad\quad + \inf_{\boldsymbol{v}^h\in L^p(I;V^h)}\{\|\boldsymbol{u}-\boldsymbol{v}^h\|_{L^p(I;V)} + \|R(\cdot;\boldsymbol{u},\boldsymbol{v}^h)\|_{L^p(I)}^{1/2}\} \\
&\quad\quad + \|(I_Q-\mathcal{P}_{Q^h})(\boldsymbol{\sigma}-\boldsymbol{\sigma}_0)\|_{L^p(I;Q)}\Big). \qquad (15.31)
\end{aligned}$$

The inequality (15.31) is the basis of convergence and error analysis for the semi-discrete solutions. Later we will specialize Theorem 15.4 to the semi-discrete approximations of Problems 15.1 with concrete choices of the potential φ and derive error estimates.

15.3 Fully Discrete Approximation

In addition to the finite dimensional spaces V^h and Q^h used in the previous section, we use a generally non-uniform partition of the time interval described in Section 7.2. Then a fully discrete approximation method based on a forward Euler scheme is the following.

Problem 15.5 *Find a displacement field* $\boldsymbol{u}^{hk} = \{\boldsymbol{u}_n^{hk}\}_{n=0}^N \subset V^h$ *and a stress field* $\boldsymbol{\sigma}^{hk} = \{\boldsymbol{\sigma}_n^{hk}\}_{n=0}^N \subset Q^h$ *such that*

$$\boldsymbol{u}_0^{hk} = \boldsymbol{u}_0^h, \ \boldsymbol{\sigma}_0^{hk} = \boldsymbol{\sigma}_0^h, \tag{15.32}$$

and for $n = 1, \dots, N$,

$$\delta_n \boldsymbol{\sigma}_n^{hk} = \mathcal{P}_{Q^h} \mathcal{E} \delta_n \boldsymbol{\varepsilon}(\boldsymbol{u}_n^{hk}) + \mathcal{P}_{Q^h} \mathcal{G}(\boldsymbol{\sigma}_{n-1}^{hk}, \boldsymbol{\varepsilon}(\boldsymbol{u}_{n-1}^{hk})), \tag{15.33}$$

$$\begin{aligned}&(\boldsymbol{\sigma}_n^{hk}, \boldsymbol{\varepsilon}(\boldsymbol{v}^h - \boldsymbol{u}_n^{hk}))_Q + j(\boldsymbol{v}^h) - j(\boldsymbol{u}_n^{hk}) \\ &\qquad \geq (\boldsymbol{f}_n, \boldsymbol{v}^h - \boldsymbol{u}_n^{hk})_V \quad \forall \, \boldsymbol{v}^h \in V^h. \end{aligned} \tag{15.34}$$

Here the element $\boldsymbol{f}_n = \boldsymbol{f}(t_n)$ is well-defined because of the assumption (15.9) which implies $\boldsymbol{f} \in W^{1,\infty}(I; V)$ and therefore, $\boldsymbol{f} \in C(\overline{I}; V)$.

We first show the unique solvability of Problem 15.5. For this, we rewrite (15.33) as

$$\boldsymbol{\sigma}_n^{hk} = \mathcal{P}_{Q^h} \mathcal{E} \boldsymbol{\varepsilon}(\boldsymbol{u}_n^{hk}) + k_n \mathcal{P}_{Q^h} \mathcal{G}(\boldsymbol{\sigma}_{n-1}^{hk}, \boldsymbol{\varepsilon}(\boldsymbol{u}_{n-1}^{hk})) + \boldsymbol{\sigma}_{n-1}^{hk} - \mathcal{P}_{Q^h} \mathcal{E} \boldsymbol{\varepsilon}(\boldsymbol{u}_{n-1}^{hk}). \tag{15.35}$$

Plugging (15.35) in (15.34), we obtain

$$\begin{aligned}&(\mathcal{E}\boldsymbol{\varepsilon}(\boldsymbol{u}_n^{hk}), \boldsymbol{\varepsilon}(\boldsymbol{v}^h - \boldsymbol{u}_n^{hk}))_Q + j(\boldsymbol{v}^h) - j(\boldsymbol{u}_n^{hk}) \\ &\qquad \geq (\mathcal{E}\boldsymbol{\varepsilon}(\boldsymbol{u}_{n-1}^{hk}) - k_n \mathcal{G}(\boldsymbol{\sigma}_{n-1}^{hk}, \boldsymbol{\varepsilon}(\boldsymbol{u}_{n-1}^{hk})) - \boldsymbol{\sigma}_{n-1}^{hk}, \boldsymbol{\varepsilon}(\boldsymbol{v}^h - \boldsymbol{u}_n^{hk}))_Q \\ &\qquad\quad + (\boldsymbol{f}_n, \boldsymbol{v}^h - \boldsymbol{u}_n^{hk})_V \quad \forall \, \boldsymbol{v}^h \in V^h. \end{aligned} \tag{15.36}$$

This is a discrete elliptic variational inequality of the second kind and has a unique solution $\boldsymbol{u}_n^{hk} \in V^h$. Once $\boldsymbol{u}_n^{hk}$ is known, we determine $\boldsymbol{\sigma}_n^{hk}$ from (15.35). Therefore, Problem 15.5 has a unique solution.

Now let us derive an error estimate for the fully discrete solution. Applying (15.35) recursively, we have

$$\boldsymbol{\sigma}_n^{hk} = \mathcal{P}_{Q^h} \mathcal{E} \boldsymbol{\varepsilon}(\boldsymbol{u}_n^{hk}) + \sum_{j=1}^n k_j \mathcal{P}_{Q^h} \mathcal{G}(\boldsymbol{\sigma}_{j-1}^{hk}, \boldsymbol{\varepsilon}(\boldsymbol{u}_{j-1}^{hk})) + \boldsymbol{\sigma}_0^{hk} - \mathcal{P}_{Q^h} \mathcal{E} \boldsymbol{\varepsilon}(\boldsymbol{u}_0^{hk}). \tag{15.37}$$

Subtracting (15.37) from (15.25) at $t = t_n$, we obtain the first error relation

$$\begin{aligned}\boldsymbol{\sigma}_n - \boldsymbol{\sigma}_n^{hk} &= (I_Q - \mathcal{P}_{Q^h})(\boldsymbol{\sigma}_n - \boldsymbol{\sigma}_0) + \mathcal{P}_{Q^h} \mathcal{E} \boldsymbol{\varepsilon}(\boldsymbol{u}_n - \boldsymbol{u}_n^{hk}) \\ &\quad + \mathcal{P}_{Q^h} \sum_{j=1}^n k_j \big[\mathcal{G}(\boldsymbol{\sigma}_{j-1}, \boldsymbol{\varepsilon}(\boldsymbol{u}_{j-1})) - \mathcal{G}(\boldsymbol{\sigma}_{j-1}^{hk}, \boldsymbol{\varepsilon}(\boldsymbol{u}_{j-1}^{hk}))\big] \\ &\quad + \mathcal{P}_{Q^h} \boldsymbol{\theta}_n^k(\mathcal{G}) + \boldsymbol{\sigma}_0 - \boldsymbol{\sigma}_0^{hk} - \mathcal{P}_{Q^h} \mathcal{E} \boldsymbol{\varepsilon}(\boldsymbol{u}_0 - \boldsymbol{u}_0^{hk}), \end{aligned} \tag{15.38}$$

where for $n = 1, \ldots, N$,

$$\boldsymbol{\theta}_n^k(\mathcal{G}) = \int_0^{t_n} \mathcal{G}(\boldsymbol{\sigma}(s), \boldsymbol{\varepsilon}(\boldsymbol{u}(s)))\, ds - \sum_{j=1}^n k_j \mathcal{G}(\boldsymbol{\sigma}_{j-1}, \boldsymbol{\varepsilon}(\boldsymbol{u}_{j-1})) \tag{15.39}$$

is a quadrature error. We have

$$\begin{aligned} \|\boldsymbol{\theta}_n^k(\mathcal{G})\|_Q &\le \sum_{j=1}^n \int_{t_{j-1}}^{t_j} \|\mathcal{G}(\boldsymbol{\sigma}(t), \boldsymbol{\varepsilon}(\boldsymbol{u}(t))) - \mathcal{G}(\boldsymbol{\sigma}_{j-1}, \boldsymbol{\varepsilon}(\boldsymbol{u}_{j-1}))\|_Q\, dt \\ &\le c \sum_{j=1}^n \int_{t_{j-1}}^{t_j} \Big(\|\boldsymbol{\sigma}(t) - \boldsymbol{\sigma}_{j-1}\|_Q + \|\boldsymbol{u}(t) - \boldsymbol{u}_{j-1}\|_V \Big)\, dt \\ &\le c\, t_n k \Big(\|\dot{\boldsymbol{\sigma}}\|_{L^\infty(I;Q)} + \|\dot{\boldsymbol{u}}\|_{L^\infty(I;V)} \Big). \end{aligned} \tag{15.40}$$

Denote

$$e_n = \|\boldsymbol{\sigma}_n - \boldsymbol{\sigma}_n^{hk}\|_Q + \|\boldsymbol{u}_n - \boldsymbol{u}_n^{hk}\|_V, \quad n = 0, \ldots, N.$$

We derive the following inequality from (15.38), using the assumptions (6.44), (6.45) and the property (15.21),

$$\begin{aligned} \|\boldsymbol{\sigma}_n - \boldsymbol{\sigma}_n^{hk}\|_Q &\le \|(I_Q - \mathcal{P}_{Q^h})(\boldsymbol{\sigma}_n - \boldsymbol{\sigma}_0)\|_Q + c\, \|\boldsymbol{u}_n - \boldsymbol{u}_n^{hk}\|_V \\ &\quad + c \sum_{j=1}^n k_j (\|\boldsymbol{\sigma}_{j-1} - \boldsymbol{\sigma}_{j-1}^{hk}\|_Q + \|\boldsymbol{u}_{j-1} - \boldsymbol{u}_{j-1}^{hk}\|_V) \\ &\quad + \|\boldsymbol{\theta}_n^k(\mathcal{G})\|_Q + c\, e_0. \end{aligned} \tag{15.41}$$

Now plugging (15.25) in (15.15) at $t = t_n$ with $\boldsymbol{v} = \boldsymbol{u}_n^{hk}$, we obtain

$$\begin{aligned} &(\mathcal{E}\boldsymbol{\varepsilon}(\boldsymbol{u}_n), \boldsymbol{\varepsilon}(\boldsymbol{u}_n - \boldsymbol{u}_n^{hk}))_Q \\ &\quad \le (\boldsymbol{f}_n, \boldsymbol{u}_n - \boldsymbol{u}_n^{hk})_V + j(\boldsymbol{u}_n^{hk}) - j(\boldsymbol{u}_n) \\ &\qquad + \Big(\int_0^{t_n} \mathcal{G}(\boldsymbol{\sigma}(s), \boldsymbol{\varepsilon}(\boldsymbol{u}(s)))\, ds, \boldsymbol{\varepsilon}(\boldsymbol{u}_n^{hk} - \boldsymbol{u}_n) \Big)_Q \\ &\qquad + (\boldsymbol{\sigma}_0 - \mathcal{E}\boldsymbol{\varepsilon}(\boldsymbol{u}_0), \boldsymbol{\varepsilon}(\boldsymbol{u}_n^{hk} - \boldsymbol{u}_n))_Q. \end{aligned}$$

Similarly, plugging (15.37) in (15.34) with any $\boldsymbol{v}^h = \boldsymbol{v}_n^h \in V^h$, we have

$$\begin{aligned} &-(\mathcal{E}\boldsymbol{\varepsilon}(\boldsymbol{u}_n^{hk}), \boldsymbol{\varepsilon}(\boldsymbol{u}_n - \boldsymbol{u}_n^{hk}))_Q \\ &\quad \le (\boldsymbol{f}_n, \boldsymbol{u}_n^{hk} - \boldsymbol{v}_n^h)_V + j(\boldsymbol{v}_n^h) - j(\boldsymbol{u}_n^{hk}) \\ &\qquad + \Big(\sum_{j=1}^n k_j \mathcal{G}(\boldsymbol{\sigma}_{j-1}^{hk}, \boldsymbol{\varepsilon}(\boldsymbol{u}_{j-1}^{hk})), \boldsymbol{\varepsilon}(\boldsymbol{v}_n^h - \boldsymbol{u}_n^{hk}) \Big)_Q \\ &\qquad + (\boldsymbol{\sigma}_0^{hk} - \mathcal{E}\boldsymbol{\varepsilon}(\boldsymbol{u}_0^{hk}), \boldsymbol{\varepsilon}(\boldsymbol{v}_n^h - \boldsymbol{u}_n^{hk}))_Q + (\mathcal{E}\boldsymbol{\varepsilon}(\boldsymbol{u}_n^{hk}), \boldsymbol{\varepsilon}(\boldsymbol{v}_n^h - \boldsymbol{u}_n))_Q. \end{aligned}$$

Adding the above two inequalities and using (15.25) at $t = t_n$, we obtain the second error relation

$$\begin{aligned}
&(\mathcal{E}\varepsilon(\boldsymbol{u}_n - \boldsymbol{u}_n^{hk}), \varepsilon(\boldsymbol{u}_n - \boldsymbol{u}_n^{hk}))_Q \\
&\quad \le (\mathcal{E}\varepsilon(\boldsymbol{u}_n - \boldsymbol{u}_n^{hk}), \varepsilon(\boldsymbol{u}_n - \boldsymbol{v}_n^h))_Q + R_n(\boldsymbol{u}_n, \boldsymbol{v}_n^h) \\
&\qquad + \left(\boldsymbol{\theta}_n^k(\mathcal{G}), \varepsilon(\boldsymbol{u}_n^{hk} - \boldsymbol{v}_n^h)\right)_Q \\
&\qquad + (\boldsymbol{\sigma}_0 - \boldsymbol{\sigma}_0^{hk} - \mathcal{E}\varepsilon(\boldsymbol{u}_0 - \boldsymbol{u}_0^{hk}), \varepsilon(\boldsymbol{u}_n^{hk} - \boldsymbol{v}_n^h))_Q,
\end{aligned}$$

where

$$R_n(\boldsymbol{u}_n, \boldsymbol{v}_n^h) = (\boldsymbol{f}_n, \boldsymbol{u}_n - \boldsymbol{v}_n^h)_V + j(\boldsymbol{v}_n^h) - j(\boldsymbol{u}_n) - (\boldsymbol{\sigma}_n, \varepsilon(\boldsymbol{u}_n - \boldsymbol{v}_n^h))_Q.$$

From this, after some tedious manipulations we obtain

$$\begin{aligned}
\|\boldsymbol{u}_n - \boldsymbol{u}_n^{hk}\|_V &\le c\left[\|\boldsymbol{u}_n - \boldsymbol{v}_n^h\|_V + |R_n(\boldsymbol{u}_n, \boldsymbol{v}_n^h)|^{1/2} + \|\boldsymbol{\theta}_n^k(\mathcal{G})\|_Q + e_0\right] \\
&\quad + c\sum_{j=1}^n k_j e_{j-1}. \qquad (15.42)
\end{aligned}$$

The inequalities (15.41) and (15.42) can be combined to yield

$$\begin{aligned}
e_n &\le c\left[\|\boldsymbol{u}_n - \boldsymbol{v}_n^h\|_V + |R_n(\boldsymbol{u}_n, \boldsymbol{v}_n^h)|^{1/2} + \|\boldsymbol{\theta}_n^k(\mathcal{G})\|_Q + e_0\right] \\
&\quad + c\,\|(I_Q - \mathcal{P}_{Q^h})(\boldsymbol{\sigma}_n - \boldsymbol{\sigma}_0)\|_Q + c\sum_{j=1}^n k_j e_{j-1}.
\end{aligned}$$

Applying Lemma 7.25 (page 163), we have

$$\max_{1\le n\le N}\left(\|\boldsymbol{\sigma}_n - \boldsymbol{\sigma}_n^{hk}\|_Q + \|\boldsymbol{u}_n - \boldsymbol{u}_n^{hk}\|_V\right) \le c\max_{1\le n\le N} g_n, \qquad (15.43)$$

where

$$\begin{aligned}
g_n &= \|\boldsymbol{u}_n - \boldsymbol{v}_n^h\|_V + |R_n(\boldsymbol{u}_n, \boldsymbol{v}_n^h)|^{1/2} + \|\boldsymbol{\theta}_n^k(\mathcal{G})\|_Q + e_0 \\
&\quad + \|(I_Q - \mathcal{P}_{Q^h})(\boldsymbol{\sigma}_n - \boldsymbol{\sigma}_0)\|_Q.
\end{aligned}$$

The quadrature error term $\|\boldsymbol{\theta}_n^k(\mathcal{G})\|_Q$ is bounded by (15.40). So we obtain from (15.43) the next result.

Theorem 15.6 *Let $(\boldsymbol{u}, \boldsymbol{\sigma}) \in W^{1,\infty}(I; V \times Q_1)$ be the solution of Problem 15.1, and let $\{(\boldsymbol{u}_n^{hk}, \boldsymbol{\sigma}_n^{hk})\}_{n=1}^N$ be the solution of the fully discrete problem 15.5. Then we have the error estimate*

$$\begin{aligned}
&\max_{1\le n\le N}\left(\|\boldsymbol{\sigma}_n - \boldsymbol{\sigma}_n^{hk}\|_Q + \|\boldsymbol{u}_n - \boldsymbol{u}_n^{hk}\|_V\right) \\
&\quad \le c\left(\|\boldsymbol{\sigma}_0 - \boldsymbol{\sigma}_0^h\|_Q + \|\boldsymbol{u}_0 - \boldsymbol{u}_0^h\|_V\right) \\
&\qquad + c\,k\left(\|\dot{\boldsymbol{\sigma}}\|_{L^\infty(I;Q)} + \|\dot{\boldsymbol{u}}\|_{L^\infty(I;V)}\right) \\
&\qquad + c\max_{1\le n\le N}\Big(\inf_{\boldsymbol{v}_n^h\in V^h}\{\|\boldsymbol{u}_n - \boldsymbol{v}_n^h\|_V + |R_n(\boldsymbol{u}_n, \boldsymbol{v}_n^h)|^{1/2}\} \\
&\qquad\quad + \|(I_Q - \mathcal{P}_{Q^h})(\boldsymbol{\sigma}_n - \boldsymbol{\sigma}_0)\|_Q\Big). \qquad (15.44)
\end{aligned}$$

The inequality (15.44) is the basis of convergence and error analysis for the fully discrete solutions.

15.4 The Signorini Contact Problem

In this section we consider the frictionless problem with Signorini's contact condition in a form with a nonzero gap function, which represents a special case of the general contact condition (15.5). The contact boundary conditions are described by the relations

$$u_\nu \le g, \quad \sigma_\nu \le 0, \quad \sigma_\nu(u_\nu - g) = 0, \quad \boldsymbol{\sigma}_\tau = \mathbf{0} \quad \text{on } \Gamma_3 \times I, \tag{15.45}$$

where g is the initial gap between the deformable body and the rigid foundation, i.e. $g(\boldsymbol{x})$ is the distance between $\boldsymbol{x} \in \Gamma_3$ and the foundation in the normal direction before the forces are applied to the body. We assume

$$g \in L^2(\Gamma_3), \quad g \ge 0 \text{ a.e. on } \Gamma_3. \tag{15.46}$$

Introduce the set of admissible displacements

$$K_g = \{\, \boldsymbol{v} \in V : v_\nu \le g \text{ a.e. on } \Gamma_3 \,\}. \tag{15.47}$$

It is straightforward to show that if $(\boldsymbol{u}, \boldsymbol{\sigma})$ is a pair of regular functions, then the contact boundary condition (15.5) holds with $U = K_g$ and

$$\varphi(\boldsymbol{v}) = \begin{cases} 0 & \text{if } \; v_\nu \le g, \\ +\infty & \text{if } \; v_\nu > g. \end{cases}$$

For this particular problem,

$$j(\boldsymbol{v}) = \begin{cases} 0 & \text{if } \boldsymbol{v} \in K_g, \\ +\infty & \text{otherwise.} \end{cases} \tag{15.48}$$

In other words, $j(\cdot)$ is the indicator function of the set K_g. Observe that j is a proper, convex and lower semicontinuous functional on V. So by Theorem 15.2 on page 322, the frictionless contact problem (15.1)–(15.4), (15.6), (15.45) has a unique weak solution $(\boldsymbol{u}, \boldsymbol{\sigma}) \in W^{1,\infty}(I; V \times Q_1)$.

Turn now to numerical approximations of the problem. We will demonstrate how Theorems 15.4 and 15.6 can be used to derive error estimates for the numerical solutions of Problem 15.1 with (15.48). In the rest of the section, the constant c may depend on the exact solution, but it is independent of the discretization parameters h and k. As in Section 7.2, we assume Ω is a polygon or polyhedron, and let $\{\mathcal{T}^h\}$ be a regular family of finite element partitions of Ω compatible to the decomposition $\Gamma = \overline{\Gamma}_1 \cup \overline{\Gamma}_2 \cup \overline{\Gamma}_3$. Let us use linear elements for the finite element space V^h and piecewise constants for Q^h, corresponding to the partition $\mathcal{T}^h$. Obviously, the condition (15.20) is satisfied.

Note that the solution satisfies

$$\boldsymbol{u} \in W^{1,\infty}(I; K_g).$$

Moreover, $j(\boldsymbol{v}) = 0$ for $\boldsymbol{v} \in K_g$.

Convergence of discrete solutions. We first consider the issue of convergence of the semi-discrete and fully discrete solutions under the basic solution regularity $(\boldsymbol{u}, \boldsymbol{\sigma}) \in W^{1,\infty}(I; K_g \times Q_1)$. For this purpose, we will need density results Theorem 7.11 (page 150) and Lemma 7.15 (page 152). We now give a convergence result.

Theorem 15.7 *Let us keep all the assumptions stated in Theorem 7.11. Let* $(\boldsymbol{u}, \boldsymbol{\sigma}) \in W^{1,\infty}(I; V \times Q_1)$ *be the solution of the Problem 15.1 with* (15.48). *Let* $(\boldsymbol{u}^h, \boldsymbol{\sigma}^h) \in W^{1,\infty}(I; K_g^h \times Q^h)$ *and* $\{(\boldsymbol{u}_n^{hk}, \boldsymbol{\sigma}_n^{hk})\}_{n=1}^N$ *be the corresponding semi-discrete and fully discrete solutions. Assume the initial gap function* $g = 0$. *If we choose the initial values* $\boldsymbol{\sigma}_0^h \in Q^h$ *and* $\boldsymbol{u}_0^h \in V^h$ *in such a way that*

$$\|\boldsymbol{\sigma}_0 - \boldsymbol{\sigma}_0^h\|_Q \to 0, \quad \|\boldsymbol{u}_0 - \boldsymbol{u}_0^h\|_V \to 0 \quad \text{as } h \to 0, \tag{15.49}$$

then we have the convergence:

$$\|\boldsymbol{\sigma} - \boldsymbol{\sigma}^h\|_{L^\infty(I;Q)} + \|\boldsymbol{u} - \boldsymbol{u}^h\|_{L^\infty(I;V)} \to 0 \quad \text{as } h \to 0. \tag{15.50}$$

$$\max_{1 \le n \le N} (\|\boldsymbol{\sigma} - \boldsymbol{\sigma}_n^{hk}\|_Q + \|\boldsymbol{u} - \boldsymbol{u}_n^{hk}\|_V) \to 0 \quad \text{as } h, k \to 0. \tag{15.51}$$

Proof. Since $j(\boldsymbol{v}) = 0$ if $\boldsymbol{v} \in K_0$, from the definition (15.30) we have, for $\boldsymbol{v}^h \in K_0$,

$$\begin{aligned} &|R(t; \boldsymbol{u}(t), \boldsymbol{v}^h)| \\ &\quad \le c\,(\|\boldsymbol{f}_0(t)\|_{L^2(\Omega)^d} + \|\boldsymbol{f}_2(t)\|_{L^2(\Gamma_2)^d} + \|\boldsymbol{\sigma}(t)\|_Q)\|\boldsymbol{u}(t) - \boldsymbol{v}^h\|_V \end{aligned}$$

for all $t \in \overline{I}$. Therefore the estimate (15.31) can be rewritten as

$$\begin{aligned} &\|\boldsymbol{\sigma} - \boldsymbol{\sigma}^h\|_{L^\infty(I;Q)} + \|\boldsymbol{u} - \boldsymbol{u}^h\|_{L^\infty(I;V)} \\ &\quad \le c\Big(\|\boldsymbol{\sigma}_0 - \boldsymbol{\sigma}_0^h\|_Q + \|\boldsymbol{u}_0 - \boldsymbol{u}_0^h\|_V \\ &\qquad + \|(I_Q - \mathcal{P}_{Q^h})(\boldsymbol{\sigma}(t) - \boldsymbol{\sigma}_0)\|_{L^\infty(I;Q)} \\ &\qquad + \inf_{\boldsymbol{v}^h \in L^\infty(I; V^h \cap K_0)} \|\boldsymbol{u} - \boldsymbol{v}^h\|_{L^\infty(I;V)}^{1/2}\Big). \end{aligned} \tag{15.52}$$

Let us apply Theorem 7.21 (page 159) to conclude the convergence (15.50). For this we need to verify Hypothesis 7.20 (page 159) with the choice $X = K_0$. By Theorem 7.11, we can take $V_0 = C^\infty(\overline{\Omega})^d \cap K_0$. Also we may take $Q_0 = H^1(\Omega)^{d\times d} \cap Q$. Then Hypothesis 7.20 is satisfied with $\alpha_1(h) \le c_1 h$ and $\alpha_2(h) \le c_2 h$. Thus under the assumption (15.49), we obtain the convergence (15.50). The convergence result (15.51) can be proven similarly, with an application of Theorem 7.23 (page 161). ■

From the proof of Theorem 15.7, we see that an essential ingredient of the convergence proof is a density result of the type described in Theorem 7.11 (page 150). If, under some assumptions, it can be shown that $K_g \cap C^\infty(\overline{\Omega})^d$ is dense in K_g, then under the same assumptions and the condition (15.49), we have the convergence results (15.50) and (15.51) under the basic solution regularity.

Error estimates. We need assumptions on the regularity of the exact solution to derive error estimates.

Theorem 15.8 *Let* $(\boldsymbol{u}, \boldsymbol{\sigma})$ *be the solution of Problem 15.1 with* (15.48), *where* g *is assumed to be concave on each line segment of* Γ_3. *Let* $(\boldsymbol{u}^h, \boldsymbol{\sigma}^h)$ *and* $\{(\boldsymbol{u}_n^{hk}, \boldsymbol{\sigma}_n^{hk})\}_{n=1}^N$ *be the corresponding semi-discrete and fully discrete solutions. Assume the solution regularity conditions* (15.18) *and*

$$\boldsymbol{u} \in C(\overline{I}; H^2(\Omega))^d, \quad \boldsymbol{u}|_{\Gamma_{3,i}} \in C(\overline{I}; H^2(\Gamma_{3,i})^d), \ 1 \le i \le i_0,$$
$$\boldsymbol{\sigma} \in C(\overline{I}; H^1(\Omega)^{d\times d}).$$

Let the initial values $\boldsymbol{u}_0^h \in V^h \cap K_g$ *and* $\boldsymbol{\sigma}_0^h \in Q^h$ *be chosen in such a way that*

$$\|\boldsymbol{\sigma}_0 - \boldsymbol{\sigma}_0^h\|_Q \le c\,h, \quad \|\boldsymbol{u}_0 - \boldsymbol{u}_0^h\|_V \le c\,h.$$

Then for $p \in [1, \infty]$, *we have the optimal order error estimates:*

$$\|\boldsymbol{\sigma} - \boldsymbol{\sigma}^h\|_{L^p(I;Q)} + \|\boldsymbol{u} - \boldsymbol{u}^h\|_{L^p(I;V)} = O(h), \tag{15.53}$$
$$\max_{1\le n\le N} (\|\boldsymbol{\sigma} - \boldsymbol{\sigma}_n^{hk}\|_Q + \|\boldsymbol{u} - \boldsymbol{u}_n^{hk}\|_V) = O(k + h). \tag{15.54}$$

Proof. For $t \in \overline{I}$, let $\Pi^h \boldsymbol{u}(t) \in V^h$ be the piecewise linear interpolant of $\boldsymbol{u}(t)$. Then we have the interpolation error estimates (7.32) and (7.34) from Section 7.2. Since g is concave on each line segment of Γ_3, we have $\Pi^h \boldsymbol{u}(t) \in K_g$, and so $j(\Pi^h \boldsymbol{u}(t)) = 0$. Certainly, $j(\boldsymbol{u}(t)) = 0$. So

$$\begin{aligned} R(t; \boldsymbol{u}(t), \Pi^h \boldsymbol{u}(t)) = & \int_\Omega \boldsymbol{f}_0(t) \cdot (\boldsymbol{u}(t) - \Pi^h \boldsymbol{u}(t))\, dx \\ & + \int_{\Gamma_2} \boldsymbol{f}_2(t) \cdot (\boldsymbol{u}(t) - \Pi^h \boldsymbol{u}(t))\, da \\ & - \int_\Omega \boldsymbol{\sigma}(t) \cdot \boldsymbol{\varepsilon}(\boldsymbol{u}(t) - \Pi^h \boldsymbol{u}(t))\, dx. \end{aligned}$$

Performing an integration by parts on the last term and using the pointwise relations (15.17), (15.19), we have

$$\begin{aligned} R(t; \boldsymbol{u}(t), \Pi^h \boldsymbol{u}(t)) &= \int_{\Gamma_3} (\boldsymbol{\sigma}(t)\boldsymbol{\nu}) \cdot (\boldsymbol{u}(t) - \Pi^h \boldsymbol{u}(t))\, da \\ &\le c\, \|\boldsymbol{\sigma}(t)\|_{H^1(\Omega)^{d\times d}} \|\boldsymbol{u}(t) - \Pi^h \boldsymbol{u}(t)\|_{L^2(\Gamma_3)^d}. \end{aligned}$$

Also, from the property of the projection,

$$\|(I_Q - \mathcal{P}_{Q^h})(\boldsymbol{\sigma}(t) - \boldsymbol{\sigma}_0)\|_Q \le c\,h\, \|\boldsymbol{\sigma}(t) - \boldsymbol{\sigma}_0\|_{H^1(\Omega)^{d\times d}}.$$

So the error estimate (15.53) follows from (15.31) with the choice $\boldsymbol{v}^h(t) = \Pi^h \boldsymbol{u}(t)$. The error estimate (15.54) can be proved similarly. ■

15.5 A Frictionless Contact Problem with Normal Compliance

In this section we consider the contact problem with the following contact boundary conditions:

$$\sigma_\nu = -\frac{1}{\mu}\,(u_\nu - g)_+^\alpha, \quad \boldsymbol{\sigma}_\tau = \mathbf{0} \quad \text{on } \Gamma_3 \times I. \tag{15.55}$$

Here $0 < \alpha \leq 1$, g is the initial gap between the elastic-viscoplastic body and the foundation and $r_+ = \max\{0, r\}$; μ is a penalization parameter which may be interpreted as the *deformability coefficient of the foundation.* We assume that

$$g \in L^2(\Gamma_3), \quad g \geq 0 \text{ a.e. on } \Gamma_3, \tag{15.56}$$

$$\mu \in L^\infty(\Gamma_3), \quad \mu \geq \mu^* > 0 \text{ a.e. on } \Gamma_3. \tag{15.57}$$

Conditions (15.55) represent an example of the normal compliance condition (5.41) for frictionless contact. Recall that the quantity $u_\nu - g$, when it is positive, represents the penetration of the body in the foundation. Therefore, in the normal compliance condition, interpenetration is allowed but penalized. Finally, Signorini's nonpenetration condition can be viewed as a limiting case of (15.55) when $\mu \to 0$, i.e. when the coefficient of deformability of the foundation tends to zero.

It is straightforward to show that if $(\boldsymbol{u}, \boldsymbol{\sigma})$ is a pair of regular functions satisfying the contact condition (15.55), then (15.5) holds with the choice $U = V$ and

$$\varphi(\boldsymbol{v}) = \frac{1}{\mu(\alpha+1)}\,(v_\nu - g)_+^{\alpha+1}.$$

For this particular contact problem,

$$j(\boldsymbol{v}) = \frac{1}{\alpha+1}\int_{\Gamma_3} \frac{1}{\mu}\,(v_\nu - g)_+^{\alpha+1}\,da. \tag{15.58}$$

Under the assumptions (15.56) and (15.57) on g and μ, we see that j is a proper, convex and lower semicontinuous functional on V. So Theorem 15.2 (page 322) assures that the frictionless contact problem (15.1)–(15.4), (15.6), (15.55) has a unique weak solution $(\boldsymbol{u}, \boldsymbol{\sigma}) \in W^{1,\infty}(I; V \times Q_1)$.

We turn now to numerical approximations of the problem and show how Theorems 15.4 and 15.6 can be used to derive error estimates for the numerical solutions of Problem 15.1 with (15.58). In the rest of the section, the constant c may depend on the exact solution, but it is independent of the discretization parameters h and k. Again, assume Ω is a polygon or polyhedron and let $\{\mathcal{T}^h\}$ be a regular family of finite element partitions of Ω compatible to the decomposition $\Gamma = \overline{\Gamma}_1 \cup \overline{\Gamma}_2 \cup \overline{\Gamma}_3$. We use linear elements for the finite element space V^h, piecewise constants for Q^h, corresponding to the partition $\mathcal{T}^h$. Again, the condition (15.20) is valid.

Convergence of discrete solutions. We first present the following convergence result.

Theorem 15.9 *Let* $(\boldsymbol{u}, \boldsymbol{\sigma}) \in W^{1,\infty}(I; V \times Q_1)$ *be the solution of Problem 15.1 with* (15.58). *Let* $(\boldsymbol{u}^h, \boldsymbol{\sigma}^h)$ *and* $\{(\boldsymbol{u}_n^{hk}, \boldsymbol{\sigma}_n^{hk})\}_{n=1}^N$ *be the corresponding semi-discrete and fully discrete solutions. Assume the initial values* $\boldsymbol{\sigma}_0^h \in Q^h$ *and* $\boldsymbol{u}_0^h \in V^h$ *are chosen in such a way that*

$$\|\boldsymbol{\sigma}_0 - \boldsymbol{\sigma}_0^h\|_Q \to 0, \quad \|\boldsymbol{u}_0 - \boldsymbol{u}_0^h\|_V \to 0 \quad \text{as } h \to 0.$$

Then we have the convergence:

$$\begin{aligned} \|\boldsymbol{\sigma} - \boldsymbol{\sigma}^h\|_{L^\infty(I;Q)} + \|\boldsymbol{u} - \boldsymbol{u}^h\|_{L^\infty(I;V)} &\to 0 \quad \text{as } h \to 0, \\ \max_{1 \le n \le N} (\|\boldsymbol{\sigma} - \boldsymbol{\sigma}_n^{hk}\|_Q + \|\boldsymbol{u} - \boldsymbol{u}_n^{hk}\|_V) &\to 0 \quad \text{as } h, k \to 0. \end{aligned}$$

Proof. Since $\alpha \in (0, 1]$, we can apply the elementary inequality (7.56) to get

$$\begin{aligned} &|j(\boldsymbol{v}^h(t)) - j(\boldsymbol{u}(t))| \\ &\quad \le c\,\|\mu\|_{L^\infty(\Gamma_3)} \int_{\Gamma_3} |v_\nu^h(t) - u_\nu(t)|\,(2\,|u_\nu(t) - g|^\alpha \\ &\qquad\qquad + |v_\nu^h(t) - u_\nu(t)|^\alpha)\,ds \\ &\quad \le c\,(\|u_\nu(t) - v_\nu^h(t)\|_{L^1(\Gamma_3)} + \|u_\nu(t) - v_\nu^h(t)\|_{L^{1+\alpha}(\Gamma_3)}^{1+\alpha}) \\ &\quad \le c\,(\|u_\nu(t) - v_\nu^h(t)\|_{L^2(\Gamma_3)} + \|u_\nu(t) - v_\nu^h(t)\|_{L^2(\Gamma_3)}^{1+\alpha}) \end{aligned}$$

for all $t \in \overline{I}$.

Hence, from the definition (15.30) of $R(\cdot; \boldsymbol{u}, \boldsymbol{v}^h)$, we have the following bound for all $t \in \overline{I}$,

$$|R(t; \boldsymbol{u}(t), \boldsymbol{v}^h(t))| \le c\,(\|\boldsymbol{u}(t) - \boldsymbol{v}^h(t)\|_V + \|\boldsymbol{u}(t) - \boldsymbol{v}^h(t)\|_V^{1+\alpha}).$$

Then as in the proof of Theorem 15.7, we can apply Theorems 7.21 and 7.23 to conclude the convergence statements. ■

Error estimates. Following the argument in the proof of Theorem 15.8, we can similarly prove the next result.

Theorem 15.10 *Let* $(\boldsymbol{u}, \boldsymbol{\sigma})$ *be the solution of Problem 15.1 with* (15.58), $(\boldsymbol{u}^h, \boldsymbol{\sigma}^h)$ *and* $\{(\boldsymbol{u}_n^{hk}, \boldsymbol{\sigma}_n^{hk})\}_{n=1}^N$ *the corresponding semi-discrete and fully discrete solutions. Assume*

$$\boldsymbol{u} \in C(\overline{I}; H^2(\Omega)^d), \quad u_\nu|_{\Gamma_{3,i}} \in C(\overline{I}; H^2(\Gamma_{3,i})), \ 1 \le i \le i_0$$

and

$$\boldsymbol{\sigma} \in C(\overline{I}; H^1(\Omega)^{d \times d}).$$

Let the initial values $\boldsymbol{u}_0^h \in V^h$ and $\boldsymbol{\sigma}_0^h \in Q^h$ be chosen in such a way that

$$\|\boldsymbol{\sigma}_0 - \boldsymbol{\sigma}_0^h\|_Q \le c\,h, \quad \|\boldsymbol{u}_0 - \boldsymbol{u}_0^h\|_V \le c\,h.$$

Then for any $p \in [1, \infty]$, we have the optimal order error estimates:

$$\|\boldsymbol{\sigma} - \boldsymbol{\sigma}^h\|_{L^p(I;Q)} + \|\boldsymbol{u} - \boldsymbol{u}^h\|_{L^p(I;V)} = O(h),$$
$$\max_{1 \le n \le N} (\|\boldsymbol{\sigma} - \boldsymbol{\sigma}_n^{hk}\|_Q + \|\boldsymbol{u} - \boldsymbol{u}_n^{hk}\|_V) = O(k + h).$$

15.6 A Convergence Result

In this section we analyze the behavior of the solution of the frictionless contact problem with normal compliance as the deformability coefficient of the foundation tends to zero. We prove that in the limit, we recover the solution of the problem of frictionless contact with a rigid foundation, i.e. to the solution of Signorini's frictionless contact problem.

For simplicity we assume the homogenous case, i.e. $\mu > 0$ is a constant. We assume that (6.44), (6.45), (15.9) and (15.46) hold and the initial data satisfy

$$\boldsymbol{u}_0 \in K_g, \qquad \boldsymbol{\sigma}_0 \in Q_1, \tag{15.59}$$
$$(\boldsymbol{\sigma}_0, \boldsymbol{\varepsilon}(\boldsymbol{v}))_Q = (\boldsymbol{f}(0), \boldsymbol{v})_V \quad \forall\, \boldsymbol{v} \in V, \tag{15.60}$$

where the set K_g and the function $\boldsymbol{f}$ are defined by (15.47) and (15.12), respectively.

Under these assumptions, it follows from Section 15.4 that the Signorini contact problem (15.1)–(15.4), (15.6), (15.45) has a unique weak solution $(\boldsymbol{u}, \boldsymbol{\sigma}) \in W^{1,\infty}(I; V \times Q_1)$, which satisfies

$$\dot{\boldsymbol{\sigma}}(t) = \mathcal{E}\boldsymbol{\varepsilon}(\dot{\boldsymbol{u}}(t)) + \mathcal{G}(\boldsymbol{\sigma}(t), \boldsymbol{\varepsilon}(\boldsymbol{u}(t))) \quad \text{a.e. } t \in I, \tag{15.61}$$
$$\boldsymbol{u}(t) \in K_g, \quad (\boldsymbol{\sigma}(t), \boldsymbol{\varepsilon}(\boldsymbol{v}) - \boldsymbol{\varepsilon}(\boldsymbol{u}(t)))_Q \ge (\boldsymbol{f}(t), \boldsymbol{v} - \boldsymbol{u}(t))_V \quad \forall\, \boldsymbol{v} \in V,\ t \in \overline{I}, \tag{15.62}$$
$$\boldsymbol{u}(0) = \boldsymbol{u}_0, \quad \boldsymbol{\sigma}(0) = \boldsymbol{\sigma}_0. \tag{15.63}$$

It also follows from the discussion in Section 15.5 that, for all $\mu > 0$, the frictionless contact problem (15.1)–(15.4), (15.6), (15.55) has a unique weak solution $(\boldsymbol{u}_\mu, \boldsymbol{\sigma}_\mu) \in W^{1,\infty}(I; V \times Q_1)$ which satisfies

$$\dot{\boldsymbol{\sigma}}_\mu(t) = \mathcal{E}\boldsymbol{\varepsilon}(\dot{\boldsymbol{u}}_\mu(t)) + \mathcal{G}(\boldsymbol{\sigma}_\mu(t), \boldsymbol{\varepsilon}(\boldsymbol{u}_\mu(t))) \quad \text{a.e. } t \in I, \tag{15.64}$$
$$(\boldsymbol{\sigma}_\mu(t), \boldsymbol{\varepsilon}(\boldsymbol{v}) - \boldsymbol{\varepsilon}(\boldsymbol{u}_\mu(t)))_Q + j_\mu(\boldsymbol{v}) - j_\mu(\boldsymbol{u}_\mu(t)) \ge (\boldsymbol{f}(t), \boldsymbol{v} - \boldsymbol{u}(t))_V \quad \forall\, \boldsymbol{v} \in V,\ t \in \overline{I}, \tag{15.65}$$
$$\boldsymbol{u}_\mu(0) = \boldsymbol{u}_0, \quad \boldsymbol{\sigma}_\mu(0) = \boldsymbol{\sigma}_0. \tag{15.66}$$

Here $j_\mu : V \to \mathbb{R}$ is the functional given by

$$j_\mu(\boldsymbol{v}) = \frac{1}{\mu(\alpha + 1)} \int_{\Gamma_3} (v_\nu - g)_+^{\alpha+1}\, da. \tag{15.67}$$

The behavior of the solution $(\boldsymbol{u}_\mu, \boldsymbol{\sigma}_\mu)$ of the problem (15.64)–(15.66) as $\mu \to 0$ is given in the following theorem.

Theorem 15.11 *Let* (6.44), (6.45), (15.9), (15.46), (15.59) *and* (15.60) *hold. Then, the solution* $(\boldsymbol{u}_\mu, \boldsymbol{\sigma}_\mu)$ *of the problem* (15.64)–(15.66) *converges to the solution* $(\boldsymbol{u}, \boldsymbol{\sigma})$ *of problem* (15.61)–(15.63), *i.e.*

$$\|\boldsymbol{u}_\mu(t) - \boldsymbol{u}(t)\|_V + \|\boldsymbol{\sigma}_\mu(t) - \boldsymbol{\sigma}(t)\|_{Q_1} \to 0 \quad \text{as } \mu \to 0, \tag{15.68}$$

for all $t \in \overline{I}$.

In addition to the mathematical interest in this result, it is important from mechanical point of view since it shows that the weak solution $(\boldsymbol{u}, \boldsymbol{\sigma})$ of Signorini's contact problem (15.1)–(15.4), (15.6), (15.45) may be obtained as the limit of the weak solution $(\boldsymbol{u}_\mu, \boldsymbol{\sigma}_\mu)$ of the frictionless contact problem with normal compliance (15.1)–(15.4), (15.6), (15.55), as the deformability coefficient of the foundation tends to zero.

The proof of Theorem 15.11 is carried out in several steps. In the rest of the section, we denote by c a strictly positive generic constant which does not depend on time or on μ, and whose value may change from place to place.

In the first step we consider the following auxiliary problem.

Problem 15.12 *Find a displacement field* $\widetilde{\boldsymbol{u}}_\mu : \overline{I} \to V$ *and a stress field* $\widetilde{\boldsymbol{\sigma}}_\mu : \overline{I} \to Q_1$ *such that*

$$\dot{\widetilde{\boldsymbol{\sigma}}}_\mu(t) = \mathcal{E}\boldsymbol{\varepsilon}(\dot{\widetilde{\boldsymbol{u}}}_\mu(t)) + \mathcal{G}(\boldsymbol{\sigma}(t), \boldsymbol{\varepsilon}(\boldsymbol{u}(t))) \quad \text{a.e. } t \in I, \tag{15.69}$$

$$(\widetilde{\boldsymbol{\sigma}}_\mu(t), \boldsymbol{\varepsilon}(\boldsymbol{v}) - \boldsymbol{\varepsilon}(\widetilde{\boldsymbol{u}}_\mu(t)))_Q + j_\mu(\boldsymbol{v}) - j_\mu(\widetilde{\boldsymbol{u}}_\mu(t)) \geq (\boldsymbol{f}(t), \boldsymbol{v} - \widetilde{\boldsymbol{u}}_\mu(t))_V \quad \forall \boldsymbol{v} \in V,\ t \in \overline{I}, \tag{15.70}$$

$$\widetilde{\boldsymbol{u}}_\mu(0) = \boldsymbol{u}_0, \quad \widetilde{\boldsymbol{\sigma}}_\mu(0) = \boldsymbol{\sigma}_0. \tag{15.71}$$

Using arguments similar to those in the proof of Theorem 15.2, we know that Problem 15.12 has a unique solution $\widetilde{\boldsymbol{u}}_\mu \in W^{1,\infty}(I;V)$, $\widetilde{\boldsymbol{\sigma}}_\mu \in W^{1,\infty}(I;Q_1)$.

Lemma 15.13 *We have the following convergence result:*

$$\|\widetilde{\boldsymbol{u}}_\mu(t) - \boldsymbol{u}(t)\|_V + \|\widetilde{\boldsymbol{\sigma}}_\mu(t) - \boldsymbol{\sigma}(t)\|_{Q_1} \to 0 \quad \text{as } \mu \to 0, \tag{15.72}$$

for all $t \in \overline{I}$.

Proof. Define a function $\boldsymbol{z} \in W^{1,\infty}(I, Q)$ by

$$\boldsymbol{z}(t) = \int_0^t \mathcal{G}(\boldsymbol{\sigma}(s), \boldsymbol{\varepsilon}(\boldsymbol{u}(s)))\, ds + \boldsymbol{\sigma}_0 - \mathcal{E}\boldsymbol{\varepsilon}(\boldsymbol{u}_0), \quad t \in \overline{I}. \tag{15.73}$$

Let $t \in \overline{I}$. It follows from (15.69)–(15.71) and (15.73) that

$$\widetilde{\boldsymbol{\sigma}}_\mu(t) = \mathcal{E}\boldsymbol{\varepsilon}(\widetilde{\boldsymbol{u}}_\mu(t)) + \boldsymbol{z}(t), \tag{15.74}$$

$$(\mathcal{E}\boldsymbol{\varepsilon}(\widetilde{\boldsymbol{u}}_\mu(t)), \boldsymbol{\varepsilon}(\boldsymbol{v}) - \boldsymbol{\varepsilon}(\widetilde{\boldsymbol{u}}_\mu(t)))_Q + (\boldsymbol{z}(t), \boldsymbol{\varepsilon}(\boldsymbol{v}) - \boldsymbol{\varepsilon}(\widetilde{\boldsymbol{u}}_\mu(t)))_Q + j_\mu(\boldsymbol{v}) - j_\mu(\boldsymbol{u}_\mu(t)) \geq (\boldsymbol{f}(t), \boldsymbol{v} - \widetilde{\boldsymbol{u}}_\mu(t))_V \quad \forall \boldsymbol{v} \in V. \tag{15.75}$$

Take $\boldsymbol{v} = \mathbf{0}$ in the inequality (15.75) to find

$$(\mathcal{E}\boldsymbol{\varepsilon}(\widetilde{\boldsymbol{u}}_\mu(t)), \boldsymbol{\varepsilon}(\widetilde{\boldsymbol{u}}_\mu(t)))_Q + j_\mu(\widetilde{\boldsymbol{u}}_\mu(t)) \\ \leq -(\boldsymbol{z}(t), \boldsymbol{\varepsilon}(\widetilde{\boldsymbol{u}}_\mu(t)))_Q + (\boldsymbol{f}(t), \widetilde{\boldsymbol{u}}_\mu(t))_V. \tag{15.76}$$

From this inequality and the positive definiteness of $\mathcal{E}$, we deduce that $\{\widetilde{\boldsymbol{u}}_\mu(t)\}_\mu$ is a bounded sequence in V. Thus, there exists $\widetilde{\boldsymbol{u}}(t) \in V$ and a subsequence of $\{\widetilde{\boldsymbol{u}}_\mu(t)\}_\mu$, again denoted by $\{\widetilde{\boldsymbol{u}}_\mu(t)\}_\mu$, such that

$$\widetilde{\boldsymbol{u}}_\mu(t) \rightharpoonup \widetilde{\boldsymbol{u}}(t) \quad \text{in } V \text{ as } \mu \to 0. \tag{15.77}$$

Using (15.76) again, we obtain

$$j_\mu(\widetilde{\boldsymbol{u}}_\mu(t)) \leq c.$$

Recalling the formula (15.67) for j_μ, by (15.77) and a lower semicontinuity argument we find

$$\int_{\Gamma_3} (\widetilde{u}_\nu(t) - g)_+^{\alpha+1}\, da = 0.$$

Hence,

$$\widetilde{\boldsymbol{u}} \in K_g. \tag{15.78}$$

Using now (15.75) and (15.67) we obtain

$$(\mathcal{E}\boldsymbol{\varepsilon}(\widetilde{\boldsymbol{u}}_\mu(t)), \boldsymbol{\varepsilon}(\boldsymbol{v}) - \boldsymbol{\varepsilon}(\widetilde{\boldsymbol{u}}_\mu(t)))_Q + (\boldsymbol{z}(t), \boldsymbol{\varepsilon}(\boldsymbol{v}) - \boldsymbol{\varepsilon}(\widetilde{\boldsymbol{u}}_\mu(t)))_Q \\ \geq (\boldsymbol{f}(t), \boldsymbol{v} - \widetilde{\boldsymbol{u}}_\mu(t))_V \quad \forall\, \boldsymbol{v} \in K_g. \tag{15.79}$$

As a consequence of (15.77)–(15.79) we deduce that

$$\widetilde{\boldsymbol{u}}(t) \in K_g, \quad (\mathcal{E}\boldsymbol{\varepsilon}(\widetilde{\boldsymbol{u}}(t)), \boldsymbol{\varepsilon}(\boldsymbol{v}) - \boldsymbol{\varepsilon}(\widetilde{\boldsymbol{u}}(t)))_Q + (\boldsymbol{z}(t), \boldsymbol{\varepsilon}(\boldsymbol{v}) - \boldsymbol{\varepsilon}(\widetilde{\boldsymbol{u}}(t)))_Q \\ \geq (\boldsymbol{f}(t), \boldsymbol{v} - \widetilde{\boldsymbol{u}}(t))_V \quad \forall\, \boldsymbol{v} \in K_g. \tag{15.80}$$

By (15.61)–(15.63) and (15.73), we infer that

$$\boldsymbol{\sigma}(t) = \mathcal{E}\boldsymbol{\varepsilon}(\boldsymbol{u}(t)) + \boldsymbol{z}(t), \tag{15.81}$$

$$\boldsymbol{u}(t) \in K_g, \quad (\mathcal{E}\boldsymbol{\varepsilon}(\boldsymbol{u}(t)), \boldsymbol{\varepsilon}(\boldsymbol{v}) - \boldsymbol{\varepsilon}(\boldsymbol{u}(t)))_Q + (\boldsymbol{z}(t), \boldsymbol{\varepsilon}(\boldsymbol{v}) - \boldsymbol{\varepsilon}(\boldsymbol{u}(t)))_Q \\ \geq (\boldsymbol{f}(t), \boldsymbol{v} - \boldsymbol{u}(t))_V \quad \forall\, \boldsymbol{v} \in K_g. \tag{15.82}$$

Since the variational inequalities (15.80) and (15.82) are uniquely solvable, we must have

$$\widetilde{\boldsymbol{u}}(t) = \boldsymbol{u}(t). \tag{15.83}$$

Using now (15.79) with $\boldsymbol{v} = \boldsymbol{u}(t)$ and the assumptions (6.44) on $\mathcal{E}$, we deduce that

$$\begin{aligned} c\,\|\widetilde{\boldsymbol{u}}_\mu(t) - \boldsymbol{u}(t)\|_V^2 &\leq (\mathcal{E}\boldsymbol{\varepsilon}(\widetilde{\boldsymbol{u}}_\mu(t)) - \mathcal{E}\boldsymbol{\varepsilon}(\boldsymbol{u}(t)), \boldsymbol{\varepsilon}(\widetilde{\boldsymbol{u}}_\mu(t)) - \boldsymbol{\varepsilon}(\boldsymbol{u}(t)))_Q \\ &= (\mathcal{E}\boldsymbol{\varepsilon}(\widetilde{\boldsymbol{u}}_\mu(t)), \boldsymbol{\varepsilon}(\widetilde{\boldsymbol{u}}_\mu(t)) - \boldsymbol{\varepsilon}(\boldsymbol{u}(t)))_Q \\ &\quad - (\mathcal{E}\boldsymbol{\varepsilon}(\boldsymbol{u}(t)), \boldsymbol{\varepsilon}(\widetilde{\boldsymbol{u}}_\mu(t)) - \boldsymbol{\varepsilon}(\boldsymbol{u}(t)))_Q \\ &\leq (\boldsymbol{z}(t), \boldsymbol{\varepsilon}(\boldsymbol{u}(t)) - \boldsymbol{\varepsilon}(\widetilde{\boldsymbol{u}}_\mu(t)))_Q + (\boldsymbol{f}(t), \widetilde{\boldsymbol{u}}_\mu(t) - \boldsymbol{u}(t))_V \\ &\quad - (\mathcal{E}\boldsymbol{\varepsilon}(\boldsymbol{u}(t)), \boldsymbol{\varepsilon}(\widetilde{\boldsymbol{u}}_\mu(t)) - \boldsymbol{\varepsilon}(\boldsymbol{u}(t)))_Q. \end{aligned}$$

By (15.77) and (15.83) we have the strong convergence of the displacement:

$$\|\widetilde{\boldsymbol{u}}_\mu(t) - \boldsymbol{u}(t)\|_V \to 0 \quad \text{as } \mu \to 0.$$

Then from the formulas (15.74) and (15.81) for $\widetilde{\boldsymbol{\sigma}}_\mu(t)$ and $\boldsymbol{\sigma}(t)$, we obtain

$$\|\widetilde{\boldsymbol{\sigma}}_\mu(t) - \boldsymbol{\sigma}(t)\|_Q \to 0 \quad \text{as } \mu \to 0.$$

Taking $\boldsymbol{v} = \boldsymbol{\phi}$, with $\boldsymbol{\phi} \in C_0^\infty(\Omega)^d$ arbitrary, in the variational inequalities (15.62) and (15.70), we find

$$\operatorname{Div} \widetilde{\boldsymbol{\sigma}}_\mu(t) = \operatorname{Div} \boldsymbol{\sigma}(t) = -\boldsymbol{f}_0(t) \quad \text{a.e. in } \Omega. \tag{15.84}$$

Thus, $\|\widetilde{\boldsymbol{\sigma}}_\mu(t) - \boldsymbol{\sigma}(t)\|_{Q_1} = \|\widetilde{\boldsymbol{\sigma}}_\mu(t) - \boldsymbol{\sigma}(t)\|_Q$ and we have the convergence result (15.68). ■

For any $\mu > 0$ we define a function $\boldsymbol{z}_\mu \in W^{1,\infty}(I; Q)$ by

$$\boldsymbol{z}_\mu(t) = \int_0^t \mathcal{G}(\boldsymbol{\sigma}_\mu(s), \boldsymbol{\varepsilon}(\boldsymbol{u}_\mu(s)))\, ds + \boldsymbol{\sigma}_0 - \mathcal{E}\boldsymbol{\varepsilon}(\boldsymbol{u}_0), \quad t \in \overline{I}. \tag{15.85}$$

The second step of the proof is the following result.

Lemma 15.14 *There exists $c > 0$ such that*

$$\|\widetilde{\boldsymbol{u}}_\mu(t) - \boldsymbol{u}_\mu(t)\|_V + \|\widetilde{\boldsymbol{\sigma}}_\mu(t) - \boldsymbol{\sigma}_\mu(t)\|_{Q_1} \le c\, \|\boldsymbol{z}_\mu(t) - \boldsymbol{z}(t)\|_Q, \tag{15.86}$$

for all $t \in \overline{I}$.

Proof. Let $t \in \overline{I}$. It follows from (15.64)–(15.66) and (15.85) that

$$\boldsymbol{\sigma}_\mu(t) = \mathcal{E}\boldsymbol{\varepsilon}(\boldsymbol{u}_\mu(t)) + \boldsymbol{z}_\mu(t), \tag{15.87}$$

$$\begin{aligned}(\mathcal{E}\boldsymbol{\varepsilon}(\boldsymbol{u}_\mu(t)), \boldsymbol{\varepsilon}(\boldsymbol{v}) - \boldsymbol{\varepsilon}(\boldsymbol{u}_\mu(t)))_Q + (\boldsymbol{z}_\mu(t), \boldsymbol{\varepsilon}(\boldsymbol{v}) - \boldsymbol{\varepsilon}(\boldsymbol{u}_\mu(t)))\\ + j_\mu(\boldsymbol{v}) - j_\mu(\boldsymbol{u}_\mu(t)) \ge (\boldsymbol{f}(t), \boldsymbol{v} - \boldsymbol{u}_\mu(t))_V \quad \forall \boldsymbol{v} \in V.\end{aligned} \tag{15.88}$$

We take $\boldsymbol{v} = \boldsymbol{u}_\mu(t)$ in (15.75), $\boldsymbol{v} = \widetilde{\boldsymbol{u}}_\mu(t)$ in (15.88), and add the two inequalities. After some algebraic manipulations with the use of the assumptions (6.44) on $\mathcal{E}$, we obtain

$$\|\widetilde{\boldsymbol{u}}_\mu(t) - \boldsymbol{u}_\mu(t)\|_V \le c\, \|\boldsymbol{z}_\mu(t) - \boldsymbol{z}(t)\|_Q.$$

Subtracting (15.87) from (15.74) and bounding, we have

$$\begin{aligned}\|\widetilde{\boldsymbol{\sigma}}_\mu(t) - \boldsymbol{\sigma}_\mu(t)\|_Q &\le \|\widetilde{\boldsymbol{u}}_\mu(t) - \boldsymbol{u}_\mu(t)\|_V + \|\boldsymbol{z}_\mu(t) - \boldsymbol{z}(t)\|_Q\\ &\le c\, \|\boldsymbol{z}_\mu(t) - \boldsymbol{z}(t)\|_Q.\end{aligned}$$

Taking $\boldsymbol{v} = \boldsymbol{\phi} \in C_0^\infty(\Omega)^d$ arbitrary in (15.65) and (15.70), we find

$$\operatorname{Div} \widetilde{\boldsymbol{\sigma}}_\mu(t) = \operatorname{Div} \boldsymbol{\sigma}_\mu(t) = -\boldsymbol{f}_0(t) \quad \text{a.e. in } \Omega.$$

Thus,

$$\|\widetilde{\boldsymbol{\sigma}}_\mu(t) - \boldsymbol{\sigma}_\mu(t)\|_{Q_1} = \|\widetilde{\boldsymbol{\sigma}}_\mu(t) - \boldsymbol{\sigma}_\mu(t)\|_Q \le c\,\|\boldsymbol{z}_\mu(t) - \boldsymbol{z}(t)\|_Q$$

and the inequality (15.86) holds. ■

We are now ready to prove Theorem 15.11.

Proof of Theorem 15.11.
Let $t \in \overline{I}$. By the norm triangle inequality, we have

$$\begin{aligned}
&\|\boldsymbol{u}_\mu(t) - \boldsymbol{u}(t)\|_V + \|\boldsymbol{\sigma}_\mu(t) - \boldsymbol{\sigma}(t)\|_{Q_1} \\
&\quad \le \|\widetilde{\boldsymbol{u}}_\mu(t) - \boldsymbol{u}_\mu(t)\|_V + \|\widetilde{\boldsymbol{\sigma}}_\mu(t) - \boldsymbol{\sigma}_\mu(t)\|_{Q_1} \\
&\qquad + \|\widetilde{\boldsymbol{u}}_\mu(t) - \boldsymbol{u}(t)\|_V + \|\widetilde{\boldsymbol{\sigma}}_\mu(t) - \boldsymbol{\sigma}(t)\|_{Q_1}.
\end{aligned}$$

Applying the inequality (15.86) and using the definitions (15.73) and (15.85) for $\boldsymbol{z}(t)$ and $\boldsymbol{z}_\mu(t)$, we obtain

$$\begin{aligned}
&\|\widetilde{\boldsymbol{u}}_\mu(t) - \boldsymbol{u}_\mu(t)\|_V + \|\widetilde{\boldsymbol{\sigma}}_\mu(t) - \boldsymbol{\sigma}_\mu(t)\|_{Q_1} \\
&\quad \le c\,\|\boldsymbol{z}_\mu(t) - \boldsymbol{z}(t)\|_Q \\
&\quad \le c \int_0^t \|\mathcal{G}(\boldsymbol{\sigma}_\mu(s), \boldsymbol{\varepsilon}(\boldsymbol{u}_\mu(s))) - \mathcal{G}(\boldsymbol{\sigma}(s), \boldsymbol{\varepsilon}(\boldsymbol{u}(s)))\|_Q ds,
\end{aligned}$$

which is bounded by

$$c \int_0^t (\|\boldsymbol{u}_\mu(s) - \boldsymbol{u}(s)\|_V + \|\boldsymbol{\sigma}_\mu(s) - \boldsymbol{\sigma}(s)\|_{Q_1})\, ds$$

with the assumptions (6.45) on $\mathcal{G}$. Therefore,

$$\begin{aligned}
&\|\boldsymbol{u}_\mu(t) - \boldsymbol{u}(t)\|_V + \|\boldsymbol{\sigma}_\mu(t) - \boldsymbol{\sigma}(t)\|_{Q_1} \\
&\quad \le c \int_0^t (\|\boldsymbol{u}_\mu(s) - \boldsymbol{u}(s)\|_V + \|\boldsymbol{\sigma}_\mu(s) - \boldsymbol{\sigma}(s)\|_{Q_1})\, ds \\
&\qquad + \|\widetilde{\boldsymbol{u}}_\mu(t) - \boldsymbol{u}(t)\|_V + \|\widetilde{\boldsymbol{\sigma}}_\mu(t) - \boldsymbol{\sigma}(t)\|_{Q_1}.
\end{aligned}$$

An application of Gronwall inequality (Lemma 7.24) yields

$$\begin{aligned}
&\|\boldsymbol{u}_\mu(t) - \boldsymbol{u}(t)\|_V + \|\boldsymbol{\sigma}_\mu(t) - \boldsymbol{\sigma}(t)\|_{Q_1} \\
&\quad \le \|\widetilde{\boldsymbol{u}}_\mu(t) - \boldsymbol{u}(t)\|_V + \|\widetilde{\boldsymbol{\sigma}}_\mu(t) - \boldsymbol{\sigma}(t)\|_{Q_1} \\
&\qquad + c \int_0^t (\|\widetilde{\boldsymbol{u}}_\mu(s) - \boldsymbol{u}(s)\|_V + \|\widetilde{\boldsymbol{\sigma}}_\mu(s) - \boldsymbol{\sigma}(s)\|_{Q_1})\, ds. \qquad (15.89)
\end{aligned}$$

On the other hand, taking $\boldsymbol{v} = \boldsymbol{u}(t)$ in (15.79) we find

$$\|\widetilde{\boldsymbol{u}}_\mu(t) - \boldsymbol{u}(t)\|_V \le c\,(\|\boldsymbol{f}(t)\|_V + \|\boldsymbol{u}(t)\|_V + \|\boldsymbol{z}(t)\|_Q).$$

From (15.74), (15.81) and (15.84) we deduce that

$$\|\widetilde{\boldsymbol{\sigma}}_\mu(t) - \boldsymbol{\sigma}(t)\|_{Q_1} \le c\,\|\widetilde{\boldsymbol{u}}_\mu(t) - \boldsymbol{u}(t)\|_V.$$

Combining the last two inequalities we obtain

$$\|\widetilde{\boldsymbol{u}}_\mu(t) - \boldsymbol{u}(t)\|_V + \|\widetilde{\boldsymbol{\sigma}}_\mu(t) - \boldsymbol{\sigma}(t)\|_{Q_1} \\ \leq c\,(\|\boldsymbol{f}(t)\|_V + \|\boldsymbol{u}(t)\|_V + \|\boldsymbol{z}(t)\|_Q). \tag{15.90}$$

Using (15.72), (15.90) and Lebesgue's convergence theorem we obtain

$$\int_0^t (\|\widetilde{\boldsymbol{u}}_\mu(s) - \boldsymbol{u}(s)\|_V + \|\widetilde{\boldsymbol{\sigma}}_\mu(s) - \boldsymbol{\sigma}(s)\|_{Q_1})\,ds \to 0 \quad \text{as } \mu \to 0. \tag{15.91}$$

Then (15.68) is a consequence of (15.72), (15.89) and (15.91). ■

15.7 Numerical Examples

To see the performance of the fully discrete scheme discussed in Section 15.3, we have checked a couple of numerical examples. We report in this section numerical results for the contact problem with normal compliance studied in Section 15.5 in one, two and three dimensional domains. Numerical examples for the Signorini contact problem considered in Section 15.4 can be found in [77, 78].

One dimensional example. The physical setting is similar to that presented in Section 11.7, see Figure 11.1 on page 246. We consider the contact problem with normal compliance with the following data:

$$\begin{aligned}
&\Omega = (0,1), \quad T = 10\,sec, \\
&\Gamma_1 = \{0\}, \quad \Gamma_2 = \emptyset, \quad \Gamma_3 = \{1\}, \quad f_0(x,t) = 10\,N/m, \\
&g = 0.25\,m, \quad \mu = 100\,m/N, \quad \alpha = 1, \\
&u_0(x) = 0\,m, \quad \sigma_0(x) = (10 - 10\,x)\,N/m, \\
&\mathcal{E} = 10\,N/m, \quad \mathcal{G}(\sigma,\varepsilon) = -\sigma + 10\,\varepsilon.
\end{aligned}$$

The exact solution of this problem is

- For $0 \leq t \leq \ln 2$ (no contact),

$$\begin{aligned}
\sigma_\mu(t,x) &= 10 - 10\,x, \\
u_\mu(t,x) &= (1 - e^{-t})\left(x - \frac{x^2}{2}\right).
\end{aligned}$$

- For $t > \ln 2$ (in contact)

$$\begin{aligned}
\sigma_\mu(t,x) &= \frac{5\,(2\,e^{-t} + 3 + 40\,\mu)}{2\,(10\,\mu + 1)} - 10\,x, \\
u_\mu(t,x) &= \frac{x^2}{2}\,(e^{-t} - 1) + x\left[\frac{2\,e^{-t} + 3 + 40\,\mu}{4\,(10\,\mu + 1)} - e^{-t}\right].
\end{aligned}$$

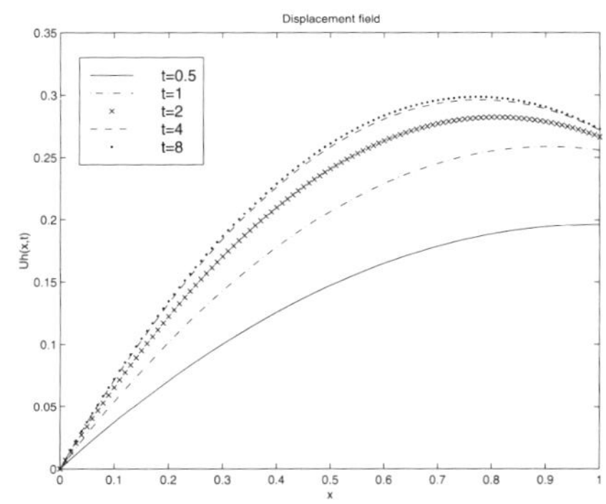

Figure 15.1: Displacement field at different times in a 1D normal compliance problem

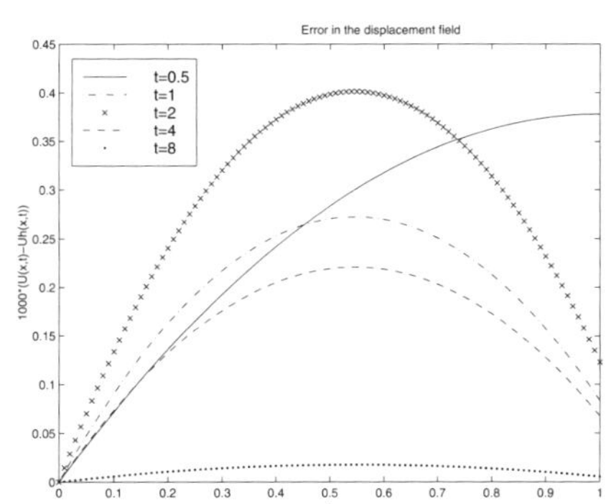

Figure 15.2: Error at different times in a 1D normal compliance problem

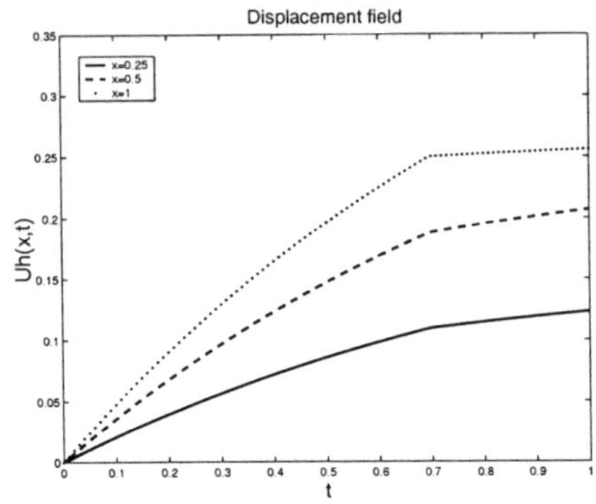

Figure 15.3: Evolution of displacements at $x = 0.25, 0.5, 1$

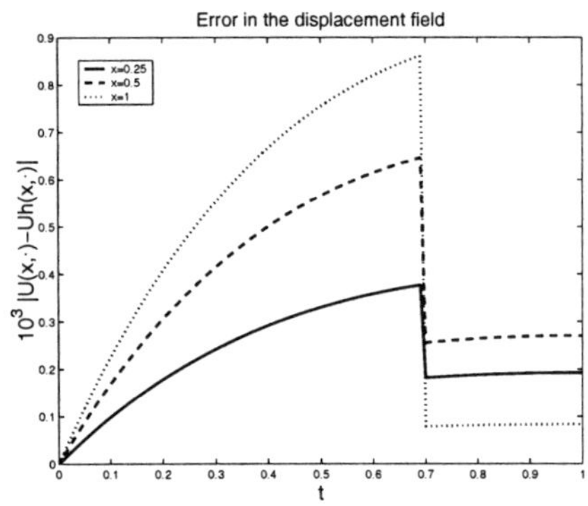

Figure 15.4: Scaled exact error at $x = 0.25, 0.5, 1$

In numerical computations, uniform partitions are used for the spatial domain $\overline{\Omega}$ and the time interval $\overline{I}$ with meshsizes h and k, respectively. Linear elements are used for V^h and piecewise constants for Q^h.

Figure 15.1 depicts the displacements at the times $t = 0.5, 1, 2, 4, 8$ sec., calculated with parameters $h = 0.01$ and $k = 0.01$. In Figure 15.2 we plot the pointwise error in the displacement field scaled by a factor 10^3.

In Figure 15.3 we show the evolution of the points $x = 0.25$, 0.5, 1. The corresponding error between the numerical solution and the exact solution is shown in Figure 15.4.

For a sequence of discretization parameters $h = k$, we have computed the corresponding numerical solutions u^{hk} and σ^{hk}, and the corresponding errors

$$\max_n \left(\|\sigma_n - \sigma_n^{hk}\|_Q + \|u_n - u_n^{hk}\|_V \right).$$

The error is plotted against the value $(k + h)$ in Figure 15.5. Linear convergence behavior is clearly observed, agreeing with the theoretical prediction made in Theorem 15.10 (page 334). In this particular example, the proportionality constant between the error and $(k + h)$ is approximately $C = 0.9557 \times 10^{-1}$.

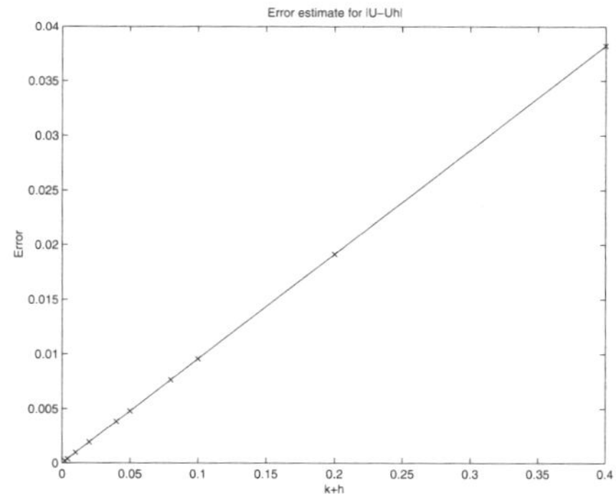

Figure 15.5: Linear convergence behavior of the fully discrete solution for a contact problem with normal compliance

Two-dimensional example. We then consider a two-dimensional contact problem with the following data:

$$\begin{aligned}
&\Omega = (0,6)\times(0,3), \quad T = 1\,sec,\\
&\Gamma_1 = 6\times[0,3], \quad \Gamma_2 = \{(0,6)\times\{3\}\}\cup\{\{0\}\times(0,3)\},\\
&\Gamma_3 = (0,6)\times\{0\},\\
&\boldsymbol{f}_0 = (0,0)N/m^2,\\
&\boldsymbol{f}_2 = \begin{cases} (0,-10\,t)\,N/m & \text{if } x_2 = 3,\\ (0,0)\,N/m & \text{otherwise,}\end{cases}\\
&g = 0\,m, \quad \mu = 100\,m^2/N, \quad \alpha = 1\\
&\boldsymbol{\sigma}_0 = \mathbf{0}\,N/m^2, \quad \boldsymbol{u}_0 = \mathbf{0}\,m.
\end{aligned}$$

The operator $\mathcal{E}$ is assumed to be the plane stress elasticity tensor, that is

$$(\mathcal{E}\boldsymbol{\tau})_{\alpha\beta} = \frac{E\nu}{1-\nu^2}\,(\tau_{11}+\tau_{22})\,\delta_{\alpha\beta} + \frac{E}{1+\nu}\,\tau_{\alpha\beta}, \quad \alpha,\beta = 1,2,$$

where E is the Young's modulus and ν is the Poisson's ratio. In this example we choose $E = 10^8 N/m^2$ and $\nu = 0.3$. We use the Perzyna's law

$$\dot{\boldsymbol{\varepsilon}} = \mathcal{E}^{-1}\dot{\boldsymbol{\sigma}} + \frac{1}{2\lambda}\,(\boldsymbol{\sigma} - \mathcal{P}_K\boldsymbol{\sigma}).$$

Thus,

$$\mathcal{G}(\boldsymbol{\sigma},\boldsymbol{\varepsilon}) = -\frac{1}{2\,\lambda}\,\mathcal{E}(\boldsymbol{\sigma} - \mathcal{P}_K\boldsymbol{\sigma}),$$

where $\lambda > 0$ is the viscosity coefficient and $\mathcal{P}_K$ is the orthogonal projection operator with respect to the inner product $(\mathcal{E}\boldsymbol{\sigma},\tau)$ over the convex subset $K\subset\mathbb{S}^2$ defined by

$$K = \{\,\boldsymbol{\tau}\in\mathbb{S}^2 : \tau_{11}^2 + \tau_{22}^2 - \tau_{11}\tau_{22} + 3\,\tau_{12}^2 \le \sigma_Y^2\,\},$$

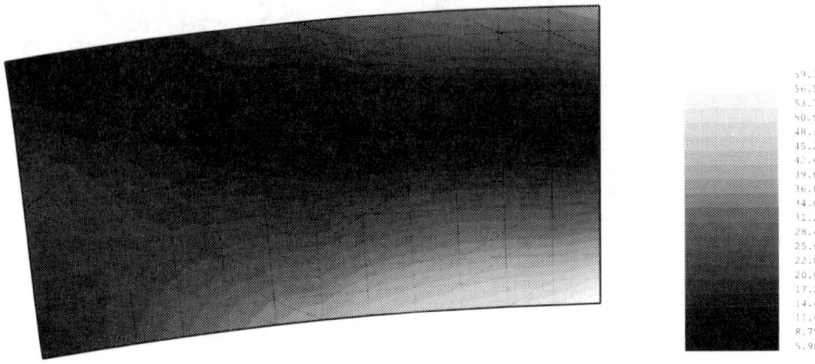

Figure 15.6: Deformed configuration and von Mises norm in a 2D normal compliance contact problem

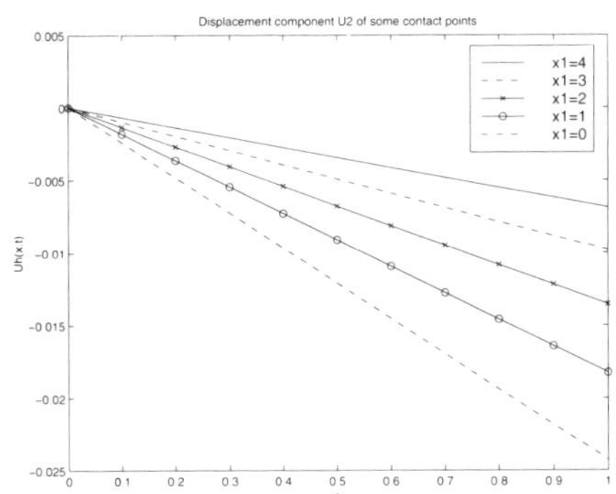

Figure 15.7: Evolution of the u_2 component in a 2D normal compliance contact problem

σ_Y being the uniaxial yield stress. For the example here, we use $\sigma_Y = \sqrt{10}\, N/m^2$ and $\lambda = 100\, N sec/m^2$.

In Figure 15.6 the deformed configuration and the von Mises norm for the stress at the final time $T = 1$ are shown, and in Figure 15.7 the time evolution of the penetration at a number of contact points is shown.

Three-dimensional example. Finally we consider a three-dimensional con-

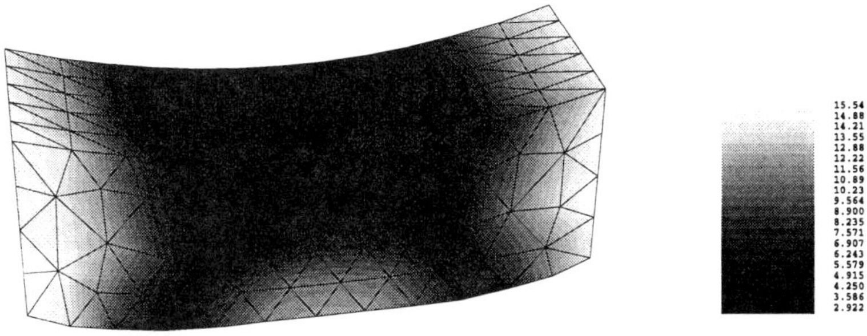

Figure 15.8: Deformed configuration and von Mises norm in a 3D normal compliance contact problem

tact problem with the following data:

$$
\begin{aligned}
&\Omega = (0,1) \times (0,3) \times (0,1), \quad T = 1sec, \\
&\Gamma_1 = [0,1] \times \{0,3\} \times [0,1], \quad \Gamma_3 = (0,1) \times (1,2) \times \{0\}, \\
&\Gamma_2 = \Gamma - (\Gamma_1 \cup \Gamma_3), \\
&\boldsymbol{f}_0 = (0,0,-10\,t)\,N/m^3, \quad \boldsymbol{f}_2 = (0,0,0)\,N/m^2, \\
&g = 0\,m, \quad \mu = 100\,m^3/N, \quad \alpha = 1 \\
&\boldsymbol{\sigma}_0 = \boldsymbol{0}\,N/m^3, \quad \boldsymbol{u}_0 = \boldsymbol{0}\,m.
\end{aligned}
$$

Here, $\mathcal{E}$ is the three-dimensional elasticity tensor,

$$
(\mathcal{E}\boldsymbol{\tau})_{ij} = \frac{E\nu}{(1+\nu)(1-2\nu)}\Big(\sum_{k=1}^{3}\tau_{kk}\Big)\delta_{ij} + \frac{E}{1+\nu}\tau_{ij}, \quad i,j = 1,2,3,
$$

where Young's modulus E and Poisson's ratio ν are taken to be $10^7 N/m^3$ and 0.3, respectively. Again we use Perzyna's law, in its three-dimensional version, with $\lambda = 100\,Nsec/m^3$. The convex subset $K \subset \mathbb{S}^3$ is

$$
\begin{aligned}
K = \{\, \boldsymbol{\tau} \in \mathbb{S}^3 : \; & (\sigma_{11} - \sigma_{22})^2 + (\sigma_{22} - \sigma_{33})^2 \\
& + (\sigma_{33} - \sigma_{11})^2 + 6\,(\sigma_{12}^2 + \sigma_{13}^2 + \sigma_{23}^2) \le \sigma_Y^2 \,\}.
\end{aligned}
$$

The uniaxial yield stress is $\sigma_Y = \sqrt{10}\,N/m^3$.

In Figures 15.8 and 15.9 we depict the deformed configuration and the von Mises norm for the stress at the final time $T = 1$, and the penetration of a number of contact nodes in time, respectively.

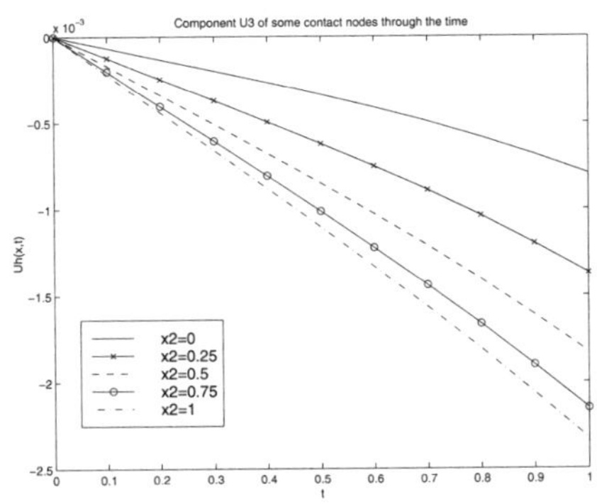

Figure 15.9: Evolution of the u_3 component in a 3D normal compliance contact problem

Chapter 16

Frictionless Contact between Two Viscoplastic Bodies

Unlike the other chapters, here we study a unilateral contact problem between two elastic-viscoplastic bodies. One purpose of the chapter is to convince the reader that analysis of contact problems between two deformable bodies does not present extra difficulties than that for contact problems between a deformable body and a rigid foundation. For the contact problem studied here, the contact is frictionless and is modeled by the classical Signorini boundary condition. We provide two different, yet equivalent, variational formulations of the model for which we prove existence and uniqueness results. We also show the continuous dependence of the solution on the data. We then turn to numerical approximations of the problem. For both spatially semi-discrete and fully discrete schemes we show the existence of unique solutions and derive error estimates. Finally, we present some numerical examples for two and three dimensional contact problems to show the performance of the numerical schemes.

16.1 Problem Statement

We consider two elastic-viscoplastic bodies whose material particles occupy open, bounded, connected domains Ω^1 and Ω^2 of $\mathbb{R}^d$. We use a superscript m to indicate that the quantity is related to the domain Ω^m, $m = 1, 2$. For each domain Ω^m, we assume its boundary Γ^m is Lipschitz continuous, and is partitioned into three parts $\overline{\Gamma_1^m}$, $\overline{\Gamma_2^m}$ and $\overline{\Gamma_3^m}$ with mutually disjoint relatively open sets Γ_1^m, Γ_2^m and Γ_3^m such that meas$\,(\Gamma_1^m) > 0$. The unit outward normal to Γ^m, defined almost everywhere, is denoted by $\boldsymbol{\nu}^m = (\nu_i^m)$. We are interested in the resulting contact process of evolution of the bodies in the time interval $\overline{I}$. Assume that the bodies are clamped on $\Gamma_1^m \times I$, and volume forces and surface tractions of densities $\boldsymbol{f}_0^m$ and $\boldsymbol{f}_2^m$ act in $\Omega^m \times I$ and $\Gamma_2^m \times I$, respectively. The two bodies are in potential contact along the common part $\Gamma_3^1 = \Gamma_3^2$, which is denoted Γ_3 below. The contact is frictionless and is modelled by the Signorini conditions on $\Gamma_3^m \times I$ in the form with a zero gap function. Finally we assume the process is quasistatic and we use (6.43) (page 135) as the viscoplastic constitutive law.

The mechanical problem we study here may be formulated as follows:

For $m = 1, 2$, find displacement fields $\boldsymbol{u}^m = (u_i^m) : \Omega^m \times \overline{I} \to \mathbb{R}^d$ and stress fields $\boldsymbol{\sigma}^m = (\sigma_{ij}^m) : \Omega^m \times \overline{I} \to \mathbb{S}^d$ such that

$$\dot{\boldsymbol{\sigma}}^m = \mathcal{E}^m \boldsymbol{\varepsilon}(\dot{\boldsymbol{u}}^m) + G^m(\boldsymbol{\sigma}^m, \boldsymbol{\varepsilon}(\boldsymbol{u}^m)) \quad \text{in } \Omega^m \times I, \tag{16.1}$$

$$\operatorname{Div} \boldsymbol{\sigma}^m + \boldsymbol{f}_0^m = \boldsymbol{0} \quad \text{in } \Omega^m \times I, \tag{16.2}$$

$$\boldsymbol{u}^m = \boldsymbol{0} \quad \text{on } \Gamma_1^m \times I, \tag{16.3}$$

$$\boldsymbol{\sigma}^m \boldsymbol{\nu}^m = \boldsymbol{f}_2^m \quad \text{on } \Gamma_2^m \times I, \tag{16.4}$$

$$\left.\begin{array}{ll} u_\nu^1 + u_\nu^2 \le 0, & \sigma_\nu^1 = \sigma_\nu^2 \le 0, \\ \sigma_\nu^1 (u_\nu^1 + u_\nu^2) = 0, & \boldsymbol{\sigma}_\tau^m = \boldsymbol{0} \end{array}\right\} \quad \text{on } \Gamma_3 \times I, \tag{16.5}$$

$$\boldsymbol{u}^m(0) = \boldsymbol{u}_0^m, \ \boldsymbol{\sigma}^m(0) = \boldsymbol{\sigma}_0^m \quad \text{in } \Omega^m. \tag{16.6}$$

Here, we use the notations

$$u_\nu^m = u_i^m \nu_i^m, \quad \sigma_\nu^m = \sigma_{ij}^m \nu_i^m \nu_j^m, \quad \boldsymbol{\sigma}_\tau^m = (\sigma_{\tau i}^m) \text{ with } \sigma_{\tau i}^m = \sigma_{ij}^m \nu_j^m - \sigma_\nu^m \nu_i^m$$

for the normal displacement, normal stress and tangential stress, respectively. Finally, $\boldsymbol{u}_0^m$ and $\boldsymbol{\sigma}_0^m$ are given initial displacement and initial stress, respectively.

We introduce the spaces

$$\begin{aligned} V^m &= \{\, \boldsymbol{v} = (v_i) : \ v_i \in H^1(\Omega^m), \ v_i = 0 \text{ on } \Gamma_1^m, \ 1 \le i \le d \,\}, \\ Q^m &= \{\, \boldsymbol{\tau} = (\tau_{ij}) : \ \tau_{ij} = \tau_{ji} \in L^2(\Omega^m), \ 1 \le i, j \le d \,\}, \\ Q_1^m &= \{\, \boldsymbol{\tau} \in Q^m : \ \operatorname{Div} \boldsymbol{\tau} \in L^2(\Omega^m)^d \,\}. \end{aligned}$$

These are Hilbert spaces with their canonical inner products. Over the space V^m, we use the inner product

$$(\boldsymbol{u}, \boldsymbol{v})_{V^m} = (\boldsymbol{\varepsilon}(\boldsymbol{u}), \boldsymbol{\varepsilon}(\boldsymbol{v}))_{Q^m} \quad \forall\, \boldsymbol{u}, \boldsymbol{v} \in V^m.$$

Since meas $(\Gamma_1^m) > 0$, Korn's inequality (7.8) (page 144) holds. Therefore, the map

$$\boldsymbol{v} \mapsto \|\boldsymbol{v}\|_{V^m} = \|\boldsymbol{\varepsilon}(\boldsymbol{v})\|_{Q^m}$$

defines a norm on V^m, which is equivalent to the norm $\|\boldsymbol{v}\|_{H^1(\Omega^m)^d}$.

We then define product spaces $\boldsymbol{V} = V^1 \times V^2$, $\boldsymbol{Q} = Q^1 \times Q^2$ and $\boldsymbol{Q}_1 = Q_1^1 \times Q_1^2$. They are all Hilbert spaces endowed with the canonical inner products denoted by $(\cdot,\cdot)_{\boldsymbol{V}}$, $(\cdot,\cdot)_{\boldsymbol{Q}}$ and $(\cdot,\cdot)_{\boldsymbol{Q}_1}$ respectively. The associated norms will be denoted by $\|\cdot\|_{\boldsymbol{V}}$, $\|\cdot\|_{\boldsymbol{Q}}$ and $\|\cdot\|_{\boldsymbol{Q}_1}$, respectively. For simplicity we shall use the notation $\boldsymbol{\varepsilon}(\boldsymbol{v}) = (\boldsymbol{\varepsilon}(\boldsymbol{v}^1), \boldsymbol{\varepsilon}(\boldsymbol{v}^2))$ for $\boldsymbol{v} = (\boldsymbol{v}^1, \boldsymbol{v}^2) \in \boldsymbol{V}$ and $\mathcal{E}\boldsymbol{\varepsilon} = (\mathcal{E}^1\boldsymbol{\varepsilon}^1, \mathcal{E}^2\boldsymbol{\varepsilon}^2)$, $\mathcal{G}(\boldsymbol{\sigma}, \boldsymbol{\varepsilon}) = (\mathcal{G}(\boldsymbol{\sigma}^1, \boldsymbol{\varepsilon}^1), \mathcal{G}(\boldsymbol{\sigma}^2, \boldsymbol{\varepsilon}^2))$ for $\boldsymbol{\varepsilon} = (\boldsymbol{\varepsilon}^1, \boldsymbol{\varepsilon}^2) \in \boldsymbol{Q}$ and $\boldsymbol{\sigma} = (\boldsymbol{\sigma}^1, \boldsymbol{\sigma}^2) \in \boldsymbol{Q}$.

In the study of the mechanical problem (16.1)–(16.6) we make the following assumptions, with $m = 1, 2$.

For the elasticity tensor $\mathcal{E}^m$, we assume

$$\left.\begin{array}{l}\text{(a) } \mathcal{E}^m : \Omega^m \times \mathbb{S}^d \to \mathbb{S}^d.\\ \text{(b) } \mathcal{E}^m_{ijkl} \in L^\infty(\Omega^m),\ 1 \le i,j,k,l \le d.\\ \text{(c) } \mathcal{E}^m \boldsymbol{\sigma} \cdot \boldsymbol{\tau} = \boldsymbol{\sigma} \cdot \mathcal{E}^m \boldsymbol{\tau} \quad \forall\, \boldsymbol{\sigma}, \boldsymbol{\tau} \in \mathbb{S}^d, \text{ a.e. in } \Omega^m.\\ \text{(d) There exists } \alpha^m > 0 \text{ such that}\\ \qquad \mathcal{E}^m \boldsymbol{\tau} \cdot \boldsymbol{\tau} \ge \alpha^m \|\boldsymbol{\tau}\|^2 \ \forall\, \boldsymbol{\tau} \in \mathbb{S}^d, \text{ a.e. in } \Omega^m.\end{array}\right\} \tag{16.7}$$

For the operator $\mathcal{G}^m$, we assume

$$\left.\begin{array}{l}\text{(a) } \mathcal{G}^m : \Omega^m \times \mathbb{S}^d \times \mathbb{S}^d \to \mathbb{S}^d.\\ \text{(b) There exists } L_{\mathcal{G}^m} > 0 \text{ such that}\\ \qquad \|\mathcal{G}^m(\boldsymbol{x}, \boldsymbol{\sigma}_1, \boldsymbol{\varepsilon}_1) - \mathcal{G}^m(\boldsymbol{x}, \boldsymbol{\sigma}_2, \boldsymbol{\varepsilon}_2)\|\\ \qquad\qquad \le L_{\mathcal{G}^m}(\|\boldsymbol{\sigma}_1 - \boldsymbol{\sigma}_2\| + \|\boldsymbol{\varepsilon}_1 - \boldsymbol{\varepsilon}_2\|)\\ \qquad \forall\, \boldsymbol{\sigma}_1, \boldsymbol{\sigma}_2, \boldsymbol{\varepsilon}_1, \boldsymbol{\varepsilon}_2 \in \mathbb{S}^d, \text{ a.e. in } \Omega^m.\\ \text{(c) For any } \boldsymbol{\sigma}, \boldsymbol{\varepsilon} \in \mathbb{S}^d,\ \boldsymbol{x} \mapsto \mathcal{G}^m(\boldsymbol{x}, \boldsymbol{\sigma}, \boldsymbol{\varepsilon}) \text{ is measurable.}\\ \text{(d) The mapping } \boldsymbol{x} \mapsto \mathcal{G}^m(\boldsymbol{x}, \boldsymbol{0}, \boldsymbol{0}) \in Q^m.\end{array}\right\} \tag{16.8}$$

For the given force densities, we assume

$$\boldsymbol{f}_0^m \in W^{1,\infty}(I; L^2(\Omega^m)^d), \quad \boldsymbol{f}_2^m \in W^{1,\infty}(I; L^2(\Gamma_2^m)^d). \tag{16.9}$$

For any $t \in \overline{I}$, let $\boldsymbol{f}(t)$ denote the element of $\boldsymbol{V}$ given by

$$\begin{aligned}(\boldsymbol{f}(t), \boldsymbol{v})_{\boldsymbol{V}} &= (\boldsymbol{f}_0^1(t), \boldsymbol{v}^1)_{L^2(\Omega^1)^d} + (\boldsymbol{f}_0^2(t), \boldsymbol{v}^2)_{L^2(\Omega^2)^d}\\ &\quad + (\boldsymbol{f}_2^1(t), \boldsymbol{v}^1)_{L^2(\Gamma_2^1)^d} + (\boldsymbol{f}_2^2(t), \boldsymbol{v}^2)_{L^2(\Gamma_2^2)^d}\end{aligned} \tag{16.10}$$

for all $\boldsymbol{v} = (\boldsymbol{v}^1, \boldsymbol{v}^2) \in \boldsymbol{V}$. The assumptions (16.9) imply

$$\boldsymbol{f} \in W^{1,\infty}(I; \boldsymbol{V}).$$

We define the set $\boldsymbol{U}$ of admissible displacement fields by

$$\boldsymbol{U} = \{\, \boldsymbol{v} = (\boldsymbol{v}^1, \boldsymbol{v}^2) \in \boldsymbol{V} : v_\nu^1 + v_\nu^2 \le 0 \text{ on } \Gamma_3 \,\}, \tag{16.11}$$

and, for all $t \in \overline{I}$, we define the set of admissible stress fields $\Sigma(t)$ by

$$\boldsymbol{\Sigma}(t) = \{\, \boldsymbol{\tau} \in (\boldsymbol{\tau}^1, \boldsymbol{\tau}^2) \in \boldsymbol{Q} : (\boldsymbol{\tau}, \boldsymbol{\varepsilon}(\boldsymbol{v}))_{\boldsymbol{Q}} \ge (\boldsymbol{f}(t), \boldsymbol{v})_{\boldsymbol{V}} \ \forall\, \boldsymbol{v} \in \boldsymbol{U} \,\}. \tag{16.12}$$

We suppose that the initial data satisfy

$$\left.\begin{array}{l}\boldsymbol{u}_0 = (\boldsymbol{u}_0^1, \boldsymbol{u}_0^2) \in \boldsymbol{U},\\ \boldsymbol{\sigma}_0 = (\boldsymbol{\sigma}_0^1, \boldsymbol{\sigma}_0^2) \in \boldsymbol{\Sigma}(0),\\ (\boldsymbol{\sigma}_0, \boldsymbol{\varepsilon}(\boldsymbol{u}_0))_{\boldsymbol{Q}} = (\boldsymbol{f}(0), \boldsymbol{u}_0)_{\boldsymbol{V}}.\end{array}\right\} \tag{16.13}$$

Using similar arguments as in Lemma 14.1 (page 310), we derive the following two weak formulations of the Signorini's frictionless problem (16.1)–(16.6).

Problem 16.1 *Find a displacement field* $\boldsymbol{u} : \overline{I} \to \boldsymbol{V}$ *and a stress field* $\boldsymbol{\sigma} : \overline{I} \to \boldsymbol{Q}_1$ *such that*

$$\dot{\boldsymbol{\sigma}}(t) = \mathcal{E}\boldsymbol{\varepsilon}(\dot{\boldsymbol{u}}(t)) + \mathcal{G}(\boldsymbol{\sigma}(t), \boldsymbol{\varepsilon}(\boldsymbol{u}(t))), \text{ a.e. } t \in I, \tag{16.14}$$

$$\boldsymbol{u}(t) \in \boldsymbol{U}, \quad (\boldsymbol{\sigma}(t), \boldsymbol{\varepsilon}(\boldsymbol{v}) - \boldsymbol{\varepsilon}(\boldsymbol{u}(t)))_{\boldsymbol{Q}} \geq (\boldsymbol{f}(t), \boldsymbol{v} - \boldsymbol{u}(t)) \quad \forall \boldsymbol{v} \in \boldsymbol{U},\ t \in \overline{I}, \tag{16.15}$$

$$\boldsymbol{u}(0) = \boldsymbol{u}_0, \qquad \boldsymbol{\sigma}(0) = \boldsymbol{\sigma}_0. \tag{16.16}$$

Problem 16.2 *Find a displacement field* $\boldsymbol{u} : \overline{I} \to \boldsymbol{V}$ *and a stress field* $\boldsymbol{\sigma} : \overline{I} \to \boldsymbol{Q}_1$ *such that*

$$\dot{\boldsymbol{\sigma}}(t) = \mathcal{E}\boldsymbol{\varepsilon}(\dot{\boldsymbol{u}}(t)) + \mathcal{G}(\boldsymbol{\sigma}(t), \boldsymbol{\varepsilon}(\boldsymbol{u}(t))) \quad \text{a.e. } t \in I, \tag{16.17}$$

$$\boldsymbol{\sigma} \in \boldsymbol{\Sigma}(t), \quad (\boldsymbol{\tau} - \boldsymbol{\sigma}(t), \boldsymbol{\varepsilon}(\boldsymbol{u}(t)))_{\boldsymbol{Q}} \geq 0 \quad \forall \boldsymbol{\tau} \in \boldsymbol{\Sigma}(t),\ t \in \overline{I}, \tag{16.18}$$

$$\boldsymbol{u}(0) = \boldsymbol{u}_0, \qquad \boldsymbol{\sigma}(0) = \boldsymbol{\sigma}_0. \tag{16.19}$$

We remark that Problems 16.1 and 16.2 are formally equivalent to the mechanical problem (16.1)–(16.6). Indeed, if $(\boldsymbol{u}, \boldsymbol{\sigma})$ represents a regular solution of the variational problem 16.1 or 16.2, then by a standard procedure, it can be shown that $(\boldsymbol{u}, \boldsymbol{\sigma})$ satisfies (16.1)–(16.6).

Well-posedness of Problems 16.1 and 16.2 is examined in the next section.

16.2 Unique Solvability and Properties of the Solution

It is easily seen that the variational problems 16.1 and 16.2 are similar to the variational problems 14.2 and 14.3, respectively. The only difference lies in the fact that here we are working on product spaces, i.e. $\boldsymbol{V} = V^1 \times V^2$, $\boldsymbol{Q} = Q^1 \times Q^2$ and $\boldsymbol{Q}_1 = Q_1^1 \times Q_1^2$. Because of this, most of the results presented in Chapter 14 may be extended in the study of Problems 16.1 and 16.2. In this section we present existence and uniqueness results as well as some properties of the solutions for Problems 16.1 and 16.2. Since the modifications are straightforward, for most of the results stated here we omit the details in the proofs.

Our first existence and uniqueness result is the following.

Theorem 16.3 *Under the assumptions* (16.7), (16.8), (16.9) *and* (16.13), *Problem 16.1 has a unique solution* $(\boldsymbol{u}, \boldsymbol{\sigma}) \in W^{1,\infty}(I; \boldsymbol{V} \times \boldsymbol{Q}_1)$.

Proof. The proof of Theorem 16.3 is carried out in several steps. It is based on similar arguments as those in the proof of Theorem 14.4. Here we sketch the main steps, for the convenience of the reader.

i) For any $\boldsymbol{\eta} \in L^\infty(I; \boldsymbol{Q})$ there exists a unique pair of functions $(\boldsymbol{u}_\eta, \boldsymbol{\sigma}_\eta) \in W^{1,\infty}(I; \boldsymbol{V} \times \boldsymbol{Q}_1)$ such that for all $t \in \overline{I}$,

$$\boldsymbol{\sigma}_\eta(t) = \mathcal{E}\boldsymbol{\varepsilon}(\boldsymbol{u}_\eta(t)) + \boldsymbol{z}_\eta(t),$$

$$\boldsymbol{u}_\eta(t) \in \boldsymbol{U},$$

$$(\boldsymbol{\sigma}_\eta(t), \boldsymbol{\varepsilon}(\boldsymbol{v}) - \boldsymbol{\varepsilon}(\boldsymbol{u}_\eta(t)))_{\boldsymbol{Q}} \geq (\boldsymbol{f}(t), \boldsymbol{v} - \boldsymbol{u}_\eta(t))_{\boldsymbol{V}} \quad \forall \boldsymbol{v} \in \boldsymbol{V},$$

where

$$z_\eta(t) = \int_0^t \eta(s)\,ds + \sigma_0 - \mathcal{E}\varepsilon(u_0).$$

Moreover, the solution satisfies

$$u_\eta(0) = u_0, \qquad \sigma_\eta(0) = \sigma_0.$$

ii) The operator $\Lambda : L^\infty(I; Q) \to L^\infty(I; Q)$ defined by

$$\Lambda\eta = \mathcal{G}(\sigma_\eta, \varepsilon(u_\eta)) \quad \forall\, \eta \in L^\infty(I; Q)$$

has a unique fixed point $\eta^* \in L^\infty(I; Q)$.

iii) The pair (u, σ) where $u = u_{\eta^*}$, $\sigma = \sigma_{\eta^*}$ is the solution of Problem 16.1 which satisfies $(u, \sigma) \in W^{1,\infty}(I; V \times Q_1)$. ■

Remark 16.4 *A careful examination of the previous arguments shows that if the space* $W^{1,\infty}$ *in the condition* (16.9) *is replaced by* $W^{1,p}$, $1 \le p < \infty$, *then Theorem 16.3 is still valid with the conclusion that Problem 16.1 has a unique solution* $(u, \sigma) \in W^{1,p}(I; V \times Q_1)$.

Let $(u, \sigma) \in W^{1,\infty}(I; V \times Q_1)$ be the solution of Problem 16.1. As in Section 8.1, we can show that

$$\operatorname{Div} \sigma^m + f_0^m = \mathbf{0} \quad \text{a.e. in } \Omega^m \times I,\ m = 1, 2, \tag{16.20}$$

and under the assumption

$$\sigma^m \nu^m \in L^\infty(I; L^2(\Gamma)),\ m = 1, 2, \tag{16.21}$$

we further have

$$\sigma^m \nu^m = f_2^m \quad \text{a.e. on } \Gamma_2^m \times I,\ m = 1, 2, \tag{16.22}$$

$$\sigma_\tau^m = \mathbf{0} \quad \text{a.e. on } \Gamma_3 \times I,\ m = 1, 2. \tag{16.23}$$

Our second existence and uniqueness result is the following.

Theorem 16.5 *Under the assumptions* (16.7), (16.8), (16.9) *and* (16.13), *Problem 16.2 has a unique solution* $(u, \sigma) \in W^{1,\infty}(I; V \times Q_1)$.

Proof. The proof of Theorem 16.5 is caried out in several steps. It is based on similar arguments as those used in the proof of Theorem 14.9, i.e.:

i) a change of variable which leads to a variational problem posed over a fixed convex set Σ_0;

ii) classical arguments on elliptic variational inequalities used to solve an auxiliary problem in which the viscoplastic part of the stress tensor is assumed to be known;

iii) a fixed point property. ■

We then consider the relation between Problem 16.1 and Problem 16.2 which represent two different variational formulations of the mechanical problem (16.1)–(16.6). We have the following equivalence result.

Theorem 16.6 *Assume that* (16.7), (16.8), (16.9) *and* (16.13) *hold and let* $\boldsymbol{u} \in W^{1,\infty}(I;\boldsymbol{V})$, $\boldsymbol{\sigma} \in W^{1,\infty}(I;\boldsymbol{Q}_1)$. *Then the couple* $(\boldsymbol{u},\boldsymbol{\sigma})$ *is a solution of the variational problem 16.1 iff it is a solution of the variational problem 16.2.*

Proof. The proof of this theorem is similar to the proof of Theorem 14.15 (page 316); in particular, properties of the projection map on the nonempty closed convex set $\boldsymbol{U} \subset \boldsymbol{V}$ are used in the proof. ■

Theorems 16.3 and 16.5 state the unique solvability of Problems 16.1 and 16.2, respectively, while Theorem 16.6 provides the equivalence of these variational problems. From these theorems we conclude that the mechanical problem (16.1)–(16.6) has a unique weak solution which solves both Problems 16.1 and 16.2.

We turn now to a continuous dependence result of the solution on the input data.

Theorem 16.7 *Assume that* (16.7), (16.8) *hold and let* $(\boldsymbol{u}_k,\boldsymbol{\sigma}_k)$ *be the solution of Problem 16.1 or 16.2 for the data* $\boldsymbol{f}_{0k}^m$, $\boldsymbol{f}_{2k}^m$, $\boldsymbol{u}_{0k}^m$, $\boldsymbol{\sigma}_{0k}^m$ *which satisfy* (16.9) *and* (16.13) *for* $k = 1,2$. *Then there exists* $c > 0$, *which depends only on* Ω, Γ_1^m, $\mathcal{E}^m$, $\mathcal{G}^m$ *and* T, *such that*

$$\begin{aligned}
&\|\boldsymbol{u}_1 - \boldsymbol{u}_2\|_{L^\infty(I;\boldsymbol{V})} + \|\boldsymbol{\sigma}_1 - \boldsymbol{\sigma}_2\|_{L^\infty(I;\boldsymbol{Q}_1)} \\
&\quad \le c\,\big(\|\boldsymbol{u}_{01} - \boldsymbol{u}_{02}\|_{\boldsymbol{V}} + \|\boldsymbol{\sigma}_{01} - \boldsymbol{\sigma}_{02}\|_{\boldsymbol{Q}_1} \\
&\qquad + \|\boldsymbol{f}_1 - \boldsymbol{f}_2\|_{L^\infty(I;\boldsymbol{V})}\big).
\end{aligned}$$

The proof of Theorem 16.7 is similar to that of Theorem 14.16. It is based on monotonicity and Gronwall type arguments. We omit the detail here.

Theorem 16.7 states that the solution of Problem 16.1 or 16.2 is Lipschitz continuous with respect to the initial displacement, the initial stress, and the force densities.

In the rest of the chapter we suppose that assumptions of Theorem 16.3 hold and we consider numerical approximations of Problem 16.1.

16.3 Semi-discrete Approximation

We will discretize the spatial domain by the finite element method. As in Section 7.2, we assume Ω^1 and Ω^2 are polygonal or polyhedral, and let $\{\mathcal{T}^h\}$ be a regular family of finite element partitions of the domain Ω in such a way that if a side of an element lies on the boundary, then the side is entirely on one of the subsets $\overline{\Gamma}_1^m$, $\overline{\Gamma}_2^m$ and $\overline{\Gamma}_3$. Corresponding to the partition $\mathcal{T}^h$, we choose $\boldsymbol{V}^h \subset \boldsymbol{V}$ to be the space of linear elements for the approximation of the displacement variable, and $\boldsymbol{Q}^h$ to be the space of piecewise constants for the approximation of the stress variable. Our discussions below is rather general and requires no specific information about the detailed structures of the finite element spaces. Then we define the discrete admissible displacement set

$$\boldsymbol{U}^h = \{\, \boldsymbol{v}^h = (\boldsymbol{v}^{1,h},\boldsymbol{v}^{2,h}) \in \boldsymbol{V}^h : v_\nu^{1,h} + v_\nu^{2,h} \le 0 \text{ on } \Gamma_3 \,\}.$$

Obviously, $\boldsymbol{U}^h \subset \boldsymbol{U}$. Let $\mathcal{P}_{\boldsymbol{Q}^h} : \boldsymbol{Q} \to \boldsymbol{Q}^h$ be the orthogonal projection defined through the relation

$$(\mathcal{P}_{\boldsymbol{Q}^h} \boldsymbol{q}, \boldsymbol{q}^h)_Q = (\boldsymbol{q}, \boldsymbol{q}^h)_Q \quad \forall \boldsymbol{q} \in \boldsymbol{Q},\ \boldsymbol{q}^h \in \boldsymbol{Q}^h.$$

We recall the property (15.21) (page 324) of an orthogonal projection operator.

Now the semi-discrete scheme is the following.

Problem 16.8 *Find a displacement field $\boldsymbol{u}^h : \overline{I} \to \boldsymbol{U}^h$ and a stress field $\boldsymbol{\sigma}^h : \overline{I} \to \boldsymbol{Q}^h$ such that*

$$\boldsymbol{u}^h(0) = \boldsymbol{u}_0^h,\ \boldsymbol{\sigma}^h(0) = \boldsymbol{\sigma}_0^h, \tag{16.24}$$

$$\dot{\boldsymbol{\sigma}}^h(t) = \mathcal{P}_{\boldsymbol{Q}^h}\mathcal{E}\boldsymbol{\varepsilon}(\dot{\boldsymbol{u}}^h(t)) + \mathcal{P}_{\boldsymbol{Q}^h}\mathcal{G}(\boldsymbol{\sigma}^h(t), \boldsymbol{\varepsilon}(\boldsymbol{u}^h(t))) \quad \text{a.e. } t \in I, \tag{16.25}$$

$$(\boldsymbol{\sigma}^h(t), \boldsymbol{\varepsilon}(\boldsymbol{v}^h - \boldsymbol{u}^h(t)))_Q \geq (\boldsymbol{f}(t), \boldsymbol{v}^h - \boldsymbol{u}^h(t))_V \ \forall \boldsymbol{v}^h \in \boldsymbol{U}^h,\ t \in I. \tag{16.26}$$

Here, $\boldsymbol{u}_0^h \in \boldsymbol{U}^h$, $\boldsymbol{\sigma}_0^h \in \boldsymbol{Q}^h$ are appropriate approximations of $\boldsymbol{u}_0$ and $\boldsymbol{\sigma}_0$.

Similar to the proof of Theorem 16.3, it can be shown that Problem 16.8 has a unique solution $(\boldsymbol{u}^h, \boldsymbol{\sigma}^h) \in W^{1,\infty}(I; \boldsymbol{U}^h \times \boldsymbol{Q}^h)$. Here we are interested in deriving estimates for the errors $\boldsymbol{u} - \boldsymbol{u}^h$ and $\boldsymbol{\sigma} - \boldsymbol{\sigma}^h$.

Theorem 16.9 *Let $(\boldsymbol{u}, \boldsymbol{\sigma}) \in W^{1,\infty}(I; \boldsymbol{U} \times \boldsymbol{Q}_1)$ be the solution of the problem 16.1, and $(\boldsymbol{u}^h, \boldsymbol{\sigma}^h) \in W^{1,\infty}(I; \boldsymbol{U}^h \times \boldsymbol{Q}^h)$ be the solution of the semi-discrete problem 16.8. Then we have the error estimate*

$$\begin{aligned}
&\|\boldsymbol{\sigma} - \boldsymbol{\sigma}^h\|_{L^\infty(I;\boldsymbol{Q})} + \|\boldsymbol{u} - \boldsymbol{u}^h\|_{L^\infty(I;\boldsymbol{V})} \\
&\quad \leq c\,(\|\boldsymbol{\sigma}_0 - \boldsymbol{\sigma}_0^h\|_{\boldsymbol{Q}} + \|\boldsymbol{u}_0 - \boldsymbol{u}_0^h\|_{\boldsymbol{V}} \\
&\qquad + c \inf_{\boldsymbol{v}^h \in L^\infty(I;\boldsymbol{U}^h)} \left\{ \|\boldsymbol{u} - \boldsymbol{v}^h\|_{L^\infty(I;\boldsymbol{V})} + |R(t; \boldsymbol{u} - \boldsymbol{v}^h)|^{1/2} \right\} \\
&\qquad + \|(I_{\boldsymbol{Q}} - \mathcal{P}_{\boldsymbol{Q}^h})(\boldsymbol{\sigma} - \boldsymbol{\sigma}_0)\|_{L^\infty(I;\boldsymbol{V})}),
\end{aligned} \tag{16.27}$$

where

$$R(t; \boldsymbol{v}) = (\boldsymbol{f}(t), \boldsymbol{v})_{\boldsymbol{V}} - (\boldsymbol{\sigma}(t), \boldsymbol{\varepsilon}(\boldsymbol{v}))_{\boldsymbol{Q}} \tag{16.28}$$

represents a residual quantity.

Proof. First we integrate the constitutive relation (16.14) and its discrete analogue (16.25), and use the initial conditions (16.16) and (16.24) to obtain

$$\boldsymbol{\sigma}(t) = \mathcal{E}\boldsymbol{\varepsilon}(\boldsymbol{u}(t)) + \int_0^t \mathcal{G}(\boldsymbol{\sigma}(s), \boldsymbol{\varepsilon}(\boldsymbol{u}(s)))\, ds + \boldsymbol{\sigma}_0 - \mathcal{E}\boldsymbol{\varepsilon}(\boldsymbol{u}_0), \tag{16.29}$$

$$\begin{aligned}
\boldsymbol{\sigma}^h(t) = {} & \mathcal{P}_{\boldsymbol{Q}^h}\mathcal{E}\boldsymbol{\varepsilon}(\boldsymbol{u}^h(t)) + \mathcal{P}_{\boldsymbol{Q}^h} \int_0^t \mathcal{G}(\boldsymbol{\sigma}^h(s), \boldsymbol{\varepsilon}(\boldsymbol{u}^h(s)))\, ds \\
& + \boldsymbol{\sigma}_0^h - \mathcal{P}_{\boldsymbol{Q}^h}\mathcal{E}\boldsymbol{\varepsilon}(\boldsymbol{u}_0^h).
\end{aligned} \tag{16.30}$$

We then subtract (16.30) from (16.29) to get the first error relation,

$$\begin{aligned}\boldsymbol{\sigma}(t)-\boldsymbol{\sigma}^h(t) &= \mathcal{P}_{Q^h}\mathcal{E}\boldsymbol{\varepsilon}(\boldsymbol{u}(t)-\boldsymbol{u}^h(t)) \\ &\quad + \mathcal{P}_{Q^h}\int_0^t \left[\mathcal{G}(\boldsymbol{\sigma}(s),\boldsymbol{\varepsilon}(\boldsymbol{u}(s))) - \mathcal{G}(\boldsymbol{\sigma}^h(s),\boldsymbol{\varepsilon}(\boldsymbol{u}^h(s)))\right] ds \\ &\quad + \boldsymbol{\sigma}_0 - \boldsymbol{\sigma}_0^h - \mathcal{P}_{Q^h}\mathcal{E}\boldsymbol{\varepsilon}(\boldsymbol{u}_0-\boldsymbol{u}_0^h) \\ &\quad + (I_Q - \mathcal{P}_{Q^h})(\boldsymbol{\sigma}(t)-\boldsymbol{\sigma}_0). \qquad (16.31)\end{aligned}$$

Now we take $\boldsymbol{v} = \boldsymbol{u}^h(t)$ in the variational inequality (16.15),

$$(\boldsymbol{\sigma}(t),\boldsymbol{\varepsilon}(\boldsymbol{u}^h(t)-\boldsymbol{u}(t)))_Q \ge (\boldsymbol{f}(t),\boldsymbol{u}^h(t)-\boldsymbol{u}(t))_V,$$

and rewrite the discrete variational inequality (16.26) in the form

$$(\boldsymbol{\sigma}^h(t),\boldsymbol{\varepsilon}(\boldsymbol{u}(t)-\boldsymbol{u}^h(t)))_Q \ge (\boldsymbol{f}(t),\boldsymbol{v}^h-\boldsymbol{u}^h(t))_V + (\boldsymbol{\sigma}^h(t),\boldsymbol{\varepsilon}(\boldsymbol{u}(t)-\boldsymbol{v}^h))_Q.$$

Adding these two inequalities, we obtain the second error relation,

$$\begin{aligned}(\boldsymbol{\sigma}(t)-\boldsymbol{\sigma}^h(t),\boldsymbol{\varepsilon}(\boldsymbol{u}(t)-\boldsymbol{u}^h(t)))_Q &\le (\boldsymbol{\sigma}(t)-\boldsymbol{\sigma}^h(t),\boldsymbol{\varepsilon}(\boldsymbol{u}(t)-\boldsymbol{v}^h))_Q \\ &\quad + R(t;\boldsymbol{u}(t)-\boldsymbol{v}^h). \qquad (16.32)\end{aligned}$$

Denote

$$e_0 = \|\boldsymbol{\sigma}_0-\boldsymbol{\sigma}_0^h\|_Q + \|\boldsymbol{u}_0-\boldsymbol{u}_0^h\|_V$$

for the initial error. We then derive the following inequality from (16.31) with the use of the assumptions (16.7) and (16.8),

$$\begin{aligned}\|\boldsymbol{\sigma}(t)-\boldsymbol{\sigma}^h(t)\|_Q &\le c\int_0^t \left(\|\boldsymbol{\sigma}(s)-\boldsymbol{\sigma}^h(s)\|_Q + \|\boldsymbol{u}(s)-\boldsymbol{u}^h(s)\|_V\right) ds \\ &\quad + c\,\|\boldsymbol{u}(t)-\boldsymbol{u}^h(t)\|_V + c\,e_0 \\ &\quad + \|(I_Q-\mathcal{P}_{Q^h})(\boldsymbol{\sigma}(t)-\boldsymbol{\sigma}_0)\|_Q. \qquad (16.33)\end{aligned}$$

Plug (16.31) into (16.32) and again use the assumptions (16.7) and (16.8). After some elementary manipulations, we have

$$\begin{aligned}\|\boldsymbol{u}(t)-\boldsymbol{u}^h(t)\|_V^2 &\le c\,\|\boldsymbol{\sigma}(t)-\boldsymbol{\sigma}^h(t)\|_Q\|\boldsymbol{u}(t)-\boldsymbol{v}^h\|_V \\ &\quad + c\,|R(t;\boldsymbol{u}(t)-\boldsymbol{v}^h)| + c\,\|\boldsymbol{u}(t)-\boldsymbol{u}^h(t)\|_V e_0 \\ &\quad + c\,\|\boldsymbol{u}(t)-\boldsymbol{u}^h(t)\|_V\int_0^t \big(\|\boldsymbol{\sigma}(s)-\boldsymbol{\sigma}^h(s)\|_Q \\ &\qquad + \|\boldsymbol{u}(s)-\boldsymbol{u}^h(s)\|_V\big)\, ds \\ &\quad + c\,\|(I_Q-\mathcal{P}_{Q^h})(\boldsymbol{\sigma}(t)-\boldsymbol{\sigma}_0)\|_Q\|\boldsymbol{u}(t)-\boldsymbol{u}^h(t)\|_V.\end{aligned}$$

Then we use the inequality

$$a\,b \le \delta\,a^2 + \frac{1}{4\,\delta}\,b^2 \quad \forall\,\delta > 0$$

to obtain

$$\begin{aligned}
&\|\boldsymbol{u}(t)-\boldsymbol{u}^h(t)\|_{\boldsymbol{V}} \\
&\quad\le \delta_0\|\boldsymbol{\sigma}(t)-\boldsymbol{\sigma}^h(t)\|_{\boldsymbol{Q}}+c\,\|\boldsymbol{u}(t)-\boldsymbol{v}^h\|_{\boldsymbol{V}}+c\,e_0 \\
&\qquad+c\,|R(t;\boldsymbol{u}(t)-\boldsymbol{v}^h)|^{1/2}+c\,\|(I_{\boldsymbol{Q}}-\mathcal{P}_{\boldsymbol{Q}^h})(\boldsymbol{\sigma}(t)-\boldsymbol{\sigma}_0)\|_{\boldsymbol{Q}} \\
&\qquad+c\int_0^t\left(\|\boldsymbol{\sigma}(s)-\boldsymbol{\sigma}^h(s)\|_{\boldsymbol{Q}}+\|\boldsymbol{u}(s)-\boldsymbol{u}^h(s)\|_{\boldsymbol{V}}\right)ds, \qquad (16.34)
\end{aligned}$$

where $\delta_0 > 0$ is arbitrary. Let δ_0 be a sufficiently small positive number. Combining (16.33) and (16.34) we obtain, for any $\boldsymbol{v}^h = \boldsymbol{v}^h(t) \in \boldsymbol{U}^h$,

$$\begin{aligned}
&\|\boldsymbol{\sigma}(t)-\boldsymbol{\sigma}^h(t)\|_{\boldsymbol{Q}}+\|\boldsymbol{u}(t)-\boldsymbol{u}^h(t)\|_{\boldsymbol{V}} \\
&\quad\le c\,\|\boldsymbol{u}(t)-\boldsymbol{v}^h(t)\|_{\boldsymbol{V}}+c\,e_0+c\,|R(t;\boldsymbol{u}(t)-\boldsymbol{v}^h(t))|^{1/2} \\
&\qquad+c\,\|(I_{\boldsymbol{Q}}-\mathcal{P}_{\boldsymbol{Q}^h})(\boldsymbol{\sigma}(t)-\boldsymbol{\sigma}_0)\|_{\boldsymbol{Q}} \\
&\qquad+c\int_0^t\left(\|\boldsymbol{\sigma}(s)-\boldsymbol{\sigma}^h(s)\|_{\boldsymbol{Q}}+\|\boldsymbol{u}(s)-\boldsymbol{u}^h(s)\|_{\boldsymbol{V}}\right)ds.
\end{aligned}$$

Then we apply Gronwall's inequality (Lemma 7.24, page 162) to obtain the estimate (16.27). ■

The estimate (16.27) is the basis for convergence analysis under the basic solution regularity $(\boldsymbol{u},\boldsymbol{\sigma}) \in W^{1,\infty}(I;\boldsymbol{U}\times\boldsymbol{Q}_1)$. For this purpose, we first quote a density result from [112, p. 141].

Lemma 16.10 *Let $d = 2$. Assume $\overline{\Gamma}_3 \cap \overline{\Gamma_1^1} = \emptyset$, $\overline{\Gamma}_3 \cap \overline{\Gamma_1^2} = \emptyset$, and there are only a finite number of boundary points of $\overline{\Gamma}_3 \cap \overline{\Gamma_2^1}$, $\overline{\Gamma}_3 \cap \overline{\Gamma_2^2}$, $\overline{\Gamma_1^1} \cap \overline{\Gamma_2^1}$, $\overline{\Gamma_1^2} \cap \overline{\Gamma_2^2}$. Let*

$$\begin{aligned}
\boldsymbol{K} = \{\, &\boldsymbol{v} = (\boldsymbol{v}^1,\boldsymbol{v}^2) \in H^1(\Omega^1)^d \times H^1(\Omega^2)^d : \\
&\boldsymbol{v}^1 = \boldsymbol{0} \text{ on } \Gamma_1^1,\ v_\nu^2 = 0 \text{ on } \Gamma_1^2,\ v_\nu^1 + v_\nu^2 \le 0 \text{ on } \Gamma_3 \,\}.
\end{aligned}$$

Then the set $\{(C^\infty(\overline{\Omega^1})^d \times (C^\infty(\overline{\Omega^2})^d)\} \cap \boldsymbol{K}$ is dense in $\boldsymbol{K}$ with respect to the norm of $H^1(\Omega^1)^d \times H^1(\Omega^2)^d$.

By a slight modification of the proof of Lemma 16.10 given in [112], we have the following density result.

Lemma 16.11 *Let $d = 2$. Assume $\overline{\Gamma}_3 \cap \overline{\Gamma_1^1} = \emptyset$, $\overline{\Gamma}_3 \cap \overline{\Gamma_1^2} = \emptyset$, and there are only a finite number of boundary points of $\overline{\Gamma}_3 \cap \overline{\Gamma_2^1}$, $\overline{\Gamma}_3 \cap \overline{\Gamma_2^2}$, $\overline{\Gamma_1^1} \cap \overline{\Gamma_2^1}$, $\overline{\Gamma_1^2} \cap \overline{\Gamma_2^2}$. Then the set $\{(C^\infty(\overline{\Omega^1})^d \times (C^\infty(\overline{\Omega^2})^d)\} \cap \boldsymbol{U}$ is dense in $\boldsymbol{U}$ with respect to the norm of $\boldsymbol{V}$.*

We have the following convergence result.

Corollary 16.12 *Assume the conditions of Lemma 16.11. Let the initial values $\boldsymbol{\sigma}_0^h \in \boldsymbol{Q}^h$ and $\boldsymbol{u}_0^h \in \boldsymbol{U}^h$ be chosen in such a way that*

$$\|\boldsymbol{\sigma}_0-\boldsymbol{\sigma}_0^h\|_{\boldsymbol{Q}} \to 0, \quad \|\boldsymbol{u}_0-\boldsymbol{u}_0^h\|_{\boldsymbol{V}} \to 0 \quad \text{as } h \to 0.$$

Then the semi-discrete method converges:

$$\|\boldsymbol{\sigma} - \boldsymbol{\sigma}^h\|_{L^\infty(I;\boldsymbol{Q})} + \|\boldsymbol{u} - \boldsymbol{u}^h\|_{L^\infty(I;\boldsymbol{V})} \to 0 \quad \text{as } h \to 0.$$

Proof. A direct estimate on the residual $R(t;\cdot)$ defined in (16.28) yields

$$|R(t;\boldsymbol{u}(t) - \boldsymbol{v}^h(t))| \le c\,(\|\boldsymbol{f}(t)\|_{\boldsymbol{V}} + \|\boldsymbol{\sigma}(t)\|_{\boldsymbol{Q}})\,\|\boldsymbol{u}(t) - \boldsymbol{v}^h(t)\|_{\boldsymbol{V}}. \tag{16.35}$$

Thus, (16.27) can be rewritten as

$$\begin{aligned}
&\|\boldsymbol{\sigma} - \boldsymbol{\sigma}^h\|_{L^\infty(I;\boldsymbol{Q})} + \|\boldsymbol{u} - \boldsymbol{u}^h\|_{L^\infty(I;\boldsymbol{V})} \\
&\quad \le c\,(\|\boldsymbol{\sigma}_0 - \boldsymbol{\sigma}_0^h\|_{\boldsymbol{Q}} + \|\boldsymbol{u}_0 - \boldsymbol{u}_0^h\|_{\boldsymbol{V}} \\
&\qquad + \|(I_{\boldsymbol{Q}} - \mathcal{P}_{\boldsymbol{Q}^h})(\boldsymbol{\sigma} - \boldsymbol{\sigma}_0)\|_{L^\infty(I;\boldsymbol{V})}) \\
&\qquad + c \inf_{\boldsymbol{v}^h \in L^\infty(I;\boldsymbol{U}^h)} \Big\{ \|\boldsymbol{u} - \boldsymbol{v}^h\|_{L^\infty(I;\boldsymbol{V})} \\
&\qquad + (\|\boldsymbol{f}\|_{L^\infty(I;\boldsymbol{V})} + \|\boldsymbol{\sigma}\|_{L^\infty(I;\boldsymbol{Q})})^{1/2}\, \|\boldsymbol{u} - \boldsymbol{v}^h\|_{L^\infty(I;\boldsymbol{V})}^{1/2} \Big\}.
\end{aligned}$$

For smooth functions, we have the finite element interpolation error estimate

$$\|\boldsymbol{v} - \Pi^h \boldsymbol{v}\|_{\boldsymbol{V}} \le c\,h\,\|\boldsymbol{v}\|_{H^2(\Omega)^d} \quad \forall\, \boldsymbol{v} \in \{(C^\infty(\overline{\Omega^1})^d \times (C^\infty(\overline{\Omega^2})^d)\} \cap \boldsymbol{U}.$$

Recalling Lemma 16.11, we can apply Theorem 7.21 (page 159) to conclude the convergence. ■

Under the additional solution regularity assumption (16.21), the bound (16.27) on the residual term can be improved. Indeed, we can use the pointwise relations (16.20), (16.22) and (16.23) to find

$$R(t;\boldsymbol{u}(t) - \boldsymbol{v}^h(t)) = \int_{\Gamma_3} \sigma_\nu^1(t)\,(v_\nu^1(t) - u_\nu^1(t) + v_\nu^2(t) - u_\nu^2(t))\,da.$$

Therefore, instead of (16.35), we have

$$|R(t;\boldsymbol{u}(t) - \boldsymbol{v}^h(t))| \le \|\sigma_\nu^1(t)\|_{L^2(\Gamma_3)} \|v_\nu(t) - u_\nu(t)\|_{L^2(\Gamma_3)},$$

where

$$\|v_\nu(t) - u_\nu(t)\|_{L^2(\Gamma_3)} = \|v_\nu^1(t) - u_\nu^1(t)\|_{L^2(\Gamma_3)} + \|v_\nu^2(t) - u_\nu^2(t)\|_{L^2(\Gamma_3)}.$$

Thus we have shown the following result.

Corollary 16.13 *Assume the additional regularity condition* (16.21). *Then*

$$\begin{aligned}
&\|\boldsymbol{\sigma} - \boldsymbol{\sigma}^h\|_{L^\infty(I;\boldsymbol{Q})} + \|\boldsymbol{u} - \boldsymbol{u}^h\|_{L^\infty(I;\boldsymbol{V})} \\
&\quad \le c\,(\|\boldsymbol{\sigma}_0 - \boldsymbol{\sigma}_0^h\|_{\boldsymbol{Q}} + \|\boldsymbol{u}_0 - \boldsymbol{u}_0^h\|_{\boldsymbol{V}}) \\
&\qquad + c \inf_{\boldsymbol{v}^h \in L^\infty(I;\boldsymbol{U}^h)} \{ \|\boldsymbol{u} - \boldsymbol{v}^h\|_{L^\infty(I;\boldsymbol{V})} + \|u_\nu - v_\nu^h\|_{L^\infty(I;L^2(\Gamma_3))}^{1/2} \} \\
&\qquad + c\,\|(I_{\boldsymbol{Q}} - \mathcal{P}_{\boldsymbol{Q}^h})(\boldsymbol{\sigma}(t) - \boldsymbol{\sigma}_0)\|_{\boldsymbol{Q}}.
\end{aligned} \tag{16.36}$$

Here the constant $c = c(\sigma_\nu^1)$ *depends on* $\|\sigma_\nu^1\|_{L^\infty(I;L^2(\Gamma_3))}^{1/2}$.

The inequality (16.36) is the basis for deriving order error estimates for the semi-discrete solutions. We present one sample result.

Assume $\boldsymbol{u} \in L^\infty(I; H^2(\Omega)^d)$, $u_\nu|_{\Gamma_{3,i}} \in L^\infty(I; H^2(\Gamma_{3,i}))$, $1 \le i \le i_0$, and $\boldsymbol{\sigma} \in L^\infty(I; H^1(\Omega)^{d\times d})$. We choose the initial values $\boldsymbol{u}_0^h \in \boldsymbol{U}^h$ and $\boldsymbol{\sigma}_0^h \in \boldsymbol{Q}^h$ in such a way that

$$\|\boldsymbol{\sigma}_0 - \boldsymbol{\sigma}_0^h\|_{\boldsymbol{Q}} + \|\boldsymbol{u}_0 - \boldsymbol{u}_0^h\|_{\boldsymbol{V}} \le c\, h.$$

Let $\Pi^h \boldsymbol{u}(t) \in \boldsymbol{V}^h$ be the piecewise linear interpolant of $\boldsymbol{u}(t)$. Then we have the interpolation error estimates (7.32) and (7.38) from Section 7.2. Hence,

$$\|\boldsymbol{u} - \Pi^h \boldsymbol{u}\|_{L^\infty(I;\boldsymbol{V})} \le c\, h\, |\boldsymbol{u}|_{L^\infty(I;H^2(\Omega)^d)},$$

$$\|u_\nu - \Pi^h u_\nu\|^{1/2}_{L^\infty(I;L^2(\Gamma_3))} \le c\, h \sum_{i=1}^{i_0} |u_\nu|^{1/2}_{L^\infty(I;H^2(\Gamma_{3,i}))}.$$

Then from (16.36), we conclude the optimal order error estimate

$$\|\boldsymbol{\sigma} - \boldsymbol{\sigma}^h\|_{L^\infty(I;\boldsymbol{Q})} + \|\boldsymbol{u} - \boldsymbol{u}^h\|_{L^\infty(I;\boldsymbol{V})} = O(h).$$

16.4 Fully Discrete Approximation

Now we consider a fully discrete approximation of Problem 16.1. We use the finite element spaces $\boldsymbol{V}^h$, $\boldsymbol{Q}^h$ and the finite element set $\boldsymbol{U}^h$ introduced in the previous section, and a general partition of the time interval described in Section 7.2. Then a fully discrete approximation method is the following.

Problem 16.14 *Find $\boldsymbol{u}^{hk} = \{\boldsymbol{u}_n^{hk}\}_{n=0}^N \subset \boldsymbol{U}^h$ and $\boldsymbol{\sigma}^{hk} = \{\boldsymbol{\sigma}_n^{hk}\}_{n=0}^N \subset \boldsymbol{Q}^h$ such that*

$$\boldsymbol{u}_0^{hk} = \boldsymbol{u}_0^h, \quad \boldsymbol{\sigma}_0^{hk} = \boldsymbol{\sigma}_0^h, \tag{16.37}$$

and for $n = 1, \ldots, N$,

$$\delta_n \boldsymbol{\sigma}_n^{hk} = \mathcal{P}_{\boldsymbol{Q}^h} \mathcal{E} \delta_n \boldsymbol{\varepsilon}(\boldsymbol{u}_n^{hk}) + \mathcal{P}_{\boldsymbol{Q}^h} \mathcal{G}(\boldsymbol{\sigma}_{n-1}^{hk}, \boldsymbol{\varepsilon}(\boldsymbol{u}_{n-1}^{hk})), \tag{16.38}$$

$$(\boldsymbol{\sigma}_n^{hk}, \boldsymbol{\varepsilon}(\boldsymbol{v}^h - \boldsymbol{u}_n^{hk}))_{\boldsymbol{Q}} \ge (\boldsymbol{f}_n, \boldsymbol{v}^h - \boldsymbol{u}_n^{hk})_{\boldsymbol{V}} \quad \forall\, \boldsymbol{v}^h \in \boldsymbol{U}^h. \tag{16.39}$$

From (16.38), we get

$$\boldsymbol{\sigma}_n^{hk} = \mathcal{P}_{\boldsymbol{Q}^h} \mathcal{E} \boldsymbol{\varepsilon}(\boldsymbol{u}_n^{hk} - \boldsymbol{u}_{n-1}^{hk}) + k_n \mathcal{P}_{\boldsymbol{Q}^h} \mathcal{G}(\boldsymbol{\sigma}_n^{hk}, \boldsymbol{\varepsilon}(\boldsymbol{u}_n^{hk})) + \boldsymbol{\sigma}_{n-1}^{hk}. \tag{16.40}$$

This is used in (16.39) to yield

$$(\mathcal{E}\boldsymbol{\varepsilon}(\boldsymbol{u}_n^{hk}), \boldsymbol{\varepsilon}(\boldsymbol{v}^h - \boldsymbol{u}_n^{hk}))_{\boldsymbol{Q}} \ge (\boldsymbol{f}_n, \boldsymbol{v}^h - \boldsymbol{u}_n^{hk})_{\boldsymbol{V}} + (\boldsymbol{\tau}_{n-1}^{hk}, \boldsymbol{\varepsilon}(\boldsymbol{v}^h - \boldsymbol{u}_n^{hk}))_{\boldsymbol{Q}}$$

for any $\boldsymbol{v}^h \in \boldsymbol{U}^h$, where

$$\boldsymbol{\tau}_{n-1}^{hk} = \mathcal{E}\boldsymbol{\varepsilon}(\boldsymbol{u}_{n-1}^{hk}) - k_n \mathcal{G}(\boldsymbol{\sigma}_{n-1}^{hk}, \boldsymbol{\varepsilon}(\boldsymbol{u}_{n-1}^{hk})) - \boldsymbol{\sigma}_{n-1}^{hk}.$$

This is a discrete elliptic variational inequality of the first kind and has a unique solution $\boldsymbol{u}_n^{hk} \in \boldsymbol{U}^h$. From (16.40) we then determine a unique $\boldsymbol{\sigma}_n^{hk} \in \boldsymbol{Q}^h$. In conclusion, Problem 16.14 has a unique solution.

Now we derive an error estimate for the fully discrete solution $(\boldsymbol{u}^{hk}, \boldsymbol{\sigma}^{hk})$. Apply (16.40) recursively to obtain

$$\boldsymbol{\sigma}_n^{hk} = \mathcal{P}_{\boldsymbol{Q}^h}\mathcal{E}\varepsilon(\boldsymbol{u}_n^{hk}) + \sum_{j=1}^{n} k_j \mathcal{P}_{\boldsymbol{Q}^h}\mathcal{G}(\boldsymbol{\sigma}_{j-1}^{hk}, \varepsilon(\boldsymbol{u}_{j-1}^{hk})) + \boldsymbol{\sigma}_0^{hk} - \mathcal{P}_{\boldsymbol{Q}^h}\mathcal{E}\varepsilon(\boldsymbol{u}_0^{hk}). \tag{16.41}$$

Subtracting (16.41) from (16.29) at $t = t_n$, we obtain the first error relation

$$\begin{aligned} \mathcal{P}_{\boldsymbol{Q}^h}\mathcal{E}\varepsilon(\boldsymbol{u}_n - \boldsymbol{u}_n^{hk}) &= \boldsymbol{\sigma}_n - \boldsymbol{\sigma}_n^{hk} + (I_{\boldsymbol{Q}} - \mathcal{P}_{\boldsymbol{Q}^h})(\boldsymbol{\sigma}_n - \boldsymbol{\sigma}_0) - \mathcal{P}_{\boldsymbol{Q}^h}\boldsymbol{\theta}_n^k(\mathcal{G}) \\ &\quad + \mathcal{P}_{\boldsymbol{Q}^h}\mathcal{E}\varepsilon(\boldsymbol{u}_n - \boldsymbol{u}_0^h) - \boldsymbol{\sigma}_0 + \boldsymbol{\sigma}_0^h \end{aligned} \tag{16.42}$$

where $\boldsymbol{\theta}_n^k(\mathcal{G})$ is the quadrature error term defined in (15.39) on page 328. Using the variational inequalities (16.39) and (16.15) at $t = t_n$ with $\boldsymbol{v} = \boldsymbol{u}_n^{hk}$, we have

$$(\boldsymbol{\sigma}_n - \boldsymbol{\sigma}_n^{hk}, \varepsilon(\boldsymbol{u}_n - \boldsymbol{u}_n^{hk}))_{\boldsymbol{Q}} \le (\boldsymbol{f}_n, \boldsymbol{u}_n - \boldsymbol{v}^h)_{\boldsymbol{V}} - (\boldsymbol{\sigma}_n^{hk}, \varepsilon(\boldsymbol{u}_n - \boldsymbol{v}^h))_{\boldsymbol{Q}}.$$

Thus we have the second error relation

$$(\boldsymbol{\sigma}_n - \boldsymbol{\sigma}_n^{hk}, \varepsilon(\boldsymbol{u}_n - \boldsymbol{u}_n^{hk}))_{\boldsymbol{Q}} \le (\boldsymbol{\sigma}_n - \boldsymbol{\sigma}_n^{hk}, \varepsilon(\boldsymbol{u}_n - \boldsymbol{v}^h))_{\boldsymbol{Q}} + R_n(\boldsymbol{u}_n - \boldsymbol{v}^h), \tag{16.43}$$

where

$$R_n(\boldsymbol{v}) = (\boldsymbol{f}_n, \boldsymbol{v})_{\boldsymbol{V}} - (\boldsymbol{\sigma}_n, \varepsilon(\boldsymbol{v}))_{\boldsymbol{Q}}. \tag{16.44}$$

Denote

$$e_n = \|\boldsymbol{\sigma}_n - \boldsymbol{\sigma}_n^{hk}\|_{\boldsymbol{Q}} + \|\boldsymbol{u}_n - \boldsymbol{u}_n^{hk}\|_{\boldsymbol{V}}, \quad n = 0, \dots, N \tag{16.45}$$

for the error of the numerical solution. From the first error relation (16.42), we then get

$$\begin{aligned} \|\boldsymbol{u}_n - \boldsymbol{u}_n^{hk}\|_{\boldsymbol{V}} &\le c\,\|\boldsymbol{\sigma}_n - \boldsymbol{\sigma}_n^{hk}\|_{\boldsymbol{Q}} + \|(I_{\boldsymbol{Q}} - \mathcal{P}_{\boldsymbol{Q}^h})(\boldsymbol{\sigma}_n - \boldsymbol{\sigma}_0)\|_{\boldsymbol{Q}} \\ &\quad + c\sum_{j=1}^{n} k_j e_{j-1} + c\, e_0 \\ &\quad + c\, k\,(\|\dot{\boldsymbol{\sigma}}\|_{L^\infty(I;\boldsymbol{Q})} + \|\dot{\boldsymbol{u}}\|_{L^\infty(I;\boldsymbol{V})}). \end{aligned} \tag{16.46}$$

Combining (16.42) and (16.43), we obtain

$$\begin{aligned} &(\boldsymbol{\sigma}_n - \boldsymbol{\sigma}_n^{hk}, \mathcal{E}^{-1}(\boldsymbol{\sigma}_n - \boldsymbol{\sigma}_n^{hk}))_{\boldsymbol{Q}} \\ &\quad \le (\boldsymbol{\sigma}_n - \boldsymbol{\sigma}_n^{hk}, \varepsilon(\boldsymbol{u}_n - \boldsymbol{v}^h))_{\boldsymbol{Q}} + R_n(\boldsymbol{u}_n - \boldsymbol{v}^h) \\ &\quad\quad + c\,\|\boldsymbol{\sigma}_n - \boldsymbol{\sigma}_n^{hk}\|_{\boldsymbol{Q}} \Big[\sum_{j=1}^{n} k_j e_{j-1} + e_0 + \|(I_{\boldsymbol{Q}} - \mathcal{P}_{\boldsymbol{Q}^h})(\boldsymbol{\sigma}_n - \boldsymbol{\sigma}_0)\|_{\boldsymbol{Q}}\Big] \\ &\quad\quad + c\, k\,(\|\dot{\boldsymbol{\sigma}}\|_{L^\infty(I;\boldsymbol{Q})} + \|\dot{\boldsymbol{u}}\|_{L^\infty(I;\boldsymbol{V})}), \end{aligned}$$

from which,

$$\|\boldsymbol{\sigma}_n - \boldsymbol{\sigma}_n^{hk}\|_{\boldsymbol{Q}} \le c\left[\|\boldsymbol{u}_n - \boldsymbol{v}^h\|_{\boldsymbol{V}} + |R_n(\boldsymbol{u}_n - \boldsymbol{v}^h)|^{1/2} + e_0\right] + c\sum_{j=1}^{n} k_j e_{j-1} + e_0 + c\,\|(I_{\boldsymbol{Q}} - \mathcal{P}_{\boldsymbol{Q}^h})(\boldsymbol{\sigma}_n - \boldsymbol{\sigma}_0)\|_{\boldsymbol{Q}} + c\,k\,(\|\dot{\boldsymbol{\sigma}}\|_{L^\infty(I;\boldsymbol{Q})} + \|\dot{\boldsymbol{u}}\|_{L^\infty(I;\boldsymbol{V})}). \tag{16.47}$$

The inequalities (16.46) and (16.47) can be used together to yield, for any $\boldsymbol{v}_n^h \in \boldsymbol{V}^h$,

$$e_n \le c\left[\|\boldsymbol{u}_n - \boldsymbol{v}_n^h\|_{\boldsymbol{V}} + |R_n(\boldsymbol{u}_n - \boldsymbol{v}_n^h)|^{1/2} + e_0\right] + c\sum_{j=1}^{n} k_j e_{j-1} + e_0 + c\,\|(I_{\boldsymbol{Q}} - \mathcal{P}_{\boldsymbol{Q}^h})(\boldsymbol{\sigma}_n - \boldsymbol{\sigma}_0)\|_{\boldsymbol{Q}} + c\,k\,(\|\dot{\boldsymbol{\sigma}}\|_{L^\infty(I;\boldsymbol{Q})} + \|\dot{\boldsymbol{u}}\|_{L^\infty(I;\boldsymbol{V})}). \tag{16.48}$$

We now apply Lemma 7.25 (page 163) to derive an error estimate from (16.48).

Theorem 16.15 *Let $(\boldsymbol{u}, \boldsymbol{\sigma}) \in W^{1,\infty}(I; \boldsymbol{U} \times \boldsymbol{Q}_1)$ be the solution of the problem 16.1, and $\boldsymbol{u}^{hk} \subset \boldsymbol{U}^h$ and $\boldsymbol{\sigma}^{hk} \subset \boldsymbol{Q}^h$ be the solution of the fully discrete problem 16.14. Then we have the error estimate*

$$\max_{1\le n\le N}\left(\|\boldsymbol{\sigma}_n - \boldsymbol{\sigma}_n^{hk}\|_{\boldsymbol{Q}} + \|\boldsymbol{u}_n - \boldsymbol{u}_n^{hk}\|_{\boldsymbol{V}}\right) \le c\,(\|\boldsymbol{\sigma}_0 - \boldsymbol{\sigma}_0^h\|_{\boldsymbol{Q}} + \|\boldsymbol{u}_0 - \boldsymbol{u}_0^h\|_{\boldsymbol{V}}) + c\max_{1\le n\le N}(\|\boldsymbol{u}_n - \boldsymbol{v}_n^h\|_{\boldsymbol{V}} + |R_n(\boldsymbol{u}_n - \boldsymbol{v}_n^h)|^{1/2}) + c\max_{1\le n\le N}\|(I_{\boldsymbol{Q}} - \mathcal{P}_{\boldsymbol{Q}^h})(\boldsymbol{\sigma}_n - \boldsymbol{\sigma}_0)\|_{\boldsymbol{Q}} + c\,k\,(\|\dot{\boldsymbol{\sigma}}\|_{L^\infty(I;\boldsymbol{Q})} + \|\dot{\boldsymbol{u}}\|_{L^\infty(I;\boldsymbol{V})}) \tag{16.49}$$

for any $\boldsymbol{v}_n^h \in \boldsymbol{U}^h$, $n = 1, \dots, N$.

Theorem 16.15 is the basis for convergence analysis and error estimation. The bound (16.35) is now modified to

$$|R_n(\boldsymbol{u}_n - \boldsymbol{v}_n^h)| \le c\,(\|\boldsymbol{f}_n\|_{\boldsymbol{V}} + \|\boldsymbol{\sigma}_n\|_{\boldsymbol{Q}})\,\|\boldsymbol{u}_n - \boldsymbol{v}_n^h\|_{\boldsymbol{V}}. \tag{16.50}$$

Then applying Theorem 7.23 (page 161), we can show the following convergence result.

Corollary 16.16 *Assume the conditions stated in Lemma 16.11. Let the initial values $\boldsymbol{\sigma}_0^{hk} \in \boldsymbol{Q}^h$ and $\boldsymbol{u}_0^{hk} \in \boldsymbol{U}^h$ be chosen in such a way that*

$$\|\boldsymbol{\sigma}_0 - \boldsymbol{\sigma}_0^{hk}\|_{\boldsymbol{Q}} + \|\boldsymbol{u}_0 - \boldsymbol{u}_0^{hk}\|_{\boldsymbol{V}} \to 0 \quad \text{as } h, k \to 0.$$

Then the fully discrete method converges:

$$\max_{1\le n\le N}\left(\|\boldsymbol{\sigma}_n - \boldsymbol{\sigma}_n^{hk}\|_{\boldsymbol{Q}} + \|\boldsymbol{u}_n - \boldsymbol{u}_n^{hk}\|_{\boldsymbol{V}}\right) \to 0 \quad \text{as } h, k \to 0.$$

Under the additional solution regularity assumption (16.21), we can improve the bound (16.50) to

$$|R_n(\boldsymbol{u}_n - \boldsymbol{v}^h)| \le \|\sigma_\nu^1(t_n)\|_{L^2(\Gamma_3)} \|v_\nu^h - u_\nu(t_n)\|_{L^2(\Gamma_3)}.$$

Thus, we obtain the following error estimate.

Corollary 16.17 *Under the additional regularity condition* (16.21) *we have*

$$\begin{aligned}
&\max_{1\le n\le N} \left(\|\boldsymbol{\sigma}_n - \boldsymbol{\sigma}_n^{hk}\|_{\boldsymbol{Q}} + \|\boldsymbol{u}_n - \boldsymbol{u}_n^{hk}\|_{\boldsymbol{V}}\right)\\
&\quad \le c\,(\|\boldsymbol{\sigma}_0 - \boldsymbol{\sigma}_0^{hk}\|_{\boldsymbol{Q}} + \|\boldsymbol{u}_0 - \boldsymbol{u}_0^{hk}\|_{\boldsymbol{V}})\\
&\qquad + c\,k\left(\|\dot{\boldsymbol{\sigma}}\|_{L^\infty(I;\boldsymbol{Q})} + \|\dot{\boldsymbol{u}}\|_{L^\infty(I;\boldsymbol{V})}\right)\\
&\qquad + c(\sigma_\nu^1) \max_{1\le n\le N} \left\{\|\boldsymbol{u}_n - \boldsymbol{v}_n^h\|_{\boldsymbol{V}} + \|(u_n)_\nu - (v_n^h)_\nu\|_{L^2(\Gamma_3)}^{1/2}\right\}\\
&\qquad + c \max_{1\le n\le N} \|(I_{\boldsymbol{Q}} - \mathcal{P}_{\boldsymbol{Q}^h})(\boldsymbol{\sigma}_n - \boldsymbol{\sigma}_0)\|_{\boldsymbol{Q}} \qquad (16.51)
\end{aligned}$$

for any $\boldsymbol{v}_n^h \in \boldsymbol{U}^h$, $n = 1, \dots, N$. *Here* $c(\sigma_\nu^1)$ *is a constant depending on* $\|\sigma_\nu^1\|_{L^\infty(I;L^2(\Gamma_3))}^{1/2}$.

The inequality (16.51) is the basis for deriving convergence order error estimates for the fully discrete solution. We present one sample result. We choose the initial values $\boldsymbol{u}_0^{hk} \in \boldsymbol{U}^h$ and $\boldsymbol{\sigma}_0^{hk} \in \boldsymbol{Q}^h$ in such a way that

$$\|\boldsymbol{\sigma}_0 - \boldsymbol{\sigma}_0^{hk}\|_{\boldsymbol{Q}} + \|\boldsymbol{u}_0 - \boldsymbol{u}_0^{hk}\|_{\boldsymbol{V}} \le c\,h.$$

For example, if the components of $\boldsymbol{\sigma}_0$ belong to $H^1(\Omega)$ and those of $\boldsymbol{u}_0$ belong to $H^2(\Omega)$, we may choose $\boldsymbol{u}_0^{hk}$ as the piecewise linear interpolant of $\boldsymbol{u}_0$ and $\boldsymbol{\sigma}_0^{hk}$ as the $L^2(\Omega)$ projection of $\boldsymbol{\sigma}_0$ onto Q^h. Then as in the case of the semi-discrete approximations discussed in the previous section, under the regularity assumptions

$$\begin{aligned}
&\boldsymbol{u} \in L^\infty(I; H^2(\Omega)^d), \quad u_\nu|_{\Gamma_{3,i}} \in L^\infty(I; H^2(\Gamma_{3,i})),\ 1 \le i \le i_0,\\
&\boldsymbol{\sigma} \in L^\infty(I; H^1(\Omega)^{d\times d}),
\end{aligned}$$

we can then derive the following optimal order error estimate from (16.51):

$$\max_{1\le n\le N} \left(\|\boldsymbol{\sigma}_n - \boldsymbol{\sigma}_n^{hk}\|_{\boldsymbol{Q}} + \|\boldsymbol{u}_n - \boldsymbol{u}_n^{hk}\|_{\boldsymbol{V}}\right) = O(k + h).$$

16.5 Numerical Examples

To show the performance of the fully discrete scheme, in this section we present some numerical examples for contact problems between two viscoplastic bodies in two and three dimensions.

Two-dimensional examples. For the first two-dimensional test problem, the data concerning the geometry are:

$$\begin{aligned}
\Omega^1 &= (0,4) \times (0,1), \\
\Omega^2 &= (0,4) \times (-1,0), \\
\Gamma_1^1 &= \{4\} \times [0,1], \\
\Gamma_1^2 &= \{4\} \times [-1,0], \\
\Gamma_3 &= (0,4) \times \{0\}, \\
\Gamma_2^1 &= \Gamma^1 \backslash (\Gamma_1^1 \cup \Gamma_3), \\
\Gamma_2^2 &= \Gamma^2 \backslash (\Gamma_1^2 \cup \Gamma_3).
\end{aligned}$$

The force densities are chosen as

$$\begin{aligned}
\boldsymbol{f}_0^1(x_1,x_2,t) &= (0,0)\, N/m^2, \\
\boldsymbol{f}_0^2(x_1,x_2,t) &= (0,0)\, N/m^2, \\
\boldsymbol{f}_2^1(x_1,x_2,t) &= \begin{cases} (0,-10\,t) N/m & \text{if } 3 \le x_1 \le 4,\ x_2 = 1, \\ (10\,t,0)\, N/m & \text{if } 0.5 \le x_2 \le 1,\ x_1 = 0, \\ 0\, N/m & \text{otherwise}, \end{cases} \\
\boldsymbol{f}_2^2(x_1,x_2,t) &= (0,0)\, N/m.
\end{aligned}$$

We use the following initial conditions

$$\boldsymbol{\sigma}_0 = \mathbf{0}\, N/m^2, \quad \boldsymbol{u}_0 = \mathbf{0}\, m.$$

We are interested in the contact evolution up to $T = 1\, sec$.

We use Perzyna's law with the same material's constants as in the two-dimensional example of Section 15.7. Thus, Young's modulus E and Poisson's ratio ν for the two viscoplastic bodies Ω^1 and Ω^2 are taken as $10^8\, N/m^2$ and 0.3, respectively. Recall that

$$\mathcal{G}(\boldsymbol{\sigma}, \boldsymbol{\varepsilon}) = -\frac{1}{2\lambda}\, \mathcal{E}(\boldsymbol{\sigma} - \mathcal{P}_K \boldsymbol{\sigma})$$

where $\lambda = 100\, N sec/m^2$, $\sigma_Y = \sqrt{10}\, N/m^2$, and

$$K = \{\, \boldsymbol{\tau} \in \mathbb{S}^2 : \tau_{11}^2 + \tau_{22}^2 - \tau_{11}\tau_{22} + 3\,\tau_{12}^2 \le \sigma_Y^2 \,\}.$$

Figure 16.1 depicts the deformed configuration and the von Mises norm for the stress at the final time $T = 1$.

In the second two-dimensional test problem we simulate the contact between two viscoplastic bodies in the setting described in Figure 16.2. We use the Perzyna law as in the previous example. We show in Figure 16.3 the deformed configuration and the von Mises norm for stress at the final time $T = 1$.

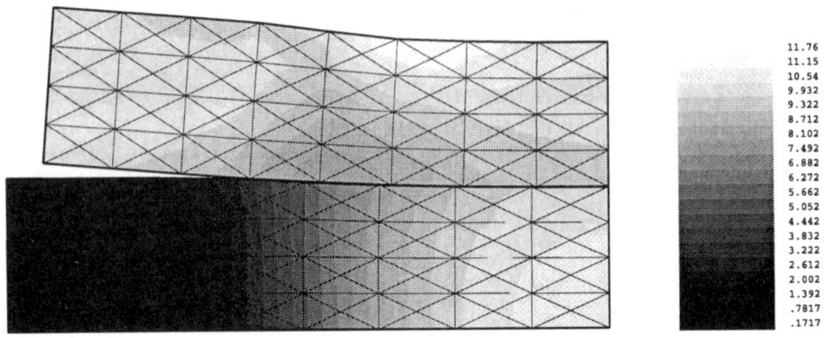

Figure 16.1: Deformed configuration and von Mises norm in the first 2D contact problem between two viscoplastic bodies

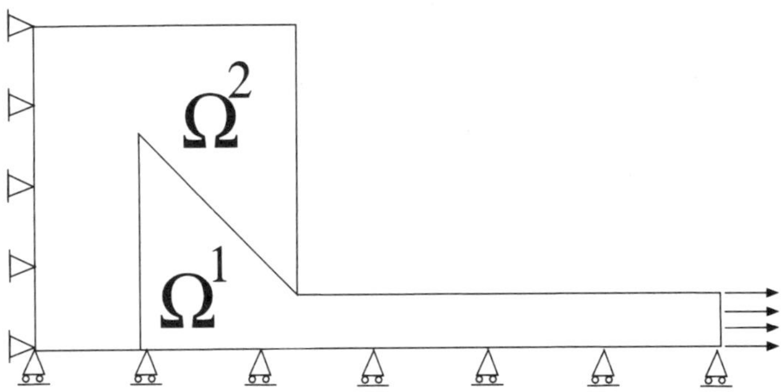

Figure 16.2: The second 2D contact problem

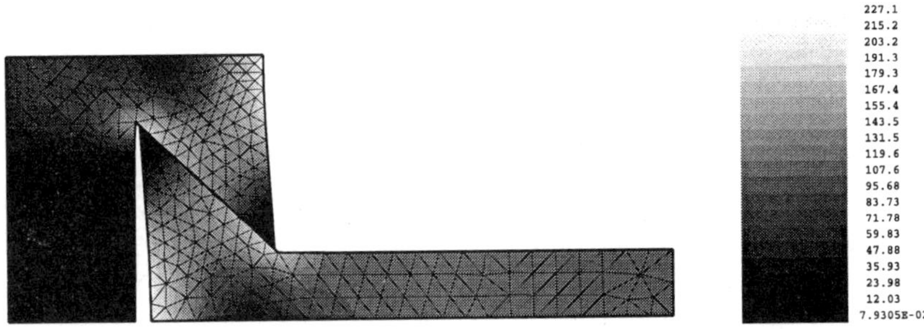

Figure 16.3: Deformed configuration and von Mises norm in the second 2D contact problem between two viscoplastic bodies

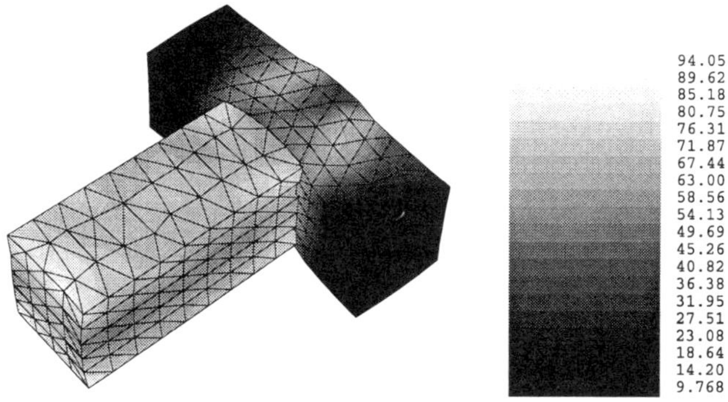

Figure 16.4: Deformed configuration and von Mises norm in a 3D contact problem between two viscoplastic bodies

Three-dimensional example. For a contact problem between two viscoplastic bodies in three dimensions, we use the following data: for the geometry,

$$\begin{aligned}
\Omega^1 &= (1,2) \times (1,4) \times (0,1),\\
\Omega^2 &= (0,3) \times (0,1) \times (0,1),\\
\Gamma_1^1 &= \emptyset,\\
\Gamma_1^2 &= \{0,3\} \times [0,1] \times [0,1],\\
\Gamma_3 &= (1,2) \times \{1\} \times (0,1),\\
\Gamma_2^1 &= \Gamma^1 \backslash (\Gamma_1^1 \cup \Gamma_3),\\
\Gamma_2^2 &= \Gamma^2 \backslash (\Gamma_1^2 \cup \Gamma_3);
\end{aligned}$$

for the force densities,

$$\begin{aligned}
\boldsymbol{f}_0^1 &= \begin{cases} (0,-100t,0)N/m^3 & \text{if } x_2 = 4,\ 1 \le x_1 \le x_2,\ 0 \le x_3 \le 1,\\ (0,0,0) & \text{otherwise,} \end{cases}\\
\boldsymbol{f}_0^2 &= (0,0,0)\,N/m^3,\\
\boldsymbol{f}_2^1 &= (0,0,0)\,N/m^2,\\
\boldsymbol{f}_2^2 &= (0,0,0)\,N/m^2;
\end{aligned}$$

and for the initial values,

$$\boldsymbol{\sigma}_0 = \mathbf{0}\,N/m^3, \quad \boldsymbol{u}_0 = \mathbf{0}\,m.$$

We simulate the contact problem up to $T = 1\,sec.$

We use Perzyna's law with the same material's constants as in the three-dimensional example of Section 15.7. Recall that $\lambda = 100\,N/m^3$, $\sigma_Y = \sqrt{10}\,N/m^3$, $E = 10^8\,N/m^3$ and $\nu = 0.3$. Figure 16.4 depicts the deformed configuration and the von Mises norm for the stress at the time $T = 1$.

Chapter 17

Bilateral Contact with Tresca's Friction Law

For the bilateral frictional contact problem studied in this chapter, the friction is modelled by Tresca's law. We provide two different, yet equivalent, variational formulations of the model for which we prove existence and uniqueness results. We also study the continuous dependence of the solution with respect to the data. We then turn to numerical approximations of the problem. For both spatially semi-discrete and fully discrete schemes we show the existence of the unique solution and present error estimates. For the fully discrete scheme, it is more delicate to prove the convergence of the numerical solution under the basic solution regularity.

17.1 Problem Statement

We assume the contact is bilateral, i.e. there is no loss of the contact during the process. Thus, the normal displacement u_ν vanishes on Γ_3 at any time. We model the friction with the Tresca friction law. The viscoplastic constitutive law of the material is (6.43) (page 135) with the assumptions (6.44) and (6.45) on the operators $\mathcal{E}$ and $\mathcal{G}$. We assume the process is quasistatic.

With the previous assumptions, the classical formulation of the mechanical problem is the following:

Find a displacement field $\boldsymbol{u} : \Omega \times \overline{I} \to \mathbb{R}^d$ and a stress field $\boldsymbol{\sigma} : \Omega \times \overline{I} \to \mathbb{S}^d$ such that

$$\dot{\boldsymbol{\sigma}} = \mathcal{E}\boldsymbol{\varepsilon}(\dot{\boldsymbol{u}}) + \mathcal{G}(\boldsymbol{\sigma}, \boldsymbol{\varepsilon}(\boldsymbol{u})) \qquad \text{in } \Omega \times I, \qquad (17.1)$$

$$\operatorname{Div} \boldsymbol{\sigma} + \boldsymbol{f}_0 = \boldsymbol{0} \qquad \text{in } \Omega \times I, \qquad (17.2)$$

$$\boldsymbol{u} = \boldsymbol{0} \qquad \text{on } \Gamma_1 \times I, \qquad (17.3)$$

$$\boldsymbol{\sigma}\boldsymbol{\nu} = \boldsymbol{f}_2 \qquad \text{on } \Gamma_2 \times I, \qquad (17.4)$$

$$\left.\begin{array}{l} u_\nu = 0, \ \|\boldsymbol{\sigma}_\tau\| \le g, \\ \|\boldsymbol{\sigma}_\tau\| < g \ \Rightarrow \ \dot{\boldsymbol{u}}_\tau = 0, \\ \|\boldsymbol{\sigma}_\tau\| = g \ \Rightarrow \ \exists \lambda \ge 0 \text{ s.t. } \boldsymbol{\sigma}_\tau = -\lambda \dot{\boldsymbol{u}}_\tau \end{array}\right\} \qquad \text{on } \Gamma_3 \times I, \qquad (17.5)$$

$$\boldsymbol{u}(0) = \boldsymbol{u}_0, \quad \boldsymbol{\sigma}(0) = \boldsymbol{\sigma}_0 \qquad \text{in } \Omega. \qquad (17.6)$$

Here $\boldsymbol{u}_0$ and $\boldsymbol{\sigma}_0$ represent the given initial displacement and initial stress. In (17.5), $g \geq 0$ is the friction bound function, i.e. the magnitude of the limiting friction traction at which slip begins. The strict inequality in (17.5) holds in the stick zone and the equality holds in the slip zone.

We use the subspace V_1 defined in (7.11) for the displacement variables and the space Q defined in (7.2) for the stress variables. Over the subspace V_1, we use the inner product of the space V. The subspace V_1 itself is a Hilbert space.

We assume that the force and the traction densities satisfy

$$\boldsymbol{f}_0 \in W^{1,\infty}(I; L^2(\Omega)^d), \quad \boldsymbol{f}_2 \in W^{1,\infty}(I; L^2(\Gamma_2)^d), \tag{17.7}$$

and the friction bound satisfies

$$g \in L^\infty(\Gamma_3), \quad g \geq 0 \text{ a.e. on } \Gamma_3. \tag{17.8}$$

We denote by $\boldsymbol{f}(t)$ the element of V_1 given by

$$(\boldsymbol{f}(t), \boldsymbol{v})_V = \int_\Omega \boldsymbol{f}_0(t) \cdot \boldsymbol{v}\, dx + \int_{\Gamma_2} \boldsymbol{f}_2(t) \cdot \boldsymbol{v}\, da, \quad \boldsymbol{v} \in V_1 \tag{17.9}$$

for all $t \in \overline{I}$ and let $j : V_1 \to \mathbb{R}_+$ be the functional

$$j(\boldsymbol{v}) = \int_{\Gamma_3} g\, \|\boldsymbol{v}_\tau\|\, da. \tag{17.10}$$

The assumptions (17.7) and (17.8) ensure that the integrals in (17.9) and (17.10) are well defined. Moreover, we have

$$\boldsymbol{f} \in W^{1,\infty}(I; V_1) \tag{17.11}$$

and

$$j(\boldsymbol{v}) \leq c\, \|g\|_{L^\infty(\Gamma_3)} \|\boldsymbol{v}\|_V \quad \forall\, \boldsymbol{v} \in V_1. \tag{17.12}$$

Denote, for all $t \in \overline{I}$, by $\Sigma(t)$ the set

$$\Sigma(t) = \{\, \boldsymbol{\tau} \in Q : (\boldsymbol{\tau}, \boldsymbol{\varepsilon}(\boldsymbol{v}))_Q + j(\boldsymbol{v}) \geq (\boldsymbol{f}(t), \boldsymbol{v})_V \ \forall\, \boldsymbol{v} \in V_1 \,\}. \tag{17.13}$$

Finally, assume that the initial data satisfy

$$\boldsymbol{u}_0 \in V_1, \quad \boldsymbol{\sigma}_0 \in \Sigma(0). \tag{17.14}$$

We have the following result.

Lemma 17.1 *If $\boldsymbol{u}$ and $\boldsymbol{\sigma}$ are regular functions satisfying* (17.2)–(17.5), *then*

$$\begin{aligned} \boldsymbol{u}(t) \in V_1,\ (\boldsymbol{\sigma}(t), \boldsymbol{\varepsilon}(\boldsymbol{v}) - \boldsymbol{\varepsilon}(\dot{\boldsymbol{u}}(t)))_Q + j(\boldsymbol{v}) - j(\dot{\boldsymbol{u}}(t)) \\ \geq (\boldsymbol{f}(t), \boldsymbol{v} - \dot{\boldsymbol{u}}(t))_V \quad \forall\, \boldsymbol{v} \in V_1, \end{aligned} \tag{17.15}$$

$$\boldsymbol{\sigma}(t) \in \Sigma(t),\ (\boldsymbol{\tau} - \boldsymbol{\sigma}(t), \boldsymbol{\varepsilon}(\dot{\boldsymbol{u}}(t)))_Q \geq 0 \quad \forall\, \boldsymbol{\tau} \in \Sigma(t) \tag{17.16}$$

for all $t \in \overline{I}$.

Proof. Let $t \in \overline{I}$ and $\boldsymbol{v} \in V_1$ be an arbitrary test function. Multiplying the equation (17.2) by $\boldsymbol{v} - \dot{\boldsymbol{u}}(t)$ and integrating by parts, we obtain

$$\begin{aligned}(\boldsymbol{\sigma}(t), \boldsymbol{\varepsilon}(\boldsymbol{v}) - \boldsymbol{\varepsilon}(\dot{\boldsymbol{u}}(t)))_Q = &\int_\Omega \boldsymbol{f}_0(t) \cdot (\boldsymbol{v} - \dot{\boldsymbol{u}}(t))\, dx \\ &+ \int_\Gamma \boldsymbol{\sigma}(t)\boldsymbol{\nu} \cdot (\boldsymbol{v} - \dot{\boldsymbol{u}}(t))\, da.\end{aligned}$$

Using the boundary conditions (17.3) and (17.4) we can replace the right hand side of the above equality by

$$(\boldsymbol{f}(t), \boldsymbol{v} - \dot{\boldsymbol{u}}(t))_V + \int_{\Gamma_3} \boldsymbol{\sigma}(t)\boldsymbol{\nu} \cdot (\boldsymbol{v} - \dot{\boldsymbol{u}}(t))\, da.$$

From the contact boundary condition (17.5) we derive the relation

$$\boldsymbol{\sigma}_\tau(t) \cdot \dot{\boldsymbol{u}}_\tau(t) = -g\, \|\dot{\boldsymbol{u}}_\tau(t)\| \quad \text{on } \Gamma_3.$$

Thus, on Γ_3,

$$\begin{aligned}\boldsymbol{\sigma}(t)\boldsymbol{\nu} \cdot (\boldsymbol{v} - \dot{\boldsymbol{u}}(t)) &= \sigma_\nu(t)\,(v_\nu - \dot{u}_\nu(t)) + \boldsymbol{\sigma}_\tau(t) \cdot (\boldsymbol{v}_\tau - \dot{\boldsymbol{u}}_\tau(t)) \\ &= \boldsymbol{\sigma}_\tau(t) \cdot \boldsymbol{v}_\tau - \boldsymbol{\sigma}_\tau(t) \cdot \dot{\boldsymbol{u}}_\tau(t) \\ &= \boldsymbol{\sigma}_\tau(t) \cdot \boldsymbol{v}_\tau + g\, \|\dot{\boldsymbol{u}}_\tau(t)\| \\ &\ge g\, (\|\dot{\boldsymbol{u}}_\tau(t)\| - \|\boldsymbol{v}_\tau\|).\end{aligned}$$

Therefore,

$$\int_{\Gamma_3} \boldsymbol{\sigma}(t)\boldsymbol{\nu} \cdot (\boldsymbol{v} - \dot{\boldsymbol{u}}(t))\, da \ge j(\dot{\boldsymbol{u}}(t)) - j(\boldsymbol{v}),$$

and we have

$$(\boldsymbol{\sigma}(t), \boldsymbol{\varepsilon}(\boldsymbol{v}) - \boldsymbol{\varepsilon}(\dot{\boldsymbol{u}}(t)))_Q \ge (\boldsymbol{f}(t), \boldsymbol{v} - \dot{\boldsymbol{u}}(t))_V + j(\dot{\boldsymbol{u}}(t)) - j(\boldsymbol{v}),$$

i.e. the inequality in (17.15) holds. The inequality (17.15) is equivalent to

$$(\boldsymbol{\sigma}(t), \boldsymbol{\varepsilon}(\dot{\boldsymbol{u}}(t)))_Q + j(\dot{\boldsymbol{u}}(t)) = (\boldsymbol{f}(t), \dot{\boldsymbol{u}}(t))_V, \tag{17.17}$$

$$(\boldsymbol{\sigma}(t), \boldsymbol{\varepsilon}(\boldsymbol{v}))_Q + j(\boldsymbol{v}) \ge (\boldsymbol{f}(t), \boldsymbol{v})_V \quad \forall\, \boldsymbol{v} \in V_1. \tag{17.18}$$

The property (17.18) indicates $\boldsymbol{\sigma}(t) \in \Sigma(t)$. For any $\boldsymbol{\tau} \in \Sigma(t)$, we subtract (17.17) from the defining inequality for $\boldsymbol{\tau} \in \Sigma(t)$ to obtain the inequality in (17.16). ■

The previous lemma leads us to consider the following weak formulations of the frictional problem (17.1)–(17.6).

Problem 17.2 *Find a displacement field* $\boldsymbol{u} : \overline{I} \to V_1$ *and a stress field* $\boldsymbol{\sigma} : \overline{I} \to Q_1$ *such that*

$$\dot{\boldsymbol{\sigma}}(t) = \mathcal{E}\boldsymbol{\varepsilon}(\dot{\boldsymbol{u}}(t)) + \mathcal{G}(\boldsymbol{\sigma}(t), \boldsymbol{\varepsilon}(\boldsymbol{u}(t))) \quad \text{a.e. } t \in I, \tag{17.19}$$

$$\begin{aligned}(\boldsymbol{\sigma}(t), \boldsymbol{\varepsilon}(\boldsymbol{v}) - \boldsymbol{\varepsilon}(\dot{\boldsymbol{u}}(t)))_Q + j(\boldsymbol{v}) - j(\dot{\boldsymbol{u}}(t)) \ge (\boldsymbol{f}(t), \boldsymbol{v} - \dot{\boldsymbol{u}}(t))_V \\ \forall\, \boldsymbol{v} \in V_1, \ \text{a.e. } t \in I,\end{aligned} \tag{17.20}$$

$$\boldsymbol{u}(0) = \boldsymbol{u}_0, \qquad \boldsymbol{\sigma}(0) = \boldsymbol{\sigma}_0. \tag{17.21}$$

Problem 17.3 *Find a displacement field* $\boldsymbol{u} : \overline{I} \to V_1$ *and a stress field* $\boldsymbol{\sigma} : \overline{I} \to Q_1$ *such that*

$$\dot{\boldsymbol{\sigma}}(t) = \mathcal{E}\boldsymbol{\varepsilon}(\dot{\boldsymbol{u}}(t)) + \mathcal{G}(\boldsymbol{\sigma}(t), \boldsymbol{\varepsilon}(\boldsymbol{u}(t))) \quad \text{a.e. } t \in I, \tag{17.22}$$

$$\boldsymbol{\sigma}(t) \in \Sigma(t), \ (\boldsymbol{\tau} - \boldsymbol{\sigma}(t), \boldsymbol{\varepsilon}(\dot{\boldsymbol{u}}(t)))_Q \geq 0 \quad \forall \boldsymbol{\tau} \in \Sigma(t), \text{ a.e. } t \in I, \tag{17.23}$$

$$\boldsymbol{u}(0) = \boldsymbol{u}_0, \qquad \boldsymbol{\sigma}(0) = \boldsymbol{\sigma}_0. \tag{17.24}$$

Well-posedness of Problems 17.2 and 17.3 is the topic of the next section.

17.2 Existence and Uniqueness Results

Our first existence and uniqueness result is the following.

Theorem 17.4 *Assume* (6.44), (6.45), (17.7), (17.8) *and* (17.14). *Then there exists a unique solution* $(\boldsymbol{u}, \boldsymbol{\sigma})$ *to Problem 17.2. Moreover, the solution satisfies*

$$\boldsymbol{u} \in W^{1,\infty}(I; V_1), \qquad \boldsymbol{\sigma} \in W^{1,\infty}(I; Q_1). \tag{17.25}$$

The proof of Theorem 17.4 is carried out in several steps, using results of evolutionary variational inequalities and Banach fixed-point theorem. Let $\boldsymbol{\eta} \in L^\infty(I; Q)$. Assume in what follows that (6.44), (6.45), (17.7), (17.8) and (17.14) hold. We define a function $\boldsymbol{z}_\eta \in W^{1,\infty}(I; Q)$ by

$$\boldsymbol{z}_\eta(t) = \int_0^t \boldsymbol{\eta}(s)\, ds + \boldsymbol{z}_0 \tag{17.26}$$

where

$$\boldsymbol{z}_0 = \boldsymbol{\sigma}_0 - \mathcal{E}\boldsymbol{\varepsilon}(\boldsymbol{u}_0). \tag{17.27}$$

Consider the following auxiliary variational problem.

Problem 17.5 *Find* $\boldsymbol{u}_\eta : \overline{I} \to V_1$ *and* $\boldsymbol{\sigma}_\eta : \overline{I} \to Q_1$ *such that*

$$\boldsymbol{\sigma}_\eta(t) = \mathcal{E}\boldsymbol{\varepsilon}(\boldsymbol{u}_\eta(t)) + \boldsymbol{z}_\eta(t) \quad \forall t \in \overline{I}, \tag{17.28}$$

$$\begin{aligned}(\boldsymbol{\sigma}_\eta(t), \boldsymbol{\varepsilon}(\boldsymbol{v}) - \boldsymbol{\varepsilon}(\dot{\boldsymbol{u}}_\eta(t)))_Q + j(\boldsymbol{v}) - j(\dot{\boldsymbol{u}}_\eta(t))\\ \geq (\boldsymbol{f}(t), \boldsymbol{v} - \dot{\boldsymbol{u}}_\eta(t))_V \quad \forall \boldsymbol{v} \in V_1, \text{ a.e. } t \in I,\end{aligned} \tag{17.29}$$

$$\boldsymbol{u}_\eta(0) = \boldsymbol{u}_0, \quad \boldsymbol{\sigma}_\eta(0) = \boldsymbol{\sigma}_0. \tag{17.30}$$

We have the following result.

Lemma 17.6 *The variational problem* 17.5 *has a unique solution* $(\boldsymbol{u}_\eta, \boldsymbol{\sigma}_\eta)$ *such that* $\boldsymbol{u}_\eta \in W^{1,\infty}(I; V_1)$ *and* $\boldsymbol{\sigma}_\eta \in W^{1,\infty}(I; Q_1)$.

Proof. Define a bilinear form $a : V_1 \times V_1 \to \mathbb{R}$ by

$$a(\boldsymbol{u}, \boldsymbol{v}) = (\mathcal{E}\boldsymbol{\varepsilon}(\boldsymbol{u}), \boldsymbol{\varepsilon}(\boldsymbol{v}))_Q \quad \forall \boldsymbol{u}, \boldsymbol{v} \in V_1.$$

It follows from the assumptions (6.44) on $\mathcal{E}$ that $a(\cdot,\cdot)$ is continuous, symmetric and V_1-elliptic. Also by the condition (17.8) we see that j is a continuous seminorm on V_1.

Let $\boldsymbol{f}_\eta : \overline{I} \to V_1$ be defined through

$$(\boldsymbol{f}_\eta(t), \boldsymbol{v})_V = (\boldsymbol{f}(t), \boldsymbol{v})_V - (\boldsymbol{z}_\eta(t), \boldsymbol{\varepsilon}(\boldsymbol{v}))_Q \quad \forall\, \boldsymbol{v} \in V_1,\ t \in \overline{I}. \tag{17.31}$$

Since $\boldsymbol{f} \in W^{1,\infty}(I;V_1)$ and $\boldsymbol{z}_\eta \in W^{1,\infty}(I;Q)$, we have $\boldsymbol{f}_\eta \in W^{1,\infty}(I;V_1)$. Moreover, since $\boldsymbol{\sigma}_0 = \mathcal{E}\boldsymbol{\varepsilon}(\boldsymbol{u}_0) + \boldsymbol{z}_0 = \mathcal{E}\boldsymbol{\varepsilon}(\boldsymbol{u}_0) + \boldsymbol{z}_\eta(0)$ by (17.27) and (17.26), the assumption $\boldsymbol{\sigma}_0 \in \Sigma(0)$ from (17.14) can be written as

$$(\mathcal{E}\boldsymbol{\varepsilon}(\boldsymbol{u}_0), \boldsymbol{\varepsilon}(\boldsymbol{v}))_Q + (\boldsymbol{z}_\eta(0), \boldsymbol{\varepsilon}(\boldsymbol{v}))_Q + j(\boldsymbol{v}) \ge (\boldsymbol{f}(0), \boldsymbol{v})_V \quad \forall\, \boldsymbol{v} \in V_1,$$

i.e.

$$a(\boldsymbol{u}_0, \boldsymbol{v}) + j(\boldsymbol{v}) \ge (\boldsymbol{f}_\eta(0), \boldsymbol{v})_V \quad \forall\, \boldsymbol{v} \in V_1.$$

Applying Theorem 4.16 (page 69) we see that there exists a unique solution $\boldsymbol{u}_\eta \in W^{1,\infty}(I;V_1)$ to the problem

$$\begin{aligned} a(\boldsymbol{u}_\eta(t), \boldsymbol{v} - \dot{\boldsymbol{u}}_\eta(t)) + j(\boldsymbol{v}) - j(\dot{\boldsymbol{u}}_\eta(t)) &\ge (\boldsymbol{f}_\eta(t), \boldsymbol{v} - \dot{\boldsymbol{u}}_\eta(t))_V \\ \forall\, \boldsymbol{v} \in V_1, \quad &\text{a.e. } t \in I, \qquad (17.32)\\ \boldsymbol{u}_\eta(0) &= \boldsymbol{u}_0. \qquad (17.33) \end{aligned}$$

Let $\boldsymbol{\sigma}_\eta \in W^{1,\infty}(I;Q)$ be given by (17.28). Then $(\boldsymbol{u}_\eta, \boldsymbol{\sigma}_\eta)$ is a solution for (17.28)–(17.30). From the inequality (17.29) we obtain $\mathrm{Div}\boldsymbol{\sigma}_\eta + \boldsymbol{f}_0 = \boldsymbol{0}$ a.e. in $\Omega \times I$. Therefore, $\boldsymbol{\sigma}_\eta \in W^{1,\infty}(I;Q_1)$. Finally, the uniqueness part of the lemma follows from the unique solvability of the variational inequality problem (17.32), (17.33). ■

We consider next an operator $\Lambda : L^\infty(I;Q) \to L^\infty(I;Q)$ defined by

$$\Lambda\boldsymbol{\eta} = \mathcal{G}(\boldsymbol{\sigma}_\eta, \boldsymbol{\varepsilon}(\boldsymbol{u}_\eta)) \quad \forall\, \boldsymbol{\eta} \in L^\infty(I;Q), \tag{17.34}$$

where $(\boldsymbol{u}_\eta, \boldsymbol{\sigma}_\eta)$ is the solution of Problem 17.5.

Lemma 17.7 *The operator Λ has a unique fixed point $\boldsymbol{\eta}^* \in L^\infty(I;Q)$.*

Proof. Let $\boldsymbol{\eta}_1, \boldsymbol{\eta}_2 \in L^\infty(I;Q)$ and denote $\boldsymbol{u}_i = \boldsymbol{u}_{\eta_i}$, $\boldsymbol{\sigma}_i = \boldsymbol{\sigma}_{\eta_i}$, and $\boldsymbol{z}_i = \boldsymbol{z}_{\eta_i}$ for $i = 1, 2$. Rewrite (17.28) and (17.29) for $\boldsymbol{u}_1$ and $\boldsymbol{u}_2$ as

$$\begin{aligned} a(\boldsymbol{u}_1, \boldsymbol{v} - \dot{\boldsymbol{u}}_1) + (\boldsymbol{z}_1, \boldsymbol{\varepsilon}(\boldsymbol{v}) - \boldsymbol{\varepsilon}(\dot{\boldsymbol{u}}_1))_Q + j(\boldsymbol{v}) - j(\dot{\boldsymbol{u}}_1) &\ge (\boldsymbol{f}, \boldsymbol{v} - \dot{\boldsymbol{u}}_1)_V, \\ a(\boldsymbol{u}_2, \boldsymbol{v} - \dot{\boldsymbol{u}}_2) + (\boldsymbol{z}_2, \boldsymbol{\varepsilon}(\boldsymbol{v}) - \boldsymbol{\varepsilon}(\dot{\boldsymbol{u}}_2))_Q + j(\boldsymbol{v}) - j(\dot{\boldsymbol{u}}_2) &\ge (\boldsymbol{f}, \boldsymbol{v} - \dot{\boldsymbol{u}}_2)_V \end{aligned}$$

for all $\boldsymbol{v} \in V_1$, a.e. on I. We take $\boldsymbol{v} = \dot{\boldsymbol{u}}_2$ in the first inequality, $\boldsymbol{v} = \dot{\boldsymbol{u}}_1$ in the second inequality, and add the two inequalities to obtain

$$\frac{1}{2}\frac{d}{dt} a(\boldsymbol{u}_1 - \boldsymbol{u}_2, \boldsymbol{u}_1 - \boldsymbol{u}_2) \le -(\boldsymbol{z}_1 - \boldsymbol{z}_2, \boldsymbol{\varepsilon}(\dot{\boldsymbol{u}}_1) - \boldsymbol{\varepsilon}(\dot{\boldsymbol{u}}_2))_Q \quad \text{a.e. on } I.$$

Let $t \in \overline{I}$. Integrate the inequality from 0 to t:

$$\begin{aligned}\frac{1}{2}\,a(\boldsymbol{u}_1(t)-\boldsymbol{u}_2(t),\boldsymbol{u}_1(t)-\boldsymbol{u}_2(t))&\\ \le -(\boldsymbol{z}_1(t)-\boldsymbol{z}_2(t),\boldsymbol{\varepsilon}(\boldsymbol{u}_1(t))-\boldsymbol{\varepsilon}(\boldsymbol{u}_2(t)))_Q&\\ +\int_0^t(\boldsymbol{\eta}_1(s)-\boldsymbol{\eta}_2(s),\boldsymbol{\varepsilon}(\boldsymbol{u}_1(s))-\boldsymbol{\varepsilon}(\boldsymbol{u}_2(s)))_Q ds.&\end{aligned}$$

Then,

$$\begin{aligned}c\,\|\boldsymbol{u}_1(t)-\boldsymbol{u}_2(t)\|_V^2 \le \|\boldsymbol{z}_1(t)-\boldsymbol{z}_2(t)\|_Q\|\boldsymbol{u}_1(t)-\boldsymbol{u}_2(t)\|_V&\\ +\int_0^t\|\boldsymbol{\eta}_1(s)-\boldsymbol{\eta}_2(s)\|_Q\|\boldsymbol{u}_1(s)-\boldsymbol{u}_2(s)\|_V ds.&\end{aligned}$$

Now

$$\boldsymbol{z}_1(t)-\boldsymbol{z}_2(t)=\int_0^t(\boldsymbol{\eta}_1(s)-\boldsymbol{\eta}_2(s))\,ds,$$

so

$$\|\boldsymbol{z}_1(t)-\boldsymbol{z}_2(t)\|_Q \le \int_0^t\|\boldsymbol{\eta}_1(s)-\boldsymbol{\eta}_2(s)\|_Q ds.$$

Therefore, we have

$$\begin{aligned}\|\boldsymbol{u}_1(t)-\boldsymbol{u}_2(t)\|_V^2 \le c\int_0^t\|\boldsymbol{\eta}_1(s)-\boldsymbol{\eta}_2(s)\|_V^2 ds&\\ +c\int_0^t\|\boldsymbol{u}_1(s)-\boldsymbol{u}_2(s)\|_V^2 ds.&\end{aligned}$$

Applying the Gronwall inequality (Lemma 7.24), we obtain

$$\|\boldsymbol{u}_1(t)-\boldsymbol{u}_2(t)\|_V^2 \le c\int_0^t\|\boldsymbol{\eta}_1(s)-\boldsymbol{\eta}_2(s)\|_Q^2 ds. \tag{17.35}$$

By the definition (17.28) for $\boldsymbol{\sigma}_1$ and $\boldsymbol{\sigma}_2$, we have

$$\begin{aligned}\boldsymbol{\sigma}_1(t)-\boldsymbol{\sigma}_2(t)&=\mathcal{E}\boldsymbol{\varepsilon}(\boldsymbol{u}_1(t)-\boldsymbol{u}_2(t))+\boldsymbol{z}_1(t)-\boldsymbol{z}_2(t)\\ &=\mathcal{E}\boldsymbol{\varepsilon}(\boldsymbol{u}_1(t)-\boldsymbol{u}_2(t))+\int_0^t(\boldsymbol{\eta}_1(s)-\boldsymbol{\eta}_2(s))\,ds.\end{aligned}$$

Then

$$\|\boldsymbol{\sigma}_1(t)-\boldsymbol{\sigma}_2(t)\|_Q^2 \le c\,\|\boldsymbol{u}_1(t)-\boldsymbol{u}_2(t)\|_V^2 + c\int_0^t\|\boldsymbol{\eta}_1(s)-\boldsymbol{\eta}_2(s)\|_Q^2 ds.$$

Using (17.35), we have

$$\|\boldsymbol{\sigma}_1(t)-\boldsymbol{\sigma}_2(t)\|_Q^2 \le c\int_0^t\|\boldsymbol{\eta}_1(s)-\boldsymbol{\eta}_2(s)\|_Q^2 ds. \tag{17.36}$$

Finally,

$$\Lambda\boldsymbol{\eta}_1(t) - \Lambda\boldsymbol{\eta}_2(t) = \mathcal{G}(\boldsymbol{\sigma}_1(t), \boldsymbol{\varepsilon}(\boldsymbol{u}_1(t))) - \mathcal{G}(\boldsymbol{\sigma}_2(t), \boldsymbol{\varepsilon}(\boldsymbol{u}_2(t))).$$

Using the assumptions (6.45) on $\mathcal{G}$, and the bounds (17.35) and (17.36), we find that

$$\|\Lambda\boldsymbol{\eta}_1(t) - \Lambda\boldsymbol{\eta}_2(t)\|_Q^2 \leq c \int_0^t \|\boldsymbol{\eta}_1(s) - \boldsymbol{\eta}_2(s)\|_Q^2 ds. \tag{17.37}$$

Lemma 17.7 follows from (17.37) and an application of Banach's fixed point theorem. ■

Now we are ready to prove Theorem 17.4.

Proof of Theorem 17.4
Existence. Let $\boldsymbol{\eta}^* \in L^\infty(I;Q)$ be the fixed point of Λ and let $(\boldsymbol{u}_{\eta^*}, \boldsymbol{\sigma}_{\eta^*})$ be the solution of Problem 17.5 for $\boldsymbol{\eta} = \boldsymbol{\eta}^*$. Using (17.28) and (17.26) we have

$$\dot{\boldsymbol{\sigma}}_{\eta^*}(t) = \mathcal{E}\boldsymbol{\varepsilon}(\dot{\boldsymbol{u}}_{\eta^*}(t)) + \boldsymbol{\eta}^*(t) \quad \text{a.e. } t \in I.$$

Since

$$\boldsymbol{\eta}^*(t) = \Lambda\boldsymbol{\eta}^*(t) = \mathcal{G}(\boldsymbol{\sigma}_{\eta^*}(t), \boldsymbol{\varepsilon}(\boldsymbol{u}_{\eta^*}(t))) \quad \text{a.e. } t \in I,$$

it follows that $(\boldsymbol{u}_{\eta^*}, \boldsymbol{\sigma}_{\eta^*})$ satisfies (17.19). Using now (17.29) and (17.30) we conclude that $(\boldsymbol{u}_{\eta^*}, \boldsymbol{\sigma}_{\eta^*})$ is a solution of Problem 17.2 with regularity (17.25).
Uniqueness. The uniqueness part follows from the uniqueness of the fixed point of the operator Λ defined by (17.34) (cf. the proof of Theorem 14.4). ■

As in previous chapters, it can be shown that

$$\operatorname{Div}\boldsymbol{\sigma} + \boldsymbol{f}_0 = \boldsymbol{0} \quad \text{a.e. in } \Omega \times I, \tag{17.38}$$

and under the assumption

$$\boldsymbol{\sigma\nu} \in L^\infty(I; L^2(\Gamma)^d), \tag{17.39}$$

we further have

$$u_\nu = 0 \text{ a.e. on } \Gamma_2 \times I, \quad \boldsymbol{\sigma\nu} = \boldsymbol{f}_2 \text{ a.e. on } \Gamma_2 \times I. \tag{17.40}$$

Our second existence and uniqueness result is the following.

Theorem 17.8 *Assume* (6.44), (6.45), (17.7), (17.8) *and* (17.14). *Then there exists a unique solution* $(\boldsymbol{u}, \boldsymbol{\sigma})$ *to Problem 17.3. Moreover, the solution satisfies* (17.25).

To prove Theorem 17.8 we first observe that the variational inequality (17.23) is posed over a time-dependent convex $\Sigma(t)$. To remove the time dependence of the constraint set, we use a change of variable similar to that used in the proof of Theorem 14.9. To this end, we define $\boldsymbol{\sigma_f} : \overline{I} \to Q$ by

$$\boldsymbol{\sigma_f} = \boldsymbol{\varepsilon}(\boldsymbol{f}).$$

Since $\boldsymbol{f} \in W^{1,\infty}(I;V_1)$, we have $\boldsymbol{\sigma_f} \in W^{1,\infty}(I;Q)$. Notice that from the definition of $\boldsymbol{\sigma_f}$, we have

$$(\boldsymbol{\sigma_f}(t), \boldsymbol{\varepsilon}(\boldsymbol{v}))_Q = (\boldsymbol{f}(t), \boldsymbol{v})_V \quad \forall \boldsymbol{v} \in V_1.$$

Then

$$\operatorname{Div} \boldsymbol{\sigma_f}(t) = -\boldsymbol{f}_0(t) \quad \text{a.e. in } \Omega \times I.$$

Thus we further have $\boldsymbol{\sigma_f} \in W^{1,\infty}(I;Q_1)$.

We can now express the set $\Sigma(t)$ as

$$\Sigma(t) = \Sigma_0 + \{\boldsymbol{\sigma_f}(t)\} \quad \forall t \in \overline{I},$$

where

$$\Sigma_0 = \{\, \boldsymbol{\tau} \in Q : \, (\boldsymbol{\tau}, \boldsymbol{\varepsilon}(\boldsymbol{v}))_Q + j(\boldsymbol{v}) \geq 0 \;\forall \boldsymbol{v} \in V_1 \,\} \tag{17.41}$$

is time independent. Let

$$\bar{\boldsymbol{\sigma}} = \boldsymbol{\sigma} - \boldsymbol{\sigma_f}, \quad \bar{\boldsymbol{\sigma}}_0 = \boldsymbol{\sigma}_0 - \boldsymbol{\sigma_f}(0), \tag{17.42}$$

and consider the following variational problem.

Problem 17.9 *Find* $\boldsymbol{u} : \overline{I} \to V_1$ *and* $\bar{\boldsymbol{\sigma}} : \overline{I} \to Q_1$ *such that*

$$\begin{aligned} \boldsymbol{\varepsilon}(\dot{\boldsymbol{u}}(t)) = \mathcal{E}^{-1}\dot{\bar{\boldsymbol{\sigma}}}(t) - \mathcal{E}^{-1}\mathcal{G}(\bar{\boldsymbol{\sigma}}(t) + \boldsymbol{\sigma_f}(t), \boldsymbol{\varepsilon}(\boldsymbol{u}(t))) \\ +\mathcal{E}^{-1}\dot{\boldsymbol{\sigma}}_{\boldsymbol{f}}(t) \quad \text{a.e. } t \in I, \end{aligned} \tag{17.43}$$

$$\bar{\boldsymbol{\sigma}}(t) \in \Sigma_0, \quad (\boldsymbol{\tau} - \bar{\boldsymbol{\sigma}}(t), \boldsymbol{\varepsilon}(\dot{\boldsymbol{u}}(t)))_Q \geq 0 \quad \forall \boldsymbol{\tau} \in \Sigma_0, \text{ a.e. } t \in I, \tag{17.44}$$

$$\boldsymbol{u}(0) = \boldsymbol{u}_0 \quad \bar{\boldsymbol{\sigma}}(0) = \bar{\boldsymbol{\sigma}}_0. \tag{17.45}$$

It is straightforward to verify the following result.

Lemma 17.10 *The couple* $(\boldsymbol{u}, \boldsymbol{\sigma})$ *is a solution to Problem* 17.3 *with regularity* $\boldsymbol{u} \in W^{1,\infty}(I;V_1)$, $\boldsymbol{\sigma} \in W^{1,\infty}(I;Q_1)$ *iff the couple* $(\boldsymbol{u}, \bar{\boldsymbol{\sigma}})$ *is a solution to Problem* 17.9 *with regularity* $\boldsymbol{u} \in W^{1,\infty}(I;V_1)$, $\bar{\boldsymbol{\sigma}} \in W^{1,\infty}(I;Q_1)$.

We turn now to the variational problem 17.9 that we solve using again a fixed point method. To this end, let $\boldsymbol{\eta} \in L^\infty(I;Q)$ and let us consider the following auxiliary problem.

Problem 17.11 *Find* $\boldsymbol{u}_\eta : \overline{I} \to V_1$ *and* $\boldsymbol{\sigma}_\eta : \overline{I} \to Q_1$ *such that*

$$\boldsymbol{\varepsilon}(\dot{\boldsymbol{u}}_\eta(t)) = \mathcal{E}^{-1}\dot{\boldsymbol{\sigma}}_\eta(t) + \boldsymbol{\eta} \quad \text{a.e. } t \in I, \tag{17.46}$$

$$\boldsymbol{\sigma}_\eta(t) \in \Sigma_0 \quad \forall t \in \overline{I}, \tag{17.47}$$

$$(\boldsymbol{\tau} - \boldsymbol{\sigma}_\eta(t), \boldsymbol{\varepsilon}(\dot{\boldsymbol{u}}_\eta(t)))_Q \geq 0 \quad \forall \boldsymbol{\tau} \in \Sigma_0, \quad \text{a.e. } t \in I, \tag{17.48}$$

$$\boldsymbol{u}_\eta(0) = \boldsymbol{u}_0, \qquad \boldsymbol{\sigma}_\eta(0) = \bar{\boldsymbol{\sigma}}_0. \tag{17.49}$$

The unique solvability of Problem 17.11 can be proved based on the following abstract result.

Theorem 17.12 *Let* $(X, \|\cdot\|_X)$ *be a real Hilbert space and let* $\varphi : X \to (-\infty, +\infty]$ *be a proper, convex, lower semicontinuous function. Then, for any* $f \in L^2(I; X)$ *and* $u_0 \in D(\varphi)$*, there exists a unique solution* $u \in H^1(I; X)$ *of the problem*

$$u(t) \in D(\partial\varphi) \quad \text{a.e. } t \in I, \tag{17.50}$$

$$\dot{u}(t) + \partial\varphi(u(t)) \ni f(t) \quad \text{a.e. } t \in I, \tag{17.51}$$

$$u(0) = u_0. \tag{17.52}$$

Theorem 17.12 is a simplified version of a more general result that can be found in [26, p. 189] or [35, p. 72]. Here $D(\varphi)$ and $D(\partial\varphi)$ denote the effective domain of φ and the domain of the subdifferential operator $\partial\varphi$, that is

$$D(\varphi) = \{\, v \in X : \varphi(v) < +\infty \,\},$$
$$D(\partial\varphi) = \{\, v \in X : \exists\, f \in X \text{ s.t. } \varphi(w) - \varphi(v) \geq (f, w - v)_X \ \forall\, w \in X \,\}.$$

We use now Theorem 17.12 to obtain the following result.

Lemma 17.13 *Problem 17.11 has a unique solution* $(\boldsymbol{u}_\eta, \boldsymbol{\sigma}_\eta)$ *with regularity* $\boldsymbol{u}_\eta \in W^{1,\infty}(I; V_1)$ *and* $\boldsymbol{\sigma}_\eta \in W^{1,\infty}(I; Q_1)$.

Proof. We observe that Σ_0 is a closed convex set in Q. Also, we have $\bar{\boldsymbol{\sigma}}_0 \in \Sigma_0$, this follows from the definition (17.42) and the assumptions (17.14) on $\boldsymbol{\sigma}_0$. Therefore, $\bar{\boldsymbol{\sigma}}_0 \in D(\psi_{\Sigma_0})$, where ψ_{Σ_0} denotes the indicator function of the set $\Sigma_0 \subset Q$. Applying Theorem 17.12 with $X = Q$, $(\boldsymbol{\sigma}, \boldsymbol{\tau})_X = (\mathcal{E}^{-1}\boldsymbol{\sigma}, \boldsymbol{\tau})_Q$, $\varphi = \psi_{\Sigma_0}$, we obtain the existence and the uniqueness of a function $\boldsymbol{\sigma}_\eta \in H^1(I; Q)$ such that

$$\boldsymbol{\sigma}_\eta(t) \in \Sigma_0 \quad \forall\, t \in \overline{I}, \tag{17.53}$$

$$(\dot{\boldsymbol{\sigma}}_\eta(t), \boldsymbol{\tau} - \boldsymbol{\sigma}_\eta(t))_X + (\mathcal{E}\boldsymbol{\eta}(t), \boldsymbol{\tau} - \boldsymbol{\sigma}_\eta(t))_X \geq 0$$
$$\forall\, \boldsymbol{\tau} \in \Sigma_0, \ \text{a.e. } t \in I, \tag{17.54}$$

$$\boldsymbol{\sigma}_\eta(0) = \bar{\boldsymbol{\sigma}}_0. \tag{17.55}$$

From (17.54) we obtain

$$(\mathcal{E}^{-1}\dot{\boldsymbol{\sigma}}_\eta(t), \boldsymbol{\tau} - \boldsymbol{\sigma}_\eta(t))_Q + (\boldsymbol{\eta}(t), \boldsymbol{\tau} - \boldsymbol{\sigma}_\eta(t))_Q \geq 0$$
$$\forall\, \boldsymbol{\tau} \in \Sigma_0, \ \text{a.e. } t \in I. \tag{17.56}$$

Taking $\boldsymbol{\tau} = \boldsymbol{0}$ and $2\,\boldsymbol{\sigma}_\eta(t)$ in (17.56), we obtain

$$(\mathcal{E}^{-1}\dot{\boldsymbol{\sigma}}_\eta(t), \dot{\boldsymbol{\sigma}}_\eta(t))_Q + (\boldsymbol{\eta}(t), \dot{\boldsymbol{\sigma}}_\eta(t))_Q = 0 \quad \text{a.e. } t \in I.$$

Then, using the assumptions (6.44) on $\mathcal{G}$, we find that

$$\|\dot{\boldsymbol{\sigma}}_\eta(t)\|_Q \leq c\, \|\boldsymbol{\eta}(t)\|_Q \quad \text{a.e. } t \in I.$$

Since $\boldsymbol{\eta} \in L^\infty(I;Q)$, we have $\boldsymbol{\sigma}_\eta \in W^{1,\infty}(I;Q)$. Moreover, from (17.53) we can deduce that

$$\operatorname{Div} \boldsymbol{\sigma}_\eta(t) = \mathbf{0} \quad \text{a.e. in } \Omega \times I.$$

Therefore, $\boldsymbol{\sigma}_\eta \in W^{1,\infty}(I;Q_1)$.

We also note that (17.56) yields

$$(\boldsymbol{\tau} - \boldsymbol{\sigma}_\eta(t), \boldsymbol{\varepsilon}_\eta(t))_Q \geq 0 \quad \forall \boldsymbol{\tau} \in \Sigma_0, \text{ a.e. } t \in I, \tag{17.57}$$

where $\boldsymbol{\varepsilon}_\eta \in L^\infty(I;Q)$ is the function given by

$$\boldsymbol{\varepsilon}_\eta = \mathcal{E}^{-1}\dot{\boldsymbol{\sigma}}_\eta + \boldsymbol{\eta}. \tag{17.58}$$

We consider the space

$$\mathcal{W} = \{\, \boldsymbol{z} \in Q : \operatorname{Div} \boldsymbol{z} = \mathbf{0} \text{ in } \Omega,\ z_\nu = 0 \text{ on } \Gamma_2 \cup \Gamma_3 \,\}.$$

For any $\boldsymbol{z} \in \mathcal{W}$, using (17.53) we obtain $\boldsymbol{\sigma}_\eta(t) \pm \boldsymbol{z} \in \Sigma_0$, $\forall t \in \overline{I}$. Taking $\boldsymbol{\tau} = \boldsymbol{\sigma}_\eta(t) \pm \boldsymbol{z}$ in (17.57), we obtain $(\boldsymbol{z}, \boldsymbol{\varepsilon}_\eta)_Q = 0\ \forall \boldsymbol{z} \in V$, a.e. on I. Since the orthogonal complement in Q of $\mathcal{W}$ is the space $\boldsymbol{\varepsilon}(V)$ (cf. Theorem 7.5, page 146), there exists $\boldsymbol{v}_\eta \in L^\infty(I;V)$ such that

$$\boldsymbol{\varepsilon}_\eta(t) = \boldsymbol{\varepsilon}(\boldsymbol{v}_\eta(t)) \quad \text{a.e. } t \in I. \tag{17.59}$$

Let us prove that $\boldsymbol{v}_\eta \in V_1$ a.e. on I. Indeed, combining (17.57) and (17.59) we obtain

$$(\boldsymbol{\tau} - \boldsymbol{\sigma}_\eta, \boldsymbol{\varepsilon}(\boldsymbol{v}_\eta))_Q \geq 0 \quad \forall \boldsymbol{\tau} \in \Sigma_0, \quad \text{a.e. on } I. \tag{17.60}$$

Let $t \in \overline{I}$ be such that

$$(\boldsymbol{\tau} - \boldsymbol{\sigma}_\eta(t), \boldsymbol{\varepsilon}(\boldsymbol{v}_\eta(t)))_Q \geq 0 \quad \forall \boldsymbol{\tau} \in \Sigma_0. \tag{17.61}$$

Since V_1 is a closed subspace of V, if $\boldsymbol{v}_\eta(t) \notin V_1$ then there exists $\tilde{\boldsymbol{\tau}} \in Q$ such that

$$(\tilde{\boldsymbol{\tau}}, \boldsymbol{\varepsilon}(\boldsymbol{v}))_Q = 0 \quad \forall \boldsymbol{v} \in V_1, \tag{17.62}$$

$$(\tilde{\boldsymbol{\tau}}, \boldsymbol{\varepsilon}(\boldsymbol{v}_\eta(t)))_Q < 0. \tag{17.63}$$

From (17.62) it follows that $\lambda\, \tilde{\boldsymbol{\tau}} \in \Sigma_0$ for all $\lambda \geq 0$. Using (17.61) we obtain

$$\lambda\, (\tilde{\boldsymbol{\tau}}, \boldsymbol{\varepsilon}(\boldsymbol{v}_\eta(t)))_Q \geq (\boldsymbol{\sigma}_\eta, \boldsymbol{\varepsilon}(\boldsymbol{v}_\eta(t)))_Q. \tag{17.64}$$

Using now (17.63) and passing to the limit $\lambda \to +\infty$ in (17.64), we reach a contradiction. Thus, $\boldsymbol{v}_\eta \in L^\infty(I;V)$ is such that $\boldsymbol{v}_\eta \in V_1$ a.e. on I, i.e. $\boldsymbol{v}_\eta \in L^\infty(I;V_1)$. Define

$$\boldsymbol{u}_\eta(t) = \int_0^t \boldsymbol{v}_\eta(s)\, ds + \boldsymbol{u}_0 \quad \forall t \in \overline{I}. \tag{17.65}$$

Since $\boldsymbol{u}_0 \in V_1$, we have $\boldsymbol{u}_\eta \in W^{1,\infty}(I; V_1)$. Moreover, from (17.58)–(17.60), (17.65), (17.53) and (17.55) it follows that $(\boldsymbol{u}_\eta, \boldsymbol{\sigma}_\eta)$ is a solution to Problem 17.11, which concludes the existence part of Lemma 17.13. The uniqueness part follows from the uniqueness of the solution to the nonlinear evolution problem (17.53)–(17.55). ■

We consider now an operator $\Lambda : L^\infty(I; Q) \to L^\infty(I; Q)$ defined by

$$\Lambda\boldsymbol{\eta} = \mathcal{E}^{-1}\dot{\boldsymbol{\sigma}}_{\boldsymbol{f}} - \mathcal{E}^{-1}\mathcal{G}(\boldsymbol{\sigma}_\eta + \boldsymbol{\sigma}_{\boldsymbol{f}}, \varepsilon(\boldsymbol{u}_\eta)), \tag{17.66}$$

where for any $\boldsymbol{\eta} \in L^\infty(I; Q)$, $(\boldsymbol{u}_\eta, \boldsymbol{\sigma}_\eta)$ denotes the solution of the variational problem 17.11.

We have the following result.

Lemma 17.14 *The operator Λ has a unique fixed point $\boldsymbol{\eta}^* \in L^\infty(I; Q)$.*

Proof. Let $\boldsymbol{\eta}_1, \boldsymbol{\eta}_2 \in L^\infty(I; Q)$ and $t \in \overline{I}$. For simplicity we denote $\boldsymbol{u}_{\eta_i} = \boldsymbol{u}_i$, $\boldsymbol{\sigma}_{\eta_i} = \boldsymbol{\sigma}_i$, $i = 1, 2$. Observe that $\boldsymbol{u}_1$ and $\boldsymbol{u}_2$ have the same initial value, and so do $\boldsymbol{\sigma}_1$ and $\boldsymbol{\sigma}_2$. From the defining inequality (17.56) for $\boldsymbol{\sigma}_1$ and $\boldsymbol{\sigma}_2$, we obtain

$$\|\boldsymbol{\sigma}_1(t) - \boldsymbol{\sigma}_2(t)\|_Q^2 \le c \int_0^t \|\boldsymbol{\eta}_1(s) - \boldsymbol{\eta}_2(s)\|_Q^2 ds.$$

Then starting with (17.46) we can derive the next inequality:

$$\|\boldsymbol{u}_1(t) - \boldsymbol{u}_2(t)\|_V^2 \le c \int_0^t \|\boldsymbol{\eta}_1(s) - \boldsymbol{\eta}_2(s)\|_Q^2 ds.$$

Now from the definition (17.66) of the operator Λ, using the assumptions (6.44) and (6.45) on $\mathcal{E}$ and $\mathcal{G}$, we have

$$\|\Lambda\boldsymbol{\eta}_1(t) - \Lambda\boldsymbol{\eta}_2(t)\|_Q^2 \le c\,(\|\boldsymbol{\sigma}_1(t) - \boldsymbol{\sigma}_2(t)\|_Q^2 + \|\boldsymbol{u}_1(t) - \boldsymbol{u}_2(t)\|_V^2).$$

Thus,

$$\|\Lambda\boldsymbol{\eta}_1(t) - \Lambda\boldsymbol{\eta}_2(t)\|_Q^2 \le c \int_0^t \|\boldsymbol{\eta}_1(s) - \boldsymbol{\eta}_2(s)\|_Q^2 ds.$$

Then the result of the lemma follows from an application of the Banach fixed-point theorem. ■

Now let us prove Theorem 17.8.

Proof of Theorem 17.8

Existence. Let $\boldsymbol{\eta}^* \in L^\infty(I; Q)$ be the fixed point of Λ and let $(\boldsymbol{u}_{\eta^*}, \boldsymbol{\sigma}_{\eta^*}) \in W^{1,\infty}(I; V_1 \times Q_1)$ be the functions given by Lemma 17.13 for $\boldsymbol{\eta} = \boldsymbol{\eta}^*$. It follows that $(\boldsymbol{u}_{\eta^*}, \boldsymbol{\sigma}_{\eta^*})$ is a solution of Problem 17.9 and, using Lemma 17.10, we obtain the existence part in Theorem 17.8.

Uniqueness. The uniqueness part in Theorem 17.8 follows from the uniqueness of the fixed point of the operator Λ defined by (17.66). Alternatively, it can be proved directly from (17.22)–(17.24), by using (6.44), (6.45) and the Gronwall inequality. ■

17.3 Some Properties of the Solution

We start this section with a study of the links between the variational problem 17.2 and 17.3 which represent two different variational formulations of the mechanical problem (17.1)–(17.6).

We have the following equivalence result.

Theorem 17.15 *Assume* (6.44), (6.45), (17.7), (17.8) *and* (17.14). *Let* $\boldsymbol{u} \in W^{1,\infty}(I; V_1)$, $\boldsymbol{\sigma} \in W^{1,\infty}(I; Q_1)$. *Then the couple* $(\boldsymbol{u}, \boldsymbol{\sigma})$ *is a solution of Problem 17.2 iff it is a solution of Problem 17.3.*

Proof. Suppose that $(\boldsymbol{u}, \boldsymbol{\sigma})$ is a solution of Problem 17.2. Taking $\boldsymbol{v} = 2\dot{\boldsymbol{u}}$ and $\boldsymbol{v} = \boldsymbol{0}$ in the inequality (17.20), we obtain

$$(\boldsymbol{\sigma}, \boldsymbol{\varepsilon}(\dot{\boldsymbol{u}}))_Q + j(\dot{\boldsymbol{u}}) = (\boldsymbol{f}, \dot{\boldsymbol{u}})_V \quad \text{a.e. on } I. \tag{17.67}$$

It is easy to see that (17.20) is equivalent to the equality (17.67) and the inequality

$$(\boldsymbol{\sigma}, \boldsymbol{\varepsilon}(\boldsymbol{v}))_Q + j(\boldsymbol{v}) \geq (\boldsymbol{f}, \boldsymbol{v})_V \quad \forall \boldsymbol{v} \in V_1, \text{ a.e. on } I.$$

Owing to the time regularity of $\boldsymbol{\sigma}$ and $\boldsymbol{f}$, this inequality holds for all $t \in \overline{I}$. Thus, $\boldsymbol{\sigma}(t) \in \Sigma(t)$ for all $t \in \overline{I}$. The inequality in (17.23) follows from the definition of $\Sigma(t)$, (17.13), and the equality (17.67). So $(\boldsymbol{u}, \boldsymbol{\sigma})$ is a solution of Problem 17.3.

Conversely, let $(\boldsymbol{u}, \boldsymbol{\sigma})$ be a solution of Problem 17.3. The subdifferentiability of j on V_1 ensures the existence of $\tilde{\boldsymbol{\tau}} : \overline{I} \to Q$ such that

$$(\tilde{\boldsymbol{\tau}}, \boldsymbol{\varepsilon}(\boldsymbol{v}) - \boldsymbol{\varepsilon}(\dot{\boldsymbol{u}}))_Q + j(\boldsymbol{v}) - j(\dot{\boldsymbol{u}}) \geq (\boldsymbol{f}, \boldsymbol{v} - \dot{\boldsymbol{u}})_V \quad \forall \boldsymbol{v} \in V_1 \tag{17.68}$$

a.e. on I. Taking $\boldsymbol{v} = 2\dot{\boldsymbol{u}}$ and $\boldsymbol{v} = \boldsymbol{0}$ in (17.68) we obtain

$$(\tilde{\boldsymbol{\tau}}, \boldsymbol{\varepsilon}(\dot{\boldsymbol{u}}))_Q + j(\dot{\boldsymbol{u}}) = (\boldsymbol{f}, \dot{\boldsymbol{u}})_V \quad \text{a.e. on } I. \tag{17.69}$$

Using (17.68) and (17.69) we deduce that $\tilde{\boldsymbol{\tau}}(t) \in \Sigma(t)$ a.e. $t \in I$. From (17.23) we find

$$(\tilde{\boldsymbol{\tau}}, \boldsymbol{\varepsilon}(\dot{\boldsymbol{u}}))_Q \geq (\boldsymbol{\sigma}, \boldsymbol{\varepsilon}(\dot{\boldsymbol{u}}))_Q \quad \text{a.e. on } I. \tag{17.70}$$

From (17.69) and (17.70) it follows that

$$(\boldsymbol{f}, \dot{\boldsymbol{u}})_V \geq (\boldsymbol{\sigma}, \boldsymbol{\varepsilon}(\dot{\boldsymbol{u}}))_Q + j(\dot{\boldsymbol{u}}) \quad \text{a.e. on } I.$$

Since $\boldsymbol{\sigma}(t) \in \Sigma(t)$ for all $t \in \overline{I}$ we have

$$(\boldsymbol{\sigma}, \boldsymbol{\varepsilon}(\dot{\boldsymbol{u}}))_Q + j(\dot{\boldsymbol{u}}) \geq (\boldsymbol{f}, \dot{\boldsymbol{u}})_V \quad \text{a.e. on } I.$$

Thus, the equality (17.67) holds and then the inequality (17.20) is valid. Therefore, $(\boldsymbol{u}, \boldsymbol{\sigma})$ is a solution of Problem 17.2. ∎

Theorems 17.4 and 17.8 state the unique solvability of Problems 17.2 and 17.3, respectively, while Theorem 17.15 expresses the equivalence of these variational problems. From these results we conclude that the mechanical problem (17.1)–(17.6) has a unique *weak solution* which solves both Problems 17.2 and 17.3.

In the following, we consider the continuous dependence of the solution on the input data.

Theorem 17.16 *Assume* (6.44), (6.45), (17.8) *and let* $(\boldsymbol{u}_i, \boldsymbol{\sigma}_i)$ *be the solution of Problem 17.2 or 17.3 for the data* $\boldsymbol{f}_{0i}$, $\boldsymbol{f}_{2i}$, $\boldsymbol{u}_{0i}$, $\boldsymbol{\sigma}_{0i}$ *which satisfy* (17.7), (17.14), $i = 1, 2$. *Then there exists a constant* $c > 0$ *depending only on* Ω, Γ_1, Γ_3, $\mathcal{E}$, $\mathcal{G}$ *and* T *such that*

$$\begin{aligned} &\|\boldsymbol{u}_1 - \boldsymbol{u}_2\|_{L^\infty(I;V)} + \|\boldsymbol{\sigma}_1 - \boldsymbol{\sigma}_2\|_{L^\infty(I;Q_1)} \\ &\quad \le c\,(\|\boldsymbol{f}_{01} - \boldsymbol{f}_{02}\|_{W^{1,\infty}(I;H)} + \|\boldsymbol{f}_{21} - \boldsymbol{f}_{22}\|_{W^{1,\infty}(I;L^2(\Gamma_2)^d)} \\ &\qquad + \|\boldsymbol{u}_{01} - \boldsymbol{u}_{02}\|_V + \|\boldsymbol{\sigma}_{01} - \boldsymbol{\sigma}_{02}\|_{Q_1}). \end{aligned} \tag{17.71}$$

Proof. Let $i = 1, 2$. By (17.20) we have

$$(\boldsymbol{\sigma}_i, \boldsymbol{\varepsilon}(\boldsymbol{v}) - \boldsymbol{\varepsilon}(\dot{\boldsymbol{u}}_i))_Q + j(\boldsymbol{v}) - j(\dot{\boldsymbol{u}}_i) \ge (\boldsymbol{f}_i, \boldsymbol{v} - \dot{\boldsymbol{u}}_i)_V \quad \forall\, \boldsymbol{v} \in V_1, \tag{17.72}$$

a.e. on I, where $\boldsymbol{f}_i$ are defined by (17.9). Integrating (17.19) and using the initial value (17.21), we get

$$\boldsymbol{\sigma}_i(t) = \mathcal{E}\boldsymbol{\varepsilon}(\boldsymbol{u}_i(t)) + \boldsymbol{z}_i(t), \tag{17.73}$$

with

$$\boldsymbol{z}_i(t) = \int_0^t \mathcal{G}(\boldsymbol{\sigma}_i(s), \boldsymbol{\varepsilon}(\boldsymbol{u}_i(s))\, ds + \boldsymbol{\sigma}_{0i} - \mathcal{E}\boldsymbol{\varepsilon}(\boldsymbol{u}_{0i}) \tag{17.74}$$

for all $t \in \overline{I}$. As in the proof of Lemma 17.6, we use the bilinear form

$$a(\boldsymbol{u}, \boldsymbol{v}) = (\mathcal{E}\boldsymbol{\varepsilon}(\boldsymbol{u}), \boldsymbol{\varepsilon}(\boldsymbol{v}))_Q$$

on the space V_1. From (17.72) and (17.73) we obtain

$$\begin{aligned} a(\boldsymbol{u}_1 - \boldsymbol{u}_2, \dot{\boldsymbol{u}}_1 - \dot{\boldsymbol{u}}_2) &\le -(\boldsymbol{z}_1 - \boldsymbol{z}_2, \boldsymbol{\varepsilon}(\dot{\boldsymbol{u}}_1) - \boldsymbol{\varepsilon}(\dot{\boldsymbol{u}}_2))_Q \\ &\quad + (\boldsymbol{f}_1 - \boldsymbol{f}_2, \dot{\boldsymbol{u}}_1 - \dot{\boldsymbol{u}}_2)_V \quad \text{a.e. on } I. \end{aligned}$$

We then integrate the previous inequality with respect to the time variable. Using (17.74), the assumptions (6.44) and (6.45) on $\mathcal{E}$ and $\mathcal{G}$, we can obtain

$$\begin{aligned} \|\boldsymbol{u}_1(t) - \boldsymbol{u}_2(t)\|_V^2 &\le c\Big(\|\boldsymbol{u}_{01} - \boldsymbol{u}_{02}\|_V^2 + \|\boldsymbol{\sigma}_{01} - \boldsymbol{\sigma}_{02}\|_Q^2 \\ &\quad + \|\boldsymbol{f}_1 - \boldsymbol{f}_2\|_{W^{1,\infty}(I;V)}^2 + \int_0^t \|\dot{\boldsymbol{f}}_1(s) - \dot{\boldsymbol{f}}_2(t)\|_V^2 ds \\ &\quad + \int_0^t (\|\boldsymbol{u}_1(s) - \boldsymbol{u}_2(s)\|_V^2 + \|\boldsymbol{\sigma}_1(s) - \boldsymbol{\sigma}_2(s)\|_Q^2)\, ds\Big) \end{aligned} \tag{17.75}$$

for all $t \in \overline{I}$. Using now (17.73)–(17.75) we obtain

$$\begin{aligned}
&\|\boldsymbol{\sigma}_1(t) - \boldsymbol{\sigma}_2(t)\|_Q^2 \\
&\quad \le c\Big(\|\boldsymbol{u}_1(t) - \boldsymbol{u}_2(t)\|_V^2 + \|\boldsymbol{u}_{01} - \boldsymbol{u}_{02}\|_V^2 + \|\boldsymbol{\sigma}_{01} - \boldsymbol{\sigma}_{02}\|_Q^2 \\
&\qquad + \int_0^t (\|\boldsymbol{u}_1(s) - \boldsymbol{u}_2(s)\|_V^2 + \|\boldsymbol{\sigma}_1(s) - \boldsymbol{\sigma}_2(s)\|_Q^2)\, ds\Big)
\end{aligned}$$

for all $t \in \overline{I}$. Thus, for any $t \in \overline{I}$,

$$\begin{aligned}
&\|\boldsymbol{u}_1(t) - \boldsymbol{u}_2(t)\|_V^2 + \|\boldsymbol{\sigma}_1(t) - \boldsymbol{\sigma}_2(t)\|_Q^2 \\
&\quad \le c\Big(\|\boldsymbol{u}_{01} - \boldsymbol{u}_{02}\|_V^2 + \|\boldsymbol{\sigma}_{01} - \boldsymbol{\sigma}_{02}\|_Q^2 + \|\boldsymbol{f}_1 - \boldsymbol{f}_2\|_{W^{1,\infty}(I;V)}^2 \\
&\qquad + \int_0^t (\|\boldsymbol{u}_1(s) - \boldsymbol{u}_2(s)\|_V^2 + \|\boldsymbol{\sigma}_1(s) - \boldsymbol{\sigma}_2(s)\|_Q^2)\, ds\Big).
\end{aligned}$$

Applying the Gronwall inequality (Lemma 7.24), we deduce that

$$\begin{aligned}
&\|\boldsymbol{u}_1 - \boldsymbol{u}_2\|_{L^\infty(I;V)} + \|\boldsymbol{\sigma}_1 - \boldsymbol{\sigma}_2\|_{L^\infty(I;Q)} \\
&\quad \le c\,(\|\boldsymbol{u}_{01} - \boldsymbol{u}_{02}\|_V + \|\boldsymbol{\sigma}_{01} - \boldsymbol{\sigma}_{02}\|_Q + \|\boldsymbol{f}_1 - \boldsymbol{f}_2\|_{W^{1,\infty}(I;V)}). \quad (17.76)
\end{aligned}$$

From the definition (17.9) of $\boldsymbol{f}(t)$, we have

$$\begin{aligned}
\|\boldsymbol{f}_1 - \boldsymbol{f}_2\|_{W^{1,\infty}(I;V)} \le c\,(&\|\boldsymbol{f}_{01} - \boldsymbol{f}_{02}\|_{W^{1,\infty}(I;H)} \\
&+ \|\boldsymbol{f}_{21} - \boldsymbol{f}_{22}\|_{W^{1,\infty}(I;L^2(\Gamma_2)^d)}).
\end{aligned}$$

Also the term $\|\boldsymbol{\sigma}_1 - \boldsymbol{\sigma}_2\|_{L^\infty(I;Q)}$ in (17.76) can be replaced by $\|\boldsymbol{\sigma}_1 - \boldsymbol{\sigma}_2\|_{L^\infty(I;Q_1)}$ since

$$\operatorname{Div} \boldsymbol{\sigma}_{0i}(t) = -\boldsymbol{f}_{0i}(t) \quad \text{a.e. in } \Omega \times \overline{I}.$$

Therefore, the inequality (17.71) follows from (17.76). ■

Remark 17.17 *The estimation* (17.71) *is a* continuous dependence result *for the solution of Problem 17.2 or Problem 17.3 with respect to the input data. We note that under the assumptions of Theorem 17.16, if* $\boldsymbol{f}_{01} = \boldsymbol{f}_{02}$ *and* $\boldsymbol{f}_{21} = \boldsymbol{f}_{22}$, *then*

$$\begin{aligned}
&\|\boldsymbol{u}_1 - \boldsymbol{u}_2\|_{L^\infty(I;V)} + \|\boldsymbol{\sigma}_1 - \boldsymbol{\sigma}_2\|_{L^\infty(I;Q_1)} \\
&\quad \le c\,(\|\boldsymbol{u}_{01} - \boldsymbol{u}_{02}\|_V + \|\boldsymbol{\sigma}_{01} - \boldsymbol{\sigma}_{02}\|_{Q_1}).
\end{aligned}$$

This is a finite time stability result *for the solution of the variational problems 17.2 and 17.3.*

In the rest of the chapter we suppose that assumptions of Theorem 17.4 hold and we consider numerical approximations of Problem 17.2.

17.4 Semi-discrete Approximation

As in Section 7.2, we assume Ω is a polygon or polyhedron, and let $\{\mathcal{T}^h\}$ be a regular family of finite element partitions of $\overline{\Omega}$ that are compatible to the boundary decomposition $\Gamma = \overline{\Gamma}_1 \cup \overline{\Gamma}_2 \cup \overline{\Gamma}_3$. We use linear elements for the displacement variable and piecewise constants for the stress field. Thus, V_1^h is the space of continuous, piecewise linear functions that are zero at the nodes on $\overline{\Gamma}_1$ and whose normal components vanish at the nodes on $\overline{\Gamma}_3$.

Let $\mathcal{P}_{Q^h} : Q \to Q^h$ be the orthogonal projection defined through the relation

$$(\mathcal{P}_{Q^h}\boldsymbol{\tau}, \boldsymbol{\tau}^h)_Q = (\boldsymbol{\tau}, \boldsymbol{\tau}^h)_Q \quad \forall\, \boldsymbol{\tau} \in Q,\ \boldsymbol{\tau}^h \in Q^h. \tag{17.77}$$

Then $\mathcal{P}_{Q^h}$ is linear and non-expansive:

$$\|\mathcal{P}_{Q^h}\boldsymbol{\tau}\|_Q \le \|\boldsymbol{\tau}\|_Q \quad \forall\, \boldsymbol{\tau} \in Q.$$

The definition of $\mathcal{P}_{V_1^h} : V_1 \to V_1^h$ is similar to that of $\mathcal{P}_{Q^h}$.

Now a semi-discrete approximation of Problem 17.2 is

Problem 17.18 *Find $\boldsymbol{u}^h : \overline{I} \to V_1^h$ and $\boldsymbol{\sigma}^h : \overline{I} \to Q^h$ such that*

$$\boldsymbol{u}^h(0) = \boldsymbol{u}_0^h, \quad \boldsymbol{\sigma}^h(0) = \boldsymbol{\sigma}_0^h, \tag{17.78}$$

and for a.e. $t \in I$,

$$\dot{\boldsymbol{\sigma}}^h(t) = \mathcal{P}_{Q^h}\mathcal{E}\boldsymbol{\varepsilon}(\dot{\boldsymbol{u}}^h(t)) + \mathcal{P}_{Q^h}\mathcal{G}(\boldsymbol{\sigma}^h(t), \boldsymbol{\varepsilon}(\dot{\boldsymbol{u}}^h(t))), \tag{17.79}$$

$$\begin{aligned}(\boldsymbol{\sigma}^h(t), \boldsymbol{\varepsilon}(\boldsymbol{v}^h) - \boldsymbol{\varepsilon}(\dot{\boldsymbol{u}}^h(t)))_Q + j(\boldsymbol{v}^h) - j(\dot{\boldsymbol{u}}^h(t))& \\ \ge (\boldsymbol{f}(t), \boldsymbol{v}^h - \dot{\boldsymbol{u}}^h(t))_V \quad \forall\, \boldsymbol{v}^h \in V_1^h.&\end{aligned} \tag{17.80}$$

Here, $\boldsymbol{u}_0^h = \mathcal{P}_{V_1^h}\boldsymbol{u}_0 \in V_1^h$, $\boldsymbol{\sigma}_0^h = \mathcal{P}_{Q^h}\boldsymbol{\sigma}_0 \in Q^h$ are orthogonal projections of $\boldsymbol{u}_0$ and $\boldsymbol{\sigma}_0$ to V_1^h and Q^h, respectively.

From the definition of $\boldsymbol{\sigma}_0^h$, we see that the discrete analog of (17.14) is valid:

$$(\boldsymbol{\sigma}_0^h, \boldsymbol{\varepsilon}(\boldsymbol{v}^h))_Q + j(\boldsymbol{v}^h) \ge (\boldsymbol{f}(0), \boldsymbol{v}^h)_V \quad \forall\, \boldsymbol{v}^h \in V_1^h.$$

This relation is needed to verify the existence of a solution of Problem 17.18.

We observe that the projection operator $\mathcal{P}_{Q^h}$ is introduced to ensure that the relation (17.79) is well-defined on the space Q^h.

Using the arguments similar to those in proving Theorem 17.4, it can be shown that Problem 17.18 has a unique solution $\boldsymbol{u}^h \in W^{1,\infty}(I; V_1^h)$ and $\boldsymbol{\sigma}^h \in W^{1,\infty}(I; Q^h)$. We have $\boldsymbol{u}^h \in C(\overline{I}; V_1^h)$ and $\boldsymbol{\sigma}^h \in C(\overline{I}; Q^h)$.

Next, we focus on the derivation of error estimates for $\boldsymbol{u} - \boldsymbol{u}^h$ and $\boldsymbol{\sigma} - \boldsymbol{\sigma}^h$. It will be convenient to use the "energy" norm

$$\|\boldsymbol{v}\|_{\mathcal{E}} = (\mathcal{E}\boldsymbol{\varepsilon}(\boldsymbol{v}), \boldsymbol{\varepsilon}(\boldsymbol{v}))_Q^{1/2}, \quad \boldsymbol{v} \in V. \tag{17.81}$$

By the assumptions (6.44) on $\mathcal{E}$, this norm is equivalent to the norm $\|\boldsymbol{v}\|_V$ over the space V.

To simplify the notation, we denote

$$\boldsymbol{s}_{\mathcal{G}}(t) = \int_0^t \mathcal{G}(\boldsymbol{\sigma}(s), \boldsymbol{\varepsilon}(\dot{\boldsymbol{u}}(s)))\, ds, \tag{17.82}$$

$$\boldsymbol{s}_{\mathcal{G}}^h(t) = \int_0^t \mathcal{G}(\boldsymbol{\sigma}^h(s), \boldsymbol{\varepsilon}(\dot{\boldsymbol{u}}^h(s)))\, ds, \tag{17.83}$$

for all $t \in \overline{I}$. We first integrate the viscoplastic constitutive relation (17.19) and its discrete analogue (17.79), and use the initial conditions (17.21) and (17.78) to obtain

$$\boldsymbol{\sigma}(t) = \mathcal{E}\boldsymbol{\varepsilon}(\boldsymbol{u}(t)) + \boldsymbol{s}_{\mathcal{G}}(t) + \boldsymbol{\sigma}_0 - \mathcal{E}\boldsymbol{\varepsilon}(\boldsymbol{u}_0), \tag{17.84}$$

$$\boldsymbol{\sigma}^h(t) = \mathcal{P}_{Q^h}\mathcal{E}\boldsymbol{\varepsilon}(\boldsymbol{u}^h(t)) + \mathcal{P}_{Q^h}\boldsymbol{s}_{\mathcal{G}}^h(t) + \boldsymbol{\sigma}_0^h - \mathcal{P}_{Q^h}\mathcal{E}\boldsymbol{\varepsilon}(\boldsymbol{u}_0^h). \tag{17.85}$$

Then we subtract (17.85) from (17.84) to get

$$\begin{aligned} \boldsymbol{\sigma}(t) - \boldsymbol{\sigma}^h(t) = {} & \mathcal{P}_{Q^h}\mathcal{E}\boldsymbol{\varepsilon}(\boldsymbol{u}(t) - \boldsymbol{u}^h(t)) + \mathcal{P}_{Q^h}\left[\boldsymbol{s}_{\mathcal{G}}(t) - \boldsymbol{s}_{\mathcal{G}}^h(t)\right] \\ & + \boldsymbol{\sigma}_0 - \boldsymbol{\sigma}_0^h - \mathcal{P}_{Q^h}\mathcal{E}\boldsymbol{\varepsilon}(\boldsymbol{u}_0 - \boldsymbol{u}_0^h) \\ & + (I_Q - \mathcal{P}_{Q^h})(\boldsymbol{\sigma}(t) - \boldsymbol{\sigma}_0), \end{aligned} \tag{17.86}$$

where $I_Q : Q \to Q$ is the identity operator.

Denote

$$e_0 = \|\boldsymbol{u}_0 - \boldsymbol{u}_0^h\|_V + \|\boldsymbol{\sigma}_0 - \boldsymbol{\sigma}_0^h\|_Q$$

for the initial error. Then we get the following inequality from (17.86), using the assumptions (6.44) and (6.45) on $\mathcal{E}$ and $\mathcal{G}$, and the non-expansiveness property of he projection operator $\mathcal{P}_{Q^h}$:

$$\begin{aligned} \|\boldsymbol{\sigma}(t) - \boldsymbol{\sigma}^h(t)\|_Q \le {} & c\,\|\boldsymbol{u}(t) - \boldsymbol{u}^h(t)\|_V + \|(I_Q - \mathcal{P}_{Q^h})(\boldsymbol{\sigma}(t) - \boldsymbol{\sigma}_0)\|_Q \\ & + c\int_0^t \left(\|\boldsymbol{\sigma}(s) - \boldsymbol{\sigma}^h(s)\|_Q + \|\boldsymbol{u}(s) - \boldsymbol{u}^h(s)\|_V\right) ds \\ & + c\,e_0. \end{aligned} \tag{17.87}$$

The previous inequality holds for all $t \in \overline{I}$.

Now plugging (17.84) in (17.20) with $\boldsymbol{v} = \dot{\boldsymbol{u}}^h(t)$, we have

$$\begin{aligned} & (\mathcal{E}\boldsymbol{\varepsilon}(\boldsymbol{u}(t)), \boldsymbol{\varepsilon}(\dot{\boldsymbol{u}}(t) - \dot{\boldsymbol{u}}^h(t)))_Q \\ & \quad \le (\boldsymbol{f}(t), \dot{\boldsymbol{u}}(t) - \dot{\boldsymbol{u}}^h(t))_V + j(\dot{\boldsymbol{u}}^h(t)) - j(\dot{\boldsymbol{u}}(t)) \\ & \qquad + \left(\boldsymbol{s}_{\mathcal{G}}(t), \boldsymbol{\varepsilon}(\dot{\boldsymbol{u}}^h(t) - \dot{\boldsymbol{u}}(t))\right)_Q \\ & \qquad + (\boldsymbol{\sigma}_0 - \mathcal{E}\boldsymbol{\varepsilon}(\boldsymbol{u}_0), \boldsymbol{\varepsilon}(\dot{\boldsymbol{u}}^h(t) - \dot{\boldsymbol{u}}(t)))_Q. \end{aligned} \tag{17.88}$$

Let $\boldsymbol{v}^h \in L^2(I; V_1^h)$ be arbitrary. We plug (17.85) in (17.80) with $\boldsymbol{v}^h = \boldsymbol{v}^h(t)$ to obtain

$$\begin{aligned} -(\mathcal{E}\boldsymbol{\varepsilon}(\boldsymbol{u}^h(t)), \boldsymbol{\varepsilon}(\dot{\boldsymbol{u}}(t) - \dot{\boldsymbol{u}}^h(t)))_Q & \\ \leq (\boldsymbol{f}(t), \dot{\boldsymbol{u}}^h(t) - \boldsymbol{v}^h(t))_V + j(\boldsymbol{v}^h(t)) - j(\dot{\boldsymbol{u}}^h(t)) & \\ + \left(\boldsymbol{s}_{\mathcal{G}}^h(t), \boldsymbol{\varepsilon}(\boldsymbol{v}^h(t) - \dot{\boldsymbol{u}}^h(t))\right)_Q & \\ + (\boldsymbol{\sigma}_0^h - \mathcal{E}\boldsymbol{\varepsilon}(\boldsymbol{u}_0^h), \boldsymbol{\varepsilon}(\boldsymbol{v}^h(t) - \dot{\boldsymbol{u}}^h(t)))_Q & \\ + (\mathcal{E}\boldsymbol{\varepsilon}(\boldsymbol{u}^h(t)), \boldsymbol{\varepsilon}(\boldsymbol{v}^h(t) - \dot{\boldsymbol{u}}(t)))_Q. & \qquad (17.89) \end{aligned}$$

The relations (17.88) and (17.89) hold for a.e. $t \in I$. Adding these relations we have

$$\begin{aligned} (\mathcal{E}\boldsymbol{\varepsilon}(\boldsymbol{u}(t) - \boldsymbol{u}^h(t)), \boldsymbol{\varepsilon}(\dot{\boldsymbol{u}}(t) - \dot{\boldsymbol{u}}^h(t)))_Q & \\ \leq (\mathcal{E}\boldsymbol{\varepsilon}(\boldsymbol{u}(t) - \boldsymbol{u}^h(t)), \boldsymbol{\varepsilon}(\dot{\boldsymbol{u}}(t) - \boldsymbol{v}^h(t)))_Q & \\ + R(t; \dot{\boldsymbol{u}}(t), \boldsymbol{v}^h(t)) + (\boldsymbol{D}(t), \boldsymbol{\varepsilon}(\dot{\boldsymbol{u}}^h(t) - \boldsymbol{v}^h(t)))_Q & \end{aligned}$$

for a.e. $t \in I$, where

$$\begin{aligned} R(t; \dot{\boldsymbol{u}}(t), \boldsymbol{v}^h(t)) = (\boldsymbol{\sigma}(t), \boldsymbol{\varepsilon}(\boldsymbol{v}^h(t) - \dot{\boldsymbol{u}}(t)))_Q + j(\boldsymbol{v}^h(t)) - j(\dot{\boldsymbol{u}}(t)) & \\ - (\boldsymbol{f}(t), \boldsymbol{v}^h(t) - \dot{\boldsymbol{u}}(t))_V & \qquad (17.90) \end{aligned}$$

and

$$\boldsymbol{D}(t) = \boldsymbol{s}_{\mathcal{G}}(t) - \boldsymbol{s}_{\mathcal{G}}^h(t) + \boldsymbol{\sigma}_0 - \boldsymbol{\sigma}_0^h - \mathcal{E}\boldsymbol{\varepsilon}(\boldsymbol{u}_0 - \boldsymbol{u}_0^h). \qquad (17.91)$$

Then we have

$$\begin{aligned} \frac{d}{dt}\|\boldsymbol{u}(t) - \boldsymbol{u}^h(t)\|_{\mathcal{E}}^2 \leq c\Big(\|\boldsymbol{u}(t) - \boldsymbol{u}^h(t)\|_{\mathcal{E}}^2 + \|\dot{\boldsymbol{u}}(t) - \boldsymbol{v}^h(t)\|_V^2 & \\ + R(t; \dot{\boldsymbol{u}}(t), \boldsymbol{v}^h(t)) + (\boldsymbol{D}(t), \boldsymbol{\varepsilon}(\dot{\boldsymbol{u}}(t) - \boldsymbol{v}^h(t)))_Q & \\ + (\boldsymbol{D}(t), \boldsymbol{\varepsilon}(\dot{\boldsymbol{u}}^h(t) - \dot{\boldsymbol{u}}(t)))_Q\Big) & \\ \leq c\Big(\|\boldsymbol{u}(t) - \boldsymbol{u}^h(t)\|_{\mathcal{E}}^2 + \|\dot{\boldsymbol{u}}(t) - \boldsymbol{v}^h(t)\|_V^2 & \\ + R(t; \dot{\boldsymbol{u}}(t), \boldsymbol{v}^h(t)) + \|\boldsymbol{D}(t)\|_Q^2 & \\ + (\boldsymbol{D}(t), \boldsymbol{\varepsilon}(\dot{\boldsymbol{u}}^h(t) - \dot{\boldsymbol{u}}(t)))_Q\Big) \quad \text{a.e. } t \in I. & \end{aligned}$$

Integrate the above inequality from 0 to t to obtain

$$\begin{aligned} \|\boldsymbol{u}(t) - \boldsymbol{u}^h(t)\|_{\mathcal{E}}^2 \leq \|\boldsymbol{u}_0 - \boldsymbol{u}_0^h\|_{\mathcal{E}}^2 + c\int_0^t \left(|R(s; \dot{\boldsymbol{u}}(s), \boldsymbol{v}^h(s))| + \|\boldsymbol{D}(s)\|_Q^2\right) ds & \\ + c\int_0^t \left(\|\boldsymbol{u}(s) - \boldsymbol{u}^h(s)\|_{\mathcal{E}}^2 + \|\dot{\boldsymbol{u}}(s) - \boldsymbol{v}^h(s)\|_V^2\right) ds & \\ + c\int_0^t \Big(\boldsymbol{D}(s), \boldsymbol{\varepsilon}(\dot{\boldsymbol{u}}^h(s) - \dot{\boldsymbol{u}}(s))\Big)_Q\, ds, & \end{aligned}$$

which holds for all $t \in \overline{I}$ due to the continuity of $\boldsymbol{u}(t)$ and $\boldsymbol{u}^h(t)$. For the last term above, we perform an integration by parts,

$$\begin{aligned}\int_0^t (\boldsymbol{D}(s), \boldsymbol{\varepsilon}(\dot{\boldsymbol{u}}^h(s) - \dot{\boldsymbol{u}}(s)))_Q\, ds \\ &= (\boldsymbol{D}(t), \boldsymbol{\varepsilon}(\boldsymbol{u}^h(t) - \boldsymbol{u}(t)))_Q - (\boldsymbol{D}(0), \boldsymbol{\varepsilon}(\boldsymbol{u}_0^h - \boldsymbol{u}_0))_Q \\ &\quad - \int_0^t (\dot{\boldsymbol{D}}(s), \boldsymbol{\varepsilon}(\boldsymbol{u}^h(s) - \boldsymbol{u}(s)))_Q\, ds,\end{aligned}$$

where

$$\dot{\boldsymbol{D}}(s) = \mathcal{G}(\boldsymbol{\sigma}(s), \boldsymbol{\varepsilon}(\dot{\boldsymbol{u}}(s))) - \mathcal{G}(\boldsymbol{\sigma}^h(s), \boldsymbol{\varepsilon}(\dot{\boldsymbol{u}}^h(s))).$$

Using the assumptions (6.45) on $\mathcal{G}$, we get the estimate

$$\begin{aligned}\int_0^t (\boldsymbol{D}(s), \boldsymbol{\varepsilon}(\dot{\boldsymbol{u}}^h(s) - \dot{\boldsymbol{u}}(s)))_Q\, ds \\ &\le c(\delta)\, \|\boldsymbol{D}(t)\|_Q^2 + \delta\, \|\boldsymbol{u}(t) - \boldsymbol{u}^h(t)\|_{\mathcal{E}}^2 + c\, e_0^2 \\ &\quad + c \int_0^t (\|\boldsymbol{u}(s) - \boldsymbol{u}^h(s)\|_{\mathcal{E}}^2 + \|\boldsymbol{\sigma}(s) - \boldsymbol{\sigma}^h(s)\|_Q^2)\, ds,\end{aligned}$$

where $\delta > 0$ is arbitrary. From the definition (17.91) of $\boldsymbol{D}(t)$, the assumptions (6.44) and (6.45) on $\mathcal{E}$ and $\mathcal{G}$, we immediately get the bound

$$\|\boldsymbol{D}(t)\|_Q \le c \int_0^t (\|\boldsymbol{u}(s) - \boldsymbol{u}^h(s)\|_{\mathcal{E}} + \|\boldsymbol{\sigma}(s) - \boldsymbol{\sigma}^h(s)\|_Q)\, ds + c\, e_0.$$

Combine the last several relations with $\delta > 0$ small enough,

$$\begin{aligned}\|\boldsymbol{u}(t) - \boldsymbol{u}^h(t)\|_{\mathcal{E}}^2 &\le c\, e_0^2 + c \int_0^t (\|\dot{\boldsymbol{u}}(s) - \boldsymbol{v}^h(s)\|_V^2 + |R(s; \dot{\boldsymbol{u}}(s), \boldsymbol{v}^h(s))|)\, ds \\ &\quad + c \int_0^t (\|\boldsymbol{u}(s) - \boldsymbol{u}^h(s)\|_{\mathcal{E}}^2 + \|\boldsymbol{\sigma}(s) - \boldsymbol{\sigma}^h(s)\|_Q^2)\, ds.\end{aligned}$$

This inequality, together with (17.87), implies

$$\begin{aligned}\|\boldsymbol{u}(t) - \boldsymbol{u}^h(t)\|_{\mathcal{E}}^2 + \|\boldsymbol{\sigma}(t) - \boldsymbol{\sigma}^h(t)\|_Q^2 \\ &\le c\, e_0^2 + c\, \|(I_Q - \mathcal{P}_{Q^h})(\boldsymbol{\sigma}(t) - \boldsymbol{\sigma}_0)\|_Q^2 \\ &\quad + c \int_0^t (\|\dot{\boldsymbol{u}}(s) - \boldsymbol{v}^h(s)\|_V^2 + |R(s; \dot{\boldsymbol{u}}(s), \boldsymbol{v}^h(s))|)\, ds \\ &\quad + c \int_0^t (\|\boldsymbol{u}(s) - \boldsymbol{u}^h(s)\|_{\mathcal{E}}^2 + \|\boldsymbol{\sigma}(s) - \boldsymbol{\sigma}^h(s)\|_Q^2)\, ds,\end{aligned}$$

for all $t \in \overline{I}$. Applying the Gronwall inequality (Lemma 7.24), we get

$$\begin{aligned}\sup_{t\in\overline{I}} &\left(\|\boldsymbol{u}(t) - \boldsymbol{u}^h(t)\|_{\mathcal{E}}^2 + \|\boldsymbol{\sigma}(t) - \boldsymbol{\sigma}^h(t)\|_Q^2\right) \\ &\le c\, e_0^2 + c \sup_{t\in\overline{I}} \|(I_Q - \mathcal{P}_{Q^h})(\boldsymbol{\sigma}(t) - \boldsymbol{\sigma}_0)\|_Q^2 \\ &\quad + c \int_I \left(\|\dot{\boldsymbol{u}}(t) - \boldsymbol{v}^h(t)\|_V^2 + |R(t; \dot{\boldsymbol{u}}(t), \boldsymbol{v}^h(t))|\right) dt.\end{aligned}$$

Notice that the term $\|\boldsymbol{u}(t) - \boldsymbol{u}^h(t)\|_{\mathcal{E}}^2$ can be replaced by $\|\boldsymbol{u}(t) - \boldsymbol{u}^h(t)\|_V^2$.

Summarizing, we have shown the following result.

Theorem 17.19 *For the semi-discrete solution of Problem 17.18, we have the error estimate*

$$\begin{aligned}\|\boldsymbol{u} - \boldsymbol{u}^h\|_{L^\infty(I;V)} &+ \|\boldsymbol{\sigma} - \boldsymbol{\sigma}^h\|_{L^\infty(I;Q)} \\ &\le c\,(\|\boldsymbol{u}_0 - \boldsymbol{u}_0^h\|_V + \|\boldsymbol{\sigma}_0 - \boldsymbol{\sigma}_0^h\|_Q) \\ &\quad + c \inf_{\boldsymbol{v}^h \in L^2(I;V_1^h)} \left\{\|\dot{\boldsymbol{u}} - \boldsymbol{v}^h\|_{L^2(I;V)} + \|R(\cdot; \dot{\boldsymbol{u}}, \boldsymbol{v}^h)\|_{L^1(I)}^{1/2}\right\} \\ &\quad + c\,\|(I_Q - \mathcal{P}_{Q^h})(\boldsymbol{\sigma} - \boldsymbol{\sigma}_0)\|_{L^\infty(I;Q)},\end{aligned} \tag{17.92}$$

where $R(\cdot;\cdot,\cdot)$ *is the residual term defined in* (17.90).

The inequality (17.92) is the basis of convergence and error analysis for the semi-discrete solutions. For convergence analysis, we use the following bound on the residual term:

$$|R(t; \dot{\boldsymbol{u}}(t), \boldsymbol{v}^h(t))| \le c\,\|\dot{\boldsymbol{u}}(t) - \boldsymbol{v}^h(t)\|_V,$$

where the constant c depends on $\|\boldsymbol{\sigma}\|_{L^\infty(I;Q)}$, $\|g\|_{L^\infty(\Gamma_3)}$, $\|\boldsymbol{f}_0\|_{L^\infty(I;L^2(\Omega)^d)}$, and $\|\boldsymbol{f}_2\|_{L^\infty(I;L^2(\Gamma_2)^d)}$. Thus

$$\|R(\cdot; \dot{\boldsymbol{u}}, \boldsymbol{v}^h)\|_{L^1(\Omega)} \le c\,\|\dot{\boldsymbol{u}} - \boldsymbol{v}^h\|_{L^2(I;V)}.$$

Hence, (17.92) implies

$$\begin{aligned}\|\boldsymbol{u} - \boldsymbol{u}^h\|_{L^\infty(I;V)} &+ \|\boldsymbol{\sigma} - \boldsymbol{\sigma}^h\|_{L^\infty(I;Q)} \\ &\le c\,(\|\boldsymbol{u}_0 - \boldsymbol{u}_0^h\|_V + \|\boldsymbol{\sigma}_0 - \boldsymbol{\sigma}_0^h\|_Q) \\ &\quad + c \inf_{\boldsymbol{v}^h \in L^2(I;V_1^h)} \left\{\|\dot{\boldsymbol{u}} - \boldsymbol{v}^h\|_{L^2(I;V)} + \|\dot{\boldsymbol{u}} - \boldsymbol{v}^h\|_{L^2(I;V)}^{1/2}\right\} \\ &\quad + c\,\|(I_Q - \mathcal{P}_{Q^h})(\boldsymbol{\sigma} - \boldsymbol{\sigma}_0)\|_{L^\infty(I;Q)}.\end{aligned}$$

We can then apply Theorem 7.21 to conclude the convergence that if $\boldsymbol{u}_0^h$ and $\boldsymbol{\sigma}_0^h$ are chosen such that

$$\|\boldsymbol{u}_0 - \boldsymbol{u}_0^h\|_V + \|\boldsymbol{\sigma}_0 - \boldsymbol{\sigma}_0^h\|_Q \to 0 \quad \text{as } h \to 0,$$

then we have the convergence

$$\|\boldsymbol{u} - \boldsymbol{u}^h\|_{L^\infty(I;V)} + \|\boldsymbol{\sigma} - \boldsymbol{\sigma}^h\|_{L^\infty(I;Q)} \to 0 \quad \text{as } h \to 0.$$

For error estimation, we need a sharper bound on the residual term. Assume the additional solution regularity (17.39). Then using the pointwise relations (17.38) and (17.40), we have

$$\begin{aligned} R(t; \dot{\boldsymbol{u}}(t), \boldsymbol{v}^h(t)) = &\int_{\Gamma_3} g\,(\|\boldsymbol{v}^h_\tau(t)\| - \|\dot{\boldsymbol{u}}_\tau(t)\|)\, da \\ &+ \int_{\Gamma_3} \boldsymbol{\sigma}(t)\boldsymbol{\nu} \cdot (\boldsymbol{v}^h(t) - \dot{\boldsymbol{u}}(t))\, da. \end{aligned}$$

So

$$|R(t; \dot{\boldsymbol{u}}(t), \boldsymbol{v}^h(t))| \le c\, \|\dot{\boldsymbol{u}}(t) - \boldsymbol{v}^h(t)\|_{L^2(\Gamma_3)}$$

a.e. $t \in I$, and the inequality (17.92) can be reduced to

$$\begin{aligned} &\|\boldsymbol{u} - \boldsymbol{u}^h\|_{L^\infty(I;V)} + \|\boldsymbol{\sigma} - \boldsymbol{\sigma}^h\|_{L^\infty(I;Q)} \\ &\quad \le c\,(\|\boldsymbol{u}_0 - \boldsymbol{u}_0^h\|_V + \|\boldsymbol{\sigma}_0 - \boldsymbol{\sigma}_0^h\|_Q) \\ &\quad\quad + c \inf_{\boldsymbol{v}^h \in L^2(I;V_1^h)} \left\{ \|\dot{\boldsymbol{u}} - \boldsymbol{v}^h\|_{L^2(I;V)} + \|\dot{\boldsymbol{u}} - \boldsymbol{v}^h\|^{1/2}_{L^2(I;L^2(\Gamma_3)^d)} \right\} \\ &\quad\quad + c\, \|(I_Q - \mathcal{P}_{Q^h})(\boldsymbol{\sigma} - \boldsymbol{\sigma}_0)\|_{L^\infty(I;Q)}. \end{aligned}$$

Assume

$$\begin{aligned} &\boldsymbol{u} \in H^1(I; H^2(\Omega)^d), \quad \dot{\boldsymbol{u}}|_{\Gamma_{3,i}} \in L^1(I; H^2(\Gamma_{3,i})^d),\ 1 \le i \le i_0, \\ &\boldsymbol{\sigma} \in L^\infty(I; H^1(\Omega)^{d\times d}). \end{aligned}$$

Then if we choose $\boldsymbol{u}_0^h$ and $\boldsymbol{\sigma}_0^h$ such that

$$\|\boldsymbol{u}_0 - \boldsymbol{u}_0^h\|_V + \|\boldsymbol{\sigma}_0 - \boldsymbol{\sigma}_0^h\|_Q \le c\, h,$$

then we have the optimal order error estimate

$$\|\boldsymbol{u} - \boldsymbol{u}^h\|_{L^\infty(I;V)} + \|\boldsymbol{\sigma} - \boldsymbol{\sigma}^h\|_{L^\infty(I;Q)} = O(h).$$

17.5 Fully Discrete Approximation

In addition to the finite element spaces V_1^h and Q^h, we need the time interval partition described in Section 7.2. Then a fully discrete approximation to be analyzed is the following.

Problem 17.20 *Find* $\boldsymbol{u}^{hk} = \{\boldsymbol{u}_n^{hk}\}_{n=0}^N \subset V_1^h$ *and* $\boldsymbol{\sigma}^{hk} = \{\boldsymbol{\sigma}_n^{hk}\}_{n=0}^N \subset Q^h$ *such that*

$$\boldsymbol{u}_0^{hk} = \boldsymbol{u}_0^h, \quad \boldsymbol{\sigma}_0^{hk} = \boldsymbol{\sigma}_0^h, \tag{17.93}$$

and for $n = 1, \ldots, N$,

$$\delta_n \boldsymbol{\sigma}_n^{hk} = \mathcal{P}_{Q^h} \mathcal{E} \boldsymbol{\varepsilon}(\delta_n \boldsymbol{u}_n^{hk}) + \mathcal{P}_{Q^h} \mathcal{G}(\boldsymbol{\sigma}_{n-1}^{hk}, \boldsymbol{\varepsilon}(\boldsymbol{u}_{n-1}^{hk})), \tag{17.94}$$

$$(\boldsymbol{\sigma}_n^{hk}, \boldsymbol{\varepsilon}(\boldsymbol{v}^h) - \boldsymbol{\varepsilon}(\delta_n \boldsymbol{u}_n^{hk}))_Q + j(\boldsymbol{v}^h) - j(\delta_n \boldsymbol{u}_n^{hk}) \\ \geq (\boldsymbol{f}_n, \boldsymbol{v}^h - \delta_n \boldsymbol{u}_n^{hk})_V \quad \forall\, \boldsymbol{v}^h \in V_1^h. \tag{17.95}$$

Here $\boldsymbol{u}_0^h = \mathcal{P}_{V^h} \boldsymbol{u}_0 \in V_1^h$, $\boldsymbol{\sigma}_0^h = \mathcal{P}_{Q^h} \boldsymbol{\sigma}_0 \in Q^h$ *are orthogonal projections of* $\boldsymbol{u}_0$ *and* $\boldsymbol{\sigma}_0$ *to* V^h *and* Q^h, *respectively.*

Let us first inductively show the unique solvability of Problem 17.20. Let $\boldsymbol{u}_{n-1}^{hk} \in V_1^h$ and $\boldsymbol{\sigma}_{n-1}^{hk} \in Q^h$ be given. We rewrite (17.94) as

$$\boldsymbol{\sigma}_n^{hk} = k_n \mathcal{P}_{Q^h} \mathcal{E} \boldsymbol{\varepsilon}(\delta_n \boldsymbol{u}_n^{hk}) + k_n \mathcal{P}_{Q^h} \mathcal{G}(\boldsymbol{\sigma}_{n-1}^{hk}, \boldsymbol{\varepsilon}(\boldsymbol{u}_{n-1}^{hk})) + \boldsymbol{\sigma}_{n-1}^{hk} \tag{17.96}$$

and use it in (17.95) to yield

$$k_n (\mathcal{E} \boldsymbol{\varepsilon}(\delta_n \boldsymbol{u}_n^{hk}), \boldsymbol{\varepsilon}(\boldsymbol{v}^h - \delta_n \boldsymbol{u}_n^{hk}))_Q + j(\boldsymbol{v}^h) - j(\delta_n \boldsymbol{u}_n^{hk}) \\ \geq (\boldsymbol{f}_n - k_n \mathcal{G}(\boldsymbol{\sigma}_{n-1}^{hk}, \boldsymbol{\varepsilon}(\boldsymbol{u}_{n-1}^{hk})) - \boldsymbol{\sigma}_{n-1}^{hk}, \boldsymbol{v}^h - \delta_n \boldsymbol{u}_n^{hk})_Q$$

for any $\boldsymbol{v}^h \in V_1^h$. This is a discrete elliptic variational inequality of the second kind with a unique solution $\delta_n \boldsymbol{u}_n^{hk} \in V_1^h$, from which we can determine $\boldsymbol{u}_n^{hk}$. Then $\boldsymbol{\sigma}_n^{hk}$ can be obtained from (17.96). Therefore Problem 17.20 has a unique solution.

In the next section, we demonstrate the convergence of the fully discrete solution under the proved solution regularity $\boldsymbol{u} \in W^{1,\infty}(I; V_1)$ and $\boldsymbol{\sigma} \in W^{1,\infty}(I; Q_1)$. Here, in the rest of the section, we derive estimates for the errors $\{\boldsymbol{u}_n - \boldsymbol{u}_n^{hk}\}_{n=1}^N$ and $\{\boldsymbol{\sigma}_n - \boldsymbol{\sigma}_n^{hk}\}_{n=1}^N$. For this purpose, we assume additionally $\boldsymbol{u} \in C^1(\overline{I}; V)$ so that (17.19) and (17.20) hold for all $t \in \overline{I}$. Similar to (17.82) and (17.83), we use the notation

$$\boldsymbol{s}_{\mathcal{G},n}^{hk} = \sum_{j=1}^{n} k_j \mathcal{G}(\boldsymbol{\sigma}_{j-1}^{hk}, \boldsymbol{\varepsilon}(\boldsymbol{u}_{j-1}^{hk})). \tag{17.97}$$

We apply (17.96) recursively to get the expression

$$\boldsymbol{\sigma}_n^{hk} = \mathcal{P}_{Q^h} \mathcal{E} \boldsymbol{\varepsilon}(\boldsymbol{u}_n^{hk}) + \mathcal{P}_{Q^h} \boldsymbol{s}_{\mathcal{G},n}^{hk} + \boldsymbol{\sigma}_0^{hk} - \mathcal{P}_{Q^h} \mathcal{E} \boldsymbol{\varepsilon}(\boldsymbol{u}_0^{hk}). \tag{17.98}$$

Then subtracting (17.98) from (17.84) at $t = t_n$, we obtain an error relation

$$\boldsymbol{\sigma}_n - \boldsymbol{\sigma}_n^{hk} = (I_Q - \mathcal{P}_{Q^h})(\boldsymbol{\sigma}_n - \boldsymbol{\sigma}_0) + \mathcal{P}_{Q^h} \mathcal{E} \boldsymbol{\varepsilon}(\boldsymbol{u}_n - \boldsymbol{u}_n^{hk}) \\ + \mathcal{P}_{Q^h} \boldsymbol{\theta}_n^{hk}(\mathcal{G}) + \boldsymbol{\sigma}_0 - \boldsymbol{\sigma}_0^{hk} - \mathcal{P}_{Q^h} \mathcal{E} \boldsymbol{\varepsilon}(\boldsymbol{u}_0 - \boldsymbol{u}_0^{hk}), \tag{17.99}$$

where

$$\boldsymbol{\theta}_n^{hk}(\mathcal{G}) = \boldsymbol{s}_{\mathcal{G}}(t_n) - \boldsymbol{s}_{\mathcal{G},n}^{hk}.$$

We denote

$$e_n = \|\boldsymbol{u}_n - \boldsymbol{u}_n^{hk}\|_V + \|\boldsymbol{\sigma}_n - \boldsymbol{\sigma}_n^{hk}\|_Q, \quad n = 0, \ldots, N$$

for the numerical solution errors.

We first present a preparatory result. Define

$$\phi_n^{hk}(\mathcal{G}) = \int_{t_n}^{t_{n+1}} \mathcal{G}(\boldsymbol{\sigma}(s), \boldsymbol{\varepsilon}(\dot{\boldsymbol{u}}(s)))\, ds - k_{n+1}\mathcal{G}(\boldsymbol{\sigma}_n^{hk}, \boldsymbol{\varepsilon}(\boldsymbol{u}_n^{hk})).$$

Then

$$\boldsymbol{\theta}_n^{hk}(\mathcal{G}) = \sum_{j=0}^{n-1} \boldsymbol{\phi}_j^{hk}(\mathcal{G}).$$

We have the following bounds on $\boldsymbol{\phi}_n^{hk}(\mathcal{G})$ and $\boldsymbol{\theta}_n^{hk}(\mathcal{G})$.

Lemma 17.21 *There exists a constant $c > 0$ such that for $n = 0, \ldots, N-1$,*

$$\|\boldsymbol{\phi}_n^{hk}(\mathcal{G})\|_Q \le c\, k_{n+1} e_n + c\, k_{n+1}^2 \left(\|\dot{\boldsymbol{u}}\|_{L^\infty(I;V)} + \|\dot{\boldsymbol{\sigma}}\|_{L^\infty(I;Q)}\right), \tag{17.100}$$

and for $n = 1, \ldots, N$,

$$\|\boldsymbol{\theta}_n^{hk}(\mathcal{G})\|_Q \le c \sum_{j=1}^{n} k_j e_{j-1} + c\, k \left(\|\dot{\boldsymbol{u}}\|_{L^\infty(I;V)} + \|\dot{\boldsymbol{\sigma}}\|_{L^\infty(I;Q)}\right). \tag{17.101}$$

Proof. We write

$$\begin{aligned}\boldsymbol{\phi}_n^{hk}(\mathcal{G}) = &\int_{t_n}^{t_{n+1}} \left[\mathcal{G}(\boldsymbol{\sigma}(s), \boldsymbol{\varepsilon}(\dot{\boldsymbol{u}}(s))) - \mathcal{G}(\boldsymbol{u}_n, \boldsymbol{\sigma}_n)\right] ds \\ &+ k_{n+1}\left[\mathcal{G}(\boldsymbol{u}_n, \boldsymbol{\sigma}_n) - \mathcal{G}(\boldsymbol{\sigma}_n^{hk}, \boldsymbol{\varepsilon}(\boldsymbol{u}_n^{hk}))\right].\end{aligned}$$

By the assumptions (6.45) on $\mathcal{G}$, it is easy to see that (17.100) holds. The inequality (17.101) follows from (17.100). ■

We obtain the following inequality from the error relation (17.99) by the use of Lemma 17.21, the assumptions (6.44) and the non-expansiveness property of the projection operator $\mathcal{P}_{Q^h}$:

$$\begin{aligned}\|\boldsymbol{\sigma}_n - \boldsymbol{\sigma}_n^{hk}\|_Q \le &\|(I_Q - \mathcal{P}_{Q^h})(\boldsymbol{\sigma}_n - \boldsymbol{\sigma}_0)\|_Q + c\,\|\boldsymbol{u}_n - \boldsymbol{u}_n^{hk}\|_V + c \sum_{j=1}^{n} k_j e_{j-1} \\ &+ c\, e_0 + c\, k\,(\|\dot{\boldsymbol{u}}\|_{L^\infty(I;V)} + \|\dot{\boldsymbol{\sigma}}\|_{L^\infty(I;Q)}).\end{aligned} \tag{17.102}$$

Now let us bound $\boldsymbol{r}_n \equiv \boldsymbol{u}_n - \boldsymbol{u}_n^{hk}$, $n = 1, \ldots, N$. For this, we plug (17.84) into (17.20) at $t = t_n$ with $\boldsymbol{w} = \delta_n \boldsymbol{u}_n^{hk}$ and get

$$\begin{aligned}(\mathcal{E}\boldsymbol{\varepsilon}(\boldsymbol{u}_n), \boldsymbol{\varepsilon}(\dot{\boldsymbol{u}}_n - \delta_n \boldsymbol{u}_n^{hk}))_Q \le\, &(\boldsymbol{f}_n, \dot{\boldsymbol{u}}_n - \delta_n \boldsymbol{u}_n^{hk})_V + j(\delta_n \boldsymbol{u}_n^{hk}) - j(\dot{\boldsymbol{u}}_n) \\ &+ \Big(\boldsymbol{s}_{\mathcal{G}}(t_n), \boldsymbol{\varepsilon}(\delta_n \boldsymbol{u}_n^{hk} - \dot{\boldsymbol{u}}_n)\Big)_Q \\ &+ (\boldsymbol{\sigma}_0 - \mathcal{E}\boldsymbol{\varepsilon}(\boldsymbol{u}_0), \boldsymbol{\varepsilon}(\delta_n \boldsymbol{u}_n^{hk} - \dot{\boldsymbol{u}}_n))_Q.\end{aligned} \tag{17.103}$$

Similarly, plugging (17.98) into (17.95) with an arbitrary $\boldsymbol{v}^h = \boldsymbol{v}_n^h \in V^h$ yields

$$-(\mathcal{E}\boldsymbol{u}_n^{hk}, \boldsymbol{v}_n^h - \delta_n \boldsymbol{u}_n^{hk})_Q \le (\boldsymbol{f}_n, \delta_n \boldsymbol{u}_n^{hk} - \boldsymbol{v}_n^h)_V + j(\boldsymbol{v}_n^h) - j(\delta_n \boldsymbol{u}_n^{hk}) + \left(\boldsymbol{s}_{\mathcal{G},n}^{hk}, \boldsymbol{v}_n^h - \delta_n \boldsymbol{u}_n^{hk}\right)_Q + (\boldsymbol{\sigma}_0^{hk} - \mathcal{E}\boldsymbol{u}_0^{hk}, \boldsymbol{v}_n^h - \delta_n \boldsymbol{u}_n^{hk})_Q. \tag{17.104}$$

In the following, it will be convenient to use the equivalent norm $\|\boldsymbol{v}\|_{\mathcal{E}}$ defined in (17.81) for the displacement variable.

Let us consider the quantity

$$A_n \equiv (\mathcal{E}\boldsymbol{\varepsilon}(\boldsymbol{r}_n), \boldsymbol{\varepsilon}(\delta_n \boldsymbol{r}_n))_Q = (\mathcal{E}\boldsymbol{\varepsilon}(\boldsymbol{u}_n - \boldsymbol{u}_n^{hk}), \boldsymbol{\varepsilon}(\delta_n \boldsymbol{u}_n - \delta_n \boldsymbol{u}_n^{hk}))_Q.$$

Immediately, we get the following lower bound:

$$A_n \ge \frac{1}{2k_n}(\|\boldsymbol{r}_n\|_{\mathcal{E}}^2 - \|\boldsymbol{r}_{n-1}\|_{\mathcal{E}}^2). \tag{17.105}$$

For any $\boldsymbol{v}_n^h \in V^h$, we write

$$\begin{aligned} A_n = {} & (\mathcal{E}\boldsymbol{\varepsilon}(\boldsymbol{u}_n), \boldsymbol{\varepsilon}(\delta_n \boldsymbol{u}_n - \dot{\boldsymbol{u}}_n))_Q + (\mathcal{E}\boldsymbol{\varepsilon}(\boldsymbol{u}_n), \boldsymbol{\varepsilon}(\dot{\boldsymbol{u}}_n - \delta_n \boldsymbol{u}_n^{hk}))_Q \\ & - (\mathcal{E}\boldsymbol{\varepsilon}(\boldsymbol{u}_n^{hk}), \boldsymbol{\varepsilon}(\delta_n \boldsymbol{u}_n - \boldsymbol{v}_n^h))_Q - (\mathcal{E}\boldsymbol{\varepsilon}(\boldsymbol{u}_n^{hk}), \boldsymbol{\varepsilon}(\boldsymbol{v}_n^h - \delta_n \boldsymbol{u}_n^{hk}))_Q. \end{aligned}$$

We then use (17.103) and (17.104) to bound the second and fourth terms,

$$\begin{aligned} A_n & \le (\mathcal{E}\boldsymbol{\varepsilon}(\boldsymbol{r}_n), \boldsymbol{\varepsilon}(\delta_n \boldsymbol{u}_n - \boldsymbol{v}_n^h))_Q + R_n(\dot{\boldsymbol{u}}_n, \boldsymbol{v}_n^h) + I_1(n) + I_2(n) + I_3(n) \\ & \le \frac{1}{2}\|\boldsymbol{r}_n\|_{\mathcal{E}}^2 + c\,\|\delta_n \boldsymbol{u}_n - \boldsymbol{v}_n^h\|_V^2 + R_n(\dot{\boldsymbol{u}}_n, \boldsymbol{v}_n^h) + I_1(n) + I_2(n) + I_3(n), \end{aligned}$$

where

$$R_n(\dot{\boldsymbol{u}}_n, \boldsymbol{v}_n^h) = (\boldsymbol{\sigma}_n, \boldsymbol{\varepsilon}(\boldsymbol{v}_n^h - \dot{\boldsymbol{u}}_n))_Q + j(\boldsymbol{v}_n^h) - j(\dot{\boldsymbol{u}}_n) - (\boldsymbol{f}_n, \boldsymbol{v}_n^h - \dot{\boldsymbol{u}}_n)_V, \tag{17.106}$$

and

$$\begin{aligned} I_1(n) & = \left(\boldsymbol{\theta}_n^{hk}(\mathcal{G}), \boldsymbol{\varepsilon}(\delta_n \boldsymbol{u}_n^{hk} - \delta_n \boldsymbol{u}_n)\right)_Q, \\ I_2(n) & = (\boldsymbol{\sigma}_0 - \boldsymbol{\sigma}_0^{hk} - \mathcal{E}\boldsymbol{\varepsilon}(\boldsymbol{u}_0 - \boldsymbol{u}_0^{hk}), \boldsymbol{\varepsilon}(\delta_n \boldsymbol{u}_n^{hk} - \delta_n \boldsymbol{u}_n))_Q, \\ I_3(n) & = \left(\boldsymbol{\theta}_n^{hk}(\mathcal{G}) + \boldsymbol{\sigma}_0 - \boldsymbol{\sigma}_0^{hk} - \mathcal{E}\boldsymbol{\varepsilon}(\boldsymbol{u}_0 - \boldsymbol{u}_0^{hk}), \boldsymbol{\varepsilon}(\delta_n \boldsymbol{u}_n - \boldsymbol{v}_n^h)\right)_Q. \end{aligned}$$

Combining the lower and upper bounds for A_n, we have

$$\|\boldsymbol{r}_n\|_{\mathcal{E}}^2 - \|\boldsymbol{r}_{n-1}\|_{\mathcal{E}}^2 \le k_n\|\boldsymbol{r}_n\|_{\mathcal{E}}^2 + c\,k_n\|\delta_n \boldsymbol{u}_n - \boldsymbol{v}_n^h\|_V^2 + 2\,k_n R_n(\dot{\boldsymbol{u}}_n, \boldsymbol{v}_n^h) + 2\,k_n(I_1(n) + I_2(n) + I_3(n)). \tag{17.107}$$

In the inequality (17.107), we change the index n to j, and sum over j from 1 to n to find

$$\|\boldsymbol{r}_n\|_{\mathcal{E}}^2 \le \|\boldsymbol{r}_0\|_{\mathcal{E}}^2 + \sum_{j=1}^n k_j \|\boldsymbol{r}_j\|_{\mathcal{E}}^2 + c \sum_{j=1}^n k_j \|\delta_j \boldsymbol{u}_j - \boldsymbol{v}_j^h\|_V^2$$
$$+ 2 \sum_{j=1}^n k_j \big[R_j(\dot{\boldsymbol{u}}_j, \boldsymbol{v}_j^h) + I_1(j) + I_2(j) + I_3(j) \big]. \tag{17.108}$$

We now write

$$\sum_{j=1}^n k_j I_1(j) = \sum_{j=1}^n \big(\boldsymbol{\theta}_j^{hk}(\mathcal{G}), \boldsymbol{\varepsilon}(\boldsymbol{r}_{j-1} - \boldsymbol{r}_j)\big)_Q$$
$$= \big(\boldsymbol{\theta}_1^{hk}(\mathcal{G}), \boldsymbol{\varepsilon}(\boldsymbol{r}_0)\big)_Q - \big(\boldsymbol{\theta}_n^{hk}(\mathcal{G}), \boldsymbol{\varepsilon}(\boldsymbol{r}_n)\big)_Q + \sum_{j=1}^{n-1} (\boldsymbol{\phi}_j^{hk}(\mathcal{G}), \boldsymbol{\varepsilon}(\boldsymbol{r}_j))_Q.$$

Using the assumptions (6.44) on $\mathcal{E}$, the estimates (17.100) and (17.101), we obtain

$$\sum_{j=1}^n k_j I_1(j) \le c\, e_0^2 + c \sum_{j=1}^{n-1} k_j (\|\boldsymbol{r}_j\|_{\mathcal{E}}^2 + \|\boldsymbol{\sigma}_j - \boldsymbol{\sigma}_j^{hk}\|_Q^2) + \frac{1}{16} \|\boldsymbol{r}_n\|_{\mathcal{E}}^2$$
$$+ c\, k^2 (\|\dot{\boldsymbol{u}}\|_{L^\infty(I;V)} + \|\dot{\boldsymbol{\sigma}}\|_{L^\infty(I;Q)})^2. \tag{17.109}$$

Similarly, we write

$$\sum_{j=1}^n k_j I_2(j) = \Big(\boldsymbol{\sigma}_0 - \boldsymbol{\sigma}_0^{hk} - \mathcal{E}\boldsymbol{\varepsilon}(\boldsymbol{u}_0 - \boldsymbol{u}_0^{hk}), \sum_{j=1}^n \boldsymbol{\varepsilon}(\boldsymbol{r}_{j-1} - \boldsymbol{r}_j)\Big)_Q$$
$$= (\boldsymbol{\sigma}_0 - \boldsymbol{\sigma}_0^{hk} - \mathcal{E}\boldsymbol{\varepsilon}(\boldsymbol{u}_0 - \boldsymbol{u}_0^{hk}), \boldsymbol{\varepsilon}(\boldsymbol{r}_0 - \boldsymbol{r}_n))_Q$$

and obtain

$$\sum_{j=1}^n k_j I_2(j) \le c\, e_0^2 + \frac{1}{16} \|\boldsymbol{r}_n\|_{\mathcal{E}}^2. \tag{17.110}$$

Finally we use the assumptions (6.44) on $\mathcal{E}$ and the estimate (17.101) to find

$$\sum_{j=1}^n k_j I_3(j) \le c\, e_0^2 + c \sum_{j=1}^{n-1} k_j (\|\boldsymbol{r}_j\|_{\mathcal{E}}^2 + \|\boldsymbol{\sigma}_j - \boldsymbol{\sigma}_j^{hk}\|_Q^2)$$
$$+ c\, k^2 (\|\dot{\boldsymbol{u}}\|_{L^\infty(I;V)} + \|\dot{\boldsymbol{\sigma}}\|_{L^\infty(I;Q)})^2$$
$$+ c \sum_{j=1}^n k_j \|\delta_j \boldsymbol{u}_j - \boldsymbol{v}_j^h\|_V^2. \tag{17.111}$$

We now combine the estimates (17.102), (17.108)–(17.111) to obtain

$$\|\boldsymbol{u}_n - \boldsymbol{u}_n^{hk}\|_{\mathcal{E}}^2 + \|\boldsymbol{\sigma}_n - \boldsymbol{\sigma}_n^{hk}\|_Q^2 \\ \le c\, J_n + c \sum_{j=1}^{n-1} k_j (\|\boldsymbol{u}_j - \boldsymbol{u}_j^{hk}\|_{\mathcal{E}}^2 + \|\boldsymbol{\sigma}_j - \boldsymbol{\sigma}_j^{hk}\|_Q^2), \qquad (17.112)$$

where

$$J_n = e_0^2 + k^2 (\|\dot{\boldsymbol{u}}\|_{L^\infty(I;V)} + \|\dot{\boldsymbol{\sigma}}\|_{L^\infty(I;Q)})^2 + \|(I_Q - \mathcal{P}_{Q^h})(\boldsymbol{\sigma}_n - \boldsymbol{\sigma}_0)\|_Q^2 \\ + \sum_{j=1}^{n} k_j \|\delta_j \boldsymbol{u}_j - \boldsymbol{v}_j^h\|_V^2 + \sum_{j=1}^{n} k_j |R_j(\dot{\boldsymbol{u}}_j, \boldsymbol{v}_j^h)|.$$

We then apply Lemma 7.25 (page 163) on (17.112) and replace $\|\cdot\|_{\mathcal{E}}$ by the equivalent norm $\|\cdot\|_V$ to obtain the following result.

Theorem 17.22 *Assume $\boldsymbol{u} \in C^1(\overline{I}; V)$. Then for the fully discrete solution of Problem 17.20, we have the error estimate*

$$\max_{1\le n\le N} \big(\|\boldsymbol{u}_n - \boldsymbol{u}_n^{hk}\|_V + \|\boldsymbol{\sigma}_n - \boldsymbol{\sigma}_n^{hk}\|_Q\big) \\ \le c\,(\|\boldsymbol{u}_0 - \boldsymbol{u}_0^{hk}\|_V + \|\boldsymbol{\sigma}_0 - \boldsymbol{\sigma}_0^{hk}\|_Q) \\ + c \max_{1\le n\le N} \|(I_Q - \mathcal{P}_{Q^h})(\boldsymbol{\sigma}_n - \boldsymbol{\sigma}_0)\|_Q \\ + c \left[\sum_{j=1}^{N} k_j \inf_{\boldsymbol{v}_j^h \in V_1^h} \big(\|\delta_j \boldsymbol{u}_j - \boldsymbol{v}_j^h\|_V^2 + |R_j(\dot{\boldsymbol{u}}_j, \boldsymbol{v}_j^h)|\big) \right]^{1/2} \\ + c\, k\, (\|\dot{\boldsymbol{u}}\|_{L^\infty(I;V)} + \|\dot{\boldsymbol{\sigma}}\|_{L^\infty(I;Q)}). \qquad (17.113)$$

The estimate (17.113) can be used for error analysis of the fully discrete solutions provided the exact solution possesses certain regularity. Suppose (17.39) is valid and

$$\boldsymbol{u} \in C^1(\overline{I}; H^2(\Omega)^d), \quad \dot{\boldsymbol{u}}|_{\Gamma_{3,i}} \in C(\overline{I}; H^2(\Gamma_{3,i})^d), \ 1 \le i \le i_0, \\ \boldsymbol{\sigma} \in C(\overline{I}; H^1(\Omega)^{d\times d}).$$

If the initial values $\boldsymbol{u}_0^h$ and $\boldsymbol{\sigma}_0^h$ are chosen such that

$$\|\boldsymbol{u}_0 - \boldsymbol{u}_0^h\|_V + \|\boldsymbol{\sigma}_0 - \boldsymbol{\sigma}_0^h\|_Q \le c\, h,$$

then it can be shown from (17.113) that

$$\max_{1\le n\le N} \big(\|\boldsymbol{u}_n - \boldsymbol{u}_n^{hk}\|_V + \|\boldsymbol{\sigma}_n - \boldsymbol{\sigma}_n^{hk}\|_Q\big) = O(h + k).$$

17.6 Convergence of the Fully Discrete Scheme

In this section, we prove the convergence of the fully discrete solution of Problem 17.2 under the basic solution regularity $(\boldsymbol{u}, \boldsymbol{\sigma}) \in W^{1,\infty}(I; V_1 \times Q)$, available from Theorem 17.4.

Notice that we can not use the estimate (17.113), because under the basic regularity condition $(\boldsymbol{u}, \boldsymbol{\sigma}) \in W^{1,\infty}(I; V_1 \times Q)$, the pointwise values $\dot{\boldsymbol{u}}_j$ and $\dot{\boldsymbol{\sigma}}_j$ that appear in (17.113) are not well-defined. Here we follow the approach developed in [96] for a convergence analysis. For this purpose we will use Lemma 7.12.

Let us again use the equivalent norm $\|\cdot\|_{\mathcal{E}}$ defined in (17.81). For the quantity $A_n = (\mathcal{E}\boldsymbol{\varepsilon}(\boldsymbol{r}_n), \boldsymbol{\varepsilon}(\delta_n \boldsymbol{r}_n))_Q$, we have the lower bound (17.105). To obtain an upper bound we begin with

$$\begin{aligned} A_n &= (\mathcal{E}\boldsymbol{\varepsilon}(\boldsymbol{u}_n), \boldsymbol{\varepsilon}(\delta_n \boldsymbol{u}_n - \delta_n \boldsymbol{u}_n^{hk}))_Q - (\mathcal{E}\boldsymbol{\varepsilon}(\boldsymbol{u}_n^{hk}), \boldsymbol{\varepsilon}(\delta_n \boldsymbol{u}_n - \boldsymbol{v}_n^h))_Q \\ &\quad - (\mathcal{E}\boldsymbol{\varepsilon}(\boldsymbol{u}_n^{hk}), \boldsymbol{\varepsilon}(\boldsymbol{v}_n^h - \delta_n \boldsymbol{u}_n^{hk}))_Q, \end{aligned}$$

where $\boldsymbol{v}_n^h \in V^h$ is arbitrary. Using (17.104) to bound the last term, we obtain

$$\begin{aligned} A_n &\le (\mathcal{E}\boldsymbol{\varepsilon}(\boldsymbol{u}_n), \boldsymbol{\varepsilon}(\delta_n \boldsymbol{u}_n - \delta_n \boldsymbol{u}_n^{hk}))_Q - (\mathcal{E}\boldsymbol{\varepsilon}(\boldsymbol{u}_n^{hk}), \boldsymbol{\varepsilon}(\delta_n \boldsymbol{u}_n - \boldsymbol{v}_n^h))_Q \\ &\quad + (\boldsymbol{f}_n, \delta_n \boldsymbol{u}_n^{hk} - \boldsymbol{v}_n^h)_V + j(\boldsymbol{v}_n^h) - j(\delta_n \boldsymbol{u}_n^{hk}) \\ &\quad + \left(\boldsymbol{s}_{\mathcal{G},n}^{hk}, \boldsymbol{\varepsilon}(\boldsymbol{v}_n^h - \delta_n \boldsymbol{u}_n^{hk})\right)_Q \\ &\quad + (\boldsymbol{\sigma}_0^{hk} - \mathcal{E}\boldsymbol{\varepsilon}(\boldsymbol{u}_0^{hk}), \boldsymbol{\varepsilon}(\boldsymbol{v}_n^h - \delta_n \boldsymbol{u}_n^{hk}))_Q. \end{aligned} \tag{17.114}$$

Now integrate (17.20) with $\boldsymbol{w} = \delta_n \boldsymbol{u}_n^{hk}$ from $t = t_{n-1}$ to t_n and use the formula (17.84) for $\boldsymbol{\sigma}(t)$ to find

$$\begin{aligned} 0 &\le \frac{1}{k_n} \int_{t_{n-1}}^{t_n} (\mathcal{E}\boldsymbol{\varepsilon}(\boldsymbol{u}(t)), \boldsymbol{\varepsilon}(\delta_n \boldsymbol{u}_n^{hk} - \dot{\boldsymbol{u}}(t)))_Q dt + j(\delta_n \boldsymbol{u}_n^{hk}) \\ &\quad - \frac{1}{k_n} \int_{t_{n-1}}^{t_n} j(\dot{\boldsymbol{u}}(t))\, dt + \frac{1}{k_n} \int_{t_{n-1}}^{t_n} \left(\boldsymbol{s}_{\mathcal{G}}(t), \boldsymbol{\varepsilon}(\delta_n \boldsymbol{u}_n^{hk} - \dot{\boldsymbol{u}}(t))\right)_Q dt \\ &\quad + (\boldsymbol{\sigma}_0 - \mathcal{E}\boldsymbol{\varepsilon}(\boldsymbol{u}_0), \boldsymbol{\varepsilon}(\delta_n \boldsymbol{u}_n^{hk} - \delta_n \boldsymbol{u}_n))_Q \\ &\quad + \frac{1}{k_n} \int_{t_{n-1}}^{t_n} (\boldsymbol{f}(t), \dot{\boldsymbol{u}}(t) - \delta_n \boldsymbol{u}_n^{hk})_V\, dt. \end{aligned}$$

We add this inequality to (17.114) and obtain

$$A_n \le R_{n,1} + R_{n,2} + R_{n,3} + R_{n,4} + R_{n,5}, \tag{17.115}$$

where

$$\begin{aligned}
R_{n,1} &= (\mathcal{E}\boldsymbol{\varepsilon}(\boldsymbol{u}_n), \boldsymbol{\varepsilon}(\delta_n\boldsymbol{u}_n - \delta_n\boldsymbol{u}_n^{hk}))_Q - (\mathcal{E}\boldsymbol{\varepsilon}(\boldsymbol{u}_n^{hk}), \boldsymbol{\varepsilon}(\delta_n\boldsymbol{u}_n - \boldsymbol{v}_n^h))_Q \\
&\quad + \frac{1}{k_n}\int_{t_{n-1}}^{t_n} (\mathcal{E}\boldsymbol{\varepsilon}(\boldsymbol{u}(t)), \boldsymbol{\varepsilon}(\delta_n\boldsymbol{u}_n^{hk} - \dot{\boldsymbol{u}}(t)))_Q dt, \\
R_{n,2} &= j(\boldsymbol{v}_n^h) - \frac{1}{k_n}\int_{t_{n-1}}^{t_n} j(\dot{\boldsymbol{u}}(t))\,dt, \\
R_{n,3} &= (\boldsymbol{f}_n, \delta_n\boldsymbol{u}_n^{hk} - \boldsymbol{v}_n^h)_V + \frac{1}{k_n}\int_{t_{n-1}}^{t_n} (\boldsymbol{f}(t), \dot{\boldsymbol{u}}(t) - \delta_n\boldsymbol{u}_n^{hk})_V\,dt, \\
R_{n,4} &= (\boldsymbol{\sigma}_0^{hk} - \mathcal{E}\boldsymbol{\varepsilon}(\boldsymbol{u}_0^{hk}), \boldsymbol{\varepsilon}(\boldsymbol{v}_n^h - \delta_n\boldsymbol{u}_n^{hk}))_Q \\
&\quad + (\boldsymbol{\sigma}_0 - \mathcal{E}\boldsymbol{\varepsilon}(\boldsymbol{u}_0), \boldsymbol{\varepsilon}(\delta_n\boldsymbol{u}_n^{hk} - \delta_n\boldsymbol{u}_n))_Q, \\
R_{n,5} &= \left(\boldsymbol{s}_{\mathcal{G},n}^{hk}, \boldsymbol{\varepsilon}(\boldsymbol{v}_n^h - \delta_n\boldsymbol{u}_n^{hk})\right)_Q + \frac{1}{k_n}\int_{t_{n-1}}^{t_n} \left(\boldsymbol{s}_{\mathcal{G}}(t), \boldsymbol{\varepsilon}(\delta_n\boldsymbol{u}_n^{hk} - \dot{\boldsymbol{u}}(t))\right)_Q dt.
\end{aligned}$$

We need to find appropriate bounds for $R_{n,i}$, $1 \le i \le 5$. First we estimate $R_{n,1}$ as follows:

$$\begin{aligned}
R_{n,1} &= \frac{1}{k_n}\Big(\mathcal{E}\boldsymbol{\varepsilon}(\boldsymbol{u}_n) - \frac{1}{k_n}\int_{t_{n-1}}^{t_n} \mathcal{E}\boldsymbol{\varepsilon}(\boldsymbol{u}(t))\,dt, \boldsymbol{\varepsilon}(\boldsymbol{r}_n - \boldsymbol{r}_{n-1})\Big)_Q \\
&\quad + \frac{1}{k_n}\int_{t_{n-1}}^{t_n} (\mathcal{E}\boldsymbol{\varepsilon}(\boldsymbol{u}(t)), \boldsymbol{\varepsilon}(\delta_n\boldsymbol{u}_n - \dot{\boldsymbol{u}}(t)))_Q dt \\
&\quad + (\mathcal{E}\boldsymbol{\varepsilon}(\boldsymbol{r}_n), \boldsymbol{\varepsilon}(\delta_n\boldsymbol{u}_n - \boldsymbol{v}_n^h))_Q - (\mathcal{E}\boldsymbol{\varepsilon}(\boldsymbol{u}_n), \boldsymbol{\varepsilon}(\delta_n\boldsymbol{u}_n - \boldsymbol{v}_n^h))_Q \\
&\le \frac{1}{k_n^2}\Big(\mathcal{E}\int_{t_{n-1}}^{t_n} (t - t_{n-1})\boldsymbol{\varepsilon}(\dot{\boldsymbol{u}}(t))\,dt, \boldsymbol{\varepsilon}(\boldsymbol{r}_n - \boldsymbol{r}_{n-1})\Big)_Q \\
&\quad + \frac{1}{k_n}\int_{t_{n-1}}^{t_n} (\mathcal{E}\boldsymbol{\varepsilon}(\boldsymbol{u}(t)), \boldsymbol{\varepsilon}(\delta_n\boldsymbol{u}_n - \dot{\boldsymbol{u}}(t)))_Q dt \\
&\quad + \frac{1}{2}\|\boldsymbol{r}_n\|_{\mathcal{E}}^2 + c\,\|\delta_n\boldsymbol{u}_n - \boldsymbol{v}_n^h\|_V^2 - (\mathcal{E}\boldsymbol{\varepsilon}(\boldsymbol{u}_n), \boldsymbol{\varepsilon}(\delta_n\boldsymbol{u}_n - \boldsymbol{v}_n^h))_Q.
\end{aligned}$$

Therefore,

$$\begin{aligned}
R_{n,1} &\le c\,\|\dot{\boldsymbol{u}}\|_{L^\infty(I_n;V)}(\|\boldsymbol{r}_n\|_{\mathcal{E}} + \|\boldsymbol{r}_{n-1}\|_{\mathcal{E}}) \\
&\quad + \frac{c}{k_n}\|\boldsymbol{u}\|_{L^\infty(I_n;V)}\int_{t_{n-1}}^{t_n} \|\delta_n\boldsymbol{u}_n - \dot{\boldsymbol{u}}(t)\|_V\,dt \\
&\quad + \frac{1}{2}\|\boldsymbol{r}_n\|_{\mathcal{E}}^2 + c\,\|\delta_n\boldsymbol{u}_n - \boldsymbol{v}_n^h\|_V^2 + c\,\|\boldsymbol{u}_n\|_V\|\delta_n\boldsymbol{u}_n - \boldsymbol{v}_n^h\|_V. \qquad (17.116)
\end{aligned}$$

Express $R_{n,2}$ as

$$R_{n,2} = \frac{1}{k_n}\int_{t_{n-1}}^{t_n} (j(\boldsymbol{v}_n^h) - j(\dot{\boldsymbol{u}}(t))\,dt.$$

Using the Lipschitz continuity of j on V_1, we find a bound for $R_{n,2}$:

$$|R_{n,2}| \le \frac{c}{k_n}\int_{t_{n-1}}^{t_n} \|\boldsymbol{v}_n^h - \dot{\boldsymbol{u}}(t)\|_V dt. \tag{17.117}$$

Rewrite $R_{n,3}$ as

$$\begin{aligned} R_{n,3} = {} & \frac{1}{k_n^2}\Big(\int_{t_{n-1}}^{t_n} (t-t_{n-1})\dot{\boldsymbol{f}}(t)\,dt, \boldsymbol{\varepsilon}(\boldsymbol{r}_{n-1}-\boldsymbol{r}_n)\Big)_Q \\ & + (\boldsymbol{f}_n, \boldsymbol{\varepsilon}(\delta_n\boldsymbol{u}_n - \boldsymbol{v}_n^h))_Q - \frac{1}{k_n}\int_{t_{n-1}}^{t_n} (\boldsymbol{f}(t), \delta_n\boldsymbol{u}_n - \dot{\boldsymbol{u}}(t))_Q dt. \end{aligned}$$

Then

$$\begin{aligned} R_{n,3} \le {} & c\,\|\dot{\boldsymbol{f}}\|_{L^\infty(I_n;V)}(\|\boldsymbol{r}_n\|_{\mathcal{E}} + \|\boldsymbol{r}_{n-1}\|_{\mathcal{E}}) + \|\boldsymbol{f}_n\|_V\|\delta_n\boldsymbol{u}_n - \boldsymbol{v}_n^h\|_V \\ & + \frac{1}{k_n}\|\boldsymbol{f}\|_{L^\infty(I_n;V)}\int_{t_{n-1}}^{t_n} \|\delta_n\boldsymbol{u}_n - \dot{\boldsymbol{u}}(t)\|_V dt. \end{aligned} \tag{17.118}$$

Rewrite $R_{n,4}$ as

$$\begin{aligned} R_{n,4} = {} & \frac{1}{k_n}(\boldsymbol{\sigma}_0 - \boldsymbol{\sigma}_0^{hk} - \mathcal{E}\boldsymbol{\varepsilon}(\boldsymbol{u}_0 - \boldsymbol{u}_0^{hk}), \boldsymbol{\varepsilon}(\boldsymbol{r}_{n-1}-\boldsymbol{r}_n))_Q \\ & - (\boldsymbol{\sigma}_0 - \mathcal{E}\boldsymbol{\varepsilon}(\boldsymbol{u}_0), \boldsymbol{\varepsilon}(\delta_n\boldsymbol{u}_n - \boldsymbol{v}_n^h))_H \\ & + (\boldsymbol{\sigma}_0 - \boldsymbol{\sigma}_0^{hk} - \mathcal{E}\boldsymbol{\varepsilon}(\boldsymbol{u}_0 - \boldsymbol{u}_0^{hk}), \boldsymbol{\varepsilon}(\delta_n\boldsymbol{u}_n - \boldsymbol{v}_n^h))_Q. \end{aligned}$$

Then

$$\begin{aligned} R_{n,4} \le {} & \frac{1}{k_n}(\boldsymbol{\sigma}_0 - \boldsymbol{\sigma}_0^{hk} - \mathcal{E}\boldsymbol{\varepsilon}(\boldsymbol{u}_0 - \boldsymbol{u}_0^{hk}), \boldsymbol{\varepsilon}(\boldsymbol{r}_{n-1}-\boldsymbol{r}_n))_Q \\ & + c\,(e_0 + \|\boldsymbol{\sigma}_0\|_Q + \|\boldsymbol{u}_0\|_V)\|\delta_n\boldsymbol{u}_n - \boldsymbol{v}_n^h\|_V. \end{aligned} \tag{17.119}$$

Express $R_{n,5}$ as

$$\begin{aligned} R_{n,5} = {} & -\big(\boldsymbol{\theta}_n^{hk}(\mathcal{G}), \boldsymbol{\varepsilon}(\boldsymbol{v}_n^h - \delta_n\boldsymbol{u}_n^{hk})\big)_Q + \big(\boldsymbol{s}_{\mathcal{G}}(t_n), \boldsymbol{\varepsilon}(\boldsymbol{v}_n^h - \delta_n\boldsymbol{u}_n^{hk})\big)_Q \\ & + \frac{1}{k_n}\int_{t_{n-1}}^{t_n} \big(\boldsymbol{s}_{\mathcal{G}}(t), \boldsymbol{\varepsilon}(\delta_n\boldsymbol{u}_n^{hk} - \delta_n\boldsymbol{u}_n) + \boldsymbol{\varepsilon}(\delta_n\boldsymbol{u}_n - \dot{\boldsymbol{u}}(t))\big)_Q\, dt. \end{aligned}$$

Integrating by parts, we obtain

$$\begin{aligned} & \frac{1}{k_n}\int_{t_{n-1}}^{t_n} \big(\boldsymbol{s}_{\mathcal{G}}(t), \boldsymbol{\varepsilon}(\delta_n\boldsymbol{u}_n - \dot{\boldsymbol{u}}(t))\big)_Q dt \\ & \qquad = \frac{1}{k_n}\int_{t_{n-1}}^{t_n} (\mathcal{G}(\boldsymbol{\sigma}(t), \boldsymbol{\varepsilon}(\dot{\boldsymbol{u}}(t))), \boldsymbol{\varepsilon}(E\boldsymbol{u}(t)))_Q dt \end{aligned}$$

where

$$E\boldsymbol{u}(t) = \boldsymbol{u}(t) - \frac{t-t_{n-1}}{k_n}\boldsymbol{u}_n - \frac{t_n - t}{k_n}\boldsymbol{u}_{n-1}, \quad t_{n-1} \le t \le t_n,\ n = 1, \dots, N$$

is the error of piecewise linear interpolation of $\boldsymbol{u}$. Also, by an elementary manipulation, we have

$$\begin{aligned}
&\frac{1}{k_n}\int_{t_{n-1}}^{t_n} \big(\boldsymbol{s}_{\mathcal{G}}(t), \boldsymbol{\varepsilon}(\delta_n \boldsymbol{u}_n^{hk} - \delta_n \boldsymbol{u}_n)\big)_Q \, dt \\
&\quad = -\frac{1}{k_n}\Big(\int_{t_{n-1}}^{t_n} (s - t_{n-1})\mathcal{G}(\boldsymbol{\sigma}(s), \boldsymbol{\varepsilon}(\dot{\boldsymbol{u}}(s)))\, ds, \boldsymbol{\varepsilon}(\delta_n \boldsymbol{u}_n^{hk} - \delta_n \boldsymbol{u}_n)\Big)_Q \\
&\qquad + \big(\boldsymbol{s}_{\mathcal{G}}(t_n), \boldsymbol{\varepsilon}(\delta_n \boldsymbol{u}_n^{hk} - \delta_n \boldsymbol{u}_n)\big)_Q .
\end{aligned}$$

Thus, we can rewrite $R_{n,5}$ as

$$\begin{aligned}
R_{n,5} = & -\big(\boldsymbol{\theta}_n^{hk}(\mathcal{G}), \boldsymbol{\varepsilon}(\boldsymbol{v}_n^h - \delta_n \boldsymbol{u}_n)\big)_Q + \frac{1}{k_n}\big(\boldsymbol{\theta}_n^{hk}(\mathcal{G}), \boldsymbol{\varepsilon}(\boldsymbol{r}_n - \boldsymbol{r}_{n-1})\big)_Q \\
& - \frac{1}{k_n^2}\Big(\int_{t_{n-1}}^{t_n} (s - t_{n-1})\mathcal{G}(\boldsymbol{u}(s), \boldsymbol{\sigma}(s))\, ds, \boldsymbol{\varepsilon}(\boldsymbol{r}_{n-1} - \boldsymbol{r}_n)\Big)_Q \\
& + \frac{1}{k_n}\int_{t_{n-1}}^{t_n} (\mathcal{G}(\boldsymbol{\sigma}(t), \boldsymbol{\varepsilon}(\dot{\boldsymbol{u}}(t))), \boldsymbol{\varepsilon}(E\boldsymbol{u}(t)))_Q \, dt \\
& + \big(\boldsymbol{s}_{\mathcal{G}}(t_n), \boldsymbol{\varepsilon}(\boldsymbol{v}_n^h - \delta_n \boldsymbol{u}_n)\big)_Q .
\end{aligned}$$

We then apply the estimates (17.101) and (17.100),

$$\begin{aligned}
R_{n,5} \le\, & c \sum_{j=1}^{n} k_j (\|\boldsymbol{u}_{j-1} - \boldsymbol{u}_{j-1}^{hk}\|_V^2 + \|\boldsymbol{\sigma}_{j-1} - \boldsymbol{\sigma}_{j-1}^{hk}\|_V^2) \\
& + c\, k^2 (\|\dot{\boldsymbol{u}}\|_{L^\infty(0,t_n;V)}^2 + \|\dot{\boldsymbol{\sigma}}\|_{L^\infty(0,t_n;Q)}^2) + \frac{1}{2}\|\boldsymbol{v}_n^h - \delta_n \boldsymbol{u}_n\|_V^2 \\
& + \frac{c}{k_n}\int_{t_{n-1}}^{t_n} \|E\boldsymbol{u}(t)\|_Q dt - \frac{1}{k_n}\big(\boldsymbol{\theta}_n^{hk}(\mathcal{G}), \boldsymbol{\varepsilon}(\boldsymbol{r}_n - \boldsymbol{r}_{n-1})\big)_Q \\
& + c\,(\|\boldsymbol{r}_n\|_{\mathcal{E}} + \|\boldsymbol{r}_{n-1}\|_{\mathcal{E}} + t_n \|\boldsymbol{v}_n^h - \delta_n \boldsymbol{u}_n\|_V),
\end{aligned} \tag{17.120}$$

where the constant c may depend on $\|\mathcal{G}(\boldsymbol{\sigma}, \boldsymbol{\varepsilon}(\boldsymbol{u}))\|_{L^\infty(I;Q)}$. Combining the lower bound (17.105) and the upper bound (17.115) for A_n, and the estimates

(17.116)–(17.120), we find that

$$\begin{aligned}\|\boldsymbol{r}_n\|_{\mathcal{E}}^2 - \|\boldsymbol{r}_{n-1}\|_{\mathcal{E}}^2 \le k_n\|\boldsymbol{r}_n\|_{\mathcal{E}}^2 + c\,k_n(\|\boldsymbol{r}_n\|_{\mathcal{E}} + \|\boldsymbol{r}_{n-1}\|_{\mathcal{E}})&\\
+ c\int_{t_{n-1}}^{t_n} \left[\|\delta_n\boldsymbol{u}_n - \dot{\boldsymbol{u}}(t)\|_V + \|\boldsymbol{v}_n^h - \dot{\boldsymbol{u}}(t)\|_V\right] dt&\\
+ c\,k_n\|\delta_n\boldsymbol{u}_n - \boldsymbol{v}_n^h\|_V^2 + c\,(1+e_0)\,\|\delta_n\boldsymbol{u}_n - \boldsymbol{v}_n^h\|_V&\\
+ 2\,(\boldsymbol{\sigma}_0 - \boldsymbol{\sigma}_0^{hk} - \mathcal{E}(\boldsymbol{u}_0 - \boldsymbol{u}_0^{hk}), \boldsymbol{r}_{n-1} - \boldsymbol{r}_n)_Q&\\
- 2\left(\boldsymbol{\theta}_n^{hk}(\mathcal{G}), \boldsymbol{r}_n - \boldsymbol{r}_{n-1}\right)_Q&\\
+ c\sum_{j=1}^{n} k_j^2(\|\boldsymbol{u}_{j-1} - \boldsymbol{u}_{j-1}^{hk}\|_V^2 + \|\boldsymbol{\sigma}_{j-1} - \boldsymbol{\sigma}_{j-1}^{hk}\|_V^2)&\\
+ c\int_{t_{n-1}}^{t_n} \|E\boldsymbol{u}(t)\|_Q dt + c\,k^3&\end{aligned}$$

where c may depend on $\|\boldsymbol{u}\|_{W^{1,\infty}(I;V)}$, $\|\boldsymbol{\sigma}\|_{W^{1,\infty}(I;Q)}$, $\|\boldsymbol{f}\|_{W^{1,\infty}(I;V)}$. Then by a series of manipulations similar to those leading to (17.112), we obtain

$$\begin{aligned}&\|\boldsymbol{u}_n - \boldsymbol{u}_n^{hk}\|_{\mathcal{E}}^2 + \|\boldsymbol{\sigma}_n - \boldsymbol{\sigma}_n^{hk}\|_Q^2\\
&\quad\le c\sum_{j=1}^{n-1} k_j(\|\boldsymbol{u}_j - \boldsymbol{u}_j^{hk}\|_V^2 + \|\boldsymbol{\sigma}_j - \boldsymbol{\sigma}_j^{hk}\|_V^2) + c\,e_0^2\\
&\qquad + c\,k + c\sum_{j=1}^{n}\int_{t_{j-1}}^{t_j} \|\delta_j\boldsymbol{u}_j - \dot{\boldsymbol{u}}(t)\|_V dt\\
&\qquad + c\sum_{j=1}^{n} k_j\|\delta_j\boldsymbol{u}_j - \boldsymbol{v}_j^h\|_V^2 + c\sum_{j=1}^{n}\int_{t_{j-1}}^{t_j} \|\boldsymbol{v}_j^h - \dot{\boldsymbol{u}}(t)\|_V dt\\
&\qquad + c\sum_{j=1}^{n}\int_{t_{j-1}}^{t_j} \|E\boldsymbol{u}(t)\|_Q dt + \|(I_Q - \mathcal{P}_{Q^h})(\boldsymbol{\sigma}_n - \boldsymbol{\sigma}_0)\|_Q^2,\end{aligned}$$

where the term $\|\boldsymbol{u}_n - \boldsymbol{u}_n^{hk}\|_{\mathcal{E}}^2$ on the left hand side can be replaced by $\|\boldsymbol{u}_n - \boldsymbol{u}_n^{hk}\|_V^2$. We then apply the technique of the proof of Theorem 17.22 to find the estimate

$$\begin{aligned}&\max_{1\le n\le N}(\|\boldsymbol{u}_n - \boldsymbol{u}_n^{hk}\|_V^2 + \|\boldsymbol{\sigma}_n - \boldsymbol{\sigma}_n^{hk}\|_Q^2)\\
&\quad\le c\left(e_0^2 + \max_{1\le n\le N}\|(I_Q - \mathcal{P}_{Q^h})(\boldsymbol{\sigma}_n - \boldsymbol{\sigma}_0)\|_Q^2 + k\right.\\
&\qquad + \sum_{j=1}^{N}\int_{t_{j-1}}^{t_j} \|\delta_j\boldsymbol{u}_j - \dot{\boldsymbol{u}}(t)\|_V dt + \sum_{j=1}^{N} k_j\|\delta_j\boldsymbol{u}_j - \boldsymbol{v}_j^h\|_V^2\\
&\qquad \left. + \sum_{j=1}^{N}\int_{t_{j-1}}^{t_j} \|\boldsymbol{v}_j^h - \dot{\boldsymbol{u}}(t)\|_V dt + \sum_{j=1}^{N}\int_{t_{j-1}}^{t_j} \|E\boldsymbol{u}(t)\|_Q dt\right). \quad (17.121)\end{aligned}$$

We now show a convergence result using (17.121).

Theorem 17.23 *Assume the conditions stated in Theorem 7.10 on the geometry of the domain* Ω. *Let* $(\boldsymbol{u}, \boldsymbol{\sigma}) \in W^{1,\infty}(I; V_1 \times Q)$ *be the solution of Problem 17.2 and* $\{(\boldsymbol{u}_n^{hk}, \boldsymbol{\sigma}_n^{hk})\}_{n=1}^N \subset V_1^h \times Q^h$ *be the solution of the corresponding fully discrete Problem 17.20. Then we have*

$$\max_{1\le n\le N}(\|\boldsymbol{u}_n - \boldsymbol{u}_n^{hk}\|_V + \|\boldsymbol{\sigma}_n - \boldsymbol{\sigma}_n^{hk}\|_Q) \to 0 \quad \text{as } h, k \to 0.$$

Proof. By Theorem 7.12 (page 150), for any $\varepsilon > 0$, there exists $\tilde{\boldsymbol{u}} \in C^\infty(\bar{I}; V_1)$ such that

$$\|\boldsymbol{u} - \tilde{\boldsymbol{u}}\|_{H^1(I;V)} \le \varepsilon.$$

Since

$$\begin{aligned}
\int_{t_{j-1}}^{t_j} \|\delta_j \boldsymbol{u}_j - \dot{\boldsymbol{u}}(t)\|_V dt &= \int_{t_{j-1}}^{t_j} \left\| \frac{1}{k_j} \int_{t_{j-1}}^{t_j} (\dot{\boldsymbol{u}}(s) - \dot{\boldsymbol{u}}(t))\, ds \right\|_V dt \\
&\le \frac{1}{k_j} \int_{t_{j-1}}^{t_j} \int_{t_{j-1}}^{t_j} \big[\|\dot{\boldsymbol{u}}(s) - \dot{\tilde{\boldsymbol{u}}}(s)\|_V + \|\dot{\boldsymbol{u}}(t) - \dot{\tilde{\boldsymbol{u}}}(t)\|_V \\
&\qquad + \|\dot{\tilde{\boldsymbol{u}}}(s) - \dot{\tilde{\boldsymbol{u}}}(t)\|_V\big]\, ds\, dt \\
&\le 2 \int_{t_{j-1}}^{t_j} \|\dot{\boldsymbol{u}}(t) - \dot{\tilde{\boldsymbol{u}}}(t)\|_V dt + k_j \int_{t_{j-1}}^{t_j} \|\ddot{\tilde{\boldsymbol{u}}}(t)\|_V dt,
\end{aligned}$$

we have

$$\sum_{j=1}^N \int_{t_{j-1}}^{t_j} \|\delta_j \boldsymbol{u}_j - \dot{\boldsymbol{u}}(t)\|_V dt \le c\,\varepsilon + k\, \|\ddot{\tilde{\boldsymbol{u}}}(t)\|_{L^1(I;V)}. \tag{17.122}$$

From

$$\|\boldsymbol{v}_j^h - \dot{\boldsymbol{u}}(t)\|_V \le \|\boldsymbol{v}_j^h - \delta_j \boldsymbol{u}_j\|_V + \|\delta_j \boldsymbol{u}_j - \dot{\boldsymbol{u}}(t)\|_V,$$

we see that

$$\begin{aligned}
\sum_{j=1}^N \int_{t_{j-1}}^{t_j} \|\boldsymbol{v}_j^h - \dot{\boldsymbol{u}}(t)\|_V dt &\le \sum_{j=1}^N k_j \|\boldsymbol{v}_j^h - \delta_j \boldsymbol{u}_j\|_V \\
&\quad + \sum_{j=1}^N \int_{t_{j-1}}^{t_j} \|\delta_j \boldsymbol{u}_j - \dot{\boldsymbol{u}}(t)\|_V dt.
\end{aligned} \tag{17.123}$$

It suffices to estimate the term $\sum_{j=1}^N k_j \|\boldsymbol{v}_j^h - \delta_j \boldsymbol{u}_j\|_V$. Noticing that

$$\boldsymbol{v}_j^h - \delta_j \boldsymbol{u}_j = \boldsymbol{v}_j^h - \delta_j \tilde{\boldsymbol{u}}_j + \frac{1}{k_j} \int_{t_{j-1}}^{t_j} (\dot{\tilde{\boldsymbol{u}}}(t) - \dot{\boldsymbol{u}}(t))\, dt,$$

we have

$$\begin{aligned}
\sum_{j=1}^N k_j \|\boldsymbol{v}_j^h - \delta_j \boldsymbol{u}_j\|_V &\le \sum_{j=1}^N k_j \|\boldsymbol{v}_j^h - \delta_j \tilde{\boldsymbol{u}}_j\|_V + \|\dot{\boldsymbol{u}} - \dot{\tilde{\boldsymbol{u}}}\|_{L^1(I;V)} \\
&\le \sum_{j=1}^N k_j \|\boldsymbol{v}_j^h - \delta_j \tilde{\boldsymbol{u}}_j\|_V + c\,\varepsilon.
\end{aligned} \tag{17.124}$$

Here $\tilde{\boldsymbol{u}}$ is such that $\tilde{\boldsymbol{u}} \in H^1(I;V_0)$ (see (7.30) on page 148 for the definition of V_0) and

$$\|\boldsymbol{u} - \tilde{\boldsymbol{u}}\|_{H^1(I;V)} \le \varepsilon. \tag{17.125}$$

It remains to bound the last term in (17.121). We have the expression

$$E\boldsymbol{u}(t) = \int_{t_{n-1}}^{t} \frac{t_n - t}{k_n} \dot{\boldsymbol{u}}(s)\, ds - \int_{t}^{t_n} \frac{t - t_{n-1}}{k_n} \dot{\boldsymbol{u}}(s)\, ds, \quad t_{n-1} \le t \le t_n.$$

Therefore,

$$\sum_{j=1}^{N} \int_{t_{j-1}}^{t_j} \|E\boldsymbol{u}(t)\|_Q dt \le \sum_{j=1}^{N} k_j \int_{t_{j-1}}^{t_j} \|\dot{\boldsymbol{u}}(s)\|_V ds \le k\, \|\dot{\boldsymbol{u}}\|_{L^1(I;V)}. \tag{17.126}$$

The bounds (17.122)–(17.126) are now used in the estimate (17.121). By the arbitrariness of $\boldsymbol{v}_j^h \in V_1^h$, $1 \le j \le N$, we get

$$\max_{1\le n\le N} (\|\boldsymbol{u}_n - \boldsymbol{u}_n^{hk}\|_V^2 + \|\boldsymbol{\sigma}_n - \boldsymbol{\sigma}_n^{hk}\|_V^2) \le c\,(e_0^2 + \varepsilon + h + k + D_{hk}(\tilde{\boldsymbol{u}})), \tag{17.127}$$

where

$$D_{hk}(\tilde{\boldsymbol{u}}) = \sum_{j=1}^{N} k_j \inf_{\boldsymbol{v}_j^h \in V_1^h} \|\delta_j \tilde{\boldsymbol{u}}_j - \boldsymbol{v}_j^h\|_V$$

and the constant c depends on $\boldsymbol{u}$, $\boldsymbol{\sigma}$, $\tilde{\boldsymbol{\sigma}}$, $\boldsymbol{f}$, $\mathcal{G}(\boldsymbol{\sigma}, \boldsymbol{\varepsilon}(\dot{\boldsymbol{u}}))$ but is independent of ε, h, k. We see that

$$\inf_{\boldsymbol{v}_j^h \in V^h} \|\delta_j \tilde{\boldsymbol{u}}_j - \boldsymbol{v}_j^h\|_V \le c\,h\, \|\delta_j \tilde{\boldsymbol{u}}_j\|_{H^2(\Omega)^d} \le \frac{c\,h}{k_j} \int_{t_{j-1}}^{t_j} \|\dot{\tilde{\boldsymbol{u}}}(t)\|_{H^2(\Omega)^d} dt,$$

and thus

$$D_{hk}(\tilde{\boldsymbol{u}}) \le c\,h\, \|\dot{\tilde{\boldsymbol{u}}}\|_{L^1(I;H^2(\Omega)^d)}.$$

So the estimate (17.127) reduces to

$$\max_{1\le n\le N} (\|\boldsymbol{u}_n - \boldsymbol{u}_n^{hk}\|_V^2 + \|\boldsymbol{\sigma}_n - \boldsymbol{\sigma}_n^{hk}\|_V^2) \le c\,(e_0^2 + \varepsilon + h + k).$$

Therefore, we have the convergence. ■

We remark here that the estimate (17.113) is used for deriving optimal order error estimates provided the exact solution has higher degree regularity, while the estimate (17.121) is suitable for convergence analysis under the basic solution regularity condition.

Chapter 18

Other Viscoplastic Contact Problems

In this chapter we consider several more mathematical models describing frictionless or frictional contact of an viscoplastic body with a foundation. For all these models we present variational formulations and establish existence and uniqueness results. The proofs are based on methods similar to those used in the previous chapters and therefore, we omit sometimes the details. Numerical approximations of some of the contact problems presented here can be provided using semi-discrete and fully discrete schemes similar to those already studied in the previous chapters.

18.1 Contact with Simplified Coulomb's Friction Law

Consider a contact condition modeled by a simplified version of Coulomb's law of dry friction when the normal stress on the contact boundary is assumed to be given (see e.g. [70, 167]). The classical formulation of the mechanical problem is the following:

Find a displacement field $\boldsymbol{u} : \Omega \times \overline{I} \to \mathbb{R}^d$ and a stress field $\boldsymbol{\sigma} : \Omega \times \overline{I} \to \mathbb{S}^d$ such that

$$\dot{\boldsymbol{\sigma}} = \mathcal{E}\boldsymbol{\varepsilon}(\dot{\boldsymbol{u}}) + \mathcal{G}(\boldsymbol{\sigma}, \boldsymbol{\varepsilon}(\boldsymbol{u})) \quad \text{in } \Omega \times I, \tag{18.1}$$

$$\operatorname{Div} \boldsymbol{\sigma} + \boldsymbol{f}_0 = \boldsymbol{0} \quad \text{in } \Omega \times I, \tag{18.2}$$

$$\boldsymbol{u} = \boldsymbol{0} \quad \text{on } \Gamma_1 \times I, \tag{18.3}$$

$$\boldsymbol{\sigma\nu} = \boldsymbol{f}_2 \quad \text{on } \Gamma_2 \times I, \tag{18.4}$$

$$\left.\begin{array}{l} \sigma_\nu = S, \quad \|\boldsymbol{\sigma}_\tau\| \le \mu\,|\sigma_\nu|, \\ \|\boldsymbol{\sigma}_\tau\| < \mu\,|\sigma_\nu| \;\Rightarrow\; \dot{\boldsymbol{u}}_\tau = \boldsymbol{0}, \\ \|\boldsymbol{\sigma}_\tau\| = \mu\,|\sigma_\nu| \;\Rightarrow\; \exists\,\lambda \ge 0 \text{ s.t. } \boldsymbol{\sigma}_\tau = -\lambda \dot{\boldsymbol{u}}_\tau \end{array}\right\} \quad \text{on } \Gamma_3 \times I, \tag{18.5}$$

$$\boldsymbol{u}(0) = \boldsymbol{u}_0, \quad \boldsymbol{\sigma}(0) = \boldsymbol{\sigma}_0 \quad \text{in } \Omega. \tag{18.6}$$

Here, $\boldsymbol{u}_0$ and $\boldsymbol{\sigma}_0$ are the given initial displacement and initial stress, S is a given normal stress and μ is the coefficient of friction.

We assume the force and traction densities satisfy conditions (17.7) (page 366) and the normal stress and the coefficient of friction satisfy

$$S \in L^\infty(\Gamma_3), \tag{18.7}$$

$$\mu \in L^\infty(\Gamma_3), \quad \mu \geq 0 \text{ a.e. on } \Gamma_3. \tag{18.8}$$

We denote by $\boldsymbol{f}(t)$ the element of V given by

$$(\boldsymbol{f}(t), \boldsymbol{v})_V = \int_\Omega \boldsymbol{f}_0(t) \cdot \boldsymbol{v}\, dx + \int_{\Gamma_2} \boldsymbol{f}_2(t) \cdot \boldsymbol{v}\, da + \int_{\Gamma_3} S v_\nu\, da \tag{18.9}$$

for all $\boldsymbol{v} \in V$ and $t \in \overline{I}$ and let $j : V \to \mathbb{R}_+$ be the functional

$$j(\boldsymbol{v}) = \int_{\Gamma_3} \mu\, |S|\, \|\boldsymbol{v}_\tau\|\, da. \tag{18.10}$$

The integrals in (18.9) and (18.10) are well defined due to the assumptions (17.7), (18.7) and (18.8).

We denote, for any $t \in \overline{I}$, by $\Sigma(t)$ the set

$$\Sigma(t) = \{\, \boldsymbol{\tau} \in Q : (\boldsymbol{\tau}, \boldsymbol{\varepsilon}(\boldsymbol{v}))_Q + j(\boldsymbol{v}) \geq (\boldsymbol{f}(t), \boldsymbol{v})_V \ \forall\, \boldsymbol{v} \in V \,\}. \tag{18.11}$$

Assume the initial data satisfy

$$\boldsymbol{u}_0 \in V, \quad \boldsymbol{\sigma}_0 \in \Sigma(0). \tag{18.12}$$

Arguments similar to those used in Lemma 17.1 lead us to consider the following weak formulations of the frictional problem (18.1)–(18.6).

Problem 18.1 *Find a displacement field* $\boldsymbol{u} : \overline{I} \to V$ *and a stress field* $\boldsymbol{\sigma} : \overline{I} \to Q_1$ *such that*

$$\begin{gathered}
\dot{\boldsymbol{\sigma}}(t) = \mathcal{E}\boldsymbol{\varepsilon}(\dot{\boldsymbol{u}}(t)) + \mathcal{G}(\boldsymbol{\sigma}(t), \boldsymbol{\varepsilon}(\boldsymbol{u}(t))) \quad \text{a.e. } t \in I, \\
(\boldsymbol{\sigma}(t), \boldsymbol{\varepsilon}(\boldsymbol{v}) - \boldsymbol{\varepsilon}(\dot{\boldsymbol{u}}(t)))_Q + j(\boldsymbol{v}) - j(\dot{\boldsymbol{u}}(t)) \geq (\boldsymbol{f}(t), \boldsymbol{v} - \dot{\boldsymbol{u}}(t))_V \\
\forall\, \boldsymbol{v} \in V, \quad \text{a.e. } t \in I, \\
\boldsymbol{u}(0) = \boldsymbol{u}_0, \qquad \boldsymbol{\sigma}(0) = \boldsymbol{\sigma}_0.
\end{gathered}$$

Problem 18.2 *Find a displacement field* $\boldsymbol{u} : \overline{I} \to V$ *and a stress field* $\boldsymbol{\sigma} : \overline{I} \to Q_1$ *such that*

$$\begin{gathered}
\dot{\boldsymbol{\sigma}}(t) = \mathcal{E}\boldsymbol{\varepsilon}(\dot{\boldsymbol{u}}(t)) + \mathcal{G}(\boldsymbol{\sigma}(t), \boldsymbol{\varepsilon}(\boldsymbol{u}(t))) \quad \text{a.e. } t \in I, \\
\boldsymbol{\sigma}(t) \in \Sigma(t), \ (\boldsymbol{\tau} - \boldsymbol{\sigma}(t), \boldsymbol{\varepsilon}(\dot{\boldsymbol{u}}(t)))_Q \geq 0 \quad \forall\, \boldsymbol{\tau} \in \Sigma(t), \ \text{a.e. } t \in I, \\
\boldsymbol{u}(0) = \boldsymbol{u}_0, \qquad \boldsymbol{\sigma}(0) = \boldsymbol{\sigma}_0.
\end{gathered}$$

We remark that Problems 18.1 and 18.2 are similar to Problems 17.2 and 17.3, the only difference being the definitions of the space V, the element $\boldsymbol{f}$ and the frictional functional j. However, we note that the regularity $\boldsymbol{f} \in W^{1,\infty}(0,T;V)$ holds and the functional j defined by (18.10) still satisfies a continuity property of the form (17.12). So with the same methods as those used in Chapter 17, we can show the following results.

Theorem 18.3 *Assume* (6.44), (6.45), (17.7), (18.7), (18.8) *and* (18.12). *Then:*

a) *There exists a unique solution* $(\boldsymbol{u}, \boldsymbol{\sigma})$ *of Problem 18.1 with regularity*

$$\boldsymbol{u} \in W^{1,\infty}(I; V), \quad \boldsymbol{\sigma} \in W^{1,\infty}(I; Q_1). \tag{18.13}$$

b) *There exists a unique solution* $(\boldsymbol{u}, \boldsymbol{\sigma})$ *to Problem 18.2 with regularity* (18.13).

c) *Let* $(\boldsymbol{u}, \boldsymbol{\sigma})$ *be a couple of functions with regularity* (18.13). *Then* $(\boldsymbol{u}, \boldsymbol{\sigma})$ *is a solution of Problem 18.1 iff it is a solution of Problem 18.2.*

Parts a) and b) of the theorem provide existence and uniqueness results in the study of the variational problems 18.1 and 18.2, respectively, while part c) expresses the equivalence of these variational problems. From these results we conclude that under the assumptions stated in the theorem, the mechanical problem (18.1)–(18.6) has a unique *weak solution* which solves both Problems 18.1 and 18.2. A continuous dependence result similar to that presented in Theorem 17.16 may be obtained for the solution of Problems 18.1 and 18.2. More details on the study of the mechanical problem (18.1)–(18.6), including numerical analysis of semi-discrete and fully discrete schemes, may be found in [43].

18.2 Contact with Dissipative Frictional Potential

In this section we extend the existence and uniqueness results presented in Chapter 17 to the case when the friction is modeled by the general velocity dependent dissipation functional used in Section 13.4. We note that in this case the nondifferentiable functional j is not a continuous semi-norm over the space of admissible velocity fields. As a result, Theorem 4.16 (page 69), which was used in the proof of Theorem 17.4, does not apply in this case. Moreover, since $j(\boldsymbol{v}) \neq 2j(\boldsymbol{v})$, it is not possible to derive dual variational formulations similar to those derived in Chapter 17. Therefore, in this section we consider only the primal variational formulation of the mechanical problem and we use Theorem 13.17 (page 299) to prove the existence of a unique weak solution to the model.

The classical formulation of the problem is the following:

Find a displacement field $\boldsymbol{u} : \Omega \times \overline{I} \to \mathbb{R}^d$ and a stress field $\boldsymbol{\sigma} : \Omega \times \overline{I} \to \mathbb{S}^d$

such that

$$\dot{\boldsymbol{\sigma}} = \mathcal{E}\boldsymbol{\varepsilon}(\dot{\boldsymbol{u}}) + \mathcal{G}(\boldsymbol{\sigma}, \boldsymbol{\varepsilon}(\boldsymbol{u})) \qquad \text{in } \Omega \times I, \tag{18.14}$$

$$\operatorname{Div} \boldsymbol{\sigma} + \boldsymbol{f}_0 = \boldsymbol{0} \qquad \text{in } \Omega \times I, \tag{18.15}$$

$$\boldsymbol{u} = \boldsymbol{0} \qquad \text{on } \Gamma_1 \times I, \tag{18.16}$$

$$\boldsymbol{\sigma\nu} = \boldsymbol{f}_2 \qquad \text{on } \Gamma_2 \times I, \tag{18.17}$$

$$\boldsymbol{u} \in U, \quad -\boldsymbol{\sigma\nu}\,(\boldsymbol{v} - \dot{\boldsymbol{u}}) \le \varphi(\boldsymbol{v}) - \varphi(\dot{\boldsymbol{u}}) \quad \forall\, \boldsymbol{v} \in U \qquad \text{on } \Gamma_3 \times I, \tag{18.18}$$

$$\boldsymbol{u}(0) = \boldsymbol{u}_0, \quad \boldsymbol{\sigma}(0) = \boldsymbol{\sigma}_0 \qquad \text{in } \Omega. \tag{18.19}$$

Here the frictional contact is modeled with the subdifferential boundary conditions (18.18) in which U represents the set of admissible contact test functions, and φ is a given function. Examples and detailed explanations of inequality problems in contact mechanics which lead to boundary conditions of this form have been presented in Section 13.4.

We suppose that $U \subset H^1(\Omega)^d$ and $\varphi : \Gamma_3 \times \mathbb{R}^d \to (-\infty, +\infty]$. To accommodate the two boundary conditions (18.16) and (18.18) on the displacement variable, we define

$$U_1 = V \cap U. \tag{18.20}$$

Let $j : U_1 \to \mathbb{R}$ be the functional

$$j(\boldsymbol{v}) = \begin{cases} \displaystyle\int_{\Gamma_3} \varphi(\boldsymbol{v})\, da & \text{if } \varphi(\boldsymbol{v}) \in L^1(\Gamma_3), \\ +\infty & \text{otherwise.} \end{cases} \tag{18.21}$$

We denote by $D(j)$ the effective domain of j, i.e.

$$D(j) = \{\, \boldsymbol{v} \in U_1 : j(\boldsymbol{v}) < +\infty \,\},$$

and we suppose in what follows that

$$U_1 \text{ is a closed subspace of } H_1 \text{ such that } C_0^\infty(\Omega)^d \subset U_1, \tag{18.22}$$

$$j \text{ is proper convex and lower semicontinuous on } U_1. \tag{18.23}$$

In U_1, we use the inner product $(\cdot, \cdot)_V$ and its induced norm.

We assume that the functions $\mathcal{E}$ and $\mathcal{G}$ satisfy the conditions (6.44) and (6.45) and the body forces and surface tractions satisfy

$$\boldsymbol{f}_0 \in H^1(I; L^2(\Omega)^d), \quad \boldsymbol{f}_2 \in H^1(I; L^2(\Gamma_2)^d). \tag{18.24}$$

Next, for each $t \in \overline{I}$, let

$$L(t, \boldsymbol{v}) = \int_\Omega \boldsymbol{f}_0(t) \cdot \boldsymbol{v}\, dx + \int_{\Gamma_2} \boldsymbol{f}_2(t) \cdot \boldsymbol{v}\, da \quad \forall\, \boldsymbol{v} \in U_1, \tag{18.25}$$

and

$$\Sigma(t) = \{\, \boldsymbol{\tau} \in Q : (\boldsymbol{\tau}, \boldsymbol{\varepsilon}(\boldsymbol{v}))_Q + j(\boldsymbol{v}) \ge L(t, \boldsymbol{v})_V \ \forall\, \boldsymbol{v} \in D(j) \,\}. \tag{18.26}$$

Finally, we assume that the initial data satisfy

$$\boldsymbol{u}_0 \in U_1, \quad \boldsymbol{\sigma}_0 \in \Sigma(0). \tag{18.27}$$

It is straightforward to show that if $(\boldsymbol{u}, \boldsymbol{\sigma})$ is a pair of sufficiently regular functions satisfying (18.15)–(18.18), then for all $t \in \overline{I}$, $\boldsymbol{u}(t) \in U_1$ and

$$\begin{aligned}&(\boldsymbol{\sigma}(t), \boldsymbol{\varepsilon}(\boldsymbol{v}) - \boldsymbol{\varepsilon}(\dot{\boldsymbol{u}}(t)))_Q + j(\boldsymbol{v}) - j(\dot{\boldsymbol{u}}(t)) \\ &\qquad \geq L(t, \boldsymbol{v} - \dot{\boldsymbol{u}}(t))_V \quad \forall\, \boldsymbol{v} \in U_1.\end{aligned}$$

This inequality, together with the constitutive relation (18.14) and the initial conditions (18.19), leads to the following variational formulation of the mechanical problem (18.14)–(18.19).

Problem 18.4 *Find a displacement field $\boldsymbol{u} : \overline{I} \to U_1$ and a stress field $\boldsymbol{\sigma} : \overline{I} \to Q_1$ such that for a.e. $t \in I$,*

$$\dot{\boldsymbol{\sigma}}(t) = \mathcal{E}\boldsymbol{\varepsilon}(\dot{\boldsymbol{u}}(t)) + \mathcal{G}(\boldsymbol{\sigma}(t), \boldsymbol{\varepsilon}(\boldsymbol{u}(t))), \tag{18.28}$$

$$\begin{aligned}&(\boldsymbol{\sigma}(t), \boldsymbol{\varepsilon}(\boldsymbol{v}) - \boldsymbol{\varepsilon}(\dot{\boldsymbol{u}}(t)))_Q + j(\boldsymbol{v}) - j(\dot{\boldsymbol{u}}(t)) \\ &\qquad \geq L(t, \boldsymbol{v} - \dot{\boldsymbol{u}}(t))_V \quad \forall\, \boldsymbol{v} \in U_1,\end{aligned} \tag{18.29}$$

and

$$\boldsymbol{u}(0) = \boldsymbol{u}_0, \quad \boldsymbol{\sigma}(0) = \boldsymbol{\sigma}_0. \tag{18.30}$$

We have the following existence and uniqueness result.

Theorem 18.5 *Assume* (6.44), (6.45), (18.22)–(18.24) *and* (18.27). *Then there exists a unique solution to Problem* 18.4. *Moreover, the solution satisfies*

$$\boldsymbol{u} \in H^1(I; U_1), \quad \boldsymbol{\sigma} \in H^1(I; Q_1). \tag{18.31}$$

In the following, we assume the conditions stated in Theorem 18.5. The proof of Theorem 18.5 is carried out in several steps. In the first step, we assume that the viscoplastic part of the stress field is a known function $\boldsymbol{\eta} \in L^2(I; Q)$. We define the function $\boldsymbol{z}_\eta \in H^1(I; Q)$ by (17.26), (17.27):

$$\begin{aligned}\boldsymbol{z}_\eta(t) &= \int_0^t \boldsymbol{\eta}(s)\, ds + \boldsymbol{z}_0, \\ \boldsymbol{z}_0 &= \boldsymbol{\sigma}_0 - \mathcal{E}\boldsymbol{\varepsilon}(\boldsymbol{u}_0).\end{aligned}$$

We consider the following auxiliary variational problem.

Problem 18.6 *Find $\boldsymbol{u}_\eta : \overline{I} \to U_1$ and $\boldsymbol{\sigma}_\eta : \overline{I} \to Q_1$ such that*

$$\boldsymbol{\sigma}_\eta(t) = \mathcal{E}\boldsymbol{\varepsilon}(\boldsymbol{u}_\eta(t)) + \boldsymbol{z}_\eta(t) \quad \forall\, t \in \overline{I}, \tag{18.32}$$

$$\begin{aligned}&(\boldsymbol{\sigma}_\eta(t), \boldsymbol{\varepsilon}(\boldsymbol{v}) - \boldsymbol{\varepsilon}(\dot{\boldsymbol{u}}_\eta(t)))_Q + j(\boldsymbol{v}) - j(\dot{\boldsymbol{u}}_\eta(t)) \geq L(t, \boldsymbol{v} - \dot{\boldsymbol{u}}(t))_V \\ &\qquad \forall\, \boldsymbol{v} \in U_1, \quad \text{a.e. } t \in I,\end{aligned} \tag{18.33}$$

$$\boldsymbol{u}_\eta(0) = \boldsymbol{u}_0, \quad \boldsymbol{\sigma}_\eta(0) = \boldsymbol{\sigma}_0. \tag{18.34}$$

We have the following result.

Lemma 18.7 *The variational problem* 18.6 *has a unique solution* $(\boldsymbol{u}_\eta, \boldsymbol{\sigma}_\eta)$ *such that* $\boldsymbol{u}_\eta \in H^1(I; U_1)$ *and* $\boldsymbol{\sigma}_\eta \in H^1(I; Q_1)$.

Proof. For $\boldsymbol{u}$, $\boldsymbol{v} \in V$, $(\boldsymbol{u}, \boldsymbol{v})_{\mathcal{E}} = (\mathcal{E}\boldsymbol{\varepsilon}(\boldsymbol{u}), \boldsymbol{\varepsilon}(\boldsymbol{v}))_Q$ defines an inner product on U_1 and its induced norm $\|\cdot\|_{\mathcal{E}}$ is equivalent to the norm $\|\cdot\|_V$ on U_1. Therefore, by the assumptions (18.22) on U_1, $(U_1, \|\cdot\|_V)$ is a real Hilbert space.

Let $\boldsymbol{f}_\eta(t)$, $t \in \overline{I}$, be the element of U_1 given by

$$(\boldsymbol{f}_\eta(t), \boldsymbol{v})_{\mathcal{E}} = L(t, \boldsymbol{v}) - (\boldsymbol{z}_\eta(t), \boldsymbol{\varepsilon}(\boldsymbol{v}))_Q. \tag{18.35}$$

Then, (18.25), (18.24) and (17.26) imply that $\boldsymbol{f}_\eta \in H^1(I; U_1)$. Moreover, using (18.35), (17.26) and (17.27) we deduce that

$$(\boldsymbol{f}_\eta(0), \boldsymbol{v})_V - (\boldsymbol{u}_0, \boldsymbol{v})_V - j(\boldsymbol{v}) = L(0, \boldsymbol{v}) - (\boldsymbol{\sigma}_0, \boldsymbol{\varepsilon}(\boldsymbol{v}))_Q - j(\boldsymbol{v}),$$

for all $\boldsymbol{v} \in D(j)$. Using the assumption (18.27), we obtain

$$(\boldsymbol{f}_\eta(0), \boldsymbol{v})_V - (\boldsymbol{u}_0, \boldsymbol{v})_V - j(\boldsymbol{v}) \le 0 \quad \forall \boldsymbol{v} \in D(j). \tag{18.36}$$

With the conditions (18.36), (18.23) and (18.27) being valid, we can use Theorem 13.17 (page 299) to deduce the existence of a unique function $\boldsymbol{u}_\eta \in H^1(I; U_1)$ such that $\boldsymbol{u}_\eta(0) = \boldsymbol{u}_0$ and

$$\begin{aligned}(\mathcal{E}\boldsymbol{\varepsilon}(\boldsymbol{u}_\eta(t)), \boldsymbol{\varepsilon}(\boldsymbol{v}) - \boldsymbol{\varepsilon}(\dot{\boldsymbol{u}}_\eta(t)))_Q + (\boldsymbol{z}_\eta(t), \boldsymbol{v} - \dot{\boldsymbol{u}}_\eta(t))_Q + j(\boldsymbol{v})\\ - j(\dot{\boldsymbol{u}}_\eta(t)) \ge L(t, \boldsymbol{v} - \dot{\boldsymbol{u}}_\eta(t)) \quad \forall \boldsymbol{v} \in U_1,\ t \in \overline{I}.\end{aligned}$$

Let $\boldsymbol{\sigma}_\eta$ be given by (18.32). Then the above inequality is identical to (18.33). By choosing $\boldsymbol{v} = \boldsymbol{u}_\eta(t) \pm \boldsymbol{\phi}$ with $\boldsymbol{\phi} \in C_0^\infty(\Omega)^d$ as test functions in (18.33), we get

$$\operatorname{Div} \boldsymbol{\sigma}_\eta + \boldsymbol{f}_0 = \boldsymbol{0} \quad \text{a.e. in } \Omega \times I.$$

Therefore, by the assumption (18.24) we see that $\boldsymbol{\sigma}_\eta \in H^1(I; Q_1)$. Moreover, since $\boldsymbol{u}_\eta(0) = \boldsymbol{u}_0$ and $\boldsymbol{z}_\eta(0) = \boldsymbol{\sigma}_0 - \mathcal{E}\boldsymbol{\varepsilon}(\boldsymbol{u}_0)$, it follows from (18.32) that $\boldsymbol{\sigma}_\eta(0) = \boldsymbol{\sigma}_0$. This concludes the existence part of Lemma 18.7. The uniqueness part follows from the uniqueness of the element $\boldsymbol{u}_\eta \in H^1(I; U_1)$, guaranteed by Theorem 13.17. ■

We now consider an operator $\Lambda : L^2(I; Q) \to L^2(I; Q)$ defined by

$$\Lambda \boldsymbol{\eta} = \mathcal{G}(\boldsymbol{\sigma}_\eta, \boldsymbol{\varepsilon}(\boldsymbol{u}_\eta)) \quad \forall \boldsymbol{\eta} \in L^2(I; Q), \tag{18.37}$$

where for any $\boldsymbol{\eta} \in L^2(I; Q)$, $(\boldsymbol{u}_\eta, \boldsymbol{\sigma}_\eta)$ denotes the solution of the variational problem 18.6. We have the following result.

Lemma 18.8 *The operator* Λ *has a unique fixed point* $\boldsymbol{\eta}^* \in L^2(I; Q)$.

Proof. Let $\boldsymbol{\eta}_1, \boldsymbol{\eta}_2 \in L^2(I;Q)$ and $t \in \overline{I}$. For the sake of simplicity we denote $\boldsymbol{z}_1 = \boldsymbol{z}_{\eta_1}$, $\boldsymbol{z}_2 = \boldsymbol{z}_{\eta_2}$, $\boldsymbol{u}_1 = \boldsymbol{u}_{\eta_1}$, $\boldsymbol{u}_2 = \boldsymbol{u}_{\eta_2}$, $\boldsymbol{\sigma}_1 = \boldsymbol{\sigma}_{\eta_1}$, $\boldsymbol{\sigma}_2 = \boldsymbol{\sigma}_{\eta_2}$. Using the defining relations (18.32) and (18.33) of Problem 18.6, we obtain

$$(\boldsymbol{u}_1, \boldsymbol{v} - \dot{\boldsymbol{u}}_1)_{\mathcal{E}} + (\boldsymbol{z}_1, \boldsymbol{\varepsilon}(\boldsymbol{v}) - \boldsymbol{\varepsilon}(\dot{\boldsymbol{u}}_1))_Q + j(\boldsymbol{v}) - j(\dot{\boldsymbol{u}}_1) \geq L(\cdot; \boldsymbol{v} - \dot{\boldsymbol{u}}_1),$$
$$(\boldsymbol{u}_2, \boldsymbol{v} - \dot{\boldsymbol{u}}_2)_{\mathcal{E}} + (\boldsymbol{z}_2, \boldsymbol{\varepsilon}(\boldsymbol{v}) - \boldsymbol{\varepsilon}(\dot{\boldsymbol{u}}_2))_Q + j(\boldsymbol{v}) - j(\dot{\boldsymbol{u}}_2) \geq L(\cdot; \boldsymbol{v} - \dot{\boldsymbol{u}}_2),$$

for all $\boldsymbol{v} \in U_1$, a.e. on I. It follows from these inequalities, after similar arguments as those used in the proof of Lemma 17.7, that

$$\|\Lambda\boldsymbol{\eta}_1(t) - \Lambda\boldsymbol{\eta}_2(t)\|_Q^2 \leq c \int_0^t \|\boldsymbol{\eta}_1(s) - \boldsymbol{\eta}_2(s)\|_Q^2 ds.$$

Iterating this inequality n times we obtain

$$\|\Lambda^n\boldsymbol{\eta}_1 - \Lambda^n\boldsymbol{\eta}_2\|_{L^2(I;Q)}^2 \leq \frac{c^n T^n}{n!}\|\boldsymbol{\eta}_1 - \boldsymbol{\eta}_2\|_{L^2(I;Q)}^2.$$

Since $c^n T^n/n! \to 0$ as $n \to \infty$, for n large enough, Λ^n is a contractive mapping on $L^2(I;Q)$. Thus, Λ has a unique fixed point in $L^2(I;Q)$. ∎

We now prove Theorem 18.5.

Proof of Theorem 18.5
Existence. Let $\boldsymbol{\eta}^* \in L^2(I;Q)$ be the fixed point of Λ. We show that $(\boldsymbol{u}_{\eta^*}, \boldsymbol{\sigma}_{\eta^*})$ is a solution for Problem 18.4. Indeed, the equality (18.28) follows from (18.32), (17.26) and (18.37), since

$$\dot{\boldsymbol{\sigma}}_{\eta^*}(t) = \mathcal{E}\boldsymbol{\varepsilon}(\dot{\boldsymbol{u}}_{\eta^*}(t)) + \dot{\boldsymbol{z}}_{\eta^*}(t),$$
$$\dot{\boldsymbol{z}}_{\eta^*}(t) = \boldsymbol{\eta}^*(t) = \Lambda\boldsymbol{\eta}^*(t) = \mathcal{G}(\boldsymbol{\sigma}_{\eta^*}(t), \boldsymbol{\varepsilon}(\boldsymbol{u}_{\eta^*}(t))),$$

a.e. $t \in I$. Inequality (18.29) follows from (18.33), and (18.30) follows from (18.34).
Uniqueness. To prove the uniqueness, let $(\boldsymbol{u}_{\eta^*}, \boldsymbol{\sigma}_{\eta^*})$ be the solution of Problem 18.4 obtained above and let $(\boldsymbol{u}, \boldsymbol{\sigma})$ be any other solution of Problem 18.4 such that $\boldsymbol{u} \in H^1(I;U_1)$, $\boldsymbol{\sigma} \in H^1(I;Q_1)$. Define $\boldsymbol{\eta} \in L^2(I;Q)$ by

$$\boldsymbol{\eta} = \mathcal{G}(\boldsymbol{\sigma}, \boldsymbol{\varepsilon}(\boldsymbol{u})), \tag{18.38}$$

and let $\boldsymbol{z}_\eta \in H^1(I;Q)$ be given by (17.26), (17.27). It follows from (18.28)–(18.30) that $(\boldsymbol{u}, \boldsymbol{\sigma})$ satisfies the variational problem (18.32)–(18.34). Since this problem has a unique solution $(\boldsymbol{u}_\eta, \boldsymbol{\sigma}_\eta)$ with regularity $\boldsymbol{u}_\eta \in H^1(I;U_1)$, $\boldsymbol{\sigma}_\eta \in H^1(I;Q_1)$, we have

$$\boldsymbol{u} = \boldsymbol{u}_\eta, \quad \boldsymbol{\sigma} = \boldsymbol{\sigma}_\eta. \tag{18.39}$$

Using now (18.37)–(18.39) we get $\Lambda\boldsymbol{\eta} = \boldsymbol{\eta}$. By the uniqueness of the fixed point of Λ we have

$$\boldsymbol{\eta} = \boldsymbol{\eta}^*.$$

Thus, $\boldsymbol{u} = \boldsymbol{u}_{\eta^*}$, $\boldsymbol{\sigma} = \boldsymbol{\sigma}_{\eta^*}$. ■

We remind the reader that numerous examples of contact and friction laws were presented in Section 13.4, which lead to an inequality of the form (18.18) such that (18.22) and (18.23) hold. So by Theorem 18.5, the corresponding boundary value contact problems for the viscoplastic materials for each of the examples have a unique weak solution.

18.3 Stress Formulation in Perfect Plasticity

In this section we consider the problem of contact with a dissipative frictional potential in the case of perfectly plastic materials. We first derive a variational formulation for the model in terms of the stress, then we establish an existence and uniqueness result.

The physical setting is similar to that in the previous section, except the viscoplastic constitutive law (18.14) is replaced by the perfectly plastic constitutive law (6.52) (page 137). More precisely, the classical formulation of the mechanical problem is the following:

Find a displacement field $\boldsymbol{u} : \Omega \times \overline{I} \to \mathbb{R}^d$ and a stress field $\boldsymbol{\sigma} : \Omega \times \overline{I} \to \mathbb{S}^d$ such that

$$\mathcal{A}\dot{\boldsymbol{\sigma}} + \partial\psi_K(\boldsymbol{\sigma}) \ni \boldsymbol{\varepsilon}(\dot{\boldsymbol{u}}) \quad \text{in } \Omega \times I, \tag{18.40}$$

$$\operatorname{Div} \boldsymbol{\sigma} + \boldsymbol{f}_0 = \boldsymbol{0} \quad \text{in } \Omega \times I, \tag{18.41}$$

$$\boldsymbol{u} = \boldsymbol{0} \quad \text{on } \Gamma_1 \times I, \tag{18.42}$$

$$\boldsymbol{\sigma\nu} = \boldsymbol{f}_2 \quad \text{on } \Gamma_2 \times I, \tag{18.43}$$

$$\boldsymbol{u} \in U, \quad -\boldsymbol{\sigma\nu} \cdot (\boldsymbol{v} - \dot{\boldsymbol{u}}) \le \varphi(\boldsymbol{v}) - \varphi(\dot{\boldsymbol{u}}) \ \forall \boldsymbol{v} \in U \quad \text{on } \Gamma_3 \times I, \tag{18.44}$$

$$\boldsymbol{u}(0) = \boldsymbol{u}_0, \quad \boldsymbol{\sigma}(0) = \boldsymbol{\sigma}_0 \quad \text{in } \Omega. \tag{18.45}$$

Here (18.40) is the perfectly plastic constitutive law in which K is the set of elastic stresses, $\mathcal{A}$ is the tensor of elastic compliance and $\partial\psi_K$ denotes the subdifferential of the indicator function $\psi_K : \mathbb{S}^d \to (-\infty, +\infty]$. As in the previous section, the inequality (18.44) models the frictional contact condition in which U represents the set of admissible test functions and φ is a given function. At the end of this section we will present a number of concrete examples of contact and friction laws which may be cast into the form (18.44) by appropriate choices of φ and U satisfying the assumptions to be formulated below.

As in Section 18.2 we suppose that $U \subset H^1(\Omega)^d$, $\varphi : \Gamma_3 \times \mathbb{R}^d \to (-\infty, +\infty]$ and we use the definitions (18.20), (18.21) for U_1 and j. We assume that (18.22) and (18.23) hold. To have a formulation in terms of the stress, we assume j is non-negative and positively homogeneous:

$$0 \le j(\boldsymbol{v}) < +\infty, \quad j(\alpha \boldsymbol{v}) = \alpha j(\boldsymbol{v}) \quad \forall \boldsymbol{v} \in U_1, \ \alpha \ge 0. \tag{18.46}$$

Next, we make the following assumptions.

$$\left.\begin{array}{l}\text{(a) } \mathcal{A} : \Omega \times \mathbb{S}^d \to \mathbb{S}^d.\\ \text{(b) } \mathcal{A}_{ijkl} \in L^\infty(\Omega), \quad 1 \le i,j,k,l \le d.\\ \text{(c) } \mathcal{A}\boldsymbol{\sigma}\cdot\boldsymbol{\tau} = \boldsymbol{\sigma}\cdot\mathcal{A}\boldsymbol{\tau} \quad \forall\,\boldsymbol{\sigma},\boldsymbol{\tau} \in \mathbb{S}^d, \text{ a.e. in } \Omega.\\ \text{(d) } \exists\,\alpha > 0 \text{ s.t. } \mathcal{A}\boldsymbol{\sigma}\cdot\boldsymbol{\sigma} \ge \alpha\,\|\boldsymbol{\sigma}\|^2\ \forall\,\boldsymbol{\sigma} \in \mathbb{S}^d, \text{ a.e. in } \Omega.\end{array}\right\} \tag{18.47}$$

$$K \subset \mathbb{S}^d \text{ is a closed convex set such that } \mathbf{0} \in K. \tag{18.48}$$

$$\boldsymbol{f}_0 \in W^{1,\infty}(I;H), \quad \boldsymbol{f}_2 \in W^{1,\infty}(I;L^2(\Gamma_2)^d). \tag{18.49}$$

For all $t \in \overline{I}$, define $\boldsymbol{f}(t) \in V$ and $\boldsymbol{\chi}(t) \in Q$ by the formulas

$$(\boldsymbol{f}(t),\boldsymbol{v})_V = \int_\Omega \boldsymbol{f}_0(t)\cdot\boldsymbol{v}\,dx + \int_{\Gamma_2} \boldsymbol{f}_2(t)\cdot\boldsymbol{v}\,da \quad \forall\,\boldsymbol{v} \in U_1, \tag{18.50}$$

$$\boldsymbol{\chi}(t) = \boldsymbol{\varepsilon}(\boldsymbol{f}(t)). \tag{18.51}$$

Note that the assumptions (18.49) imply

$$\boldsymbol{\chi} \in W^{1,\infty}(I;Q) \tag{18.52}$$

We assume the initial stress satisfies

$$\boldsymbol{\sigma}_0 = \boldsymbol{\chi}(0). \tag{18.53}$$

Let $\mathcal{K}$ denote the set

$$\mathcal{K} = \{\,\boldsymbol{\tau} \in Q : \boldsymbol{\tau}(\boldsymbol{x}) \in K \text{ a.e. } \boldsymbol{x} \in \Omega\,\}. \tag{18.54}$$

We assume that there exists $\delta > 0$ such that

$$\boldsymbol{\chi}(t) + \boldsymbol{\tau} \in \mathcal{K} \quad \forall\,t \in \overline{I},\ \forall\,\boldsymbol{\tau} \in Q \text{ with } \|\boldsymbol{\tau}\|_Q \le \delta. \tag{18.55}$$

This is known as a "safe loading" assumption in the literature ([127, 156, 203]).

Let $\Sigma(t)$ be the subset of Q defined by

$$\Sigma(t) = \{\,\boldsymbol{\tau} \in Q : (\boldsymbol{\tau},\boldsymbol{\varepsilon}(\boldsymbol{v}))_Q + j(\boldsymbol{v}) \ge (\boldsymbol{f}(t),\boldsymbol{v})_V\ \forall\,\boldsymbol{v} \in U_1\,\}. \tag{18.56}$$

We now derive a variational formulation of the mechanical problem (18.40)–(18.45). To this end, we assume that $(\boldsymbol{u},\boldsymbol{\sigma})$ are regular functions satisfying (18.40)–(18.45). Let $\boldsymbol{v} \in U_1$ be arbitrary, $t \in \overline{I}$. Using standard arguments we obtain

$$(\boldsymbol{\sigma}(t),\boldsymbol{\varepsilon}(\boldsymbol{v}) - \boldsymbol{\varepsilon}(\dot{\boldsymbol{u}}(t)))_Q + j(\boldsymbol{v}) - j(\dot{\boldsymbol{u}}(t)) \ge (\boldsymbol{f}(t),\boldsymbol{v} - \dot{\boldsymbol{u}}(t))_V. \tag{18.57}$$

Due to the positive homogeneity of $j(\cdot)$, the inequality (18.57) is equivalent to the equality

$$(\boldsymbol{\sigma}(t),\boldsymbol{\varepsilon}(\dot{\boldsymbol{u}}(t)))_Q + j(\dot{\boldsymbol{u}}(t)) = (\boldsymbol{f}(t),\dot{\boldsymbol{u}}(t))_V \tag{18.58}$$

and the inequality

$$(\boldsymbol{\sigma}(t), \boldsymbol{\varepsilon}(\boldsymbol{v}))_Q + j(\boldsymbol{v}) \geq (\boldsymbol{f}(t), \boldsymbol{v})_V \quad \forall \boldsymbol{v} \in U_1.$$

Thus, in particular, $\boldsymbol{\sigma}(t) \in \Sigma(t)$. The subdifferential inclusion (18.40) implies that $\boldsymbol{\sigma}(t)$ belongs to $\mathcal{K}$ defined in (18.54). We conclude that

$$\boldsymbol{\sigma}(t) \in \mathcal{K} \cap \Sigma(t). \tag{18.59}$$

By the equality (18.58) and the definition (18.56) of $\Sigma(t)$, we find

$$(\boldsymbol{\tau} - \boldsymbol{\sigma}(t), \boldsymbol{\varepsilon}(\dot{\boldsymbol{u}}(t)))_Q \geq 0 \quad \forall \boldsymbol{\tau} \in \Sigma(t).$$

Now from the subdifferential inclusion (18.40) it follows that

$$(\mathcal{A}\dot{\boldsymbol{\sigma}}(t), \boldsymbol{\tau} - \boldsymbol{\sigma}(t))_Q \geq (\boldsymbol{\tau} - \boldsymbol{\sigma}(t), \boldsymbol{\varepsilon}(\dot{\boldsymbol{u}}(t)))_Q \quad \forall \boldsymbol{\tau} \in \mathcal{K}.$$

Therefore,

$$(\mathcal{A}\dot{\boldsymbol{\sigma}}(t), \boldsymbol{\tau} - \boldsymbol{\sigma}(t))_Q \geq 0 \quad \forall \boldsymbol{\tau} \in \mathcal{K} \cap \Sigma(t). \tag{18.60}$$

Combining (18.45), (18.59) and (18.60) we obtain the following variational formulation in terms of the stress field for the problem (18.40)–(18.45).

Problem 18.9 *Find a stress field* $\boldsymbol{\sigma} : \overline{I} \to Q_1$ *such that*

$$\boldsymbol{\sigma}(t) \in \mathcal{K} \cap \Sigma(t) \quad \forall t \in \overline{I}, \tag{18.61}$$

$$(\mathcal{A}\dot{\boldsymbol{\sigma}}(t), \boldsymbol{\tau} - \boldsymbol{\sigma}(t))_Q \geq 0 \quad \forall \boldsymbol{\tau} \in \mathcal{K} \cap \Sigma(t), \quad \text{a.e. } t \in I, \tag{18.62}$$

$$\boldsymbol{\sigma}(0) = \boldsymbol{\sigma}_0. \tag{18.63}$$

The main result in this section is the following.

Theorem 18.10 *Assume* (18.47)–(18.49), (18.53) *and* (18.55). *There exists a unique solution* $\boldsymbol{\sigma}$ *to Problem 18.9 with* $\boldsymbol{\sigma} \in H^1(I; Q)$, $\operatorname{Div} \boldsymbol{\sigma} \in W^{1,\infty}(I; H)$.

The proof of Theorem 18.10 is carried out in several steps. In the first step we consider a regularized version of Problem 18.9, which depends on a viscosity parameter. For $\mu > 0$, define an operator $\mathcal{G}_\mu : Q \to Q$ by

$$\mathcal{G}_\mu(\boldsymbol{\tau}) = \frac{1}{\mu}(\boldsymbol{\tau} - \mathcal{P}_\mathcal{K}\boldsymbol{\tau}), \quad \boldsymbol{\tau} \in Q, \tag{18.64}$$

where $\mathcal{P}_\mathcal{K} : Q \to \mathcal{K}$ is the orthogonal projection on the non-empty closed subset $\mathcal{K} \subset Q$.

We consider the following auxiliary problem.

Problem 18.11 *Find* $\boldsymbol{\sigma}_\mu : \overline{I} \to Q_1$ *such that*

$$\boldsymbol{\sigma}_\mu(t) \in \Sigma(t) \quad \forall t \in \overline{I}, \tag{18.65}$$

$$(\mathcal{A}\dot{\boldsymbol{\sigma}}_\mu(t), \boldsymbol{\tau} - \boldsymbol{\sigma}_\mu(t))_Q + (\mathcal{G}_\mu(\boldsymbol{\sigma}_\mu(t)), \boldsymbol{\tau} - \boldsymbol{\sigma}_\mu(t))_Q \geq 0$$
$$\forall \boldsymbol{\tau} \in \Sigma(t), \quad \text{a.e. } t \in I, \tag{18.66}$$

$$\boldsymbol{\sigma}_\mu(0) = \boldsymbol{\sigma}_0. \tag{18.67}$$

Clearly, Problem 18.11 represents the variational formulation in terms of the stress of the contact problem in which the constitutive law (18.40) is replaced by the regularized Perzyna law (6.47) (page 136) with $\mathcal{E}^{-1} = \mathcal{A}$.

We have the following existence and uniqueness result.

Lemma 18.12 *Problem 18.11 has a unique solution* $\boldsymbol{\sigma}_\mu \in H^1(I; Q_1)$.

Proof. Let $\boldsymbol{u}_0 \in U_1$. From the definition of $\boldsymbol{\sigma}_0$ (cf. (18.53) and (18.51)) it follows that $\boldsymbol{\sigma}_0 \in \Sigma(0)$. Then, we apply Theorem 18.5 to obtain the existence and the uniqueness of a couple of functions $(\boldsymbol{u}_\mu, \boldsymbol{\sigma}_\mu)$ such that

$$\boldsymbol{u}_\mu \in H^1(I; U_1), \quad \boldsymbol{\sigma}_\mu \in H^1(I; Q_1), \tag{18.68}$$

$$\boldsymbol{\varepsilon}(\dot{\boldsymbol{u}}_\mu(t)) = \mathcal{A}\dot{\boldsymbol{\sigma}}_\mu(t) + \mathcal{G}_\mu(\boldsymbol{\sigma}_\mu(t)) \quad \text{a.e. } t \in I, \tag{18.69}$$

$$\begin{aligned}&(\boldsymbol{\sigma}_\mu(t), \boldsymbol{\varepsilon}(\boldsymbol{v}) - \boldsymbol{\varepsilon}(\dot{\boldsymbol{u}}_\mu(t)))_Q + j(\boldsymbol{v}) - j(\dot{\boldsymbol{u}}_\mu(t)) \\ &\quad \geq (\boldsymbol{f}(t), \boldsymbol{v} - \dot{\boldsymbol{u}}_\mu(t))_V \quad \forall \boldsymbol{v} \in U_1, \text{ a.e. } t \in I,\end{aligned} \tag{18.70}$$

$$\boldsymbol{u}_\mu(0) = \boldsymbol{u}_0, \quad \boldsymbol{\sigma}_\mu(0) = \boldsymbol{\sigma}_0. \tag{18.71}$$

Due to the positive homogeneity of j, the variational inequality (18.70) is equivalent to

$$(\boldsymbol{\sigma}_\mu(t), \boldsymbol{\varepsilon}(\dot{\boldsymbol{u}}_\mu(t)))_Q + j(\dot{\boldsymbol{u}}_\mu(t)) = (\boldsymbol{f}(t), \dot{\boldsymbol{u}}_\mu(t))_V, \quad \text{a.e. } t \in I,$$
$$(\boldsymbol{\sigma}_\mu(t), \boldsymbol{\varepsilon}(\boldsymbol{v}))_Q + j(\boldsymbol{v}) \geq (\boldsymbol{f}(t), \boldsymbol{v})_V \quad \forall \boldsymbol{v} \in U_1, \text{ a.e. } t \in I.$$

Thus, $\boldsymbol{\sigma}_\mu(t) \in \Sigma(t)$ for all $t \in \overline{I}$ and

$$(\boldsymbol{\tau} - \boldsymbol{\sigma}_\mu(t), \boldsymbol{\varepsilon}(\dot{\boldsymbol{u}}_\mu))_Q \geq 0 \quad \forall \boldsymbol{\tau} \in \Sigma(t), \text{ a.e. } t \in I.$$

This inequality can be rewritten in the form (18.66) after we replace $\boldsymbol{\varepsilon}(\dot{\boldsymbol{u}}_\mu(t))$ by the right hand side of (18.69). Hence, $\boldsymbol{\sigma}_\mu$ is a solution of Problem 18.11.

The uniqueness part follows from standard arguments, by the assumptions (18.47) on $\mathcal{A}$ and the monotonicity of the operator $\mathcal{G}_\mu$. ■

We will then derive a priori estimates on the solution $\boldsymbol{\sigma}_\mu$ of Problem 18.11 and take the limit $\mu \to 0$. In the rest of this section we denote by c a strictly positive generic constant which does not depend on t or μ and whose value may change from place to place. We denote by $\mathcal{F}_\mu : Q \to \mathbb{R}_+$ the functional

$$\mathcal{F}_\mu(\boldsymbol{\tau}) = \frac{1}{2\mu}\|\boldsymbol{\tau} - \mathcal{P}_\mathcal{K}\boldsymbol{\tau}\|_Q^2, \quad \boldsymbol{\tau} \in Q. \tag{18.72}$$

Lemma 18.13 *The sequence* $\{\boldsymbol{\sigma}_\mu\}$ *is bounded in* $L^\infty(I; Q)$.

Proof. Let $\mu > 0$ and denote

$$\overline{\boldsymbol{\sigma}}_\mu = \boldsymbol{\sigma}_\mu - \boldsymbol{\chi}. \tag{18.73}$$

We take $\boldsymbol{\tau} = \boldsymbol{\chi}(t)$ in (18.66) to obtain

$$(\mathcal{A}\dot{\overline{\boldsymbol{\sigma}}}_\mu(t), \overline{\boldsymbol{\sigma}}_\mu(t))_Q + (\mathcal{G}_\mu(\boldsymbol{\sigma}_\mu(t)), \overline{\boldsymbol{\sigma}}_\mu(t))_Q \leq -(\mathcal{A}\dot{\boldsymbol{\chi}}(t), \overline{\boldsymbol{\sigma}}_\mu(t))_Q \quad \text{a.e. } t \in I. \tag{18.74}$$

Note that the operator $\mathcal{G}_\mu$ defined by (18.64) is the Gâteaux-derivative of the convex function $\mathcal{F}_\mu$ given by (18.72) (for a proof, see [96, Lemma 8.6]). Therefore,

$$\mathcal{F}_\mu(\boldsymbol{\tau}) - \mathcal{F}_\mu(\boldsymbol{\sigma}) \geq (\mathcal{G}_\mu(\boldsymbol{\sigma}), \boldsymbol{\tau} - \boldsymbol{\sigma})_Q \quad \forall\, \boldsymbol{\sigma}, \boldsymbol{\tau} \in Q. \tag{18.75}$$

By the safe loading assumption (18.55), $\boldsymbol{\chi}(t) \in \mathcal{K}$. Hence, $\mathcal{F}_\mu(\boldsymbol{\chi}(t)) = 0$ for all $t \in \overline{I}$. We choose $\boldsymbol{\tau} = \boldsymbol{\chi}(t)$, $\boldsymbol{\sigma} = \boldsymbol{\sigma}_\mu(t)$ in (18.75), and use the definition (18.73) of $\overline{\boldsymbol{\sigma}}_\mu$ to find

$$(\mathcal{G}_\mu(\boldsymbol{\sigma}_\mu(t)), \overline{\boldsymbol{\sigma}}_\mu(t))_Q \geq (\mathcal{F}_\mu(\boldsymbol{\sigma}_\mu(t)) \quad \forall\, t \in \overline{I}.$$

Then from (18.74) we have

$$(\mathcal{A}\dot{\overline{\boldsymbol{\sigma}}}_\mu(t), \overline{\boldsymbol{\sigma}}_\mu(t))_Q + \mathcal{F}_\mu(\boldsymbol{\sigma}_\mu(t)) \leq -(\mathcal{A}\dot{\boldsymbol{\chi}}(t), \overline{\boldsymbol{\sigma}}_\mu(t))_Q \quad \text{a.e. } t \in I. \tag{18.76}$$

Now $\mathcal{F}_\mu(\boldsymbol{\sigma}_\mu(t)) \geq 0$, we obtain

$$\frac{1}{2}\frac{d}{dt}(\mathcal{A}\overline{\boldsymbol{\sigma}}_\mu(t), \overline{\boldsymbol{\sigma}}_\mu(t))_Q \leq -(\mathcal{A}\dot{\boldsymbol{\chi}}(t), \overline{\boldsymbol{\sigma}}_\mu(t))_Q \quad \text{a.e. } t \in I.$$

Notice that $\overline{\boldsymbol{\sigma}}_\mu(0) = \mathbf{0}$, since $\boldsymbol{\sigma}_\mu(0) = \boldsymbol{\sigma}(0) = \boldsymbol{\chi}(0)$. Integrate the above inequality from $t = 0$ to $t = s$ and use the assumptions (18.47) on $\mathcal{A}$ to obtain

$$\begin{aligned}
\|\overline{\boldsymbol{\sigma}}_\mu(s)\|_Q^2 &\leq c \int_0^s \|\dot{\boldsymbol{\chi}}(t)\|_Q \|\overline{\boldsymbol{\sigma}}_\mu(t)\|_Q \, dt \\
&\leq c \int_0^s \|\dot{\boldsymbol{\chi}}(t)\|_Q^2 dt + c \int_0^s \|\overline{\boldsymbol{\sigma}}_\mu(t)\|_Q^2 dt \quad \forall\, s \in \overline{I}.
\end{aligned}$$

By the Gronwall inequality, we obtain

$$\|\overline{\boldsymbol{\sigma}}_\mu\|_{L^\infty(I;Q)} \leq c\, \|\dot{\boldsymbol{\chi}}\|_{L^2(I;Q)}.$$

Thus, $\{\overline{\boldsymbol{\sigma}}_\mu\}$ and then $\{\boldsymbol{\sigma}_\mu\} = \{\overline{\boldsymbol{\sigma}}_\mu + \boldsymbol{\chi}\}$ is bounded in $L^\infty(I;Q)$. ∎

Lemma 18.14 *The sequence $\{\mathcal{G}_\mu(\boldsymbol{\sigma}_\mu)\}$ is bounded in $L^1(I;Q)$.*

Proof. We integrate (18.74) on I,

$$\int_I (\mathcal{A}\dot{\overline{\boldsymbol{\sigma}}}_\mu(t), \overline{\boldsymbol{\sigma}}_\mu(t))_Q dt + \int_I (\mathcal{G}_\mu(\boldsymbol{\sigma}_\mu(t)), \overline{\boldsymbol{\sigma}}_\mu(t))_Q dt \leq -\int_I (\mathcal{A}\dot{\boldsymbol{\chi}}(t), \overline{\boldsymbol{\sigma}}_\mu(t))_Q dt.$$

Now

$$\begin{aligned}
\int_I (\mathcal{A}\dot{\overline{\boldsymbol{\sigma}}}_\mu(t), \overline{\boldsymbol{\sigma}}_\mu(t))_Q dt &= \frac{1}{2}\,(\mathcal{A}\overline{\boldsymbol{\sigma}}_\mu(T), \overline{\boldsymbol{\sigma}}_\mu(T))_Q - \frac{1}{2}\,(\mathcal{A}\overline{\boldsymbol{\sigma}}_\mu(0), \overline{\boldsymbol{\sigma}}_\mu(0))_Q \\
&= \frac{1}{2}\,(\mathcal{A}\overline{\boldsymbol{\sigma}}_\mu(T), \overline{\boldsymbol{\sigma}}_\mu(T))_Q \\
&\geq 0,
\end{aligned}$$

and

$$-\int_I (\mathcal{A}\dot{\boldsymbol{\chi}}(t), \overline{\boldsymbol{\sigma}}_\mu(t))_Q dt \le c\,\|\dot{\boldsymbol{\chi}}\|_{L^2(I;Q)} \|\overline{\boldsymbol{\sigma}}_\mu\|_{L^2(I;Q)}.$$

So by the boundedness of $\{\|\overline{\boldsymbol{\sigma}}_\mu\|_{L^2(I;Q)}\}$ from Lemma 18.13, we obtain

$$\int_I (\mathcal{G}_\mu(\boldsymbol{\sigma}_\mu(t)), \overline{\boldsymbol{\sigma}}_\mu(t))_Q \, dt \le c. \tag{18.77}$$

Let $\boldsymbol{\tau} \in Q$ such that $\|\boldsymbol{\tau}\|_Q \le \delta$. By the safe loading assumption (18.55) we have $\mathcal{F}_\mu(\boldsymbol{\chi}(t) + \boldsymbol{\tau}) = 0$ for all $t \in \overline{I}$. Then applying the inequality (18.75) with $\boldsymbol{\tau}$ and $\boldsymbol{\sigma}$ replaced by $\boldsymbol{\chi}(t) + \boldsymbol{\tau}$ and $\boldsymbol{\sigma}_\mu(t)$, respectively, we obtain

$$(\mathcal{G}_\mu(\boldsymbol{\sigma}_\mu(t)), \boldsymbol{\tau})_Q \le (\mathcal{G}_\mu(\boldsymbol{\sigma}_\mu(t)), \overline{\boldsymbol{\sigma}}_\mu(t))_Q \quad \forall\, t \in \overline{I}.$$

Thus for all $t \in \overline{I}$,

$$\begin{aligned} \|\mathcal{G}_\mu(\boldsymbol{\sigma}_\mu(t))\|_Q &= \frac{1}{\delta} \sup_{\|\boldsymbol{\tau}\|_Q \le \delta} (\mathcal{G}_\mu(\boldsymbol{\sigma}_\mu(t)), \boldsymbol{\tau})_Q \\ &\le \frac{1}{\delta} (\mathcal{G}_\mu(\boldsymbol{\sigma}_\mu(t)), \overline{\boldsymbol{\sigma}}_\mu(t))_Q. \end{aligned}$$

Then by (18.77) we see that $\{\mathcal{G}_\mu(\boldsymbol{\sigma}_\mu)\}$ is bounded in $L^1(I;Q)$. ■

Lemma 18.15 *The sequence $\{\dot{\boldsymbol{\sigma}}_\mu\}$ is bounded in $L^2(I;Q)$.*

Proof. We introduce the set

$$\Sigma_0 = \{\, \boldsymbol{\tau} \in Q : \ (\boldsymbol{\tau}, \boldsymbol{\varepsilon}(\boldsymbol{v}))_Q + j(\boldsymbol{v}) \ge 0 \ \forall\, \boldsymbol{v} \in U_1 \,\}.$$

Then we can write

$$\Sigma(t) = \Sigma_0 + \{\boldsymbol{\chi}(t)\} \quad \forall\, t \in \overline{I}.$$

Let $\boldsymbol{\tau} \in \Sigma_0$ be arbitrary and use $\boldsymbol{\tau} + \boldsymbol{\chi}(t)$ as the test function in (18.66) to obtain

$$\begin{aligned} (\mathcal{A}\dot{\overline{\boldsymbol{\sigma}}}_\mu(t), \boldsymbol{\tau} - \overline{\boldsymbol{\sigma}}_\mu(t))_Q &+ (\mathcal{A}\dot{\boldsymbol{\chi}}(t), \boldsymbol{\tau} - \overline{\boldsymbol{\sigma}}_\mu(t))_Q \\ &+ (\mathcal{G}_\mu \boldsymbol{\sigma}_\mu(t), \boldsymbol{\tau} - \overline{\boldsymbol{\sigma}}_\mu(t))_Q \ge 0 \quad \text{a.e. } t \in I, \end{aligned} \tag{18.78}$$

where $\dot{\boldsymbol{\sigma}}_\mu(t)$ was replaced by $\dot{\overline{\boldsymbol{\sigma}}}_\mu(t) + \dot{\boldsymbol{\chi}}(t)$. Since $\overline{\boldsymbol{\sigma}}_\mu(t) \in \Sigma_0 \ \forall\, t \in \overline{I}$, we can take $\boldsymbol{\tau} = \overline{\boldsymbol{\sigma}}_\mu(t \pm h)$ in (18.78) with $h > 0$. Divide the resulting inequalities by h and pass to the limit $h \to 0$. We obtain

$$\begin{aligned} (\mathcal{A}\dot{\overline{\boldsymbol{\sigma}}}_\mu(t), \dot{\overline{\boldsymbol{\sigma}}}_\mu(t))_Q &+ (\mathcal{A}\dot{\boldsymbol{\chi}}(t), \dot{\overline{\boldsymbol{\sigma}}}_\mu(t))_Q \\ &+ (\mathcal{G}_\mu(\boldsymbol{\sigma}_\mu(t)), \dot{\overline{\boldsymbol{\sigma}}}_\mu(t))_Q = 0 \quad \text{a.e. } t \in I. \end{aligned}$$

We integrate this equality over I. Using the regularity $\boldsymbol{\chi} \in W^{1,\infty}(I;Q)$ and the boundedness of $\{\|\mathcal{G}_\mu(\boldsymbol{\sigma}_\mu)\|_{L^1(I;Q)}\}$ from Lemma 18.14, we have

$$\|\dot{\overline{\boldsymbol{\sigma}}}_\mu\|^2_{L^2(I;Q)} + \int_I (\mathcal{G}_\mu(\boldsymbol{\sigma}_\mu(t)), \dot{\boldsymbol{\sigma}}_\mu(t))_Q \, dt \le c + c\, \|\dot{\overline{\boldsymbol{\sigma}}}_\mu\|_{L^2(I;Q)}. \tag{18.79}$$

Since $\mathcal{G}_\mu$ is the Gâteaux-derivative of the convex function $\mathcal{F}_\mu$,

$$\begin{aligned}\int_I (\mathcal{G}_\mu(\boldsymbol{\sigma}_\mu(t)), \dot{\boldsymbol{\sigma}}_\mu(t))_Q \, dt &= \int_I \frac{d}{dt} \mathcal{F}_\mu(\boldsymbol{\sigma}_\mu(t)) \, dt \\ &= \mathcal{F}_\mu(\boldsymbol{\sigma}_\mu(T)) - \mathcal{F}_\mu(\boldsymbol{\sigma}_\mu(0)).\end{aligned}$$

Notice that $\boldsymbol{\sigma}_\mu(0) = \boldsymbol{\sigma}_0 \in \mathcal{K}$ by the safe loading assumption (18.55) and therefore we have $\mathcal{F}_\mu(\boldsymbol{\sigma}_\mu(0)) = 0$. Thus,

$$\int_I (\mathcal{G}_\mu(\boldsymbol{\sigma}_\mu(t)), \dot{\boldsymbol{\sigma}}_\mu(t))_Q \, dt = \mathcal{F}_\mu(\boldsymbol{\sigma}_\mu(T)) \ge 0.$$

The inequality (18.79) then implies

$$\|\dot{\overline{\boldsymbol{\sigma}}}_\mu\|^2_{L^2(I;Q)} \le c + c\, \|\dot{\overline{\boldsymbol{\sigma}}}_\mu\|_{L^2(I;Q)}.$$

Therefore, $\{\|\dot{\overline{\boldsymbol{\sigma}}}_\mu\|_{L^2(I;Q)}\}$ is bounded, and since $\dot{\boldsymbol{\sigma}}_\mu = \dot{\overline{\boldsymbol{\sigma}}}_\mu + \dot{\boldsymbol{\chi}}$, $\{\|\dot{\boldsymbol{\sigma}}_\mu\|_{L^2(I;Q)}\}$ is also bounded. ■

We now prove Theorem 18.10.

Proof of Theorem 18.10

Existence. By Lemmas 18.13 and 18.15 we deduce that there exists a subsequence of $\{\boldsymbol{\sigma}_\mu\}$, still denoted by $\{\boldsymbol{\sigma}_\mu\}$, and an element $\boldsymbol{\sigma} \in H^1(I;Q)$ such that as $\mu \longrightarrow 0$ we have

$$\boldsymbol{\sigma}_\mu \rightharpoonup^* \boldsymbol{\sigma} \quad \text{in } L^\infty(I;Q), \tag{18.80}$$

$$\dot{\boldsymbol{\sigma}}_\mu \rightharpoonup \dot{\boldsymbol{\sigma}} \quad \text{in } L^2(I;Q). \tag{18.81}$$

Let us show that $\boldsymbol{\sigma}$ is the solution of Problem 18.9. First, we note that (18.80) and (18.81) imply, for a further subsequence $\{\boldsymbol{\sigma}_\mu\}$ if necessary,

$$\boldsymbol{\sigma}_\mu(t) \to \boldsymbol{\sigma}(t) \quad \text{strongly in } Q, \ \forall\, t \in \overline{I}. \tag{18.82}$$

Since $\boldsymbol{\sigma}_\mu(t) \in \Sigma(t)$ for all $t \in \overline{I}$, we have

$$\boldsymbol{\sigma}(t) \in \Sigma(t) \quad \forall\, t \in \overline{I}.$$

We derive from (18.75) that

$$\mathcal{F}_\mu(\boldsymbol{\sigma}(t)) \le \mathcal{F}_\mu(\boldsymbol{\sigma}_\mu(t)) - (\mathcal{G}_\mu(\boldsymbol{\sigma}(t)), \boldsymbol{\sigma}_\mu(t) - \boldsymbol{\sigma}(t))_Q, \quad \forall\, t \in \overline{I}.$$

We integrate the inequality over I and use the definitions of $\mathcal{F}_\mu$ and $\mathcal{G}_\mu$ to obtain

$$\int_I \|\boldsymbol{\sigma}(t) - \mathcal{P}_{\mathcal{K}}\boldsymbol{\sigma}(t)\|_Q^2 dt \le \mu \int_I \mathcal{F}_\mu(\boldsymbol{\sigma}_\mu(t))\, dt \\ - \int_I (\boldsymbol{\sigma}(t) - \mathcal{P}_{\mathcal{K}}\boldsymbol{\sigma}(t), \boldsymbol{\sigma}_\mu(t) - \boldsymbol{\sigma}(t))_Q\, dt. \qquad (18.83)$$

We now integrate (18.76) over I and use the boundedness of $\{\|\boldsymbol{\sigma}_\mu\|_{L^\infty(I;Q)}\}$ to obtain

$$\int_I \mathcal{F}_\mu(\boldsymbol{\sigma}_\mu(t))\, dt \le c.$$

So in the limit $\mu \to 0+$, the first term on the right hand side of (18.83) tends to 0, and so does the second term because of the weak convergence $\boldsymbol{\sigma}_\mu \rightharpoonup \boldsymbol{\sigma}$ in $L^2(I;Q)$. Thus, from (18.83) we obtain

$$\int_I \|\boldsymbol{\sigma}(t) - \mathcal{P}_{\mathcal{K}}\boldsymbol{\sigma}(t)\|_Q^2 dt = 0.$$

Therefore, $\boldsymbol{\sigma}(t) = \mathcal{P}_{\mathcal{K}}\boldsymbol{\sigma}(t) \in \mathcal{K}$ a.e. $t \in \overline{I}$. Since $\boldsymbol{\sigma}$ is continuous in t,

$$\boldsymbol{\sigma}(t) \in \mathcal{K} \quad \forall\, t \in \overline{I}.$$

Hence (18.61) is satisfied.

Second, using (18.66) and (18.75) we have

$$(\mathcal{A}\dot{\boldsymbol{\sigma}}_\mu(t), \boldsymbol{\tau} - \boldsymbol{\sigma}_\mu(t))_Q + \mathcal{F}_\mu(\boldsymbol{\tau}) - \mathcal{F}_\mu(\boldsymbol{\sigma}_\mu(t)) \ge 0 \quad \forall\, \boldsymbol{\tau} \in \Sigma(t), \text{ a.e. } t \in I.$$

Since $\mathcal{F}_\mu(\boldsymbol{\sigma}_\mu(t)) \ge 0$, we then have

$$(\mathcal{A}\dot{\boldsymbol{\sigma}}_\mu(t), \boldsymbol{\tau} - \boldsymbol{\sigma}_\mu(t))_Q \ge 0 \quad \forall\, \boldsymbol{\tau} \in \mathcal{K} \cap \Sigma(t), \text{ a.e. } t \in I. \qquad (18.84)$$

Let $\boldsymbol{\tau} = \boldsymbol{\tau}(t)$ in (18.84) with elements $\boldsymbol{\tau} \in L^2(I;Q)$ such that $\boldsymbol{\tau} \in \mathcal{K} \cap \Sigma(t)$ a.e. $t \in I$, and integrate the result over I to obtain

$$\int_I (\mathcal{A}\dot{\boldsymbol{\sigma}}_\mu(t), \boldsymbol{\tau}(t))_Q dt \ge \int_I (\mathcal{A}\dot{\boldsymbol{\sigma}}_\mu(t), \boldsymbol{\sigma}_\mu(t))_Q dt.$$

Letting $\mu \to 0+$ in this inequality and using the weak convergence (18.80), (18.81), by a lower-semicontinuity argument we deduce that

$$\int_I (\mathcal{A}\dot{\boldsymbol{\sigma}}(t), \boldsymbol{\tau})_Q\, dt \ge \int_I (\mathcal{A}\dot{\boldsymbol{\sigma}}(t), \boldsymbol{\sigma}(t))_Q\, dt.$$

We can then apply Lebesgue's Theorem 2.23 (page 38) to localize the previous inequality to deduce (18.62).

Finally, since $\boldsymbol{\sigma}_\mu(0) = \boldsymbol{\sigma}_0$, by the convergence (18.82), we get the initial condition (18.63).

We conclude that $\boldsymbol{\sigma} \in H^1(I;Q)$ is a solution of Problem 18.9. Notice that $\boldsymbol{\sigma}(t) \in \Sigma(t)$ implies $\operatorname{Div} \boldsymbol{\sigma}(t) = -\boldsymbol{f}_0(t)$ a.e. in Ω, for all $t \in \overline{I}$. With the regularity (18.49) on $\boldsymbol{f}_0$, we find $\operatorname{Div} \boldsymbol{\sigma} \in W^{1,\infty}(I, H)$.
Uniqueness. To prove uniqueness, let $\{\boldsymbol{\sigma}_i\}$, $i = 1, 2$, be two solutions of Problem 18.9, then

$$
\begin{aligned}
&(\mathcal{A}\dot{\boldsymbol{\sigma}}_1(t), \boldsymbol{\tau} - \boldsymbol{\sigma}_1(t))_Q \geq 0, \\
&(\mathcal{A}\dot{\boldsymbol{\sigma}}_2(t), \boldsymbol{\tau} - \boldsymbol{\sigma}_2(t))_Q \geq 0
\end{aligned}
$$

for all $\boldsymbol{\tau} \in \mathcal{K} \cap \Sigma(t)$, a.e. $t \in I$. We take $\boldsymbol{\tau} = \boldsymbol{\sigma}_2(t)$ in the first inequality, $\boldsymbol{\tau} = \boldsymbol{\sigma}_1(t)$ in the second inequality and add the two inequalities to obtain

$$
\begin{aligned}
\frac{1}{2}\frac{d}{dt}&(\mathcal{A}(\boldsymbol{\sigma}_1(t) - \boldsymbol{\sigma}_2(t)), \boldsymbol{\sigma}_1(t) - \boldsymbol{\sigma}_2(t))_Q \\
&= (\mathcal{A}\dot{\boldsymbol{\sigma}}_1(t) - \mathcal{A}\dot{\boldsymbol{\sigma}}_2(t), \boldsymbol{\sigma}_1(t) - \boldsymbol{\sigma}_2(t))_Q \\
&\leq 0
\end{aligned}
$$

a.e. $t \in I$. Integrate the inequality over $[0, s]$ and use the initial condition to obtain

$$
(\mathcal{A}(\boldsymbol{\sigma}_1(s) - \boldsymbol{\sigma}_2(s)), \boldsymbol{\sigma}_1(s) - \boldsymbol{\sigma}_2(s))_Q \leq 0 \quad \forall\, s \in \overline{I}.
$$

Thus, $\boldsymbol{\sigma}_1 = \boldsymbol{\sigma}_2$ and the solution is unique. ∎

We end this section by presenting a number of friction laws which lead to inequalities of the form (18.44) and for which conditions (18.22), (18.23) and (18.46) hold. So Theorem 18.10 can be applied to the weak formulations in terms of the stress for the corrsponding contact boundary value problems for perfectly plastic materials.

Example 18.16 *Bilateral contact with Tresca's friction law.*

We assume the contact is maintained on the contact boundary and the friction bound is prescribed,

$$
\left.
\begin{aligned}
&u_\nu = 0, \quad \|\boldsymbol{\sigma}_\tau\| \leq g, \\
&\|\boldsymbol{\sigma}_\tau\| < g \ \Rightarrow\ \dot{\boldsymbol{u}}_\tau = \boldsymbol{0}, \\
&\|\boldsymbol{\sigma}_\tau\| = g \ \Rightarrow\ \exists\, \lambda \geq 0 \text{ s.t. } \boldsymbol{\sigma}_\tau = -\lambda \dot{\boldsymbol{u}}_\tau
\end{aligned}
\right\}
\quad \text{on } \Gamma_3 \times I. \tag{18.85}
$$

Here g represents the friction bound, i.e., the magnitude of the limiting friction traction at which slip begins. We assume $g \in L^\infty(\Gamma_3)$ and $g \geq 0$ a.e. on Γ_3. The contact described with (18.85) is bilateral, i.e., there is no loss of contact during the process. The set of admissible test functions U is given by

$$
U = \{\, \boldsymbol{v} \in H^1(\Omega)^d : v_\nu = 0 \text{ on } \Gamma_3 \,\}.
$$

It is straightforward to show that if $(\boldsymbol{u}, \boldsymbol{\sigma})$ is a pair of regular functions satisfying (18.85) then

$$
\boldsymbol{\sigma}\boldsymbol{\nu} \cdot (\boldsymbol{v} - \dot{\boldsymbol{u}}) \geq g\, \|\dot{\boldsymbol{u}}_\tau\| - g\, \|\boldsymbol{v}_\tau\| \quad \forall\, \boldsymbol{v} \in U,
$$

a.e. on $\Gamma_3 \times I$. Thus, (18.44) holds with the choice $\varphi(\boldsymbol{v}) = g\,\|\boldsymbol{v}_\tau\|$ and we obtain from (18.20) and (18.21) that

$$\begin{aligned} U_1 &= V_1, \\ j(\boldsymbol{v}) &= \int_{\Gamma_3} g\,\|\boldsymbol{v}_\tau\|\,da \quad \forall\,\boldsymbol{v} \in V_1. \end{aligned}$$

The assumptions (18.22), (18.23) and (18.46) are obviously satisfied. Therefore, Theorem 18.10 yields the existence and uniqueness of a function $\boldsymbol{\sigma}$ which represents a weak solution in terms of the stress for the mechanical problem (18.40)–(18.43), (18.45), (18.85). The existence of a weak solution in terms of the velocity for the same problem is proved in [2], where the space $BD(\Omega)$ of bounded deformations is used in the derivation and analysis of the weak formulation.

Example 18.17 *Frictionless contact with normal damped response.*

We consider a contact problem similar to that proposed in [70, p. 147]. It consists of the following contact boundary conditions:

$$\left.\begin{aligned} & g_1 \le \sigma_\nu \le g_2, \\ & g_1 < \sigma_\nu < g_2 \;\Rightarrow\; \dot{u}_\nu = 0, \\ & \sigma_\nu = g_1 \;\Rightarrow\; \dot{u}_\nu \ge 0, \\ & \sigma_\nu = g_2 \;\Rightarrow\; \dot{u}_\nu \le 0, \\ & \boldsymbol{\sigma}_\tau = \boldsymbol{0} \end{aligned}\right\} \quad \text{on } \Gamma_3 \times I. \tag{18.86}$$

Here g_1 and g_2 are prescribed functions in $L^\infty(\Gamma_3)$ such that $g_1 \le 0 \le g_2$, a.e. on Γ_3. Note that the contact conditions included in (18.86) lead to a multivalued relation between the normal stress σ_ν and the normal velocity $\dot{u}_\nu$. Recalling (5.45) (page 107), we can view (18.86) as a frictionless contact condition with normal damped response.

It is straightforward to show that if $(\boldsymbol{u}, \boldsymbol{\sigma})$ is a pair of regular functions satisfying (18.86), then the boundary condition (18.44) holds with

$$U = H^1(\Omega)^d, \quad \varphi(\boldsymbol{v}) = g_2 v_\nu^- - g_1 v_\nu^+.$$

Here, r^+ and r^- represent the magnitudes of the positive and negative parts of r: $r^+ = \max\{r, 0\}$, $r^- = \max\{-r, 0\}$. Then

$$\begin{aligned} U_1 &= V, \\ j(\boldsymbol{v}) &= \int_{\Gamma_3} (g_2 v_\nu^- - g_1 v_\nu^+)\,da \quad \forall\,\boldsymbol{v} \in V. \end{aligned}$$

The assumptions (18.22), (18.23) and (18.46) are easy to verify. Hence, we may apply Theorem 18.10 to the mechanical problem (18.40)–(18.43), (18.45), (18.86) and conclude that it has a unique solution for the weak formulation in terms of the stress.

Example 18.18 *Normal damped response with Tresca's friction law.*

This is a combination of (18.85) and (18.86) above. The contact boundary conditions are the following:

$$\left.\begin{array}{l} g_1 \le \sigma_\nu \le g_2, \\ g_1 < \sigma_\nu < g_2 \;\Rightarrow\; \dot{u}_\nu = 0, \\ \sigma_\nu = g_1 \;\Rightarrow\; \dot{u}_\nu \ge 0, \\ \sigma_\nu = g_2 \;\Rightarrow\; \dot{u}_\nu \le 0, \\ \|\boldsymbol{\sigma}_\tau\| \le g, \\ \|\boldsymbol{\sigma}_\tau\| < g \;\Rightarrow\; \dot{\boldsymbol{u}}_\tau = \boldsymbol{0}, \\ \|\boldsymbol{\sigma}_\tau\| = g \;\Rightarrow\; \exists\lambda \ge 0 \text{ s.t. } \boldsymbol{\sigma}_\tau = -\lambda\dot{\boldsymbol{u}}_\tau \end{array}\right\} \quad \text{on } \Gamma_3 \times I. \qquad (18.87)$$

Here g_1, g_2 and g are prescribed functions from $L^\infty(\Gamma_3)$ such that $g \ge 0$, $g_1 \le 0 \le g_2$, a.e. on Γ_3. The boundary conditions (18.87) imply (18.44) with

$$U = H^1(\Omega)^d, \quad \varphi(\boldsymbol{v}) = g_2 v_\nu^- - g_1 v_\nu^+ + g\,\|\boldsymbol{v}_\tau\|.$$

Then for this problem,

$$\begin{aligned} U_1 &= V, \\ j(\boldsymbol{v}) &= \int_{\Gamma_3} (g_2 v_\nu^- - g_1 v_\nu^+ + g\,\|\boldsymbol{v}_\tau\|)\, da \quad \forall\, \boldsymbol{v} \in V. \end{aligned}$$

The assumptions (18.22), (18.23) and (18.46) are satisfied. We conclude from Theorem 18.10 that the mechanical problem (18.40)–(18.43), (18.45) and (18.87) has a unique weak stress solution.

We end this section by noticing that, following [96, Chapter 13], it is possible to provide numerical analysis of semi-discrete and fully discrete schemes in solving Problem 18.9.

18.4 A Frictionless Contact Problem for Materials with Internal State Variable

Most of the results concerning the contact problems for elastic-visco-plastic materials of the form (6.43) (page 135) can be extended to the case of materials with internal state variable modelled by a constitutive law of the form (6.54), (6.58). To illustrate this, in this section we consider a frictionless contact problem for such kind of materials. The contact is modeled with the general normal compliance condition (5.41), already used in Chapter 11. The classical formulation of the model for this process is the following:

Find a displacement field $\boldsymbol{u} : \Omega \times \overline{I} \to \mathbb{R}^d$, a stress field $\boldsymbol{\sigma} : \Omega \times \overline{I} \to \mathbb{S}^d$ and an internal state variable field $\boldsymbol{\beta} : \Omega \times \overline{I} \to \mathbb{R}^m$ such that

$$\dot{\boldsymbol{\sigma}} = \mathcal{E}(\boldsymbol{\varepsilon}(\dot{\boldsymbol{u}})) + \mathcal{G}(\boldsymbol{\sigma}, \boldsymbol{\varepsilon}(\boldsymbol{u}), \boldsymbol{\beta}) \quad \text{in } \Omega \times I, \tag{18.88}$$

$$\dot{\boldsymbol{\beta}} = \varphi(\boldsymbol{\sigma}, \boldsymbol{\varepsilon}(\boldsymbol{u}), \boldsymbol{\beta}) \quad \text{in } \Omega \times I, \tag{18.89}$$

$$\text{Div}\, \boldsymbol{\sigma} + \boldsymbol{f}_0 = \boldsymbol{0} \quad \text{in } \Omega \times I, \tag{18.90}$$

$$\boldsymbol{u} = \boldsymbol{0} \quad \text{on } \Gamma_1 \times I, \tag{18.91}$$

$$\boldsymbol{\sigma\nu} = \boldsymbol{f}_2 \quad \text{on } \Gamma_2 \times I, \tag{18.92}$$

$$-\sigma_\nu = p_\nu(u_\nu - g) \quad \text{on } \Gamma_3 \times I, \tag{18.93}$$

$$\boldsymbol{\sigma}_\tau = \boldsymbol{0} \quad \text{on } \Gamma_3 \times I, \tag{18.94}$$

$$\boldsymbol{u}(0) = \boldsymbol{u}_0, \quad \boldsymbol{\sigma}(0) = \boldsymbol{\sigma}_0, \quad \boldsymbol{\beta}(0) = \boldsymbol{\beta}_0 \quad \text{in } \Omega. \tag{18.95}$$

Here $\boldsymbol{\beta}$ denotes an internal state variable whose evolution is governed by the differential equation (18.89) in which φ is a nonlinear function to be described below. In (18.95), $\boldsymbol{u}_0$, $\boldsymbol{\sigma}_0$ and $\boldsymbol{\beta}_0$ represent the initial displacement, the initial stress and the initial internal state variable, respectively.

We recall that in the normal compliance contact condition (18.93), g represents the gap between the potential contact surface and the foundation, measured along the outward normal direction, and p_ν is a prescribed function such that $p_\nu(r) = 0$ for $r \le 0$. The quantity $u_\nu - g$, when positive, represents the penetration of the body in the foundation.

To obtain a variational formulation of the problem (18.88)–(18.95), in addition to the spaces V, Q and Q_1, we introduce the space

$$Y = \{\, \boldsymbol{\alpha} = (\alpha_k) : \alpha_k \in L^2(\Omega),\ 1 \le k \le m \,\}$$

for internal state variables. This is a real Hilbert space with its canonical inner products.

In the study of the mechanical problem (18.88)–(18.95), we assume conditions (6.44), (6.64), (6.65) for the constitutive functions $\mathcal{E}$, $\mathcal{G}$ and φ, respectively. We assume the normal compliance function p_ν satisfies:

$$\left.\begin{array}{l}
\text{(a) } p_\nu : \Gamma_3 \times \mathbb{R} \to \mathbb{R}_+.\\
\text{(b) There exists an } L_\nu > 0 \text{ such that}\\
\qquad |p_\nu(\boldsymbol{x}, r_1) - p_\nu(\boldsymbol{x}, r_2)| \le L_\nu |r_1 - r_2|\\
\qquad \forall\, r_1, r_2 \in \mathbb{R}, \text{ a.e. } \boldsymbol{x} \in \Gamma_3.\\
\text{(c) } (p_\nu(\boldsymbol{x}, r_1) - p_\nu(\boldsymbol{x}, r_2))\,(r_1 - r_2) \ge 0\\
\qquad \forall\, r_1, r_2 \in \mathbb{R}, \text{ a.e. } \boldsymbol{x} \in \Gamma_3.\\
\text{(d) The mapping } \boldsymbol{x} \mapsto p_\nu(\boldsymbol{x}, r) \text{ is Lebesgue measurable}\\
\qquad \text{on } \Gamma_3 \text{ for any } r \in \mathbb{R}.\\
\text{(e) The mapping } \boldsymbol{x} \mapsto p_\nu(\boldsymbol{x}, r) = 0 \text{ for } r \le 0.
\end{array}\right\} \tag{18.96}$$

Let us make some comments on the assumptions (18.96). Condition (b) requires the function p_ν to grow at most linearly; condition (c) states that the reaction of the obstacle increases with the penetration; condition (e) says that when there is no penetration (i.e. $u_\nu \le g$) then there is no reaction of the obstacle ($\sigma_\nu = 0$) and, finally, condition (d) represents a technical condition imposed from mathematical reasons. We observe that the assumptions (18.96) are satisfied by the functions p_ν defined in (5.42) and (5.44).

We also suppose that the forces and the tractions have the regularity

$$\boldsymbol{f}_0 \in W^{1,\infty}(I;H), \quad \boldsymbol{f}_2 \in W^{1,\infty}(I;L^2(\Gamma_2)^d) \tag{18.97}$$

and the gap function g is such that

$$g \in L^2(\Gamma_3), \quad g \geq 0 \text{ a.e. on } \Gamma_3. \tag{18.98}$$

Next, using Riesz Representation Theorem, we can define $\boldsymbol{f}(t) \in V$ by

$$(\boldsymbol{f}(t), \boldsymbol{v})_V = \int_\Omega \boldsymbol{f}_0(t) \cdot \boldsymbol{v}\, dx + \int_{\Gamma_2} \boldsymbol{f}_2(t) \cdot \boldsymbol{v}\, da \quad \forall\, \boldsymbol{v} \in V,\ t \in \overline{I}. \tag{18.99}$$

By the regularity conditions (18.97), the integrals in the above formula are well-defined and we have

$$\boldsymbol{f} \in W^{1,\infty}(I;V). \tag{18.100}$$

Let $j : V \times V \to \mathbb{R}$ be the functional

$$j(\boldsymbol{u}, \boldsymbol{v}) = \int_{\Gamma_3} p_\nu(u_\nu - g)\, v_\nu\, da \quad \forall\, \boldsymbol{u}, \boldsymbol{v} \in V. \tag{18.101}$$

Due to the assumptions (18.96) and (18.98), the integral in (18.101) is well-defined.

Finally, we assume that the initial data satisfy the conditions

$$\boldsymbol{u}_0 \in V, \quad \boldsymbol{\sigma}_0 \in Q_1, \quad \boldsymbol{\beta}_0 \in Y, \tag{18.102}$$

$$(\boldsymbol{\sigma}_0, \boldsymbol{\varepsilon}(\boldsymbol{v}))_Q + j(\boldsymbol{u}_0, \boldsymbol{v}) = (\boldsymbol{f}(0), \boldsymbol{v})_V \quad \forall\, \boldsymbol{v} \in V. \tag{18.103}$$

We note that the relation (18.103) represents a compatibility condition at the initial time.

Now assume $(\boldsymbol{u}, \boldsymbol{\sigma}, \boldsymbol{\beta})$ are regular functions satisfying (18.90)–(18.94). Then by a standard procedure, it is straightforward to see that for all $t \in \overline{I}$,

$$\boldsymbol{u}(t) \in V, \quad (\boldsymbol{\sigma}(t), \boldsymbol{\varepsilon}(\boldsymbol{v}))_Q + j(\boldsymbol{u}(t), \boldsymbol{v}) = (\boldsymbol{f}(t), \boldsymbol{v})_V \quad \forall\, \boldsymbol{v} \in V.$$

This relation, together with (18.88), (18.89), and the initial condition (18.95), leads to the following variational formulation of the mechanical problem (18.88)–(18.95).

Problem 18.19 *Find a displacement field* $\boldsymbol{u} : \overline{I} \to V$, *a stress field* $\boldsymbol{\sigma} : \overline{I} \to Q_1$ *and an internal state variable field* $\boldsymbol{\beta} : \overline{I} \to Y$ *such that*

$$\dot{\boldsymbol{\sigma}}(t) = \mathcal{E}\boldsymbol{\varepsilon}(\dot{\boldsymbol{u}}(t)) + \mathcal{G}(\boldsymbol{\sigma}(t), \boldsymbol{\varepsilon}(\boldsymbol{u}(t)), \boldsymbol{\beta}(t)) \quad \text{a.e. } t \in I, \tag{18.104}$$

$$\dot{\boldsymbol{\beta}}(t) = \varphi(\boldsymbol{\sigma}(t), \boldsymbol{\varepsilon}(\boldsymbol{u}(t)), \boldsymbol{\beta}(t))) \quad \text{a.e. } t \in I, \tag{18.105}$$

$$(\boldsymbol{\sigma}(t), \boldsymbol{\varepsilon}(\boldsymbol{v}))_Q + j(\boldsymbol{u}(t), \boldsymbol{v}) = (\boldsymbol{f}(t), \boldsymbol{v})_V \quad \forall\, \boldsymbol{v} \in V,\ t \in \overline{I}, \tag{18.106}$$

$$\boldsymbol{u}(0) = \boldsymbol{u}_0, \quad \boldsymbol{\sigma}(0) = \boldsymbol{\sigma}_0, \quad \boldsymbol{\beta}(0) = \boldsymbol{\beta}_0. \tag{18.107}$$

Well-posedness of Problem 18.19 is provided in the following theorem.

Theorem 18.20 *Assume* (6.44), (6.64), (6.65), (18.96)–(18.98), (18.102) *and* (18.103). *Then there exists a unique solution* $(\boldsymbol{u}, \boldsymbol{\sigma}, \boldsymbol{\beta})$ *to Problem 18.19. Moreover, the solution satisfies*

$$\boldsymbol{u} \in W^{1,\infty}(I;V), \quad \boldsymbol{\sigma} \in W^{1,\infty}(I;Q_1), \quad \boldsymbol{\beta} \in W^{1,\infty}(I;Y). \tag{18.108}$$

We conclude that the mechanical problem (18.88)–(18.95) has a unique *weak solution* with regularity (18.108).

The proof of Theorem 18.20 is carried out in several steps. Assume in what follows that (6.44), (6.64), (6.65), (18.96)–(18.98), (18.102) and (18.103) hold. For any $\boldsymbol{\eta} \in L^\infty(I; Q \times Y)$, we denote by $\boldsymbol{z}_\eta = (\boldsymbol{z}_\eta^1, \boldsymbol{z}_\eta^2) \in W^{1,\infty}(I; Q \times Y)$ the function given by

$$\boldsymbol{z}_\eta(t) = \int_0^t \boldsymbol{\eta}(s)\, ds + \boldsymbol{z}_0, \tag{18.109}$$

where

$$\boldsymbol{z}_0 = (\boldsymbol{\sigma}_0 - \mathcal{E}\boldsymbol{\varepsilon}(\boldsymbol{u}_0), \boldsymbol{\beta}_0). \tag{18.110}$$

We consider the following variational problem.

Problem 18.21 *Find a displacement field* $\boldsymbol{u}_\eta : \overline{I} \to V$ *and a stress field* $\boldsymbol{\sigma}_\eta : \overline{I} \to Q_1$ *such that for all* $t \in \overline{I}$,

$$\boldsymbol{\sigma}_\eta(t) = \mathcal{E}\boldsymbol{\varepsilon}(\boldsymbol{u}_\eta(t)) + \boldsymbol{z}_\eta^1(t), \tag{18.111}$$

$$(\boldsymbol{\sigma}_\eta(t), \boldsymbol{\varepsilon}(\boldsymbol{v}))_Q + j(\boldsymbol{u}_\eta(t), \boldsymbol{v}) = (\boldsymbol{f}(t), \boldsymbol{v})_V \quad \forall\, \boldsymbol{v} \in V. \tag{18.112}$$

Clearly, we are solving the problem for the displacement and the elastic part of the stress field. We have the following existence and uniqueness result.

Lemma 18.22 *There exists a unique solution to Problem 18.21 such that* $\boldsymbol{u}_\eta \in W^{1,\infty}(I;V)$, $\boldsymbol{\sigma}_\eta \in W^{1,\infty}(I;Q_1)$. *Moreover, the solution satisfies*

$$\boldsymbol{u}_\eta(0) = \boldsymbol{u}_0, \quad \boldsymbol{\sigma}_\eta(0) = \boldsymbol{\sigma}_0. \tag{18.113}$$

Proof. Let $a : V \times V \to \mathbb{R}$ denote the bilinear form given by

$$a(\boldsymbol{u}, \boldsymbol{v}) = (\mathcal{E}\boldsymbol{\varepsilon}(\boldsymbol{u}), \boldsymbol{\varepsilon}(\boldsymbol{v}))_Q \quad \forall\, \boldsymbol{u}, \boldsymbol{v} \in V.$$

Due to the assumptions (6.44) on $\mathcal{E}$, we know that $a(\cdot,\cdot)$ is continuous and coercive on V,

$$|a(\boldsymbol{u}, \boldsymbol{v})| \le c\, \|\boldsymbol{u}\|_V \|\boldsymbol{v}\|_V \quad \forall\, \boldsymbol{u}, \boldsymbol{v} \in V, \tag{18.114}$$

$$a(\boldsymbol{v}, \boldsymbol{v}) \ge \alpha_0 \|\boldsymbol{v}\|_V^2 \quad \forall\, \boldsymbol{v} \in V. \tag{18.115}$$

From its definition (18.101), we see that $j(\cdot,\cdot)$ satisfied the following inequalities:

$$j(\boldsymbol{u}_1,\boldsymbol{u}_1-\boldsymbol{u}_2)-j(\boldsymbol{u}_2,\boldsymbol{u}_1-\boldsymbol{u}_2)\geq 0 \quad \forall\,\boldsymbol{u}_1,\boldsymbol{u}_2\in V, \tag{18.116}$$

$$j(\boldsymbol{u}_1,\boldsymbol{v})-j(\boldsymbol{u}_2,\boldsymbol{v})\leq c\,\|\boldsymbol{u}_1-\boldsymbol{u}_2\|_V\|\boldsymbol{v}\|_V \quad \forall\,\boldsymbol{u}_1,\boldsymbol{u}_2,\boldsymbol{v}\in V, \tag{18.117}$$

where the assumptions (18.96) on the normal compliance function p_ν are used.

Using Riesz Representation Theorem we may define an operator $B : V \to V$ and an element $\boldsymbol{f}_\eta(t)\in V$ by the relations

$$(B\boldsymbol{u},\boldsymbol{v})_V = a(\boldsymbol{u},\boldsymbol{v})+j(\boldsymbol{u},\boldsymbol{v}), \tag{18.118}$$

$$(\boldsymbol{f}_\eta(t),\boldsymbol{v})_V = (\boldsymbol{f}(t),\boldsymbol{v})_V-(\boldsymbol{z}^1_\eta(t),\boldsymbol{\varepsilon}(\boldsymbol{v}))_Q, \tag{18.119}$$

for all $\boldsymbol{u},\boldsymbol{v}\in V$, $t\in\overline{I}$. Let $\boldsymbol{u}_1,\boldsymbol{u}_2\in V$. Then

$$\begin{aligned}(B\boldsymbol{u}_1-B\boldsymbol{u}_2,\boldsymbol{u}_1-\boldsymbol{u}_2)_V &= a(\boldsymbol{u}_1-\boldsymbol{u}_2,\boldsymbol{u}_1-\boldsymbol{u}_2)\\ &\quad + j(\boldsymbol{u}_1,\boldsymbol{u}_1-\boldsymbol{u}_2)-j(\boldsymbol{u}_2,\boldsymbol{u}_1-\boldsymbol{u}_2).\end{aligned}$$

Using the inequalities (18.115) and (18.116), we deduce

$$(B\boldsymbol{u}_1-B\boldsymbol{u}_2,\boldsymbol{u}_1-\boldsymbol{u}_2)_V\geq \alpha_0\|\boldsymbol{u}_1-\boldsymbol{u}_2\|^2_V. \tag{18.120}$$

We also have

$$(B\boldsymbol{u}_1-B\boldsymbol{u}_2,\boldsymbol{v})_V = a(\boldsymbol{u}_1-\boldsymbol{u}_2,\boldsymbol{v})+j(\boldsymbol{u}_1,\boldsymbol{v})-j(\boldsymbol{u}_2,\boldsymbol{v})$$

for all $\boldsymbol{v}\in V$. Using the inequalities (18.114) and (18.117), we obtain

$$(B\boldsymbol{u}_1-B\boldsymbol{u}_2,\boldsymbol{v})_V\leq c\,\|\boldsymbol{u}_1-\boldsymbol{u}_2\|_V\|\boldsymbol{v}\|_V \quad \forall\,\boldsymbol{v}\in V.$$

Thus,

$$\|B\boldsymbol{u}_1-B\boldsymbol{u}_2\|_V\leq c\,\|\boldsymbol{u}_1-\boldsymbol{u}_2\|_V. \tag{18.121}$$

The inequalities (18.120) and (18.121) show that B is a strongly monotone Lipschitz continuous operator on V. Therefore, by Corollary 4.4 (page 63), there exists a unique element $\boldsymbol{u}_\eta(t)\in V$ such that

$$B\boldsymbol{u}_\eta(t)=\boldsymbol{f}_\eta(t) \quad \forall\, t\in\overline{I}. \tag{18.122}$$

Define $\boldsymbol{\sigma}_\eta(t)$ by (18.111). Then the equality (18.122) can be rewritten in the form (18.112). Choosing $\boldsymbol{v}=\pm\boldsymbol{\phi}$, $\boldsymbol{\phi}\in C_0^\infty(\Omega)^d$ arbitrary, in (18.112), we have

$$(\boldsymbol{\sigma}_\eta(t),\boldsymbol{\varepsilon}(\boldsymbol{\phi}))_Q=(\boldsymbol{f}(t),\boldsymbol{\phi})_V \quad \forall\, t\in\overline{I},$$

and then

$$\operatorname{Div}\boldsymbol{\sigma}_\eta(t)+\boldsymbol{f}_0(t)=\boldsymbol{0} \quad \text{in } \Omega,\ \forall\, t\in\overline{I}.$$

Therefore, from the regularity assumption (18.97) on $\boldsymbol{f}_0$, we have $\boldsymbol{\sigma}_\eta(t)\in Q_1$ for all $t\in\overline{I}$.

Now, let $t_1, t_2 \in \overline{I}$. We derive from the equation (18.122), with the use of the inequality (18.120) and the definition (18.119) of $\boldsymbol{f}_\eta$, that

$$\|\boldsymbol{u}_\eta(t_1) - \boldsymbol{u}_\eta(t_2)\|_V \le c\big(\|\boldsymbol{f}(t_1) - \boldsymbol{f}(t_2)\|_V + \|\boldsymbol{z}^1(t_1) - \boldsymbol{z}^1(t_2)\|_Q\big).$$

Then for $\boldsymbol{\sigma}_\eta$ defined in (18.111), we have

$$\|\boldsymbol{\sigma}_\eta(t_1) - \boldsymbol{\sigma}_\eta(t_2)\|_Q \le c\big(\|\boldsymbol{u}_\eta(t_1) - \boldsymbol{u}_\eta(t_2)\|_V + \|\boldsymbol{z}^1(t_1) - \boldsymbol{z}^1(t_2)\|_Q\big).$$

Since $\boldsymbol{f} : \overline{I} \to V$ and $\boldsymbol{z}^1 : \overline{I} \to Q$ are Lipschitz continuous, we deduce that $\boldsymbol{u}_\eta \in W^{1,\infty}(I;V)$ and $\boldsymbol{\sigma}_\eta \in W^{1,\infty}(I;Q)$. Since $\operatorname{Div} \boldsymbol{\sigma}_\eta = -\boldsymbol{f}_0 \in W^{1,\infty}(I;H)$, we further have $\boldsymbol{\sigma}_\eta \in W^{1,\infty}(I;Q_1)$. This concludes the existence part of Lemma 18.22. The uniqueness part follows from the unique solvability of the time-dependent nonlinear equation (18.122).

Finally, the assumptions (18.102) and (18.103) on the initial data and the definitions (18.109) and (18.110) for $\boldsymbol{z}_\eta(t)$ and $\boldsymbol{z}_0$ imply that $(\boldsymbol{u}_0, \boldsymbol{\sigma}_0)$ satisfies (18.111), (18.112) at $t = 0$. We then obtain the initial condition (18.113) from the unique solvability of the system (18.111), (18.112) for any $t \in \overline{I}$. ∎

Define $\boldsymbol{\beta}_\eta \in W^{1,\infty}(I;Y)$ by

$$\boldsymbol{\beta}_\eta = \boldsymbol{z}_\eta^2 \tag{18.123}$$

and define $\Lambda : L^\infty(I;Q \times Y) \to L^\infty(I;Q \times Y)$ by

$$\Lambda\boldsymbol{\eta} = \Big(G(\boldsymbol{\sigma}_\eta, \boldsymbol{\varepsilon}(\boldsymbol{u}_\eta), \boldsymbol{\beta}_\eta), \varphi(\boldsymbol{\sigma}_\eta, \boldsymbol{\varepsilon}(\boldsymbol{u}_\eta), \boldsymbol{\beta}_\eta)\Big) \tag{18.124}$$

for any $\boldsymbol{\eta} \in L^\infty(I;Q \times Y)$.

We have the following result.

Lemma 18.23 *The operator Λ has a unique fixed point $\boldsymbol{\eta}^* \in L^\infty(I, Q \times Y)$.*

Proof. Let $\boldsymbol{\eta}_1, \boldsymbol{\eta}_2 \in L^\infty(I;Q\times Y)$ and let $t \in \overline{I}$. We use the simplified notations $\boldsymbol{u}_i = \boldsymbol{u}_{\eta_i}$, $\boldsymbol{\sigma}_i = \boldsymbol{\sigma}_{\eta_i}$, $\boldsymbol{\beta}_i = \boldsymbol{\beta}_{\eta_i}$, $\boldsymbol{z}_i = (\boldsymbol{z}_i^1, \boldsymbol{z}_i^2) = \boldsymbol{z}_{\eta_i}$ for $i = 1, 2$. By (18.111) and (18.112) we have

$$\boldsymbol{\sigma}_i(t) = \mathcal{E}\boldsymbol{\varepsilon}(\boldsymbol{u}_i(t)) + \boldsymbol{z}_i^1(t), \tag{18.125}$$

$$(\boldsymbol{\sigma}_i(t), \boldsymbol{\varepsilon}(\boldsymbol{v}))_Q + j(\boldsymbol{u}_i(t), \boldsymbol{v}) = (\boldsymbol{f}(t), \boldsymbol{v})_V \quad \forall\, \boldsymbol{v} \in V, \tag{18.126}$$

for $i = 1, 2$. From (18.125) and (18.126) we can get the following relation:

$$\begin{aligned}
&(\mathcal{E}\boldsymbol{\varepsilon}(\boldsymbol{u}_1(t)) - \mathcal{E}\boldsymbol{\varepsilon}(\boldsymbol{u}_2(t)), \boldsymbol{\varepsilon}(\boldsymbol{u}_1(t)) - \boldsymbol{\varepsilon}(\boldsymbol{u}_2(t)))_Q \\
&\quad + j(\boldsymbol{u}_1(t), \boldsymbol{u}_1(t) - \boldsymbol{u}_2(t)) - j(\boldsymbol{u}_2(t), \boldsymbol{u}_1(t) - \boldsymbol{u}_2(t)) \\
&\quad = (\boldsymbol{z}_1^1(t) - \boldsymbol{z}_2^1(t), \boldsymbol{\varepsilon}(\boldsymbol{u}_1(t)) - \boldsymbol{\varepsilon}(\boldsymbol{u}_2(t)))_Q.
\end{aligned} \tag{18.127}$$

Using the assumptions (6.44) on $\mathcal{E}$ and the inequality (18.116), we deduce that

$$\|\boldsymbol{u}_1(t) - \boldsymbol{u}_2(t)\|_V \le c\, \|\boldsymbol{z}_1(t) - \boldsymbol{z}_2(t)\|_{Q\times Y}.$$

Then from the definitions (18.125) for $\boldsymbol{\sigma}_1$ and $\boldsymbol{\sigma}_2$, and (18.123) for $\boldsymbol{\beta}_1$ and $\boldsymbol{\beta}_2$, we obtain

$$\|\boldsymbol{\sigma}_1(t) - \boldsymbol{\sigma}_2(t)\|_Q + \|\boldsymbol{\beta}_1(t) - \boldsymbol{\beta}_2(t)\|_Y \le c\,\|\boldsymbol{z}_1(t) - \boldsymbol{z}_2(t)\|_{Q\times Y}.$$

Using the definition (18.109) for $\boldsymbol{z}_1$ and $\boldsymbol{z}_2$, we have

$$\|\boldsymbol{z}_1(t) - \boldsymbol{z}_2(t)\|_{Q\times Y} \le \int_0^t \|\boldsymbol{\eta}_1(s) - \boldsymbol{\eta}_2(s)\|_{Q\times Y} ds.$$

Therefore,

$$\begin{aligned}&\|\boldsymbol{u}_1(t) - \boldsymbol{u}_2(t)\|_V + \|\boldsymbol{\sigma}_1(t) - \boldsymbol{\sigma}_2(t)\|_Q + \|\boldsymbol{\beta}_1(t) - \boldsymbol{\beta}_2(t)\|_Y \\ &\quad \le c \int_0^t \|\boldsymbol{\eta}_1(s) - \boldsymbol{\eta}_2(s)\|_{Q\times Y} ds. \qquad (18.128)\end{aligned}$$

Recall the definition of the operator Λ, (18.124), using the assumptions (6.64) and (6.65) on $\mathcal{G}$ and φ, and the estimate (18.128), we obtain

$$\|\Lambda\boldsymbol{\eta}_1(t) - \Lambda\boldsymbol{\eta}_2(t)\|_{Q\times Y} \le c \int_0^t \|\boldsymbol{\eta}_1(s) - \boldsymbol{\eta}_2(s)\|_{Q\times Y} ds.$$

Reiterate this inequality n times,

$$\|\Lambda^n\boldsymbol{\eta}_1 - \Lambda^n\boldsymbol{\eta}_2\|_{L^\infty(I;Q\times Y)} \le \frac{c^n T^n}{n!}\|\boldsymbol{\eta}_1 - \boldsymbol{\eta}_2\|_{L^\infty(I;Q\times Y)}.$$

Thus, for n sufficiently large, Λ^n is a contraction on the Banach space $L^\infty(I; Q\times Y)$. So by Corollary 1.63 (page 23), Λ has a unique fixed point $\boldsymbol{\eta}^* \in L^\infty(I; Q\times Y)$. ■

Let us now prove Theorem 18.20.

Proof of Theorem 18.20.
Existence. Let $\boldsymbol{\eta}^* \in L^\infty(I; Q\times Y)$ be the fixed point of Λ and let $(\boldsymbol{u}_{\eta^*}, \boldsymbol{\sigma}_{\eta^*})$ represent the solution of Problem 18.21 for $\boldsymbol{\eta} = \boldsymbol{\eta}^*$. We denote by $\boldsymbol{\beta}_{\eta^*}$ the function given by (18.123) for $\boldsymbol{\eta} = \boldsymbol{\eta}^*$. Let us show that $(\boldsymbol{u}_{\eta^*}, \boldsymbol{\sigma}_{\eta^*}, \boldsymbol{\beta}_{\eta^*})$ is a solution of Problem 18.21. Indeed, the equalities (18.104) and (18.105) follow from (18.111), (18.109), (18.123) and (18.124) since

$$\begin{aligned}&\dot{\boldsymbol{\sigma}}_{\eta^*}(t) = \mathcal{E}\boldsymbol{\varepsilon}(\dot{\boldsymbol{u}}_{\eta^*}(t)) + \dot{\boldsymbol{z}}^1_{\eta^*}(t),\\ &\dot{\boldsymbol{z}}^1_{\eta^*}(t) = {\boldsymbol{\eta}^*}^1(t) = \mathcal{G}(\boldsymbol{\sigma}_{\eta^*}(t), \boldsymbol{\varepsilon}(\boldsymbol{u}_{\eta^*}(t)), \boldsymbol{\beta}_{\eta^*}(t)),\\ &\dot{\boldsymbol{\beta}}_{\eta^*}(t) = \dot{\boldsymbol{z}}^2_{\eta^*}(t) = {\boldsymbol{\eta}^*}^2(t) = \varphi(\boldsymbol{\sigma}_{\eta^*}(t), \boldsymbol{\varepsilon}(\boldsymbol{u}_{\eta^*}(t)), \boldsymbol{\beta}_{\eta^*}(t)),\end{aligned}$$

a.e. $t \in I$. Next, (18.106) is a direct consequence of (18.112). The initial conditions (18.107) follow from (18.113), (18.123), (18.109) and (18.110). Finally, the regularity (18.108) follow from Lemma 18.22 and (18.123).
Uniqueness. The uniqueness part follows from similar arguments as those used in proving Theorem 14.4. It is a consequence of the uniqueness of the fixed point of the operator Λ defined in (18.124). ■

Bibliographical Notes

The first existence and uniqueness result in the study of quasistatic problems with viscoplastic materials of the form (6.43) (page 135) in the case when a full coupling of the stress and the strain is involved in $\mathcal{G}$ was obtained in [120]. This result was extended in [115] to the the study of dynamic problems. Other references on this topic include [114, 193]. A survey of various functional and numerical methods in the study of such kind of viscoplastic materials is given in [121]. In all these works only classical displacement-traction boundary conditions were considered.

The study of contact problems involving viscoplastic constitutive laws of the form (6.43) is more recent. Most of the results presented in Section 14.1 were obtained in [194]. A fixed point method similar to that used in the proof of Theorem 14.4, as well as in many other places in this book, was already employed in [145] in the study of a displacement-traction problem with viscoplastic materials with hardening. An existence result for the Signorini frictionless problem involving Maxwell-Norton materials was recently obtained in [23]. There, the constitutive law is of the form (6.43) with $\mathcal{G}$ independent of ε. Since the corresponding constitutive function $\mathcal{G}$ is not Lipschitz continuous, the existence of the solution is obtained using different arguments, based on a time-regularization method.

General frictionless contact problems of the form (15.1)–(15.6) were considered in [183] and the numerical analysis of such kind of problems, provided in Sections 15.2–15.6 was obtained in [46]. The convergence result in Theorem 15.11 was obtained in [195]. The numerical results in Section 15.7 were presented following [77, 78].

Chapter 16 was written based on the papers [78, 100, 181]. Theorems 16.3, 16.5 –16.7 were obtained in [181], Theorems 16.9 and 16.15 in [100] and the numerical results presented in Section 16.3 were obtained in [78]. Recently, error analysis for a non-conforming finite element method in solving the frictionless contact problems with viscoplastic materials of the form (6.43) have been considered in [83, 84].

The variational analysis of the bilateral contact with Tresca's friction law presented in Chapter 17 was obtained in [5] and the numerical approximation of this problem, presented in Sections 17.4–17.6, was written following the ideas in [43]. There, semi-discrete and fully discrete approximations of an abstract nonlinear evolutionary system was considered, with emphasis in the study of frictional contact problems for viscoplastic materials.

Theorem 18.3 was derived in [6] from an abstract existence and uniqueness result and Theorem 18.5 was obtained in [196], among other results. Section 18.3 was written following [4]. There, an existence and uniqueness result for abstract differential inclusions is obtained and applications in the study of frictional contact problems with perfect plastic materials are provided. More details in the study of displacement-traction boundary value problems with perfectly plastic materials can be found in [12, 127, 203, 204] and, more recently, in [20, 133]. Results concerning the analysis of regularity of weak solutions in the study of perfectly plastic materials can be found in the monograph [87]. The frictionless contact problem (18.88)–(18.95) involving viscoplastic materials with internal state variable has been considered in [81]. There Theorem 18.20 was obtained, error estimates for a fully discrete scheme to solve the problem were derived, and numerical examples were provided.

We end these comments with the remark that quasistatic contact problems with variants of Coulomb law of friction for viscoplastic materials of the form (6.43) lead to substantial mathematical difficulties and remain to be open problems. Also, we notice that despite the fact that considerable progress has been made in the variational and numerical analysis of dynamic contact problems with viscoelastic bodies, there are few results in the study of dynamic contact problems involving viscoplastic bodies and many problems on this topic still remain open.

Bibliography

[1] R.A. Adams, *Sobolev Spaces*, Academic Press, New York, 1975.

[2] A. Amassad, M. Shillor and M. Sofonea, A quasistatic contact problem for an elastic perfectly plastic body with Tresca's friction, *Nonlinear Analysis TMA* **35** (1999), 95–109.

[3] A. Amassad, M. Shillor and M. Sofonea, A quasistatic contact problem with slip dependent coefficient of friction, *Math. Meth. Appl. Sci.* **22** (1999), 267–284.

[4] A. Amassad, M. Shillor and M. Sofonea, A nonlinear evolution inclusion in perfect plasticity with friction, *Acta Mathematica Universitatis Comenianae* **LXX** (2001), 1–14.

[5] A. Amassad and M. Sofonea, Analysis of a quasistatic viscoplastic problem involving Tresca friction law, *Discrete and Continuous Dynamical Systems* **4** (1998), 55–72.

[6] A. Amassad and M. Sofonea, Analysis of some nonlinear evolution systems arising in rate-type viscoplasticity, in W. Chen and S. Hu, eds., *Dynamical System and Differential Equations*, an added volume to *Discrete and Continuous Dynamical Systems*, 1998, 58–71.

[7] L.-E. Andersson, A quasistatic frictional problem with normal compliance, *Nonlinear Analysis TMA* **16** (1991), 347–370.

[8] L.-E. Andersson, A global existence result for a quasistatic contact problem with friction, *Advances in Mathematical Sciences and Applications* **5** (1995), 249–286.

[9] L.-E. Andersson, A quasistatic frictional problem with a normal compliance penalization term, *Nonlinear Analysis TMA* **37** (1999), 689–705.

[10] L.-E. Andersson, Existence results for quasistatic contact problems with Coulomb friction, *Appl. Math. Optim.* **42** (2000), 169–202.

[11] S.S. Antman, *Nonlinear Problems of Elasticity*, Springer-Verlag, New York, 1995.

[12] G. Anzellotti and S. Luckhaus, Dynamical evolution of elasto-perfectly plastic bodies, *Appl. Math. Optim.* **15** (1987), 121–140.

[13] J.F. Archard, Contact and rubbing of flat surfaces, *J. Appl. Phys.* **24** (1953), 981–988.

[14] K. Atkinson and W. Han, *Theoretical Numerical Analysis: A Functional Analysis Framework*, Springer-Verlag, New York, TAM, Volume 39, 2001.

[15] B. Awbi, O. Chau and M. Sofonea, Variational analysis of a frictional contact problem for viscoelastic bodies, *Intern. Math. Journal* **1** (2002), 333–348.

[16] B. Awbi, El H. Essoufi and M. Sofonea, A viscoelasstic contact problem with normal damped response and friction, *Annales Polonici Mathematici* **LXXV** (2000), 233–246.

[17] B. Awbi, M. Rochdi and M. Sofonea, Abstract evolution equations for viscoelastic frictional contact problems, *Journal of Applied Mathematics and Physics* (*ZAMP*) **50** (1999), 1–18.

[18] B. Awbi, M. Shillor and M. Sofonea, Dual formulation of a quasistatic viscoelastic contact problem with Tresca's friction law, *Applicable Analysis* **79** (2001), 1–20.

[19] I. Babuška and A.K. Aziz, Survey lectures on the mathematical foundations of the finite element method, in A.K. Aziz, ed., *The Mathematical Foundations of the Finite Element Method with Applications to Partial Differential Equations*, Academic Press, New York, 1972, 3–359.

[20] I. Babuška and P. Shi, Regularity of solution to a one-dimensional plasticity model, *Mathematical Modelling and Numerical Analysis* (M^2AN) **32** (1998), 521–537.

[21] I. Babuška and T. Strouboulis, *The Finite Element Method and its Reliability*, Oxford University Press, 2001.

[22] C. Baiocchi and A. Capelo, *Variational and Quasivariational Inequalities: Applications to Free-Boundary Problems*, John Wiley, Chichester, 1984.

[23] P. Baral and P. Quintanela, Existence of solution for a contact problem of Signorini type in Maxwell-Norton materials, *preprint.*

[24] M. Barboteu, W. Han and M. Sofonea, Numerical analysis of a bilateral frictional contact problem for linearly elastic materials, to appear in *IMA Journal of Numerical Analysis.*

[25] M. Barboteu, W. Han and M. Sofonea, A frictionless contact problem for viscoelastic materials, *Journal of Applied Mathematics* **2** (2002), 1–21.

[26] V. Barbu, *Nonlinear Semigroups and Differential Equations in Banach Spaces*, Editura Academiei, Bucharest-Noordhoff, Leyden, 1976.

[27] V. Barbu, *Optimal Control of Variational Inequalities*, Pitman, Boston, 1984.

[28] G. Bayada, M. Chambat and A. Lakhal, A dynamic contact problem with regularized friction law: application to impact problems, *Math. Comput. Modelling* **28** (1998), 67–85.

[29] F. Ben Belgacem, P. Hild and P. Laborde, The mortar finite element method for contact problems, *Math. Comput. Modelling* **28** (1998), 263–271.

[30] F. Ben Belgacem, P. Hild and P. Laborde, Extension of the mortar finite element method to a variational inequality modeling unilateral contact, *Math. Models. Methods Appl. Sci* **9** (1999), 287–303.

[31] D. Braess, *Finite Elements: Theory, Fast Solvers, and Applications in Solid Mechanics*, Cambridge University Press, Cambridge, 1997.

[32] S.C. Brenner and L.R. Scott, *The Mathematical Theory of Finite Element Methods*, Springer-Verlag, New York, 1994.

[33] H. Brézis, Equations et inéquations non linéaires dans les espaces vectoriels en dualité, *Ann. Inst. Fourier* **18** (1968), 115–175.

[34] H. Brézis, Problèmes unilatéraux, *J. Math. Pures et Appl.* **51** (1972), 1–168.

[35] H. Brézis, *Opérateurs maximaux monotones et semi-groupes de contractions dans les espaces de Hilbert*, Mathematics Studies, North Holland, Amsterdam, 1973.

[36] H. Brézis, *Analyse fonctionnelle—Théorie et applications*, Masson, Paris, 1987.

[37] M. Burguera and J.M. Viaño, Numerical solving of frictionless contact problems in perfect plastic bodies, *Comput. Methods Appl. Mech. Engrg.* **120** (1995), 303–322.

[38] M. Campillo and I.R. Ionescu, Initiation of antiplane shear instability under slip dependent friction, *Journal of Geophysical Research* **102 B9** (1997) 363–371.

[39] O. Chau, W. Han and M. Sofonea, Analysis and aproximation of a viscoelastic contact problem with slip dependent friction, *Dynamics of Continuous, Discrete and Impulsive Systems* **8** (2001), 153–174.

[40] O. Chau, W. Han and M. Sofonea, A dynamic frictional contact problem with normal damped response, *Acta Applicandae Mathematicae* **71** (2002), 159–178.

[41] O. Chau, D. Motreanu and M. Sofonea, Quasistatic frictional problems for elastic and viscoelastic materials, to appear in *Applications of Mathematics.*

[42] O. Chau, L. Selmani and M. Sofonea, Analysis of a quasistatic displacement-traction problem for viscoelastic materials, *Mathematical and Computer Modelling* **32** (2000), 453–464.

[43] J. Chen, W. Han and M. Sofonea, Numerical analysis of a nonlinear evolutionary system with applications in viscoplasticity, *SIAM Journal of Numerical Analysis* **38** (2000), 1171–1199.

[44] J. Chen, W. Han and M. Sofonea, Numerical analysis of a class of evolution systems arising in viscoplasticity, *Computational and Applied Mathematics* **19** (2000), 279–306.

[45] J. Chen, W. Han and M. Sofonea, Numerical analysis of a quasistatic problem of sliding frictional contact with wear, *Methods and Applications of Analysis* **7** (2000), 687–704.

[46] J. Chen, W. Han and M. Sofonea, Numerical analysis of a contact problem in rate-type viscoplasticity, *Numer. Funct. Anal. and Optimiz.* **22** (2001), 505–527.

[47] P.G. Ciarlet, *The Finite Element Method for Elliptic Problems*, North Holland, Amsterdam, 1978.

[48] P.G. Ciarlet, *Mathematical Elasticity, Volume I: Three Dimensional Elasticity*, Studies in Mathematics and its Applications, Vol. 20, North-Holland, Amsterdam, 1988.

[49] P.G. Ciarlet, Basic error estimates for elliptic problems, in P.G. Ciarlet and J.-L. Lions, eds., *Handbook of Numerical Analysis*, Vol. II, North-Holland, Amsterdam, 1991, 17–351.

[50] C. Ciulcu, D. Motreanu and M. Sofonea, Analysis of an elastic contact problem with slip dependent coefficient of friction, *Mathematical Inequalities & Applications* **4** (2001), 465–479.

[51] M. Cocu, Existence of solutions of Signorini problems with friction, *Int. J. Engng. Sci.* **22** (1984), 567–581.

[52] M. Cocu, E. Pratt and M. Raous, Existence d'une solution du problème quasistatique de contact unilatéral avec frottement non local, *C. R. Acad. Sci. Paris,* **320**, Série I (1995), 1413–1417.

[53] M. Cocu, E. Pratt and M. Raous, Formulation and approximation of quasistatic frictional contact, *Int. J. Engng. Sci.* **34** (1996), 783–798.

[54] M. Cocu and J.M. Ricaud, Existence results for a class of implicit evolution inequalities and application to dynamic unilateral contact problems with friction, *C. R. Acad. Sci. Paris* **329**, Série I (1999), 839–844.

[55] M. Cocu and J.M. Ricaud, Analysis of a class of implicit evolution inequalities associated to dynamic contact problems with friction, *Int. J. Engrg. Sci.* **328** (2000), 1534–1549.

[56] P. Coorevits, P. Hild and J.P. Pelle, A posteriori error estimates for unilateral contact with matching and nonmatching meshes, *Comput. Methods Appl. Mech. Engrg.* **186** (2000), 65–83.

[57] C. Corneschi, T.-V. Hoarau-Mantel and M. Sofonea, A quasistatic contact problem with slip dependent coefficient of friction for elastic materials, *Journal of Applied Analysis* **8** (2002), 59–80.

[58] N. Cristescu, *Dynamic Plasticity*, North-Holland, Amsterdam, 1967.

[59] N. Cristescu and I. Suliciu, *Viscoplasticity*, Martinus Nijhoff Publishers, Editura Tehnica, Bucharest, 1982.

[60] A. Curnier, A theory of friction, *Int. J. Solids Structures* **20** (1984), 637–647.

[61] I. Demkowicz and J.T. Oden, On some existence and uniqueness results in contact problems with non local friction, *Nonlinear Analysis, Theory, Methods & Applications* **6** (1982), 1075–1093.

[62] C. Do, On the dynamic deformation of a bar against an obstacle, in S. Nemat-Nasser, ed., *Variational Mathods in the Mechanics of Solids*, Pergamon, Oxford, 1980, 237–241.

[63] I. Doghri, *Mechanics of Deformable Solids*, Springer, Berlin, 2000.

[64] P. Doktor, On the density of smooth functions in certain subspaces of Sobolev space, *Comment. Math. Univ. Carolinae* **14** (1973), 609–622.

[65] S. Drabla, M. Rochdi and M. Sofonea, On a frictionless contact problem for elastic-viscoplastic materials with internal state variables, *Mathematical and Computer Modelling* **26** (1997), 31–47.

[66] S. Drabla, M. Sofonea and B. Teniou, Analysis of some frictionless contact problems for elastic bodies, *Ann. Pol. Mat.* **LXIX** (1998), 75–88.

[67] A.D. Drozdov, *Finite Elasticity and Viscoelasticity—A Course in the Nonlinear Mechanics of Solids*, World Scientific, Singapore, 1996.

[68] D.C. Drucker, Elementary results of dimensional analysis for friction and wear in steady state sliding, *Wear* **40** (1976), 129–133.

[69] G. Duvaut, Loi de frottement non locale, *J. Méc. Thé. Appl.* Special issue (1982), 73–78.

[70] G. Duvaut and J.-L. Lions, *Inequalities in Mechanics and Physics*, Springer-Verlag, Berlin, 1976.

[71] C. Eck and J. Jarušek, Existence results for the static contact problem with Coulomb friction, *Math. Meth. Appl. Sci.* **8** (1998), 445–468.

[72] C. Eck and J. Jarušek, Existence results for the semicoercive static contact problem with Coulomb friction, *Nonlinear Analysis TMA* **42** (2000), 961–976.

[73] I. Ekeland and R. Temam, *Convex Analysis and Variational Problems*, North-Holland, Amsterdam, 1976.

[74] L.C. Evans, *Partial Differential Equations*, AMS Press, Providence, 1999.

[75] R.S. Falk, Error estimates for the approximation of a class of variational inequalities, *Mathematics of Computation* **28** (1974), 963–971.

[76] E. Fernández-Cara and C. Moreno, Critical point approximation through exact regularization, *Mathematics of Computation* **50** (1988), 139–153.

[77] J.R. Fernández-García, W. Han, M. Shillor and M. Sofonea, Numerical analysis and simulations of quasistatic frictionless contact problems, *Int. J. Appl. Math. and Comp. Sci.* **11** (2001), 205–222.

[78] J.R. Fernández-García, W. Han, M. Shillor and M. Sofonea, Recent advances on variational and numerical analysis in contact mechanics, *Proc. MTNS 2000*, Perpignan, France, 2000, published on CD-ROM.

[79] J.R. Fernández–García, W. Han and M. Sofonea, Numerical simulations for a frictional viscoelastic problem with normal compliance, *preprint.*

[80] J.R. Fernández–García, W. Han and M. Sofonea, Numerical simulations for in the study of a viscoelastic contact problem with normal damped response, *preprint.*

[81] J.R. Fernández-García, W. Han, M. Sofonea and J.M. Viaño, Variational and numerical analysis of a frictionless contact problem for elastic-viscoplastic materials with internal state variable, *Quart. J. Mech. Appl. Math.* **54** (2001), 501–522.

[82] J.R. Fernández-García, M. Sofonea and J.M. Viaño, A frictionless contact problem for elastic-viscoplastic materials with normal compliance, *Numerische Mathemätik* **90** (2002), 689–719.

[83] J.R. Fernández–García, P. Hild and J.M. Viaño, The mortar finite element method for elastic-viscoplastic contact, *Proc. MTNS 2000*, Perpignan, France, 2000, published on CD-ROM.

[84] J.R. Fernández–García, P. Hild and J.M. Viaño, Numerical approximation of the elastic-viscoplastic contact problem with non-matching meshes, to appear in *Numerische Mathemätik.*

[85] G. Fichera, Problemi elastostatici con vincoli unilaterali. II. Problema di Signorini con ambique condizioni al contorno, *Mem. Accad. Naz. Lincei, S. VIII, Vol. VII, Sez. I,* **5** (1964) 91–140.

[86] A. Friedman, *Variational Principles and Free-boundary Problems*, John Wiley, New York, 1982.

[87] M. Fuchs and G. Seregin, *Variational Methods for Problems from Plasticity Theory and for Generalized Newtonian Fluids*, Lecture Notes in Mathematics 1749, Springer-Verlag, Berlin, Heidelberg, 2000.

[88] P. Germain and P. Muller, *Introduction à la mécanique des milieux continus*, Masson, Paris, 1980.

[89] H. Giesekus, An alternative approach to the linear theory of viscoelasticity and some characteristic effects being distinctive of the type of material, *Rheol. Acta* **34** (1995), 2–11.

[90] R. Glowinski, *Numerical Methods for Nonlinear Variational Problems*, Springer-Verlag, New York, 1984.

[91] R. Glowinski, J.-L. Lions, and R. Trémolières, *Numerical Analysis of Variational Inequalities*, North-Holland, Amsterdam, 1981.

[92] A. Guran, ed., *Provceedings of the First International Symposium on Impact and Friction of Solids, Structures and Intelligent Machines*, World Scientific, Singapore, 2000.

[93] A. Guran, F. Pfeiffer and K. Popp, eds., *Dynamics with Friction: Modeling, Analysis and Experiment, Part I*, World Scientific, Singapore, 1996.

[94] M.E. Gurtin, *An Introduction to Continuum Mechanics*, Academic Press, New York, 1981.

[95] W. Han, On the numerical approximation of a frictional contact problem with normal compliance, *Numer. Funct. Anal. and Optimiz.* **17** (1996), 307–321.

[96] W. Han and B.D. Reddy, *Plasticity: Mathematical Theory and Numerical Analysis*, Springer-Verlag, New York, 1999.

[97] W. Han and B.D. Reddy, Convergence of approximations to the primal problem in plasticity under conditions of minimal regularity, *Numerische Mathematik* **87** (2000), 283–315.

[98] W. Han, M. Shillor and M. Sofonea, Variational and numerical analysis of a quasistatic viscoelastic problem with normal compliance, friction and damage, *Journal of Computational and Applied Mathematics*, **137** (2001), 377–398.

[99] W. Han and M. Sofonea, Analysis and numerical approximation of an elastic frictional contact problem with normal compliance, *Applicationes Mathematicae* **26** (1999), 415–435.

[100] W. Han and M. Sofonea, Numerical analysis of a frictionless contact problem for elastic-viscoplastic materials, *Comput. Methods Appl. Mech. Engrg.* **190** (2000), 179–191.

[101] W. Han and M. Sofonea, Evolutionary Variational inequalities arising in viscoelastic contact problems, *SIAM Journal of Numerical Analysis* **38** (2000), 556–579.

[102] W. Han and M. Sofonea, On numerical approximation of a frictionless contact problem for elastic-viscoplastic materials, in *Integral Methods and Engineering*, B. Bertram, C. Constanda and A. Struthers Eds., Chapman & Hall/CRC Reserch Notes in Mathematics, Vol. 418 (2000), 173–178.

[103] W. Han and M. Sofonea, Numerical analysis of a quasistatic viscoelastic problem with friction and damage, in *Advanced in Scientific Computation*, Conference in Honor of Professor H.-C. Huang on the occasion of his retirement, Z.C. Shi, M. Mu, W.M. Xue and J. Zou, eds., Science Press, 2000, 51–60.

[104] W. Han and M. Sofonea, Time-dependent variational inequalities for viscoelastic contact problems, *Journal of Computational and Applied Mathematics* **136** (2001), 369–387.

[105] W. Han and L.H. Wang, Non-conforming finite element analysis for a plate contact problem, to appear in *SIAM Journal on Numerical Analysis.*

[106] J. Haslinger and I. Hlaváček, Contact between elastic bodies–I. Continuous problem, *Apl. Mat.* **25** (1980), 324–347.

[107] J. Haslinger and I. Hlaváček, Contact between elastic bodies–II. Finite element analysis, *Apl. Mat.* **26** (1981), 263–290.

[108] J. Haslinger and I. Hlaváček, Contact between elastic bodies–III. Dual finite element analysis, *Apl. Mat.* **26** (1981), 321–344.

[109] J. Haslinger, I. Hlaváček, J. Nečas, Numerical methods for unilateral problems in solid mechanics, *Handbook of Numerical Analysis, Vol IV*, P.G. Ciarlet and J.L. Lions, eds., North-Holland, Amsterdam, 1996, 313–485.

[110] P. Hild, Numerical implementation of two nonconforming finite element methods for unilateral contact, *Comput. Methods Appl. Mech. Engrg.* **184** (2000), 99–123.

[111] J.-B. Hiriart-Urruty and C. Lemaréchal, *Convex Analysis and Minimization Algorithms, I, II*, Springer-Verlag, Berlin, 1993.

[112] I. Hlaváček, J. Haslinger, J. Nečas, and J. Lovíšek, *Solution of Variational Inequalities in Mechanics*, Springer-Verlag, New York, 1988.

[113] I. Hlaváček and J. Lovíšek, A finite element analysis for the Signorini problem in plane elastostatics, *Aplikace Matematiky* **22** (1977), 215–227.

[114] I.R. Ionescu, Error estimates of an Euler's method for a quasistatic elastic-viscoplastic problem, *Zeitschrift fur Angewandte Mathematik und Mechanik* (*ZAMM*) **70** (1990), 173–180.

[115] I.R. Ionescu, Dynamic processes for a class of elastic-viscoplastic materials, *Studii si Cercetari Matematice* **44** (1992), 113–125.

[116] I.R. Ionescu and Q.-L. Nguyen, Dynamic contact problems with slip dependent friction in viscoelasticity, *Int. J. Appl. Math. Comput. Sci.* **12** (2002), to appear.

[117] I.R. Ionescu and J.C. Paumier, Friction dynamique avec coefficient dépandant de la vitesse de glissement, *C. R. Acad. Sci. Paris* **316**, Série I (1993), 121–125.

[118] I.R. Ionescu and J.-C. Paumier, On the contact problem with slip rate dependent friction in elastodynamics, *European J. of Mech., A—Solids* **13** (1994), 556–568.

[119] I.R. Ionescu and J.-C. Paumier, On the contact problem with slip displacement dependent friction in elastostatics, *Int. J. Engng. Sci.* **34** (1996), 471–491.

[120] I.R. Ionescu and M. Sofonea, Quasistatic processes for elastic-viscoplastic materials, *Quarterly of Applied Mathematics* **46** (1988), 220–243.

[121] I.R. Ionescu and M. Sofonea, *Functional and Numerical Methods in Viscoplasticity*, Oxford University Press, Oxford, 1993.

[122] J. Jarušek, Contact problem with given time-dependent friction force in linear viscoelasticity, *Comment. Math. Univ. Carolinae* **31** (1990), 257–262.

[123] J. Jarušek, Dynamic contact problems for bodies with a singular memory, *Boll. Unione Mat. Ital.* **7** (1995), 581–592.

[124] J. Jarušek, Dynamic contact problems with given friction for viscoelastic bodies, *Czechoslovak Mathematical Journal* **46** (1996), 475–487.

[125] J. Jarušek, Remark to dynamic contact problems for bodies with a singular memory, *Comment. Math. Univ. Carolinae* **39** (1998), 545–550.

[126] J. Jarušek and C. Eck, Dynamic contact problems with small Coulomb friction for viscoelastic bodies. Existence of solutions, *Mathematical Models and Methods in Applied Sciences* **9** (1999), 11–34.

[127] C. Johnson, Existence theorems in plasticity, *J. Math. Pures et Appl.* **55** (1976), 431–444.

[128] C. Johnson, *Numerical Solutions of Partial Differential Equations by the Finite Element Method*, Cambridge University Press, Cambridge, 1987.

[129] K.L. Johnson, *Contact Mechanics*, Cambridge University Press, Cambridge, 1987.

[130] J.J. Kalker, *Three-dimensional Elastic Bodies in Rolling Contact*, Kluwer Academic Publishers, Dordrecht, 1990.

[131] H. Kardestuncer and D.H. Norrie, eds., *Finite Element Handbook*, McGraw-Hill Book Company, New York, 1987.

[132] A.M. Khludnev and J. Sokolowski, *Modelling and Control in Solid Mechanics*, Birkhäuser-Verlag, Basel, 1997.

[133] A.M. Khludnev and J. Sokolowski, On solvability of boundary value problems in elastoplasticity, *Control and Cybernetics* **27** (1998), 311–330.

[134] N. Kikuchi and J.T. Oden, *Contact Problems in Elasticity: A Study of Variational Inequalities and Finite Element Methods*, SIAM, Philadelphia, 1988.

[135] J.U. Kim, A one-dimensional dynamic contact problem in linear viscoelasticity, *Mathematical Methods in the Applied Sciences* **13** (1990), 55–79.

[136] D. Kinderlehrer and G. Stampacchia, *An Introduction to Variational Inequalities and their Applications*, Academic Press, New York, 1980.

[137] A. Klarbring, A. Mikelič and M. Shillor, Frictional contact problems with normal compliance, *Int. J. Engng. Sci.* **26** (1988), 811–832.

[138] A. Klarbring, A. Mikelič and M. Shillor, A global existence result for the quasistatic frictional contact problem with normal compliance, in G. Del Piero and F. Maceri, eds., *Unilateral Problems in Structural Analysis*, Vol. **4**, Birkhäuser, Boston, 1991, 85–111.

[139] Y. Komura, Nonlinear semigroups in Hilbert spaces, *Journal of the Mathematical Society of Japan* **19** (1967), 493–507.

[140] M. Krizek and P. Neittaanmaki, *Mathematical and Numerical Modelling in Electrical Engineering: Theory and Applications*, Kluwer Academic Publishers, Dordrecht, 1996.

[141] K.L. Kuttler, Dynamic friction contact problem with general normal and friction laws, *Nonlinear Analysis TMA* **28** (1997), 559–575.

[142] K.L. Kuttler and M. Shillor, Set-valued pseudomonotone maps and degenerate evolution inclusions, *Communications in Contemporary Mathematics* **1** (1999), 87–123.

[143] K.L. Kuttler and M. Shillor, Dynamic bilateral contact with discontinuous friction coefficient, *Nonlinear Analysis TMA* **45** (2001), 309–327.

[144] K.L. Kuttler and M. Shillor, Dynamic contact with normal compliance wear and discontinuous friction coefficient, *preprint.*

[145] P. Laborde, On visco-plasticity with hardening, *Numerical Functional Analysis and Optimization* **1** (1979), 315–339.

[146] J. Lemaître and J.-L. Chaboche, *Mechanics of Solids Materials*, Cambridge University Press, Cambridge, 1990.

[147] F. Léné, Sur les matériaux élastiques à énergie de déformation non quadratique, *Journal de Mécanique* **13** (1974), 499–534.

[148] J.-L. Lions and E. Magenes, *Problèmes aux limites non-homogènes I*, Dunod, Paris, 1968.

[149] G.I. Marchuk, Splitting and alternating direction methods, in P.G. Ciarlet and J.-L. Lions, eds., *Handbook of Numerical Analysis*, Vol. I, North-Holland, Amsterdam, 1990, 197–464.

[150] G.I. Marchuk and V.V. Shaidurov, *Difference Methods and Their Extrapolations*, Springer-Verlag, New York, 1983.

[151] D. Maugis, *Contact, Adhesion and Rupture of Elastic Solids*, Springer-Verlag, Berlin, Heidelberg, 2000.

[152] J.A.C. Martins and J.T. Oden, Existence and uniqueness results for dynamic contact problems with nonlinear normal and friction interface laws, *Nonlinear Analysis TMA* **11** (1987), 407–428.

[153] A. Matei, V.V. Motreanu and M. Sofonea, On the Signorini frictionless contact problem for linear viscoelastic materials, to appear in *Applicable Analysis.*

[154] S.J. Matysiak, ed., Contact Mechanics, Special issue of *Journal of Theoretical and Applied Mechanics* **39**(3) (2001).

[155] J.J. Moreau, Proximité et dualité dans un espace hilbertien, *Bulletin de la Société Mathématique de France* **93** (1965), 273–283.

[156] J.J. Moreau, Application of convex analysis to the treatment of elastoplastic systems, in P. Germain and B. Nayroles, eds., *Applications of Methods of Functional Analysis to Problems in Mechanics*, Lecture Notes in Math, Springer-Verlag, Berlin, 1976.

[157] J.J. Moreau, Application of convex analysis to some problems of dry friction in H. Zorski, ed., *Trends in Applications of Pure Mathematics to Mechanics*, Vol. II, Pitman, London, 1979.

[158] D. Motreanu and M. Sofonea, Evolutionary variational inequalities arising in quasistatic frictional contact problems for elastic materials, *Abstract and Applied Analysis* **4** (1999), 255–279.

[159] D. Motreanu and M. Sofonea, Quasivariational inequalities and applications in frictional contact problems with normal compliance, *Adv. Math. Sci. Appl.* **10** (2000), 103–118.

[160] J. Nečas, *Les méthodes directes en théorie des équations elliptiques*, Academia, Praha, 1967.

[161] J. Nečas and I. Hlaváček, *Mathematical Theory of Elastic and Elastico-Plastic Bodies: An Introduction*, Elsevier Scientific Publishing Company, Amsterdam, Oxford, New York, 1981.

[162] Q.S. Nguyen, *Stability and Nonlinear Solid Mechanics*, John Wiley & Sons, LTD, Chichester, 2000.

[163] J.T. Oden and J.A.C. Martins, Models and computational methods for dynamic friction phenomena, *Computer Methods in Applied Mechanics and Engineering* **52** (1985), 527–634.

[164] J.T. Oden and E.B. Pires, Nonlocal and nonlinear friction laws and variational principles for contact problems in elasticity, *J. Appl. Mech.* **50** (1983), 67–76.

[165] J.T. Oden and J.N. Reddy, *An Introduction to the Mathematical Theory of Finite Elements*, John Wiley, New York, 1976.

[166] M. Ohnaka, Nonuniformity of the constitutive law parameters for shear rupture and quasistatic nucleation to dynamic rupture: A physical model of earthquake generation model, in *Earthquake Prediction: The Scientific Challange*, Irvine, California, U.S. Acad. of Sci., 1996.

[167] P.D. Panagiotopoulos, *Inequality Problems in Mechanics and Applications*, Birkhäuser, Boston, 1985.

[168] A. Papo, Rheological models for gypsum plaster pastes, *Rheol. Acta* **27** (1988), 320–325.

[169] P. Perzyna, The constitutive equations for rate sensitive plastic materials, *Quarterly of Applied Mathematics* **20** (1963), 321–332.

[170] A. Quarteroni and A. Valli, *Numerical Approximation of Partial Differential Equations*, Springer-Verlag, Berlin, 1994.

[171] E. Rabinowicz, *Friction and Wear of Materials*, Wiley, New York, 1965.

[172] E. Rabinowicz, *Friction and Wear of Materials*, 2nd edition, Wiley, New York, 1995.

[173] M. Raous, P. Chabrand and F. Lebon, Numerical methods for frictional contact problems and applications *J. Theor. Appl. Mech.*, special issue, Suppl. N1 **7** (1988), 111–128.

[174] M. Raous, M. Jean and J.J. Moreau, eds., *Contact Mechanics*, Plenum Press, New York, 1995.

[175] B.D. Reddy, *Introductory Functional Analysis with Applications to Boundary Value Problems and Finite Elements*, Springer, New York, 1998.

[176] R. Rocca, Existence of a solution for a quasistatic contact problem with local friction, *C. R. Acad. Sci. Paris* **328**, Série I (1999), 1253–1258.

[177] R. Rocca and M. Cocu, Existence and approximation of a solution to quasistatic problem with local friction, to appear in *Int. J. Engrg. Sci.*

[178] M. Rochdi, M. Shillor and M. Sofonea, Quasistatic viscoelastic contact with normal compliance and friction, *Journal of Elasticity* **51** (1998), 105–126.

[179] M. Rochdi, M. Shillor and M. Sofonea, A quasistatic contact problem with directional friction and damped response, *Applicable Analysis* **68** (1998), 409–422.

[180] M. Rochdi, M. Shillor and M. Sofonea, Analysis of a quasistatic viscoelastic problem with friction and damage, *Adv. Math. Sci. Appl.* **10** (2000), 173–189.

[181] M. Rochdi and M. Sofonea, On frictionless contact between two elastic-viscoplastic bodies, *Q. J. Mech. Appl. Math.* **50** (1997), 481–496.

[182] M. Rochdi and M. Sofonea, On the existence and behaviour of the solution for a class of nonlinear evolution systems, *Rev. Roumaine Math. Pures Appl.* **42** (1997), 659–667.

[183] M. Rochdi, M. Sofonea and I. Rosca, On existence, behaviour, and numerical approach of the solution for contact problems in rate-type viscoplasticity, in A.B. Sabir et al, eds., *Proc. ESDA '96*, Montpellier, France, 1996, Vol. 1, 87–92.

[184] D. Roylance and S.S. Wang, Penetration mechanics of textile structures: influence of nonlinear viscoelastic relaxation, *Polym. Eng. Sci.* **18** (1978), 1068–1072.

[185] J.F. Rodrigues, *Obstacle Problems in Mathematical Physics*, North-Holland, Amsterdam, 1987.

[186] J. Rojek, J.J. Telega, Contact problems with friction, adhesion and wear in orthopaedic biomechanics. I: General developments, *J. Theor. Appl. Mech.* **39** (2001), 655–677.

[187] J. Rojek, J.J. Telega, S. Stupkiewicz, Contact problems with friction, adhesion and wear in orthopaedic biomechanics. II: Numerical implementation and application to implanted knee joints, *J. Theor. Appl. Mech.* **39** (2001), 679–706.

[188] M. Shillor, ed., Recent advances in contact mechanics, Special issue of *Math. Comput. Modelling* **28**(4–8) (1998).

[189] M. Shillor, Quasistatic problems in contact mechanics, *Int. J. Appl. Math. Comput. Sci.* **11** (2001), 189–204.

[190] M. Shillor and M. Sofonea, A quasistatic viscoelastic contact problem with friction, *Int. J. Engng. Sci.* **38** (2000), 1517–1533.

[191] A. Signorini, Sopra a une questioni di elastostatica, *Atti della Società Italiana per il Progresso delle Scienze*, 1933.

[192] T.L. Smith, Physical properties of polymers—an introductory discussion, *Polym. Eng. Sci.* **13** (1973), 161–175.

[193] M. Sofonea, Some remarks on the behaviour of the solution in dynamic process for rate-type models, *Zeitschrift fur Angewandte Mathematik und Physik (ZAMP)* **41** (1990), 656–668.

[194] M. Sofonea, On a contact problem for elastic-viscoplastic bodies, *Nonlinear Analysis TMA* **29** (1997), 1037–1050.

[195] M. Sofonea, A convergence result in the study of frictionless viscoplastic contact problems, *Rev. Roum. Math. Pures Appl.* **45** (2000), 343–351.

[196] M. Sofonea and M. Shillor, Variational analysis of quasistatic viscoplastic contact problems with friction, *Communications in Applied Analysis* **5** (2001), 135–151.

[197] M. Sofonea and J.M. Viaño, eds., Mathematical Modelling and Numerical Analysis in Solid Mechanics, Special issue of *Int. J. Appl. Math. Comput. Sci.* **12** (1) (2002), to appear.

[198] I. Stakgold, *Green's Functions and Boundary Value Problems*, John Wiley & Sons, Inc., 1979.

[199] G. Strang and G. Fix, *An Analysis of the Finite Element Method*, Prentice-Hall, Englewood Cliffs, NJ, 1973.

[200] J.C. Strikwerda, *Finite Difference Schemes and Partial Differential Equations*, Wadsworth & Brooks/Cole Advanced Books & Software, 1989.

[201] N. Strömberg, *Continuum Thermodynamics of Contact, Friction and Wear*, Thesis No. 491, Department of Mechanical Engineering, Linköping Institute of Technology, Linköping, Sweden, 1995.

[202] N. Strömberg, L. Johansson and A. Klarbring, Derivation and analysis of a generalized standard model for contact friction and wear, *Int. J. Solids Structures* **33** (1996), 1817–1836.

[203] P. Suquet, Evolution problems for a class of dissipative materials, *Quarterly of Applied Mathematics* **38** (1981), 391–414.

[204] P. Suquet, Sur les équations de la plasticité: existyence et régularité des solutions, *Journal de Mécanique* **20** (1981), 3–39.

[205] P. Suquet, *Plasticité et homogéneisation*, Thesis, University of Paris VI, 1982.

[206] B. Szabó and I. Babuška, *Finite Element Analysis*, John Wiley, Inc., New York, 1991.

[207] W. Szczepinski, ed., *Experimental Methods in Mechanics of Solids*, Elsevier, Amsterdam, 1990.

[208] J.J. Telega, Limit analysis theorems in the case of Signorini's boundary conditions, *Arch. Mech.* **37** (1985), 549–562.

[209] J.J. Telega, Topics on unilateral contact problems of elasticity and inelasticity, in J.J. Moreau and P.D. Panagiotopoulos, eds. *Nonsmooth Mechanics and Applications*, Springer-Verlag, Wien, 1988, 340–461.

[210] J.J. Telega, Quasi-static Signorini's contact problem with friction and duality, *International Series of Numerical Mathematics* **101** (1991), 199–214.

[211] R. Temam and A. Miranville, *Mathematical Modeling in Continuum Mechanics*, Cambridge University Press, Cambridge, 2001.

[212] V. Thomée, Finite difference methods for linear parabolic equations, in P.G. Ciarlet and J.-L. Lions, eds., *Handbook of Numerical Analysis, I*, North-Holland, Amsterdam, 1990, 5–196.

[213] J.M. Viaño, Análisis de un método numérico con elementos finitos para problemas de contacto unilateral sin rozamiento en elasticidad: Aproximación y resolución de los problemas discretos, *Rev. Internac. Mtod. Numr. Clc. Dise. Ingr.* **2** (1986), 63–86.

[214] R.L. Wheeden and A. Zygmund, *Measure and Integral: An Introduction to Real Analysis*, Marcel Dekker, Inc., New York, 1977.

[215] P. Wriggers and P.D. Panagiotopoulos (Eds.), *New Developments in Contact Problems*, Springer-Verlag, Wien, New York, 1999.

[216] E. Zeidler, *Nonlinear Functional Analysis and its Applications.* III*: Variational Methods and Optimization*, Springer-Verlag, New York, 1986.

[217] E. Zeidler, *Nonlinear Functional Analysis and its Applications.* IV*: Applications to Mathematical Physics*, Springer-Verlag, New York, 1988.

[218] E. Zeidler, *Nonlinear Functional Analysis and its Applications,* II/A*: Linear Monotone Operators*, Springer-Verlag, New York, 1990.

[219] E. Zeidler, *Nonlinear Functional Analysis and its Applications,* II/B*: Nonlinear Monotone Operators*, Springer-Verlag, New York, 1990.

[220] E. Zeidler, *Applied Functional Analysis: Applications of Mathematical Physics*, Springer-Verlag, New York, 1995.

[221] E. Zeidler, *Applied Functional Analysis: Main Principles and Their Applications*, Springer-Verlag, New York, 1995.

[222] A. Zmitrowicz, A thermodynamical model of contact, friction and wear: I. Governing equation, II. Constitutive equations for materials and linearized theories, III. Constitutive equations for friction, wear and friction heat, *Wear* **114** (1987), 135–168, 169–197, 199–221.

[223] A. Zmitrowicz, *On the Thermodynamics of Contact, Friction and Wear*, Publications of Institute Maszyn Przeplywowych, PAN, Gdansk, 287/1212, 1989.

[224] A. Zmitrowicz, *Variational Formulation of Contact, Friction and Wear Problems*, Publications of Institute Maszyn Przeplywowych, PAN, Gdansk, 504/1463, 1999.

[225] A. Zmitrowicz, Strong and weak formulation of contact problems with various laws of friction frictional heat and wear, *Machine Dynamics Problems* **24** (2000) 209–222.

Index

Titles in This Series

30 **Weiman Han and Mircea Sofonea,** Quasistatic Contact Problems in Viscoelasticity and Viscoplasticity, 2002

29 **Shuxing Chen and S.-T. Yau, Editors,** Geometry and Nonlinear Partial Differential Equations, 2002

28 **Valentin Afraimovich and Sze-Bi Hsu,** Lectures on Chaotic Dynamical Systems, 2002

27 **M. Ram Murty,** Introduction to p-adic Analytic Number Theory, 2002

26 **Raymond Chan, Yue-Kuen Kwok, David Yao, and Qiang Zhang, Editors,** Applied Probability, 2002

25 **Donggao Deng, Daren Huang, Rong-Qing Jia, Wei Lin, and Jian Zhong Wong, Editors,** Wavelet Analysis and Applications, 2002

24 **Jane Gilman, William W. Menasco, and Xiao-Song Lin, Editors,** Knots, Braids, and Mapping Class Groups—Papers Dedicated to Joan S. Birman, 2001

23 **Cumrun Vafa and S.-T. Yau, Editors,** Winter School on Mirror Symmetry, Vector Bundles and Lagrangian Submanifolds, 2001

22 **Carlos Berenstein, Der-Chen Chang, and Jingzhi Tie,** Laguerre Calculus and Its Applications on the Heisenberg Group, 2001

21 **Jürgen Jost,** Bosonic Strings: A Mathematical Treatment, 2001

20 **Lo Yang and S.-T. Yau, Editors,** First International Congress of Chinese Mathematicians, 2001

19 **So-Chin Chen and Mei-Chi Shaw,** Partial Differential Equations in Several Complex Variables, 2001

18 **Fangyang Zheng,** Complex Differential Geometry, 2000

17 **Lei Guo and Stephen S.-T. Yau, Editors,** Lectures on Systems, Control, and Information, 2000

16 **Rudi Weikard and Gilbert Weinstein, Editors,** Differential Equations and Mathematical Physics, 2000

15 **Ling Hsiao and Zhouping Xin, Editors,** Some Current Topics on Nonlinear Conservation Laws, 2000

14 **Jun-ichi Igusa,** An Introduction to the Theory of Local Zeta Functions, 2000

13 **Vasilios Alexiades and George Siopsis, Editors,** Trends in Mathematical Physics, 1999

12 **Sheng Gong,** The Bieberbach Conjecture, 1999

11 **Shinichi Mochizuki,** Foundations of p-adic Teichmüller Theory, 1999

10 **Duong H. Phong, Luc Vinet, and Shing-Tung Yau, Editors,** Mirror Symmetry III, 1999

9 **Shing-Tung Yau, Editor,** Mirror Symmetry I, 1998

8 **Jürgen Jost, Wilfrid Kendall, Umberto Mosco, Michael Röckner, and Karl-Theodor Sturm,** New Directions in Dirichlet Forms, 1998

7 **D. A. Buell and J. T. Teitelbaum, Editors,** Computational Perspectives on Number Theory, 1998

6 **Harold Levine,** Partial Differential Equations, 1997

5 **Qi-keng Lu, Stephen S.-T. Yau, and Anatoly Libgober, Editors,** Singularities and Complex Geometry, 1997

4 **Vyjayanthi Chari and Ivan B. Penkov, Editors,** Modular Interfaces: Modular Lie Algebras, Quantum Groups, and Lie Superalgebras, 1997

3 **Xia-Xi Ding and Tai-Ping Liu, Editors,** Nonlinear Evolutionary Partial Differential Equations, 1997

2.2 **William H. Kazez, Editor,** Geometric Topology, 1997

2.1 **William H. Kazez, Editor,** Geometric Topology, 1997

TITLES IN THIS SERIES

1 **B. Greene and S.-T. Yau, Editors,** Mirror Symmetry II, 1997